# Topics in Biodiversity and Conservation

Volume 20

Springer's book series, Topics in Biodiversity and Conservation, brings together some of the most exciting and topical papers in biodiversity and conservation research. The result is a series of useful themed collections covering issues such as the diversity and conservation of specific habitats or groups of organisms, and the key dilemma of resource use versus conservation.

Based on Springer's popular journal, Biodiversity and Conservation, the series provides access to selected peer-reviewed papers which represent the cutting edge of current research to provide a valuable overview of progress in each topic addressed. With their diversity of case studies and depth of investigation, these collections will be of particular interest for courses including biodiversity and/or conservation issues, and to advanced students and researchers working in related fields.

More information about this series at https://link.springer.com/bookseries/7488

Florencia Montagnini
Editor

# Biodiversity Islands: Strategies for Conservation in Human-Dominated Environments

Springer

*Editor*
Florencia Montagnini
School of the Environment,
The Forest School
Yale University
New Haven, CT, USA

ISSN 1875-1288 ISSN 1875-1296 (electronic)
Topics in Biodiversity and Conservation
ISBN 978-3-030-92233-7 ISBN 978-3-030-92234-4 (eBook)
https://doi.org/10.1007/978-3-030-92234-4



This Springer imprint is published by the registered company Springer Nature Switzerland AG
The registered company address is: Gewerbestrasse 11, 6330 Cham, Switzerland

*This book is dedicated to my father.*

# Foreword

It is unfortunate that a book on this topic needed to be written, but it is on a highly relevant subject given the worldwide destruction of natural habitats and the loss of so many biological species. As we, at present, face the sixth great species extinction, the remaining biodiversity survive in small patches or islands, which have become extremely important for the survival of the species they contain or the migrant birds they host. As someone who has worked a lifetime in botanical gardens, I am very aware of the small patches of original vegetation that they often contain and their value for the pollinators that visit the flowers or the occasional visit of migratory birds as they pause on their journey. At the Royal Botanical Gardens at Kew in London, there is rare-listed hoverfly and an endangered species of lichen among other species preserved in this biodiversity island. In my field work in the highly fragmented Atlantic rainforest of Brazil and Argentina, we are still finding new and undescribed species of plants in the small remnants of the original forest. It is fortunate that still many species of animals and plants survive in these often small islands, making them extremely important, and a book drawing attention to them is most welcome.

Until relatively recently, much more attention was given to marine islands following the work of MacArthur and Wilson in 1967 and because of their much-threatened biodiversity, but now there is a growing realisation of the importance of human-made islands on the mainland. The creation of biodiversity islands has been the topic of important research in the Biological Dynamics of Forest Fragments Project near Manaus, Brazil. I have spent many hours identifying plant species for this project. The original name of the project "Minimum Critical Size of Ecosystems Project" indicates its original research purpose to provide data about the minimum area needed to preserve a functioning area of rainforest to assist in the establishment of reserves and conservation areas. This book clearly demonstrates that today there are biodiversity islands of many different sizes, shapes and purposes.

This book treats a great variety of different types of biodiversity islands, all of which are areas of high biodiversity surrounded by highly degraded or intensely used landscapes that act as refuges for the surviving species of the original ecosystem. The many examples given here clearly show the critical importance of biodiversity islands for conservation, restoration and sustainable management of several productive agroforestry systems. It is good to be taken around the world with examples of biodiversity islands in both the tropical and the temperate regions. I like the fact that these examples include not only areas of pristine natural habitats such as the Monteverde Cloud Forest in Costa Rica or the forest islands in the Paraguayan Chaco but also several examples from highly managed islands in agroforestry and regenerative agriculture systems. Some of the examples of the policies and political motivations given in various chapters should be helpful to anyone involved in the creation or management of a biodiversity island. Several chapters here show examples of harmonising food production with conservation. This unity of purpose is important and is far more likely to be of long-term success than placing conservation and agriculture in separate camps. Several chapters show the importance of alternative ways to produce food from more integrated management systems that also preserve biodiversity. The social, ecological, ethical and economic benefits of such systems are clearly outlined in several of the chapters.

I congratulate the editor of this book for gathering together such a varied and useful compilation of the ongoing work on biodiversity islands. This will be of considerable use to people involved in the design of future biodiversity islands because it has much to say about the motivations and politics and also about their size and spatial distribution whether from fragments of the original vegetation or from restoration of degraded and intensely used areas. It will be a most useful tool for both conservation and restoration. My hope is that this will be used by conservation organisations, local communities and indigenous peoples to create effective islands of biodiversity in many different ecosystems of the world and for many more creative types of management.

Lyme Regis, United Kingdom
e-mail: siriain01@yahoo.co.uk

Professor Sir Ghillean T. Prance FRS, VMH

# Preface

A biodiversity island is an area of high biodiversity located within ecologically degraded, human-dominated landscapes. Biodiversity in the "islands" exceeds the surrounding landscape biodiversity baseline. These biodiversity islands thereby act as ecological refuges, promoting restoration and conservation in altered ecosystems prevalent today throughout the globe.

Biodiversity islands can provide food, water, fuels, and fibers, as well as genetic, medicinal, biochemical, and ornamental resources, pollination services, biological pest control, and maintenance of life cycles of migratory species. These landscapes hold promise for protecting a multitude of plant and animal species for present and future generations. The presence of biodiversity islands spread over a large area can decrease the chances of habitat loss from fire, disease, and other disturbances.

Biodiversity islands can exist within a wide range of human-dominated landscapes, including forest, agricultural, and urban settings, and can vary in scale from square meters to thousands of square kilometers. Design strategies for biodiversity islands depend on the spatial distribution of reserves throughout the landscape, the degree of site degradation, the species present, and their locations within the urban to rural spectrum.

This book is intended to provide an overview for the identification and establishment of biodiversity islands, presenting examples and case studies where the biodiversity islands approach is being used in a variety of locations and contexts worldwide. This book will contribute to design parameters on appropriate sizing and spatial distribution of biodiversity islands to be effective in conservation and regeneration across the landscape, using integrated landscape management approaches.

The chapters discuss current challenges faced today by biodiversity conservation researchers, practitioners, and policy makers and propose innovative approaches to tackle them. Contributors are an assemblage of researchers, academicians, and practitioners from biodiversity conservation, environmental management, forestry, agroecology, agroforestry, and related fields who approach the issues from unique perspectives.

This book comprises five parts: **Part I, Introduction**, establishes the framework for understanding the complexities of biodiversity islands and the variety of strategies that can be used to establish them. The Introduction defines the term "biodiversity islands" and their size, location, and distribution in the landscape; stresses their many ecological, social, and economic benefits; and discusses potential limitations of the use of this framework along with ways to overcome them. **Part II, Biodiversity Islands Establishment and Management: Challenges and Alternatives,** shows how design strategies may depend on landscape use within the matrix of habitat fragmentation, with integrated landscape management (ILM), including sustainable agriculture, agroforestry, and community-led action, providing a framework for implementation. **Part III, Biodiversity Islands Across the Globe: Case Studies,** shows how varied agroecological strategies were applied in the formation or conservation of biodiversity islands in human-dominated landscapes in Paraguay, Peru, Costa Rica, Colombia, Great Britain, Argentina, Panama, and the USA. The variety of case studies from different types of landscapes from several regions of the world reveals the role biodiversity islands play in conserving local flora and fauna that have been largely diminished by anthropogenic activities, while providing cultural connections to nature and supplying ecosystem services that make biodiversity islands advantageous to farmers and nearby communities. **Part IV, Safeguarding the Environmental, Economic, and Social Benefits of Biodiversity Islands,** further details the economic, social, political, and cultural aspects of the establishment and persistence of biodiversity islands in anthropogenic landscapes, emphasizing how community-led action contributes to their development and subsequent management, with examples from Puerto Rico, Ecuador, Brazil, India, the USA, Panama, and Ethiopia. **Part V, Conclusions**, summarizes the lessons learned while compiling this volume and lays out the pending challenges and potential solutions ahead.

One late summer afternoon, about 2 years ago, while relaxing in the porch of a house in suburban/rural Northford, Connecticut, a fox ran across the garden, apparently not feeling too threatened by our presence. When wondering where this small animal was coming from, and where did it go when it finally ran away, Kjell E Berg suggested that the water reservoir located about 100 meters from the house was a nice undisturbed forest that perhaps was functioning as a biodiversity island. Soon the idea of digging more into the concept grew in all directions; the next day, Brett Levin at Yale enthusiastically took it as his own project, and soon we wrote the introductory chapter of this book among the three of us.

Other ideas followed as we developed a website: https://biodiversityislands.org/ and led a meeting session called "Biodiversity Islands: Pockets of Protected Land in Human Dominated Environments" at a IUFRO (International Union of Forest Research Organizations) conference in Posadas, Misiones, Argentina, in October 2018. The structure and contents of this book further developed as we met and held conversations with students, colleagues, and friends whose enthusiasm, energy, and joyful attitude made this book possible from start to end. The more than a 100 authors who contributed chapters for this book drove the rest of the way with their dynamism, dedication, and persistence. Numerous colleagues and friends also helped with their intellectual input and moral support.

There was a total of 105 contributors from 11 countries (32 Argentina; 2 Brazil; 1 Canada; 14 Colombia; 2 Costa Rica; 10 Mexico; 4 Panama; 11 Paraguay; 3 Peru; 23 USA; 2 UK). Different chapters report research, case studies, and experiences from 14 countries: Argentina, Brazil, Colombia, Costa Rica, Ecuador, Ethiopia, India, Mexico, Panama, Paraguay, Peru, Puerto Rico, the UK, and the USA. Thus, the book includes examples of biodiversity islands from tropical as well as temperate regions, ranging from natural habitats to agroforestry and regenerative agriculture systems, and from relatively small to large geographic areas of the world.

A holistic, multidisciplinary perspective was taken in approaching each theme, encompassing factors and variables from multiple disciplines. The contributing authors present views from the academic, practitioner, and policy-making perspectives, offering alternatives and suggestions for promoting strategies that support biodiversity conservation through intentionally designed frameworks for sustainable forest landscapes. With the current worldwide trend of habitat destruction and the need to preserve biodiversity and its values, this book is an essential tool as it provides suggestions and concrete examples that can be used by a variety of stakeholders in various settings throughout the world. This book is useful to researchers, farmers, foresters, landowners, land managers, city planners, and policy makers alike.

New Haven, CT, USA — Florencia Montagnini

# Acknowledgments

Many chapter contributors acted as independent reviewers of other colleagues' chapters. In addition, other external reviewers generously gave their time to read and offer useful suggestions to improve the chapters. There was a total of 55 reviewers from the academic as well as from the practitioner's realms. The following is a list of chapter reviewers: Oscar J Abelleira, Dara Albrecht, Victor Arroyo-Rodriguez, Gary Bentrup, Kjell E Berg, Robert Bushbacher, Jonathan Cornelius, Sara del Fierro, Beatriz Eibl, Alberto Esquivel, Ben Everett-Lane, Glenn Galloway, Sergius Gandolfi, Eva Garen, Libertario González, Heather Griscom, David Hawksworth, Karen Kainer, Keith Kirby, B. Mohan Kumar, Rafaela Laino Guanes, Ariel Lugo, Brett Levin, Philip Marshall, Paula Meli, Zoyla Mireya Clavo Peralta, Irene Montes-Londoño, Gabriela Morales-Nieves, Mathew Moran, Carlos Navarro, Quint Newcomer, Fernando Niella, Joseph Orefice, Alison Ormbsy, Nahuel Pachas, Pablo Peri, Daniel Piotto, Julio Prieto, Neptali Ramírez-Marcial, Juan Rivero de Aguilar, Carmen María Rojas González, Ricardo Rozzi, Rocío Santos-Gally, John Schelhas, Sara Scherr, Emily Sigman, Jacob Slusser, Ryan Smith, Rosina Soler, Eric Toensmeier, Mateo Vega, Zoe Volenec, Sheila Ward, Catherine Watson, and Gustavo Zuleta. Many thanks to them for their generosity and dedication.

Sara del Fierro and Ryan Smith, both from Yale University's School of the Environment (YSE), performed multiple roles as most efficient, dynamic, and enthusiastic book editors, assistants, reviewers, and co-authors. Dara Albrecht and Ben Everett-Lane, both at Yale College, majoring in environmental science, generously volunteered their time as dedicated, energetic, and passionate editors, assistants, reviewers, and co-authors. They all made the task feel more important and appreciated, and their contributions and collegiality are immensely appreciated.

Financial, logistical, and administrative support from Yale School of the Environment (YSE) made this book possible.

Finally, this book was written to soothe the grief of losing Sunset, constant and faithful companion whose energy, strength, and perseverance were always contagious and made the ride through life smooth and enjoyable for so many years.

Northford and New Haven, CT, USA
June 12, 2021

Florencia Montagnini

# Contents

# About the Editor

**Florencia Montagnini** has over 30 years of experience researching and teaching in topics on sustainability of managed ecosystems in the tropics, such as forest, tree plantations, and agroforestry systems, with a special emphasis on Latin America. Her work as a scientific advisor and consultant has also taken her to Africa and South East Asia. Her research encompasses sustainable land-use systems that integrate ecological principles with economic, social, and political factors; the principles and applications of forest landscape restoration; the reforestation of degraded lands with native species; identification and quantification of ecological services (biodiversity, carbon sequestration, and watershed protection); organic farming using indigenous resources; biodiversity conservation in human-dominated landscapes; and biodiversity islands. She received her BS in agronomy from the National University of Rosario, Argentina; her master's degree in ecology from the Venezuelan Institute of Scientific Research (IVIC), Caracas, Venezuela; and her PhD in ecology from the University of Georgia. Since 1989, she has worked as a professor and researcher at the Yale School of the Environment, as well as the Tropical Agriculture Research and Higher Education Center (CATIE). She has written 11 books and over 250 scientific articles about the ecology of tropical forests, agroforestry systems, native species reforestation, and forest landscape restoration.

# Part I
# Introduction

# Chapter 1
# Introduction. Biodiversity Islands: Strategies for Conservation in Human-Dominated Environments

**Florencia Montagnini, Brett Levin, and Kjell E. Berg**

**Abstract** This chapter serves to conceptualize, identify and promote implementation of the framing tool we term Biodiversity Islands: ecological refuges where plants and animals can thrive without major interference from human activity, thereby providing ecological, economic, and social benefits at the ecosystem, landscape, and global levels. Design strategies for these biodiversity islands depend on their purpose, as well as on the spatial distribution of reserves throughout the landscape, degree of landscape degradation, species present, and location within the urban-rural spectrum. Biodiversity islands can exist in a range of human-dominated landscapes (e.g., agricultural, wetland, urban) and can range from square meters to many square kilometers in size. In promoting biodiversity islands, we encourage the application of integrated landscape management and inclusive community led action, and discuss possible financial and non-monetary incentives for their implementation and protection. This book presents examples of biodiversity islands from throughout the world and discusses potential difficulties with their use. We expect this book to be useful to researchers, farmers, foresters, landowners, land managers, city planners, and policy makers.

**Keywords** Agroforestry · Biodiversity conservation · Community action · Degraded landscapes · Ecosystem services · Landscape restoration

---

F. Montagnini (✉)
School of the Environment, The Forest School, Yale University, New Haven, CT, USA
e-mail: florencia.montagnini@yale.edu

B. Levin
Dietrick Institute for Applied Insect Ecology, Ventura, CA, USA

K. E. Berg
Bioren Systems, Hampden, MA, USA



F. Montagnini (ed.), *Biodiversity Islands: Strategies for Conservation in Human-Dominated Environments*, Topics in Biodiversity and Conservation 20,
https://doi.org/10.1007/978-3-030-92234-4_1

## 1.1 Introduction

The concept of biodiversity islands helps focus and frame the importance of saving intact sections of land where plants and animals can thrive without major degenerative interference from human activity. A biodiversity island is an area of high biodiversity within ecologically degraded or threatened, human-dominated landscapes. Building upon the foundations of island biogeography theory (MacArthur and Wilson 1967), small and large biodiversity islands act as ecological refugia, protected areas, or reserves within the landscape, with biodiversity in the “island” greatly exceeding the surrounding landscape’s biodiversity baseline.

Historically, the refugia concept was used to describe climatically stable areas in which species survived past Quaternary glacial–interglacial oscillations. The framework of “Anthropocene refugia” extends this concept in recognition of the so-called “Anthropocene” era in which human activities are the dominant driving force on ecosystems and are major factors limiting species distributions. “Anthropocene refugia” then refer to areas that provide spatial and temporal protection from human activities and that will remain suitable for a given taxonomic unit in the long-term. This framework focuses on developing appropriate conservation strategies for wildlife taxa around the world (Monsarrat et al. 2019). As we refer to them here, biodiversity islands go a step further in proposing a framework to contribute to the protection of a multitude of plant and animal species and to the provision of ecological, social, and economic benefits for present and future generations.

The size, configuration, and position of biodiversity islands in the landscape may be mostly opportunistic, determined by their purpose, as well as by patterns of human settlement, development, and utilization or extraction of natural resources. When newly established, planned and designed, their characteristics, i.e. size and configuration may vary according to various scientific guidelines (Laurance 2008; Laurance et al. 2018). Matrix type and quality are also important determinants of taxonomic diversity (Boesing et al. 2018). Biodiversity islands may act as buffer zones between areas of greater human impact or degradation. Multiple biodiversity islands spread over a large area in an optimal configuration can decrease chances of biodiversity loss through creation of repopulation reserves and biological corridors (Harvey et al. 2008).

Biodiversity islands can also be designed and established to serve restoration purposes, as in applied nucleation/island tree planting strategies used to accelerate recovery of degraded tropical forests and pastures (Holl 2002; Holl et al. 2017; Santos-Gally et al. 2019). Depending on their design and specific considerations for their management and preservation, biodiversity islands can effectively serve to protect all levels of biological diversity, including genetic variability, species diversity, functional diversity, as well as ecosystems and landscape diversity (Eibl et al. 2022; Niella et al. 2022; Santos-Gally and Boege 2022).

A key challenge in designing and managing biodiversity islands, as explained in chapters of this book authored by Arroyo-Rodríguez et al., Clavo Peralta et al., Kirby, and a few others, is the question of priorities and tradeoffs inherent in most conservation approaches. Designing a protected area with emphasis on one or few target species may not favor other species with different habitat requirements. Deciding on a specific type of biodiversity island design without considering the broader landscape, or the value of alternative land uses can undermine wider conservation objectives. Biodiversity islands must be always designed and managed with careful attention to both broad and local contexts and objectives.

Biodiversity islands are particularly valuable in areas of biodiversity hotspots. Biodiversity hotspots are regions that contain significantly high concentrations of plant or animal endemism and experience high rates of habitat loss. Because of those features, attention is needed to protect their biodiversity. The 36 global hotspots currently identified represent just 2.3% of Earth's land surface, but they contain around 50% of the world's endemic plant species and 42% of all terrestrial vertebrate species (Mittermeier et al. 2011; Conservation International 2014). Integrating biodiversity islands within these biodiversity hotspots would allow multiple small sites on a relatively small area of land to contribute to significant beneficial biodiversity outcomes.

Although we emphasize tropical forest landscapes, biodiversity is also highly threatened in other types of ecosystems. Temperate grasslands ecosystems, for instance, are especially threatened because they have been long preferred for economic uses and human settlement. Another good example from temperate regions are ancient woodland islands in the British countryside, as described by Kirby in this volume. Wetlands and drylands are at increasing risk as well. In drylands, there are fewer species adapted to the harsh environmental conditions, and these are especially important and difficult to substitute, in contrast with humid forests where many ecological functions are provided by several different species. Other non-forest ecosystems, such as well-designed and managed agricultural systems, grazing areas, non-native tree cover, 'green' infrastructure and eco-friendly built structures, can also improve ecosystem services such as watershed functions, sedimentation control, and carbon sequestration and storage. Therefore biodiversity islands are just as important in non-forest ecosystems, as shown in respective chapters in this book (example Laino et al., Vásquez et al.).

Recognizing, developing, and protecting biodiversity correlates with an increase in economic benefit due to a greater presence of natural capital. Natural capital is the stock of natural assets that make human life possible. These include the environmentally derived provisions necessary for fresh air, food, clean water, medicine, warmth, and shelter. Many of the benefits that may arise through the development of biodiversity islands could be captured by natural capital accounting by placing a monetary value on the environmental or ecosystem services provided. The estimated total value of ecosystem services worldwide has been estimated to be between $125 and $145 trillion USD per year, offering an enormous contribution to human wellbeing (Costanza et al. 2014). Biodiversity plays an important role within this valuation by supplying provisions of food and water, fuels and fibers, genetic resources, medicinal and other biochemical supplies, ornamental products, pollination services, biological control, and habitat for the maintenance of life cycles of migratory species (Kumar 2010). Assessing economic value for biodiversity can guide strategies and policies for preserving ecosystem integrity over time (Pimentel et al. 1997).

While it is important to recognize these economic benefits as a tool for evaluating environmental gain through the development of biodiversity islands, it is also important to recognize the non-monetary benefits. These can take the form of cultural, religious, spiritual, educational, and experiential value within the landscape, in addition to the inherent value of the biodiversity itself. One may view biodiversity islands as natural capital "banks" within the landscape, which preserve and store these important resources not only for their present value but also for the future. For example, a natural area or piece of fallow land that may otherwise be disregarded or developed can be protected as a biodiversity island to provide value to a farm, community, or region for generations to come.

Although the threats to biodiversity are dramatic and broad-ranging, binding international agreements for biodiversity conservation have thus far been largely unsuccessful. The Convention on Biological Diversity (CBD) established a strategic action plan for 2002 to 2010 which ran short of reaching its goals. The 2010 Aichi Biodiversity Targets which aimed to reduce loss of natural habitats and expand nature reserves and other effective area-based conservation reserves from 10% to 17% of the world's land by 2020, have also fallen behind (Watts 2018). Following the 14th Meeting of the Conference of the Parties to the Convention on Biological Diversity (CBD COP 14 Egypt) in 2018 there was a unanimous decision to accelerate action to achieve the Aichi Biodiversity Targets by 2020 at the global, regional, national and subnational levels. The meeting also developed a comprehensive and participatory process for the post-2020 global biodiversity framework to be agreed upon in Beijing at the next Conference of Parties (COP 15) in 2020 (State Information Service 2018).

The post-2020 global biodiversity framework applies a "theory of change" approach, a strategic planning framework used to help plan, implement and evaluate the impacts of the actions taken. It provides a powerful tool for organizing measurable goals and solutions, and for evaluating both short-and long-term impacts in a consistent, meaningful and transparent structure (https://www.cbd.int/doc/c/efb0/1f84/a892b98d2982a829962b6371/wg2020-02-03-en.pdf). In this context, it becomes imperative to increase efforts to promote biodiversity conservation strategies, such as biodiversity islands, that can be effective at the local, regional and global levels.

The interaction between climate change and habitat isolation (mainly caused by habitat loss; Fahrig 2003) adds another significant threat to biodiversity conservation. Restoration of degraded ecosystems across the world can work towards addressing climate change and biodiversity loss (Lovejoy and Hannah 2019). For example, favoring natural regeneration and reforestation of degraded lands can contribute to carbon sequestration, as well as to recovery of habitats and biodiversity. The work of recognizing and developing biodiversity islands provides one step in the restoration process, which can contribute to reversing isolation and increasing biodiversity worldwide.

Natural resource managers urgently need a synthesis of our rapidly growing understanding of these issues in order to promote strategies for biodiversity conservation in human-dominated landscapes. Building on existing literature on relevant conceptual aspects, as well as on case studies and experiences with biodiversity conservation throughout the world, this book serves to identify and promote strategies for framing and planning biodiversity islands within a landscape. We advocate for the adoption of biodiversity islands in human modified landscapes, whether private, public, rural, or urban. We present examples of biodiversity islands at several different scales throughout the world and discuss potential difficulties with the use of this framework. We expect this book to be useful to researchers, farmers, foresters, landowners, land managers, city planners, and policymakers.

## 1.2 Design Considerations: Configuration, Size, and Location in the Landscape

Design strategies for biodiversity islands depend on many factors including their purpose, patch dynamics and spatial distribution of reserves throughout the landscape, the degree of site degradation, species present, and their locations within the urban to rural spectrum. Our understanding of the relationship between spatial distribution of reserves throughout fragmented landscapes and biodiversity conservation has evolved significantly in recent years. The Theory of Island Biogeography, developed by MacArthur and Wilson (1967), first framed the dynamics of population ecology and speciation across fragmented landscapes of marine islands through time around the factors of island size, distance from other islands, and methods of species dispersal and immigration. This framework has since been applied numerous times to terrestrial ecosystems worldwide to understand population dynamics and diversity in areas where reserves act as "islands" throughout fragmented landscapes.

For example, Diamond (1975) recognized patch dynamics as an essential component in reserve design. A minimum dynamic area must provide recolonization sources for relevant species diversity to prevent extinction (Pickett and Thompson 1978). Open areas, such as pastures and annual croplands, decrease landscape connectivity and resource availability for forest species and recolonization (Dunning et al. 1992).

More recently however, habitat fragmentation research has gone beyond island biogeographic theory (Laurance 2008). Island biogeography provides few predictions about how community composition in fragments should change over time, and which species would be most vulnerable. In addition, edge effects can be an important driver of local species extinctions and ecosystem change. The matrix of modified vegetation surrounding fragments can strongly influence fragment connectivity, affecting the demography, genetics, and survival of local populations. Most fragmented landscapes are also altered by other anthropogenic changes, such as hunting, logging, fires, and pollution, which can interact synergistically with habitat fragmentation (Laurance 2008).

The "island" metaphor has been long used because of its simplicity in communicating the concept to a wide range of stakeholders. However, the concept is complex and multifaceted (Haila 2002), and varies by location and habitat, as different contributors of this book demonstrate in their respective chapters (Kirby 2022; Santos-Gally and Boege 2022; Montes-Londoño et al. 2022; Negret et al. 2022; Soler et al. 2022; and others).

The scale of biodiversity islands can range from square meters to thousands of square kilometers (Table 1.1). Examples presented as case studies in this book range from small persisting biodiverse land patches in the British countryside as described by Kirby, to relatively larger protected areas in tropical forest locations in Costa Rica, Peru, Ecuador as shown in contributions by Newcomer et al., Clavo Peralta et al., Esbach et al. respectively, and several in between. There has been significant debate around promoting increased biodiversity through few large reserves versus more numerous small reserves (Simberloff and Abele 1982; Tjørve 2010). Research as to whether single large or several smaller reserves (SLOSS) are superior, also known as "the SLOSS debate," led to the development of the Biological Dynamics of Forest Fragments Project (formerly the Minimum Critical Size of Ecosystem Project), which monitors features of multiple size fragmentation regimes in Amazon forests over time (Lovejoy and Bierregaard 1990; Laurance et al. 2018).

This unresolved debate is critical for prioritizing conservation actions, particularly for increasingly disturbed biodiversity hotspots. Recent research results suggest that small patches have greater importance for biodiversity conservation than previously anticipated (Arroyo-Rodríguez et al. 2009; Hernández-Ruedas et al. 2014; Volenec and Dobson 2019; Wintle et al. 2019). Results from the Lacandona rainforest of Mexico, where the effect of forest patch size on species density of different taxonomic groups was examined, support this assessment (Arroyo-Rodríguez et al. 2022).

Recent study further supports the "habitat amount hypothesis," which emphasizes total habitat area within a landscape (Watling et al. 2020). This theory argues that the species density (i.e., number of species in plots of fixed size) is more strongly and positively related to habitat amount (e.g., forest) in the landscape than to the size of the patch in which the plot is located (Fahrig 2013). This suggests that the greater the total area of reserves within a fragmented landscape, the greater the potential biodiversity, whether it be numerous small reserves, fewer large reserves, or any mixture of these sizing parameters (Fahrig 2013, 2017).

In addition, many remaining forest patches are non-randomly distributed and strongly correlated with topography, and soil types and quality due to human disturbances. Many fragments are located at inaccessible or less desirable areas at higher elevations and on steeper slopes, with varying slope aspects. The spatial distribution of forest fragments influences tropical tree conservation, as fragment location is the main driver of tree species maintenance within landscapes (Liu and Slik 2014).

**Table 1.1** Examples of different types of existing biodiversity islands throughout the world

| Name | Scale | Location | Landscape | Species | Objective | Source |
|---|---|---|---|---|---|---|
| Islands of nature | Several square kilometers | Worldwide | Indigenous territories | Native forest species | Sustainable food production and biodiversity | United Nations IPBES (2019) |
| Private Forest reserves | Hectares | Organic yerba mate farms, Misiones, Argentina, and other, in many locations worldwide | Agriculture and Forest | Native forest species | Sources of pollinators, biological pest control, water protection, ecotourism | Montagnini et al. (2011), Montagnini et al. (2022) |
| Islands of resilience | Hectares | Santa Fe Province, Argentina | Extensive industrial agriculture | Exotic and native species | Restoration of biodiversity | Libertario González, pers. comm. (June 2019) and González et al. (2022) |
| Dayak gardens | Hectares | Indonesia | Managed forest | Native species | Long term production of crops, fruits, and timber | Peters (2018) |
| Forest windbreaks | Hectares | Several locations worldwide | Agricultural | Native and exotic species | Windbreak, connectivity | Montagnini and del Fierro (2022) |
| Hutan Desa | Hectares | Indonesia | Native managed forests and agricultural | Native and production species | Protection of community resources | Moeliono et al. (2015) |
| Church forests | 3–300 ha | Ethiopia | Deforested land | Native species | Fertile oasis | Abbott (2019) and Baez Schon et al. (2022) |
| Religious sites and sacred groves | Vary | Several locations worldwide (see text) | Native forests, lakes, rivers | Native species | Spiritual, conservation | Several authors, see text |
| Native tree Species Islands | Square meters | Los Tuxtlas, Veracruz, Mexico | Degraded pastures restored to Silvopastoral systems | Native tree species | Pasture restoration and productivity | Santos-Gally et al. (2019) and Santos-Gally and Boege (2022) |

(continued)

**Table 1.1** (continued)

| Name | Scale | Location | Landscape | Species | Objective | Source |
|---|---|---|---|---|---|---|
| Patches of trees (nucleation) | Square meters | Latin America / tropics | Degraded agricultural or pasture | Native tree species | Landscape restoration | Holl (2002) and Holl et al. (2017) |
| Isolated trees in pastures | Square meters | Valle del Cauca, Colombia, and many other locations | Degraded pastures | Pioneer native tree species | Catalyze (jump start) succession | Esquivel and Calle (2002), others |
| Edible Forest garden | Square meters | New England, USA | Urban and suburban | Native and exotic species | Sustainable food production and biodiversity | Toensmeier and Bates (2013) and Toensmeier (2022) |

The edge effect within the matrix influences a set of environmental conditions that may be favorable to different species and highly detrimental to others (Tuff et al. 2016; Laurance et al. 2018). For example, Pfeifer et al. (2017) demonstrates that edge effects can be both positive and negative, with highly endangered species tending to show stronger negative responses to edges. Overall, increasing vegetation cover types in the matrix is an efficient conservation strategy for maintaining higher biodiversity levels in fragmented landscapes (Boesing et al. 2018). Also, a certain amount of disturbance may provide additional niches for greater biodiversity to be housed (Fahrig 2017).

In human managed landscapes, an ideal biodiversity friendly landscape could be an area with significant continuous forest cover in a single large reserve, along with sections of fragmented landscapes of different sized patches (Melo et al. 2013). Land-use intensification can be mitigated by maintaining isolated trees, living fences, and a limited number of open areas within the heterogeneous anthropogenic matrix. Local efforts to avoid chronic disturbances and decrease further intensification may include protection against excessive hunting, unsustainable logging, and mismanaged firewood extraction (Fig. 1.1).

**Fig. 1.1** Fragmented landscape on the road to Meru District, near Nairobi, Kenya, where the World Agroforestry Center promotes tree planting and agroforestry systems among small farmers and farmers' associations to improve livelihoods, enhancing soil fertility, biodiversity and other ecosystem services (www.worldagroforestry.org). (Photo: F. Montagnini)

Computational methods have been developed for understanding patterns of biodiversity throughout reserve networks. Site-selection algorithms provide computational methods to identify reserves containing as many species as possible to maximize the representation of species (Cabeza and Moilanen 2001). Part of the method utilizes irreplaceability indices and vulnerability indices for determining species' conservation value. Such innovations may prove valuable with greater technological advancement through time.

Other factors to take into consideration in the design of biodiversity islands are the degree of degradation and the species present in the targeted landscape. Monitoring of key plant and animal species is important to understanding how such population dynamics reflect the baseline of biodiversity within the area prior to degradation. After determining the degradation status, it is possible to decide the most feasible and appropriate restoration pathway (Clewell et al. 2004). Strategies should reflect past land use history and presence of damaging invasive species, when applicable.

When biodiversity islands are designed and established to serve restoration purposes, as in applied nucleation or tree island planting strategies, several factors can influence their design and effectiveness. This approach is based on nucleation theory (Yarranton and Morrison 1974), a natural recovery process where pioneer shrubs and trees establish patchily and facilitate recruitment via enhanced seed dispersal and improved establishment conditions. Patches spread outward clonally and/or by facilitating the colonization of later-successional species. The use of "tree islands" or applied nucleation has shown a great deal of promise as a restoration approach, given that it simulates a natural recovery pattern and reduces tree planting costs. Recently this strategy has been tested to discern to what extent restored plant communities are similar to reference forests (Holl 2002; Holl et al. 2017). It has also been successfully applied to restore and add biodiversity to tropical pastures (Santos-Gally et al. 2019; Santos-Gally and Boege 2022), as well as to restore degraded riparian and open forests in temperate regions. These nucleation strategies provide clusters or islands of habitat, which can help to restore forests and increase their connectivity.

## 1.3 Biodiversity Islands in Rural and Urban Settings

Biodiversity islands may exist anywhere within the rural to urban spectrum. In human-dominated landscapes, the degree and matrix of fragmentation is context specific, determined by various social and ecological dimensions. Whether urban or rural, the protection of natural areas and restoration efforts are both needed for further creation of biodiversity islands. The first and most crucial step in doing so is recognizing and protecting these biodiversity islands within the landscape as valuable assets.

Creation of government administered parks, private reserves, and voluntarily conserved lands allow for large biodiversity islands to emerge within rural, cultivated contexts. In addition, many farmers deliberately leave portions of original forest cover in their properties to serve as a refuge for pollinators, to provide a source of beneficial insects and birds that can contribute to pest control, and to protect water sources. For example, in Misiones, Argentina, several organic farms feature these relatively large areas that sometimes also serve ecotourism purposes (Montagnini et al. 2011) (Table 1.1). Farmers may also have motivations such as personal conviction along with ethical, cultural, or spiritual drivers that lead them to create and protect reserves in their private lands.

In urban and built environments, including buildings, housing, bridges, roads, airports and other, biodiversity islands such as urban green infrastructure, greenways, habitat networks, parks, and planted trees, can provide important habitat for wildlife and other species and offer other services, such as enhanced climate resilience. Communities can play a leading role in these efforts to ensure that they are initiated and managed sustainably and locally.

In both rural and urban settings, relict areas that have since been engulfed by development can function as biodiversity islands. For example, the old Governor's House in the center of Chennai (Madras), India has ancient forest remnants where even a new fungus on a new insect species has been collected (David Hawksworth, personal communication, October 2019). Similarly, the old farm houses that were abandoned as they were surrounded by the extensive monoculture agriculture systems that now dominate the landscape in the "pampas" or Argentinian plains are now covered by diverse vegetation that has repopulated the area and harbors rich fauna. Currently, researchers from INTA[1] are studying the biodiversity of these "resilience islands", as they call them, and are proposing ways to integrate them in the productive landscapes of the region (González et al. 2022).

### *1.3.1 The Contribution of Sustainable Agriculture to Biodiversity Islands*

Sustainable agricultural management techniques geared toward harmonizing ecosystem productivity and biodiversity conservation can contribute to mitigating or reversing detrimental effects of human impacts on landscapes (Montagnini and Berg 2019). Incorporation of biodiversity friendly land uses into actively managed buffer zones or biological corridors can contribute to the long-term conservation value of protected areas (DeFries et al. 2007; Harvey et al. 2008; Chazdon et al. 2009; Clavo Peralta et al. 2022; Giraldo et al. 2022; Laino et al. 2022, among others). In landscapes lacking protected areas or intact forests, agriculture, agroforestry,

[1] The National Institute for Agricultural Technology (Instituto Nacional de Tecnología Agropecuaria, INTA) is an Argentine federal extension agency in charge of the generation, adaptation and diffusion of technologies, knowledge and learning procedures for agriculture, forest and agro-industrial activities within an ecologically clean environment.

remnant vegetation, plantations, wetlands and managed forest patches provide critical habitats and refugia for biodiversity (Harvey et al. 2006, 2008; Harvey and González 2007; Bhagwat et al. 2008; Chazdon et al. 2009; Levin 2022a).

For example, the planting of diverse flowering insectary hedgerows may serve as an optimal buffer zone between land sparing areas and agricultural areas on a farm. Insectary hedgerows can offer diverse sources of pollen and nectar, creating a balanced on-farm insect ecology inclusive of beneficial predatory insects that help to manage pests (Chandler et al. 1998; Landis et al. 2000). Other interventions may include agroforestry systems (AFS), which may contribute to higher species diversity within a relatively homogeneous landscape (Galluzzi et al. 2010; Udawatta et al. 2019; Montagnini 2020; Montagnini and del Fierro 2022).

In recent years, the general consensus on the escalating rates of biodiversity loss in agricultural systems has given rise to an increase in the number of international, national and local actions related to biodiversity management in agricultural systems. A few notable initiatives among these include the World Project for Farm Conservation at Bioversity International (formerly IPGRI, The International Plant Genetic Resources Institute); the People, Land Management and Environmental Change Project (PLEC); the network for Development and Conservation of Farm Biodiversity (CBDC); the International Center for Tropical Agriculture (CIAT); the Institute for Soil Biology and Fertility (TSBF); the World Project for Pollinators supported by the FAO; and other projects supported by the Global Environment Facility (GEF), among several others (Jarvis et al. 2011; Montagnini et al. 2022).

Many national and regional rural development programs also recognize the importance of biodiversity and thus include biodiversity conservation goals, for example, SANBIO in South Africa (https://www.nepadsanbio.org/) which is working in several African countries on projects to improve local livelihoods. In addition, some large World Bank-funded projects, and the Green Climate Fund pipeline project for landscape restoration (https://www.greenclimate.fund/) that currently focus more on carbon, could be (and in some cases are) incorporating explicit biodiversity objectives.

### *1.3.2 The Role of Agroforestry in Biodiversity Islands*

Agroforestry systems (AFS), which combine trees and crops on the same land, can increase productivity in the short and long term while promoting biodiversity and bringing social, environmental and economic benefits to the farmer and society (Montagnini and Metzel 2017; Montagnini 2020; Montagnini and del Fierro 2022). AFS are often important components of buffer zones of protected areas. Thus, they can be a great tool for biodiversity islands. The most common AFS are shaded annual and perennial crops, silvopastoral systems, live fences, and windbreaks. The characteristics of the systems vary greatly according to their design, objectives, species involved, and regions. Caution should therefore be taken when deciding on design and management of AFS to be included as component parts of biodiversity islands.

Several examples of indigenous multistrata AFS that are closely integrated within the forest landscape show remarkable biodiversity (Redford and Padoch 1992). These types of multistrata AFS may constitute a kind of biodiversity island on their own. Alternatively, they may be integrated within biodiversity islands or may be used to create buffer zones. Other types of multistrata AFS, including homegardens and shaded perennial crops such as cacao, coffee and yerba mate, can also contribute significantly to biodiversity conservation (Fig. 1.2) (Montagnini and Berg 2019; Montagnini 2020; Montagnini and del Fierro 2022).

Likewise, silvopastoral systems (SPS) can be easily integrated with other landscape level strategies, such as connectivity corridors, in order to conserve biodiversity and enhance other environmental services within agricultural landscapes (Fig. 1.3). In many regions where cattle production predominates, several unprotected forest remnants of high conservation value are often embedded within a matrix of cattle grazing areas made up of pasture monocultures with few trees. In

**Fig. 1.2** Homegarden and nursery in Nkenlikok, Cameroon. The World Agroforestry Centre introduced agroforestry methods to rural farmers in the central African country some 20 years ago as part of projects intended to ensure smallholder households increase their use of trees in agricultural landscapes to improve food security, nutrition, income, health, shelter, social cohesion, energy resources and environmental sustainability (Montagnini and Metzel 2017). Several traditionally used tree and other plant species are reproduced in farmers' nurseries aiding to in-situ conservation of local genetic resources. (Photo: F. Montagnini)

**Fig. 1.3** A traditional silvopastoral system in Indonesia: buffaloes grazing under rubber trees near Lubuk Beringin Village, Bungo Regency, Jambi Province, Sumatra. The first village forest (Hutan Desa) with 2356 ha was inaugurated and designated in 2009 under the management of Lubuk Beringin village administration. Under the Forestry Minister Regulation No. p49/Menhut-II/2008, village communities can be granted legal right to manage state forests for their own prosperity (World Agroforestry Centre (ICRAF), Bogor (Indonesia) 2010). Livestock in small farms in the village include goats, sheep, chickens and buffalo. (Photo: F. Montagnini)

such areas, SPS with complex vegetation structures can support important levels of biodiversity (Harvey et al. 2005, 2006; Sáenz et al. 2007) and provide ecosystem services such as natural pest management, carbon sequestration, water and soil conservation, nutrient cycling, hydrological protection, and crop pollination (Chazdon et al. 2009; Calle et al. 2010). Hence it is possible to enhance biodiversity by strategically placing elements such as live fences, scattered trees, riparian buffers, and connectivity corridors within the landscape (Esquivel and Calle 2002; Murgueitio et al. 2011; Calle and Holl 2019; Santos-Gally et al. 2019; Calle et al. 2022; Giraldo et al. 2022; Santos-Gally and Boege 2022) (Table 1.1).

Windbreaks and living fences are examples of AFS that connect portions of remnant forests in agricultural landscapes (Francesconi et al. 2011). Windbreaks tend to be favored by farmers and can be instrumental in biodiversity conservation and landscape connectivity in fragmented areas. Windbreaks therefore have the potential to play similar roles in the connectivity of biodiversity islands across a landscape (Montagnini 2020; Montagnini and del Fierro 2022).

The low intensive management multistrata AFS with native species have the greatest potential to harbor high amounts of biodiversity. Biodiversity decreases as management intensity increases in more homogeneous tree crop/animal combinations. However, even the less heterogeneous types of AFS provide greater biodiversity than would otherwise be realized in conventional monoculture agriculture or in degraded landscapes. In addition, farmers value and protect AFS because of their contributions to their livelihoods, thus ensuring the conservation of services they provide, including biodiversity (Montagnini 2020; Montagnini and del Fierro 2022).

### *1.3.3 Biodiversity Islands in Urban Environments*

Urban areas are highly modified and complex landscapes, within which green or open areas are valuable for human well-being as well as wildlife (Pickett et al. 2004). Urban Green Infrastructure (UGI) provides important contributions to the sustainability of urban systems and offers a nature-based solution to reduce the impact of rapid urbanization, a consequence of the increased rates of rural to urban migrations happening today worldwide. Urban species diversity and richness and the use of green spaces for biodiversity conservation are of paramount importance for developing climate-resilient cities (Biazen 2015). The UGI may include several types of vegetation assemblages in parks, playgrounds, sidewalk tree plantings and even cemeteries when specific efforts are made to increase their aesthetic value to visitors (https://www.nationalgeographic.com/animals/2019/10/cemeteries-home-to-diverse-plants-animals/).

Indicators of the size of the UGI, such as the amount of green space per person, are used by funding institutions such as the Interamerican Development Bank (IADB, https://www.iadb.org) as criteria to decide on financing specific urban projects. For example, the Emergent and Sustainable Cities IADB initiative has designated the city of Rosario, Argentina as the greenest city of the country, with an average of 12 $m^2$ of green space per inhabitant. In contrast, in the capital city of Buenos Aires, there are only 3.5 green $m^2$ per person (https://buenavibra.es/por-el-mundo/destinos/argentina/rosario-fue-declarada-la-ciudad-mas-ecologica-de-argentina/). The city of Rosario, which has a population of about 1.2 million covers a total of 178 $Km^2$, with a total of 2070 streets, avenues and short alleys (Fig. 1.4). There are a total of 250 green municipal areas, with the Independence Park (Parque de la Independencia) of a total of 126 ha standing out for its floristic richness. The growth of urban vegetation has increased over the last three decades thanks to tree planting as well as the creation of new green areas (https://fcagr.unr.edu.ar/?p=13370).

**Fig. 1.4** View of downtown Rosario, Argentina showing flowering trees and other vegetation of the abundant urban green infrastructure. (Photo: "Carácter" by Sebastian Infante; www.sebastianinfante.com). This picture won the first prize in a contest organized in November 2019 by Aehgar, Asociación Empresaria Hotelera Gastronómica y Afines Rosario (Hotels, Gastronomy and Related Empresarial Association of Rosario)

Cities provide habitats for a rich and diverse range of plants and animals. Enhancement of biodiversity in urban ecosystems can have a positive impact on the quality of life and education of urban dwellers and thus facilitate the preservation of biodiversity in natural ecosystems. An essential first step to managing biodiversity in urban environments more effectively is to gain a full understanding of the interplay between landscape (matrix effects) and local factors (patch effects) that affect urban biodiversity (Savard et al. 2000).

Many cities have a network of habitat fragments or urban greenways comprising areas of semi-natural habitats, secondary succession, ruderal and pioneer environments and open areas. These habitats may be important features for biodiversity both as stable and as transient habitats (McIntyre et al. 2001), and may also be valuable for their possible function as corridors and stepping stones to facilitate species dispersal.

In urban landscape planning, urban greenways and wildlife corridors are increasingly advocated for in order to encourage animals and plants to move around urban areas, thereby preserving or enhancing urban biodiversity. Urban planning needs to consider the design and establishment of greenways to serve as dispersal route ways as well as habitat, providing a chain of different habitats that permeate the urban environment. City planners can have a positive impact on urban biodiversity by slowing the pace of redevelopment (Angold et al. 2006).

The establishment of parks and the integration of ecological landscape design bring increased biodiversity to suburban and urban settings. Actively designing or setting aside areas within an urban park to serve as biodiversity islands can contribute significantly to biodiversity conservation. Some community led efforts, sustained and managed locally, can also integrate biodiversity islands and make substantial contributions to biodiversity conservation. For example, in Philadelphia, the long-established Fairmont Park contains a wide range of species and habitats such as pollinator and butterfly gardens within the extensive urban forests following the Schuylkill River and Wissahickon Creek (https://myphillypark.org). Similarly, the Urban Oases in the New Haven Watershed program in New Haven, CT bring together a variety of community partners to establish a network of green spaces for wildlife and community members at parks throughout the city (http://newhavenwildliferefuge.org/).

Planting trees and increasing vegetation in cities can create small scale biodiversity islands. These efforts also have social and educational value. For example, the Urban Resources Initiative, a university non-profit collaboration based in New Haven, CT, USA, is one of many likeminded organizations that plants trees while working with the local community, thereby increasing habitat for biodiversity (https://uri.yale.edu).

Similarly, in Philadelphia, PA, USA, Philly Peace Park is an example of a neighborhood-managed ecology campus championing sustainable production of food while fostering education and community involvement (www.PhillyPeacePark.org). As another example, the small town of Salliqueló in the Buenos Aires Province in Argentina has carried out a tree-planting program called "Nace un niño, nace un árbol" ("A child is born, a tree is born") for 24 years, with over 3000 trees planted so far (cypnoticias.com.ar. Salliqueló, una ciudad que piensa en verde).

In urban settings, gardens can play a variety of roles as biodiversity islands at different scales, including at the landscape level, contributing to conservation of forest habitat and creating new habitats for other species. In many cases this is attained by improving connectivity between remaining natural forest fragments within city limits, thereby creating "stepping stones" across the urban landscape and reducing the edge-effect at forest boundaries (Negret et al. 2022; Toensmeier 2022). Urban green areas potentially serve as venues for promoting not only local biodiversity conservation but also novel species assemblages that are different from those found in pristine areas (Soler et al. 2022).

## 1.4 Landscape Management Approaches for the Implementation of Biodiversity Islands

Biodiversity islands can exist in a wide range of human-dominated landscapes. They may be actively implemented as part of a complex landscape design that may include agriculture, forests, plantations, wetlands or other uses, or they may be the result of a passive management practice, i.e., as part of the landscapes that were left untouched for practical or other reasons. The method of implementation is important, as it will determine the characteristics, position in the landscape, and management of the biodiversity island.

Land management decision-making may take many different forms. Public, private, and communal land each involve different stakeholders as key decision makers who may choose to implement biodiversity islands. An integrated landscape approach is needed to understand the effects of landscape structure and dynamics on conservation of biodiversity, provision of ecosystem services, and sustainability of rural livelihoods (Tscharntke et al. 2005; Chazdon et al. 2009). Integrated landscape management (ILM) is an approach inclusive of various stakeholders that can be instrumental in furthering the creation of biodiversity islands. ILM supports engagement across sectors and scales, increases broad multilateral coordination, and ensures harmonization of planning, implementation and monitoring processes at the landscape, regional and national levels (Thaxton et al. 2016).

With the ILM approach, landscapes can be designed and managed to secure a future for both biodiversity and rural livelihoods. This rationale can be used in planning strategies to promote the adoption and development of biodiversity islands. Landowners can integrate small scale land sparing to set aside pieces of the property as untouched natural settings that can act as biodiversity islands. On the other hand, when coexistence is identified as an end goal, land sharing may allow wildlife to thrive in human-dominated landscapes (Crespin and Simonetti 2019). While land sparing (e.g., protected areas) is needed to protect biodiversity, land sharing (e.g., agroforestry systems) may contribute to the maintenance of ecological services. The debate between land sparing and sharing continues, with several authors advocating for the land sparing alternative for the sake of conservation (Phalan 2018; Phalan et al. 2011). For a given area, the question could be resolved with consideration of site-specific conditions of habitat disturbance and development, e.g., conventional pastures versus degraded lands versus improved sustainable cattle raising versus reserves. Both land sparing and land sharing as complementary strategies can provide valuable protection of species diversity over time (Grass et al. 2019).

Strategies to promote biodiversity islands at the landscape level may include mitigating threats to biodiversity loss in agricultural landscapes; protecting, diversifying, and sustainably managing tree cover within an agricultural matrix, including all types of agroforestry systems (AFS); promoting and conserving indigenous, traditional, and ecologically based agricultural practices; and restoring degraded lands (Harvey et al. 2006, 2008; Chazdon et al. 2009).

The notion of a "landscape approach" is not new, but in recent years it has gained in importance and is a major topic of national and international policy discourse (Denier et al. 2015). The landscape may be the most appropriate scale for action between national and local scales. A landscape approach, using ILM, can allow stakeholders to decide on land and water use in such a manner that community, commercial and conservation interests are more balanced and sustainable. ILM arose from diverse, innovative strategies – from indigenous territorial development, to integrated watershed management, to landcare. It involves new levels of collaboration through place-based partnerships that are inclusive of communities, governments, businesses, and land managers (Montagnini et al. 2022).

Many examples of applications of ILM with a strong biodiversity focus have been documented recently (Denier et al. 2015). For example, in northwest Vietnam, the Australian Centre for International Agricultural Research (AGIAR) and the Consortium of International Agricultural Research Centers (CGIAR) have introduced agroforestry techniques into monocropped landscapes at the farm level, while also providing techniques for the expansion of AFS to the landscape level. The techniques include "training of trainers," championing farmers, organizing farmer field days and setting up community tree nurseries. Agroforestry was scaled up in collaboration with provincial governments and local farmers. The project not only diversifies income at the farm level, but also provides important environmental services at the landscape scale, such as reduced pressure on forests for timber, reduced soil erosion and protection against storms (http://www.worldagroforestry.org/project/agroforestry-livelihoods-smallholder-farmers-northwest-viet-nam).

The landscape approach to development and conservation is proliferating worldwide, particularly in the past five years (Sara Scherr, EcoAgriculture Partners, personal communication, January 2020). ILM initiatives in Africa, Latin America/Caribbean and South/Southeast Asia have inspired the creation of new partnerships that emphasize biodiversity (Denier et al. 2015). International conservation-oriented organizations working in this field include, among others, the African Wildlife Foundation (https://www.awf.org/), which uses a landscape approach to conservation to improve the livelihoods of local people, and the International Union for Conservation of Nature (IUCN)'s Livelihoods and Landscapes Strategies, which address human and environmental needs in large areas of land with a special emphasis on the sustainable use of forests (https://www.iucn.org/asia/thailand/countries/thailand/livelihoods-and-landscapes-strategies).

The new alliances inspired by the ILM approach promote collaboration in research, in the co-design of conservation programs and policies, and in management of human-modified landscapes. This can be accomplished using participatory and multidisciplinary approaches in research and management (Chazdon et al. 2009; Esbach et al. 2022; Newcomer et al. 2022; Painter et al. 2022; Vásquez et al. 2022).

In addition, socioeconomic, legal, and political actions can be used to promote biodiversity islands at the landscape level. These may include, for example, the use of economic instruments, such as payments for ecosystem services, as discussed in Sect. 1.6 of this chapter, or improving environmental laws and enforcement to reduce deforestation, regulate logging, conserve on-farm tree cover, and reduce agrochemical use (Sheban 2022; Sigman 2022). Action can also be taken to promote ecologically sustainable production systems such as agroforestry or to leverage local and regional political support for existing initiatives for biodiversity protection (Newcomer et al. 2022).

## 1.5 How Community Led Action Can Advance the Development of Biodiversity Islands

Community led action can contribute to the development of biodiversity islands. Motivations for such action may be inherent within cultures or learned and applied with ethical, philosophical, or scientific motivations. Land access and land tenure provide the basis for community led efforts, while other powerful instruments and legal tools to be used in service of conservation include education, political engagement, and conservation easements. Community developed biodiversity islands may be governed as public lands, commons, cooperatives, or through tactics of community based conservation, in which social complexities are appreciated and democratic decision-making processes are used. Examples of grassroots community action for the advancement of conservation practices are numerous, diverse, and worldwide (Otto et al. 2013; Levin 2022b; Morales-Nieves 2022). Both non-profit and for-profit organizations are capable of further expanding community led action for the development of biodiversity islands.

The types of legal frameworks, methods of enforcement, modes of implementation, and levels of community engagement that can be used to protect and manage biodiversity islands are site and context dependent (Levin 2022b). Conservation easements are one such tool that may be advocated for by a community for the development of biodiversity islands. Conservation easements are voluntary legal agreements between landowners and a land trust or government agency that protect conservation values on a property by permanently limiting uses of the land and offering tax incentives to the land owner. A land trust is a non-profit organization that acquires land or conservation easements through support from donations or government funding. In many instances, a land trust can also act as a conservation organization to help draft, implement, and ensure compliance with the easement.

Conservation easements may include land uses such as recreation, forestry, agroforestry, or agriculture. These easements, which provide conservation value in addition to opportunities for revenue generation, are known as working land conservation easements. Throughout the world, communities can support and develop local land trusts to assist in the development of biodiversity islands by supporting landowners in putting their land into conservation. One of many such examples is the Harrison Family of North Branford, CT, who recently transferred ownership of their 8.5 ha property to the North Branford Land Conservation Trust (NBLCT) (www.NBLandTrust.org). The family donated the land, although it was prime for real estate development, so that it could be protected and preserved in its natural state in perpetuity (Figs. 1.5 and 1.6) (Totoket Times 2018). As such, the Harrison Farm, now under the care and maintenance of NBLCT, serves as a true biodiversity island in an otherwise suburban landscape.

Religious sites and sacred groves are other examples where community led action protects biodiversity rich sites on a landscape in ways that can be considered biodiversity islands. Examples include the Mizorom sacred groves in Northeastern India, sacred pools protected by Tchabè communities in Central Benin, sacred cacao

**Fig. 1.5** A section of forest recently donated to the North Branford Land Conservation Trust (NBLCT) by the Harrison Family and protected in perpetuity by a conservation easement. (Photo: F. Montagnini)

groves of the Maya, and other sacred groves in Zimbabwe, Ghana, Thailand, China, and Nepal (Gómez-Pompa et al. 1990; Gadgil et al. 1993; Bhagwat and Rutte 2006; Ceperley et al. 2010). One such successful example is church forests in Ethiopia (Table 1.1), which are small pockets of protected forest surrounding orthodox churches. They number over 35,000 and are spread all over the country. These small but fertile oases, which range from 3 to 300 ha, are some of the last remnants of the tall, lush native forests that once covered Ethiopia, and which, along with their biodiversity, have all disappeared (Abbott 2019; Baez Schon et al. 2022).

Other community protected sites include the village forests in Indonesia, known as Hutan Desa, which are legally recognized for the ecosystem services and benefits to society they provide. Their management and protection are guided by traditional communal governance as well (Moeliono et al. 2015). Although the original purpose in most cases was not to create biodiversity islands, they may be developed as a consequence of such community actions.

There are numerous examples of organized community efforts to support biodiversity outcomes that align with the creation of biodiversity islands (UNDP 2006; Levin 2022b). From the Zapatistas farmers movement in southern Mexico (Esteva 1999), to community restoration projects in Northern Ethiopia (Ethiopian Biodiversity Institute 2014; Sigman 2022), to the newly emerging global concept of Ecosystem Restoration Camps (https://www.ecosystemrestorationcamps.org/), rural

**Fig. 1.6** The Harrison Family during a ribbon cutting ceremony at the newly created Harrison Farm Preserve, celebrating the transfer of ownership from the family to North Branford Land Conservation Trust (NBLCT). (Photo: F. Montagnini)

community action provides alternative methods for enhancing biodiversity outcomes in rural settings (Nazarea et al. 2014).

Non-profit organizations such as EcoAgriculture Partners (https://ecoagriculture.org/) and The Forests Dialogue (https://theforestsdialogue.org/), aim to help engage communities for biodiversity conservation in rural settings through the ILM approach previously described. For example, in their new initiative, "1000 Landscapes for 1 Billion People. A Radical Collaboration for Resilient Communities and Restored Nature", EcoAgriculture supports locally led initiatives, for example, via their Landscape Action Platform co-designed with communities (Concept Note for Discussion. EcoAgriculture Partners and Rainforest Alliance, September 2019).

Numerous other international non-profit organizations, such as Commonland (https://www.commonland.com/), Conservation International (https://www.conservation.org/), Solidaridad (https://www.solidaridadnetwork.org/), World Wildlife Fund (https://www.worldwildlife.org/), and Wetlands International (https://www.wetlands.org/), just to name a few, promote practical strategies for local use of natural resources in harmony with biodiversity conservation in many of their projects around the world. Likewise, networks of landscape initiatives, for example, the International Model Forest Network (https://imfn.net/), and the Satoyama Initiative (https://satoyama-initiative.org) work with and connect member organizations worldwide to promote ways for local communities to manage their natural environment for production activities, to nurture their traditions and culture, and to maintain ecosystems and biodiversity.

In suburban settings, parks, home gardens, community garden projects, and ecovillages also act as biodiversity islands. In the urban context, the rise of urban forestry, parks, urban community gardens, and educational centers which support these ends continue to grow in popularity. Along this wide spectrum of population density, there is great potential for the continuation of community led action to promote and sustain biodiversity islands.

## 1.6 The Way Forward: Valuing and Financing Biodiversity Conservation

Biodiversity conservation is in great peril due to significant gaps in what is available and what is needed for financing conservation in a feasible and tangible way. According to current estimates, for example, conserving biodiversity to meet the Aichi targets and other conservation efforts worldwide would require between $150 and $440 billion (CBD High-Level Panel 2014; Credit Suisse et al. 2014); with some estimates as high as $977 billion (Talberth and Gray 2012). Meanwhile, only about $30 billion has been spent on biodiversity in the 2010 decade (i.e., only 3+ billion dollars/year) via different mechanisms, such as biodiversity-relevant taxes, PES programs, biodiversity offsets, and bilateral biodiversity-relevant overseas development assistance (OECD 2018). Other studies including other instruments estimate spending between $39 and $49 billion annually (OECD 2019), with some as high as $50 billion (Parker et al. 2012).

There are many specific examples throughout the world that serve to illustrate the urgent need to find alternative feasible ways to finance conservation efforts such as biodiversity islands. Biodiversity alone is not easy to finance; however, it can be financed through strategies that incorporate other valued resources or ecosystem services such as water or carbon (Sheban 2022). When businesses and communities recognize these integrated benefits as essential to their own supply chains, resource bases, and human health, financing becomes more tangible.

The valuation of ecosystem services is a first step in financing biodiversity conservation. Assessment of ecosystem services varies among spatial scales, landscapes, and interactions with stakeholders. Proper valuation is critical for developing programs that seek to compensate farmers and other actors for the environmental services and benefits they provide, including biodiversity (Pagiola et al. 2017). Such programs may include Payments for Environmental Services (PES), which encourage projects that enhance conservation, restoration, production, and rural development via compensation (Montagnini and Finney 2011).

Recent decades have witnessed a considerable increase in PES programs that pay or compensate land users financially for land management practices intended to provide or ensure ecosystem services, with over 550 active programs around the globe and an estimated US$36–42 billion in annual transactions (Salzman et al. 2018). The biodiversity and habitat PES sector uses offsets directed to ensure habitat and biodiversity protection. This sector remains the least developed in terms of geographical scope and is most challenging for countries to put in place. Mitigation credit banks are growing but primarily in developed countries, with a few examples also in developing countries. With transactions estimated at US$3.6 billion per year, compensatory mitigation banking continues to grow. In developing countries, mitigation carried out directly by the party producing the impact or by a subcontractor, known as 'permittee-responsible mitigation', is the most commonly found option for compliance, although many allow developers to pay a compensation fee in lieu of an offset, which is generally used to fund conservation projects carried out by the public sector or an NGO. Voluntary biodiversity offsets are a recent policy development and remain small (Salzman et al. 2018).

Unlike in water PES for which the beneficiaries are straightforward and local, the beneficiaries of biodiversity are often widespread, and the specific benefits are indirect or non-material. Institutions comparable to water utilities that can collect fees on behalf of many beneficiaries do not exist, and common metrics are difficult to determine. As a result, biodiversity PES programs remain limited (Salzman et al. 2018).

Compared to PES systems that include only one environmental service, systems that incorporate bundling or layering of multiple services can make sustainable land uses more attractive to farmers and reduce perverse incentives (Montagnini and Finney 2011). These valuation systems that incorporate bundling of multiple services may protect biodiversity even when PES are given for the provision of some other ecosystem service such as water or carbon. For example, various land use types can be assigned an "Environmental Service Index" based on tree cover and habitat type, bundling biodiversity and carbon (Montagnini and Finney 2011). Under the provision of external funding, a payment can be estimated for each land use type to compensate farmer participation and thus encourage establishment of biodiversity islands on farms. Bundling services can protect biodiversity islands much like the market mechanisms supporting the protection of the Panama Canal watershed (Adamowicz et al. 2019).

Perhaps one of the greatest opportunities for biodiversity conservation is to incorporate biodiversity objectives into the new generation of climate change action programs, for which funds are increasingly available. While the language currently places the emphasis on climate change mitigation and adaptation with terms such as "climate-smart agriculture" or "climate-smart landscapes" many of these programs can and do incorporate multiple ecological objectives, beyond just climate.

In many jurisdictions, taxes levied on activities detrimental to ecosystems worldwide can be used to share the costs of environmental services such as carbon storage and biodiversity. As an example, Sweden recently introduced emissions taxes on airline travel to discourage the use of fossil fuels (Nordic Business Insider 2018). A fair and equitable distribution could divide the proceeds proportionately between countries that are exposed to the air travel emissions. These funds could be used to finance biodiversity islands, preserves and carbon sequestration in affected countries.

In contrast to the taxed approach, industries may recognize the importance of self regulation to address environmental degradation of their practices. Continuing with the example of the aviation industry, a global climate agreement has been reached, CORSIA (Carbon Offsetting and Reduction Scheme for International Aviation), which focuses on investing in emission reductions in unrelated fields to compensate for airline emissions that exceed the 2020 Aichi Targets (https://www.iata.org/policy/environment/Pages/corsia.aspx). World ecosystem services could be greatly enhanced if these types of funds and efforts to offset excessive emissions focused on supporting carbon sequestration and biodiversity protection through biodiversity islands and preserves.

Likewise, worldwide investments aligned with positive environmental outcomes are gaining popularity. For example, Norway's $1 Trillion pension fund has committed to zero deforestation in its public procurements (Rainforest Foundation Norway 2016). Similar initiatives could be applied in other parts of the world, perhaps with a focus on a more equitable allocation of emissions tax revenues towards protecting natural resources.

The integration of biodiversity goals in agricultural and other supply chains (including even products for high fashion, sophisticated consumer markets, government procurement, etc.) is gaining momentum as well (see examples in Montagnini and Metzel 2017; Montes-Londoño et al. 2022; Newcomer et al. 2022; Sheban 2022). In this context, the Coalition for Private Investment in Conservation (CPIC) works to support and scale biodiversity-friendly private investment (http://cpicfinance.com/). In its Statement of Intent, CPIC affirms the need for "investment opportunities that provide measurable, science-based conservation benefits and social impact to participating communities and to biodiversity, while delivering at-scale financial returns for investors." To address the current gap in low investment, inadequate returns, and lack of coordinated knowledge around scalable investment, CPIC calls for "a concerted, systematic effort focused on creating investment products that provide a conservation and financial bottom line..." (cpicfinance.com/wp-content/uploads/2017/05/05_10_17_CPIC-Statement-of-Intent-Final.pdf). The Statement is endorsed by numerous NGOs and public and private entities who have agreed to be part of this collaborative effort.

Because it takes advantage of smaller parcels or fragments of land, the concept of biodiversity islands may be more amenable to funding allocations by governments and policymakers than more conventional sparing approaches that dedicate large areas of land to biodiversity conservation. In situations when governments feel pressured to develop lands to satisfy economic needs of the population, dedicating relatively smaller areas such as biodiversity islands to conservation, instead of designating them for strictly economic uses, may seem more feasible (Montagnini et al. 2022).

## 1.7 Conclusions: Potential for Recognizing and Integrating Biodiversity Islands in Human-Dominated Landscapes

Based on the evidence and experiences presented in this book, there is great potential for recognizing and further integrating biodiversity islands in human-dominated landscapes to achieve positive economic, ecological, and social outcomes. Design strategies for biodiversity islands are dependent on landscape use within the matrix of habitat fragmentation. Integrated landscape management (ILM), including elements of sustainable agriculture, agroforestry and community led action, may provide the guiding framework for implementation of biodiversity islands in complex landscape matrices involving a variety of stakeholders. Land use systems made up of complex assemblages embedded in a forestry matrix, as is the case in many traditional indigenous sacred sites and agroforestry systems, can be considered biodiversity islands as well.

In more degraded landscapes, mixed production systems can act as buffer zones surrounding preserved sites, protecting biodiversity islands or serving as biodiversity islands themselves. Restructuring financial systems to value strategies that protect ecosystem services can enable individuals, businesses, and communities to carry out actions to support the establishment and preservation of biodiversity islands. While financial incentives are crucial, still, other motivations such as personal conviction along with ethical, cultural, or spiritual drivers may prevail in fostering actions towards the creation and protection of biodiversity islands over time.

As with other conservation and restoration initiatives, there are some inherent potential difficulties in the use of this framework. The configuration, design, and other factors previously discussed influence the effectiveness of biodiversity islands. For example, sometimes biodiversity islands may be created to protect target species, disregarding other species with less mobility or specific resource requirements.

In addition, not all biodiversity islands have the same conservation value. It would be useful to prioritize to some degree the conservation value of different biodiversity islands. For instance, large reserves may not have the same value as isolated trees in urban areas or living fences in agricultural lands. Conservation plans can be misguided without this prioritization, e.g. a landowner may cut his or her forest, maintaining some isolated trees or living fences because the latter are also classified as biodiversity islands. In fact, biodiversity islands should be prioritized within each context where they may be applied, whether that be urban, rural, wetlands, etc. A forest in a city cannot be compared to a forest in a rural setting. This prioritization should be done by the respective local planners in each specific case, e.g., municipal authorities, local or international NGOs, park and recreation agencies, etc., with consideration of the location's specific biodiversity goals and constraints.

Moreover, depending on the social and economic context, emphasizing biodiversity conservation may be prone to criticism by those who place higher value on the provision of certain services for society. On the other hand, biodiversity friendly land uses, such as biodiversity islands themselves or agroforestry systems, may be seen by many as enough of a contribution to biodiversity conservation, thereby excluding or replacing other necessary efforts.

Another challenge of biodiversity islands is that because of their tendency to be small and few within a larger landscape, climate change may pose a serious problem to their viability as biodiversity habitat. Parks and protected areas are already facing this challenge as the habitat for species of critical interest shifts with climate change.

Finally, many may argue in favor of the growing trend of recognizing the rights of nature independent of the services it provides, for which biodiversity islands may be used to protect nature regardless of its purpose or potential uses (Prieto Méndez 2013). The growing international movement around the rights of nature provides an alternative justification for biodiversity islands based on the inalienable right of nature to exist, which goes beyond the more anthropocentric rationale based on the ecosystem services they provide (https://therightsofnature.org/what-is-rights-of-nature/).

The purpose of this book is to present examples and case studies where the biodiversity islands approach is being used in a variety of locations and contexts worldwide. This book contributes to informing design parameters on appropriate sizing and spatial distribution of biodiversity islands in order to be effective in conservation and regeneration across the landscape, using integrated landscape management approaches. The book is organized in five parts: **I. Introduction; II. Biodiversity Islands Establishment and Management: Challenges and Alternatives; III. Biodiversity Islands across the Globe: Case Studies; IV. Safeguarding the Environmental, Economic, and Social Benefits of Biodiversity Islands;** and **V. Conclusions**.

The contributing authors present views from the academic, the practitioner and the policymaker perspectives, offering alternatives and suggestions for promoting strategies that support biodiversity conservation through intentionally designed frameworks for sustainable human-dominated landscapes. This book provides suggestions and concrete examples that can be used by a variety of stakeholders in various settings throughout the world. In all, this book covers examples and case studies from 16 countries, including 10 from Latin America (Argentina, Brazil, Colombia, Costa Rica, Ecuador, Mexico, Panama, Paraguay, Peru and Puerto Rico), as well as from the USA, Sweden, Finland, Germany, France, India, Indonesia, Cameroon, Kenya, Ethiopia and others. This book is useful to researchers, farmers, foresters, landowners, land managers, city planners, and policy makers alike.

**Acknowledgements** The authors thank Rafael Herrera, Joseph Orefice, Victor Arroyo-Rodríguez, Sara Scherr and two anonymous reviewers who offered constructive and useful criticism on this chapter. A Conference Session entitled: Biodiversity Islands: Pockets of Protected Land in Human Dominated Environments, that the authors led as part of the IUFRO Conference "Adaptive Management for Forested Landscapes in Transformation", Posadas, Misiones, Argentina, October 1, 2018, inspired and initiated this chapter as well as other contributions contained in this book.

## References


Abbott A (2019) Biodiversity thrives in Ethiopia's church forests. Nature 565:548–549

Adamowicz W, Calderon-Etter L, Entem A, Fenichel EP, Hall JS, Lloyd-Smith P, Ogden FL, Regina JA, Rad MR, Stallard RF (2019) Assessing ecological infrastructure investments. Proc Nat Acad Sci. https://doi.org/10.1073/pnas.1802883116

Angold PG, Sadler JP, Hill MO, Pullin A, Rushton S, Austin K, Small E, Wood B, Wadsworth R, Sanderson R, Thompson K (2006) Biodiversity in urban habitat patches. Sci Total Environ 360: 196–204

Arroyo-Rodríguez V, Pineda E, Escobar F, Benítez-Malvido J (2009) Value of small patches in the conservation of plant-species diversity in highly fragmented rainforest. Conserv Bio 23:729–739

Arroyo-Rodríguez V, Arasa-Gisbert R, Arce-Peña N, Cervantes-López MJ, Cudney-Valenzuela SJ, Galán-Acedo C, Hernández-Ruedas MA, San-José M (2022) The importance of small rainforest patches for biodiversity conservation: a multi-taxonomic assessment in the Lacandona region, Mexico. In: Montagnini F (ed) Biodiversity islands: strategies for conservation in human-dominated environments. Series: Topics in biodiversity and conservation. Springer, Cham, pp 41–60

Baez Schon M, Woods CL, Cardelús CL (2022) Sacred church forests in northern Ethiopia: biodiversity and Cultural Islands. In: Montagnini F (ed) Biodiversity islands: strategies for conservation in human-dominated environments, Series: Topics in Biodiversity and Conservation. Springer, Cham, pp 531–549

Bhagwat SA, Rutte C (2006) Sacred groves: potential for biodiversity management. Frontiers Ecol Environ 4(10):519–524

Bhagwat SA, Willis KJ, Birks HJB, Whittaker RJ (2008) Agroforestry: a refuge for tropical biodiversity? Trends Ecol Evol 23:261–267

Biazen M (2015) The value of urban green infrastructure and its environmental response in urban ecosystem: a systematic review. Inter J Environ Sci 4(2):89–101

Boesing AL, Nichols E, Metzger JP (2018) Biodiversity extinction thresholds are modulated by matrix type. Ecography 41(9):1520–1533

Cabeza M, Moilanen A (2001) Design of reserve networks and the persistence of biodiversity. Trends Ecol Evol 16(5):242–248

Calle A, Holl KD (2019) Riparian forest recovery following a decade of cattle exclusion in the Colombian Andes. For Ecol Manag 452. https://doi.org/10.1016/j.foreco.2019.117563

Calle Z, Guariguata MR, Giraldo E, Chará J (2010) La producción de maracuyá (Passiflora edulis) en Colombia: perspectivas para la conservación del hábitat a través del servicio de polinización. Interciencia 35(3):207–212

Calle Z, Molina CCH, Molina DCH, Molina DEJ, Molina EJJ, Murgueitio B, Murgueitio A, Murgueitio E (2022) A highly productive Biodiversity Island within a monoculture landscape: El Hatico nature reserve (Valle del Cauca, Colombia). In: Montagnini F (ed) Biodiversity islands: strategies for conservation in human-dominated environments. Series: Topics in biodiversity and conservation. Springer, Cham, pp 279–304

CBD High-Level Panel (2014) Resourcing the Aichi Biodiversity Targets: an assessment of benefits, investments and resource needs for implementing the strategic plan for biodiversity 2011–2020. Second report of the high-level panel on global assessment of resources for implementing the strategic plan for biodiversity 2011–2020. Montreal, Canada https://www.cbd.int/financial/hlp/doc/hlp-02-report-en.pdf

Ceperley N, Montagnini F, Natta A (2010) Significance of sacred sites for riparian forest conservation in Central Benin. Bois Forêts Tropiques 303:5–23

Chandler J, Corbett A, Lamb C, Long R, Reberg-Horton C, Stimmann M (1998) Beneficial insects move from flowering plants to nearby crops. Calif Agric 52:23–26

Chazdon RL, Harvey C, Komar O, Griffith DM, Ferguson BG, Martínez-Ramos M, Morales H, Nigh R, Soto-Pinto L, van Bruegel M, Philpott S (2009) Beyond reserves: a research agenda for conserving biodiversity in human-modified tropical landscapes. Biotropica 41(2):142–153

Clavo Peralta ZM, Vela Alvarado JW, Alvez-Valles CM (2022) Biodiversity islands and dominant species in agricultural landscapes of the Western South Amazon, Peru. In: Montagnini F (ed) Biodiversity islands: strategies for conservation in human-dominated environments. Series: Topics in biodiversity and conservation. Springer, Cham, pp 207–235

Clewell A, Aronson J, Winterhalder K (2004) The SER international primer on ecological restoration. Society for Ecological Restoration International Science & Policy Working Group (Version 2: October, 2004) https://www.ctahr.hawaii.edu/LittonC/PDFs/682_SERPrimer.pdf

Conservation International (2014) Hotspots: targeted investment in nature's most important places. https://www.conservation.org/How/Pages/Hotspots.aspx. Retrieved 1 Sep 2018

Costanza R, de Groot R, Sutton P, van der Ploeg S, Anderson SJ, Kubiszewski I, Farber S, Turner RK (2014) Changes in the global value of ecosystem services. Glob Environ Chang 26:152–158

Credit Suisse, World Wildlife Fund, McKinsey & Company (2014) Conservation finance: moving beyond donor funding toward an investor-driven approach. https://www.credit-suisse.com/media/cc/docs/responsibility/conservation-finance-en.pdf. Retrieved 1 Nov 2018

Crespin SJ, Simonetti JA (2019) Reconciling farming and wild nature: integrating human–wildlife coexistence into the land-sharing and land-sparing framework. Ambio 48:131–138. https://doi.org/10.1007/s13280-018-1059-2

DeFries R, Hansen A, Turner BL, Reid R, Liu JG (2007) Land use change around protected areas: management to balance human needs and ecological function. Ecol Appl 17:1031–1038

Denier L, Scherr S, Shames S, Chatterton P, Hovani L, Stam N (2015) The little sustainable landscapes book. Global Canopy Programme, Oxford

Diamond JM (1975) The island dilemma: lessons of modern biogeographic studies for the design of natural reserves. Biol Conserv 7(2):129–146

Dunning JB, Danielson BJ, Pulliam HR (1992) Ecological processes that affect populations in complex landscapes. Oikos 65:169–175

Eibl B, Montagnini F, López MA, Lavignolle P, Cortes J, De La Vega M, López LN, Dummel C, Küppers G (2022) Conservation, connectivity and registration of seed areas in remnants of natural reserves in the province of Misiones, Argentina. In: Montagnini F (ed) Biodiversity islands: strategies for conservation in human-dominated environments. Series: Topics in biodiversity and conservation. Springer, Cham, pp 157–181

Esbach MS, Puri M, Botero-Arias R, Loiselle BA (2022) Beyond the island: safeguarding biodiversity requires integrated approaches. In: Montagnini F (ed) Biodiversity islands: strategies for conservation in human-dominated environments. Series: Topics in biodiversity and conservation. Springer, Cham, pp 551–568

Esquivel MJ, Calle Z (2002) Árboles aislados en potreros como catalizadores de la sucesión en la Cordillera Occidental Colombiana. Agrofor Amér 9(33–34):43–47

Esteva G (1999) The Zapatistas and people's power. Cap Class 23(2):153–182

Ethiopian Biodiversity Institute (2014) Government of the Federal Democratic Republic of Ethiopia: Ethiopia's fifth national report to the convention on biological diversity, Ethiopian Biodiversity Institute Addis Ababa, Ethiopia. 86 pp

Fahrig L (2003) Effects of habitat fragmentation on biodiversity. Ann Rev Ecol Evol Syst 34:487–515. https://doi.org/10.1146/annurev.ecolsys.34.011802.132419

Fahrig L (2013) Rethinking patch size and isolation effects: the habitat amount hypothesis. J Biogeogr 40(9):1649–1663

Fahrig L (2017) Ecological responses to habitat fragmentation per se. Ann Rev Ecol Evol Syst 48: 1–23

Francesconi W, Montagnini F, Ibrahim M (2011) Living fences as linear extensions of forest remnants: a strategy for restoration of connectivity in agricultural landscapes. In: Montagnini F, Finney C (eds) Restoring degraded landscapes with native species in Latin America. Nova Science Publishers, New York, pp 115–126

Gadgil M, Berkes F, Folke C (1993) Indigenous knowledge for biodiversity conservation. Ambio 22:151–156

Galluzzi G, Eyzaguirre P, Negri V (2010) Home gardens: neglected hotspots of agro-biodiversity and cultural diversity. Biodivers Conserv 19(13):3635–3654

Giraldo LP, Chará J, Calle Z, Blanco JF, Chará-Serna AM (2022) Riparian corridors: longitudinal biodiversity islands in farming landscapes. In: Montagnini F (ed) Biodiversity islands: strategies for conservation in human-dominated environments. Series: Topics in biodiversity and conservation. Springer, Cham, pp 139–156

Gómez-Pompa A, Flores JS, Fernández MA (1990) The sacred cacao groves of the Maya. Lat Amer Antiq 1(3):247–257

González LH, Neffeng G, Benedetto V, Sánchez Miñarro M, García A, Biasatti R, Rimoldi P, Alesio C, Paiz D (2022) The anthropization of the natural landscape in the Pampa region in Santa Fe province, Argentina. Resilience of the environment, opportunity for a change of the Agri-food paradigm. In: Montagnini F (ed) Biodiversity islands: strategies for conservation in human-dominated environments. Series: Topics in biodiversity and conservation. Springer, Cham, pp 353–386

Grass I, Loos J, Baensch S, Batáry P, Librán-Embid F, Ficiciyan A, Klaus F, Riechers M, Rosa J, Tiede J et al (2019) Land-sharing/−sparing connectivity landscapes for ecosystem services and biodiversity conservation. Peo and Nat 1:262–272

Haila Y (2002) A conceptual genealogy of fragmentation research: from island biogeography to landscape ecology. Eco App 12(2):321–334

Harvey CA, González J (2007) Agroforestry systems conserve species rich but modified assemblages of tropical birds and bats. Biodiv Conserv 16:2257–2292

Harvey CA, Villanueva C, Villacís J, Chacón M, Muñoz D, López M, Ibrahim M, Taylor R, Martínez JL, Navas A, Sáenz J, Sánchez D, Medina A, Vílchez S, Hernández B, Pérez A, Ruiz F, López F, Lang I, Kunth S, Sinclair FL (2005) Contribution of live fences to the ecological integrity of agricultural landscapes in Central America. Agric Ecosyst Environ 111:200–230

Harvey CA, Medina A, Sánchez D, Vílchez S, Hernández B, Saenz J, Maes J, Casanoves F, Sinclair FL (2006) Patterns of animal diversity associated with different forms of tree cover retained in agricultural landscapes. Ecol Appl 16:1986–1999

Harvey CA, Komar O, Chazdon R, Ferguson BJ, Finegan B, Griffith DM, Martínez-Ramos M, Morales H, Nigh R, SotoPinto L, Van Breugel M, Wishnie M (2008) Integrating agricultural landscapes with biodiversity conservation in the Mesoamerican hotspot. Conserv Bio 22(1): 8–15

Hernández-Ruedas MA, Arroyo-Rodríguez V, Meave JA, Martínez-Ramos M, Ibarra-Manríquez G, Martínez E, Jamangapé G, Melo FPL, Santos BA (2014) Conserving tropical tree diversity and forest structure: the value of small rainforest patches in moderately-managed landscapes. PLoS One 9:e98931

Holl KD (2002) Tropical moist forest restoration. In: Perrow MR, Davy AJ (eds) Handbook of ecological restoration, vol 2. Cambridge University Press, Cambridge, pp 539–558

Holl KD, Leighton Reid J, Chaves-Fallas JM, Oviedo-Brenes F, Zahawi RA (2017) Local tropical forest restoration strategies affect tree recruitment more strongly than does landscape forest cover. J Appl Ecol 54:1091–1099

Jarvis DI, Padoch C, Cooper HD (2011) Manejo de la agrobiodiversidad en los ecosistemas agrícolas. Bioversity International, Rome

Kirby K (2022) Islands of trees in Long-fragmented landscapes in Great Britain. In: Montagnini F (ed) Biodiversity islands: strategies for conservation in human-dominated environments. Series: Topics in biodiversity and conservation. Springer, Cham, pp 337–352

Kumar P (ed) (2010) The economics of ecosystems and biodiversity ecological and economic foundations. Earthscan, London/Washington. 422 pp

Laino R, Musalem K, Laino L, Caballero A, Bueno D, Aranda L, Esquivel A, Ferreira Riveros M, Romero Nardelli L, Cantero C, Irala R (2022) Island forests among savannahs: conservation of biodiversity in rangelands of the Paraguayan humid Chaco. In: Montagnini F (ed) Biodiversity islands: strategies for conservation in human-dominated environments. Series: Topics in Biodiversity and Conservation. Springer, Cham, pp 185–205

Landis DA, Wratten SD, Gurr GM (2000) Habitat management to conserve natural enemies of arthropod pests in agriculture. Ann Rev Entomo 45(1):175–201

Laurance WF (2008) Theory meets reality: how habitat fragmentation research has transcended island biogeographic theory. Biol Conserv 141(7):1731–1744

Laurance WF, Camargo JL, Fearnside PM, Lovejoy TE, Williamson GB, Mesquita RC, Meyer CF, Bobrowiec PE, Laurance SG (2018) An Amazonian rainforest and its fragments as a laboratory of global change. Biol Rev 93(1):223–247

Levin B (2022a) Regenerative agricultural systems as biodiversity islands. In: Montagnini F (ed) Biodiversity islands: strategies for conservation in human-dominated environments. Series: Topics in biodiversity and conservation. Springer, Cham, pp 61–88

Levin B (2022b) The role of community led action for the creation of biodiversity islands. In: Montagnini F (ed) Biodiversity islands: strategies for conservation in human-dominated environments. Series: Topics in biodiversity and conservation. Springer, Cham, pp 487–504

Liu J-J, Slik JWF (2014) Forest fragment spatial distribution matters for tropical tree conservation. Biolog Cons 171:99–106

Lovejoy TE, Bierregaard RO (1990) Central Amazonian forests and the minimum critical size of ecosystems project. In: Gentry AH (ed) Four neotropical rainforests. Yale University Press, New Haven, pp 60–71

Lovejoy TE, Hannah L (eds) (2019) Biodiversity and climate change transforming the biosphere. Yale University Press, New Haven. 416 pp

MacArthur RH, Wilson EO (1967) The theory of island biogeography. Princeton University Press, Princeton, p 203

McIntyre NE, Rango J, Fagan WF, Faeth SH (2001) Ground arthropod community structure in a heterogeneous urban environment. Landsc Urban Plan 52:257–274

Melo FP, Arroyo-Rodríguez V, Fahrig L, Martínez-Ramos M, Tabarelli M (2013) On the hope for biodiversity-friendly tropical landscapes. Trends Ecol Evol 28(8):462–468

Mittermeier RA, Turner WR, Larsen FW, Brooks TM, Gascon C (2011) Global biodiversity conservation: the critical role of hotspots. In: Zachos FE, Habel JC (eds) Biodiversity hotspots. Springer, Verlag Berlin Heidelberg, pp 3–22

Moeliono M, Mulyana A, Adnan H, Manalu P, Yuliani L (2015) Village forests (hutan desa): empowerment, business or burden? World Agroforestry Centre-ICRAF, Southeast Asia Regional Office, Bogor

Monsarrat S, Jarvi S, Svenning J-C (2019) Anthropocene refugia: integrating history and predictive modelling to assess the space available for biodiversity in a human-dominated world. Phil Trans of the R Soc B 374:20190219

Montagnini F (2020) The contribution of agroforestry to restoration and conservation: biodiversity islands in degraded landscapes. In: Gupta SR, Dagar JC, Teketay D (eds) Agroforestry for degraded landscapes: recent advances and emerging challenges, vol 1. Springer Nature, Singapore, pp 445–479. https://doi.org/10.1007/978-981-15-4136-0_15

Montagnini F, Berg KE (2019) Feeding the world. In: Carver F, Gossington H, Manuel C (eds) Sustainable development goals. Transforming our world. Witan Media Ltd/United Nations Association, Painswick/London, pp 98–100. https://www.sustainablegoals.org.uk/feeding-the-world/

Montagnini F, del Fierro S (2022) Functions of agroforestry systems as biodiversity islands in productive landscapes. In: Montagnini F (ed) Biodiversity islands: strategies for conservation in human-dominated environments. Series: Topics in biodiversity and conservation. Springer, Cham, pp 89–116

Montagnini F, Finney C (2011) Payments for environmental services in Latin America as a tool for restoration and rural development. Ambio 40:285–297

Montagnini F, Metzel R (2017) The contribution of agroforestry to sustainable development goal 2: end hunger, achieve food security and improved nutrition, and promote sustainable agriculture. In: Montagnini F (ed) Integrating landscapes: agroforestry for biodiversity conservation and food sovereignty, Advances in Agroforestry 12. Springer, Cham, pp 11–45

Montagnini F, Eibl B, Barth SR (2011) Organic yerba mate: an environmentally, socially and financially suitable agroforestry system. Bois For Tropiq 308:59–74. https://drflorenciamontagnini.wordpress.com/organic-yerba-mate-an-environmentally-socially-and-financially-suitable-agroforestry-system/

Montagnini F, Smith RTW, Everett-Lane B, del Fierro S, Albrecht D (2022) Conclusions: lessons learned and pending challenges. In: Montagnini F (ed) Biodiversity islands: strategies for conservation in human-dominated environments. Series: Topics in biodiversity and conservation. Springer, Cham, pp 677–703

Montes-Londoño I, Calle A, Montes-Calderon A, Montes-Botero O (2022) Hacienda Pinzacuá: an island of regenerative agriculture in the piedmont of the Colombian Central Andes. In: Montagnini F (ed) Biodiversity islands: strategies for conservation in human-dominated environments. Series: Topics in biodiversity and conservation. Springer, Cham, pp 305–335

Morales-Nieves GM (2022) Priorities, perspectives and use of a community forest by surrounding residents in Mayagüez, Puerto Rico: protecting the forest for its services. In: Montagnini F (ed) Biodiversity islands: strategies for conservation in human-dominated environments. Series: Topics in biodiversity and conservation. Springer, Cham, pp 505–530

Murgueitio RE, Calle Z, Uribe F, Calle A, Solorio B (2011) Native trees and shrubs for the productive rehabilitation of cattle ranching lands. For Ecol Manag 261:1654–1663

Nazarea VD, Rhoades RE, Andrews-Swann JE (eds) (2014) Seeds of resistance, seeds of hope: place and agency in the conservation of biodiversity. University of Arizona Press, Tucson, AZ. 288 pp

Negret HRC, Negret R, Montes-Londoño I (2022) Residential garden Design for Urban Biodiversity Conservation: Experience from Panama City, Panama. In: Montagnini F (ed) Biodiversity islands: strategies for conservation in human-dominated environments. Series: Topics in biodiversity and conservation. Springer, Cham, pp 387–417

Newcomer Q, Camacho Céspedes F, Stallcup L (2022) The Monteverde Cloud Forest: evolution of a biodiversity island in Costa Rica. In: Montagnini F (ed) Biodiversity islands: strategies for conservation in human-dominated environments. Series: Topics in biodiversity and conservation. Springer, Cham, pp 237–278

Niella F, Rocha P, Tuzinkievicz A, Buchweis R, Bulman C, Thalmayr P, González J, Montagnini F, Sharry S (2022) Native multipurpose species in the Yaboti Biosphere Reserve, Misiones-Argentina. In: Montagnini F (ed) Biodiversity islands: strategies for conservation in human-dominated environments. Series: Topics in biodiversity and conservation. Springer, Cham, pp 461–483

Nordic Business Insider (2018) Norwegian Air axes two Stockholm-US direct routes. https://nordic.businessinsider.com. Retrieved 1 Dec 2018

OECD (2018) Tracking economic instruments and finance for biodiversity. OECD Publishing, Paris. https://www.oecd.org/environment/resources/Tracking-Economic-Instruments-and-Finance-for-Biodiversity.pdf

OECD (2019) Biodiversity: finance and the economic and business case for action. Report prepared for the G7 Environment Ministers' meeting, 5–6 May 2019. Available at: https://www.oecd.org/environment/resources/biodiversity/G7-report-Biodiversity-Finance-and-the-Economic-and-Business-Case-for-Action.pdf

Otto J, Zerner C, Robinson J, Donovan R, Lavelle M, Villarreal R, Salafsky N, Alcorn J, Seymour F, Kleyneyer C, Pearl M (2013) Natural connections: perspectives in community-based conservation. Island Press, Washington DC

Pagiola S, Honey-Rosés J, Freire-González J (2017) Assessing the permanence of land use change induced by payments for environmental services: evidence from Nicaragua. World Bank eLibrary. https://doi.org/10.1596/27537

Painter K, Buschbacher R, Souto Silva LC, Costa e Silva E (2022) Agroecology and forest conservation in three types of land reform community in the Cacao Region of Bahia, Brazil. In: Montagnini F (ed) Biodiversity islands: strategies for conservation in human-dominated environments. Series: Topics in biodiversity and conservation. Springer, Cham, pp 569–599

Parker C, Cranford M, Oakes N, Leggett M (2012) The little biodiversity finance book, 3rd edn. The Global Canopy Programme, Oxford. 101 pp

Peters CM (2018) Managing the wild. Stories of people and plants and tropical forests. Yale University Press, New Haven. 208 pp

Pfeifer M, Lefebvre V, Peres CA, Banks-Leite C, Wearn OR, Marsh CJ, Butchart SHM, Arroyo-Rodríguez V, Barlow J, Cerezo A et al (2017) Creation of forest edges has a global impact on forest vertebrates. Nature 551:187–191

Phalan BT (2018) What have we learned from the land sparing-sharing model? Sustainability 10(1760):doi:10.3390/su10061760

Phalan B, Onial M, Balmford A, Green RE (2011) Reconciling food production and biodiversity conservation: land sharing and land sparing compared. Science 333(6047):1289–1291

Pickett ST, Thompson JN (1978) Patch dynamics and the design of nature reserves. Biol Conserv 13(1):27–37

Pickett STA, Cadenasso ML, Grove JM (2004) Resilient cities: meaning, models, and metaphor for integrating the ecological, socio-economic, and planning realms. Landsc Urban Plan 69:369–384

Pimentel D, Wilson C, McCullum C, Huang R, Dwen P, Flack J, Tran Q, Saltman T, Cliff B (1997) Economic and environmental benefits of biodiversity. BioSci 47(11):747–757

Prieto Méndez JM (2013) Derechos de la naturaleza. Fundamento, contenido y exigibilidad jurisdiccional. Corte Constitucional del Ecuador, CEDEC, Quito. 280 pp

Rainforest Foundation Norway (2016) Norwegian state commits to zero deforestation. https://www.regnskog.no/en/news/norwegian-state-commits-to-zero-deforestation-1. Retrieved 7 Feb 2018

Redford KH, Padoch C (eds) (1992) Conservation of Neotropical forests. Working from traditional resource use. Columbia University Press, New York, p 475

Sáenz JC, Villatoro F, Ibrahim M, Fajardo D, Pérez M (2007) Relación entre las comunidades de aves y la vegetación en agropaisajes dominados por la ganadería en Costa Rica, Nicaragua y Colombia. Agrofor Amér 45:37–48

Salzman J, Bennett G, Carroll N, Goldstein A, Jenkins M (2018) The global status and trends of payments for ecosystem services. Nat Sustain 1:136–144

Santos-Gally R, Boege K (2022) Biodiversity islands. Native tree islands within silvopastoral systems in a neotropical region. In: Montagnini F (ed) Biodiversity islands: strategies for conservation in human-dominated environments. Series: Topics in biodiversity and conservation. Springer, Cham, pp 117–138

Santos-Gally R, Boege K, Fornoni J, Domínguez CA (2019) Cómo transitar hacia una ganadería tropical sostenible. Oikos. web.ecologia.unam.mx

Savard JPL, Clergeau P, Mennechez G (2000) Biodiversity concepts and urban ecosystems. Land Urb Plan 48:131–142

Sheban K (2022) Preserving biodiversity in Appalachian mixed mesophytic forest through permit-based harvest of ginseng and other forest botanicals. In: Montagnini F (ed) Biodiversity islands: strategies for conservation in human-dominated environments. Series: Topics in biodiversity and conservation. Springer, Cham, pp 601–628

Sigman E (2022) Safeguarding the benefits of biodiversity islands in northern Ethiopia in the midst of political change. In: Montagnini F (ed) Biodiversity islands: strategies for conservation in human-dominated environments. Series: Topics in biodiversity and conservation. Springer, Cham, pp 647–674

Simberloff D, Abele LG (1982) Refuge design and island biogeographic theory: effects of fragmentation. Amer Nat 120(1):41–50

Soler R, Benítez J, Solá F, Lencinas MV (2022) Biodiversity islands at the world's southernmost city: plant, bird and insect conservation in urban forests and peatlands of Ushuaia, Argentina. In: Montagnini F (ed) Biodiversity islands: strategies for conservation in human-dominated environments. Series: Topics in biodiversity and conservation. Springer, Cham, pp 419–437

State Information Service (2018) COP14 decisions, political declaration unanimously adopted-Minister. http://sis.gov.eg/Story/136497/COP14-decisions%2c-political-declaration-unanimously-adopted%2D%2Dminister?lang=en-us. Retrieved 7 Feb 2019

Talberth J, Gray E (2012) Global costs of achieving the Aichi biodiversity targets a scoping assessment of anticipated costs of achieving targets 5, 8, and 14. Center for Sustainable Economy, Washington, DC https://sustainable-economy.org/wp-content/uploads/Aichi-58-and-14-Final-Report.pdf

Thaxton M, Forster T, Hazlewood P, Mercado L, Neely C, Scherr SJ, Wertz L, Wood S, Zandri E (2016) Landscape partnerships for sustainable development: achieving the SDGs through integrated landscape management. White paper produced by The landscapes for people, food, and nature initiative task force on the sustainable development goals. www.peoplefoodandnature.org. 28 pp

Tjørve E (2010) How to resolve the SLOSS debate: lessons from species-diversity models. J Theor Biol 264(2):604–612

Toensmeier E (2022) Paradise lot: a temperate-climate urban agroforestry biodiversity Island. In: Montagnini F (ed) Biodiversity islands: strategies for conservation in human dominated-environments. Series: Topics in biodiversity and conservation. Springer, Cham, pp 439–459

Toensmeier E, Bates J (2013) Paradise lot; two plant geeks, one tenth of an acre, and the making of an edible garden oasis in the city. Chelsea Green Publishing, White River Junction, VT

Totoket Times (2018) Press release from the North Branford conservation trust. celebrating a new chapter and leaving a legacy. Totoket. Times 25(17):14

Tscharntke T, Klein AM, Kruess M, Steffan-Dewenter I, Thies C (2005) Landscape perspectives on agricultural intensification and biodiversity—ecosystem service management. Ecol Lett 8:857–874

Tuff KT, Tuff T, Davies KF (2016) A framework for integrating thermal biology into fragmentation research. Ecol Lett 19(4):361–374

Udawatta RP, Rankoth LM, Jose S (2019) Agroforestry and biodiversity. Sustainability 11:2879. https://doi.org/10.3390/su11102879

UNDP (2006) Community action to conserve biodiversity. linking biodiversity conservation with poverty reduction. Case studies from Latin America and the Caribbean. UNDP, New York. http://www.undp.org/content/undp/en/home/librarypage/environment-energy/ecosystems_and_biodiversity/community-action-to-conserve-biodiversity.html. Retrieved 7 Feb 2019

United Nations IPBES (2019) Global assessment report on biodiversity and ecosystem services. https://sdg.iisd.org/news/ipbes-previews-2019-global-assessment-report-on-biodiversity/

Vásquez V, Barber C, Dguidegue Y, Caughlin TT, García R, Metzel R (2022) Farmer perceptions of tropical dry forest restoration practices on the Azuero Peninsula of Panama – implications for increasing biodiversity in a human-dominated landscape. In: Montagnini F (ed) Biodiversity islands: strategies for conservation in human-dominated environments. Series: Topics in biodiversity and conservation. Springer, Cham, pp 629–646

Volenec ZM, Dobson AP (2019) Conservation value of small reserves. Conserv Bio. https://doi.org/10.1111/cobi.13308

Watling J, Arroyo-Rodríguez V, Pfeifer M, Baeten L, Banks-Leite C, Cisneros L, Fang R, Hamel-Leigue C, Lachat T, Leal I, Lens L, Possingham H, Raheem D, Ribeiro D, Slade E, Urbina-Cardona N, Wood E, Fahrig L (2020) Support for the habitat amount hypothesis from a global synthesis of species density studies. Ecol Lett 23(4):674–681. https://doi.org/10.1111/ele.13471. (Advances in Agroforestry 12. Cham: Springer) pp 179–209

Watts J (2018) Stop biodiversity loss or we could face our own extinction, warns UN. https://www.theguardian.com/environment/2018/nov/03/stop-biodiversity-loss-or-we-could-face-our-own-extinction-warns-un. Retrieved 27 Nov 2018

Wintle BA, Kujala H, Whitehead A, Cameron A, Veloz A, Kukkala A, Moilanen A, Gordon A, Lentini PE, Cadenhead NCR, Bekessy SA (2019) Global synthesis of conservation studies reveals the importance of small habitat patches for biodiversity. PNAS 116:909–914

World Agroforestry Centre (ICRAF), Bogor (Indonesia) (2010) Lubuk Beringin village forest. The first in Indonesia. SEA Regional Office. Contributor(s): Rewards for Use of and Shared Investment in Pro-Poor Environmental Services (RUPES). http://www.worldagroforestry.org/publication/lubuk-beringin-village-forest-first-indonesia

Yarranton GA, Morrison RG (1974) Spatial dynamics of a primary succession: nucleation. J Ecol 62:417–428

# Part II
# Biodiversity Islands Establishment and Management: Challenges and Alternatives

# Chapter 2
# The Importance of Small Rainforest Patches for Biodiversity Conservation: A Multi-taxonomic Assessment

**Víctor Arroyo-Rodríguez, Ricard Arasa-Gisbert, Norma P. Arce-Peña, Martín J. Cervantes-López, Sabine J. Cudney-Valenzuela, Carmen Galán-Acedo, Manuel A. Hernández-Ruedas, Miriam San-José, and Lenore Fahrig**

**Abstract** Tropical forests are being rapidly deforested worldwide. The remaining forest is distributed in different-sized forest patches, but the species preservation value of small patches remains debated. Some studies suggest that edge effect can decrease forest-specialist species diversity, particularly in small patches, which are expected to be mainly occupied by a few disturbance-adapted species. We tested this hypothesis by sampling plants, dung beetles, amphibians, reptiles, birds, and mammals in the fragmented Lacandona rainforest in Mexico. We separately evaluated forest-specialist and habitat generalist species. As positive patch size effects on species richness can be simply related to the sample area effect (i.e. larger samples have a higher chance of holding more species), we assessed the effect of patch size on the number of species of each group in samples of constant size (species density). We also evaluated whether species density is lower in forest patches than in continuous forest sites. We found that patch area was generally a poor predictor of

---

V. Arroyo-Rodríguez (✉)
Laboratorio de Ecología de Paisajes Fragmentados, Instituto de Investigaciones en Ecosistemas y Sustentabilidad, Universidad Nacional Autónoma de México, Morelia, Michoacán, Mexico

Escuela Nacional de Estudios Superiores, Universidad Nacional Autónoma de México, Mérida, Yucatán, Mexico

R. Arasa-Gisbert · N. P. Arce-Peña · M. J. Cervantes-López · S. J. Cudney-Valenzuela · M. A. Hernández-Ruedas · M. San-José
Laboratorio de Ecología de Paisajes Fragmentados, Instituto de Investigaciones en Ecosistemas y Sustentabilidad, Universidad Nacional Autónoma de México, Morelia, Michoacán, Mexico

C. Galán-Acedo
Escuela Nacional de Estudios Superiores, Universidad Nacional Autónoma de México, Morelia, Michoacán, Mexico

L. Fahrig
Geomatics and Landscape Ecology Laboratory, Department of Biology, Carleton University, Ottawa, Canada

 
F. Montagnini (ed.), *Biodiversity Islands: Strategies for Conservation in Human-Dominated Environments*, Topics in Biodiversity and Conservation 20,
https://doi.org/10.1007/978-3-030-92234-4_2

species density, and that in most study groups the density of species did not differ between continuous forest and forest patches. Remarkably, most results were independent of habitat specialization. These findings add to the increasing evidence that, on a per-sample area basis, small patches are valuable biodiversity islands for conservation of forest-specialist species and are not the near-exclusive habitat of generalist species. Our results indicate a need to redress the neglect of small patches in conservation plans, even for forest-specialist species in fragmented rainforests, in order to help maintain species diversity.

**Keywords** Biodiversity-friendly landscapes · Habitat fragmentation · Habitat loss · Landscape configuration · Landscape restoration

## 2.1 Introduction

Deforestation is causing the net annual loss of more than 12 million hectares of forests globally (Hansen et al. 2013). The remaining forest is distributed in millions of different-sized forest patches, most of them very small, with the global meanpatch size ranging between 13 and 17 ha (Taubert et al. 2018). This situation threatens biodiversity maintenance since forest loss is a major cause of species loss (Newbold et al. 2016; Watling et al. 2020). Because forest area can have an effect on species loss at a local scale, some studies suggest that small forest patches can be of relatively low conservation value because: (1) they cannot maintain viable populations of forest-specialist species (i.e. those that use forest interior as their primary habitat); (2) populations in small patches are more susceptible to human-caused disturbances (e.g. hunting, logging, livestock incursion, wildfire); and (3) most forest-specialist species are negatively impacted by edge effects (i.e. biotic and abiotic changes at forest edges) and should therefore be more prone to extinction in small patches, which have a high edge-to-area ratio (Diamond 1975, 1976; Willis 1984; Laurance et al. 2002; Banks-Leite et al. 2010; Fletcher et al. 2018; Phalan 2018). Thus, smaller patches are expected to have a lower number of species than large ones.

The lower number of species in small patches can, however, be caused by the sample area effect, i.e. small samples (patches) can contain a lower number of individuals, so they have a higher chance of containing a lower species richness as well (Fahrig 2013; Chase et al. 2018). Thus, to better understand the impact of patch size on species diversity, the number of species needs to be measured in same-sized samples (i.e. species density; Fahrig 2013). Small patches may be edge-affected habitats of ideal conditions for a few generalists and/or invasive species (i.e. those that use resources from forest interior, forest edges, regenerating forest stands, and agricultural lands), but of less favorable conditions for most forest-specialist species (Laurance et al. 2002; Banks-Leite et al. 2010; Tabarelli et al. 2012; Fletcher et al. 2018). Therefore, the species density should be positively related to patch size. Similarly, compared to samples in continuous forest sites, species density should be lower in forest patches, especially when considering patches of $<100$ ha, as edge effects can penetrate 500 m or more into forest patches (Laurance et al. 2002; Harper et al. 2005; Pfeifer et al. 2017).

In contrast to this hypothesis, evidence indicates that on a per-sample area basis, small patches can have as high or higher value for conservation of forest species as large patches (Arroyo-Rodríguez et al. 2020; Fahrig 2020; Watling et al. 2020). Among other factors (reviewed by Fahrig 2020), this may be related to the fact that most species are able to use resources from multiple patches in the landscape rather than being constrained by the resources available in a single patch. This is plausible if forest edges and the surrounding anthropogenic matrix do not represent a barrier to the movement of individuals in fragmented landscapes, as demonstrated in several studies (Mendenhall et al. 2013; Ferreira et al. 2018; Galán-Acedo et al. 2019). Therefore, the number of species in a given site (species density) may depend more on the forest cover available in the local landscape surrounding the site than on the size of the patch in which the site is located (Fahrig 2013). A recent global meta-analysis supports this hypothesis (Watling et al. 2020). However, considerable skepticism remains about the value of small patches in biodiversity conservation, especially in tropical regions, so this controversial topic requires further evaluation (Fletcher et al. 2018).

Here, we report results of studies in which we compiled data on the seed rain, saplings, trees, dung beetles, amphibians, reptiles, birds, bats, medium- and large-bodied terrestrial mammals, and arboreal mammals recorded in the interior of old-growth forest sites (forest sites, hereafter). The area of study included different-sized forest patches and continuous forest sites in the Lacandona rainforest, Mexico (Fig. 2.1). These data were collected as part of several of our previous studies on the same subject (Table 2.1). In particular, we assessed, for the first time: (1) the effect of patch size on species density (i.e. species richness in samples of fixed size); and (2) the difference in species density between forest patches and continuous forest sites. These analyses were done by separately assessing forest-specialist and forest-generalist species. Such an evaluation is important in the context of biodiversity island design, as our findings will indicate if forest patches of any size can be beneficial to the protection of forest biodiversity.

## 2.2 Methods

### 2.2.1 Study Area

The Mexican portion of the Lacandona rainforest (16°10′56″N - 90°53′21″W) is located in southeastern Chiapas State, Mexico (Fig. 2.1). The climate is warm with a mean annual temperature of 24 °C and humid with a mean annual rainfall of 2143 mm. This region preserves one of the largest rainforest tracts in Mexico and represents a priority area for biodiversity conservation in Mesoamerica (Arriaga et al. 2000). Within this region, the Montes Azules Biosphere Reserve preserves 331,200 ha of contiguous, largely undisturbed forest. Nevertheless, deforestation outside this reserve has resulted in the loss of more than 45% of original forest cover, due to its conversion to cattle pastures, annual crops, and oil-palm plantations

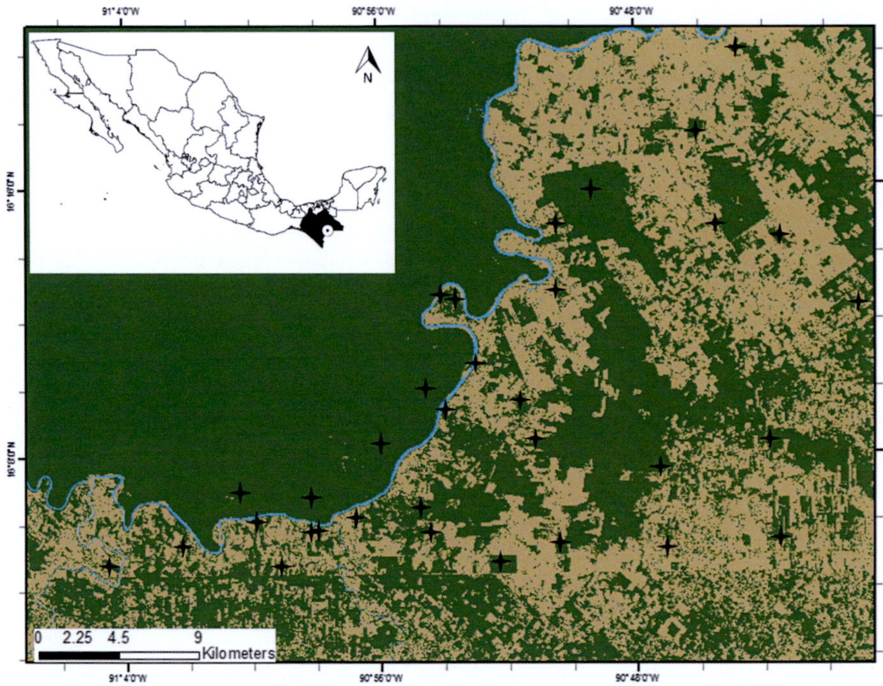


**Fig. 2.1** Location of the Lacandona rainforest in southeastern Mexico. We sampled continuous forest sites within the Montes Azules Biosphere Reserve, and forest patches in the Marqués de Comillas municipality (black stars). Forest cover is indicated with dark green, and the anthropogenic matrix with brown. The blue line is the Lacantún River. Source: Digitized by C. Galán-Acedo using Landsat images and the CLASlite software

(Carabias et al. 2015; Fig. 2.2a and b). Such forest loss has been particularly notable in the Marqués de Comillas region, which is separated from the Biosphere Reserve by the Lacantún River (Figs. 2.1 and 2.2a).

### 2.2.2 *Study Forest Sites*

As part of previous independent studies of different taxonomic groups, we have data from 12 to 26 forest sites depending on the taxa: 0 to 4 sites within the Montes Azules Biosphere Reserve, and 5 to 24 forest patches within the Marqués de Comillas municipality (Table 2.1; Fig. 2.1). Continuous forest sites refer to circular areas of 100 ha each within the Montes Azules Biosphere Reserve (continuous forest sites hereafter), which were spaced 4 km apart on average and at least 1 km from the nearest edge of the Lacantún river (Fig. 2.1). Forest patches within the Marqués de Comillas municipality were spaced at least 1500 m apart, and forest patch size ranges differed among studies (Table 2.1). All sites were located in lowland areas

**Table 2.1** Summary of study designs. Rows refer to independent studies of different biological groups in forest patches (FP) and continuous forest sites (CF) in the Lacandona rainforest, Mexico

| Biological group | Response variable[a] | Sampling period | # forest sites | Patch size (ha) | Sampling method in each site | Refs[b] |
|---|---|---|---|---|---|---|
| Seeds | Species density | Feb. 2015 – Feb. 2016 | 19 FP + 1 CF | 2.8–129 | 9 seed traps (0.5 $m^2$ each) in a grid of 8 × 8 m | 1 |
| Saplings | Species density | Jan. – May 2018 | 19 FP + 1 CF | 2.8–157 | 25 plots (8 $m^2$ each) in a grid of 120 × 120 m | 2 |
| Trees | Species density | Apr. – Jun. 2015 | 19 FP + 1 CF | 2.8–129 | 10 parallel 50 × 2-m plots (plot distance = 10 m) | 3 |
| Dung beetles | Species density* | Jul. – Sep. 2012 | 21 FP + 3 CF | 2.8–129 | 4 baited pitfall traps per transect. Number of transects proportional to patch area | 4 |
| Amphibians & reptiles | Species density | May – Sep. 2012 | 9 FP + 3 CF | 2.8–91.9 | Visual encounters in 6 non-fixed transects | 5 |
| Amphibians & reptiles | Species density | May – Nov. 2018 | 5 FP | 1.7–69.8 | Visual encounters in 6 non-fixed transects | 6 |
| Birds | Species density* | May – Aug. 2012 | 17 FP + 3 CF | 2.8–129 | Point counts (number proportional to patch size) | 7 |
| Bats | Species density | May – Sep. 2012 | 12 FP + 3 CF | 2.8–91.9 | 5 mist nests located 50 m between each other | 8 |
| Terrestrial mammals[c] | Species density | Apr. – Aug. 2011 | 24 FP + 4 CF | 2.8–129 | Camera traps (150 camera trap nights per site) | 9 |
| | Species density | Apr. – Aug. 2017 | 24 FP + 4 CF | 2.8–129 | Camera traps (150 camera trap nights per site) | 10 |
| Arboreal mammals | Species density | May 2018 – May 2019 | 19 FP + 1 CF | 2.8–157 | Camera traps on 5 focal trees | 11 |

[a]In all cases but two (indicated with *), we calculated the number of species (i.e. count response variable) recorded in samples of fixed size (species density). The responses marked with asterisk were recorded with a sampling effort that increased with patch size. In those cases, species density was calculated as the mean number of species per transect (dung beetles) or point count (birds), so they are continuous response variables

[b]References: 1. San-José et al. (2020); 2. R. Arasa-Gisbert (unpubl. data); 3. M.A. Hernández-Ruedas (unpubl. data); 4. Sánchez-de-Jesús et al. (2015); 5. Russildi et al. (2016); 6. Cervantes-López (unpubl. data); 7. Carrara et al. (2015); 8. Arroyo-Rodríguez et al. (2016); 9. Garmendia et al. (2013); 10. N. Arce-Peña (unpubl. data); 11. S. Cudney-Valenzuela (unpubl. data)

[c]Terrestrial mammals were sampled in the same sites but in different time periods. We calculated the accumulated number of species sampled at each site across time periods

**Fig. 2.2** Pictures from the study landscape showing the Lacantun river (panel **a**), which separates the continuous forest of the Montes Azules Biosphere Reserve (left side) and the Marqués de Comillas region (right side). This latter region is highly heterogeneous, as the remaining forest patches are surrounded by a matrix of cattle pastures (panels **a** and **b**) and crops (e.g. oil-palm plantations in panel **c**), with different treed elements (e.g. living fences and isolated standing trees in panels **a** and **b**). (Photos: Víctor Arroyo-Rodríguez)

(100–200 meters above sea level (m.a.s.l.)) with similar soil and weather conditions to avoid the potential confounding effects of these variables. Also, because most of the region was rapidly deforested during the 1980s, we assume that most patches have a similar age (30–40 years).

### 2.2.3 *Study Taxa*

We surveyed seeds, saplings, trees, dung beetles, amphibians, reptiles, birds, bats, medium- to large-bodied terrestrial mammals, and arboreal mammals (Table 2.1). All surveys were done within forest interior (>50 m from the nearest forest edge). The sampling methods and efforts are detailed in their original studies and summarized and referenced in Table 2.1, but a brief overview is given here. Note that all taxa but two (i.e. birds and dung beetles) were sampled in sample areas of constant size.

#### 2.2.3.1 Seeds, Saplings and Trees

We sampled seeds, saplings and trees in the center of each forest site ($n = 20$ sites in all cases; Table 2.1). To sample seeds, we placed 9 seed traps (trap area = 0.5 $m^2$, 1-mm nylon mesh) in a grid of 3 × 3 traps separated by 4 m (total sampling effort = 20 × 9 = 180 traps). Traps were hung approximately 90 cm above the ground, and trap contents were recovered every 15 days for one year (see San-José et al. 2020). Saplings were recorded in 25 circular plots (1.60-m radius, 8-$m^2$ each; total sampling area per site = 200 $m^2$) in a grid of 5 × 5 plots (inter-plot distance = 30 m). We counted and identified all tree saplings (excluding palms and lianas) with ≥30 cm height and < 1 cm of diameter at breast height (DBH). Trees were sampled using a modification of Gentry's (1982) protocol; at each site, we recorded only tree species (including palms) with DBH ≥ 1 cm in ten parallel 50 × 2-m transects (i.e. 0.1 ha per patch), separated by 10 m.

#### 2.2.3.2 Dung Beetles

We collected dung beetles in 24 forest sites (Table 2.1) using baited pitfall traps (i.e. 1 L plastic containers buried with the top edge leveled with the soil surface). We baited traps with 25 g of a mixture of human and pig excrement (7:3). We placed four traps separated by 50 m along transects. Sampling effort increased with patch area (i.e. one transect every 20 ha, and five transects in continuous forest sites). When sampling ≥2 transects in a site, we separated them by ≥150 m. In total, we used 196 pitfall traps, and we collected, identified, and counted all individuals and species (Sánchez-de-Jesús et al. 2016).

#### 2.2.3.3 Amphibians and Reptiles

We sampled amphibians and reptiles as part of two studies in 12 and 5 forest sites (Table 2.1) but using the same protocol. In particular, we used visual encounter surveys (Crump and Scott 1994). We surveyed all forest sites (continuous forest sites and forest patches) using the same methods and a similar sampling effort in all sample sites independently of their size. In particular, we divided each sample site in six sections of similar size, and then sampled all amphibians and reptiles using non-fixed transects (one transect per section). During each visit we sampled a different section, totaling 6 days per forest site. Each visit included a day (10:00 to 13:00 h) and night (19:00 to 22:00 h) sampling period (3 h each). All individuals were identified *in situ* using the field guides of Lee (2000) and Campbell (1998) and released in the same place where they were captured (see Russildi et al. 2016).

#### 2.2.3.4 Birds

We sampled birds in 20 forest sites (Table 2.1) using unlimited radius point counts (Bibby et al. 2000). We sampled each site three times (once per month) following a randomly selected order. We distributed point counts by dividing each site in three sections of similar size, and during each visit, we sampled a different section from 5: 30 to 10:30 h. Point counts were separated 200 m from each other. The number of point counts increased with the size of the patches and sections (i.e. 1 to 5 points per section depending on its size). In total, we surveyed 130 point counts in patches and 72 in continuous forest sites. In each point count we recorded all birds seen or heard during a 15 min period, considering only those perched on trees, on the floor, feeding or using other resources of the study forest (see Carrara et al. 2015).

#### 2.2.3.5 Bats

We sampled bats in 15 forest sites placing 5 mist nests (12 $\times$ 2.5 m) 50 m apart in natural linear openings (e.g. animal trails, streams) at the center of each site (Table 2.1). Mist nets were kept open for 5 h after sunset and checked every 30 min (Kunz 1982). Each site was sampled for six non-consecutive nights (n = 90 nights). We did not sample bats during the full moon because of their reduced activity. We placed captured bats in cloth bags (30 $\times$ 40 cm) for later identification using a field key (Medellín et al. 2008), and we marked them on the right tibia to avoid repeated counts. We released them within 2 h of capture at the capture site (see Arroyo-Rodríguez et al. 2016).

#### 2.2.3.6 Medium- and Large-Bodied Terrestrial Mammals

Terrestrial mammals were sampled as part of two studies in the same 28 forest sites and using exactly the same sampling protocol (Table 2.1). In each forest site, we placed a camera trap in an apparently ‘high quality’ location, i.e. with evident signs of use by mammals (e.g. area with several mammal footprints). The camera was left for 30 days, with a recovery time of 30 seconds per picture. Cameras were serviced once a month (e.g. change of batteries, download pictures). After 30 days, we moved the camera to a different ‘high quality’ location within the same forest site, but 100 to 1000 m apart, as done in previous studies (Saito and Koike 2013). In total, we registered 150 camera trap nights per site, with five sample locations in each site (Garmendia et al. 2013). For the analyses described below, we combined the information from the two studies by accumulating the species sampled in each site after the two sampling periods.

#### 2.2.3.7 Arboreal Mammals

We also used camera traps to record arboreal mammals in 20 forest sites (Table 2.1). At the geographical center of each forest site, we selected five trees (four reached the canopy and one the understory) with good climbing conditions (i.e. branches $\geq$20 cm wide, preferably hardwood species, with adequate architecture to install a camera trap facing other main branches). We placed one camera trap on one of the five trees per patch and we changed the location of the cameras, rotating them once a month among the five trees, except from October to December when they remained on the same tree (7387 total camera trap nights). We serviced cameras (e.g. change of batteries, download pictures) every time we changed their location. We used bait (tuna, peanut butter with oatmeal, and a banana) on one tree per site to increase the probability of photo-capturing arboreal mammals.

### 2.2.4 *Data Analyses*

We separately assessed responses of forest-specialist and generalist species, which were classified based on scientific literature (Garmendia et al. 2013; Hernández-Ruedas et al. 2014; Carrara et al. 2015; Sánchez-de-Jesús et al. 2016; Arroyo-Rodríguez et al. 2016; Russildi et al. 2016; IUCN 2019), and the assistance of local experts. All analyses were done with R 3.0.1 (R Core Team 2013).

#### 2.2.4.1 Effect of Patch Size on Response Variables

Our response variables were estimated on a per-sample area basis: i.e. species density. This was already the case when sampling effort was the same in each patch (i.e. seed rain, saplings, trees, amphibians, reptiles, bats, medium- and large-bodied terrestrial mammals, and arboreal mammals). When sampling effort increased with patch size (i.e. dung beetles and birds, Table 2.1), we calculated the mean number of species per transect (dung beetles) or per point count (birds).

After a visual inspection of the scatter plots, we discovered that species density was not related to patch size in most species groups (i.e. null patch size effect), and when related, patch size seemed to be linearly related to species density (Table 2.2). Therefore, to statistically test these relationships we used generalized linear models (GLM) and compared those models to the null model (including only the intercept). Such a comparison was done following an information-theoretic criteria. In particular, we estimated the Akaike Information Criterion corrected for small samples (AICc) and selected the model with lower AICc. When the linear and null models showed similar empirical support (i.e. $\Delta$AICc <2), we selected the simplest (null) model. We tested GLMs with a Poisson error distribution for count response variables (see column of response variables in Table 2.1), and we corrected for over-dispersion by including a quasi-poisson error distribution when the ratio between residual deviance to degrees of freedom was higher than 1. For responses variables that were means (dung beetles and birds; Table 2.1), we used a Gaussian error distribution (Crawley 2007).

#### 2.2.4.2 Differences Between Continuous Forest and Forest Patches

Differences in all response variables per unit area (as in Sect. 2.2.4.1) between continuous and fragmented forests were also tested with GLMs, but only for those taxa for which we had enough replicates of continuous forest ($n \geq 3$ sites). As described in Sect. 2.2.4.1, we fitted a Poisson error distribution to count response variables (number of species) and a Gaussian error distribution to continuous response variables (mean number of species). Here, we also corrected the models for over-dispersion, when needed.

## 2.3 Results

### *2.3.1 Effect of Patch Size on the Number of Species*

In total, we sampled 111 species of tree seeds, 161 tree saplings, 152 trees, 43 dung beetles, 26 amphibians, 42 reptiles, 84 birds, 34 bats, 21 terrestrial mammals, and 15 arboreal mammals. Patch size was generally a poor predictor of species density,

**Table 2.2** Effect of forest patch size on the species density of different taxa in the Lacandona rainforest, Mexico

| Taxon | Habitat | Model | Coef | AICc | $\Delta_i$ | $w_i$ |
|---|---|---|---|---|---|---|
| Seeds | FS | **Null** | | **92.54** | **0.0** | **0.75** |
| | | GLM | 0.0015 | 94.74 | 2.2 | 0.25 |
| | FG | **Null** | | **87.17** | **0.0** | **0.74** |
| | | GLM | 0.0015 | 89.37 | 2.2 | 0.25 |
| Saplings | FS | **Null** | | **111.61** | **0.0** | **0.77** |
| | | GLM | −0.0004 | 114.01 | 2.4 | 0.23 |
| | FG | **Null** | | **143.32** | **0.0** | **0.60** |
| | | GLM | 0.0014 | 144.12 | 0.8 | 0.40 |
| Trees | FS | **Null** | | **120.95** | **0.0** | **0.73** |
| | | GLM | 0.0010 | 122.95 | 2.0 | 0.27 |
| | FG | **Null** | | **132.09** | **0.0** | **0.74** |
| | | GLM | 0.0007 | 134.19 | 2.1 | 0.26 |
| Dung beetles | FS | **GLM** | **−0.062** | **100.95** | **0.0** | **0.92** |
| | | Null | | 105.95 | 5 | 0.08 |
| | FG | GLM | −0.040 | 102.69 | 0.0 | 0.55 |
| | | **Null** | | **103.09** | **0.4** | **0.45** |
| Amphibians | FS | **Null** | | **58.5** | **0.0** | **0.64** |
| | | GLM | 0.006 | 59.6 | 1.1 | 0.36 |
| | FG | GLM | −0.008 | 61.1 | 0.0 | 0.53 |
| | | **Null** | | **61.3** | **0.2** | **0.47** |
| Reptiles | FS | **Null** | | **52.56** | **0.0** | **0.70** |
| | | GLM | −0.005 | 54.26 | 1.7 | 0.30 |
| | FG | **Null** | | **64.16** | **0.0** | **0.53** |
| | | GLM | −0.006 | 64.36 | 0.2 | 0.47 |
| Birds | FS | **Null** | | **11.27** | **0.0** | **0.74** |
| | | GLM | −0.003 | 13.37 | 2.1 | 0.26 |
| | FG | **GLM** | **−0.028** | **57.80** | **0.0** | **0.85** |
| | | Null | | 61.3 | 3.5 | 0.15 |
| Bats | FS | **Null** | | **48.56** | **0.0** | **0.81** |
| | | GLM | 0.0002 | 51.46 | 2.9 | 0.19 |
| | FG | **Null** | | **51.95** | **0.0** | **0.81** |
| | | GLM | −0.0011 | 54.85 | 2.9 | 0.19 |
| Terrestrial mammals | FG | **Null** | | **110.48** | **0.0** | **0.67** |
| | | GLM | 0.0026 | 111.88 | 1.4 | 0.33 |
| Arboreal mammals | FG | **Null** | | **87.55** | **0.0** | **0.77** |
| | | GLM | −0.0006 | 89.95 | 2.4 | 0.23 |

We separately assessed forest-specialist (FS) and forest-generalist (FG) species. We compared the generalized linear model (GLM) with the null model (which included only the intercept) using information-theoretic criteria (AICc: Akaike Information Criterion corrected for small samples; $\Delta_i$: difference in AICc between the best model and model i; $w_i$: Akaike weights). The selected model in each case is in boldface. Note that when the linear and null models showed similar empirical support (i.e. ΔAICc <2), we selected the simplest (i.e. null) model

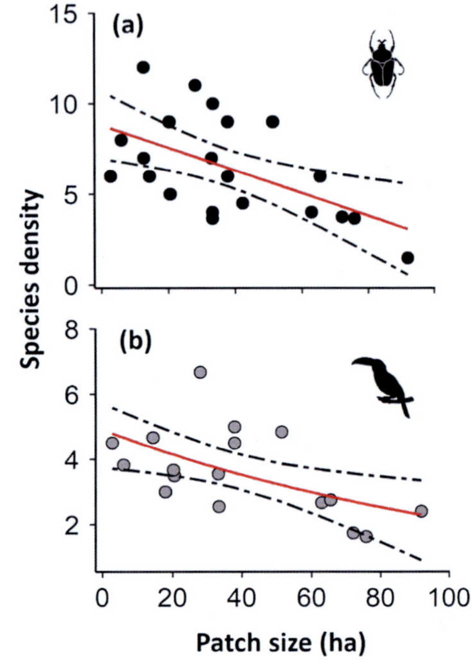


**Fig. 2.3** The most important effects of patch size on species density of forest-specialist dung beetles (black dots) and generalist birds (gray dots) in the Lacandona rainforest, Mexico. Only models with empirical support are shown (see Table 2.2). The red line shows the predicted estimates from the best fitting models and the dotted lines indicate 95% confidence intervals

with the null model being the best fitting model in 16 of 18 assessments (89%; Table 2.2). Only the mean number of specialist dung beetles and generalist birds were linearly and negatively related to patch size (Fig. 2.3).

### *2.3.2 Continuous Forest vs Forest Patches*

In 8 out of 11 cases (72%), the species density was similar in forest patches and continuous forest sites (Table 2.3). The density of terrestrial mammals (all generalist species) and the densities of forest-specialist amphibians and reptiles were 1.3 to 1.7 times higher in continuous forest than in forest patches (Table 2.3).

## 2.4 Discussion

This study demonstrates the high conservation value of small forest patches for biodiversity conservation in the Lacandona rainforest – a species-rich but vanishing tropical region in southeastern Mexico. We predicted that if small patches are of low quality for forest-specialist species, the density of forest-specialist species should be positively related to patch size and be higher in continuous forest than forest patches.

**Table 2.3** Differences in the mean number of forest-specialist and generalist species between continuous forest sites (CF) and forest patches (FP) in the Lacandona rainforest, Mexico, for biological groups sampled in at least three continuous forest sites (see Table 2.1)

| Study taxa | CF | FP | t | p |
|---|---|---|---|---|
| *Forest-specialist species* | | | | |
| Dung beetles | 3.9 ± 0.1 | 6.5 ± 2.8 | 1.69 | 0.09 |
| Amphibians | 6.0 ± 4.2 | 4.4 ± 1.6 | −2.5 | **0.01** |
| Reptiles | 7.0 ± 2.2 | 4.0 ± 1.3 | −2.56 | **0.01** |
| Birds | 0.5 ± 0.1 | 0.8 ± 0.3 | 0.6 | 0.55 |
| Bats | 8.0 ± 3.0 | 5.3 ± 1.4 | −1.69 | 0.09 |
| *Generalist species* | | | | |
| Dung beetles | 2.7 ± 0.2 | 5.8 ± 2.6 | 2.02 | 0.06 |
| Amphibians | 3.3 ± 1.5 | 5.5 ± 2.0 | 1.49 | 0.14 |
| Reptiles | 5.0 ± 1.7 | 7.0 ± 2.2 | 1.21 | 0.23 |
| Birds | 1.6 ± 0.2 | 3.6 ± 1.3 | 1.69 | 0.09 |
| Bats | 9.7 ± 3.2 | 8.2 ± 1.3 | −0.8 | 0.43 |
| Terrestrial mammals | 15.0 ± 1.6 | 11.6 ± 2.8 | −2.26 | **0.03** |

Mean (±SD) values are indicated in each forest type. Significant differences are highlighted with boldface (t = t value in GLM tests, p = probability)

Conversely, if generalist species can use resources from forest interiors, forest edges, and even from some matrix types, we expected no effect of patch size on generalists, and a similar density of generalist species in both forest types.

Our findings support our predictions for generalist species, but not for specialist ones. Patch area is generally a poor predictor of species density. In fact, where there was a difference (specialist dung beetles and generalist birds), species density was higher in small than large patches. Also, in all but three study groups (generalist terrestrial mammals, and specialist amphibians and reptiles), the density of species did not differ between continuous forest sites and forest patches. Importantly, most findings were similar for forest-specialist and habitat generalist species. Therefore, our findings add to an increasing number of studies (e.g. Fahrig 2020; Watling et al. 2020) indicating that, on a per-area basis, a unit of habitat in a small forest patch may not only have a conservation value similar to that of large ones, but may be even greater.

The fact that species density is rarely related to patch area and that it does not usually differ between continuous forest and forest patches can be related to several non-exclusive factors. First, in this region, forest patches are embedded in a highly heterogeneous matrix with many treed elements (i.e. living fences, isolated trees, tree crops; Fig. 2.2a–c), and this spatial context can reduce edge influence on forest patches (Harper et al. 2005; Arroyo-Rodríguez et al. 2017). Second, these treed elements in the matrix can also provide important supplementary resources to forest species (Asensio et al. 2009; Mendenhall et al. 2013; Hernández-Ordóñez et al. 2015; Ferreira et al. 2018; Galán-Acedo et al. 2019), and can therefore prevent the loss of species in small patches (reviewed by Dunning et al. 1992, Arroyo-Rodríguez et al. 2020). In fact, there is evidence that species-area relationships are shallower (less extinction driven) where matrix quality is higher (Reider et al. 2018). Third, forest-specialist and generalist species can use these treed elements to move across the landscape (Mendenhall et al. 2013; Galán-Acedo et al. 2019), allowing individuals to use resources from multiple patches rather than being constrained to the resources available in a single patch (Fahrig 2013, 2020). This is particularly likely in Lacandona's moderately-deforested region, where mean (±SD) inter-patch isolation distance (edge-to-edge) in 1000-m radius landscapes is relatively short (99.8 ± 104.9 m; San-José et al. 2020). Therefore, as predicted by the 'habitat amount hypothesis' (Fahrig 2013) and supported by empirical evidence (Watling et al. 2020), species density in this region may depend more on forest cover in the local landscape surrounding the site than on the size of the patch in which the sample site is located. Although we did not test the effect of forest cover on species density, our database does not show any relation between forest cover and patch size (Arroyo-Rodríguez et al. 2013). Thus, our findings are, to some extent, consistent with this hypothesis, which predicts that if habitat amount (i.e. forest cover) remains constant, species density should also be independent of patch size (see Fig. 7b in Fahrig 2013). These results provide potential for a new approach to future research in the region.

Interestingly, our findings were independent of the degree of habitat specialization of the species. This contradicts the idea that small patches are mainly valuable for conserving generalist species, but not for habitat specialist species (Diamond 1976; Willis 1984; Tabarelli et al. 2012; Fletcher et al. 2018). In her recent review, Fahrig (2020) also found no support for this assumption, possibly because, "as long as patches are close enough together, persistence across multiple patches could occur by frequent immigration and/or by individual space use that incorporates multiple patches" (Fahrig 2020). In fact, as stated above, individuals from many forest specialist species have been recorded moving through the anthropogenic matrix, using several small patches as part of their home ranges (Asensio et al. 2009; Mendenhall et al. 2013; Ferreira et al. 2018; Galán-Acedo et al. 2019). Therefore, the high conservation value of small forest patches is not limited to generalist species, as these small patches can also contain forest specialists.

However, it is important to consider that the density of forest-specialist amphibians and reptiles is lower in forest patches than continuous forest. This finding supports previous studies in the region showing that some amphibian and reptile species from continuous forest sites are very rare or absent in forest patches (Hernández-Ordóñez et al. 2014, Russildi et al. 2016). These included amphibians such as *Craugastor alfredi*, *Smilisca cyanosticta*, *Smilisca cyanosticta*, and *Bolitoglossa mexicana*, and reptiles such as *Celestus rozellae*, *Anolis capito* and *Porthidium nasutum* (Fig. 2.4). Currently there is scarce information on the life history traits that limit the persistence of these forest-specialist species in forest patches, but they could have high population variability and restricted habitat needs (Gibbs 1998). Thus, in agreement with recent proposals to design biodiversity-friendly landscapes (Arroyo-Rodríguez et al. 2020), a mixed strategy of preserving many small patches and a few large patches would maximize biodiversity conservation in the region.

**Fig. 2.4** Examples of generalist terrestrial mammals (**a–c**), and forest-specialist reptiles (**d–e**) and amphibians (**f–g**) that are very rare or absent in all studied forest patches, despite the fact our sampling effort was higher at forest patches than continuous forest (see Table 2.1). Species names = *Panthera onca* (**a**), *Puma concolor* (**b**), *Tapirus bairdii* (**c**), *Anolis capito* (**d**), *Porthidium nasutum* (**e**), *Smilisca cyanosticta* (**f**) and *Bolitoglossa mexicana* (**g**). Photographers: Norma Arce-Peña (pictures **a–c**), and Martín J. Cervantes-López (pictures **d–g**)

The density of medium- and large-sized terrestrial mammals was also lower in forest patches than continuous forest. This is surprising because they are all generalist species. This decrease of medium- and large-sized mammal density might be related to higher hunting pressures in forest patches than in the continuous forest of the reserve. Hunting is a well-known driver of terrestrial mammal elimination in human-modified landscapes (Lamb et al. 2017; Benítez-López et al. 2019; Osuri et al. 2020; Deere et al. 2020). In fact, our database, which includes large-bodied vertebrates such as the Jaguar (*Panthera onca*), Cougar (*Puma concolor*), Baird's tapir (*Tapirus bairdii*), Collared peccary (*Pecari tajacu*), and White-tailed deer (*Odocoileus virginianus*) (Fig. 2.4; see the complete list of species in Garmendia et al. 2013), are all frequently hunted in human-modified tropical forests, resulting in "empty" or "half-empty" forests (e.g. Peres 2001; Peres and Palacios 2007; Deere et al. 2020). Therefore, we speculate that hunting is a major driver of mammal "defaunation" in forest patches, although this remains to be tested. If this hypothesis is correct, in addition to preventing forest loss in the region, the conservation of terrestrial mammals will likely require enforcing strict controls on hunting, as has already been highlighted for other fragmented regions (e.g. Deere et al. 2020).

## 2.5 Final Remarks and Conservation Implications

Given the accelerated deforestation and fragmentation of natural forests worldwide, understanding the importance of forest patch size in preserving biodiversity has never been so urgent. Our findings suggest that patch size is a poor predictor of species density in fragmented tropical forests. This does not mean that forest loss does not have negative effects on biodiversity. Forest loss is a well-known driver of population decline and species extinction (Fahrig 2003; Watling et al. 2020), and thus, preventing deforestation in all landscapes, including those where little or no deforestation has yet occurred, should be a top priority, especially in the tropics (Peres 2005; Gibson et al. 2011; Phalan 2018; Edwards et al. 2019; Walker et al. 2019; Arroyo-Rodríguez et al. 2020).

In tropical regions that have been already deforested and fragmented, an important question is whether preservation and restoration strategies should be focused on large (Diamond 1975) or small patches (Fahrig 2020). The present study supports recent conclusions that preservation of all forests is important to biodiversity, especially preservation of small patches in human-dominated areas (Deane and He 2018; Belder et al. 2019; Wintle et al. 2019; Fahrig 2020). This is important since conservation strategies usually prioritize large patches. In the Lacandona rainforest, for instance, payments for ecosystem services are only given to landowners that preserve forest areas of >100 ha (CONAFOR 2020), leaving smaller patches open to future destruction. This threatens biodiversity maintenance in the region, because based on our and previous findings, small patches have, on a per-sample area basis, as high or higher value for biodiversity conservation as large patches and are not the near-exclusive habitat of generalist species. Thus, an urgent priority for biodiversity conservation is to redress the neglect of small forest patches in conservation plans within this and other biodiversity hotspots.

## 2.6 Conclusion

Several previous studies have suggested that small forest patches are less suitable to forest-specialist species due to their high edge-to-area ratio, which amplifies detrimental edge effects. This has led to a neglect of small forest patches as key biodiversity islands that contribute to the preservation of species diversity. Our results show that, independently of the taxa and habitat specialization, there are equivalent levels of species density across different-sized forest patches, and species density does not differ between continuous forest and forest patches. This contradicts the notion that forest-specialist species are unable to survive in small patches, likely because they are able to move across the landscape to use resources from several forest patches and even from the anthropogenic matrix. These findings have critical applied implications in the context of biodiversity islands. As deforestation expands, forest patches are becoming increasingly prevalent and smaller, particularly in rainforest landscapes. While prior work has been done to encourage landowners to preserve large forest patches, a greater attempt should be made to conserve small land patches. These small patches not only increase the amount of habitat for forest species, but they can also be a keystone of species movement, which is why their preservation should be considered a priority towards the conservation of the region's habitat. In examining the different factors that contribute to biodiversity island design, small forest patches should be given as much consideration as larger patches because they can often have the same or greater benefit to the wellbeing of the regional environment and the species within it.

**Acknowledgments** We are indebted to A. Garmendia, E. Carrara, H. A. Sánchez-de-Jesús, C. Rojas, M. Solà, I. M. González-Perez, D. Jiménez-González and G. Russildi for collecting some of the data. This research was supported by Rufford small grants (projects 16237-1, 22049-1, 23706-1, 24642-1), SEP-CONACyT (project 2015-253946), Programa de Apoyo a Proyectos de Investigación e Innovación Tecnológica (PAPIIT), DGAPA-UNAM (projects IA-203111, IB-200812, RR-280812, and IN-204215), and Idea Wild. CG-A thanks DGAPA-UNAM for her postdoctoral scholarship. We thank the landowners in the Marqués de Comillas region for allowing us to collect data on their properties, as well as the Montes Azules Biosphere Reserve and the National Commission of Natural Protected Areas. We especially thank Audon Jamangapé, Adolfo Jamangapé, Gilberto Jamangapé and Marta Aguilar for their invaluable field assistance and accommodation in the Marqués de Comillas region. Zoe Volenec, Ben Everett-Lane and an anonymous reviewer made valuable comments on the manuscript that helped us to improve it.

## References

Arriaga L, Espinoza J, Aguilar C, Martínez E, Gómez L, Loa E (2000) Regiones terrestres prioritarias de México. CONABIO, Mexico City

Arroyo-Rodríguez V, González-Perez IM, Garmendia A, Solà M, Estrada A (2013) The relative impact of forest patch and landscape attributes on black howler monkey populations in the fragmented Lacandona rainforest, Mexico. Landsc Ecol 28:1717–1727

Arroyo-Rodríguez V, Rojas C, Saldaña-Vázquez RA, Stoner KE (2016) Landscape composition is more important than landscape configuration for phyllostomid bat assemblages in a fragmented biodiversity hotspot. Biol Conserv 198:84–92

Arroyo-Rodríguez V, Saldaña-Vázquez RA, Fahrig L, Santos BA (2017) Does forest fragmentation cause an increase in forest temperature? Ecol Res 32:81–88

Arroyo-Rodríguez V, Fahrig L, Tabarelli M, Watling JI, Tischendorf L, Benchimol M, Cazetta E, Faria D, Leal IR, Melo FPL, Morante-Filho JC, Santos BA, Arasa-Gisbert R, Arce-Peña N, Cervantes-López MJ, Cudney-Valenzuela S, Galán-Acedo C, San-José M, Vieira ICG, Slik

JWF, Nowakowski J, Tscharntke T (2020) Designing optimal human-modified landscapes for forest biodiversity conservation. Ecol Lett 23:1404–1420
Asensio N, Arroyo-Rodríguez V, Dunn JC, Cristóbal-Azkarate J (2009) Conservation value of landscape supplementation for howler monkeys living in forest patches. Biotropica 41:768–773
Banks-Leite C, Ewers RM, Metzger JP (2010) Edge effects as the principal cause of area effects on birds in fragmented secondary forest. Oikos 119:918–926
Belder DJ, Pierson JC, Ikin K, Blanchard W, Westgate MJ, Crane M, Lindenmayer DB (2019) Is bigger always better? Influence of patch attributes on breeding activity of birds in box-gum grassy woodland restoration planting. Biol Conserv 236:134–152
Benítez-López A, Santini L, Schipper AM, Busana M, Huijbregts MAJ (2019) Intact but empty forests? Patterns of hunting-induced mammal defaunation in the tropics. PLoS Biol 17: e3000247
Bibby CJ, Burgess ND, Hill DA, Mustoe S (2000) Bird census techniques. Academic, London
Campbell J (1998) Amphibians and reptiles of Northern Guatemala, the Yucatan, and Belize, Animal Natural History series. University of Oklahoma Press, Norman
Carabias J, de la Maza J, Cadena R (2015) Conservación y desarrollo sustentable en la Selva Lacandona. 25 años de actividades y experiencias. Natura y Ecosistemas Mexicanos A.C, Mexico City
Carrara E, Arroyo-Rodríguez V, Vega-Rivera JH, Schondube JE, de Freitas SM, Fahrig L (2015) Impact of landscape composition and configuration on forest specialist and generalist bird species in the fragmented Lacandona rainforest, Mexico. Biol Conserv 184:117–126
Chase JM, McGill BJ, McGlinn DJ, May F, Blowes SA, Xiao X, Knight TM, Purschke O, Gotelli NJ (2018) Embracing scale-dependence to achieve a deeper understanding of biodiversity and its change across communities. Ecol Lett 21:1737–1751
CONAFOR (2020) Apoyos CONAFOR. Comisión Nacional Forestal, Mexico. https://www.gob.mx/conafor/acciones-y-programas/apoyos-conafor
Crawley MJ (2007) Statistical modelling in the R book. Wiley, Chichester
Crump ML, Scott NJ Jr (1994) Visual encounter surveys. In: Heyer WR, Donnelly MA, McDiarmid RW, Hayek LC, Foster MC (eds) Measuring and monitoring biological diversity: standard methods for amphibians. Smithsonian Institution Press, Washington, pp 84–92
Deane DC, He F (2018) Loss of only the smallest patches will reduce species diversity in most discrete habitat networks. Glob Change Biol 24:5802–5814
Deere NJ, Guillera-Arroita G, Platts PJ, Mitchell SL, Baking E, Bernard H, Haysom JK, Reynolds G, Seaman DJI, Davies ZG, Struebig MJ (2020) Implications of zero-deforestation commitments: forest quality and hunting pressure limit mammal persistence in fragmented tropical landscapes. Conserv Lett 13:e12701
Diamond JM (1975) The island dilemma: lessons of modern biogeographic studies for the design of natural reserves. Biol Conserv 7:129–146
Diamond JM (1976) Island biogeography and conservation: strategy and limitations. Science 193: 1027–1029
Dunning JB, Danielson BJ, Pulliam HR (1992) Ecological processes that affect populations in complex landscapes. Oikos 65:169–175
Edwards DP, Socolar JB, Mills SC, Burivalova Z, Pin Koh L, Wilcove DS (2019) Conservation of tropical forests in the Anthropocene. Curr Biol 29:R1008–R1020
Fahrig L (2003) Effects of habitat fragmentation on biodiversity. Annu Rev Ecol Evol Syst 34:487–515
Fahrig L (2013) Rethinking patch size and isolation effects: the habitat amount hypothesis. J Biogeogr 40:1649–1663
Fahrig L (2020) Why do several small patches hold more species than few large patches? Glob Ecol Biogeogr 29:615–628
Ferreira AS, Peres CA, Bogoni JA, Cassano CR (2018) Use of agroecosystem matrix habitats by mammalian carnivores (Carnivora): a global-scale analysis. Mammal Rev 48:312–327

Fletcher RJ, Didham RK, Banks-Leite C, Barlow J, Ewers RM, Rosindell J, Holt RD, Gonzalez A, Pardini R, Damschen EI, Melo FPL, Ries L, Prevedello JA, Tscharntke T, Laurance WF, Lovejoy T, Haddad NM (2018) Is habitat fragmentation good for biodiversity? Biol Conserv 226:9–15

Galán-Acedo C, Arroyo-Rodríguez V, Andresen E, Arregoitia LV, Vega E, Peres CA, Ewers RM (2019) The conservation value of human-modified landscapes for the world's primates. Nat Commun 10:152

Garmendia A, Arroyo-Rodríguez V, Estrada A, Naranjo E, Stoner KE (2013) Landscape and patch attributes impacting medium- and large-sized terrestrial mammals in a fragmented rain forest. J Trop Ecol 29:331–344

Gentry AH (1982) Patterns of Neotropical plant species diversity. In: Hecht MK, Wallace B, Prance GT (eds) Evolutionary biology. Plenum Press, New York, pp 1–84

Gibbs JP (1998) Distribution of woodland amphibians along a forest fragmentation gradient. Landsc Ecol 13:263–268

Gibson L, Lee TM, Koh LP, Brook BW, Gardner TA, Barlow J, Peres CA, Bradshaw CJA, Laurance WF, Lovejoy TE, Sodhi NS (2011) Primary forests are irreplaceable for sustaining tropical biodiversity. Nature 478:378–383

Hansen MC, Potapov PV, Moore R, Hancher M, Turubanova SA, Tyukavina A, Thau D, Stehman SV, Goetz SJ, Loveland TR, Kommareddy A, Egorov A, Chini L, Justice CO, Townshend JRG (2013) High-resolution global maps of 21st-century forest cover change. Science 342:850–853

Harper KA, MacDonald SE, Burton PJ, Chen J, Brosofske KD, Saunders SC, Euskirchen ES, Roberts D, Jaiteh MS, Esseen P (2005) Edge influence on forest structure and composition in fragmented landscapes. Conserv Biol 19:768–782

Hernández-Ordóñez O, Martínez-Ramos M, Arroyo-Rodríguez V, González-Hernández A, González-Zamora A, Zárate DA, Reynoso VH (2014) Distribution and conservation status of amphibian and reptile species in the Lacandona rainforest, Mexico: an update after 20 years of research. Trop Conserv Sci 7:1–25

Hernández-Ordóñez O, Urbina-Cardona N, Martínez-Ramos M (2015) Recovery of amphibian and reptile assemblages during old-field succession of tropical rainforests. Biotropica 47:377–388

Hernández-Ruedas MA, Arroyo-Rodríguez V, Meave JA, Martínez-Ramos M, Ibarra-Manríquez G, Martínez E, Jamangapé G, Melo FPL, Santos BA (2014) Conserving tropical tree diversity and forest structure: the value of small rainforest patches in moderately-managed landscapes. PLoS One 9(6):e98931

IUCN (2019) The IUCN red list of threatened species. Version 2019-3. http://www.iucnredlist.org. Downloaded on 10 December 2019

Kunz TH (1982) Ecology of bats. Plenum Publishing Corporation, New York

Lamb CT, Mowat G, McLellan BN, Nielsen SE, Boutin S (2017) Forbidden fruit: human settlement and abundant fruit create an ecological trap for an apex omnivore. J Anim Ecol 86:55–65

Laurance WF, Lovejoy TE, Vasconcelos HL, Bruna EM, Didham RK, Stouffer PC, Gascon C, Bierregaard RO, Laurance SG, Sampaio E (2002) Ecosystem decay of Amazonian forest fragments: a 22-year investigation. Conserv Biol 16:605–618

Lee JC (2000) A field guide to the amphibians and reptiles of the Maya world: the lowlands of Mexico, Northern Guatemala, and Belize. Cornell University Press, Ithaca

Medellín R, Arita H, Sánchez O (2008) Identificación de los murciélagos de México. Clave de Campo. Asociación Mexicana de Mastozoología, México D.F

Mendenhall CD, Kappel CV, Ehrlich PR (2013) Countryside biogeography. In: Levin SA (ed) Encyclopedia of biodiversity, 2nd edn. Academic, Waltham, pp 347–360

Newbold T, Hudson LN, Arnell AP, Contu S, De Palma A, Ferrier S, Hill SLL, Hoskins I, Lysenko AJ, HRP P, Burton VJ, CWT C, Emerson S, Gao D, Hale GP, Hutton J, Jung M, Sanchez-Ortiz K, Simmons BI, Whitmee S, Zhang H, Scharlemann JPW, Purvis A (2016) Has land use pushed terrestrial biodiversity beyond the planetary boundary? A global assessment. Science 353:288–291

Osuri AM, Mendiratta U, Naniwadekar R, Varma V, Naeem S (2020) Hunting and forest modification have distinct defaunation impacts on tropical mammals and birds. Front For Glob Change 2:87

Peres CA (2001) Synergistic effects of subsistence hunting and habitat fragmentation on Amazonian forest vertebrates. Conserv Biol 15:1490–1505

Peres CA (2005) Why we need megareserves in Amazonia. Conserv Biol 19:728–733

Peres CA, Palacios E (2007) Basin-wide effects of game harvest on vertebrate population densities in Amazonian forests: implications for animal-mediated seed dispersal. Biotropica 39:304–315

Pfeifer M, Lefebvre V, Peres CA, Banks-Leite C, Wearn OR, Marsh CJ, Butchart SHM, Arroyo-Rodríguez V, Barlow J, Cerezo A, Cisneros L, D'Cruze N, Faria D, Hadley A, Harris SM, Klingbeil BT, Kormann U, Lens L, Medina-Rangel GF, Morante-Filho JC, Olivier P, Peters SL, Pidgeon A, Ribeiro DB, Scherber C, Schneider-Maunoury L, Struebig M, Urbina-Cardona N, Watling JI, Willig MR, Wood EM, Ewers RM (2017) Creation of forest edges has a global impact on forest vertebrates. Nature 551:187–191

Phalan BT (2018) What have we learned from the land sparing-sharing model? Sustainability 10: 1760

R Core Team (2013) R version 3.0. 1. The R core team, 2001e2013. http://www.rproject.org

Reider IJ, Donnelly MA, Watling JI (2018) The influence of matrix quality on species richness in remnant forest. Landsc Ecol 33:1147–1157

Russildi G, Arroyo-Rodríguez V, Hernández-Ordóñez O, Pineda E, Reynoso VH (2016) Species- and community-level responses to habitat spatial changes in fragmented rainforests: assessing compensatory dynamics in amphibians and reptiles. Biodivers Conserv 25:375–392

Saito M, Koike F (2013) Distribution of wild mammal assemblages along an urban-rural-forest landscape gradient in warm temperate East Asia. PLoS One 8(5):e65464

Sánchez-de-Jesús H, Arroyo-Rodríguez V, Andresen E, Escobar F (2016) Forest loss and matrix composition are the major drivers shaping dung beetle assemblages in a fragmented rainforest. Landsc Ecol 31:843–854

San-José M, Arroyo-Rodríguez V, Meave JA (2020) Regional context and dispersal mode drive the impact of landscape structure on seed dispersal. Ecol Appl 30:e02033

Tabarelli M, Peres CA, Melo FP (2012) The 'few winners and many losers' paradigm revisited: emerging prospects for tropical forest biodiversity. Biol Conserv 155:136–140

Taubert F, Fischer R, Groeneveld J, Lehmann S, Müller MS, Rödig E, Wiegand T, Huth A (2018) Global patterns of tropical forest fragmentation. Nature 554:519–522

Walker RT, Simmons C, Arima E, Galvan-Miyoshi Y, Antunes A, Waylen M, Irigaray M (2019) Avoiding Amazonian catastrophes: prospects for conservation in the 21st century. One Earth 1: 202–215

Watling JI, Arroyo-Rodríguez V, Pfeifer M, Baeten L, Banks-Leite C, Cisneros LM, Fang R, Hamel-Leigue AC, Lachat T, Leal IR, Lens L, Possingham HP, Raheem DC, Ribeiro DB, Slade EM, Urbina-Cardona JN, Wood EM, Fahrig L (2020) Support for the habitat amount hypothesis from a global synthesis of species density studies. Ecol Lett 23:674–681

Willis EO (1984) Conservation, subdivision of reserves, and the antidismemberment hypothesis. Oikos 42:396–398

Wintle BA, Kujala H, Whitehead A, Cameron A, Veloz S, Kukkala A, Moilanen A, Gordon A, Lentini PE, Cadenhead NCR, Bekessy SA (2019) Global synthesis of conservation studies reveals the importance of small habitat patches for biodiversity. Proc Natl Acad Sci U S A 116: 909–914

# Chapter 3
# Regenerative Agriculture as Biodiversity Islands

**Brett Levin**

**Abstract** When the amount of biological diversity in an agricultural system is significantly higher than the baseline biodiversity of the surrounding area, the agricultural system itself may be recognized as a biodiversity island. Regenerative agricultural systems, which build and maintain fertility through time, may increase and maintain biodiversity as an integrated component of food production. Increases in biodiversity within an agricultural system can span all biological taxonomic kingdoms and vast numbers of classes and species within each. As such, regenerative agricultural management techniques geared toward harmonizing agricultural productivity and biodiversity conservation can contribute to mitigating or reversing detrimental effects of human impacts on landscapes. Greater diversity through intercropping, companion planting, combinations of perennial and annuals crops, cover cropping, hedgerows and diverse edge plantings, reduced agrochemical use, silvopasture with rotational grazing, and selection of rare, heirloom, underutilized, or diverse genetics allows for biodiversity to harmonize with agricultural production. In landscapes lacking protected areas or intact ecosystems, habitat restoration and preservation within agricultural systems can enable both farm productivity and biodiversity to increase. An integration of restoration and agriculture through farmer managed natural regeneration, rewilding, and incorporation of traditional ecological knowledge as operational management approaches within a regenerative agricultural framework may also achieve such ends. Much of the origins of regenerative agriculture emerged from indigenous practice of food production and traditional ecological knowledge that maintains biodiversity. Examples of regenerative agriculture as biodiversity islands, where farm productivity and improved biodiversity are achieved, span a multitude of crops, regions, and cultures throughout the world.

**Keywords** Agroforestry · Cover cropping · Intercropping · Habitat restoration · Hedgerows · Reduced agrochemical use · Silvopasture · Traditional ecological knowledge

---

B. Levin (✉)
Dietrick Institute for Applied Insect Ecology, Ventura, CA, USA
e-mail: Brett.Levin@aya.yale.edu



F. Montagnini (ed.), *Biodiversity Islands: Strategies for Conservation in Human-Dominated Environments*, Topics in Biodiversity and Conservation 20,
https://doi.org/10.1007/978-3-030-92234-4_3

## 3.1 Introduction

Historically, expansion of agriculture has contributed to biodiversity loss globally. Conversion of native ecosystems to less diverse agriculturally productive landscapes tends to follow trends in global human population (Crist et al. 2017). Generally, a positive feedback loop exists related to population growth and ecosystem loss; as human populations increase, more land is converted to agriculture, providing the calories necessary for further population growth and greater land conversion (Henderson and Loreau 2019, Gustafson et al. 2020). While the technological and operational achievements of modern industrial agriculture provide relatively inexpensive calories to significant populations on a daily basis, the negative social costs in pollution, farmworker exploitation and health, and biodiversity loss are extensive. Even with considerable advancements and gains in yield per hectare through genetic selection and operational improvements, global biodiversity loss trends continue in response to agricultural expansion (Newbold et al. 2016). An inverse relationship typically emerges between increased crop specialization with production to achieve economies of scale and decreased landscape biodiversity and ecological function (Klasen et al. 2016). From deforestation for cattle ranching in the Amazon to commodity crop expansion in sub-Saharan Africa and corn and soybean production throughout the Midwestern United States, the global trends of rising global population linked with agricultural expansion and biodiversity loss continue (Fearnside 2017).

With the increase of such landscape degradation, ecosystem loss, and loss of species diversity throughout the globe, an urgency of design and communication frameworks is needed to protect biodiversity within the agricultural matrix (Kidd et al. 2019). Biodiversity islands, which may be defined as areas of high biodiversity nested within ecologically degraded, human-dominated landscapes, are one such instrument (Montagnini et al. 2022). The number of species in a biodiversity island is greater than the biodiversity of the surrounding human managed landscape. As such, biodiversity islands can take many forms, such as parklands surrounded by urban sprawl, conserved forestland surrounded by degraded and overgrazed pasture, or riparian corridors in a monocropped agricultural matrix (Montagnini et al. 2022). These ecological refugia can provide social, economic, and environmental value through time. Agricultural systems which utilize the practices of regenerative agriculture may harmonize ecosystem and agricultural productivity with biodiversity conservation, thus operating as biodiversity islands within the landscape.

Regenerative agriculture is an emerging term with a variety of definitions stemming from a diversity of land use approaches, ecological and social contexts, and lineages of agricultural practice. Consensus regarding an exact scientific definition for regenerative agriculture poses a challenge and frameworks for socio-economic and social implementation are sparse within the academic literature (Schreefel et al. 2020). In recent years, the term regenerative agriculture has gained popularity, differentiating itself from organic, conventional, conservation, or sustainable agriculture. Explicit practice-based definitions of regenerative agriculture may by limiting, given the system's approach and broad range of contexts towards which regenerative agricultural principles may be applied (Soloviev and Landua 2016). Regenerative agriculture draws from centuries of indigenous and traditional agricultural practices and decades of scientific study and applied research on organic farming, soil health, agroecology, permaculture, holistic management, and agroforestry around the globe. Generally, regenerative agriculture is a system of farming principles and practices that increase biodiversity, enrich soils, improve watersheds, and enhance ecosystem services (White 2020). They aim to capture carbon in soil and aboveground biomass, reversing current global trends of atmospheric carbon dioxide accumulation while offering increased yields, resilience to climate instability, and higher health and vitality for farming and ranching communities (www.regenerativeagriculturedefinition.com). The social aspects of agricultural production are also addressed by regenerative agriculture, in which production supports just and reciprocal relationships amongst all stakeholders. While a sustainable system maintains itself through time, a regenerative system builds and enhances ecological and social functioning, recognizing whole systems rather than reductionist viewpoints (Gibbons 2020). The definition and practice of regenerative agriculture continues to evolve.

This chapter focuses on regenerative agricultural systems which support wildlife, biodiversity conservation, and a diversity of genetic resources harmonized with farm productivity. Such agricultural methods may take many forms, from land sparing and land sharing, through traditional cultivation methods, and various other working-lands, agroecological management, and operational techniques (Perfecto et al. 2009; Gliessman 2016; Altieri 2018; Wagner 2020). Through time, as agricultural practices enhance fertility, sequester carbon, improve soil structure and water holding capacity, and reduce agrochemical inputs, farm biodiversity may increase as well (Toensmeier 2016; Rhodes 2017; Meena et al. 2020). When such biodiverse agricultural areas are within ecologically degraded human dominated surroundings, they act as biodiversity islands within the landscape. While general practices are described in this chapter, frameworks for implementation must consider social, economic, environmental, and cultural circumstances of each location. The following techniques and considerations described are useful for farmers, policy makers, researchers, and decision makers in landscape management.

## 3.2 Regenerative Management Increasing Biodiversity

Regenerative agricultural systems may be designed and managed to increase the on-farm presence of cultivated and wild species from numerous taxonomic kingdoms. This can include systems for water catchment, roads and pathway placement, and crop selection as well as more specific practices such as intercropping, polycultures, agroforestry, insectary hedgerows, reduced agrochemical use, and habitat restoration. These practices may also be interwoven with cultural and social restoration and the use of traditional ecological knowledge to fully foster regeneratively managed agricultural systems as part of the development of biodiversity islands within a landscape.

### 3.2.1 System Design and Management Plans

System design and management plans for increasing biodiversity through regenerative agriculture are highly specific to the region, social context, diversity of crop selection, and particular biophysical attributes of the site, such as geology, soil, and hydrology (Cabeza and Moilanen 2001; Mendenhall et al. 2014). The degree of existing ecological degradation, surrounding patch dynamics, and associated population ecology as it relates to island biogeography are also important factors determining species migration and baseline site biodiversity (Tavares et al. 2019). In severely degraded sites or existing monoculture industrialized agricultural systems, significant changes in cultural and management practices may be necessary to increase on-farm biodiversity. Yield of singular specialized crops may need to decrease to achieve greater on-farm biodiversity, while greater diversified crop yields, species abundance, and provision of ecosystem services can result (Altieri 2015). In many indigenous, traditional, and agroecological agricultural systems, management and design integrate biodiversity into production. A key principle of such agroecological management is designing agricultural ecosystems to mimic the function of local ecosystems through productive and diverse native species or agronomic crop analogs (Gliessman 2016; Altieri 2018). Integrating such practice through improved agricultural methods promotes habitat for a broad range of microbial, animal, plant, and fungi communities (Altieri 1999; Benayas and Bullock 2012). Sustainable intensification of agriculture through the application of agroecological principles can also increase trophic complexity, niche formation, and the biodiversity potential of the agroecosystem (Liere et al. 2017; Atkinson and Watson 2019).

Accordingly, system design and associated management for improved soil health can greatly increase biodiversity potential (Wagg et al. 2014). Terrestrial ecosystem functioning and biodiversity are controlled largely by soil microbial dynamics and soil health, whereas soil health is the capacity of soil to function as an essential living system, within ecosystem and land-use boundaries, to sustain plant and animal productivity, maintain or enhance water and air quality, and promote plant and animal life (Giller et al. 1997; Lehmann et al. 2020). Management practices that enhance soil health are key indicators of ecosystem productivity and associated biodiversity (Brussaard et al. 2007).

From the physical and biophysical perspective, proper system design is essential in developing biodiverse and productive agricultural systems. Well-designed systems can be more productive, pest resistant, and water efficient, and they conserve and cycle nutrients more effectively (Doré et al. 2011; Ching 2018). Darren Dougherty's Regrarian's Platform®, built upon P.A. Yeoman's Keyline Scale of Permanence, offers one framework for considering factors according to the amount of time and effort required for farm modification by humans (Yeomans 1958). The consideration of these factors, including climate, geography, water, access, forestry, buildings, fences, soil, economy, and energy, can lead to the lasting commercial and ecological viability of the agricultural system (Chabay et al. 2015, www.Regrarians.org).

Long-term improved biodiversity can also correlate directly with other farm system design considerations that utilize an ecosystem approach (Dominati et al. 2019). For example, effective context-specific water management is extremely important for landscape productivity and improved biodiversity potential. When culturally, ecologically, and financially appropriate, earthworks designed and implemented to optimize on-farm water retention can increase the capacity for onsite cultivation and associated biodiversity (Socci et al. 2019). Keyline design, developed by agricultural innovator P.A. Yeomans, involves processes such as bringing water from valleys to ridgelines, to capture, slow, spread, store, and integrate water on the farm (Yeomans 1958). Utilizing such water management approaches, rainwater can become a tool for soil building, increased biomass accumulation, and increased biodiversity, rather than an erosive and potentially polluting force.

Planting arrangement and species selection are also important design considerations. For example, several vegetation strata, including low-lying groundcovers, understory herbs, low to mid story shrubs, and trees, some of which may be nitrogen-fixing, reduce dependency on agricultural inputs, enhancing synergisms among both biological and biophysical system functioning (Nair 2017). As these strategically designed biodiverse agricultural systems develop, in-situ mulching, improved nutrient cycling, increased water retention, more buffered temperatures, reduced soil evaporative loss, and increased biological control may result (Schroth et al. 2001; Lichtfouse 2018). Use of plants and animal breeds adapted to local conditions can also reduce dependence on foreign inputs and further increase nutrient cycling, soil fertility, and add biodiversity to the system (Enri et al. 2017; Jose et al. 2019). Regenerative agriculture may harness such principles, design strategies, and management techniques to achieve improved biodiversity outcomes for the farm system. If located within a degraded landscape with low biodiversity, such practices may allow for a biodiversity island to emerge.

### *3.2.2 Intercropping and Polycultures*

Polycultures and intercropping involve the cultivation of multiple plant species in mixtures, typically in two or more parallel rows, though they may be planted in complex assemblages. These practices increase biodiversity and are utilized throughout the world for several agricultural benefits including diversified yields, improved biological pest control, weed control, reduced wind erosion, and improved water infiltration (McLaughlin and Mineau 1995; Corrado et al. 2019). Compared to a monocrop system, an intercropped system is inherently more biodiverse, given the increased number of plant species and varieties in cultivation, either planted at the same time or successionally on the same land. Food web interactions and habitat complexity also increase (Moss et al. 2020). Synergistic relationships of vertebrate, invertebrate, and microbial communities support and harbor more complex, resilient, and biodiverse farm ecosystems (Jackson et al. 2007; Zhao et al. 2019). Intercropping can consist of annuals, perennials, or combinations of the two.

The selection of intercropped species requires knowledge of which species grow well together. The degree of biodiversity improvements will depend on the species selected and cultivation strategy. In some instances, weeds are left to grow as a trap crop for insect pests, increasing invertebrate diversity (Reddy 2017). In other instances, diversified polycultures are intentionally planted from nursery stock or seeded in situ. Studies indicate that intercropping increases invertebrate abundance compared to monocrop systems (Cárcamo and Spence 1994; Tilman 2020). Similarly, in terms of microbial biodiversity, intercropping of diverse landscapes and tree species results in greater soil microbiological diversity (Lacombe et al. 2009; Chen et al. 2020). Overall, through intercropping, biodiversity of plants, soil, and animal life can increase and contribute to the creation of biodiversity islands in otherwise degraded landscapes.

### *3.2.3 Agroforestry*

Agroforestry systems (AFS), an intensive land management system that optimizes benefits from the biological interactions created when trees and/or shrubs are deliberately combined with crops and/or livestock, can increase farm productivity while supporting biodiversity and providing social and economic benefits for farmers (Leakey 1999; Jose 2009, 2012; Montagnini and Metzel 2017; Udawatta et al. 2019; Montagnini 2020, https://www.aftaweb.org/). AFS are heterogeneous in their design, management, and species composition, and therefore have diverse values in terms of restoration, productivity, and conservation.

Within AFS, the highest amounts of biodiversity are typically found within successional AFS, home gardens, forest gardens, and other complex multi-strata systems (Huang et al. 2002). Other simpler AFS with fewer species of plants or animals will typically foster less biodiversity, though they may still be considered a biodiversity island if the surrounding landscape is degraded. In most instances, the AFS will be the most biodiverse cultivated system of the human dominated landscape (Schroth et al. 2004; McNeely and Schroth 2006). Riparian corridors, living fences, windbreaks, and perimeter hedges may also provide connectivity in fragmented agricultural landscapes and help bring greater biodiversity to a farm (Jose 2012). Perennial crops under shade or silvopastoral systems (SPS) with scattered trees for shade can also provide greater biodiversity and ecological benefits than monocultures or degraded fallow lands, though providing less biodiversity than complex multistrata AFS (Leakey 1999). To favor biodiversity restoration and conservation, AFS should increase their structural complexity in terms of strata and number of species (Montagnini and del Fierro 2022). Management and regenerative agricultural practices that employ agroforestry systems can increase biodiversity and aid in the development of agricultural lands as biodiversity islands within the landscape.

Correspondingly, successional agroforestry systems established through enrichment planting can mimic natural regeneration to produce biodiverse and productive food systems (Young 2017). Successional agroforestry systems consist of multi-strata multifunctional species assemblages that collectively appear to have a similar structure to native forests. They can include both introduced and native species. Native species may emerge from the existing seedbank where seeds are otherwise unavailable. Tree-growth and crop productivity are achieved by management that promotes functional characteristics of key natural successional stages of the native landscape. As stands mature, unique habitats emerge, creating the conditions for greater biodiversity and opportunities for the establishment of a greater variety of successional productive species.

Many native or imported species cannot be planted in open plantations because seedlings are shade tolerant and otherwise will not germinate. As such, enrichment plantings of food bearing species within degraded landscapes may produce highly diverse agroforestry systems (Montagnini et al. 1997). Such practices are not new. For centuries, numerous indigenous cultures recognized the resiliency and food bearing potential of biodiverse successional agroforestry systems as forest analogues (Bertsch 2017).

### *3.2.4 Cover Cropping*

Cover cropping, an agricultural technique where pure or mixed stands of perennial or annual herbaceous plants are grown to cover soil and improve fertility, has also been shown to increase farm biodiversity. As a tool for regenerative agriculture, cover cropping legumes, cereals, and other plant mixtures can improve soil structure, soil fertility, pest management, and biodiversity. Moreover, cover cropping improves water holding capacity and infiltration, reduces soil erosion, adds organic matter to soil horizons, cycles nutrients, nourishes the soil food web, reduces weed competition, and aids in the regulation of soil temperatures (Altieri 2015).

Cover crop management practices vary significantly depending on regional climate, species selection, tillage and clipping frequency, and time of seeding (Finch and Sharp 1976). Rye (*Secale cereale* L.), clovers (*Trifolium* spp.), vetches (*Vicia* spp.), alfalfa (*Medicago sativa*), and leguminous *Pueraria, Stylosanthes*, and *Centrosema* are commonly planted as cover crops (Altieri 1995). One study indicated the planting of leguminous cover crops (*Mucuna pruriens var. utilis*) increases soil macrofauna and nematofauna in maize cultivation (Blanchart et al. 2006). In rubber plantations, Kuzdu, *Pueraria phaseoloides* is a nitrogen fixing legume commonly used as a cover crop. Wild peanuts (*Arachis pintoi*) are commonly used as cover crops for coffee and silvopastural systems in Central and South America (De La Cruz et al. 1994).

A practice used from tropical to temperate systems, cover crops also promote invertebrate diversity, increase populations of beneficial parasitoids, and can improve biological pest control (Altieri and Schmidt 1985). As a tool in regenerative agriculture, cover cropping provides multiple benefits and increases belowground and aboveground biodiversity of the farm system. As such, cover cropping can enhance the capacity of regenerative agricultural systems as biodiversity islands.

### 3.2.5 *Insectary Hedgerows*

Planting a broad range of flowering perennial and annual species in hedgerows throughout a farm can harbor a diverse and balanced insect ecology while greatly reducing pest pressure on crops and maintaining on-farm biodiversity in regenerative agricultural systems (Long et al. 1998; Landis et al. 2000). These insectary hedgerows include plants that attract both pests and associated beneficial predatory insects, providing a breeding ground for beneficial insect populations to increase and expand into cultivated spaces. Flowering species which bloom in succession throughout the growing season should be included to ensure that nectar is available to support beneficial insects throughout the growing season (Holland 2019).

Within insectary hedgerows, plants can perform different functions related to biological pest control and on-farm biodiversity. Plants that are more attractive to a pest than the crop plant may be monitored as indicators of pest populations and developing pest pressures. For example, pole beans are more attractive to spider mites than tomato, pepper, cucumber, or strawberry. As such, indication of spider mites on pole beans can allow farmers to control outbreaks before significant crop damage occurs. Similarly, trap crops are plants which are more attractive to pests than the commercial crop, taking most of the pest damage and sparing the desired crop (Parker et al. 2016). Used in conjunction with one another, a monitoring plant can also act as a trap plant.

When beneficial insects become established by feeding on the pests, these trap crops can become banker crops providing a food source for the increase of beneficial insect populations (Miller et al. 2017). Banker plants attract and host pests and are used as an insectary to grow more beneficial insects (Balzan 2017). For example, fast growing cereal grasses such as ryegrass can be used to attract aphids that become a food source for aphid predators and parasites. In each of these instances, the simple increase of plant and arthropod diversity in the system, through the planting of insectary hedgerows, promotes great biodiversity and establishment of the agricultural system as a biodiversity island in an otherwise degraded, human dominated landscape.

### 3.2.6 *Reduced Agrochemical Use*

Another important aspect of regenerative agriculture as it relates to biodiversity is the reduction of agrochemical use. Use of pesticides, herbicides, and conventional fertilization all may contribute towards decreases in biodiversity (Benton et al. 2003; Mandal et al. 2020). Agrochemical use may also negatively affect nutrient cycling, the soil food web, decomposition of soil organic matter, beneficial insects and natural predator populations. Excessive use of agrochemicals may also increase $NO_2$ and other greenhouse gas emissions, thus affecting air quality and farmworker health, which are antithetical to the outcomes and principles of regenerative agriculture (Kimbrell 2002).

Insect populations have significantly decreased in recent years, largely attributable to increased pesticide use (Sánchez-Bayo and Wyckhuys 2019). In many instances, the effects of these chemical applications go beyond their point of use and can be associated with decreased biodiversity in the broader ecosystem. Runoff of excess nitrogen and phosphorus in fertilizers enters waterways and reduces aquatic biodiversity (Ali et al. 2011).

Fortunately, as described in this section, biological practices for pest management, weed abatement, and fertility are feasible and can increase on farm biodiversity without agrochemical use (Jørgensen and Kudsk 2006). Such regenerative practices can also decrease the costs of inputs through time by improving in situ nutrient cycling (Coleman et al. 1983). Agrochemical applications contradict the biological practices of regenerative agriculture particularly related to biodiversity. Therefore, their use should be minimized in the establishment and maintenance of agricultural systems as biodiversity islands.

### *3.2.7 Habitat Restoration Within Regenerative Agriculture*

Another method to increase on-farm biodiversity is through the restoration of habitat and ecosystems within low diversity farm systems. Establishing areas of natural vegetation on farms allows the landscape to fulfill greater ecological function and provides additional ecosystem services simultaneously with agricultural production. In degraded lands, restoration of habitat towards these ends directly relates to the ecological objectives and goals of regenerative agriculture. The co-benefits of on-farm habitat restoration include production of nonagricultural products, habitat for various life forms, prevention of soil erosion through runoff and wind, increased carbon sequestration, and increased water infiltration and watershed health (Benton et al. 2003). The intentional integration of habitat restoration within the landscape is therefore a strategy a farmer may choose to implement as part of a regenerative agricultural system. Examples of biodiversity enhancing on-farm habitat restoration include farmer managed natural regeneration, successional agroforestry systems which integrate native species, and rewilding of farmlands.

Farmer managed natural regeneration increases biodiversity and farm productivity by allowing the existing on-farm sources of regeneration to germinate, grow, and compete with other vegetation. Through observation and selection of which useful species emerge, one can manage, tend, and harvest from more diversified farm ecosystems. This can be achieved by allowing natural regeneration to take place in fields or on selected patches within the farm (Wintle et al. 2019) Valuable species are selected to persist, thus creating a low-cost and biodiverse foundation from which productive agricultural systems may emerge. For example, in the Sahel region of Niger, rather than weeding all species, farmers may select specific species to remain in the fields. By caring for these naturally regenerating drought tolerant species, greater diversity and yields result. These practices have been a contributing factor in a low-cost option for increased diversity and indigenous genetics of gardens and agriculture throughout the region (Reij and Garrity 2016).

Allowing succession to occur in a slightly more hands-off approach may be known as rewilding or natural regeneration. Allocating land for rewilding, some areas of crop cultivation may be lost, but the trade-off results in greater diversity, pollination, and other ecosystem services (Navarro and Pereira 2015). These areas may also be seeded with a diversity of desired annual and perennial species, with minimal continued management.

The beneficial outcomes of natural regeneration on sections of farmland are particularly clear in certain grazing systems. For example, The Knepp Wildland Project in the United Kingdom originally utilized a traditional pastureland. As cattle were removed from sections of the farm, those areas underwent rapid natural regeneration. In some areas, existing seed banks were able to emerge and other areas were seeded with desirable species. After tree establishment, cattle were reintroduced to sections of the farm, where they had access to increased forage and greater shade, functioning as a silvopastoral system (Tree 2019). This integration of rewilding and natural regeneration provided habitat for a vastly greater number of local species, while still providing farm yields. The farm was transformed into a biodiversity island within the landscape.

### 3.2.8 Silvopasture and Rotational Grazing

Silvopasture with rotational grazing is another management strategy which can improve agricultural biodiversity as part of a regenerative agricultural system (Jose et al. 2019). Silvopasture is a type of agroforestry system consisting of the intentional combination of trees, forage plants and livestock together as an integrated, intensively-managed production system (https://www.aftaweb.org/). Silvopasture can provide profitable opportunities for tree growers, forest landowners, and livestock producers through the integration of what are typically separate production of tree crops and livestock. The benefits of rotational grazing are site and context specific but can include improved forage production, soil health, fertility, soil carbon storage, drought resistance, weed control, human and animal relationships, animal welfare, an extended grazing season, reduced forage waste, and reduced parasite problems (Orefice and Carroll 2017; Jose et al. 2019). Combined with silvopasture, additional economic benefits of tree production may emerge such as reduced fertilizer requirements, improved yields, increased weight gain, and reduced fodder needs (Gabriel 2018). Compared to monoculture tree cultivation or continuous grazing, silvopasture with rotational grazing can greatly increase biodiversity (McDermott and Rodewald 2014). Silvopastoral systems can harbor a high diversity of cultivated species in addition to a wide range of arboreal wildlife habitat.

As for rotational grazing, the diversity, quality, and longevity of forage species can increase when adequate rest is given to the grazed area, when compared to a continuous grazing system. Such outcomes are dependent on stocking rate, paddock size, longevity of grazing, and regional climatic and biophysical factors of the farm (Gabriel 2018). If managed optimally for the appropriate context of the farm, rotational grazing provides opportunities for a greater variety of forage species to persist and for greater profitability (Orefice et al. 2019). Additionally, greater farm biodiversity may be present in rare, native, or unique livestock breeds as well as the incorporation of different species including goats, chickens, ducks, pigs, cattle, buffalo, and others (Gabriel 2018). With proper management, silvopasture and rotational grazing can allow for greater biodiversity to emerge (McAdam and McEvoy 2009).

With both silvopasture and rotational grazing, one should learn the benefits as well as the risks before adopting the practices. Integration of multi-species grazing schemes may increase parasite loads if not managed properly. In certain areas, legislation prevents grazing on lands used for the cultivation of food crops within a specified time period preceding harvest in order to prevent contamination risks. Transition of land into silvopasture or grazing areas without proper management can damage soil, cause erosion, and eliminate opportunities for natural regeneration when appropriate or desired (Orefice et al. 2019). The complexity and diversity of approaches for integrating silvopasture and rotational grazing depends on farm location and larger holistic framework of farm context (Savory and Butterfield 1998). If properly applied and managed, silvopasture with rotation grazing is an agricultural practice which may increase on-farm biodiversity and allow for a farm to become a biodiversity island within a degraded landscape.

### *3.2.9 The Use of Rare, Heirloom, and Underutilized Species and Cultivars*

When unique, rare, and diverse species of plants and livestock are cultivated in regenerative agricultural systems, these farm systems can serve as biodiversity islands within a human dominated and degraded landscape. Greater crop diversity of cultivated species increases the overall biodiversity of the agricultural system and can allow for increased food security, decreased pest pressures, more resilience to climate change, and enhanced connection between cultures and locally produced foods (Smith et al. 2008; Chateil et al. 2013; Gaudin et al. 2015).

Rare, heirloom, regional and family cultivars of fruit and vegetable crops were once commonplace globally, though an inverse relationship tends to exist between industrialized agriculture expansion and landrace presence and diversity (Nazarea

2005). Fortunately, farms, organizations, and community groups are working to continue to keep such species alive while increasing genetic diversity through time (Abebe 2005). Regenerative agriculture systems such as urban community gardens and homegardens commonly cultivate genetically diverse and heirloom crops (Bardhan et al. 2012; Redondo-Brenes and Montagnini 2010). Conservation of on-farm crop diversity is extremely important to both biodiversity and the cultures from which these crops arose (Brush 2000). In situ and ex situ methods of conservation allow for genetic resources to be preserved through time while expanding crop diversity (Swanson and Goeschl 2000). More diverse crops have the potential to support greater soil life diversity and insect diversity, with differing phenologies, nutrient requirements, decomposition rates, and structure (Redlich et al. 2018). When in urban environments, these systems can be important refugia for biodiversity, as well as places where people can connect with nature (Toensmeier 2022). When rare, heirloom, or underutilized species are cultivated, seeds saved and passed on through time, greater genetic diversity can persist in the human dominated landscape.

Maintaining the diversity of plants in cultivation can take many forms. Rowen White, through her work as an indigenous seed breeder, cofounder of the Sierra Seeds Cooperative, and chair of the board of The Seed Savers Exchange, emphasizes the genetic, cultural, and historical importance of seed saving (White et al. 2018). The Felix Gillet Institute, founded by the late Amigo Bob Cantisano, explores neglected homesteads and agricultural sites seeking surviving heirloom varieties of fruit and nut trees throughout California (https://felixgillet.org/). Plant explorers such as David Fair-Child traveled the globe bringing diverse and underutilized species into cultivation (Fair-Child 1939). Plant breeders such as the famed Luther Burbank utilize innovative breeding techniques to greatly expand favorable characteristics and useful varieties and diversity of plants in cultivation (Burbank 1915). Through seed saving, recovery of heirloom varieties, exploration of underutilized species, and innovative breeding for new genetics, the diversity of cultivated plants can expand, even when challenged by economic forces spurring an opposite trend (Nazarea 2005). Regenerative agriculture can utilize the broad range of cultivar diversity to keep the genetics, stories, and species diversity alive. In doing so, regenerative agricultural sites can continue to develop as biodiversity islands within the landscape.

## 3.3 Social Dimensions of Regenerative Agriculture as Biodiversity Islands

In attaining greater biodiversity in regenerative agricultural systems, social and cultural factors should not be overlooked. Restorative action can go beyond practices of cultivation and ecological management to address restoration and regeneration of community and human relationships. Respect for the cultural origins of regenerative agriculture and the historical ecology of cultivated lands, farmworker health, empowerment, right-livelihoods, and the affordability and access to regeneratively cultivated foods by consumers must integrate with enhanced biodiversity for the agricultural system to be truly regenerative. The potential to address these social considerations is a key task for regenerative agriculture into the future.

### *3.3.1 Traditional Ecological Knowledge: The Roots of Regenerative Agriculture*

Traditional Ecological Knowledge (TEK) or Ecological Indigenous Knowledge (EIK) and the historical ecology of cultivated lands may contribute significantly towards the development and continuation of food production systems which act as biodiversity islands within a landscape. Much of regenerative agriculture is built upon this knowledge. When not already in place, indigenous, knowledge-holding representatives should hold positions of authority and decision-making within agricultural organizations. Land return to indigenously managed lands is another approach towards empowerment and social restoration in regenerative agriculture. Many indigenous communities inhabit areas where the diversity of plant and animal species have been utilized for thousands of years (Rocha et al. 2017). This can be seen in many terraced landscapes throughout the globe (Fig. 3.1).

Most inhabited places of the earth have an associated historical ecology, although in many instances this knowledge has been deeply eroded due to various social and economic factors (Balée 1998). Where traditional cultures remain, often certain members of the indigenous culture still hold knowledge of traditions and practices related to cultivating, managing, processing, and consuming diverse, native species (Berkes et al. 1994). For example, there is extensive knowledge of Native American management and cultivation of California's pre-colonial landscape (Anderson 2013). In areas where such knowledge is nearly lost from collective memory or culture, there is significant opportunity in the rediscovery and re-empowerment of such traditional knowledge as a fundamental component of productive and biodiverse regenerative agricultural systems of the future.

Technical and scientific knowledge of such traditional ecological knowledge may provide medicinal, nutritional, and otherwise valuable products. Among many others, guayusa, yerba mate, and cacao have well-documented examples of indigenous knowledge being integrated with scientific techniques to develop modern cultivation practices. In instances where local knowledge is used to gain insights into cultivation, harvesting, and processing, and bringing new products to market, one must be cautious to avoid exploitation and acknowledge the social responsibility within regenerative agriculture. For example, Yoco (*Paullinia yoco*), a vine that is wild-harvested for its caffeinated bark by the Secoya people, has had populations greatly reduced in recent years. A Yale researcher began working to restore populations of Yoco by designing systems with the community to enrich the forests

**Fig. 3.1** A village outside Muktinath, Nepal. Irrigated terrace agriculture with incorporation of various annual grains and tree crops, adding to the biodiversity of the landscape. Techniques of traditional and regenerative agriculture allow for subsistence farming, organic nutrient cycling, and efficient use of water to transform otherwise inhospitable terrain into a biodiverse landscape. (Photo: Brett Levin)

with more yoco while monitoring key environmental outcomes of these systems. (Fig. 3.2) The project looks very carefully under which ecological conditions the yoco can be established, such as light, soils, and species assemblages. This connection between traditional, technical, and community empowerment is an example of how regenerative agriculture can enhance biodiversity through the incorporation of traditional ecological knowledge. These productive sites emerge as biodiversity islands within landscapes that are rapidly being cleared due to logging and conversion of forest to pasture.

Correspondingly, ethnobiology and ethnophenology address the human and cultural component of how species and genetics are selected through time (Nabhan 2016). Ethnobiology explores the complex interactions among cultures, their languages and resource management practices with genes, foods, medicines, habitats, and landscapes for addressing critical links between culture, cultivation, and ecological diversity. Ethnophenology refers to the cultural perception of the timing of recurrent natural history events and environmental conditions in the selection and managing of specific species. For example, records from the early 1900s provide anecdotal evidence that for the Hidatsa people of the Missouri River, the sunflower seed was always the first seed planted in the spring based on observations of the melting of ice along the banks of the Missouri River around April. This was followed by planting of corn in May based on the observation of the emergence of leaves of

**Fig. 3.2** Expanding the traditional Secoya people cultivation through forest enrichment with Yoco in a multilayered, biodiverse forest system in the Amazon of Ecuador. (Photo: Luke Weiss)

the wild gooseberry bushes. These strong observations of environmental cues are learned by cultures with longstanding connections to land.

Through the incorporation of greater biodiversity into farm systems, regenerative agriculture can learn and build upon these traditions to provide a more perceptive, inclusive, and harmonious approach to cultivation, rather than mandate by strict agronomic management procedures (Nabhan 2014; Albuquerque and de Sousa 2016). Regenerative agriculture farms may become locations of applied practice towards cultural restoration and biodiversity enhancement in addition to agronomic cultivation.

### *3.3.2 Farmworker Health, Empowerment, and Right-Livelihood*

Farmworker health, empowerment, and right-livelihood are other important social matters that must be considered as part of biodiversity-enhancing regenerative agricultural systems. Is an agricultural system that enhances biodiversity but harms and exploits its workers truly regenerative? Farmworker health issues caused by exposure to toxic agrochemicals is rampant in industrialized agriculture (Salzman

2018; Saxton 2021). By implementing biodiversity-enhancing agricultural systems, farmworker health can improve (Afshari et al. 2021). Opportunities in farmworker training and empowerment may also be integrated into biodiversity-friendly agricultural systems (Braun and Duveskog 2011). One such opportunity may be the training of farmworkers to recognize, monitor, and collect data on soil health and pest pressures, which can allow for more targeted biological approaches towards cultivation and pest management. A survey of Farmer Field School (FFS) programs, which employ local knowledge sharing and training in pest management, not only reduced pesticide use and improved associated farmworker health, but proved economically advantageous (Rejesus and Jones 2020).

Additionally, workers should be paid a living wage for their time and energy if the agricultural system is to be recognized as regenerative. Worker-ownership and cooperative business structures can provide long-term equity, wealth building and authority to all levels of the organization while biodiversity practices are implemented (Alkon and Guthman 2017). Such approaches are utilized in coffee plantations throughout El Salvador and farmworker cooperatives of the United States (Bacon et al. 2008; Gray 2013). Sylvanaqua farms is a unique example in the Washington DC area focused on both the social and ecological components of regenerative agriculture (https://www.sylvanaqua.com/). Overall, for a farm system that enhances biodiversity to be recognized as regenerative, it must also consider the health and wellbeing of the farmers who work the land.

### 3.3.3 *Affordability and Access to Regenerative Agriculture*

The lack of affordability and access of regeneratively grown crops and associated products is another often overlooked social dimension of regenerative agriculture. If only the wealthy can afford regenerative agriculture, is it truly regenerative? When on-farm biodiversity improvements cause prices to increase, low-income consumers become excluded from the market. Similarly, when the negative social costs of degradative agriculture are not factored into pricing, prices remain artificially low (Pascual and Perrings 2007).

Such issues can be addressed through numerous progressive and grassroots strategies. Removal of existing governmental subsidies that support degradative practices is essential. Support of new subsidies for regenerative practices advance affordability. Negative social costs such as biodiversity loss, soil loss, greenhouse gas emissions, water quality degradation, and effects of industrial agriculture on human health must be factored into pricing (Mouysset et al. 2015). Payments for ecosystem services can also reduce prices of food and products to consumers that are grown regeneratively (Lankoski 2016). Additionally, with grassroots efforts to develop local regenerative agricultural systems and community supported agriculture projects as biodiversity islands, low-income access and affordability can improve. Examples are numerous and worldwide, spanning urban to rural areas lacking access to regeneratively grown produce (Adam 2006; Duchemin et al. 2008; Lovell 2010).

Overall, a transition from degenerative agricultural practices to regenerative practices requires a cultural shift to one that sees natural systems as essential, valuable, and inherently interconnected. Regenerative agricultural practices that increase biodiversity should also improve social and human wellbeing. Much of the origins of regenerative agriculture emerged from indigenous practices of food production and traditional ecological knowledge that maintains biodiversity. Recognizing, appreciating, and empowering this history is an essential component of the story of regenerative agriculture that is commonly appropriated, dismissed, or ignored. For agriculture to be truly regenerative, it must use a systems approach and consider impacts to the interrelated human systems that make cultivation, distribution, and food access possible.

## 3.4 Regenerative Agriculture in the Modern Age

As regenerative agriculture continues to expand, it is important to highlight some of the key organizations currently shaping the conversation. Future considerations regarding funding, monitoring of environmental outcomes, and education are essential for the continued widespread adoption of regenerative agriculture which may act as biodiversity islands.

### *3.4.1 Organizations Supporting Regenerative Agriculture for Biodiversity Conservation*

The promotion and adoption of regenerative agriculture continues to expand in various sectors. Organizations and projects supporting the advancement of biodiverse regenerative agriculture are diverse and worldwide. Investment entities, farms, service organizations, consumer packaged goods manufacturers, and nonprofit organizations are rapidly expanding their language and practices surrounding regenerative agriculture (http://www.ethansoloviev.com/regenerative-agriculture-industry-map/). The International Federation of Organic Agriculture Movements (IFOAM), the National Center for Appropriate Technology (NCAT) and their ATTRA program, Ecological Farming Association (Ecofarm), The Tropical Agricultural Research and Higher Education Center (CATIE), The Society of Ethnobiology, and countless other organizations support research and communication of these ideas. Some additional examples are reviewed in greater detail below to highlight the diversity of geography and scope of work from which supporting organizations exist.

SOCLA (Sociedad Cientifica de Agricultura Latino Americana de Agroecologia, Latin America Scientific Society of Agroecology) is a network of researchers, professors, extensionists, and other professionals which promotes agroecological and regenerative practices to confront the crisis of industrial agriculture. SOCLA promotes agroecology as a scientifically justifiable and sustainable rural development strategy in Latin America. To accomplish its objectives SOCLA organizes scientific congresses, holds short training courses in various countries, produces publications on key issues, and maintains working groups that provide information, analysis and technical advice to numerous civil and farmers organizations involved in agroecology in the region. Recently, the North American SOCLA chapter was launched. This work promotes the integration of greater biodiversity within regenerative agricultural systems (https://www.socla.co/).

Rodale Institute has been a leader in organic and regenerative agriculture since its founding in 1947 in Kutztown, Pennsylvania. As a pioneer in this field, they support new farmers, contribute valuable research, and educate consumers regarding the benefits of organic products. A key component of their work encourages biodiversity through regenerative agriculture. Rodale recognizes that a rich mix of microorganisms, plants, and animals on the farm creates healthy soil, strong crops, and resilient natural systems that don't require chemical intervention to manage pests and diseases. This knowledge is shared broadly through their public, outreach and education, in addition to being applied on their own agricultural land. In 2018, Rodale Institute helped spearhead a new, holistic, high-bar standard for agriculture certification. Regenerative Organic Certification, or ROC, is overseen by the Regenerative Organic Alliance, a non-profit made up of experts in farming, ranching, soil health, animal welfare, as well as farmer and worker fairness (https://rodaleinstitute.org/). The certification consists of three pillars: soil health, animal welfare, and social fairness. Attaining certification supports approaches to land management and associated processes that contribute to the health of ecosystems and human communities.

The Land Institute, founded by Wes Jackson in the 1970s, has been working to develop perennial grain crops and support biodiverse polycultures. Located in Salinas, Kansas, they have recently had success with the development of a new species of perennial grain called Kernza, which has the potential to transform much of the world's grain production into perennial agriculture, thereby contributing to soil protection and the preservation of waterways. The Land Institute is also developing a crop protection program that relies on biological control using natural enemies (https://landinstitute.org/our-work/ecological-intensification/).

The Al Bahaya project in Saudi Arabia is transforming a barren desert into productive savanna grasslands and agroforestry systems using regenerative agriculture techniques and extensive stone terracing to capture water. Choosing the appropriate species has been essential. When the system was first established, irrigation was utilized and later it was cut off. For 31 months there was no precipitation. After it finally rained, the species were able to recover and begin a biological cascade towards rejuvenation. (https://www.youtube.com/watch?v=T39QHprz-x8). Here, once the system was established, regenerative agricultural processes utilize the resources available, no matter how scarce, to build biodiversity and productive agricultural systems. Such considerations are the basis of the future of regenerative agriculture as biodiversity islands within degraded landscapes.

The Savanna Institute, a nonprofit organization located in Wisconsin, is a leader in temperate agroforestry research, laying the groundwork for widespread agroforestry in the Midwest US. Working in collaboration with farmers and scientists, the Savanna Institute is developing perennial food and fodder crops within multifunctional polyculture systems, grounded in ecology, and inspired by the savanna biome, with an emphasis on tree crops. Chestnuts and hazelnuts tend to be the backbone of The Savanna Institute's diverse agroforestry systems and they strategically enact their mission via research, education, and outreach (http://savannainstitute.org/).

Numerous family farms with goals of integrating biodiversity and food production also continue to emerge. New Forest Farm is a diverse restoration agriculture research site in southwestern Wisconsin, USA. Located on a former cornfield, through the efforts of Mark Shephard, the land has been transformed into a biodiverse perennial agriculture ecosystem. Utilizing innovative water management techniques, various trees, shrubs, vines, canes, grasses, forbs and fungi have been planted, organized to optimize yield and efficiency in harvesting and management. Woody crops include hazelnuts, chestnuts, walnuts, and apples (Shepard 2013). The diverse plantings and biology present within New Forest Farm make it a biodiversity island within the surrounding vast expanse of monoculture corn and soy production. Polyface farm is another example of a biodiverse regenerative agriculture family farm. Spearheaded by the Salatin family, the operation produces pastured poultry and a broad range of crops focusing on soil health, community health, and the continued improvement of the land base. Through time, measured improvements in biodiversity have resulted (Salatin 2010). Such operations as biodiversity islands within the landscape integrate old farm knowledge with new innovations, paving the future of a new, regenerative, and biodiverse agricultural paradigm.

The Savory Institute and Holistic Management International both promote, advocate, and teach about regenerative agriculture through holistic rangeland management and holistic decision making (https://holisticmanagement.org/, https://savory.global/). Holistic management was born from the work of Allan Savory, a Zimbabwean ecologist. Properly managed livestock are the ecological foundation of the holistic context. The general objectives are to help ranchers and land stewards strengthen local economies, improve local food quality, heal the environment, improve wildlife habitats, and enhance community. The teachings train farmers to recognize their goals, plan appropriately based on specific contexts, and manage livestock to mimic natural ecological patterning of mob grazing while improving soil carbon sequestration and overall rangeland biodiversity as compared to conventional grazing and cattle raising operations.

These organizations are a small sampling of many more groups focused on advancing biodiversity through regenerative agriculture. It is also important to recognize the millions of smallholders practicing similar techniques and sharing traditional knowledge throughout the world. As awareness and interest continues to grow for increasing biodiversity in degraded landscapes while producing food, one can expect the influence of these bodies to continue to expand and new organizations to continue to emerge.

### 3.4.2 *Considering the Future: Funding, Monitoring, and Education*

Alongside private sector approaches, governments can continue to support and grow programs for agricultural practices that encourage farmers to increase farm biodiversity. Governments can work towards goals of increased agricultural biodiversity in the same way successful widespread adoption of organic programs in Europe took place. This was achieved through increased funding of training programs, offsetting certification costs, and improving the quality of government advisory services, all of which have proven highly effective (Mills et al. 2020). In the United States, the Department of Agriculture and the Natural Resources Conservation Service currently have several financial incentives for farmers to adopt practices such as riparian corridors, windbreaks, and hedgerows (Duru et al. 2015). The Environmental Quality Incentives Program (EQIP) aids agricultural producers through technical and financial support through public funding to address natural resource degradation and to improve the environment through increased water and air quality, conserved ground and surface water, increased soil health, reduced soil erosion, improved or created wildlife habitat, and mitigation against increasing weather volatility through public funding. Of these conservation practices, many contribute to the development of agricultural biodiversity islands within a landscape (https://www.nrcs.usda.gov/wps/portal/nrcs/main/national/programs/financial/eqip/). Though well-funded with a budget of $1.75 billion in fiscal years 2019 and 2020, $1.8 billion in fiscal year 2021, $1.85 billion in fiscal year 2022 and $2.025 billion in fiscal year 2023, there remains opportunity for greater financing of biodiversity enhancing conservation practices (https://www.fb.org/market-intel/eqip-and-csp-conservation-programs-in-the-2018-farm-bill). This type of financial assistance can be greatly expanded upon, and include all the previously mentioned practices, which can increase farm resilience, yields, and on-farm biodiversity. Within the United States, this can be addressed through a revision of funding priorities federally in the Farm Bill, and locally through state action and cooperative extensions.

Additionally, as private funding and markets for payments for ecosystem services and carbon sequestration in agriculture continue to advance, it is important to consider the potential to integrate biodiversity within such projects. Many carbon offset projects focus solely on biomass production and carbon sequestration. Focusing on the maximum biomass growth possible to obtain as many carbon credits as possible may place higher value on fast growing species than native, bio-regionally appropriate food-bearing species. In these instances, where projects focus on biomass generation for either carbon or bioenergy, biodiversity can decrease through time rather than improve (Abreu et al. 2017). By incorporating some of the practices mentioned above, these projects can have mutually beneficial outcomes of biomass production, sequestration, and improved biodiversity outcomes.

Advancements in monitoring of biodiversity coupled with carbon sequestration and other ecosystem services may provide another significant increase in the adoption of regenerative agriculture. When the benefits and positive impacts of these practices are measurable with greater certainty, value can be associated with such practices, and the positive social benefits can be attributed to individual farms and farmers. The externalities of any farm, positive or negative, influence the rest of the landscape. When such externalities are properly monitored and valued, society is more able to perceive those benefits, which in turn makes regenerative agriculture more attractive. This opens further opportunity for community engagement, investment, funding, and more widespread adoption of biodiverse regenerative farming, sparking the development of biodiversity islands throughout degraded landscapes (https://www.regen.network/, https://www.ecosystemmarketplace.com/).

Online educational opportunities for learning regenerative agricultural practices that enhance biodiversity outcomes have grown significantly. Reports, podcasts, webinars, workshops, conferences, virtual university extension programs which are now widely available for free, provide information that is both conceptual and specific for bioregional applied practice. Many examples of such media can be found on websites and platforms such as, https://attra.ncat.org/multimedia/, https://ecoagriculture.org/, https://imfn.net/, https://satoyama-initiative.org/, https://www.nature.org/en-us/, https://www.conservation.org/, https://www.worldwildlife.org/, https://onehealthinitiative.com/, and others mentioned in Chap. 1 Sect. 1.5 of this volume.

Additionally, there are a growing number of technical and scientific publications accessible to a broad audience, such as *Working with Nature: Resource Management for Sustainability* (Jordan 1998), *Tomorrow's Biodiversity* (Shiva 2000), *Call of the Reed Warbler* (Massy 2017), *Growing a Revolution* (Montgomery 2018), and *Reclaiming the Commons* (Shiva 2020). Such resources and writings are inspiring a new generation of educators, policy makers, and farmers to engage in the work of developing biodiverse regenerative agricultural systems which may act as biodiversity islands within the landscape.

Regenerative agriculture emerged from traditional knowledge and ecological observations through time. While conducted mostly by indigenous people and smallholders throughout the world, over the past century, writers and practitioners worldwide have continued to advance the science and practice of regenerative agriculture in the western paradigm. Such notable proponents include Amigo Bob Cantisano, Bill Mollison, Christine Jones, Cyril G Hopkins, Darren Dougherty, David Montgomery, Edward Faulkner, Eric Toesnmeier, Ethan Soloviev, Eve Balfour, Everette "Deke" Dietrick, F.H. King, Gabe Brown, J. Russell Smith, J.I. Rodale, Joel Salatin, John Jeavons, John Kempf, John Lundgren, Judith Schwartz, Miguel Altieri, Leah Penniman, Mark Shepard, Masanobu Fukuoka, Newman Turner, P.A. Yeomans, Reginaldo Haslett-Marroquin, Richard Perkins, Rudolph Steiner, Sir Albert Howard, Thomas Barrett, Vandana Shiva, William Albrecht, Wendell Berry, and many others. Through an ever-growing application of scientific, philosophical, ethical, and on-the-ground practice, the role and impact of biodiverse regenerative agriculture continues to expand, increasing the development of biodiversity islands in degraded lands.

## 3.5 Conclusions

Biodiversity enhancing regenerative agricultural practices can be applied wherever agriculture is practiced. Throughout the twentieth century, a shift in agricultural production to large scale, industrial, monoculture production with the use of agrochemicals became commonplace. While providing cheap calories, this agriculture has a myriad of negative consequences for both humans and the environment. This chapter has described successful examples and techniques for the advancement of a different approach towards agriculture, where agrochemical use is limited or eliminated, diverse genetics are utilized, and design and agricultural techniques are examined to enhance biodiversity within agricultural systems.

Much of regenerative agriculture emerged from indigenous land use systems. Building from this knowledge as science, diverse agroecosystems can continue to spread throughout the globe with great success. A diversity of organizations and case studies were presented to highlight the scope, scale, and diversity of regenerative agriculture as it contributes to biodiversity islands in a landscape. If managed following the approaches described throughout this chapter, farms can become biodiversity islands within the matrix of degraded landscapes.

Financial, cultural, and ecological opportunities abound in the transition of degraded lands into agriculturally based biodiversity islands. As adoption of biodiversity into agricultural lands becomes more commonplace and more farms throughout the landscape harbor greater levels of biodiversity, one can envision an agricultural future in which biodiversity islands are the norm. As such, speciation and species preservation may continue to increase through time in harmony with human habitation and agricultural production, and the biodiversity island which is planet earth may flourish towards a greater bounty and beauty that is evident in the diversity of life.

## References

Abebe T (2005) Diversity in homegarden agroforestry systems of southern Ethiopia. PhD thesis, Wageningen University, Wageningen

Abreu RC, Hoffmann WA, Vasconcelos HL, Pilo NA, Rossatto DR, Durigan G (2017) The biodiversity cost of carbon sequestration in tropical savanna. Sci Adv 3(8):e1701284. https://doi.org/10.1126/sciadv.1701284

Adam KL (2006) Community supported agriculture. ATTRA-National Sustainable Agriculture Information Service, Butte. http://attra.ncat.org/

Afshari M, Karimi-Shahanjarini A, Khoshravesh S, Besharati F (2021) Effectiveness of interventions to promote pesticide safety and reduce pesticide exposure in agricultural health studies: a systematic review. PLoS One 16(1):e0245766
Albuquerque UP, de Sousa DCP (2016) Ethnobiology and biodiversity conservation. In: Albuquerque UP, de Sousa DCP (eds) Introduction to ethnobiology. Springer, Cham, pp 227–232
Ali M, Khan SJ, Aslam I, Khan Z (2011) Simulation of the impacts of land-use change on surface runoff of Lai Nullah Basin in Islamabad, Pakistan. Landsc Urban Plan 102(4):271–279
Alkon A, Guthman J (eds) (2017) The new food activism: opposition, cooperation, and collective action. University of California Press, Berkeley
Altieri MA (1995) Biodiversity and biocontrol: Lessons from insect pest management. Academic Press Adv in Plant Path 11(10):191–209
Altieri MA (1999) The ecological role of biodiversity in agroecosystems. In: Altieri MA (ed) Invertebrate biodiversity as bioindicators of sustainable landscapes. Elsevier, Amsterdam, pp 19–31
Altieri MA (2015) Agroecology, key concepts, principles and practices. Penang, Third World Network (TWN)-SOCLA, Jutaprint
Altieri MA (2018) Agroecology: the science of sustainable agriculture. CRC Press, Boca Raton
Altieri MA, Schmidt LL (1985) Cover crop manipulation in northern California orchards and vineyards: effects on arthropod communities. Biol Agric Hortic 3(1):1–24
Anderson MK (2013) Tending the wild: native American knowledge and the management of California's natural resources. University of California Press, Berkeley
Atkinson D, Watson CA (eds) (2019) The science beneath organic production. Wiley, New York
Bacon C, Mendez VE, Gliessman S, Goodman D, Fox J (eds) (2008) Confronting the coffee crisis: fair trade, sustainable livelihoods and ecosystems in Mexico and Central America. MIT Press, Cambridge
Balée W (1998) Historical ecology: premises and postulates. In: Balée W (ed) Advances in historical ecology. Columbia University Press, New York, pp 13–29
Balzan MV (2017) Flowering banker plants for the delivery of multiple agroecosystem services. Arthropod Plant Interact 11(6):743–754
Bardhan S, Jose S, Biswas S, Kabir K, Rogers W (2012) Homegarden agroforestry systems: an intermediary for biodiversity conservation in Bangladesh. Agrofor Syst 85(1):29–34
Benayas JMR, Bullock JM (2012) Restoration of biodiversity and ecosystem services on agricultural land. Ecosystems 15(6):883–899
Benton TG, Vickery JA, Wilson JD (2003) Farmland biodiversity: is habitat heterogeneity the key? Trends Ecol Evol 18(4):182–188
Berkes F, Folke C, Gadgil M (1994) Traditional ecological knowledge, biodiversity, resilience and sustainability. In: Berkes F, Folke C, Gadgil M (eds) Biodiversity conservation. Springer, Dordrecht, pp 269–287
Bertsch A (2017) Indigenous successional agroforestry: integrating the old and new to address food insecurity and deforestation. In: Montagnini F (ed) Integrating landscapes: agroforestry for biodiversity conservation and food sovereignty, Advances in agroforestry 12. Springer, Cham, pp 65–178
Blanchart E, Villenave C, Viallatoux A, Barthès B, Girardin C, Azontonde A, Feller C (2006) Long-term effect of a legume cover crop (Mucuna pruriens var. utilis) on the communities of soil macrofauna and nematofauna, under maize cultivation, in southern Benin. Eur J Soil Biol 42:S136–S144
Braun A, Duveskog D (2011) The farmer field school approach – history, global assessment and success stories, Background paper for the IFAD rural poverty report. International Fund for Agricultural Development, Rome
Brush SB (ed) (2000) Genes in the field: on-farm conservation of crop diversity. Boca Raton, International Development Research Centre and International Plant Genetic Resources Institute/ Lewis Publishers

Brussaard L, De Ruiter PC, Brown GG (2007) Soil biodiversity for agricultural sustainability. Agric Ecosyst Environ 121(3):233–244

Burbank L (1915) Luther Burbank: his methods and discoveries and their practical application, vol 12. Luther Burbank Press, New York and London

Cabeza M, Moilanen A (2001) Design of reserve networks and the persistence of biodiversity. Trends Ecol Evol 16(5):242–248

Cárcamo HA, Spence JR (1994) Crop type effects on the activity and distribution of ground beetles (Coleoptera: Carabidae). Environ Entomol 23(3):684–692

Chabay I, Frick M, Helgeson J (eds) (2015) Land restoration: reclaiming landscapes for a sustainable future. Academic, New York

Chateil C, Goldringer I, Tarallo L, Kerbiriou C, Le Viol I, Ponge JF, Salmon S, Gachet S, Porcher E (2013) Crop genetic diversity benefits farmland biodiversity in cultivated fields. Agric Ecosyst Environ 171:25–32

Chen XD, Dunfield KE, Fraser TD, Wakeli SA, Richardson AE, Condron LM (2020) Soil biodiversity and biogeochemical function in managed ecosystems. Soil Res 58(1):1–20

Ching LL (2018) Agroecology for sustainable food systems, Environment and development series, 19. Third World Network, Penang

Coleman D, Reid C, Cole C (1983) Biological strategies of nutrient cycling in soil systems. Adv Ecol Res 13:1–55

Corrado C, Elena T, Giancarlo R, Stefano C (2019) The role of agrobiodiversity in sustainable food systems design and management. In: Nandwani D (ed) Genetic diversity in horticultural plants. Springer, Cham, pp 245–271

Crist E, Mora C, Engelman R (2017) The interaction of human population, food production, and biodiversity protection. Science 356(6335):260–264

De La Cruz R, Suárez S, Ferguson JE (1994) The contribution of Arachis pintoi as a ground cover in some farming systems of Tropical America. CIAT Publication, Cali

Dominati EJ, Maseyk FJ, Mackay AD, Rendel JM (2019) Farming in a changing environment: increasing biodiversity on farm for the supply of multiple ecosystem services. Sci Total Environ 662:703–713

Doré T, Makowski D, Malézieux E, Munier-Jolain N, Tchamitchian M, Tittonell P (2011) Facing up to the paradigm of ecological intensification in agronomy: revisiting methods, concepts and knowledge. Eur J Agron 34(4):197–210

Duchemin E, Wegmuller F, Legault AM (2008) Urban agriculture: multi-dimensional tools for social development in poor neighbourhoods. Field Actions Sci Rep 2(1):1–8. https://doi.org/10.5194/facts-2-1-2009

Duru M, Therond O, Martin G, Martin-Clouaire R, Magne MA, Justes E, Journet EP, Aubertot JN, Savary S, Bergez JE, Sarthou JP (2015) How to implement biodiversity-based agriculture to enhance ecosystem services: a review. Agron Sustain Dev 35(4):1259–1281

Enri SR, Probo M, Farruggia A, Lanore L, Blanchetete A, Dumont B (2017) A biodiversity-friendly rotational grazing system enhancing flower-visiting insect assemblages while maintaining animal and grassland productivity. Agric Ecosyst Environ 241:1–10

Fair-Child D (1939) The world was my garden: travels of a plant explorer. Soil Sci 48(4):356

Fearnside P (2017) Business as usual: a resurgence of deforestation in the Brazilian Amazon. Yale Environ 360:1–6

Finch CU, Sharp WC (1976) Cover crops in California orchards and vineyards. US Department of Agriculture, Washington, DC

Gabriel S (2018) Silvopasture: a guide to managing grazing animals, forage crops, and trees in a temperate farm ecosystem. Chelsea Green Publishing, White River Junction

Gaudin AC, Tolhurst TN, Ker AP, Janovicek K, Tortora C, Martin RC, Deen W (2015) Increasing crop diversity mitigates weather variations and improves yield stability. PLoS One 10(2): e0113261

Gibbons LV (2020) Regenerative—the new sustainable? Sustainability 12(13):5483

Giller KE, Beare MH, Lavelle P, Izac AM, Swift MJ (1997) Agricultural intensification, soil biodiversity and agroecosystem function. Appl Soil Ecol 6(1):3–16

Gliessman S (2016) Transforming food systems with agroecology. Agroecol Sustain Food Syst 40(3):187–189. https://doi.org/10.1080/21683565.2015.1130765

Gray M (2013) Labor and the locavore: the making of a comprehensive food ethic. University of California Press, Berkeley

Gustafson JP, Raven PH, Ehrlich PR (2020) Population, agriculture, and biodiversity: problems and prospects. University of Missouri Press, Columbia

Henderson K, Loreau M (2019) An ecological theory of changing human population dynamics. People Nat 1(1):31–43

Holland JM (2019) Contribution of hedgerows to biological control. In: Dover JW (ed) The ecology of hedgerows and field margins. Routledge, Abingdon, pp 123–146

Huang W, Luukkanen O, Johanson S, Kaarakka V, Räisänen S, Vihemäki H (2002) Agroforestry for biodiversity conservation of nature reserves: functional group identification and analysis. Agrofor Syst 55(1):65–72

Jackson LE, Pascual U, Brussaard L, de Ruiter P, Bawa KS (2007) Biodiversity in agricultural landscapes: investing without losing interest. Agric Ecosyst Environ 121(3):193–195

Jordan CF (1998) Working with nature: resource management for sustainability. Taylor & Francis, Amsterdam

Jørgensen LN, Kudsk P (2006) Twenty years' experience with reduced agrochemical inputs: effect on farm economics, water quality, biodiversity and environment. In: Home-Grown Cereals Authority (HGCA) agronomist conference, Grantham, UK

Jose S (2009) Agroforestry for ecosystem services and environmental benefits: an overview. Agrofor Syst 76(1):1–10

Jose S (2012) Agroforestry for conserving and enhancing biodiversity. Agrofor Syst 85(1):1–8

Jose S, Walter D, Kumar BM (2019) Ecological considerations in sustainable silvopasture design and management. Agrofor Syst 93(1):317–331

Kidd LR, Garrard GE, Bekessy SA, Mills M, Camilleri AR, Fidler F, Fielding KS, Gordon A, Gregg EA, Kusmanoff AM, Louis W (2019) Messaging matters: a systematic review of the conservation messaging literature. Biol Conserv 236:92–99

Kimbrell A (ed) (2002) The fatal harvest reader: the tragedy of industrial agriculture. Island Press, Washington, DC

Klasen S, Meyer KM, Dislich C, Euler M, Faust H, Gatto M, Hettig E, Melati DN, Jaya INS, Otten F, Pérez-Cruzado C (2016) Economic and ecological trade-offs of agricultural specialization at different spatial scales. Ecol Econ 122:111–120

Lacombe S, Bradley RL, Hamel C, Beaulieu C (2009) Do tree-based intercropping systems increase the diversity and stability of soil microbial communities? Agric Ecosyst Environ 131(1–2): 25–31

Landis DA, Wratten SD, Gurr GM (2000) Habitat management to conserve natural enemies of arthropod pests in agriculture. Annu Rev Entomol 45(1):175–201

Lankoski J (2016) Alternative payment approaches for biodiversity conservation in agriculture. https://www.oecd-ilibrary.org/content/paper/5jm22p4ptg33-en

Leakey RR (1999) Agroforestry for biodiversity in farming systems. In: Collins WW, Calvin OQ (eds) Biodiversity in agroecosystems. Lewis Publishers, New York, pp 127–145

Lehmann J, Bossio DA, Kögel-Knabner I, Rillig MC (2020) The concept and future prospects of soil health. Nat Rev Earth Environ. https://doi.org/10.1038/s43017-020-0080-8

Lichtfouse E (ed) (2018) Sustainable agriculture reviews 31: biocontrol, vol 31. Springer, New York

Liere H, Jha S, Philpott SM (2017) Intersection between biodiversity conservation, agroecology, and ecosystem services. Agroecol Sustain Food Syst 41(7):723–760

Long R, Corbett A, Lamb C, Reberg-Horton C, Chandler J, Stimmann M (1998) Beneficial insects move from flowering plants to nearby crops. Calif Agric 52(5):23–26

Lovell ST (2010) Multifunctional urban agriculture for sustainable land use planning in the United States. Sustainability 2(8):2499–2522

Mandal A, Sarkar B, Mandal S, Vithanage M, Patra AK, Manna MC (2020) Impact of agrochemicals on soil health. In: Prasad MNV (ed) Agrochemicals detection, treatment and remediation. Butterworth-Heinemann, Oxford, pp 161–187

Massy C (2017) Call of the reed warbler: a new agriculture—a new earth. University of Queensland Press, St. Lucia, p 569

McAdam JH, McEvoy PM (2009) The potential for silvopastoralism to enhance biodiversity on grassland farms in Ireland. In: McAdam JH, McEvoy PM (eds) Agroforestry in Europe. Springer, Dordrecht, pp 343–356

McDermott ME, Rodewald AD (2014) Conservation value of silvopastures to Neotropical migrants in Andean forest flocks. Biol Conserv 175:140–147

McLaughlin A, Mineau P (1995) The impact of agricultural practices on biodiversity. Agric Ecosyst Environ 55(3):201–212

McNeely JA, Schroth G (2006) Agroforestry and biodiversity conservation–traditional practices, present dynamics, and lessons for the future. Biodivers Conserv 15(2):549–554

Meena RS, Kumar S, Datta R, Lal R, Vijayakumar V, Brtnicky M, Sharma MP, Yadav GS, Jhariya MK, Jangir CK, Pathan SI (2020) Impact of agrochemicals on soil microbiota and management: a review. Land 9(2):34

Mendenhall CD, Karp DS, Meyer CF, Hadly EA, Daily GC (2014) Predicting biodiversity change and averting collapse in agricultural landscapes. Nature 509(7499):213–217

Miller TP, Rebek EJ, Schnelle MA (2017) Banker plants for control of greenhouse pests. Oklahoma Cooperative Extension Service, Oklahoma State University, Stillwater. https://extension.okstate.edu/

Mills J, Ingram J, Dibari C, Merante P, Karaczun Z, Molnar A, Sánchez B, Iglesias A, Ghaley BB (2020) Barriers to and opportunities for the uptake of soil carbon management practices in European sustainable agricultural production. Agroecol Sustain Food Syst 44(9):1185–1211

Montagnini F (2020) The contribution of agroforestry to restoration and conservation: biodiversity islands in degraded landscapes. In: Dagar JC, Gupta SR, Teketay D (eds) Agroforestry for degraded landscapes: recent advances and emerging challenges. Springer, New York, pp 445–479. https://doi.org/10.1007/978-981-15-4136-0_15

Montagnini F, del Fierro S (2022) Functions of agroforestry systems as biodiversity islands in productive landscapes. In: Montagnini F (ed) Biodiversity Islands: strategies for conservation in human-dominated environments. Topics in Biodiversity and Conservation, Springer, Cham, pp 89–116

Montagnini F, Metzel R (2017) The contribution of agroforestry to sustainable development goal 2: end hunger, achieve food security and improved nutrition, and promote sustainable agriculture. In: Montagnini F (ed) Integrating landscapes: agroforestry for biodiversity conservation and food sovereignty, Advances in agroforestry 12. Springer, Cham, pp 11–45

Montagnini F, Eibl B, Grance L, Maiocco D, Nozzi D (1997) Enrichment planting in overexploited subtropical forests of the Paranaense region of Misiones. Argent For Ecol Manage 99(1–2): 237–246

Montagnini F, Levin B, Berg KE (2022) Introduction. Biodiversity Islands: strategies for conservation in human-dominated environments. In: Montagnini F (ed) Biodiversity Islands: strategies for conservation in human-dominated environments, Topics in Biodiversity and Conservation. Springer, Cham, pp 1–35

Montgomery D (2018) Growing a revolution: bringing our soil back to life. W.W. Norton and Company Inc., New York

Moss C, Lukac M, Harris F, Outhwaite CL, Scheelbeek PF, Green R, Berstein FM, Dangour AD (2020) The effects of crop diversity and crop type on biological diversity in agricultural landscapes: a systematic review protocol [version 2; peer review: 1 approved, 2 approved with reservations]. Wellcome Open Res 4(101). https://doi.org/10.12688/wellcomeopenres.15343.2

Mouysset L, Doyen L, Pereau JC, Jiguet F (2015) Benefits and costs of biodiversity in agricultural public policies. Eur Rev Agric Econ 42(1):51–76

Nabhan GP (2014) Food security, biodiversity and human health: ethnobiology as a predictive science. J Ethnobiol 34(1):7–11

Nabhan GP (2016) Enduring seeds: native American agriculture and wild plant conservation. University of Arizona Press, Tucson

Nair PK (2017) Managed multi-strata tree+ crop systems: an agroecological marvel. Front Environ Sci. https://doi.org/10.3389/fenvs.2017.00088

Navarro LM, Pereira HM (eds) (2015) Rewilding European landscapes. Springer, Cham

Nazarea VD (2005) Heirloom seeds and their keepers: marginality and memory in the conservation of biological diversity. University of Arizona Press, Tucson

Newbold T, Hudson LN, Arnell AP, Contu S, De Palma A, Ferrier S, Hill SL, Hoskins AJ, Lysenko I, Phillips HR, Burton VJ (2016) Has land use pushed terrestrial biodiversity beyond the planetary boundary? A global assessment. Science 353(6296):288–291

Orefice JN, Carroll J (2017) Silvopasture—it's not a load of manure: differentiating between silvopasture and wooded livestock paddocks in the northeastern United States. J For 115(1): 71–72

Orefice J, Smith RG, Carroll J, Asbjornsen H, Howard T (2019) Forage productivity and profitability in newly-established open pasture, silvopasture, and thinned forest production systems. Agrofor Syst 93:51–65

Parker JE, Crowder DW, Eigenbrode SD, Snyder WE (2016) Trap crop diversity enhances crop yield. Agric Ecosyst Environ 232:254–262

Pascual U, Perrings C (2007) Developing incentives and economic mechanisms for in situ biodiversity conservation in agricultural landscapes. Agric Ecosyst Environ 121(3):256–268

Perfecto I, Vandermeer JH, Wright AL (2009) Nature's matrix: linking agriculture, conservation and food sovereignty. Routledge, London

Reddy P (2017) Intercropping. In: Reddy P (ed) Agro-ecological approaches to pest management for sustainable agriculture. Springer, Singapore, pp 109–131

Redlich S, Martin EA, Steffan-Dewenter I (2018) Landscape-level crop diversity benefits biological pest control. J Appl Ecol 55(5):2419–2428

Redondo-Brenes A, Montagnini F (2010) Contribution of homegardens, agrosilvopastoral systems, and other human dominated land-use types to the avian diversity of a biological corridor in Costa Rica. Adv Environ Res 2(2):111–148

Reij C, Garrity D (2016) Scaling up farmer-managed natural regeneration in Africa to restore degraded landscapes. Biotropica 48(6):834–843

Rejesus RM, Jones MS (2020) Perspective: enhancing economic evaluations and impacts of integrated pest management Farmer Field Schools (IPM-FFS) in low-income countries. Pest Manag Sci 76(11):3527–3536

Rhodes CJ (2017) The imperative for regenerative agriculture. Sci Prog 100(1):80–129

Rocha P, Niella F, Keller H, Montagnini F, Metzel R, Eibl B, Kornel J, Romero F, López L, Araujo J, Barquinero J (2017) Ecological indigenous (EIK) and scientific (ESK) knowledge integration as a tool for sustainable development in indigenous communities. Experience in Misiones, Argentina. In: Montagnini F (ed) Integrating landscapes: agroforestry for biodiversity conservation and food sovereignty. Springer, Cham, pp 235–260

Salatin J (2010) Sheer ecstasy of being a lunatic farmer. Chelsea Green Publishing, White River Junction

Salzman T (2018) The farmworker health landscape: barriers to improving health and safety in US agriculture. Doctoral dissertation, Tufts University, Medford/Somerville

Sánchez-Bayo F, Wyckhuys KA (2019) Worldwide decline of the entomofauna: a review of its drivers. Biol Conserv 232:8–27

Savory A, Butterfield J (1998) Holistic management: a new framework for decision making. Island press, Washington, DC

Saxton DI (2021) The Devil's fruit: farmworkers, health, and environmental justice. Rutgers University Press, New Brunswick
Schreefel L, Schulte RPO, de Boer IJM, Schrijver AP, van Zanten HHE (2020) Regenerative agriculture – the soil is the base. Glob Food Sec 26:100404. https://doi.org/10.1016/j.gfs.2020.100404
Schroth G, Lehmann J, Rodrigues MRL, Barros E, Macêdo JL (2001) Plant-soil interactions in multistrata agroforestry in the humid tropics. Agrofor Syst 53(2):85–102
Schroth G, Izac AMN, Vasconcelos HL, Gascon C, da Fonseca GA, Harvey CA (eds) (2004) Agroforestry and biodiversity conservation in tropical landscapes. Island Press, New York
Shepard M (2013) Restoration agriculture. Acres U.S.A., Austin
Shiva V (2000) Tomorrow's biodiversity. Thames & Hudson, London
Shiva V (2020) Reclaiming the commons: biodiversity, indigenous knowledge, and the rights of mother earth. Synergetic Press, Santa Fe
Smith RG, Gross KL, Robertson GP (2008) Effects of crop diversity on agroecosystem function: crop yield response. Ecosystems 11(3):355–366
Socci P, Errico A, Castelli G, Penna D, Preti F (2019) Terracing: from agriculture to multiple ecosystem services. In: Oxford research encyclopedia of environmental science. Oxford University Press, Oxford
Soloviev ER, Landua G (2016) Levels of regenerative agriculture. https://www.terra-genesis.com/wp-content/uploads/2017/03/Levels-of-Regenerative-Agriculture-1. Accessed Online March 22 2021
Swanson T, Goeschl T (2000) Optimal genetic resource conservation: in situ and ex situ. In: Genes in the field: on-farm conservation of crop diversity. International Development Research Centre and International Plant Genetic Resources Institute/Lewis Publishers, Boca Raton, pp 165–191
Tavares PD, Uzêda MC, Pires ADS (2019) Biodiversity conservation in agricultural landscapes: the importance of the matrix. Floresta e Ambiente. https://doi.org/10.1590/2179-8087.066417
Tilman D (2020) Resource competition and community structure (vol. 17). Princeton University Press. Princeton, New Jersey
Toensmeier E (2016) The carbon farming solution: a global toolkit of perennial crops and regenerative agriculture practices for climate change mitigation and food security. Chelsea Green Publishing, White River Junction
Toensmeier E (2022) Paradise lot: a temperate-climate urban agroforestry Biodiversity Island. In: Montagnini F (ed) Biodiversity Islands: strategies for conservation in human-dominated environments. Topics in Biodiversity and Conservation. Springer, Cham, pp 439–459
Tree I (2019) Wilding: returning nature to our farm. New York Review of Books, London
Udawatta R, Rankoth L, Jose S (2019) Agroforestry and biodiversity. Sustainability 11(10):2879
Wagg C, Bender SF, Widmer F, Van Der Heijden MG (2014) Soil biodiversity and soil community composition determine ecosystem multifunctionality. Proc Nat Acad Sci 111(14):5266–5270
Wagner DL (2020) Insect declines in the anthropocene. Annu Rev Entomol 65:457–480
White C (2020) Why regenerative agriculture? Am J Econ Soc 79(3):799–812
White R, Charlebois E, Martin S (2018) Seed "rematriation". Panel discussion at the intertribal food sovereignty summit. Pequot Museum and Research Center, 21–23 August 2018, Ledyard, CT, USA
Wintle BA, Kujala H, Whitehead A, Cameron A, Veloz S, Kukkala A, Moilanen A, Gordon A, Lentini PE, Cadenhead NC, Bekessy SA (2019) Global synthesis of conservation studies reveals the importance of small habitat patches for biodiversity. Proc Nat Acad Sci 116(3):909–914
Yeomans PA (1958) The challenge of landscape: the development and practice of Keyline. Keyline Pub. Pty, Sydney
Young KJ (2017) Mimicking nature: a review of successional agroforestry systems as an analogue to natural regeneration of secondary forest stands. In: Montagnini F (ed) Integrating landscapes: agroforestry for biodiversity conservation and food sovereignty. Springer, Cham, pp 179–209
Zhao ZH, Hui C, Reddy GV, Ouyang F, Men XY, Ge F (2019) Plant species richness controls arthropod food web: evidence from an experimental model system. Ann Entomol Soc Am 112(1):27–32

# Chapter 4
# Functions of Agroforestry Systems as Biodiversity Islands in Productive Landscapes

**Florencia Montagnini and Sara del Fierro**

**Abstract** Given their ability to harmonize productivity with environmental functions, agroforestry systems (AFS) are an important strategy for conservation within human managed landscapes. AFS are heterogeneous in their design, management, and species composition, with consequences for their restoration, conservation, and productivity functions. AFS can function as biodiversity islands or can be incorporated into existing biodiversity islands as buffer zones because they can be integrated into already productive landscapes. This chapter provides an overview of the various ecological, social, and economic benefits of the main types of AFS systems and their applications as and within biodiversity islands. It also discusses the use of incentives to support and promote AFS in order to safeguard the contributions they provide to landscape biodiversity and rural communities.

**Keywords** Buffer zones · Certification · Connectivity · Markets · Organic farming · Payments for ecosystem services (PES)

## 4.1 Introduction

Sustainable agricultural management that seeks to balance ecosystem productivity and conservation can play an important role in mitigating or reversing detrimental human induced effects on landscapes. Agroforestry systems (AFS), which combine trees and crops or pasture on the same land, can increase productivity in the short and long term, while also supporting biodiversity and providing social and economic benefits for farmers (Jose 2009, 2012; Montagnini et al. 2015; Montagnini and Metzel 2017; Montagnini 2020; Udawatta et al. 2019).

F. Montagnini (✉)
School of the Environment, The Forest School, Yale University, New Haven, CT, USA
e-mail: florencia.montagnini@yale.edu

S. del Fierro
United States Department of Agriculture (USDA), Washington, DC, USA



F. Montagnini (ed.), *Biodiversity Islands: Strategies for Conservation in Human-Dominated Environments*, Topics in Biodiversity and Conservation 20,
https://doi.org/10.1007/978-3-030-92234-4_4

A growing body of evidence supports the overall assertion that floral, faunal, and soil microbial diversity are significantly greater in AFS as compared to monocultures, adjacent crop lands, and even some forests (Bhagwat et al. 2008; Udawatta et al. 2019). Among the soil organisms, arbuscular mycorrhizal fungi, bacteria, and enzyme activities have been found to be significantly greater in AF than in conventional agriculture (Udawatta et al. 2019). Agroforestry can also create spatially concentrated high biodiversity near trees due to favorable soil-plant-water-microclimate conditions. The greater biodiversity in AFS has been attributed to their more favorable microclimate and soil conditions, and to their heterogeneity in comparison with monocrops.

AFS are used in a variety of contexts and systems, serving various needs and functions and incorporating different crops and local practices to meet those needs. The specific characteristics of each AFS vary strongly according to system design, objectives and species involved, with strong differences both within and between ecological regions. Their locations, and within them the context-specific economic, social, and political factors, can influence the practices used, along with their productivity, sustainability, and environmental services (Montagnini and Metzel 2017; Montagnini 2020). Therefore, functions of AFS, including their role in biodiversity conservation and restoration, can vary widely, however, they tend to enhance conservation of biodiversity when compared to monoculture systems.

As the agricultural frontier continues to advance and landscapes become further fragmented, driven by the need to supply food and resources to a growing (in number and area) human population, measures to preserve biodiversity are urgently needed (Montagnini and Berg 2019). One such measure is the promotion of biodiversity islands, protected and managed areas of high biological diversity, within otherwise degraded human-dominated landscapes, where plants and animals can thrive without major degenerative interference from human activity (Montagnini and Berg 2019; Montagnini et al. 2022). Since AFS are often important components among land uses in buffer zones of protected areas, they work well as parts of biodiversity islands, creating a smoother transition to areas of greater human impact. AFS can also serve as biodiversity islands themselves within cultivated landscapes, consisting of higher species compositions and providing greater species habitat than otherwise conventional systems.

This chapter provides an overview of AFS systems, drawing from studies around the world. Since biodiversity in AFS varies strongly according to system characteristics, the following section discusses features and biodiversity contributions of the most frequent AFS: multistrata systems including homegardens and successional AFS, perennial crops under shade, silvopastoral systems, and living fences and windbreaks. For each type, the role of local and indigenous knowledge and several of their ecological, social, and economic benefits are highlighted. The chapter then offers ways to consider AFS as components of biodiversity islands in human-dominated landscapes. The chapter ends by discussing incentive systems that can be incorporated to promote and sustain AFS in order to safeguard the contributions they provide to both biodiversity and rural communities.

## 4.2 Predominant Types of Agroforestry Systems and Their Benefits

The predominant types of AFS used globally are multistrata AFS, perennial crops under shade, silvopastoral systems, living fences, windbreaks and riparian buffers. The main features of these major types, as well as of sub-types within them, along with their ecological, economic, social, and cultural benefits, are summarized in Table 4.1.

### *4.2.1 Multistrata Agroforestry Systems*

#### 4.2.1.1 Homegardens

Homegardens, also called 'forest gardens,' are defined as intimate, multistory combinations of various trees and crops, sometimes in association with domestic animals, around homesteads (Nair and Kumar 2006). In homegardens, plants are generally categorized into three main groups: cultivated, protected or spared. Cultivated plants are those that are sown or planted by the owner, while protected plants are transplanted or spontaneous plants that are encouraged by the farmer. Spared plants are those that spontaneously grow in the garden and are not removed (Blanckaert et al. 2004). Each design and management strategy can contribute to the presence and preservation of species of interest to the farmer and to landscape biodiversity (Fig. 4.1).

Food plants, including crops and trees, are often the most abundant species found in homegardens. Homegardens are also valuable sources of income via cash crops (Nair and Kumar 2006). Species complexity in homegardens is a result of deliberate and meticulous selection and management by farmers to grow and steward the products they consider important for their subsistence and livelihood. Different factors, such as remoteness from urban centers, management, and modernization, influence biodiversity in homegardens throughout the world, as seen from examples from Kerala, India (Peyre et al. 2006), southern Ethiopia (Abete et al. 2006), the Peruvian Amazon (Wezel and Ohl 2006), and Meso America (Kumar and Nair 2006; Montagnini 2006).

Homegardens span a wide range of climatic and geographic conditions, including areas of typically lower plant diversity. Table 4.1 shows cases from Puebla, Mexico, eastern Cuba, and the Mexican Plateau where homegardens in particularly semi-arid contexts provide culturally important and sustenance species in areas that would otherwise have lower biodiversity and fewer economic options.

Homegardens also serve as sites for domestication of useful species in several regions. Cases from Pará, Brazil, Copán, Honduras, and the Tehuacán-Cuicatlán valley in central Mexico demonstrate the use of homegardens in these regions to serve as "banks" for ancient and potential future crop species. They also serve as

**Table 4.1** Summary of types of AFS, their ecological and economic/cultural/social benefits, and case studies cited in the text.

| Type of system | Ecological benefits | Economic/cultural/social benefits | Case studies cited |
|---|---|---|---|
| Homegardens | Increased plant diversity in arid to semi-arid areas where low diversity is expected | Production of culturally important food crops (chilli (*Capsicum* spp.), tomato (*Lycopersicum esculentum)*, and several Cactaceae and Solanaceae species) in arid to semi-arid areas | Puebla, Mexico (Blanckaert et al. 2004) |
| | Reduced climatological impacts; increased soil fertility | Production of medicinal and food plants | Mexican plateau (Terrones Rincón et al. 2011) |
| | Increased species diversity, higher ecosystem productivity | Production of fruit trees in semi-arid area where other options are more difficult and less viable | Eastern Cuba (Wezel and Bender 2003) |
| | Conservation of crop species | Serve as crop species "banks", validation facilities for farmer decision-making, crop improvement and propagation | Pará, Brazil (Callo-Concha and Denich 2011) |
| | Serve as gene bank, preserving species not found elsewhere, including edible vegetable species and fruits, such as the chayo (*Cnidoscolus chaymansa*) | Serve as research fields for new varieties and cultivars and gene banks for edible vegetable and fruit species | El Camalote, Honduras (House and Ochoa 1998) |
| | Increased species diversity and conservation | Larger, better quality fruits, such as highly consumed fruit, tempesquistle (*Sideroxylon palmei*) | Tehuacán-Cuicatlán, Mexico (González Soberanis and Casas 2004) |
| | Serve as refuge for wildlife during disturbances like fire and for critical food source post-fire | Production of food sources for wildlife; food and cash income for farmers | Petén, Guatemala (Griffith 2000) |
| Successional AFS | Improved soil fertility; bees attracted for pollination | Preservation of indigenous system using distinct phases: *Milpa* (cultivated maize field), *arbusto* (shrub with planting), and *acahual* (fallow shrub), then return to *Selva Alta* (high forest); readily harvestable crops; honey bees | Chiapas, Mexico (Diemont et al. 2006; Diemont et al. 2011) |

(continued)

**Table 4.1** (continued)

| Type of system | Ecological benefits | Economic/cultural/social benefits | Case studies cited |
|---|---|---|---|
| | Managed gardens mirror natural succession stages and functions | Production of fruit trees | Dayak people of Indonesia (Peters 2018) |
| | Cultivated crops serve ecological functions of natural plants; restoration of degraded lands | Increase in agricultural production; reduced risk of drought-related harvest loss | NE Brazil (Schulz 2011). |
| | Attract wildlife for seed dispersal, increased soil fertility | Attract wildlife for hunting | Worldwide (Bertsch 2017) |
| Perennial crops under shade: Coffee | Display high biodiversity and high structural complexity; provide habitats; serve as perches/nesting sites; provide food sources; improve local microclimates; increase sustainability | Provision of coffee crop, product diversification, decreased economic risks, decreased need of external inputs | Mexico (Davidson 2005) |
| | Seed deposition; serve as buffer zones and biological corridors | Provision of coffee crop, product diversification, decreased economic risks, decreased need of external inputs | Mexico (Moguel and Toledo 1999) |
| | Increased butterfly species richness | Provision of coffee crop, product diversification, decreased economic risks, decreased need of external inputs | Chiapas, Mexico (Mas and Dietsch 2003) |
| | Increased plant diversity, better soil cover, soil conservation | Competitive coffee yields in intermediate intensity managed systems | Costa Rica (Rossi et al. 2011). |
| | Provide habitat for migrating birds | Provision of coffee crop, higher sustainability, product diversification, decreased need of external inputs | Neotropics (Somarriba et al. 2004). |
| | Increased bird diversity | Provision of coffee crop, higher sustainability, product diversification, decreased need of external inputs | Colombia, worldwide (Chait 2015; several others). |
| | Increased diversity of arthropods, mammals, amphibians and reptiles | Provision of coffee crop, higher sustainability, product diversification, decreased need of external inputs | Neotropics, worldwide (Mas and Dietsch 2003; Teodoro et al. 2011; Rossi et al. 2011; Chait 2015). |

(continued)

**Table 4.1** (continued)

| Type of system | Ecological benefits | Economic/cultural/social benefits | Case studies cited |
|---|---|---|---|
| | Increased diversity of pollinators | Increases in yield, weight of the bean, and quality of the coffee | Colombia, worldwide (Chait 2015) |
| Perennial crops under shade: Cacao | Conservation of Atlantic forest biodiversity | Provision of cacao crop, higher sustainability, product diversification, decreased need of external inputs | Cabruca systems in Bahia and Espírito Santo in eastern Brazil (Rolim and Chiarello 2004). |
| | Maintained species richness of trees, fungi, invertebrates, and vertebrates | Increased yield, higher sustainability, product diversification, decreased need of external inputs | Indonesia (Clough et al. 2011) |
| | Landscape restoration; increased water infiltration capacity; soil and water conservation; expanded areas with biological significance; enhanced biological corridors | Increased resilience of productive systems in the face of climate change; greater food sovereignty and employment opportunities | Cacao Alliance participants in El Salvador (Montagnini and Metzel 2017) |
| | Increased plant diversity, better soil cover, soil conservation | Improved livelihoods via stronger resilience to pests and diseases | Indonesia (Roshetko et al. 2016) |
| Perennial crops under shade: Yerba mate | Maintained bird species abundance (relative to adjacent forest reserve), including globally threatened species | Provision of yerba mate crop | Paraguay (Cockle et al. 2005) |
| SPS | Increased bird fauna habitat | Provision of cattle products, higher sustainability, product diversification, improved cattle comfort and productivity | Cordoba, Colombia (Múnera et al. 2009) |
| | Preservation of native species; nitrogen fixation, soil conservation, and natural biological control of pests | Production of timber, edible fruits for cattle | Neotropics (Calle et al. 2017) |
| | Provide resources for native wildlife; reduce risk of local extinction | Provision of cattle products, higher sustainability, product diversification, improved cattle comfort and productivity | Neotropics (Rivera et al. 2013; Montoya-Molina et al. 2016) |

(continued)

**Table 4.1** (continued)

| Type of system | Ecological benefits | Economic/cultural/social benefits | Case studies cited |
|---|---|---|---|
| Intensive SPS | Greatest richness in bird species of land uses; three times as many bird species as pasture systems without trees | Increased cattle productivity, higher sustainability, product diversification, improved cattle comfort and productivity | El Hatico, Colombia (Fajardo et al. 2009). |
| | Increase in abundance and diversity of birds; general plant and animal diversity; improved natural biological control; more rare/endangered species sightings | Increased cattle productivity, higher sustainability, product diversification, improved cattle comfort and productivity | La Vieja, Colombia (Calle et al. 2009). |
| Living fences | Promote biodiversity, increase connectivity among forest patches, serve as biological corridors | Protect agricultural plots and cattle; provide fuelwood, fruits, fodder, and shade for cattle | Neotropics (Harvey et al. 2005, 2008; Francesconi et al. 2011a, b; Ibrahim et al. 2011) |
| | Higher species richness of butterflies than in pastures with high or low tree densities | Divide, separate, and protect agricultural plots | Esparza, Costa Rica (Tobar et al. 2007) |
| | Preferred habitat and corridors for mantled howler monkeys | Divide, separate, and protect agricultural plots | Esparza, Costa Rica (Rosales and Sáenz 2007) |
| Windbreaks | Maintain spatial organization of landscape; contribute to fauna species mobility, serve as part of biological corridors | Serve as windbreaks to protect crops, provide tree products | Chaco, Argentina (Tamashiro 2018) |

research sites for validating and testing new varieties and methods, and as venues for developing products, such as fruits, that are larger or have better quality than in the wild (Table 4.1).

Homegardens are human impacted areas due to their proximity to homes. This makes them unique in that their biodiversity is often linked to human habitation—more so than other agroforestry systems that are practiced in remote locations where biodiversity can be more "wild" and less novel. Through the vast potential of species diversity of food plants and the associated habitat generated from such species, homegardens may harbour greater amounts of biological diversity than otherwise conventional agricultural systems or degraded landscapes, allowing for homegardens to function as biodiversity islands in the landscape.

**Fig. 4.1** Homegarden in Embu district, Kenya, featuring coffee, manioc, plantain, sugar cane, and a number of fruit, fuelwood, and timber species in the upper strata. The World Agroforestry Center has been studying and promoting homegardens and other agroforestry systems since the 1990s in collaboration with local organizations, based on ethnobotanical surveys and small farmer's needs, using local as well as exotic tree species for fodder, soil fertility and soil erosion control. (http://old.worldagroforestry.org/Units/Library/Books/Book%2006/html/14.5_kari_kefri_icraf_agrpfore.htm?n=135). Photo: F. Montagnini

#### 4.2.1.2 Successional Agroforestry Systems

In restoration ecology, successional processes are often manipulated to restore devastated landscapes (Hobbs et al. 2007). Successional AFS (SAFS) attempt to replicate the spatial and temporal dynamics of forest succession, with species assemblages that are planted, maintained, and modified over time to mirror the successional stages of secondary forest development, for achieving both ecosystem conservation or restoration as well as for production of subsistence and marketable goods (Fig. 4.2). Successional agroforests can be similar to natural forest ecosystems as they are characterized by high plant diversity and managed in accordance with the natural succession of species (Peneireiro 1999; Micollis et al. 2016).

Such systems have been practiced by indigenous peoples for centuries. The Lacandon Mayan of Chiapas, Mexico and the Kenyah Dayak of Kalimantan, Indonesia have long managed homegardens as SAFS according to traditional indigenous practices (Table 4.1). In many indigenous SAFS, plant species are selected not

**Fig. 4.2** In this system, indigenous people from the Napo Province in the Ecuadorian Amazon introduce useful species such as fruit, medicinal, and food species, which occupy consecutive successional stages, thereby restoring and enriching the biodiversity of the disturbed forests. This management practice can be considered a type of successional agroforestry system. Photo: F. Montagnini

only to provide fruit and products, including honey, for personal consumption and soil fertility, but also to attract wildlife, which deposit waste thus increasing soil fertility, bring new seed sources, and can be hunted for food (Bertsch 2017).

Many of these SAFS practices from traditional, indigenous knowledge have been adapted to modern agroecosystems (Schulz 2011; Young 2017). In northeastern Brazil, Nicaragua, and Belize, farmers integrate and enhance natural succession within their AFS, for instance by planting locally adapted edible plants with similar functional characteristics as plants of the same successional level within the native ecosystem, beginning with plants that augment organic material, and then slowly integrating plants of higher successional levels. Highly degraded areas have been regenerated with this method, resulting in an approximately four-fold increase of agricultural production, while also reducing the risk of drought-related harvest loss via crop diversification and the use of perennial plants (Schulz 2011).

### 4.2.2 *Perennial Crops Under Shade in Agroforestry Systems*

AFS of perennial crops under shade are broadly used throughout the tropics, with cacao and coffee being the most frequent worldwide (Montagnini et al. 1992; Beer et al. 1998; Chait 2015). The environmental value of shade trees is provided by their forest-like structure (Perfecto et al. 2005). Shade trees also have social and economic value in reducing the vulnerability of households to climatic stress, pest outbreaks, falling prices and food insecurity (Tscharntke et al. 2011). Enriching the diversity of natural shade trees, for instance by planting leguminous species, can also provide additional positive impacts such as increased soil fertility (Montagnini et al. 1992; Beer et al. 1998). Reducing pesticide spraying protects the functional agrobiodiversity of the system, including organisms that provide biological control against pests and diseases and pollinators that enhance cacao and coffee yield. From a landscape perspective, natural forest maintained alongside agroforestry increases the diversity of functionally important organisms (Tscharntke et al. 2011).

In AFS of perennial crops under shade, the amount of crop shading cover is a proxy of agricultural intensification (Beer et al. 1998). Shade trees provide long-term resistance and resilience in the presence of unmanageable pest pressure, vulnerability to changing climate and difficulties in rejuvenating the perennial crops. Shade removal, although it may increase short-term yield gains, may compromise this long-term resilience.

#### 4.2.2.1 Coffee

Coffee (*Coffea arabica*) is of paramount economic importance in more than 50 countries worldwide, and although it has potential to influence biodiversity conservation over large areas, its cultivation can also be a cause of deforestation if done poorly (Somarriba et al. 2004). Many regions where coffee is cultivated fall within areas identified as mega-diversity sites (Somarriba et al. 2004). In some areas, the landscape has been so severely degraded that the only remaining tree cover is in coffee plantations. For example, in El Salvador, the most densely populated country in Latin America with less than 10% of its natural forests remaining, the vast majority of remaining tree cover is associated with shade coffee (F. Montagnini, personal observations 2002; Blackman et al. 2012).

Traditional, indigenous shaded coffee AFS harbor relatively high biological diversity and provide high structural complexity (Moguel and Toledo 1999). These systems illustrate valuable features of AFS: providing habitat for a variety of species, serving as perches and nesting sites, providing food resources, and improving local microclimates that are amenable for a wide variety of species (Davidson 2005). Shaded coffee AFS can also serve as sites for seed deposition and germination, act as buffer zones and biological corridors, and serve as a refuge for wildlife from surrounding areas that have been deforested (Moguel and Toledo 1999).

Coffee AFS play an important role in providing habitat for–and increasing the local diversity of–birds, arthropods, mammals, and, to a lesser extent, amphibians and reptiles. In Chiapas, Mexico, Costa Rica, and elsewhere in the Neotropics, several studies have found an inverse relationship between the diversity of certain species and the intensity of the management of the coffee AFS (Table 4.1). In experimental coffee AFS at CATIE,[1] Costa Rica, intermediate management intensity produced competitive coffee yields, and organically managed plots had high herbaceous diversity and were as productive as chemically managed plots, suggesting that AFS can balance agricultural productivity while maintaining a significant number of herbaceous species (Rossi et al. 2011).

#### 4.2.2.2 Cacao

Cacao (*Theobroma cacao*) AFS can conserve natural resources and improve small farmers' livelihoods and self-sufficiency, by offering a varied production of food and cash crops (Cerda et al. 2014; Gross et al. 2016). Studies from Bahia and Espírito Santo, Brazil, and Indonesia, have demonstrated that in addition to the economic benefit they provide, cacao AFS conserve local biodiversity. In Indonesia, the species richness of trees, fungi, invertebrates, and vertebrates did not decrease with increased cacao yield, indicating that moderate shade and adequate labor can be combined with a complex habitat structure to provide both high biodiversity and high yields (Clough et al. 2011).

As with coffee and other crops, cacao's benefits to biodiversity greatly depend on design and management of the AFS. Even though moderate shade levels rarely reduce cacao or coffee yield, farmers in many parts of the world are converting shaded cacao and coffee systems into unshaded monocultures to increase short-term income (Tscharntke et al. 2011). However, benefiting from the long-term advantages of shaded cacao agroforestry does not necessarily exclude intermediate levels of intensification. For example, in Sulawesi, Indonesia, it has been shown that reducing canopy cover from 80% to 40% can double the income of local farmers with only minor changes in biodiversity and associated ecosystem services (Tscharntke et al. 2011). In another study in southern Cameroon of a project where increased use of fungicides and the expansion of cultivated area aimed to reduce rural poverty, overall plant diversity decreased only slightly with management intensification (Gockowski et al. 2010). In another example in Sulawesi, Indonesia, the transformation of the conventional cacao cultivation systems to cacao AFS has improved livelihoods for small farmers by increasing diversity in the cacao AFS, which enhanced yields that were previously poor due to pests and diseases (Roshetko et al. 2016).

[1] Centro Agronómico Tropical de Investigación y Enseñanza, Tropical Agriculture Research and Higher Education Center, Turrialba, Costa Rica.

**Fig. 4.3** Agroforestry system of organic yerba mate (*Ilex paraguariensis*) and timber trees in Misiones, Argentina, subtropical Atlantic Forest region. Organic yerba mate, grown generally under several species of native trees and shrubs of timber, fruit or other uses, can get price surpluses, which has led to an increased interest in organic farming and in yerba mate cultivation under shade in recent years (Montagnini et al. 2011; Eibl et al. 2017). Photo: F. Montagnini

#### 4.2.2.3 Yerba Mate

Yerba mate, *Ilex paraguariensis*, is a native tree from South America whose leaves are used to prepare an infusion or tea of popular local consumption, with a market expanding internationally due to its nutritious and energizing properties (Montagnini et al. 2011; Eibl et al. 2015, 2017). Yerba mate trees are usually grown in monocultures with conventional management, resulting in decreased plant productivity and soil erosion in the long term. Since the yerba mate tree grows naturally in subtropical forest and is shade tolerant, however, it is adequate for growing under the canopy of other tree species in AFS (Fig. 4.3).

Yerba mate grows naturally in the Atlantic forest of southeastern Brazil, northeastern Argentina, and eastern Paraguay, a region that is one of the world's biodiversity hotspots, with about 1–8% of all species worldwide, and high rates of plant, insect and mammal endemism (Myers et al. 2000; Calmon et al. 2011). The Atlantic forest is one of the most highly impacted rainforest areas in the world, where over five centuries of deforestation resulted in a ~ 84% loss of area, with deforestation continuing currently at a rate of 20,000 ha per year (Ribeiro et al. 2009; SOS Mata Atlântica and INPE 2014). Agriculture, cattle-ranching, and industry have replaced much of the Atlantic forest, and its diverse fauna is threatened by high grade logging, hunting, habitat loss, and habitat fragmentation (Cockle et al. 2005; Brewer 2011).

Yerba mate is grown by small or medium to large farmers as family businesses, farmers' cooperatives, or large-scale enterprises, both for local consumption and for export. Yerba mate cultivation thus expands a whole range of systems, from extensive monocultures, to AFS with 1–2 tree species for shade, to more complex, multistrata systems in the case of most of the organic yerba mate AFS (Ilany et al. 2010; Montagnini et al. 2011; Eibl et al. 2015, 2017) (Fig. 4.3).

In a study in Paraguay, bird species' presence and abundance were compared between a forest reserve and an adjacent plantation of shade-grown yerba mate, where some of the forest understory and trees were removed and yerba mate was planted below the tree canopy. Of the 145 species that were regularly recorded in the forest, 66%, including five globally threatened species, were also recorded in the yerba mate AFS. Within the yerba mate AFS, higher tree density did not lead to a greater abundance of birds. Yerba mate AFS under native trees could therefore be used to rehabilitate cleared land and allow recolonization by Atlantic forest bird species (Cockle et al. 2005).

In another study in the 'San Rafael' Reserve and its buffer area (Itapua Department, Paraguay), researchers analyzed the potential benefits of forests with shade-grown yerba mate for birds, amphibian and reptiles, comparing species richness and composition between three environments: forest with shade-grown yerba mate, forest edge and monoculture crop plantations. Their results suggest that forests with yerba mate plantations maintain high bird species richness, with its species composition differing significantly from edges and croplands (Cabral et al. 2020).

Yerba mate AFS can also boost the biodiversity conservation capabilities of forest fragments through increasing connectivity. Currently, with the expansion of the yerba mate cultivation area and an increased interest in growing it in AFS due to potential price surpluses, it would be interesting to further ascertain the role of yerba mate AFS on restoring and conserving biodiversity, as compared to other cultivating systems (Montagnini et al. 2011; Montagnini 2020).

### 4.2.3 Silvopastoral Systems

Silvopastoral systems (SPS) can incorporate more sustainable and biodiversity friendly approaches into livestock rearing operations. Silvopastoral systems, which involve the combination of trees with pastures and livestock, are classified by the functions and configuration or structure of trees within the system. SPS may include dispersed trees in pastures, live fences in pastures, fodder banks, tree alley pasture systems, and pastures with windbreaks (Pezo and Ibrahim 1999). Because they are more structurally complex than grass monocultures, SPS have important benefits for biodiversity. Many types of SPS (e.g., high density trees in pastures and live multi-strata fences) have levels of species richness comparable to those of early secondary forest, and networks of live fences in pastures are important for landscape

**Fig. 4.4** Silvopastoral systems using planted hybrid pines (*Pinus taeda* x *Pinus caribaea*), Braford cattle (hybrids of Brahman x Hereford), with *Brachiaria bryzantha* grass in Misiones, Argentina. This highly technified type of system based on exotic species of animals, trees and grasses is less biodiverse than more traditional systems based on local species of plants and animals, however, it is more structurally complex and therefore more biodiverse than grass monocultures. Photo: F. Montagnini

connectivity (Harvey et al. 2005; Francesconi et al. 2011a, b; Ibrahim et al. 2011). For example, in a comparison of bird diversity among different land uses in Cordoba, Colombia, the SPS had the greatest total number of bird species, followed by old fallows, forest fragments, and pastures with low tree density (Múnera et al. 2009).

The biodiversity benefits of SPS depend on the system components and management, with larger biodiversity present in the more complex systems, such as SPS with natural regenerating trees in pastures, than in the fodder banks or planted timber trees in pastures (Fig. 4.4). In particular, using native species as part of SPS confers several advantages to biodiversity conservation (Montagnini and Finney 2011; Murgueitio et al. 2011; Montagnini et al. 2013; Santos-Gally and Boege 2022). Native trees and palms play important roles in tropical livestock systems by providing direct benefits through production of timber and edible fruits for the cattle, and indirect benefits through nitrogen fixation, soil conservation, and natural biological control of pests, as well as by providing resources for wildlife (Rivera et al. 2013; Montoya-Molina et al. 2016; Calle et al. 2017). Endangered or vulnerable tree and palm species that are deliberately added to cattle ranching systems may have a lower risk of local extinction.

#### 4.2.3.1 Intensive Silvopastoral Systems

Intensive Silvopastoral Systems (ISPS) are agroforestry arrangements that combine high-density cultivation of fodder shrubs (4000–40,000 plants per ha) with improved tropical grasses, and trees or palms at densities of 100–600 individuals per ha. ISPS were initially developed in Colombia and have expanded to Mexico and Brazil, among other countries (Murgueitio et al. 2009, 2011; Chará et al. 2017). Several agroecological principles and strategies are applied in managing ISPS, including: (i) use of several layers of vegetation (herbs, shrubs, trees, and palms) to maximize energy transformation; (ii) reduced dependency on agrochemical inputs and energy, emphasizing interactions and synergisms among biological components to enhance recycling and biological control; and (iii) incorporation of biodiversity into the system components and its surroundings (Chará et al. 2017). Under these conditions, biodiversity restoration and conservation are enhanced in ISPS relative to other SPS, and even more so to conventional pastures.

In ISPS, the canopy cover, tree diversity, and structural complexity of vegetation contribute to their functions as habitat, refuge and food resources (Chará et al. 2015). In various studies from Colombia, examples of ISPS had greater richness of birds and other animal and plant species, provided natural biological control, and resulted in increased sightings of rare and endangered species (Table 4.1; Fajardo et al. 2009; Calle et al. 2009).

At the landscape level, ISPS can contribute to connectivity among patches of forest as well as to the recovery of strategic sites for the provision of environmental services (Calle et al. 2012, 2022). Several examples have been documented where the movement of organisms has been facilitated by ISPS. A matrix permeable to bird movement can avoid the collapse of small populations of wildlife that are isolated in forest fragments (Chará et al. 2015).

Because ISPS intensify production in high yielding systems on less land, they can help to avoid deforestation. At the same time, ISPS are a form of production that utilize multi-functional landscapes of high value for biodiversity without sacrificing productivity and economic feasibility (Chará et al. 2015). In this way, they incorporate the benefits of both land sparing and land sharing in a single productive and biodiversity friendly system.

### *4.2.4 Living Fences and Windbreaks*

#### 4.2.4.1 Living Fences

The principal role of living fences is to divide, separate, and protect agricultural plots or cattle. They also provide several services and products—including fuelwood, fruits, fodder, and shade for cattle—and a major environmental function: promotion of biodiversity (Harvey et al. 2005, 2008; Francesconi et al. 2011a, b; Ibrahim et al. 2011). In Esparza, Costa Rica, species richness of butterflies was found to be higher in multi-strata live fences than in pastures with high and low tree densities (Tobar et al. 2007; De Tobar and Ibrahim 2010). In this same area, Rosales and Sáenz (2007), found that Mantled Howler Monkeys (*Alouatta palliata*) preferred riparian forest and forest fragments for daily activities and also used living fences to move within and across pastured areas.

Living fences promote bird abundance and diversity by providing habitat and can be used by generalist and savanna specialist species. At the landscape level, living fences can provide effective connectivity among forest patches (Francesconi et al. 2011a, b; Francesconi and Montagnini 2015).

Like other AFS, their structure and composition are important factors influencing their usage by bird species. The presence of birds in living fences could be improved by increasing tree diversity and allowing trees to grow to mature stages or to develop broad crowns (Francesconi et al. 2011a, b). However, some of these features, such as having larger trees in the fences, may not be as practical for the farmer as it may be harder to manage the wire, and the trees maybe more difficult to prune. Thus living fence design should recognize the tradeoffs between their productive and conservation functions.

#### 4.2.4.2 Windbreaks

As they are often the only arboreal component of an agricultural landscape, windbreaks and hedges play important roles in providing habitats and resources for animals and other plants. Like living fences, windbreaks and hedges also function as natural corridors for animal movements across landscapes (Harvey et al. 2005, 2008). Windbreaks tend to be favored by farmers and can be instrumental in biodiversity conservation and landscape connectivity in fragmented areas. For example, forest windbreaks in the perimeter of agricultural fields are frequently used in the Chaco region of Argentina, where about 32% of the original forest remains on average. These windbreaks range from 30 to 50 m wide and 1700 m long, representing just 5% of the forest area but providing up to 40% connectivity among forest fragments. Windbreaks therefore help to maintain the spatial organization of the landscape and can contribute to the mobility of different species among forest fragments (Tamashiro 2018). Like living fences, windbreaks have a variety of agricultural functions and their design and management for environmental functions must be compatible with their agricultural use.

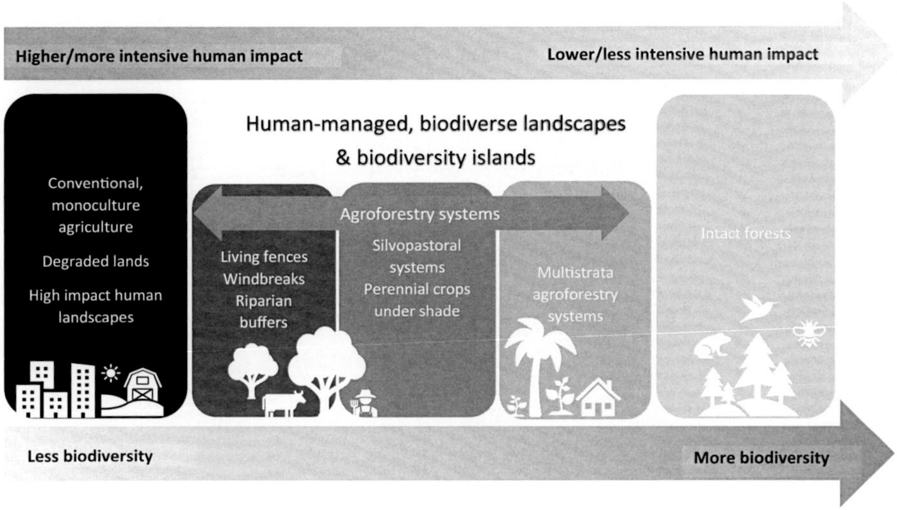


**Fig. 4.5** Schematic representation of landscapes along a spectrum of high to low human impact and low to high biodiversity

## 4.3 Applying Agroforestry Systems as and Within Biodiversity Islands

As demonstrated by the different types of AFS and examples of their applications in a variety of contexts, the contributions of AFS to biodiversity in managed landscapes depend on the type of AFS, its management, component species, and position within the landscape matrix (Montagnini 2020). The low-intensive management, multistrata AFS with native species has the greatest potential to harbor the largest biodiversity (Figs. 4.1, 4.2, and 4.3). As management intensity increases, biodiversity decreases accordingly (Fig. 4.4). However, even the less heterogeneous AFS provide greater biodiversity than would otherwise be realized in conventional monoculture agriculture or degraded landscapes. In addition, farmers value and protect AFS for their contributions to their livelihoods, thereby also ensuring the maintenance of ecosystem services they provide, including biodiversity. Figure 4.5 illustrates the location of AFS along this spectrum of high to low human impact and low to high biodiversity. The diagram points to intact forests as, generally, the land use supporting the highest biodiversity. However, certain types of AFS can sometimes have higher diversity of species than intact forests nearby, because the AFS can have open areas that attract species that would not use the forest, for instance, some species of birds and butterflies (Múnera et al. 2009).

Because AFS can be incorporated into a range of managed landscapes, they have great potential to function as biodiversity islands or as component parts of biodiversity islands. Building upon the foundations of island biogeography, biodiversity islands act as ecological refugia, protected areas, or reserves within a landscape and can exist in a wide range of human dominated landscapes (MacArthur and Wilson 1967; Tjørve 2010). They may be actively implemented as part of a complex landscape that includes several land uses (e.g., agriculture, forest plantations, etc.) or they may be part of a passive management practice (i.e., areas left untouched for practical or economic reasons). Biodiversity islands size, configuration, and position within the landscape may vary according to different features, including patterns of human settlement, development, and utilization of natural resources (Laurance 2008). The method of implementation is important, as it will determine the characteristics, position in the landscape, and management of the biodiversity island and surrounding landscapes.

Living fences and windbreaks function as biodiversity islands within cultivated areas, providing, as described earlier, habitat and resources for wildlife and other plant species. Like other biodiversity islands, they can serve as biological corridors, providing landscape connectivity within fragmented areas. Even relatively small patches of land, when strategically located around perimeters throughout a cultivated landscape, can provide important services to biodiversity.

Similarly, homegardens and successional AFS can also serve as important biodiversity islands within suburban or urban landscapes (Negret et al. 2022; Soler et al. 2022; Toensmeier 2022). Though their relative area may be small, the presence of these AFS within developed landscapes can provide otherwise missing ecological and biological functions.

AFS are also important components among land uses in buffer zones of protected areas. Biodiversity islands may therefore include a buffer zone that incorporates AFS to transition from the preserved "island" to areas of greater human impact or degradation. AFS may uniquely occupy these buffer zones since they are a source of productivity, but still provide ecological benefits, offering a more gradual transition for wildlife from the more protected areas. Multiple biodiversity islands spread over a large area in an optimal configuration can decrease chances of biodiversity loss through creation of repopulation reserves and biological corridors.

To ensure the effectiveness of different types of AFS as part of biodiversity islands, land users should take a landscape approach, considering both the prevalent land uses and the natural ecosystems in the region of study. The prevalent land uses in a landscape can be arranged along a continuum of successional stages, from the earliest stages of succession (degraded lands) to more mature stages (forests), with AFS lying in between these two extremes. Landowners can integrate small scale land sparing to set aside pieces of the property as untouched natural settings to act as biodiversity islands. Alternatively, or in addition, they can explore land sharing by using AFS to incorporate biodiversity conservation and food production on the same land. Both land sparing and land sharing can provide valuable protection of species diversity through time (Phalan et al. 2011). In either instance, outcomes result in a greater biodiversity within the biodiversity islands as compared to otherwise degraded landscapes.

Deciding on the design and management of AFS within a landscape must, however, be done appropriately and cautiously. For example, AFS promotion may be inappropriately used to justify forest cutting and to advance the agricultural frontier, citing the role that AFS plays in restoring or preserving biodiversity. AFS should instead be used to compensate for biodiversity loss by restoring and preserving biodiversity in regions where the landscape has already been converted to agriculture or otherwise degraded. In addition, it must be noted that the type of biodiversity provided by AFS might not be the desired or natural biodiversity of the region. Increasing overall biodiversity (through novel habitats in AFS) could be in fact detrimental to some types of native biodiversity. The type of AFS, its intensity, and use of native versus new crop species all impact the type of biodiversity to be promoted or restored.

The same judiciousness should be taken when planning and designing AFS as component parts of biodiversity islands. When compared to a completely unnatural system (like crop monoculture), AFS stand out for their biodiversity. However, compared to an intact native forest, AFS biodiversity may have less benefits, although it may still be better than the alternative monocultures or degraded landscapes. While AFS should not be a replacement for intact forests in terms of biodiversity value, they are a useful tool for conservation in increasingly human impacted landscapes.

## 4.4 Supporting and Promoting Agroforestry Systems as Biodiversity Islands with Incentive Programs

Given the ecological and social benefits of the various types of AFS presented in this chapter, several studies have looked at programs or designs that have helped to incentivize and expand their applications. In several examples worldwide, certification schemes that guarantee higher quality and ecologically-sound management have helped to facilitate the sale of AFS products to specialty export markets (Montagnini and Metzel 2017; Rocha et al. 2017). In combination with direct sale made possible by aggregating the harvest through producer cooperatives, certification holds the potential to help sustain the livelihoods of family farmers confronting an evolving market. By compensating farmers for the extra labor required to produce certified organic, biodiversity friendly products, certification can bridge the gap between financial and biodiversity benefits. Organic AFS are also more biodiversity friendly since the lack of pesticide and herbicide use favors both plant and animal diversity (Montagnini et al. 2011; Rossi et al. 2011).

Several commodities grown as perennial crops in AFS such as coffee, cacao, yerba mate, guayusa, and açaí have been able to achieve price surpluses that can serve as an incentive for the farmer to turn to certified organic or biodiversity friendly products (Montagnini and Metzel 2017; Rocha et al. 2017). In Pará, Brazil, açaí (*Euterpe oleraceae*) is harvested from forests as well as from homegardens in the estuary of the Amazon River. When marketed well, local producer associations and cooperatives facilitate the collective sale of açaí, offering alternative points of sale that may recognize the higher quality product, in response to the demand from export markets that value traditional production (Pepper and de Freitas Navegantes Alves 2017). In Argentina, although some times yerba mate production may not be very attractive due to price instabilities, organic yerba mate producers can get substantial price surpluses on their product, thereby increasing interest in organic farming and in yerba mate cultivation under shade in recent years (Montagnini et al. 2011; Eibl et al. 2017).

Using a similar model that promotes a product that has a favorable niche in the market, biodiverse cacao AFS have often been incorporated into restoration and rural development projects in Latin America and beyond (Cerda et al. 2014; Gross et al. 2016). This is especially true where the resulting biodiversity friendly product can obtain higher market prices, as is the case with wild and cultivated cacao in the Amazon region of Bolivia (Rocha et al. 2017). In El Salvador, the Cacao Alliance seeks to position the country as an exclusive origin for high quality fine aromatic cacao in the profitable gourmet segments in international markets (Montagnini and Metzel 2017). Cacao AFS generate social and environmental benefits such as: (a) restoring productive landscapes through increased vegetative cover; (b) increasing water infiltration capacity; (c) increasing size and quality of areas with restored biological significance where there has been reduction of ecological niches due to habitat fragmentation; and d) improving connections between biological corridors (Frank Sullyvan Cardoza Ruiz, Cacao Alliance, El Salvador, personal communication, September 2016).

As cattle ranching is expected to continue being an important land use in many regions, the use of payments for ecosystem services (PES) mechanisms can be one way to provide incentives for farmers to make their cattle ranching activities more environmentally friendly. A recent project in Latin America has examined whether PES has increased the adoption of SPS on cattle farms in Esparza (Costa Rica), Matiguás (Nicaragua), and Quindío (Colombia) (Ibrahim et al. 2011). An environmental service index (ESI) was developed to determine the level of PES, with birds as the primary indicator of biodiversity. The number of bird species observed in pastures with high tree densities or multistrata live fences was higher than that in degraded pastures and grass monocultures and was comparable to the number of species in riparian and secondary forest (Sáenz et al. 2007). The percentage of tree cover and the number of tree species were the two most important parameters that explained variation in bird species on different land uses.

Before the PES project began, farmers managed the pastures with the use of herbicides to control weeds, which was associated with high mortality of saplings and juvenile stages of native tree species (Ibrahim and Camargo 2001). With the implementation of PES, the use of herbicides was reduced significantly. In addition to managing natural regeneration to increase tree cover in pastures, private farmers in Costa Rica and Colombia were trained and supported to produce plants of focal tree species of interest for conservation. These plants were sold to many cattle farmers receiving PES and were planted along live fence lines and riparian forest that were fenced off to keep cattle away from the water sources.

Similarly, the *Mainstreaming Biodiversity into Sustainable Cattle Ranching* project (MBSCR) promotes the planting of 50 focal species of native trees and palms of global conservation concern in cattle farms in five regions in Colombia (Calle et al. 2015, 2017). In this context, focal species are native trees and palms that can be incorporated directly into SPS, live fences, or riparian buffers to enhance biodiversity and environmental services in cattle dominated landscapes. To achieve this goal, the MBSCR project uses a short-term PES to partially offset investment costs in land uses that are compatible with biodiversity. In recognition of the special effort that must be made to adopt focal species, farmers eligible for PES receive an additional bonus for planting and caring for these native species on their farms.

These programs and systems have been used in various contexts to provide support and incentives to farmers. As agricultural activities and human settlements continue to expand, such incentives may play an increasing role in aiding farmers and other land-users in incorporating AFS and other biodiversity friendly practices onto their productive lands or into their neighborhoods, in order to create biodiversity islands that foster connectivity and biodiversity. Figure 4.6 provides a conceptual diagram that summarizes the elements contributing to AFS, the predominant types of AFS, and their primary ecological and economic or social benefits.

## 4.5 Conclusions

With increasing threats to natural ecosystems worldwide due to human population pressures, and consequent biodiversity losses and fragmentation, new strategies are needed to restore, conserve, and connect ecosystems and landscapes. AFS present a compromise that meet both the needs of biodiversity conservation and sustainable productivity. Multistrata systems, including homegardens and successional AFS, hold the highest biodiversity, when compared to more simple AFS. The more simplified AFS designs, such as some types of perennial crops under shade or SPS with only few trees species for shade, still can provide more ecological benefits than monocultures or degraded lands, though they may offer lower biodiversity than the more complex multistrata systems. Living fences, windbreaks and hedges along the perimeters of cultivated land can provide connectivity in fragmented agricultural landscapes. To favor biodiversity restoration and conservation, AFS should increase their structural complexity in terms of number of species and strata.

Biodiversity islands can contribute to the protection of plants and wildlife in human managed landscapes and AFS have great potential to be incorporated into their designs, because of their intermediate role between intensively managed land and intact, protected land. AFS can be integrated as components of biodiversity islands, especially in buffer zones of protected areas, to smooth transitions from higher to lower impact areas. Alternatively, more complex AFS can also constitute biodiversity islands themselves, for instance among more intensively cultivated land or within human settled landscapes, where they can provide both economic and social functions along with their biological and ecological roles. Promotion of AFS should not be used to justify conversion of forests to agricultural systems, but rather, can be planned within a broader strategy that seeks to maintain or restore areas of natural forest within cultivated landscapes.

Like other conservation endeavors, financial incentives are often needed to promote and sustain AFS in order to provide a competitive advantage over more intensive agriculture. AFS inherently provide economic benefits because of the agricultural products they provide, but additional incentives may help to bolster their biodiversity and ecological contributions. Payments for Ecosystem Services (PES), certification schemes such as organic certification, and niche marketing, have been used successfully in this regard and may be integrated into AFS to promote its usage and to encourage more favorable environmental practices, including pesticide-free management and/or the use of native species. These types of incentives may support the increased use of AFS as and within biodiversity islands, in turn providing services for biodiversity and rural communities alike.

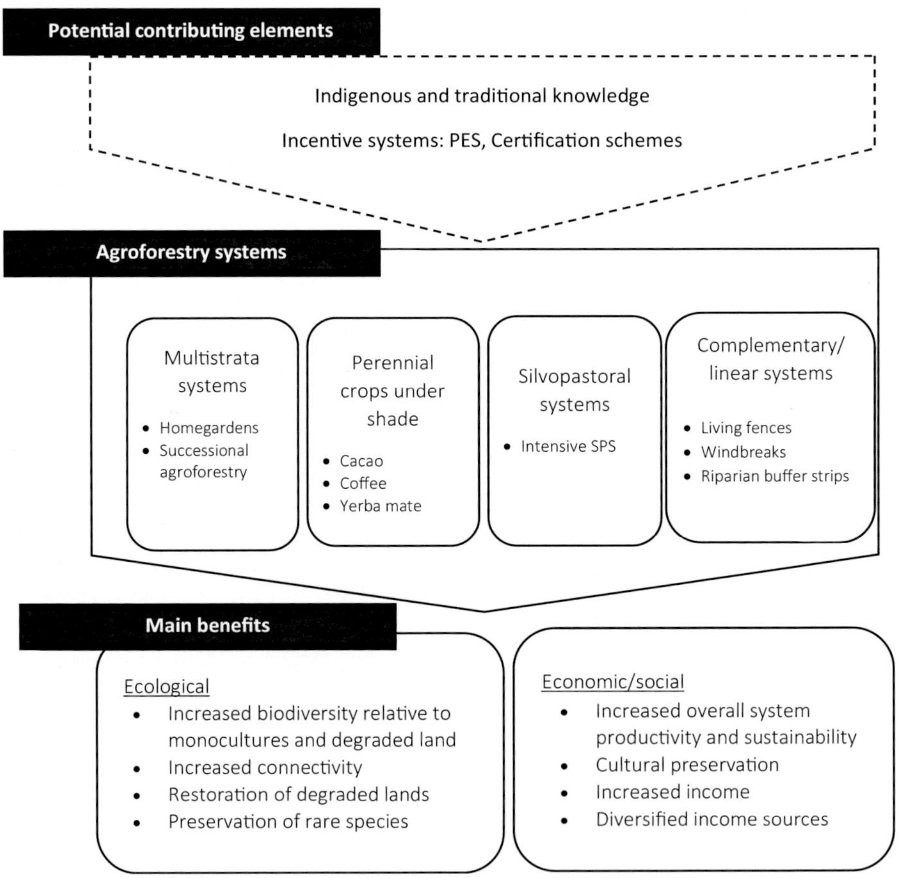


**Fig. 4.6** Schematic representation of elements contributing to AFS, the predominant types of AFS, and their primary ecological and economic/social benefits

**Acknowledgements** Thanks to Brett Levin (Yale University School of Forestry and Environmental Studies) who provided valuable assistance in the preparation of this chapter, and to Kjell Berg for his continued input and encouragement. Brett Levin and Joseph Orefice provided useful comments and insights as reviewers.

## References

Abete T, Wiersum KF, Bongers F, Sterck F (2006) Diversity and dynamics in homegardens of southern Ethiopia. In: Kumar BM, Nair PKR (eds) Tropical homegardens: a time-tested example of sustainable agroforestry. Advances in agroforestry 3. Springer, Dordrecht, 377 pp, pp 123–142

Beer J, Muschler R, Kass D, Somarriba E (1998) Shade management in coffee and cacao plantations. Agrofor Syst 38:134–164

Bertsch A (2017) Indigenous successional agroforestry: integrating the old and new to address food insecurity and deforestation. In: Montagnini F (ed) Integrating landscapes: agroforestry for biodiversity conservation and food sovereignty, Advances in agroforestry 12. Springer, Cham, pp 165–178

Bhagwat SA, Willis KJ, Birks HJB, Whittaker RJ (2008) Agroforestry: a refuge for tropical biodiversity? Trends Ecol Evol 23(5):261–267

Blackman A, Ávalos-Sartorio B, Chow F (2012) Land cover change in agroforestry: shade coffee in El Salvador. Land Econ 88(1):75–101

Blanckaert I, Swennen RL, Paredes Flores M, Rosas López R, Lira Saade R (2004) Floristic composition, plant uses and management practices in homegardens of San Rafael Coxcatlán, Valley of Tehuacán, Mexico. J Arid Environ 57:39–62

Brewer MD (2011) Strategies for cost-effective native species restoration in the sub-tropical Atlantic Forest of southern Brazil. In: Montagnini F, Finney C (eds) Restoring degraded landscapes with native species in Latin America. Nova Science Publishers, New York, pp 173–195

Cabral H, Coronel-Bejarano D, Ruiz R, Canete L, Britez E, Rojas V (2020) Living in the shadows: diversity of amphibians, reptiles and birds in shade-grown yerba mate (Ilex Paraguariensis) plantations. Austral Ecol:1–12. https://doi.org/10.1111/aec.12942

Calle A, Montagnini F, Zuluaga AF (2009) Farmer's perceptions of Silvopastoral system promotion in Quindío, Colombia. Bois For Trop 300(2):79–94

Calle Z, Murgueitio E, Chará J (2012) Integrating forestry, sustainable cattle ranching, and landscape restoration. Unasylva 63:31–40

Calle Z, Hernández M, Murgueitio E, Giraldo AM, Uribe F, Zuluaga AF (2015) Especies focales del Proyecto Ganadería Colombiana Sostenible. Carta Fedegán 148:54–60

Calle Z, Giraldo AM, Cardozo A, Galindo A, Murgueitio E (2017) Enhancing biodiversity in neotropical silvopastoral systems: use of indigenous trees and palms. In: Montagnini F (ed) Integrating landscapes: agroforestry for biodiversity conservation and food sovereignty, Advances in agroforestry 12. Springer, Cham, pp 417–438

Calle Z, Hernando C, José E, José J, Hernán C, Murgueitio B, Murgueitio A, Murgueitio E (2022) A highly productive biodiversity island: El Hatico Nature Reserve (Valle Del Cauca, Colombia). In: Montagnini F (ed) Biodiversity islands: strategies for conservation in human-dominated environments. Topics in Biodiversity and Conservation. Springer, Cham, pp 279–304

Callo-Concha D, Denich M (2011) Operationalizing environmental services of agroforestry systems: functional biodiversity assessment in Tomé-Açú, Pará, Brazil. In: Montagnini F, Francesconi W, Rossi E (eds) Agroforestry as a tool for landscape restoration. Nova Science Publishers, New York, pp 143–156

Calmon M, Brancalion PH, Paese A, Aronson J, Castro P, da Silva SC, Rodrigues RR (2011) Emerging threats and opportunities for large-scale ecological restoration in the Atlantic Forest of Brazil. Restor Ecol 19(2):154–158

Cerda R, Deheuvels O, Calvache R, Niehaus L, Saenz Y, Kent J, Vilchez S, Villota A, Martinez C, Somarriba E (2014) Contribution of cocoa agroforestry systems to family income and domestic consumption: looking toward intensification. Agrofor Syst. https://doi.org/10.1007/s10457-014-9691-8

Chait G (2015) Café en Colombia: servicios ecosistémicos, conservación de la biodiversidad. In: Montagnini F, Somarriba E, Murgueitio E, Fassola H, Eibl B (eds) Sistemas Agroforestales. Funciones productivas, socioeconómicas y ambientales, Serie Técnica Informe Técnico 402. CATIE/Fundación CIPAV, Turrialba/Cali, pp 349–361

Chará J, Camargo JC, Calle Z, Bueno L, Murgueitio E, Arias L, Dossman M, Molina EJ (2015) Servicios ambientales de sistemas silvopastoriles intensivos: mejoramiento del suelo y restauración ecológica. In: Montagnini F, Somarriba E, Murgueitio E, Fassola H, Eibl B (eds) Sistemas agroforestales: funciones productivas, socioeconómicas y ambientales, Serie técnica informe técnico No. 402. CATIE/Editorial CIPAV, Turrialba/Cali, pp 331–347

Chará JD, Rivera J, Barahona R, Murgueitio E, Deblitz C, Reyes E, Martins Mauricio R, MolinaJJ FM, Zuluaga AF (2017) Intensive silvopastoral systems: economics and contribution to climate change mitigation and public policies. In: Montagnini F (ed) Integrating landscapes: agroforestry for biodiversity conservation and food sovereignty, Advances in agroforestry 12. Springer, Cham, pp 395–416

Clough Y, Barkmann J, Juhrbandt J, Kessler M, Cherico Wangera T, Anshary A, Buchorig D, Cicuzza D, Darrasi K, Dwi Putrak D, Erasmi S, Pitopang M, Schmidt C, Schulze CH, Seidell D, Steffan-Dewenter I, Stenchly K, Vidal S, Weista M, Wielgoss AW, Tscharntke T (2011) Combining high biodiversity with high yields in tropical agroforests. Proc Natl Acad Sci USA 108(20):8311–8316

Cockle KL, Leonard ML, Bodrati AA (2005) Presence and abundance of birds in an Atlantic forest reserve and adjacent plantation of shade-grown yerba mate, in Paraguay. Biodivers Conserv 14: 3265–3288

Davidson S (2005) Shade coffee agro-ecosystems in Mexico: a synopsis of the environmental services and socio-economic considerations. J Sustain For 21(1):81–95

De Tobar L, Ibrahim M (2010) ¿Las cercas vivas ayudan a la conservación de la diversidad de mariposas en paisajes agropecuarios? Rev Biol Trop 58(1):447–463

De Tobar L, Ibrahim M, Casasola F (2007) Diversidad de mariposas en un paisaje agropecuario del Pacífico central de Costa Rica. Agroforestería en las Américas 45:58–65

Diemont SAW, Martin JF, Levy-Tacher SI (2006) Lacandon Maya forest management: restoration of soil fertility using native tree species. Ecol Eng 28:205–212

Diemont SAW, Bohn JL, Rayome DD (2011) Comparisons of Mayan forest management, restoration, and conservation. For Ecol Manag 261:1696–1705

Eibl B, Montagnini F, López M, Montechiesi R, Barth S, Esterche E (2015) Ilex paraguariensis A. St.-Hil., yerba mate orgánica bajo dosel de especies nativas maderables, una propuesta de producción sustentable. In: Montagnini F, Somarriba E, Murgueitio E, Fassola H, Eibl B (eds) Sistemas agroforestales: funciones productivas, socioeconómicas y ambientales, Serie técnica informe técnico No. 402. CATIE/Editorial CIPAV, Turrialba/Cali, pp 153–177

Eibl B, Montagnini F, López M, López LN, Montechiesi R, Esterche E (2017) Organic yerba mate, *Ilex paraguariensis*, in association with native tree species: a sustainable production alternative. In: Montagnini F (ed) Integrating landscapes: agroforestry for biodiversity conservation and food sovereignty, Advances in agroforestry 12. Springer, Cham, pp 261–281

Fajardo ND, González RJ, Neira LA (2009) Sistemas ganaderos amigos de las aves. In: Murgueitio E, Cuartas C, Naranjo J (eds) Ganadería del futuro: Investigación para el desarrollo, 2da edn. Fundación CIPAV, Cali, pp 171–203

Francesconi W, Montagnini F (2015) Los SAF como estrategia para favorecer la conectividad funcional del paisaje fragmentado. In: Montagnini F, Somarriba E, Murgueitio E, Fassola H, Eibl B (eds) Sistemas agroforestales: funciones productivas, socioeconómicas y ambientales, Serie técnica informe técnico No. 402. CATIE/Editorial CIPAV, Turrialba/Cali, pp 363–379

Francesconi W, Montagnini F, Ibrahim M (2011a) Living fences as linear extensions of forest remnants: a strategy for restoration of connectivity in agricultural landscapes. In: Montagnini F, Finney C (eds) Restoring degraded landscapes with native species in Latin America. Nova Science Publishers, New York, pp 115–126

Francesconi W, Montagnini F, Ibrahim M (2011b) Using bird distribution to evaluate the potential of living fences to restore landscape connectivity in pasturelands. In: Montagnini F, Francesconi W, Rossi E (eds) Agroforestry as a tool for landscape restoration. Nova Science Publishers, New York, pp 133–142

Gockowski J, Tchatat M, Dondjang JP, Hietet G, Fouda T (2010) An empirical analysis of the biodiversity and economic returns to cocoa agroforests in Southern Cameroon. J Sustain For 29(6):638–670. https://doi.org/10.1080/10549811003739486

González-Soberanis C, Casas A (2004) Traditional management and domestication of tempesquistle, Sideroxylon palmeri (Sapotaceae) in the Tehuacan-Cuicatlán Valley, Central México. J Arid Environ 59:245–258

Griffith DM (2000) Agroforestry: a refuge for tropical biodiversity after fire. Conserv Biol 14:325–326

Gross L, Castro-Tanzi S, Scherr SJ (2016) Connecting farms to landscapes for biodiversity conservation. The CAMBio project in Central America. Ecoagriculture discussion paper no. 15, Ecoagriculture Partners, Washington DC, 31 pp

Harvey C, Villanueva C, Villacís J, Chacón M, Muñoz D, López M, Ibrahim M, Gómez R, Taylor R, Martínez J, Navas A, Sáenz J, Sánchez D, Medina A, Vílches S, Hernández B, Pérez A, Ruíz F, López F, Lang I, Kunth S, Sinclair F (2005) Contribution of live fences to the ecological integrity of agricultural landscapes. Agric Ecosyst Environ 111:200–230

Harvey CA, Komar O, Chazdon R, Ferguson BG, Finegan B, Griffith DM, Martínez-Ramos M, Morales H, Nigh R, Soto-Pinto L, Van Breugel M, Wishnie M (2008) Integrating agricultural landscapes with biodiversity conservation in the Mesoamerican hotspot. Cons Biol 22(1):8–15

Hobbs RJ, Walker LR, Walker J (2007) Integrating restoration and succession. In: Walker LR, Walker J, Hobbs RJ (eds) Linking restoration and ecological succession. Springer, New York, pp 168–179

House PH, Ochoa L (1998) La diversidad de especies útiles en diez huertos en la aldea de Camalote, Honduras. In: Lok R (ed) Huertos caseros tradicionales de América Central: características, beneficios e importancia, desde un enfoque multidisciplinario. Centro Agronómico Tropical de Investigación y Enseñanza (CATIE), Turrialba, pp 61–84

Ibrahim M, Camargo JC (2001) ¿Cómo aumentar la regeneración de árboles maderable en potreros? Agroforestería en las Américas 8(32):35–41

Ibrahim M, Casasola F, Villanueva C, Murgueitio E, Ramírez E, Sáenz S, Sepúlveda C (2011) Payment for environmental services as a tool to encourage the adoption of silvo-pastoral systems and restoration of agricultural landscapes dominated by cattle in Latin America. In: Montagnini F, Finney C (eds) Restoring degraded landscapes with native species in Latin America. Nova Science Publishers, New York, pp 197–219

Ilany T, Ashton M, Montagnini F, Martinez C (2010) Using agroforestry to improve soil fertility: effects of intercropping on Ilex paraguariensis (yerba mate) plantations with Araucaria angustifolia. Agrofor Syst 80(3):399–409

Jose S (2009) Agroforestry for ecosystem services and environmental benefits: an overview. Agrofor Syst 76:1–10

Jose S (2012) Agroforestry for conserving and enhancing biodiversity. Agrofor Syst 85(1):1–8

Kumar BM, Nair PKR (2006) (eds) Tropical Homegardens: a time-tested example of sustainable agroforestry, Advances in agroforestry 3. Springer, Dordrecht, 377 pp

Laurance WF (2008) Theory meets reality: how habitat fragmentation research has transcended island biogeographic theory. Biol Conserv 141(7):1731–1744

MacArthur RH, Wilson EO (1967) The theory of island biogeography. Princeton University Press, Princeton, p 203

Mas AH, Dietsch T (2003) An index of management intensity for coffee agroecosystems to evaluate butterfly species richness. Ecol Appl 13(5):1491–1501

Miccolis A, Mongeli Peneireiro F, Rodrigues Marques H, Mascia Vieira DL, Francia Arco-Verde-M, Rigon Hoffmann M, Rehder T, Barbosa Pereira AV (2016) Agroforestry Systems for agroecological restoration: how to reconcile conversation with production, options for the Cerrado and the Caatinga. Brasília: Instituto Sociedade, População e Natureza – ISPN/World Agroforestry Centre – ICRAF, 240p. http://old.worldagroforestry.org/downloads/Publications/PDFS/B19034.pdf

Moguel P, Toledo V (1999) Biodiversity conservation in traditional coffee systems of Mexico. Cons Bio 13(1):11–21

Montagnini F (2006) Homegardens of Mesoamerica: biodiversity, food security, and nutrient management. In: Kumar BM, Nair PKR (eds) Tropical homegardens: a time-tested example of sustainable agroforestry, Advances in agroforestry, vol 3. Springer, Dordrecht, pp 61–84

Montagnini F (2020) The contribution of agroforestry to restoration and conservation: biodiversity islands in degraded landscapes. In: Dagar JC, Gupta SR, Teketay D (eds) Agroforestry for degraded landscapes: recent advances and emerging challenges, vol 1. Springer, Singapore, pp 445–479. https://doi.org/10.1007/978-981-15-4136-0_15

Montagnini F, Berg KE (2019) Feeding the world. In: Carver F, Gossington H, Manuel C (eds) Sustainable development goals. Transforming our world. Witan Media/United Nations Association – UK, Painswick/London, pp 98–100. https://www.sustainablegoals.org.uk/feeding-the-world/

Montagnini F, Finney C (2011) Payments for environmental services in Latin America as a tool for restoration and rural development. Ambio 40:285–297

Montagnini F, Metzel R (2017) The contribution of agroforestry to sustainable development goal 2: end hunger, achieve food security and improved nutrition, and promote sustainable agriculture. In: Montagnini F (ed) Integrating landscapes: agroforestry for biodiversity conservation and food sovereignty, Advances in agroforestry 12. Springer, Cham, pp 11–45

Montagnini F et al (1992) Sistemas agroforestales. Principios y aplicaciones en los trópicos, 2d. ed. Organización para Estudios Tropicales (OTS). San José, 622 pp. http://www.ots.ac.cr/images/downloads/information-resources/library/sistemasagroforestales.pdf

Montagnini F, Eibl B, Barth SR (2011) Organic yerba mate: an environmentally, socially and financially suitable agroforestry system. Bois For Tropiques 308:59–74. https://drflorenciamontagnini.wordpress.com/organic-yerba-mate-an-environmentally-socially-and-financially-suitable-agroforestry-system/

Montagnini F, Ibrahim M, Murgueitio Restrepo E (2013) Silvopastoral systems and mitigation of climate change in Latin America. Bois For Tropiques 316(2):3–16

Montagnini F, Somarriba E, Murgueitio E, Fassola H, Eibl B (eds) (2015) Sistemas agroforestales: funciones productivas, socioeconómicas y ambientales. Serie técnica informe técnico No. 402. CATIE/Editorial CIPAV, Turrialba/Cali, 454pp

Montagnini F, Levin B, Berg KE (2022) Introduction. Biodiversity islands: strategies for conservation in human-dominated environments. In: Montagnini F (ed) Biodiversity islands: strategies for conservation in human-dominated environments, Topics in Biodiversity and Conservation. Springer, Cham, pp 1–35

Montoya-Molina S, Giraldo-Echeverri C, Montoya-Lerma J, Chará J, Escobar F, Calle Z (2016) Land sharing vs land sparing in the dry Caribbean lowlands: a dung beetles' perspective. Appl Soil Ecol 98:204–212

Múnera E, Bock BC, Bolívar Vergara DM, Botero Botero JA (2009) Composición y estructura de la avifauna en diferentes hábitats en el Departamento de Córdoba, Colombia. In: Murgueitio E, Cuartas C, Naranjo J (eds) Ganadería del futuro: Investigación para el desarrollo, 2da Ed. Fundación CIPAV, Cali, pp 206–225

Murgueitio E, Cuartas C, Naranjo J (eds) (2009) Ganadería del futuro: Investigación para el desarrollo, 2da Ed. Fundación CIPAV (Centro para la Investigación en Sistemas Silvopastoriles de Producción Agropecuaria), Cali, p 490

Murgueitio E, Calle Z, Uribe F, Calle A, Solorio B (2011) Native trees and shrubs for the productive rehabilitation of tropical cattle ranching lands. For Ecol Manag 261(10):1654–1663
Myers N, Mittermeier RA, Mittermeier CG, Fonseca G, Kent J (2000) Biodiversity hotspots for conservation priorities. Nature 403:858–863
Nair PKR, Kumar BM (2006) Introduction. In: Kumar BM, Nair PKR (eds) Tropical homegardens: a time-tested example of sustainable agroforestry, Advances in agroforestry, vol 3. Springer, Dordrecht, pp 1–10
Negret HRC, Negret R, Montes-Londoño I (2022) Residential garden design for urban biodiversity conservation: experience from Panama City, Panama. In: Montagnini F (ed) Biodiversity islands: strategies for conservation in human-dominated environments. Topics in Biodiversity and Conservation. Springer, Cham, pp 387–417
Peneireiro FM (1999) Sistemas Agroflorestais Dirigidos pela Sucessão Natural: um Estudo de Caso. (Agroforestry systems driven by natural succession: a case study). Universidade de São Paulo, Escola Superior de Agricultura "Luiz de Queiroz". Master's dissertation, 138 pp
Pepper LG, De Freitas Navegantes Alves L (2017) Small-scale Açaí in the global market: adding value to ensure sustained income for forest farmers in the Amazon Estuary. In: Montagnini F (ed) Integrating landscapes: agroforestry for biodiversity conservation and food sovereignty, Advances in agroforestry 12. Springer, Cham, pp 211–234
Perfecto I, Vandermeer J, Mas AA, Soto Pinto L (2005) Biodiversity, yield, and shade coffee certification. Ecol Econ 54:435–446
Peters CM (2018) Managing the wild. Stories of people and plants and tropical forests. Yale University Press, New Haven, p 208
Peyre A, Guidal A, Wiersum KF, Bongers F (2006) Homegardens dynamics in Kerala, India. In: Kumar BM, Nair PKR (eds) Tropical homegardens: a time-tested example of sustainable agroforestry, Advances in agroforestry, vol 3. Springer, Dordrecht, pp 87–104
Pezo D, Ibrahim M (1999) Sistemas Silvopastoriles, 2nd ed. Proyecto Agroforestal CATIE/GTZ, Módulo de Enseñanza Agroforestal No. 2. CATIE, Turrialba, Costa Rica, 275pp
Phalan B, Onial M, Balmford A, Green RE (2011) Reconciling food production and biodiversity conservation: land sharing and land sparing compared. Science 333(6047):1289–1291
Ribeiro MC, Metzger JP, Martensen AC, Ponzoni FJ, Hirota MM (2009) The Brazilian Atlantic Forest: how much is left, and how is the remaining forest distributed? Implications for conservation. Biol Conserv 142:1141–1153
Rivera L, Armbrecht I, Calle Z (2013) Silvopastoral systems and ant diversity conservation in a cattle-dominated landscape of the Colombian Andes. Agric Ecosyst Environ 181:188–194
Rocha P, Niella F, Keller H, Montagnini F, Metzel R, Eibl B, Kornel J, Romero F, López L, Araujo J, Barquinero J (2017) Ecological indigenous (EIK) and scientific (ESK) knowledge integration as tool for sustainable development in indigenous communities. experience in Misiones, Argentina. In: Montagnini F (ed) Integrating landscapes: agroforestry for biodiversity conservation and food sovereignty, Advances in agroforestry 12. Springer, Cham, pp 235–260
Rolim SG, Chiarello AG (2004) Slow death of Atlantic forest trees in cocoa agroforestry in southeastern Brazil. Biodivers Conserv 13:2679–2694
Rosales M, Sáenz JC (2007) Uso de coberturas de vegetación por tropas de monos congos (Alouatta palliata) y carablanca (Cebus capucinus) en un agropaisaje y percepciones de finqueros con respecto a la conservación de los primates en Esparza, Costa Rica. In: III Congreso Iberoamericano Sobre Desarrollo y Ambiente (CISDA), Heredia, Costa Rica
Roshetko JM, Purwanto E, Moeliono M, Widayati A, Purnomosidhi P, Mahrizal, Wau D, Perdana A, Martini E, Gaol A, Paramita E, Dahlia L, Suyanto KN, Manurung G, Yuliani L, Rohadi D, Manalu P, Umar A, Millang S (2016) Agroforestry and Forestry in Sulawesi: linking knowledge with action. Annual progress report year 5 (April 2015–March 2016). World Agroforestry Centre (ICRAF) Southeast Asia Regional Program, Center for International Forestry Research; Bogor, Indonesia, Operation Wallacea Trust, Bogor, Indonesia, Faculty of Forestry, Hasanuddin University, Makassar, Indonesia, 137 pp
Rossi E, Montagnini F, de Melo E (2011) Effects of management practices on coffee productivity and herbaceous species diversity in agroforestry systems in Costa Rica. In: Montagnini F,

Francesconi W, Rossi E (eds) Agroforestry as a tool for landscape restoration. Nova Science Publishers, New York, pp 115–132

Sáenz JC, Villatoro F, Ibrahim M, Fajardo D, Pérez M (2007) Relación entre las comunidades de aves y la vegetación en agropaisajes dominados por la ganadería en Costa Rica, Nicaragua y Colombia. Agroforestería en las Américas 45:37–48

Santos-Gally R, Boege K (2022) Biodiversity islands. Native tree islands within silvopastoral systems in a neotropical region. In: Montagnini F (ed) Biodiversity islands: strategies for conservation in human-dominated environments, Topics in Biodiversity and Conservation. Springer, Cham, pp 117–138

Schulz J (2011) Imitating natural ecosystems through successional agroforestry for the regeneration of degraded lands – a case study of smallholder agriculture in northeastern Brazil. In: Montagnini F, Francesconi W, Rossi E (eds) Agroforestry as a tool for landscape restoration. Nova Science Publishers, New York, pp 3–17

Soler R, Benítez J, Solá F, Lencinas MV (2022) Biodiversity islands at the world's southernmost city: plant, bird and insect conservation in urban forests and peatlands of Ushuaia, Argentina. In: Montagnini F (ed) Biodiversity islands: strategies for conservation in human-dominated environments, Topics in Biodiversity and Conservation. Springer, Cham, pp 419–437

Somarriba E, Harvey CA, Samper M, Anthony F, Gonzalez J, Staver C, Rice RA (2004) Biodiversity conservation in neotropical coffee (Coffea Arabica) plantations. In: Schroth G, da Fonseca GAB, Harvey CA, Gascon C, Vasconcelos HL, Izac MN (eds) Agroforestry and biodiversity conservation in tropical landscapes. Island Press, Washington, D.C., pp 198–226

SOS Mata Atlântica, INPE (2014) Atlas dos remanescentes florestais da Mata Atlântica: período 2012–2013. Instituto Nacional de Pesquisas Espaciais http://www.inpe.br/ https://www.sosma.org.br/

Tamashiro S (2018) Corredores verdes reducen impactos del desmonte. Retrieved 17 October 2018, from http://curi.nearural.com/ampliar.php?mkt_hm=6&id=40360&utm_source=email_marketing&utm_admin=8097&utm_medium=email&utm_campaign=#.W4l9KnmQBck.email

Teodoro AV, Munoz A, Tscharntke T (2011) Early succession arthropod community changes on experimental passion fruit plant patches along a land-use gradient in Ecuador. Agric Ecos Environ 140(1–2):14–19

Terrones Rincón TDRL, Hernández Martínez MA, Ríos Ruiz SA, Martínez Ayala C (2011) Non-wood products from native multipurpose trees from agroforestry homegardens in the semiarid Mexican Plateau. In: Montagnini F, Francesconi W, Rossi E (eds) Agroforestry as a tool for landscape restoration. Nova Science Publishers, New York, pp 85–97

Tjørve E (2010) How to resolve the SLOSS debate: lessons from species-diversity models. J Theor Biol 264(2):604–661

Toensmeier E (2022) Paradise lot: a temperate urban multistrata agroforestry island of biodiversity. In: Montagnini F (ed) Biodiversity islands: strategies for conservation in human dominated environments. Topics in Biodiversity and Conservation, Springer, Cham, pp 439–459

Tscharntke T, Clough Y, Bhagwat SA, Buchori D, Faust H, Hertel D, Holscher D, Juhrbandt J, Kessler M, Perfecto I, Scherber C, Schroth G, Veldkamp E, Wanger TC (2011) Multifunctional shade-tree management in tropicalagroforestry landscapes – a review. J Appl Ecol 48:619–629

Udawatta RP, Rankoth LM, Jose S (2019) Agroforestry and biodiversity. Sustainability 11:2879. https://doi.org/10.3390/su11102879

Wezel A, Bender S (2003) Plant species diversity of homegardens of Cuba and its significance for household food supply. Agrofor Syst 57:39–49

Wezel A, Ohl J (2006) Homegarden plant diversity in relation to remoteness from urban centers: a case study from the Peruvian Amazon. In: Kumar BM, Nair PKR (eds) Tropical homegardens: a time-tested example of sustainable agroforestry, Advances in agroforestry 3. Springer, Dordrecht, pp 159–184

Young K (2017) Mimicking nature: a review of successional agroforestry systems as an analogue to natural regeneration of secondary forest stands. In: Montagnini F (ed) Integrating landscapes: agroforestry for biodiversity conservation and food sovereignty, Advances in agroforestry 12. Springer, Cham, pp 179–209

# Chapter 5
# Biodiversity Islands: The Role of Native Tree Islands Within Silvopastoral Systems in a Neotropical Region

**Rocio Santos-Gally and Karina Boege**

**Abstract** Neotropical rainforests have lost about 35% of their native vegetation due to deforestation for agricultural purposes and/or livestock grazing. In addition, many forest remnants are immersed in a grassland matrix with low productivity, due to soil degradation and erosion, high temperature and low humidity. Under such a scenario, silvopastoral systems (SPS) can be a sustainable alternative to increase the profitability of livestock production, allowing areas not suitable for livestock to be restored for biodiversity conservation. In this chapter we review the advantages of SPS from their prehistoric appearance to the more innovative intensive SPS (iSPS), and propose that the inclusion of native tree islands (as ecological restoration plots) can further increase the recovery of biodiversity in tropical regions. For the implementation of these native tree islands, we review different approaches for ecological restoration, from passive ones, such as exclusion of cattle in pastures, to the most labor-intensive ones, such as planting seedlings within cattle farms. We also discuss the relevance of considering the different components of biological diversity (genetic, functional and phylogenetic diversity) for species selection during these ecological restoration efforts and highlight the use of phylogenetic diversity as a useful predictor of functional diversity and ecological dynamics. Finally, we present an example of a recent intervention in a tropical region of Mexico including the implementation of islands of native trees taking phylogenetic diversity into account. Our proposal is to conserve and restore, not only the species but the unique evolutionary history of the great biodiversity accumulated in the tropics.

**Keywords** Agroforestry · Biodiversity conservation · Ecosystem functioning · Functional traits · Livestock · Phylogenetic diversity

---

R. Santos-Gally (✉)
CONACYT-Instituto de Ecología, Universidad Nacional Autónoma de México, México, CDMX, Mexico
e-mail: rsantos@iecologia.unam.mx

K. Boege
Instituto de Ecología, Universidad Nacional Autónoma de México, México, CDMX, Mexico



F. Montagnini (ed.), *Biodiversity Islands: Strategies for Conservation in Human-Dominated Environments*, Topics in Biodiversity and Conservation 20,
https://doi.org/10.1007/978-3-030-92234-4_5

## 5.1 Introduction

We are currently facing a global biodiversity loss of a magnitude not seen before during human history. To date, 35% of the original vegetation cover around the world has been lost, and of the remaining forested areas about 82% have been degraded by human activities (Watson et al. 2018). Furthermore, it is estimated that every year between 11,000 and 58,000 plant species go extinct (Ceballos et al. 2017). In the case of vertebrate animals, more than 50% of the species are endangered (Hoffmann et al. 2010; Dirzo et al. 2014; Ceballos et al. 2017), and populations of most representative insect orders are decreasing in numbers (Hoffmann et al. 2010; Dirzo et al. 2014). For plants, the situation is critical as well, particularly in the tropics, as most of the world's tropical tree species now are considered to be threatened (Steege et al. 2015). Last, but not least important, our reduced knowledge regarding microorganism diversity does not even allow us to estimate how many species are being lost at this taxonomic level.

A generalized decrease in biodiversity across the world has led to a striking reduction of ecosystem functionality and different ecological processes providing climate regulation, water availability, nutrient cycling, soil fertility and disease control (Haines-Young and Potschin 2010; Segan et al. 2016). The loss of these ecosystem services has affected human socioeconomic resilience, compromising natural resources and livelihood for future generations. Food and water provisioning, resilience to extreme climatic events, incidence of pest and zoonotic diseases, are just some processes already affected by anthropogenic activities in the last century (Haines-Young and Potschin 2010; Watson et al. 2018).

In both terrestrial and aquatic ecosystems, the main drivers of species extinction afecting ecosystem processes are habitat loss, the spread of invasive species, acute climate change, widespread environmental pollution, and the over-exploitation of species (Dirzo et al. 2014; Bellard et al. 2016; Román-Palacios and Wiens 2020). In particular, the expansion of farmland has led to the loss of ~100 million ha of tropical forests around the world in just over three decades (AGAL 2008; Lewis et al. 2015). Tropical regions in Latin America have been severely affected in the last decades, due to the expansion of agricultural and livestock activities, with the loss of between 3.8 (Achard et al. 2014) and 4.88 (Baccini et al. 2012) million ha/year of wet and dry forests, with a mean annual deforestation rate of 0.49% (Achard et al. 2014). For example, as a result of an aggressive policy promoting the introduction of African improved grasses and cattle species for extensive livestock production, Mexico has lost up to 80% of the original tropical forest cover in the last 50 years (Challenger and Soberón 2008).

Extensive cattle ranching requires large extensions of pastures, and a heavy investment in herbicides and pesticides to maintain high productivity, which makes it economically challenging and, in the long run, inefficient (Arellano et al. 2018). Without any tree cover, open pastures promote a significant increase in local temperatures, the loss of water sources, soil compaction, a reduction of available nutrients in the system and an overall biodiversity loss at all taxonomic levels, from soil microbes to large vertebrates at the top of food chains. Hence, this productive activity causes a continuous deforestation of tropical forests (Steinfeld et al. 2006).

The maintenance of healthy and sustainable socio-ecological systems requires a generalized transformation process and reconsideration of the current ways of carrying out agricultural, forestry and livestock activities. Different alternatives should aim for sustainable food production systems encompassing the conservation of biodiversity, ecosystem functioning and ecological processes, incorporating the values, knowledge and interests of local producers and the enhancement of local markets. In this sense, agroforestry and silvopastoral systems (SPS) play an important role in advancing the interactive trajectories of socio-economic and environmental changes (Altieri 1999; Montagnini 2017; Chará et al. 2019; Calle 2020). Through so-called land sharing, these systems optimize ecological processes by increasing functional trait diversity (e.g. incorporating nitrogen fixing species) that in turn delivers ecosystem services to replace external inputs (e.g. fertilizers), turning fragmented and less-productive landscapes into a biodiverse production matrix (Perfecto and Vandermeer 2010; Cardinale et al. 2012; Calle et al. 2013; Kremen and Merenlender 2018). For example, the combination of different vegetation strata promotes the use of solar energy in the conversion of biomass, which can be used in animal feed, and can also add nutrients to the soil through leaf decomposition, and water filtration by roots (Cardinale et al. 2012). In this sense, agroecology and its associated systems can be considered an inextricable component of the biodiversity conservation agenda.

In this chapter we review how SPS in the neotropics can be a sustainable alternative for livestock production. We highlight the productive and economic benefits derived from SPS, especially those where the use of high protein forage species is intensified, and how, ideally, such systems can liberate fragile areas within farms for ecological restoration. We discuss different approaches to ecological restoration (from cattle exclusion and natural regeneration to plantations), both in un-productive abandoned pastures, and within cattle ranches. As SPS prove their productive efficiency and producers become more sensitive to increasing biodiversity in their land, the discussion of which species are the most suitable for the proper functioning of the ecosystem becomes crucial. Therefore, we emphasize the importance of considering genetic, functional and phylogenetic diversity when selecting plant species for ecological restoration.

For example, previous studies suggest that an increase in genetic diversity is associated with a higher ability to adapt and respond to environmental changes (Kettenring et al. 2014). Phylogenetic diversity – a measure of the amount of evolutionary history represented within a community – has also been suggested to increase ecosystem resilience and productivity due to a broader diversity of traits that represent the functional diversity of the ecosystem (Cadotte et al. 2008, 2013). Community diversity is quantified through different measures of phylogenetic and functional diversity, which are based on the distance that separates the species from their most recent common ancestor. Conserving and/or restoring biodiversity, taking into account these evolutionary relationships, means that we are protecting not only ecological processes but also their evolutionary history.

## 5.2 Silvopastoral Systems

### *5.2.1 SPS as a Sustainable Alternative for Livestock Production*

Silvopastoral systems are characterized by intentional and/or spontaneous arrangements that combine trees or shrubs, herbs and grasses for nutrition, fiber, medicine, and energy for people and animals. This alternative way of feeding livestock has a long history related to human use of available resources in particular ecosystems (Jose et al. 2019). Among the oldest examples is the *Dehesa* landscape from the Iberian Peninsula (also called *Montado* in Portugal), Italy, France and Morocco, which includes the combination of different trees (*Quercus ilex*, *Q. suber*, *Q. rubur*, *Argania spinosa, Platanus orientalis*, etc.) and grasses with a variety of livestock types (cows, sheep, pigs, horses, goats) (Fig. 5.1). On a historic timescale, use of *Dehesas* likely began in the mid-late Neolithic (ca. 4500–3300 BC), when humans used management interventions such as thinning to establish pastures underneath Mediterranean tree canopies to feed livestock (López-Sáez et al. 2007, 2014). Paleoecological records show a transition from a Mediterranean vegetation, dominated by oaks and junipers in the tree layer (*Q. ilex*-type, *Juniperus*-type), and olives (*Olea*-type), lentisk (*Pistaccia lentiscus*-type) and gum rockrose (*Cystus ladanifer*-type) in the shrub strata, towards an increase of heliophilous taxa, such as ribwort plantain (*Plantago lanceolata*-type), nettle (*Urtica dioica* type), Amaranthaceas and

**Fig. 5.1** Silvopastoral systems in the South of Europe and North Africa, (**a**) SPS in France with *Quercus robur*, grasses and *Aubrac* cattle, (**b**) *Quercus suber*, herbs and grasses with black Iberian pigs in the Sierra Norte of Seville, Spain; (**c**) Moroccan mountain goats with *Argania spinosa*; **d**) France, *Platanus orientalis*, herbs and grasses with horses. (Photos: R. Santos-Gally)

other nitrophilic species (those adapted to growth in nitrogen-rich soils), such as *Rumex*-type and *Aster*-type, associated with the presence of domestic animals and agriculture. In addition, archaeological records suggest the use of acorns to produce flour, which could have been used to supplement the human diet (Cerrillo et al. 2005) providing an additional benefit. The perpetuation of the *Dehesa* landscape since prehistoric times is an example showing that anthropic intervention in ecosystems can be sustainable, ensuring productive benefits while conserving biodiversity and climate regulation (Garrido et al. 2017; Ferraz-de-Oliveira et al. 2016). In fact, the *Dehesa* has been considered a traditional cultural ecosystem or semi-natural ecosystem (according to the European Union legal context) and serves as a model ecosystem in ecological restoration (Gann et al. 2019).

Neotropical livestock farming, since its inception in the sixteenth century, also originally occurred under the shelter of trees (Murgueitio and Ibrahim 2008; Guevara et al. 2018; Jose et al. 2019), ensuring a sustainable productive system preserving local biodiversity. It was not until the last century that a mistaken conception of incompatibility between trees and grasses permeated among ranchers, basically due to the introduction of heliophilous grasses of the genera *Cynodon*, *Megathyrsus*, *Brachiaria*, *Urochloa*, *Pennisetum*, *Dichanthium*, *Cenchrus*, *Bothriochloa* and others which were well adapted to browsing by large herbivores and to fire. Their use promoted what we know today as extensive cattle ranches without any tree cover.

Nevertheless, the presence of trees within the productive landscape is still common in tropical farms (Kú et al. 1999; Guevara et al. 2018). These trees have multiple uses as a supply of fruit, medicine, forage, shade, wood, live fences, windbreaks and river corridors. The presence of trees is also evidence of the importance of traditional ecological knowledge and that their presence is of utmost importance for the management and conservation of natural resources (Raymond et al. 2010; Guevara et al. 2018; Tarbox et al. 2020). This knowledge should be an ally in the adoption and expansion of silvopastoral practices. Furthermore, synergies between local and scientific ecological knowledge can promote co-produced innovations for the transition towards more sustainable practices (Reed et al. 2008).

### 5.2.2 *Innovations of Intensive Silvopastoral Systems for Livestock Sustainability in Latin America*

Intensive silvopastoral systems (iSPS) are a novel agroecological alternative consisting of a well-designed combination of different vegetation strata such as grasses, herbs, shrubs, trees and/or palms. This species arrangement increases biodiversity at different levels and therefore, functional traits that underpin different functions in ecosystemic processes and their benefits for productive systems such as nutrient recycling, soil fertility enhancement, increased primary productivity, and climate buffering. In particular, highly proteinic forage plants, with functional traits such as high leaf N and P content for direct browsing by livestock are a key element in iSPS. Species currently used for the shrub strata within Latin American iSPS are *Leucaena leucocephala*, *Tithonia diversifolia* and *Guazuma ulmifolia* (Murgueitio et al. 2015). Planted in high densities (more than 10,000 $ha^{-1}$) within the paddocks, *L. leucocephala* facilitates high fixation and transfer of nitrogen, while *T. diversifolia* favors the solubilization of phosphorous in acid soils thus benefitting the associated pastures (Ojeniyi et al. 2012; González 2013; Bacab 2013). These species also provide high protein content for cattle, which translates into higher productivity. The result of this arrangement is a more productive and diverse system, with greater species richness and more diverse functional traits, although biodiversity is not necessarily maximized.

Intensive rotation of livestock among paddocks practiced with permanent or mobile electric fences is essential to allow forage plant and grass recovery. It is also fundamental to ensure a supply of good quality water through fixed or mobile drinkers and to provide mineral salt. Live fences often divide paddocks and are used to limit the area where the cows graze (Murgueitio et al. 2019). In addition, arrangements in the tree stratum may include timber, fruit and forage species which serve as food and shade for livestock, as well as isolated or small groups (i.e. "vegetation islands") of native trees providing ecological services such as pollination, nutrient recycling, water filtration, fruit and seed dispersion, and carbon fixation. Depending on the ecoregion and productive interest, the number of adult trees can range from 100 to 600 $ha^{-1}$ (Chará et al. 2019). The rotation of livestock and the retention of forest cover, through living fences and trees in paddocks, play a key role for ecological restoration (Martínez-Ramos et al. 2016). The first helps reduce soil compaction and improves nutrient recycling, thus facilitating the appearance of new propagules. The latter act as biological corridors, to facilitate the movement of pollinators, seed dispersers, herbivores and other animals (Omeja et al. 2016).

The different components of the iSPS (Fig. 5.2) increase biodiversity and its associated functional traits. Ideally, these systems allow a transformation of livestock farming landscapes of homogeneous vegetation to complex networks of vegetation, where the areas adjacent to the pastures are restored to form biological corridors, and the points where the productive zones join can form native tree islands (see green squares in Fig. 5.2). Native tree islands become a source of propagules, nesting sites, biological pest control and a sink for soil nutrients, and the increase in these ecological processes is linked to environmental, productive and economic improvements. Cattle is fed better thanks to efficient and quality grazing resulting from high-protein forage. The animals suffer less heat stress, as the temperature in wooded paddocks can decrease between 4–8 °C in reference to open pasture areas, and distances to search for water or food are reduced through access to mobile drinkers and more forage biomass. All this can result in an increase of 5 to up to 10 times the amount of meat production and up to an additional 80% in the volume of milk produced in comparison with conventional pastures (Murgueitio and Giraldo 2009; Navas 2010; González 2013; Bacab et al. 2013; Chará et al. 2019; Jose et al. 2019; Murgueitio et al. 2019). In addition, intensive rotation of livestock results in an

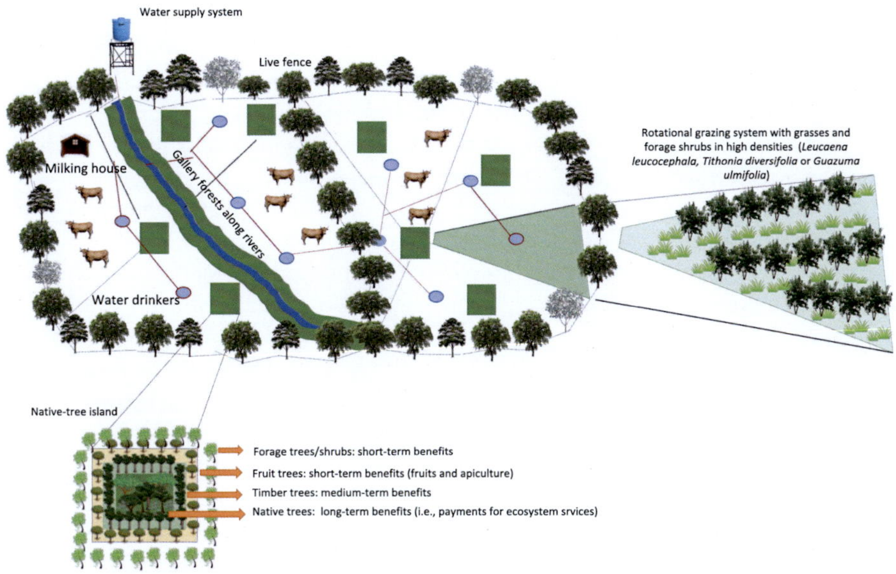


**Fig. 5.2** Scheme of a cattle ranch with intensive silvopastoral systems, native-tree islands (green squares), a riparian vegetation corridor, a water system for drinking fountains and living fences. (Modified from Santos-Gally et al. 2019)

increase of stocking rate per ha of four to five times higher than what is achieved in extensive cattle ranching (Murgueitio et al. 2019). Expenses on external inputs, such as fertilizers, can be reduced to zero due to the increased nitrogen fixation and other nutrients contributed by forage shrubs (González 2013, Murgueitio et al. 2019). Forage biomass is also increased by up to 47% compared to that within a monoculture of pastures, therefore reducing the need for food supplementation by more than half (González 2013; Calle et al. 2013).

An iSPS, when well designed and managed, is a habitat with a more complex structural vegetation than the treeless grass monocultures of extensive pastures, and therefore, becomes a refuge enabling increased diversity of arthropofauna, local and migratory birds, reptiles and small mammals (Harvey and Haber 1998; Giraldo et al. 2010; Rivera et al. 2013; Omeja et al. 2016). The increased biodiversity of iSPS allows some of the ecological processes of natural habitats to be restored, such as the acceleration of manure recycling by different types of dung beetles. Some iSPS have been shown to conserve 60.7% of forest beetle species, thus reducing laying sites of blood-eating cattle flies which can negatively impact weight gain by cattle (Murgueitio and Giraldo 2009; Giraldo et al. 2018). Additional ecological processes are restored such as fruit and seed dispersal, pollination and nutrient recycling. In this way, iSPS could also present a greater diversity of functional traits, although not necessarily maximizing phylogenetic diversity.

At an interspecific level, functional traits determine the adaptive attributes by which an organism can survive in a given environment (response traits) and their contribution to ecosystem functioning and services (effect traits) (Violle et al. 2007; Cardinale et al. 2012; Srivastava et al. 2012). For example, in tropical forests, whether roots are deep or superficial has implications for resource competition, as species with deep roots can survive better during long periods of water stress (Paz et al. 2015). Species with pivotal deep roots contribute to soil development, groundwater and streamflow regulation, soil carbon storage and moisture content in the atmosphere (Pierret et al. 2016). Other functional traits, such as leaves and roots with lower C:N ratios, influence organic matter decomposition rates (these being faster when the C:N ratios are lower) and thus soil microbiota and nutrient cycling. In this context, the inclusion of native-tree islands within SPS could significantly increase phylogenetic diversity and hence the number of functional groups in a particular landscape.

Furthermore, the change of diet resulting from the increased diversity of tannin-rich species modifies the digestive processes of cows, reducing the amount of methane and nitrous oxide emitted into the atmosphere, and thus their contribution of greenhouse gas emissions (Rivera et al. 2015). Compared with conventional extensive livestock, the iSPS generates an increase in the environmental services that result from the greater diversity in functional traits of the different species that are found within it. Therefore, an ecological restoration program that prioritizes phylogenetic diversity also prevents the erosion of evolutionary history, when species with distant recent common ancestors are purposely chosen, as in the hypothesized case of islands of native vegetation within an iSPS (as shown in Fig. 5.2).

In general, iSPS allow intensifying production within the parts of the property that are most suitable for livestock, generating greater productivity and profitability, ideally freeing up areas that have a vocation other than that of livestock production. These areas are ideal to fulfill ecological restoration processes, becoming forests fragments used for conservation purposes and the recovery of ecosystem services in the long term (Perfecto and Vandermeer 2010; Giraldo et al. 2018). In the following sections we discuss different implementation experiences and challenges related with the establishment of native tree species either within the degraded pastures and iSPS, or in adjacent areas released from cattle ranching and designated for conservation. We emphasize the importance of considering plant phylogenetic diversity as a novel tool for ecological restoration in these areas.

## 5.3 Experiences of Pasture Restoration in Dry and Humid Neotropical Forests

Different restoration experiences of tropical forests in abandoned cattle pastures have shown the complex and context-dependent factors influencing the success of such efforts (Martínez-Garza et al. 2016). Many empirical studies provide different approaches to using tree islands or forest patches of native vegetation within cattle ranches to recover biodiversity and ecosystem services (Aide et al. 2000; Griscom et al. 2005; Martínez-Ramos et al. 2012; González-Tokman et al. 2018). Whereas most studies have been performed in abandoned pastures (Aide et al. 1994, 2000; Martínez-Ramos et al. 2012), a few experiences show that even in active livestock pastures, vegetation restoration is possible and is a viable way of achieving more sustainable cattle ranching practices (Martínez-Garza and Howe 2003; de la Peña-Domene et al. 2013; Calle et al. 2017). Next, we describe some key lessons from experiences with ecosystem restoration in Latin America and Mexico in particular, providing important details and challenges that must be considered when establishing native tree islands within livestock ranches.

### 5.3.1 Intensity and Costs of Interventions

Depending on the site conditions, the intensity of soil degradation present in restoration areas (Chazdon 2003), and on the available economic resources, restoration interventions in degraded pastures can range from minimal, low-cost approaches to very intensive and costly plantations (Chazdon 2003; Griscom et al. 2005; Martínez-Garza et al. 2016). Minimal intensity interventions consist of the exclusion of particular areas from cattle, allowing natural vegetative succession to occur (Hobbs and Norton 1996; Aide et al. 2000). Studies of natural succession in abandoned pastures in neotropical dry and humid forests show that, if propagule species are available nearby, cattle exclusion and fire protection allow the establishment of early and intermediate stages of forest succession in a range of 25–40 years, with species richness levels equivalent to mature forests, although species enrichment can be required to achieve the same species composition (Aide et al. 1994, 2000).

Similarly, in a tropical dry forest of Panama, Griscom et al. (2005) report that, when excluding cattle from abandoned pastures, up to 67 species (mostly pioneer species) were established in experimental plots, with significantly greater stem and basal areas than plants established where cattle was not excluded. In Mexico, chronosequence studies of abandoned pastures in both tropical dry and humid forests show that after 20 years of cattle exclusion, different biotic communities including plants, amphibians, reptiles, bats and birds can reach the same diversity and structure as mature forests, although functional traits may take longer to recover (Martínez-Ramos et al. 2012). However, these forest succession trends are strongly influenced by the level of land degradation and distance to propagule sources (Chazdon 2003; Martínez-Ramos et al. 2012). Although these studies suggest that restoration of tropical forests is possible by excluding cattle and protecting target areas from fires, recovering ecosystem services often takes much longer. Hence, incorporating vegetation islands into sustainable cattle ranching initiatives may require some degree of additional intervention.

An intermediate level of intervention involves the exclusion of cattle from the areas to be restored, together with actions accelerating secondary succession. These actions include the enrichment of restoration areas with pioneer species (Martínez-

Ramos and García-Orth 2007), the removal of competitor species such as grasses, vines or ferns (Martínez-Ramos et al. 2012, González-Tokman et al. 2018), and the protection of seeds to reduce seed predation (Martínez- Ramos 2012). For example, in Southeastern Mexico, the enrichment of areas with early successional stages ("acahuales") with the pioneer species *Ochroma pyramidale* was reported to accelerate the re-establishment of different late-successional species, increasing the availability of organic matter in the soils, and the eradication of fern species arresting natural succession (Levy et al. 2016). Other experiences have also included the protection of isolated trees within pastures or even the introduction of artificial bird perches to stimulate seed dispersal from forest fragments (Shiels and Walker 2003; Laborde et al. 2008; Martínez-Garza et al. 2016). The presence of isolated fig trees in livestock pastures in Los Tuxtlas, México, for instance, has been shown to promote the establishment up to 73 native tree and shrub species (Laborde et al. 2008).

The most intensive level of intervention to restore ecosystem functioning is the establishment of tree plantations of a diverse array of native plant species within areas excluded from cattle (Martínez-Gaza et al. 2016). This strategy, although more costly, may accelerate the processes of natural secondary succession, the recovery of soil fertility, nutrient cycling and availability of different niches for higher trophic levels (del-Val et al. 2016), facilitating the recovery of species interactions and ecological processes. The establishment of plantations requires adequate species selection, depending on the purposes of the desired ecological restoration and the availability of native propagules in the surrounding areas. For example, if the objective of the plantation is to accelerate the process of natural succession, planting a combination of pioneer and late successional tree species, in particular those dispersed by animals, may further enhance the recruitment of new species (Martínez-Garza et al. 2013; de la Peña-Domene et al. 2013). If long-term carbon fixation is of interest, species with high wood density and greater water conduction efficiency are recommended (Martínez-Cabrera et al. 2009). Furthermore, species selection may determine the success of these efforts, as not all tree species are adapted to the same soil characteristics and climatic conditions. Which species to include in a native vegetation island should consider not only the local availability of seeds or propagules, but also previous experience and local knowledge on their propagation, phenology, germination techniques or ability to be propagated via stakes.

### *5.3.2 Technical and Biological Factors Influencing the Establishment of Vegetation Islands*

The technical and biological solutions are complex and need to be adjusted for each case, considering the costs, suitable species and the design of the areas to be recovered. Equally important to consider are the primary objectives for which those vegetation islands or ecological restoration plots are created. Are landholders interested in recovering native biodiversity for ethical, cultural and/or economic reasons? Are they aware of the abilities of vegetation islands to improve ecosystem services, thereby enhancing the productivity of their cattle ranches? Are they interested in including timber or fruit tree species to diversify the products coming from their ranches? Addressing this diversity of interests is necessary for species selection and may require the inclusion of timber or fruit trees in vegetation islands within livestock pastures (Minnemeyer et al. 2011; Calle et al. 2012). For example, in Colombia, the successful improvement of pastures combines the establishment of high densities of forage species for cattle, such as *Guazuma ulmifolia*, with strips (instead of vegetation islands) of native timber species such as *Cordia gerascanthus* and *Tabebuia rosea* and the endangered *Pachira quinata* (Calle et al. 2012).

Once species are selected, choosing the right propagation techniques is also relevant. Some species can be established vegetatively through stakes, whereas others perform better when planted as seeds or saplings. For example, in Chiapas, Mexico, a restoration experiment demonstrated that plant survivorship and performance was increased when plants were planted as seeds or saplings, rather than propagated from stakes (Douterlungne and Ferguson 2016). This was probably due to greater survival of plants with pivotal roots, in contrast with stakes that need to develop their roots.

Overall, these different ecological restoration efforts in tropical forests can provide guidelines for the implementation of native-tree islands within cattle pastures and conservation areas, in which the costs and work implicit at different level of intensity described here must be taken into account. Most of these experiences however, have been performed in abandoned pastures, without the challenge of livestock exerting a continuous pressure on the excluded plots. Some strategies to deal with this pressure are a) excluding livestock from plots with long-lasting living fences, and b) planting buffers of forage plants to prevent livestock from entering the plantations.

## 5.4 Relevance of Incorporating Phylogenetic Diversity in Ecological Restoration

Intensive silvopastoral systems that contain vegetation islands can protect natural ecosystems, recover biodiversity, and restore ecological functioning in landscapes modified and fragmented by productive activities. Therefore, these systems are useful for reducing biodiversity loss, which can lead to ecosystem degradation and loss of ecological benefits (Perfecto and Vandermeer 2010). In addition, when deforestation is followed by agriculture and livestock grazing, the potential for forest regeneration is decreased. Ecological restoration projects with native tree species in SPS in Colombia have used the focal species approach for plant species selection (Calle et al. 2017). This includes the selection of those plant species most threatened by degrading processes such as habitat loss and fragmentation, climate change, alteration of biochemical cycles, or species introductions, in turn facilitating the conservation of less vulnerable species (Lambeck 1997). It has been proposed that the specific management and landscape restoration requirements for certain species could be representative of the requirements of other species of the same or different taxonomic groups (Lindenmayer et al. 2002). The use of vulnerable or endangered species in SPS, as for example, *Mimosa trianae* and *Swietenia macrophylla* in Colombia (Calle et al. 2017) has contributed to their conservation, with parallel benefits such as the provision of habitat for birds and insects or improved micro-habitats for the establishment of other species. Other restoration efforts have shown that the introduction of greater plant species diversity or greater plant functional trait diversity can enhance particular ecosystem processes and the associated ecosystem services for human well-being (Benayas et al. 2009; Doherty et al. 2011; Montoya et al. 2012).

There is sufficient evidence that biodiversity and the diverse functional traits within a given environment determine the stability of ecosystem functions through time (Cardinale et al. 2012). Furthermore, a congruence between phylogenetic diversity and functional trait diversity has been reported in different ecological communities (Cadotte 2019), indicating that those communities with species dispersed along the branches of the phylogenetic tree of life, i.e. the tree-like representation of the phylogenetic relationships that describes the evolutionary history of Earth's species, tend to be more functionally diverse.

Phylogenetic diversity can be viewed as a proportion of the diversity of a group of interest that is represented within a specific ecological community. If the composition of these species includes lineages with distant common ancestors, then there is a greater probability of finding different functional traits between the species due to a greater degree of independent evolutionary histories. Phylogenetically distant species tend to have dissimilar traits, whereas closely related species tend to have similar traits and ecological niches, competing for the same resources and responding similarly to natural selection agents (Felsenstein 1985; Díaz et al. 2013). For example, under current climate change conditions, tropical dry forests face an increase in temperature and drought (IPCC 2007; Aguirre-Gutiérrez et al. 2020). In some tropical forest communities in West Africa, an increase in the composition of drought tolerant species with low ratios of leaf mass to sapwood mass (LM:SM) and higher photosynthetic rates has been observed (Fauset et al. 2012). A shift towards greater species homogeneity in landscapes due to changing environmental conditions has been demonstrated at functional and phylogenetic levels (Aguirre-Gutiérrez et al. 2020), highlighting the need to cover the different components of diversity for more effective ecosystem conservation. Such homogenization of species composition at different levels, including genetic variation, phylogenetic diversity, and in particular, functional traits, can reduce the diversity of species responses to environmental fluctuations, overall reducing community resilience to changing climates.

Hence, species diversity and species functional traits are important to biotic interactions and dynamics of communities (Navarro-Cano et al. 2014). Given the intra and interspecific variation in functional traits (morphological, physiological, structural, behavioral, biochemical, etc.) their quantification to assess their value to ecosystem processes is not a simple task. In addition, measuring certain traits such as the structure and size of tree roots is also challenging. Therefore, phylogenetic diversity has the added practical advantage of being a useful proxy for functional diversity.

Recently, Cadotte et al. (2019) found that the degree of correlation between phylogenetic and functional diversity reported in 36 studies was positive with values ranging from 0.4 to 1.0. The tendency for phylogenetically distant species to be functionally disparate allows for their coexistence, improving their efficiency in the use of limiting resources and buffering harsh abiotic conditions (Navarro-Cano et al. 2014). Studies on plant assemblages of high phylogenetic diversity have shown higher productivity in terms of plant biomass (Cadotte 2013), higher soil microbial productivity (Navarro-Cano et al. 2014), or decreased plant-pathogen infection range (Gilbert and Webb 2007). Phylogenetic and functional diversity are thus two metrics that can be used to restore diverse, functional and resilient communities (Forest et al. 2007; Verdú et al. 2012; Hipp et al. 2015; Thévenin et al. 2018; Cadotte et al. 2019).

Intraspecific genetic variation in plants is another form of diversity that can modify ecosystem functioning. This may be relevant to ecological restoration for its ability to modify soil properties, microbiota composition, and ultimately the role of plant-soil feedbacks in determining plant performance (Schweitzer 2018). Indeed, a comparison of different genotypes of *Pinus pinaster* from three different geographic regions showed that genetic variation determines the phylogenetic structure of rhizosphere microbial communities. This, in turn, modulated the enzymatic processes related to the C, N and P cycles (Pérez-Izquierdo et al. 2019). Thus, plant genotypes affect ecosystem properties and services through the different impacts that they can have on nutrient availability and soil fertility.

## 5.5 Phylogenetic Diversity in Native Tree Islands Within Silvopastoral Systems

Genetic, functional and phylogenetic diversity are associated with increases in the abilities of communities to adapt and respond to changes in the environment, as well as the resilience and productivity of the ecosystem (Díaz et al. 2013). To test our hypothesis that high phylogenetic diversity in native-tree islands within iSPS can enhance the ecological outcomes of restoration efforts, thereby promoting a high diversity of functional traits, we implemented an iSPS with native-tree islands within the pastures. These vegetation islands had different degrees of phylogenetic diversity, allowing for future studies to test how different plant communities influence ecological processes and the recovery of ecosystem services. Because they were only recently established (2018–2019) and data collection is in process, no results regarding the influence of phylogenetic diversity on ecosystem processes and services has yet been obtained, but here it is described how native tree islands were implemented to inspire further restoration efforts within iSPS elsewhere.

We implemented a native-tree island experiment in the Mexican humid tropics (18.552469 N, −94.998846 W) (Santos-Gally et al. 2019) within extensive cattle pastures in a region where a high evergreen rainforest was gradually deforested starting in the 1940s. As a result, by 1990 the region consisted mainly of a patchwork of extensive cattle pastures with different species of native grasses (*Andropogon sp.*, *Digitaria sp.*, *Lasiacis divaricata*, *Oplismenus burmannii*, *Panicum* and *Setaria sp.*), some introduced species (*Cynodon plectostachyus*. *C. dactylon*, *Brachiaria sp.*), and few scattered native trees. The soil has a clay to clay-loam texture and is acidic to moderately acidic (pH 5.2–5.7), probably due to the high Fe values. The land use is mostly semi-extensive livestock farming with 0.9 cows/ha, mainly dual-purpose zebu and zebu-swiss cow breeds. Within a 100 ha livestock farm, we chose an area of 10 ha to develop an iSPS including ecological restoration with scattered native-tree islands. Because soils are acidic to moderately acidic, we chose *Tithonia diversifolia, Leucaena diversifolia* and *Guazuma ulmifolia* as the species for the high-density shrub stratum. In the design of the selection of forage species, we considered the rancher's preferences, the presence of the species in natural habitats, preference as a fodder species by the cows, high protein content (20–28%), as well as the species' ability to fix nitrogen and solubilize phosphorus in the soil (*L. diversifolia* and *T. diversifolia*, respectively).

Within the cattle pasture, we established native-tree islands which had either high or low phylogenetic diversity (HPD and LPD, measured as the sum of all the branches that comprise each group in a time-calibrated phylogenetic tree of all angiosperm families), with 30 plant species per island, belonging to 25 and 10 different plant families, respectively, according to the Angiosperm Phylogeny Group (APG III) (Stevens 2001) classification (N = 11 islands per treatment). With the intention of accelerating plant succession in the plant community assemblage, a greater number of tolerant species than pioneer species was selected. Within each treatment, 20 species were shade-tolerant while 10 species were pioneers. Twelve species were shared between treatments (six tolerant and six pioneers) and 36 species were not shared between treatments (28 shade-tolerant and eight pioneers). In total we selected 48 native tree species. To measure the phylogenetic diversity of each group, we used a time-calibrated phylogenetic tree reconstruction of all angiosperms families and some gymnosperm families as an outgroup (Fiz-Palacios et al. 2011). Two phylogenetic trees were reconstructed according to the families included in each treatment. Phylogenetic diversity was measured for each treatment using the package *picante* (Kembel et al. 2010) in R (R Team 2018) and calculated as the sum of all branches spanning the group of species of interest (Faith 1992). The HPD treatment, composed of species from 25 angiosperm families, had a phylogenetic diversity that was three times greater than the LPD treatment, with species from 10 families.

Seeds were collected from at least 3 to 5 mother trees for both phylogenetic groups, for the purpose of increasing intraspecific genetic variation. Seeds were germinated in a mixture of black soil and soil from around the parent tree in an effort to incorporate parent tree's soil microbial community. Seedlings were planted in the experimental islands (plots) of the two phylogenetic groups between August 2018

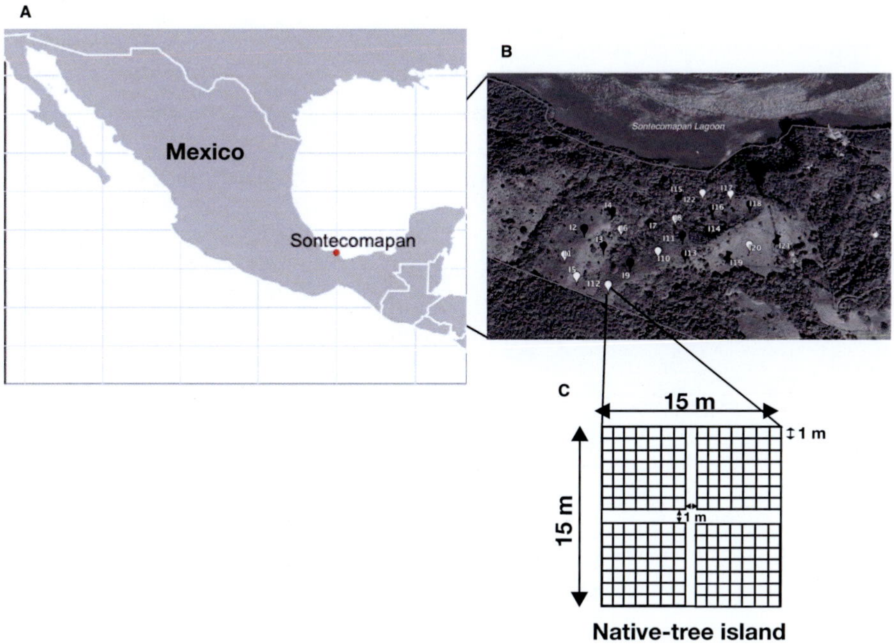


**Fig. 5.3** (**a**) Map of Mexico showing the location of the study site in Sontecomapan, Veracruz; (**b**) distribution of native-tree islands in Rancho Los Amigos, black symbols represent islands with high phylogenetic diversity, white symbols those with low phylogenetic diversity. (**c**) Sketch of a native-tree island representing how saplings where planted

and January 2019. The seedlings were planted once they had reached an appropriate size for planting, determined by the phenology of the species, their germination and growth rate. Grass was controlled several times by clearing around the saplings. The location and type (HPD or LPD) of each 15 $m^2$ biodiversity island within the 10-hectare paddock was assigned at random (Fig. 5.3). The islands were excluded from cattle with two barbed wire lines.

Physicochemical and biological soil parameters including measurements of arthropods and microbes, as well as the diversity of pollinator and invertebrate species in the plant stratum, were recorded. These measurements were taken prior to the establishment of the vegetation islands to determine responses to high and low phylogenetic diversity treatments in biodiversity islands over the 10-year study. These measurements will be compared 3, 6 and 9 years after the establishment of the vegetation islands. In addition, the survival of seedlings, competition, litter decomposition, growth rate, plant palatability to herbivores, seed dispersion and pollination diversity will be compared between treatments.

Most of the results of this ecological restoration experiment within a silvopastoral system will be obtained in coming years. Nonetheless, we present some early results from the implementation stage. First, we found that the seeds of the species that were used to implement the biodiversity islands had an average germination success of 32% (Santos-Gally et al. 2019). Survival of saplings for both HPD and LPD plots was greater after the rainy season (96%) than after the dry season (69%). Because we expect HPD islands will contain a greater diversity of functional traits, allowing greater complementarity in the use of resource when compared to LPD islands, we predict that HPD islands will present higher productivity, better decomposition of organic matter and greater availability of nutrients, which in turn will be facilitated by a greater diversity of microorganisms in the soil (Cadotte et al. 2008; Cadotte 2013). If the diversity of reproductive and dispersal traits increases with phylogenetic diversity, we expect that the diversity of pollinator and disperser species will increase in islands with high phylogenetic diversity. Furthermore, if structural diversity of vegetation increases in islands with HPD, then we expect greater diversity in the recruitment of seedlings on these islands.

Previous studies have shown that seedling survival is frequently affected by environmental variability such as that which occurs during a prolonged dry season (Martínez-Garza et al. 2013), as well as by competition with grasses (González-Tokman et al. 2018) or, in the case of restoration attempts within production paddocks, livestock pressure. To avoid this pressure, we recommend isolating restoration areas with an electric fence and connecting plots with corridors. Such corridors could have the additional advantage of serving to separate paddocks and provide shade for cattle (e.g. paddock division in Fig. 5.2), as well as potentially functioning as corridors for some small vertebrates. In our case, the control of the exotic grass *Brachiaria brizantha* CIAT 6780, was an important two-pronged problem. First, because it is known that its root system competes efficiently for water resources (Guenni et al. 2002), and second, once established as a pasture, its eradication requires continuous clearing. Sometimes if the exotic grasses turn out to be a very aggressive competitor with seedlings, the use of graminicides is also suggested at this stage.

## 5.6 Conclusions

Intensive silvopastoral systems play a crucial role as part of the biodiversity conservation agenda, especially in Neotropical regions such as Colombia, Brazil and Mexico. Native tree islands within iSPS increase native biodiversity, serving as a source of propagules, nesting sites, biological pest control and contributing to soil nutrient recycling. The iSPS enriched with tree islands and/or corridors can be envisioned as a mosaic of forest fragments and grasslands that allow for increased vegetation cover, reestablishing ecological and evolutionary processes, and improving ecosystem services in Neotropical regions. Hence, establishing iSPS with islands of phylogenetically and functionally diverse native vegetation can contribute to the stability, resilience, and functioning of tropical ecosystems.

In our experience, the implementation of native tree islands within productive pasturelands as a form of active restoration has posed several challenges. For example, the costs related to seed collection, germination and transplanting seedlings, as well as the establishment of a nursery, are significant and may be unaffordable to producers. In addition, native tree islands need to be protected from cattle and competition from grasses such as *Brachiaria*. Once these challenges are overcome, native tree islands can present an opportunity for the integration of livestock production and biodiversity conservation in human-modified landscapes.

**Acknowledgements** J.R. Pérez-Ishiwara, S. Sinaca-Colin assisted with different aspects of the work. Special thanks for F. Serrano of LANCIS- IE-UNAM for the development of tools for plant mapping within the vegetation islands, and to M. Baltazar with computing assistance. This work was supported by funding from the Universidad Nacional Autónoma de México (PAPIIT IN-202419).

## References

Achard F, Beuchle R, Mayaux P, Stibig H-J, Bodart C, Brink A, Donnay F, Lupi A, Carboni S, Desclee B, Donnay F, Eva HD, Lupi A, Rasi R, Seliher R, Simonetti D (2014) Determination of tropical deforestation rates and related carbon losses from 1990 to 2010. Glob Chang Biol 20: 2540–2554

AGAL (2008) Livestock policy briefs 03: cattle ranching and deforestation. FAO-ONU, Rome. http://www.fao.org/ag/againfo/resources/en/pubs_sap.html

Aguirre-Gutiérrez J, Malhi Y, Lewis SL, Fauset S, Adu-Bredu S, Affum-Baffoe K, Baker TR, Gvozdevaite A, Hubau W, Moore S, Peprah T, Ziemińska K, Phillips OL, Oliveras (2020) Long-term droughts may drive drier tropical forests towards increased functional, taxonomic and phylogenetic homogeneity. Nat Commun 11:3346. https://doi.org/10.1038/s41467-020-16973-4

Aide TM, Cavelier J (1994) Barriers to lowland tropical forest restoration in the Sierra Nevada de Santa Marta, Colombia. Restor Ecol 2:219–229

Aide TM, Zimerman JK, Pascarella JB, Rivera LW, Humfredo M-V (2000) Forest regeneration in a chronosequence of tropical abandoned pastures: implications for restoration ecology. Restor Ecol 8:328–338

Altieri MA (1999) Applying agroecology to enhance productivity of peasant farming systems in Latin America. Env Dev Sust 1:197–217

Arellano L, López-Ortiz S, Pérez-Hernández J, Alonso G, Torres-Rivera JA, Huerta C, Cruz-Rosales M (2018) La Ganaderia Bovina Veracruzana. Hacia la sutentabilidad. In: Halffter G, Cruz M, Huerta C. (Comps.) Ganadería sustentable en el Golfo de México. Instituto de Ecología, A.C., México pp. 189–212

Bacab HM, Madera NB, Solorio FJ, Vera F, Marrufo DF (2013) Los sistemas silvopastoriles intensivos con Leucaena leucocephala: una opción para la ganadería tropical. Avances Inv Agrop 17:67–81

Baccini A et al (2012) Estimated carbon dioxide emissions from tropical deforestation improved by carbon-density maps. Nature Clim Change 2:182–185

Bellard C, Cassey P, Blackburn TM (2016) Alien species as a driver of recent extinctions. Biol Lett 12:20150623

Benayas JMR, Newton AC, Diaz A, Bullock JM (2009) Enhancement of biodiversity and ecosystem services by ecological restoration: a meta-analysis. Science 325:1121–1124

Cadotte MW, Cardinale BJ, Oakley TH (2008) Evolutionary history and the effect of biodiversity on plant productivity. Proc Natl Acad Sci U S A 105:17012–17017. https://doi.org/10.1073/pnas.0805962105
Cadotte MW (2013) Experimental evidence that evolutionarily diverse assemblages result in higher productivity. Proc Natl Acad Sci U S A 110:8996–9000
Cadotte MW, Carboni M, Si XF, Tatsumi S (2019) Do traits and phylogeny support congruent community diversity patterns and assembly inferences? J Ecol 107:2065–2077
Calle A (2020) Can short-term payments for ecosystem services deliver long-term tree cover change? Ecosyst Serv 42:101084
Calle Z, Murgueitio E, Chará J (2012) Integrating forestry, sustainable cattle ranching and landscape restoration. Unasylva 239:31–40
Calle Z, Murgueitio E, Chará J, Molina CH, Zuluaga AF, Calle A (2013) A strategy for scaling up intensive silvopastoral systems in Colombia. J Sustain Forest 32:677–693
Calle Z, Giraldo AM, Cardozo A, Galindo A, Murgueitio E (2017) Enhancing biodiversity in Neotropical silvopastoral systems: use of indigenous trees and palms. In: Montagnini F (ed) Integrating landscapes: agroforestry for biodiversity conservation and food sovereignty. Advances agroforestry, vol 12. Springer, Dordrecht, pp 417–438
Cardinale B, Duffy JE, Gonzalez A, Hooper DU, Perrings C, Venail P (2012) Biodiversity loss and its impact on humanity. Nature 486:59–67
Ceballos G, Ehrlich PR, Dirzo R (2017) Biological annihilation via the ongoing sixth mass extinction signaled by vertebrate population losses and declines. Proc Natl Acad Sci U S A 114:E6089–E6096
Cerrillo E, Pastor Villegas J, Prada Gallardo A, González Cordero A, Morales Muñiz A, López Sáez JA, López García P, Arnanz A, Juan-Treserras J, Matamala JC, Garrido García JA (2005) Bases económicas y ambientales para el estudio de las comunidades neolíticas del centro-oeste peninsular: perspectivas desde el yacimiento de Los Barruecos. In: Arias Cabal P., Ontañón Peredo, R. García-Moncó, C. (Eds.), III Congreso del Neolítico en la Península Ibérica. Monografías del Instituto Internacional de Investigaciones Prehistóricas de Cantabria 1, Universidad de Cantabria, Santander. 1056 pp
Challenger A, Soberón J (2008) Los ecosistemas terrestres. In Capital natural de México. Vol. I: Conocimiento actual de la biodiversidad. Conabio, México, pp 87–108
Chará J, Reyes E, Peri P, Otte J, Arce E, Schneider F (2019) Silvopastoral systems and their contribution to improved resource use and sustainable development goals: evidence from Latin America. FAO, CIPAV and Agri Benchmark, Cali, p 60
Chazdon RL (2003) Tropical forest recovery: legacies of human impact and natural disturbances. Perspect Plant Ecol Evol Syst 6:51–71
de la Peña-Domene M, Martínez-Garza C, Howe HF (2013) Early recruitment dynamics in tropical restoration. Ecol Appl 23:1124–1134
del-Val E, Martínez-Garza C, Boege K, Juan-Baeza I, Solis L (2016) Restauración de poblaciones de invertebrados e interacciones bióticas en selvas estacionales de Jalisco y Morelos. In Ceccon E. and Martínez-Garza C. (coord.) Experiencias mexicanas en la restauración de los ecosistemas. Universidad Nacional Autónoma De México. ISBN: 978–607–02-8157-0
Díaz S, Purvis A, Cornelissen JHC, Mace GM, Donoghue MJ, Ewers RM, Jordano P, Pearse WD (2013) Functional traits, the phylogenyof function, and ecosystem service vulnerability. Ecol Evol 3:2958–2975
Dirzo R, Young HS, Galetti M, Ceballos G, Isaac NJB, Collen B (2014) Defaunation in the Anthropocene. Science 345:401–406
Douterlungne D, Ferguson BG (2016) Estrategias para el establecimiento de árboles en pastizales para la restauración de la selva húmeda en Chiapas. In Ceccon E. and Martínez-Garza (coord.) Experiencias mexicanas en la restauración de los ecosistemas. Universidad Nacional Autónoma de México. ISBN: 978–607–02-8157-0
Doherty J, Callaway JC, Zedler JB (2011) Diversity–function relationships changed in a long-term restoration experiment. Ecol Appl 21:2143–2155

Faith DP (1992) Conservation evaluation and phylogenetic diversity. Biol Conserv 61:1–10
Fauset S, Baker TR, Lewis SL, Feldpausch TR, Affum-Baffoe K, Foli EG, Hamer KC, Swaine MD (2012) Drought-induced shifts in the floristic and functional composition of tropical forests in Ghana. Ecol Lett 15:1120–1129
Felsenstein J (1985) Phylogenies and the comparative method. Am Nat 125:1–15
Ferraz-de-Oliveira MI, Azeda C, Pinto-Correia T (2016) Management of Montados and Dehesas for high nature value: an interdisciplinary pathway. Agr Syst 90:1–6
Fiz-Palacios O, Schneider H, Heinrichs J, Savolainen V (2011) Diversification of land plants: insights from a family-level phylogenetic analysis. BMC Evol Biol 11:341
Forest F, Grenyer R, Rouget M, Davies TJ, Cowling RM, Faith DP, Balmford A, Manning JC, Procheş Ş, van der Bank M, Reeves G, Hedderson TAJ, Savolainen V (2007) Preserving the evolutionary potential of floras in biodiversity hotspots. Nature 445:757–760
Gann G, McDonald T, Walder B, Aronson J, Nelson CR, Eisenberg C et al (2019) International principles and standards for the practice of ecological restoration. Restor Ecol 27:S3–S46
Garrido P, Elbakidze M, Angelstam P, Plieninger T, Pulido F, Moreno G (2017) Stakeholder perspectives of wood-pasture ecosystem services: a case study from Iberian dehesas. Land Use Policy 60:324–333
Gilbert GS, Webb CO (2007) Phylogenetic signal in plant pathogen-host range. Proc Natl Acad Sci U S A 104:4979–4983
Giraldo C, Escobar F, Montoya S (2018) Escarabajos del estiércol en paisajes ganaderos de Colombia. Fundación CIPAV, Cali
Giraldo C, Escobar F, Chará J, Calle Z (2010) The adoption of silvopastoral systems promotes the recovery of ecological processes regulated by dung beetles in the Colombian Andes. Insect Conserv Divers 4:115–122. https://doi.org/10.1111/j.1752-4598.2010.00112.x
González JM (2013) Costos y beneficios de un sistema silvopastoril intensivo (SSPI), con base en Leucaena leucocephala. Estudio de caso en el municipio de Tepalcatepec, Michoacán. México Avances Inv Agrop 17:35–50
González-Tokman D, Barradas VL, Boege K, Domínguez CA, del-Val E, Saucedo E, Martínez-Garza C (2018) Mulching maximizes growth rates but not survival of 11 tree species in restoration plantings in a Mexican dry forest. Restor Ecol 26:642–649
Griscom HP, Ashton PMS, Berlyn GP (2005) Seedling survival and growth of native tree species in pastures: implications for dry tropical forest rehabilitation in Central Panama. For Ecol Manage 218:306–318
Guenni O, Marín D, Baruch Z (2002) Responses to drought of five Brachiaria species. I. Biomass production, leaf growth, root distribution, water use and forage quality. Plant and Soil 243:229–241
Guevara S, Laborde J, Sánchez-Ríos G (2018) El paisaje de la ganadería: una experiencia tropical. in: Halffter G., Cruz M., Huerta C. (Comps.) Ganadería sustentable en el Golfo de México. Instituto de Ecología, A.C., México, 432 pp
Haines-Young R, Potschin M (2010) The links between biodiversity, ecosystem services and human Well-being. In: Raffaelli D, Frid C (eds) Ecosystem ecology: a new synthesis. Ecological Reviews Series, Cambridge University Press, Cambridge
Harvey CA, Haber WA (1998) Remnant trees and the conservation of biodiversity in costa Rican pastures. Agr Syst 44:37–68. https://doi.org/10.1023/A:1006122211692
Hipp AL, Larkin DJ, Barak RS, Bowles ML, Cadotte MW, Jacobi SK, Lonsdorf E, Scharenbroch BC, Williams E, Weiher E (2015) Phylogeny in the service of ecological restoration. Am J Bot 102:647–648
Hobbs, RJ Norton, DA (1996) Towards a conceptual framework for restoration ecology. Restor Ecol 4: 93–110
Hoffmann M, Hilton-Taylor C, Angulo A, Bohm M, Brooks TM, Butchart SH, Carpenter KE, Chanson J, Collen B, Cox NA et al (2010) The impact of conservation on the status of the world's vertebrates. Science 630:1503–1509

IPCC (2007) Climate change 2007: impacts, adaptation and vulnerability. Fourth Assessment Report, Geneva

Jose S, Walter D, Mohan Kumar B (2019) Ecological considerations in sustainable silvopasture design and management. Agr Syst 93:317–331

Kettenring KM, Mercer KL, Reinhardt Adams C, Hines J (2014) Application of genetic diversity-ecosystem function research to ecological restoration. J Appl Ecol 51:339–348

Kembel SW, Cowan PD, Helmus MR, Cornwell WK, Morlon H, Ackerly DD, Blomberg SP, Webb CO (2010) Picante: tools for integrating phylogenies and ecology. Bioinformatics 26:1463–1464

Kremen C, Merenlender AM (2018) Landscapes that work for biodiversity and people. Science 362:6020

Kú VJ, Ramírez AL, Jiménez FG, Alayón AJ, Ramírez CL (1999) Árboles y arbustos para la producción animal en el trópico mexicano. En: Sánchez MD, Rosales MM editores. Agroforestería para la producción animal en América Latina. Roma, Italia. 231–258 pp

Laborde J, Guevara S, Sánchez-Ríos G (2008) Tree and shrub seed dispersal in pastures: the importance of rainforest trees outside forest fragments. Écoscience 15:6–16

Lambeck RJ (1997) Focal species: a multi-species umbrella for nature conservation. Conserv Biol 11:849–856

Levy-Tacher SI, Aguirre-Rivera JR, Vleut I, Román-Dañobeytia F, Perales-Rivera H, Zúñiga-Morales, J, et al. (2016) Experiencias y perspectivas para la rehabilitación ecológica en zonas de amortiguamiento de las áreas naturales protegidas Montes Azules (Chiapas) y Calakmul (Campeche). In Ceccon E. and Martínez-Garza C. (coord.) Experiencias mexicanas en la restauración de los ecosistemas. Universidad Nacional Autónoma de México. ISBN: 978–607–02-8157-0

Lewis SL, Edwards DP, Galbraith D (2015) Increasing human dominance of tropical forests. Science 349:827–832

Lindenmayer DB, Manning AD, Smith PL, Possingham HP, Fischer J, Oliver I, McCarthy MA (2002) The focal-species approach and landscape restoration: a critique. Conserv Biol 16:33–345

López-Sáez JA, López-García P, López-Merino L, Cerrillo E, González A, Prada A (2007) Origen prehistórico de la dehesa en Extremadura: Una perspectiva paleoambiental. Rev Estudios Extremeños 63:493–509

López-Sáez JA, Abel-Schaad D, Pérez-Díaz S, Blanco-González A, Alba-Sánchez F, Dorado M, Ruiz-Zapata B, Gil-García MJ, Gómez-González C, Franco-Múgica F (2014) Vegetation history, climate and human impact in the Spanish central system over the last 9000 years. Quat Int 353:98–122

Martínez-Cabrera HI, Jones CS, Espino S, Schenk HJ (2009) Wood anatomy and wood density in shrubs: responses to varying aridity among transcontinental transects. Am J Bot 96:1388–1398

Martínez-Garza C, Howe HF (2003) Restoring tropical diversity: beating the time tax on species loss. J Appl Ecol 40:423–429

Martínez-Garza C, Bongers F, Poorter L (2013) Are functional traits good predictors of species performance in restoration plantings in tropical abandoned pastures? Forest Ecol Manag 303: 35–45

Martínez-Garza C, Campo J, Ricker M, Tobón W (2016) Effect of initial soil properties on six year growth of 15 tree species in tropical restoration plantings. Ecol Evol 6:8686–8694

Martínez-Ramos M, García-Orth X (2007) Sucesión ecológica y restauración de las selvas húmedas. B Soc Bot Mex 80:69–84

Martínez-Ramos M, Barraza L, Balvanera P, Benítez Malvid J, Bongers F et al (2012) Manejo de bosques tropicales: bases científicas para la conservación, restauración y aprovechamiento de ecosistemas en paisajes rurales. Invest Ambiental 4:107–126

Martínez-Ramos M, Pingarroni A, Rodríguez-Velázquez R, Toledo-Chelala L, Zermenño-Hernaández I, Bongers F (2016) Natural forest regeneration and ecological restoration in human-modified tropical landscapes. Biotropica 48:745–757

Minnemeyer S, Laestadius L, Sizer N, Saint-Laurent C, Potapov P (2011) A world of opportunity. Washington, DC, World Resources Institute. Available at: www.wri.org/restoringforests

Montagnini F (Ed.) (2017) Integrating landscapes: agroforestry for biodiversity conservation and food sovereignty. Advances in Agroforestry,12. Springer, Dordrecht

Montoya D, Rogers L, Memmott J (2012) Emerging perspectives in the restoration of biodiversity-based ecosystem services. Trends Ecol Evol 27:666–672

Murgueitio E, Ibrahim M (2008) Ganadería y Medio Ambiente en américa Latina. In: Murgueitio E, Cuartas C, Naranjo J (eds) Ganadería del futuro: Investigación para el desarrollo. Fundación CIPAV, Cali, Colombia, p 490

Murgueitio E, Giraldo C (2009) Sistemas silvopastoriles y control de parásitos. Rev Carta Fedegan 115:60–63

Murgueitio E, Chará J, Barahona R, Rivera JE (2019) Development of sustainable cattle rearing in silvopastoral systems in Latin America. Cuban J Agric Sci 53:1–7

Murgueitio E, Barahona R, Chará J, Flores MX, Mauricio RM, Molina JJ (2015) Los Sistemas silvopastoriles intensivos en América Latina alternativa sostenible para enfrentar el cambio climático en la ganadería. Cuban J Agric Sci 49:541–554

Navas A (2010) Importance of silvopastoral systems on caloric stress reduction in tropical livestock productions. Ver Med Vet 19:113–122

Navarro-Cano JA, Goberna M, Valiente-Banuet A, Montesinos-Navarro A, García C, Verdú M (2014) Plant phylodiversity enhances soil microbial productivity in facilitation-driven communities. Oecologia 174:909–920

Ojeniyi SO, Odedina SA, Agbede TM (2012) Soil productivity improving attributes of Mexican sunflower (Tithonia diversifolia) and siam weed (Chromolaena odorata). Emirates J Food Agr 24:243–247

Omeja PAO, Lawes MJ, Corriveau A, Valenta KV, Sarkar D, Paim FP, Chapman CAC (2016) Recovery of tree and mammal communities during large-scale forest regeneration in Kibale National Park, Uganda. Biotropica 48:770–779

Paz H, Pineda-García F, Pinzón-Pérez L (2015) Root depth and morphology in response to soil drought: comparing ecological groups along secondary succession in a tropical dry forest. Oecologia 179:551–561

Pérez-Izquierdo L, Zabal-Aguirre M, González-Martínez SC, Buée M, Verdú M, Rincón A, Goberna M (2019) Plant intraspecific variation modulates nutrient cycling through its below-ground rhizo-spheric microbiome. J Ecol 107:1594–1605

Perfecto I, Vandermeer J (2010) The agroecological matrix as alternative to the landsparing/ agriculture intensification model. Proc Natl Acad Sci U S A 107:5786–5791

Pierret A, Maeght JL, Clément C, Montoroi JP, Hartmann C, Gonkhamdee S (2016) Understanding deep roots and their functions in ecosystems: an advocacy for more unconventional research. Ann Bot 118:621–635

R Team (2018) R: a language and environment for statistical computing. R Foundation for Statistical Computing, Vienna, Austria

Raymond C, Fazey I, Reed M, Stringer L, Robinson G, Evely A (2010) Integrating local and scientific knowledge for environmental management. J Environ Manage 91:1766–1777

Reed MS, Dougill AJ, Baker T (2008) Participatory indicator development: what can ecologists and local communities learn from each other? Ecol Appl 18:1253–1269

Rivera LF, Armbrecht I, Calle Z (2013) Silvopastoral systems and ant diversity conservation in a cattle-dominated landscape of the Colombian Andes. Agric Ecosyst Environ 181:188–194

Rivera JE, Molina IC, Donney's G, Villegas G, Chará J, Barahona R (2015) Dinámica de fermentación y producción de metano en dietas de sistemas silvopastoriles intensivos con Leucaena leucocephala y sistemas convencionales orientados a la producción de leche. Livest. Res. Rural Dev. 27: Article #76. Available: http://www.lrrd.org/lrrd27/4/rive27076.html

Román-Palacios C, Wiens JJ (2020) Recent responses to climate change reveal the drivers of species extinction and survival. Proc Natl Acad Sci U S A 117:4211–4217

Santos-Gally, R., Boege, K., Fornoni, F. & Domínguez, C. 2019. Ganadería sostenible en la región de Los Tuxtlas, Veracruz, México: el equilibrio entre la producción y la conservación de la biodiversidad. Chará, J., Peri, P. & Rivera, J., (eds). ActasIX Congreso Sistemas Silvopastoriles. Aporte a los objetivos de desarrollo sostenible. Cali, Colombia, p. 318.

Schweitzer JA, Van Nuland M, Bailey JK (2018) Intraspecific plant–soil feedbacks link ecosystem ecology and evolutionary biology. In: Ohgushi T, Wurst S, Johnson S (eds) Aboveground–belowground community ecology. Springer Nature Switzerland AG, Cham, Suiza, pp 69–84

Segan DB, Murray KA, Watson JEM (2016) A global assessment of current and future biodiversity vulnerability to habitat loss–climate change interactions. Global Ecol Conserv 5:12–21

Shiels AB, Walker LR (2003) Bird perches increase forest seeds on Puerto Rican landslides. Restor Ecol 11:457–465

Srivastava DS, Cadotte MW, MacDonald AAM, Marushia RG, Mirotchnick N (2012) Phylogenetic diversity and the functioning of ecosystems. Ecol Lett 15:637–648

Steege NC, Pitman A, Killeen TJ, Laurance WF, Peres CA et al (2015) Estimating the global conservation status of more than 15,000 Amazonian tree species. Science 1:1–10

Steinfeld H, Gerber P, Wassenaar T, Castel V, Rosales M, De Haasn C (2006) Livestock's long shadow. Environmental issues and options. Food and Agriculture Organization of the United Nations, Rome, p 390

Stevens PF (2001 onwards) Angiosperm Phylogeny Website. Version 14, July 2017 [and more or less continuously updated since] http://www.mobot.org/MOBOT/research/APweb/

Tarbox BC, Swisher M, Calle Z, Wilson CH, Flory SL (2020) Decline in local ecological knowledge in the Colombian Andes may constrain silvopastoral tree diversity. Restor Ecol. http://onlinelibrary.wiley.com/doi/10.1111/rec.13153/suppinfo

Thévenin C, Mouchet M, Robert A, Kerbiriou C, Sarrazin F (2018) Reintroductions of birds and mammals involve evolutionarily distinct species at the regional scale. Proc Natl Acad Sci U S A 115:3404–3409

Verdú M, Gómez-Aparicio L, Valiente-Banuet V (2012) Phylogenetic relatedness as a tool in restoration ecology: a meta-analysis. Proc Biol Sci 279:176–1767

Violle C, Navas ML, Vile D, Kazakou E, Fortunel C, Hummel I, Garnier E (2007) Let the concept of trait be functional! Oikos 116:882–892

Watson JEM, Evans T, Venter O et al (2018) The exceptional value of intact forest ecosystems. Nat Ecol Evol 2:599–610. https://doi.org/10.1038/s41559-018-0490-x

# Chapter 6
# Riparian Forests: Longitudinal Biodiversity Islands in Agricultural Landscapes

**Lina Paola Giraldo, Julián Chará, Zoraida Calle D, and Ana M. Chará-Serna**

**Abstract** Riparian buffers safeguard the only remaining forest fragments in many agricultural landscapes of the Colombian Andean region. These linear landscape elements contribute to the conservation of terrestrial biodiversity in agricultural landscapes by providing shelter, reproduction sites, food, and connectivity for arthropods, amphibians, mammals and birds. Thus, riparian buffers play a critical role as biodiversity islands. In addition, forested riparian buffers protect aquatic environments and water quality by reducing the input of pollutants from catchment areas, improving physical habitat with shade, and adding allochthonous materials that provide the main source of energy for stream ecosystems. This chapter summarizes the results of research conducted during the past two decades by the Center for Research in Sustainable Agricultural Systems (CIPAV) in the Central Andes coffee-growing region of Colombia. These studies highlight the critical role of forested riparian buffers for conservation and ecosystem services. We provide a synthesis of lessons learned on the effects of both cattle grazing and riparian forest cover on stream ecosystems. This body of research also demonstrates that streams protected by riparian forests support complex and biodiverse macroinvertebrate assemblages and may respond positively to the ecological restoration of riparian strips. The chapter concludes with recommendations for restoring and protecting riparian buffers from agricultural practices, partially through incentives to landowners. These insights have emerged from decades of research and institutional experience on riparian restoration initiatives.

L. P. Giraldo (✉)
Centro para la Investigación en Sistemas Sostenibles de Producción Agropecuaria (CIPAV), Cali, Colombia

Instituto de Biología, Universidad de Antioquia, Medellín, Colombia
e-mail: lina@fun.cipav.org.co

J. Chará · Zoraida Calle D · A. M. Chará-Serna
Centro para la Investigación en Sistemas Sostenibles de Producción Agropecuaria (CIPAV), Cali, Colombia
e-mail: julian@fun.cipav.org.co; zoraida@fun.cipav.org.co; ana@fun.cipav.org.co

 
F. Montagnini (ed.), *Biodiversity Islands: Strategies for Conservation in Human-Dominated Environments*, Topics in Biodiversity and Conservation 20,
https://doi.org/10.1007/978-3-030-92234-4_6

**Keywords** Andean streams · Colombian Andes · Ecological restoration · Riparian biodiversity · Riparian buffers · La Vieja River

## 6.1 Introduction

Forested riparian buffers (also known as riparian forests) are strips of vegetation that grow along rivers and streams, and around springs and wetlands. These longitudinal elements, distributed along the water network, act as biodiversity islands, connecting forest fragments, conserving water sources, and providing environmental benefits to adjacent agroecosystems (Ericsson and Stevens 1996; Naiman et al. 2000; Schroth et al. 2004; Lees and Peres 2007; Palmer et al. 2014; Luke et al. 2019).

In Colombia, forested riparian buffers, which have an average width of 24 m, are often the only patches of woody vegetation that remain in many agricultural landscapes. Although forested riparian buffers often occupy small land areas, they make a disproportionate contribution to the landscape-scale conservation of birds, arthropods, and other organisms that provide essential services such as biological pest control, seed dispersal, pollination and carbon sequestration (Schroth et al. 2004; Marczak et al. 2010).

However, despite their importance, these forest strips are being destroyed and replaced with pastures or cropland, which has had negative effects on aquatic environments, water quality, and terrestrial biodiversity (Braccia and Voshell 2007; Chará et al. 2007; Riseng et al. 2011; Skłodowski et al. 2014). This is the case of the Central Andes coffee-growing region of Colombia, where many forests, including riparian corridors, were replaced with coffee or banana plantations and pastures during the second half of the twentieth century (Sadeghian et al. 1999). In some areas of this region, land cover transformation has been successfully reversed through restoration projects carried out by local farmers and the Center for Research in Sustainable Agricultural Systems (CIPAV, an autonomous Colombian organization with 35 years of experience in research, training and outreach on sustainable agricultural production systems; Calle 2020). Several of the restoration projects developed by CIPAV in this coffee-growing region have focused on the implementation of environmentally friendly agroforestry and silvopastoral systems and the release of riparian areas for forest restoration (Calle 2020).

This chapter synthesizes the findings of research conducted for two decades along with restoration projects in this coffee-growing region. These studies evaluated the role of riparian forests in the protection of terrestrial biodiversity and aquatic environments by monitoring the results of several restoration initiatives focused on these key landscape elements. The chapter ends with a synthesis of the lessons learned from these studies, together with recommendations that can be applied to riparian restoration.

### 6.1.1 Study Area

The region of central-western Colombia known as the "Eje Cafetero" (coffee-growing region) includes a variety of ecosystems of the Central Andes (Cordillera Central), from lowland rainforests to snow-capped mountains. Its high biodiversity and unique ecosystems are threatened by landscape transformation to establish pastures or crops. For this reason, it has been recognized internationally as a conservation priority (CARDER-FONADE 2002; Uribe-Gómez 2008). In the 1990s, many coffee growers in the region eliminated their plantations as a result of the economic instability triggered by the fall of international coffee prices, making pastures for bovine livestock become the dominant land use at elevations between 1200 and 1800 m (Sadeghian et al. 1999). The adoption and management of livestock grazing systems have changed the species composition and vegetation structure of riparian buffers in this region (Chará-Serna et al. 2015; Fig. 6.1).

CIPAV studies were done at La Vieja river basin in the Central Andes coffee-growing region (Fig. 6.2). These watersheds are located on rolling hills and valleys, at altitudes between 900 and 2400 m, with average annual rainfall of 1900–2600 mm and mean temperature between 12 and 24 °C, varying with elevation. Rainfall exhibits a bimodal seasonality, with two annual periods of high precipitation (April–May and October–November). The studies focused on the effects of agricultural systems (mainly cattle ranching) on headwater micro-basins (<100 ha), the role of forested riparian buffers in mitigating these effects on aquatic environments, and the conservation of terrestrial biodiversity.

**Fig. 6.1** Riparian forest in a cattle ranching landscape, coffee-growing region of Colombia. (Photo: Julián Chará)

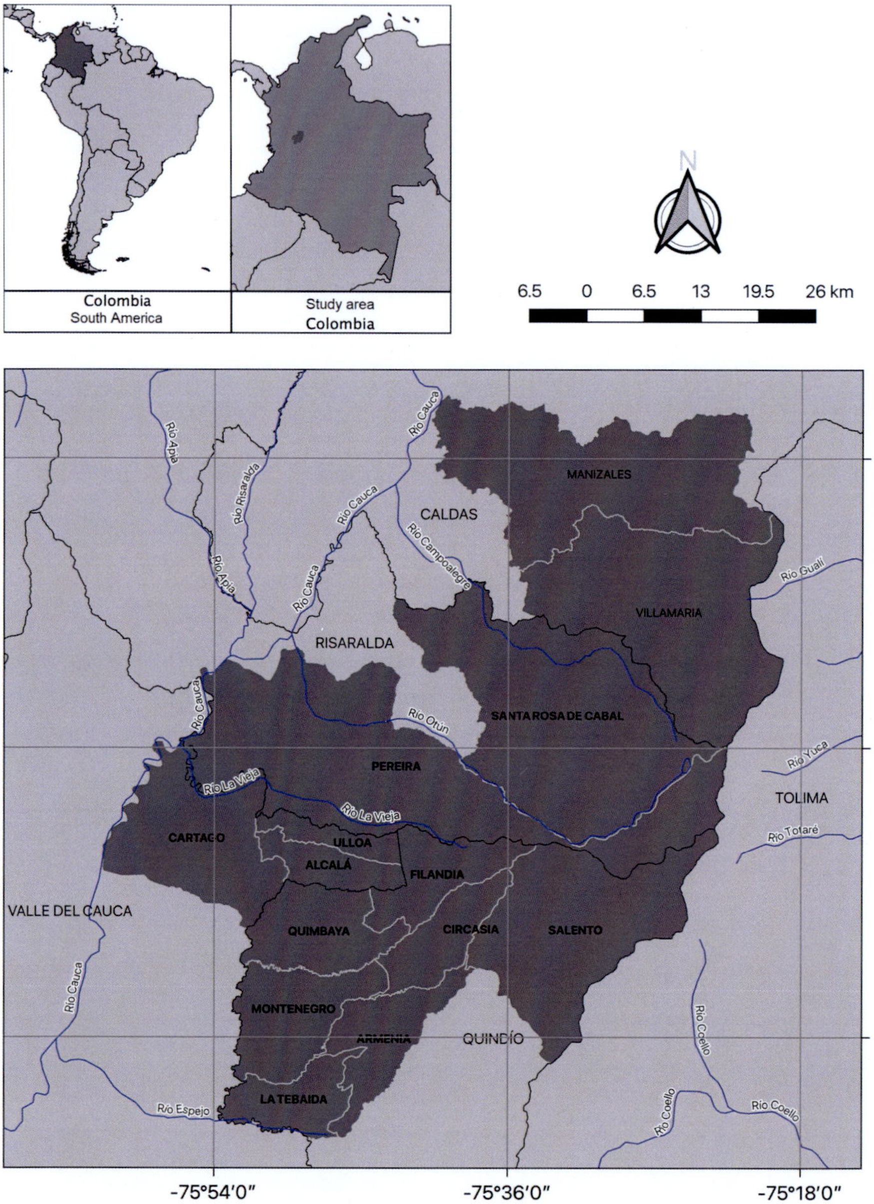


**Fig. 6.2** Study area in the Central Andes coffee-growing region, Colombia. (Map: Julián Mendivil)

Close to 20% of the land in La Vieja river basin is covered by secondary, mature and riparian forests. The most species-rich botanical families in this area are Lauraceae, Rubiaceae, Moraceae, Euphorbiaceae and Fabaceae. Abundant species in riparian forests include *Ocotea* sp., *Calliandra pittieri*, *Miconia* sp., *Cordia alliodora*, *Guadua angustifolia*, *Cupania americana*, *Sorocea trophoides*, *Oreopanax cecropifolius*, *Piper crassinervium*, *Anacardium excelsum*, *Cecropia angustifolia*, *Croton magdalenensis*, *Heliconia platystachys*, *Brosimum alicastrum*, *Aiphanes horrida*, *Cinnamomum triplinerve*, *Guarea guidonia* and *Urera caracasana* (Méndez and Calle 2010). A large portion of this species richness is confined to riparian forests, in contrast with the agricultural matrix, where the diversity of woody plants is very low (Méndez and Calle 2010).

## 6.2 Riparian Forests and Biodiversity

Riparian forests link terrestrial and aquatic ecosystems through processes that occur at different spatial scales. At the watershed scale, these forests contribute to ecological functions that depend on species movement or landscape connectivity. On a local scale, riparian forests provide organic matter and shade to streams, maintain slope stability, and protect stream beds, thereby determining habitat quality and aquatic biodiversity.

### *6.2.1 Terrestrial Biodiversity (Regional or Watershed Scale)*

Riparian forests protect some of the most biodiverse and dynamic ecosystems on the planet (Naiman et al. 2005). Their vegetation may include unique species assemblages and sustain animal populations that depend on these ecosystems for shelter, reproduction, food and passage (Moore and Richardson 2003; Sabo et al. 2005). Studies on mammals, birds and dung beetles have shown that forested riparian buffers support more terrestrial biodiversity than the surrounding agricultural matrix (Fajardo et al. 2009; Gray et al. 2014; Zimbres et al. 2017; Luke et al. 2019). These ecosystems are biodiversity islands because they are often the last forest remnants in agricultural landscapes. In this context, their conservation and restoration help mitigate biodiversity loss and habitat fragmentation (Lees and Peres 2007). Due to their linear configuration, riparian forests act as biological corridors, connecting forest patches in fragmented landscapes, and facilitating migration and dispersal of birds, mammals, reptiles, amphibians, insects and other organisms that provide key ecosystem services (Naiman et al. 2005; Medina et al. 2007; Gray et al. 2017).

Studies conducted in agricultural landscapes of the Colombian coffee-growing region have found that compared to all other types of ecosystems, forested riparian buffers and forest fragments have the most structurally complex and species-rich vegetation (Table 6.1). Out of 390 woody plant species known to exist in the agricultural basin of La Vieja river, 278 species (71%) were found in riparian forests (Calle and Méndez 2009), where the dominant species include *Guadua angustifolia* (bamboo), *Cupania americana*, *Sorocea trophoides*, *Oreopanax cecropifolius*, *Piper crassinervium*, *Anacardium excelsum* and *Cecropia angustifolia*.

**Table 6.1** Bird, woody plant and dung beetle species richness in riparian forests and other land uses in cattle farms of the Central Andes coffee-growing region (CSCRP: Colombian Sustainable Cattle Ranching Project) (http://ganaderiacolombianasostenible.co)

| Land use | Woody plant species richness (Calle and Méndez 2009) | Bird species richness (Fajardo et al. 2009) | Woody plant species richness (CSCRP) |
|---|---|---|---|
| Riparian forest | 278 | 103 | 183 |
| Secondary and mature forest | 264 | 92 | 199 |
| Secondary growth areas | – | 88 | – |
| Bamboo forest (*Guadua angustifolia*) | 89 | – | – |
| Agriculture | 86 | – | – |
| Scattered trees in paddocks | – | – | 15 |
| Live fences | – | – | 25 |
| Intensive silvopastoral systems | – | – | 18 |
| Enhanced treeless pasture | – | 45 | 1 |
| Natural treeless pasture | 102 | 38 | – |

More recent studies done in this coffee-growing region by the Colombian Sustainable Cattle Ranching Project (http://ganaderiacolombianasostenible.co), found a high diversity of plants in riparian forests (unpublished data; Table 6.1). Some threatened and scarce tree species were found in riparian forests, including *Cedrela odorata*, *Swietenia macrophylla*, *Podocarpus oleifolius*, *Anacardium excelsum* and *Astronium graveolens*.

Additionally, compared to other landscape elements and land uses in the region, riparian forests in this basin showed the highest bird species richness (103 of the 229 species in the landscape, ≈45%), and contained most of the 41 bird species of global conservation concern recorded in the region (Fajardo et al. 2009) (Table 6.1). Endemic and nearly endemic bird species observed in this study include the 'grayish piculet' (*Picumnus granadensis*), 'flamerumped tanager' (*Ramphocelus flammigerus*), 'apical flycatcher' (*Myiarchus apicalis*), 'crested ant-tanager' (*Habia cristata*), 'bar-crested antshrike' (*Thamnophilus multistriatus*), 'scrub tanager' (*Tangara vitriolina*) and 'grasshopper sparrow' (*Ammodramus savannarum*).

### *6.2.2 Aquatic Biodiversity (Local Scale)*

Forested riparian buffers mitigate the impact of agricultural activities on aquatic ecosystems through different mechanisms. Riparian vegetation filters and retains sediments, organic matter, nutrients, chemical substances and pathogens released from the catchment area, preventing them from entering aquatic ecosystems. Furthermore, tree shade reduces fluctuations in water temperature, and the roots of dense

**Fig. 6.3** Stream protected by a riparian forest on a cattle farm in the coffee-growing region, Colombia. (Photo: Carlos Pineda)

vegetation stabilize riverbanks, protecting them from erosion (Osborne and Kovacic 1993; Mingoti and Vettorazzi 2011; Schilling and Jacobson 2014; Tanaka et al. 2016) (Fig. 6.3). Together, these mechanisms enhance hydrological regulation, improve water quality and contribute to the conservation of aquatic biodiversity.

Studies done in agricultural landscapes of the Central Andes coffee-growing region have shown that headwater streams protected with riparian forests often contain a considerable diversity of aquatic macroinvertebrates (Chará et al. 2007; Giraldo et al. 2014; Villada et al. 2017; Ramírez et al. 2018). This biodiversity is also related to water quality and characteristics of the streambed such as the abundance of stones (Table 6.2). Macroinvertebrate orders like Ephemeroptera, Plecoptera and Trichoptera (also known as EPT taxa) play an important role in processing leaf litter contributed by the riparian vegetation to the aquatic environment and are particularly sensitive to habitat alteration. Therefore, they are considered bioindicators of conserved ecosystems. Up to 77% of the families and 42% of the genera of Trichoptera reported for Colombia were found to be associated with forested riparian buffers (Ascúntar et al. 2014).

Recent studies of small streams protected by riparian forests within agricultural landscapes have expanded the known distributions of several species of the orders Trichoptera, Plecoptera and Coleoptera in Colombia (Zúñiga et al. 2014, 2015; González-Córdoba et al. 2015, 2016). Additionally, research in these small ecosystems has resulted in the discovery and description of new aquatic insect species for the country (Molineri et al. 2016). These findings support the value of small streams as unexpected reservoirs of biodiversity in agricultural landscapes.

**Table 6.2** Mean values of physical and biological variables in watersheds of the Central Andes coffee-growing region of Colombia. The impact estimate is the arithmetic difference between watersheds with forested and pasture-dominated riparian buffers. Based on Chará et al. (2007), Giraldo et al. (2014), Villada et al. (2017), Ramírez et al. (2018) and summarized in Giraldo (2019)

| Variable | Watersheds with forested riparian buffers n = 24 | Watersheds with pasture-dominated riparian buffers n = 30 | Impact estimate (%) |
|---|---|---|---|
| Width of streambed (bank to bank) (m) | 2.3 | 4.2 | 82.6 (+) |
| Depth (cm) | 17.5 | 13.1 | 25.1 (−) |
| % of rocks | 67 | 13 | 80.5 (−) |
| % of mud | 18 | 61 | 238.8 (+) |
| *Macroinvertebrates* | | | |
| Mean abundance | 751.8 | 2811.2 | 274 (+) |
| Richness | 83 | 72 | 13.2 (−) |
| % EPT[a] | 36.1 | 4.2 | 88.3 (−) |
| % Diptera | 25.5 | 42.5 | 66.6 (+) |
| % Mollusca | 9.7 | 42.2 | 335 (+) |
| *Water quality* | | | |
| Temperature (°C) | 18.4 | 21.8 | 18.4 (+) |
| Total solids (mg $L^{-1}$) | 85.5 | 146.8 | 71.6 (+) |
| Total suspended solids (mg $L^{-1}$) | 9.7 | 139 | 1332.9 (+) |
| BOD $_{5\text{-}20°\,C}$ (mg. $L^{-1}O_2$) | 2.3 | 6.2 | 169.5 (+) |
| Ammonia nitrogen (mg.$L^{-1}$ N-$NH_3$) | 0.47 | 0.67 | 42.5 (+) |
| Dissolved oxygen (mg.$L^{-1}$) | 6.1 | 4.3 | 29.5 (−) |
| Fecal coliforms (MPN. 100 $mL^{-1}$) | 1596.3 | 36200.6 | 2167.7 (+) |

[a]Ephemeroptera, Plecoptera, Trichoptera
*BOD* biochemical oxygen demand, *MPN* most probable number

Similar studies have shown that the elimination of riparian forests often triggers severe changes in the composition of the aquatic fauna, such as a loss of diversity of sensitive EPT taxa, and an increase in the abundance of groups that are tolerant to organic pollution, such as Diptera and Mollusca (Chará et al. 2007; Giraldo et al. 2014; Ramírez et al. 2018) (Table 6.2). Agricultural practices have also been shown to affect aquatic macroinvertebrates indirectly, by increasing nitrogen concentrations and reducing the width of forested riparian strips. These alterations reduce habitat quality for aquatic fauna by limiting the availability of coarse substrates within stream channels (Chará-Serna et al. 2015).

Riparian forests provide large quantities of coarse particulate organic material (leaves, flowers, fruits, branches) to the aquatic environment. These organic inputs are the primary source of energy for food webs in forested headwater streams, where the closed canopy limits light availability and primary productivity (Hynes 1975; Vannote et al. 1980; Wallace et al. 1997). The processing of allochthonous inputs of organic matter initiates the transfer of energy that then flows through the aquatic food web in these ecosystems (Jones 1997; Wallace et al. 1997; Lamberti and Gregory 2006; Aldridge et al. 2009; Yoshimura 2012). Coarse plant material also maintains aquatic biodiversity by providing habitats and physical structures that are used by fauna (Suurkuukka et al. 2014). In a comparative study of protected and unprotected streams, Giraldo (2019) found a greater abundance (584 vs. 40 individuals), richness (10 vs. 5 genera) and biomass (3.6 vs. 0.35 g) of leaf-processing macroinvertebrates per grasp in streams with riparian forests.

In addition, riparian vegetation offers shelter for emerging adult stages of aquatic insects, which may not be able to fly far and constitute an important food source for birds, bats, amphibians and other insectivorous species. Riparian forests provide habitat for other species of arthropods, mollusks, crustaceans, and small fish that are consumed by terrestrial organisms, contributing to the transfer of energy between terrestrial and aquatic ecosystems (Paetzold et al. 2005).

## 6.3 Impacts of the Loss of Riparian Forests on Streams

Even though riparian forests perform essential functions related to the protection of water quality and biodiversity, they have been highly impacted around the world (Kuglerová et al. 2014). Farmers often remove riparian forests to establish pastures and crops because the riparian forest soils are richer in nutrients than the surrounding areas (Naiman et al. 2005). Replacing native forests with pastures or crops leads to several negative effects, including increased inputs of sediments and pollutants to water sources, reduced water regulation capacity, and biodiversity loss (Duehr and Siepker 2006; Chará et al. 2007; Lorion and Kennedy 2009; Turunen et al. 2019). Table 6.2 summarizes several differences in physical, biological and water quality variables between water sources with and without the protection of the riparian vegetation. Unprotected sites tend to have shallower water, lower proportions of coarse substrates, lower species richness, higher abundance of organisms and higher values in parameters such as water temperature, solids, biochemical oxygen demand (BOD), nitrogen, and fecal coliforms.

The following list presents some lessons learned about the effects of livestock and riparian forests on headwaters of the Central Andes coffee-growing region (described in detail in Giraldo (2019), based on studies of water quality, habitat quality, aquatic macroinvertebrates and the flow of coarse particulate organic matter (Chará et al. 2007, 2011; Camargo et al. 2011; Giraldo et al. 2014; Chará-Serna et al. 2015; Galindo et al. 2017; Ramírez et al. 2018; Giraldo 2019).

- Cattle grazing in catchment areas causes undesirable effects such as soil compaction, reduced infiltration capacity, varying degrees of erosion, and the loss of forests that protect streams. Comparative studies of soils under two types of riparian vegetation carried out in the area have found that soils had lower apparent density (0.7 vs. 0.9 g/cm$^3$), higher total porosity (70% vs. 60%) and lower susceptibility to compaction (85% vs. 88.3%) in bamboo (*Guadua angustifolia*) riparian forests than in pastures (Camargo et al. 2011).
- The degradation or removal of riparian forests reduces canopy cover and shade. Vegetation structure and composition become simplified as plant covers dominated by grasses and pioneer shrubs (mostly Piperaceae and Melastomataceae) replace more diverse woody vegetation.
- Although woody plants from nearby areas continue to disperse their seeds to degraded riparian strips, the vigorous growth of grasses may temporarily inhibit the establishment of trees and shrubs. In sites without restoration treatments, pastures can cover up to 52% of the area (Galindo et al. 2017).
- The loss of riparian forests and their buffering services amplifies the negative effects of grazing on watersheds. Without shade, water temperature, organic matter, nutrients and pathogens increase while dissolved oxygen decreases (Chará et al. 2007; Giraldo et al. 2014). Each of these changes implies a loss of water quality with negative consequences for nearby human populations in addition to local species.
- The removal of riparian woody vegetation facilitates the direct access of cattle to streambeds. Without the strong roots that stabilize stream margins, cattle trampling rapidly deteriorates banks and slopes.
- Damage to the banks accelerates erosion and sedimentation of the streambed and changes channel morphology. The average width of the bed in unprotected streams is 5.4 m, compared to 2.2 m in sites protected by riparian forests (Chará et al. 2007).
- Streams where riparian forests have been eliminated and cattle have direct access to the channel tend to be shallower than protected streams, with a significant fraction of coarse substrates being replaced by fine sediments such as silt and sand. In cattle areas, up to 100% of the riverbed of unprotected streams can become covered by very fine substrates (Giraldo et al. 2014).
- The loss of riparian forest reduces the inputs of wood, litter, and other coarse materials in streams. Fallen organic matter forms important microhabitats such as pools and small turbulences, provides colonization substrates for organisms, and is an essential source of energy for macroinvertebrates. Pools occupy a smaller proportion of the area in streams impacted by livestock activities than in those protected by riparian forests (13% vs. 46%, respectively; Chará et al. 2007).
- Lower water quality and modified physical conditions of streams cause changes in macroinvertebrate communities. In these circumstances, groups that tolerate habitat degradation, such as mollusks (Physidae) and dipterans (mainly of the Chironomidae and Simuliidae families) tend to increase in abundance and dominance, but the overall richness of species, families, and orders tends to decrease.

- On average, streams protected by riparian forests receive 7.3 times more leaf litter per year than unprotected streams (9150 kg vs. 1255 kg per hectare of woody vegetation, respectively; Giraldo 2019).
- Leaves form the largest proportion of litter that enters forested streams. However, in buffer strips covered by grasses, a qualitative change occurs in the composition of accumulated material when the inputs of wood, flowers and fruits are lost. Due to the lack of logs that form pools and structures that retain materials on the stream bed, the rate of storage can be four times lower in streams with grasses (Giraldo 2019).
- Although the studied watersheds are immersed in agricultural landscapes and occupy relatively small areas (<100 ha), their aquatic ecosystems harbor impressive biodiversity, represented mainly by macroinvertebrates. The conservation of these watersheds is essential to protect this biota. Small watersheds are also the main sources of water for rural communities, so their conservation is also critical from a public health perspective.

## 6.4 Restoration of Riparian Forests in Agricultural Landscapes

Riparian restoration should start by guaranteeing the protection of existing forest remnants and improving connectivity between the upper and lower sections of watersheds. When restoring riparian forests, it is useful to define a set of clear objectives or reference conditions in terms of ideal forest structure and composition. The selection of restoration techniques will depend on the specific conditions of each site, including the characteristics of the remnant vegetation, soil conditions, and the proximity to seed sources (e.g., other forest fragments).

The restoration of riparian areas in agricultural landscapes should be prioritized, facilitating the gradual reestablishment of woody vegetation and its associated ecological functions in terrestrial and aquatic ecosystems (Meli et al. 2019). Certain characteristics of restored riparian forests, such as their width, length, and vegetation structure, determine the magnitude of the environmental benefits of restoration, such as reductions in nutrient cycling rates, the protection of aquatic environments, and landscape-scale biodiversity conservation.

Fencing riparian strips is one of the most popular methods to initiate the restoration of riparian forests in agricultural landscapes. Fencing has been shown to enhance the natural regeneration of riparian vegetation by preventing the access of cattle to riparian areas. For example, riparian strips that had been protected from grazing during the last 10 to 14 years in La Vieja river basin had recovered a similar assemblage of dominant species as reference forest ecosystems in the area, including common species of trees such as *Cupania americana*, *Inga edulis*, *Cecropia angustifolia* and *Croton magdalenensis*, and some locally threatened species such as *Anacardium excelsum*, *Oreopanax cecropifolius*, *Trichilia pallida*, *Aiphanes horrida*, *Nectandra turbacensis*, *Ocotea macropoda* and *Machaerium capote*

**Table 6.3** Vegetation structure in reference sites and riparian buffers undergoing restoration at La Vieja river basin (Calle and Holl 2019)

| | Reference | Restored |
|---|---|---|
| Average tree species density | 8.5 species ha$^{-1}$ | 19 species ha$^{-1}$ |
| Density of tree stems | 300 stems ha$^{-1}$ | 750 stems ha$^{-1}$ |
| Basal area | 8 m$^2$ ha$^{-1}$ | 14.6 m$^2$ ha$^{-1}$ |
| Average canopy cover | Not available | 89% |
| Grass cover | Not available | < 5% |

(Calle and Holl 2019). Additionally, restored areas had more species, a higher density of tree stems, higher canopy cover and lower grass cover than reference forests (Table 6.3).

Different factors may slow down or prevent the spontaneous regeneration of woody species in fenced riparian buffers. A frequent issue in deforested cattle ranching watersheds is the uncontrolled growth of grasses on the riparian strips. Dense grass growth may inhibit the regeneration of shrubs and trees, even after the removal of grazing, temporarily halting secondary succession. Techniques of assisted natural regeneration, such as the periodic control of competing plants and the enrichment planting of pioneer trees, can be used to accelerate forest recovery. Fast-growing shrubs can be planted to shade out grasses, slow their growth and facilitate the regeneration of woody plants, offsetting the inhibitory effects of grass growth. For example, Galindo et al. (2017) studied the effect of *Tithonia diversifolia* and *Piper auritum* planted at high-density to shade the grasses and facilitate the establishment of native trees. After 15 months, *T. diversifolia* was able to reduce grass cover by 81% and enhanced the survivorship of native trees planted underneath.

Riparian restoration efforts can also have beneficial effects on aquatic environments. A recent study of several cattle ranching watersheds in the Andean region of Colombia showed that the early growth of native vegetation in riparian strips enhances the chemical and biological properties of aquatic ecosystems (Giraldo et al. 2020). The biochemical oxygen demand (BOD), which measures organic water pollution, was significantly lower in the studied streams 36 months after the beginning of riparian restoration activities. Similarly, restored streams showed an increase in dissolved oxygen, as well as a decrease in turbidity and fecal coliforms. Regarding the composition of biological communities, the relative abundance of tolerant aquatic insects of the family Chironomidae (Diptera) significantly decreased through the 3 years of sampling, whereas the abundance of the family Hydropsychidae (Trichoptera) showed moderate increases. Together these bio-indicators suggest that the stream ecosystem is being restored.

## 6.5 Conclusions and Recommendations

Given the strategic value of riparian forests, all initiatives to protect and restore these landscape elements and their ecosystems should be designed with the complete basin in mind. The following recommendations are based on lessons learned through various studies carried out by CIPAV in Colombian cattle ranching landscapes (Murgueitio and Ibrahim 2009; Chará and Giraldo 2011; Calle et al. 2012; Chará et al. 2018; Giraldo 2019).

### *6.5.1 Watershed Scale*

- Conservation initiatives for riparian areas and aquatic environments should prioritize landscape elements that can gradually reestablish some ecological functions of forests. Tree coverage should be increased not only in stream banks, but in the entire catchment areas.
- The adoption of silvopastoral systems, agroforestry systems, strategically placed live fences, windbreaks, and other tree-based elements promote the recovery of soil, biodiversity, hydrologic regulation, microclimate, natural biological pest control, and carbon sequestration.
- Agrochemicals should be reduced gradually and finally eliminated in pastures and cropland. Veterinary medicines should be used more sparingly to reduce water pollution.
- Solid wastes and urine from livestock facilities should be treated with agroecological practices that enhance nutrient recycling in farming systems, such as water decontamination with biodigesters and the production of compost.

### *6.5.2 Stream Segment Scale*

- A necessary first step when restoring forest cover in heavily eroded or degraded riparian strips is the restriction of livestock access to the streams.
- Drinking stations must be provided at each paddock to prevent cattle from entering into streams.
- One key action when conserving riparian forests within agricultural landscapes is fencing of existing forest fragments and guaranteeing their effective protection.
- After fencing the riparian strip, the process of assisted natural regeneration involves controlling competing plants and enrichment planting with native species. Together, these actions promote the recovery of aquatic environments.
- Riparian forest restoration accelerates the recovery of aquatic biodiversity and key ecological functions, such as the processing and transfer of leaf litter within the ecosystem.

- An important principle that should guide the restoration of riparian forests is the provision of heterogeneous leaf litter to the aquatic ecosystem. Planted and regenerating species should provide thick and thin, small and large leaves, and both fast and slow decaying litter. Since the organisms that contribute to the fragmentation of leaf litter require some palatable, nutrient dense food resources, tree species that provide such attractive resources for aquatic organisms should be included in the restoration treatments.
- Trees that contribute significant amounts of litter to the soil of the riparian areas will be attractive to edaphic macrofauna and will accelerate the recovery of the stream segment's hydrological properties.
- The richness and singularity of entomofauna in Andean stream ecosystems prove that forested riparian buffers function as biodiversity islands and support efforts to restore these landscape elements.

### *6.5.3 Social Issues*

- The main environmental service provided by riparian forests in headwaters is the maintenance of water quality for rural populations, urban areas, and agriculture. Clear measures are required to achieve their conservation and recovery in order to ensure their productive use.
- The vulnerability of riparian forests is a direct result of conservation and management decisions made by landowners. Preserving these landscape elements, their ecosystems, and their myriad services requires an increasing willingness of farmers to adopt environmentally friendly and sustainable agricultural practices and undertake the restoration of areas critical for biodiversity. Colombia and other Latin American countries need clear policies for headwater conservation. These policies must guarantee the sustainable use of resources and the continuity of environmental services.
- Incentives such as payment for environmental services and discounts in property taxes motivate farmers to conserve and restore forest remnants adjacent to aquatic environments.

**Acknowledgements** The authors gratefully acknowledge the support of Minciencias and the Colombian Autonomous Fund for Science, Technology and Innovation Francisco José de Caldas through contract 80740-006-2020, and the support to L.P. Giraldo through the scholarship for doctoral studies, call 567.

## References

Aldridge KT, Brookes JD, Ganf GG (2009) Rehabilitation of stream ecosystem functions through the reintroduction of coarse particulate organic matter. Restor Ecol 17:97–106

Ascúntar O, Zuñiga MDC, Giraldo LP, Ramírez Y, Chará J, Ramos B (2014) Contribución al conocimiento de los tricópteros adultos (Insecta: Trichoptera) de los andes orientales de Colombia. In: La biodiversidad sensible: patrimonio natural irreemplazable. IV Congreso Colombiano de Zoología. Diciembre 1–5 de 2014. Libro de resúmenes. Asociación Colombiana de Zoología. Disponible en línea: www.congresocolombianodezoologia.org; www.aczcolombia.org

Braccia A, Voshell R (2007) Benthic macroinvertebrate responses to increasing levels of cattle grazing in blue Ridge Mountain streams, Virginia, USA. Environ Monit Assess 131:185–200

Calle A (2020) Can short-term payments for ecosystem services deliver long-term tree cover change? Ecosyst Serv 42:101084

Calle A, Holl KD (2019) Riparian forest recovery following a decade of cattle exclusion in the Colombian Andes. For Ecol Manag 452:117563

Calle Z, Méndez LE (2009) Estructura y Composición de la Vegetación Arbórea en el Agropaisaje del Río La Vieja. In: Rodriguez JM, Camargo JC, Niño J, Pineda AM, Arias LM, Echeverry MA, Miranda CL (eds) Valoración de la Biodiversidad en la Ecorregión del Eje Cafetero. CIEBEREG, Pereira, pp 185–192

Calle Z, Murgueitio E, Chará JD (2012) Integrating forestry, sustainable cattle ranching and landscape restoration. Unasylva 239:31–40

Camargo JC, Chará JD, Giraldo LP, Chará-Serna AM, Pedraza GX (2011) Beneficios de los corredores ribereños de *Guadua angustifolia* en la protección de ambientes acuáticos en la Ecorregión Cafetera de Colombia. 1. Efectos sobre las propiedades del suelo. Recursos Naturales y Ambiente 61:53–59

CARDER-FONADE (Ministerio del Medio Ambiente) (2002) Ecorregión Eje cafetero: un territorio de oportunidades. Corporación Alma Mater, Pereira

Chará JD, Giraldo C (2011) Servicios ambientales de la biodiversidad en paisajes agropecuarios. Fundación CIPAV, Cali, 76p

Chará JD, Pedraza G, Giraldo LP, Hincapié D (2007) Efecto de corredores ribereños sobre el estado de quebradas en la zona ganadera del río La Vieja, Colombia. RAFA 45:72–78

Chará JD, Giraldo LP, Chará-Serna AM, Pedraza GX (2011) Beneficios de los corredores ribereños de *Guadua angustifolia* en la protección de ambientes acuáticos en la Ecorregión Cafetera de Colombia. 2. Efectos sobre la escorrentía y captura de nutrientes. Revista Recursos Naturales y Ambiente 61:60–66

Chará J, Reyes E, Peri P, Otte J, Arce E, Schneider F (2018) Silvopastoral systems and their contribution to improved resource use and sustainable development goals: evidence from Latin America. FAO, CIPAV. Editorial CIPAV, Cali, 58p. Licence: CC BY-NC-SA 3.0 IGO. isbn: 978-958-9386-84-2

Chará-Serna AM, Chará JD, Giraldo LP, Zúñiga MDC, Allan JD (2015) Understanding the impacts of agriculture on Andean stream ecosystems of Colombia: a causal analysis using aquatic macroinvertebrates as indicators of biological integrity. Freshwater Sci 34:727–740

Duehr JP, Siepker MJ (2006) Relation of riparian buffer strips to in-stream habitat, macroinvertebrates and fish in a small Iowa stream. J Iowa Acad Sci 113:49–55

Ericsson A, Stevens V (1996) Fringe benefits: a land owner's guide to the value and stewardship of riparian habitat. Forest Renewal BC. Fraser River Action Plan. Environment Canada, 28pp. www.landstewardship.org

Fajardo D, Johnston-González R, Neira L, Chará J, Murgueitio E (2009) Influencia de sistemas silvopastoriles en la diversidad de aves en la cuenca del río La Vieja, Colombia. Recursos Naturales y Ambiente 58:9–16

Galindo V, Calle Z, Chará J, Armbrecht I (2017) Facilitation by pioneer shrubs for the ecological restoration of riparian forests in the Central Andes of Colombia. Restor Ecol 25:731–737

Giraldo LP (2019) Efecto de la vegetación ribereña sobre la dinámica de la materia orgánica particulada gruesa e insectos fragmentadores en quebradas de paisajes ganaderos (Eje Cafetero de Colombia). PhD thesis, Universidad de Antioquia, Medellín, Colombia

Giraldo LP, Chará J, Zúñiga MDC, Chará-Serna AM, Pedraza G (2014) Impacto del uso del suelo agropecuario sobre macroinvertebrados acuáticos en pequeñas quebradas de la cuenca del río La Vieja (Valle del Cauca, Colombia). Rev Biol Trop 62:203–219

Giraldo LP, Chará J, Chará-Serna AM, Ramírez YP (2020) Restauración de corredores ribereños en paisajes ganaderos de la zona andina colombiana: Efectos tempranos sobre el ambiente acuático. Revista Acad Colomb Ciencias Exactas 44:652–664

González-Córdoba M, Zúñiga MDC, Manzo V (2015) Riqueza genérica y distribución de Elmidae (Insecta: Coleoptera, Byrrhoidea) en el departamento del Valle del Cauca, Colombia. Biota Colombiana 16:51–74

González-Córdoba M, Zúñiga MDC, Manzo V, Giraldo LP, Chará J (2016) Notelmis Hinton y Onychelmis Hinton (Coleoptera: Elmidae: Elminae) de Colombia: Claves taxonómicas ilustradas. Boletín del Museo de Entomología de la Universidad del Valle 16:1–17

Gray CL, Slade EM, Mann DJ, Lewis OT (2014) Do riparian reserves support dung beetle biodiversity and ecosystem services in oil palm-dominated tropical landscapes? Ecol Evol 4: 1049–1060

Gray REJ, Slade EM, Chung AYC, Lewis OT (2017) Riparian reserves in oil palm plantations may provide movement corridors for invertebrates. bioRxiv:204990. https://doi.org/10.1101/204990

Hynes H (1975) The stream and its valley. Verh Int Ver Theor Angew Limnol 19:1–15

Jones JB (1997) Benthic organic matter storage in streams: influence of detrital import and export, retention mechanisms, and climate. In: Webster JR, Meyer JL (eds) Stream organic matter budgets. J North Am Benthol Soc 16:109–119

Kuglerová L, Agren A, Jansson R, Laudon H (2014) Towards optimizing riparian buffer zones: ecological and biogeochemical implications for forest management. For Ecol Manag 334:74–84

Lamberti GA, Gregory SV (2006) CPOM transport, retention, and measurement. In: Hauer FR, Lamberti GA (eds) Methods in stream ecology, 2nd edn. Academic, San Diego, pp 273–292

Lees AC, Peres CA (2007) Conservation value of remnant riparian forest corridors of varying quality for Amazonian birds and mammals. Conserv Biol 22:439–449

Lorion C, Kennedy B (2009) Relationships between deforestation, riparian forest buffers and benthic macroinvertebrates in neotropical headwater streams. Freshw Biol 54:165–180

Luke SH, Slade EM, Gray CL, Annammala KV, Drewer J, Williamson J, Agama AL, Ationg M, Mitchell S, Vairappan CS, Struebig MJ (2019) Riparian buffers in tropical agriculture: scientific support, effectiveness and directions for policy. J Appl Ecol 56:85–92

Marczak LB, Sakamaki T, Turvey SL, Deguise I, Wood SLR, Richardson JS (2010) Are forested buffers an effective conservation strategy for riparian fauna? An assessment using meta-analysis. Ecol Appl 20:126–134

Medina A, Harvey CA, Merlo DS, Vílchez S, Hernández B (2007) Bat diversity and movement in an agricultural landscape in Matiguás, Nicaragua. Biotropica 39:120–128

Meli P, Calle A, Calle Z, Ortiz-Arrona CI, Sirombra M, Brancalion PHS (2019) Riparian-forest buffers: bridging the gap between top-down and bottom-up restoration approaches in Latin America. Land Use Policy 87:104085. https://doi.org/10.1016/j.landusepol.2019.104085

Méndez LE, Calle Z (2010) Plantas de la Cuenca media del río La Vieja. CIPAV y CIEBEREG, Cali

Mingoti R, Vettorazzi CA (2011) Relative reduction in annual soil loss in microwatersheds due to the relief and forest cover. Eng Agric 31:1202–1211

Molineri C, Zúñiga MDC, Ramos BC, Giraldo LP, Cardona W (2016) Three new species of Leptohyphes Eaton (Ephemeroptera: Leptohyphidae) from Colombia. Iheringia, Série Zoologia 106:e2016024

Moore RD, Richardson J (2003) Progress towards understanding the structure, function and ecological significance of small stream channels and their riparian zone. Can J For Res 33: 1349–1351

Murgueitio E, Ibrahim M (2009) Ganadería y Medio Ambiente en América Latina. In: Murgueitio E, Cuartas C, Naranjo J (eds) Ganadería del futuro: Investigación para el desarrollo. Fundación CIPAV, Cali, pp 19–39

Naiman RJ, Bilby RE, Bisson PA (2000) Riparian ecology and management in the Pacific coastal rain forest. Bioscience 50:996–1011

Naiman RJ, Décamps H, McClain ME (2005) Riparia: ecology, conservation and management of streamside communities. Elsevier Academic Press, San Diego

Osborne LL, Kovacic DA (1993) Riparian vegetated buffer strips in water-quality restoration and stream management. Freshw Biol 29:243–258

Paetzold A, Schubert CJ, Tockner K (2005) Aquatic-terrestrial linkages along braided-river: riparian arthropods feeding on aquatic insects. Ecosystems 8:748–759

Palmer MA, Hondula KL, Koch BJ (2014) Ecological restoration of streams and rivers: shifting strategies and shifting goals. Annu Rev Ecol Evol Syst 45:247–269

Ramírez YP, Giraldo LP, Zúñiga MDC, Ramos BC, Chará J (2018) Influencia de la ganadería en los macroinvertebrados acuáticos en microcuencas de los Andes centrales de Colombia. Rev Biol Trop 66:1244–1257

Riseng CM, Wiley MJ, Black RW, Munn MD (2011) Impacts of agricultural land use on biological integrity: a causal analysis. Ecol Appl 21:3128–3146

Sabo JL, Sponseller R, Dixon M, Gade D, Harms T, Heffernan J, Jani A, Katz G, Soykan C, Watts J, Welter J (2005) Riparian zones increase regional species diversity by harboring different, not more species. Ecology 86:56–62

Sadeghian S, Rivera J, Gómez M (1999) Impacto de sistemas de ganadería sobre las características físicas, químicas y biológicas de suelos en los Andes de Colombia. In: Sánchez M, Rosales M (eds) Agroforestería para la Producción Animal en América Latina, Producción y Sanidad Animal No. 143. FAO, Roma, pp 123–142

Schilling KE, Jacobson P (2014) Effectiveness of natural riparian buffers to reduce subsurface nutrient losses to incised streams. Catena 114:140–148

Schroth G, da Fonseca GAB, Harvey CA, Gascon C, Vasconcelos HL, Izac AM (2004) Agroforestry and biodiversity conservation in tropical landscapes. Island Press, Washington, DC

Skłodowski M, Kiedrzynska E, Kiedrzynski M, Urbaniak M, Zielinska KM, Kurowski JK, Zalewski M (2014) The role of riparian willows in phosphorus accumulation and PCB control for lotic water quality improvement. Ecol Eng 70:1–10

Suurkuukka H, Virtanen R, Suorsa V, Soininen J, Paasivirta L, Muotka T (2014) Woodland key habitats and stream biodiversity: does small-scale terrestrial conservation enhance the protection of stream biota? Biol Conserv 170:10–19

Tanaka MO, Souza ALT, Moschini LE, Oliveira AK (2016) Influence ofwatershed land use and riparian characteristics on biological indicators of stream water quality in southeastern Brazil. Agric Ecosyst Environ 216:33–339

Turunen J, Markkula J, Rajakallio M, Aroviita J (2019) Riparian forests mitigate harmful ecological effects of agricultural diffuse pollution in medium-sized streams. Sci Total Environ 649:495–503

Uribe-Gómez F (2008) La Dimensión Ambiental en la Construcción de la Ecorregión Eje Cafetero. In: Rodriguez JM, Camargo JC, Niño J, Pineda AM, Arias LM, Echeverry MA, Miranda CL (eds) Valoración de la Biodiversidad en la Ecorregión del Eje Cafetero. CIEBEREG, Pereira, pp 21–38

Vannote RL, Minshall GW, Cummins KW, Sedell JR, Cushing CE (1980) The river continuum concept. Can J Fish Aquat Sci 37:130–137

Villada S, Triana LA, Días L (2017) Grupos funcionales alimentarios de insectos acuáticos en quebradas andinas afectadas por agricultura y minería. Caldasia 39:370–387

Wallace JB, Eggert SL, Meyer JL, Webster JR (1997) Multiple trophic levels of a forest stream linked to terrestrial litter inputs. Science 277:102–104

Yoshimura M (2012) Effects of forest disturbances on aquatic insect assemblages. J Entomol Sci 15:145–154

Zimbres B, Peres CA, Machado RB (2017) Terrestrial mammal responses to habitat structure and quality of remnant riparian forests in an Amazonian cattle-ranching landscape. Biol Conserv 206:283–292

Zúñiga MDC, Giraldo LP, Calero H, Ramírez YP, Chará JD (2014) Anacroneuria caraca stark y A. jewetti stark (Insecta: Plecoptera: Perlidae): Primeros registros para los Andes Orientales y el pie de monte de la orinoquía Colombiana. Boletín del Museo de Entomología de la Universidad del Valle 15:12–19

Zúñiga MDC, Giraldo LP, Ramírez YP, Chará JD, Ramos BC (2015) Neoatriplectides (Trichoptera: Atriplectididae) en Colombia:Notas sobre su taxonomía, ecología y distribución en el Neotrópico. Rev Colomb Entomol 41:149–152

# Chapter 7
# Conservation and Registration of Seed Sources in Reserve Remnants in the Province of Misiones, Argentina

**Beatriz I. Eibl, Florencia Montagnini, Lucas N. López, Héctor F. Romero, Claudio J. Dummel, Guillermo Küppers, Miguel A. López, Pedro Lavignolle, Jorge Cortes, and Marcelo A. De La Vega**

**Abstract** The Upper Parana Atlantic Forest, a biodiversity hotspot of highly threatened biodiversity, extends to northeastern Argentina in the province of Misiones. In Argentina, national and provincial laws promote the restoration of degraded forest areas. This is a crucial process necessary to the establishment of biodiversity islands, which requires plant propagation material of native species to ensure their viability and diversity. Since 2018, a registry of forest areas and seed trees has been implemented nationwide to serve as a repository of biodiversity and plant propagation material, controlled by the National Seed Institute (INASE) Regulation n°318/18, elaborated in scientific/technical cooperation with the School of Forestry of the National University of Misiones. In these registered areas, which can be in private, public, or protected lands, propagation material is certified, with priority given to arboreal species and rare, endemic, threatened and/or vulnerable species. Each species is registered in every site where it is documented in order to record as much genetic variability as possible. Propagation material certification establishes a protocol that allows native species to be included, under certified control, in sustainable, productive, and multiple-purpose land use systems. To date, more than 1000 registered seed trees have been recorded in more than 54 hectares of

B. I. Eibl (✉) · L. N. López · H. F. Romero · C. J. Dummel · G. Küppers · M. A. López
Facultad de Ciencias Forestales, Universidad Nacional de Misiones, Eldorado, Misiones, Argentina
e-mail: hcr_lopez@yahoo.ar

F. Montagnini
School of the Environment, The Forest School, Yale University, New Haven, CT, USA
e-mail: florencia.montagnini@yale.edu

P. Lavignolle · J. Cortes · M. A. De La Vega
Instituto Nacional de Semillas (INASE), Dirección de Certificación y Control, Buenos Aires, Argentina
e-mail: plavig@inase.gov.ar; jcortes@inase.gov.ar; madelavega@inase.gov.ar

 
F. Montagnini (ed.), *Biodiversity Islands: Strategies for Conservation in Human-Dominated Environments*, Topics in Biodiversity and Conservation 20,
https://doi.org/10.1007/978-3-030-92234-4_7

forest remnants. Registered seed trees facilitate restoration of degraded areas, increasing site diversity via dispersion by birds, small animals and wind. The certification of native species for multiple use contributes to their conservation and to their use for restoration, while also generating resources for owners, helping to conserve biodiversity.

**Keywords** Atlantic forest · Diversity · Germination · Natural reserves · Production · Propagation material · Restoration

## 7.1 Introduction

The eco-region of the Atlantic Forest in South America is an internationally recognized biodiversity hotspot due to its biologically rich terrestrial ecosystems (Galindo-Leal and Câmara 2003; Mittermeier et al. 2011). In Argentina, the Atlantic Forest, locally known as the Upper Parana Atlantic Forest, emerges from the coast and stretches into the interior, sharing portions with Brazil and Paraguay (Di Bitetti et al. 2003). The part of the Upper Parana Atlantic Forest in the Province of Misiones, known as the Selva Misionera or Selva Paranaense, contains the greatest biodiversity in Argentina, with dense multi-stratified vegetation in the forests and isolated trees in open fields. Many previously unfound species continue to be discovered in the forest and added to the local biodiversity lists, proving it has the ideal habitat conditions to support a variety of life (Keller 2007).

Restoration and conservation practices in this region of the Atlantic Forest can help protect one of the last relicts of native forest and therefore conserve native species. In this region, restored areas of protective forests on the borders of streams and slopes provide connectivity between different areas of restored forests in the basins of the Uruguay, Iguazú and Paraná rivers, which contributes to the biodiversity of the region. Additionally, because the Upper Parana Atlantic Forest shares its atmosphere, biodiversity, and surface and underground water with Brazil and Paraguay, its biodiversity is connected to those nations, so protecting the Argentinian section of forest will contribute to the protection of land in those nations as well. However, the advancement of the agricultural frontier has caused changes in land use, population growth, timber extraction from natural forests and increased isolation of protected areas, which has compromised the integrity of the genetic and species diversity of these regions and the durability of these natural formations (Young 2003; Gibbs et al. 2010; Mittermeier et al. 2011; Armenteras et al. 2017). The protection of the Atlantic Forest is crucial to the environmental health of the region, which is why the Argentinian government has made previous efforts to conserve the land.

The government of Argentina has passed laws promoting restoration and conservation efforts, and restoration efforts are ongoing. Strategies for biodiversity conservation in this region must include a series of actions applied by public institutions that complement each other in order to provide an enforced, comprehensive environmental protection plan. For instance, the Provincial Laws XVI n°60 "Green

Corridor" (Ley Provincial XVI n°60 2010) and n°29 (Ley Provincial n°29 2010) of Protected Natural Areas present an opportunity to maintain natural forest connectivity in the Province of Misiones by conserving forest in public areas, private parks and nature reserves.

Conserving biodiversity can also be a profitable renewable resource through the use of local species for economic purposes. Therefore, following the principles of its conservation and rational use, biodiversity can be considered as an asset for economic and social development purposes, providing additional benefits toward conservation practices (Ugarte-Guerra et al. 2010).

However, several factors must be considered when conserving land and establishing biodiversity islands. The degree of degradation and the species present in the targeted landscape are important factors to evaluate when designing biodiversity islands. After determining the degradation status, it is possible to determine the most feasible and appropriate restoration pathway that will maximize the efficiency and impact of conservation efforts (Montagnini et al. 2022). In Argentina, the national and provincial laws establish and promote the restoration of degraded areas, which requires propagation material of native species including seeds, fruits, cuttings, etc. As part of our previous work on restoration of degraded lands with native tree species that can be useful for productive purposes, a list of plant species was compiled. The creation of this database facilitated the selection of species most compatible with the productive systems in the region (including plantations, agroforestry, and silvopastoral systems) (Eibl et al. 2015b). These specifically-chosen plant species provide long-term products and store atmospheric carbon, in addition to the advantages of diversity conservation via native forests strips and patches.

In this chapter, we describe the methodology, procedures and contributions of a successful registry and certification system of seed sources on land areas in need of restoration and conservation of species diversity in the region. The registry allows for the propagation of genetic material used in restoration and/or enrichment plans to be traced and monitored to evaluate its success and potential use in domestication programs.

## 7.2 Legal Framework Supporting Tree Planting in Argentina

In Argentina, the National Law on Minimum Budgets for the Environmental Protection of Native Forests, n°26,331 (Ley Nacional n°26331 2007). classifies native forests into three management categories: conservation, sustainable management and potential conversion to other uses. Forests that fall in the first two categories must be restored with native species, which may be required by the governing agency in each province. In the case of Misiones, the governing agency is the Ministry of Ecology and Renewable Natural Resources (MEyRNR). In addition to government regulation, National Law n°27,487 (Ley Nacional n°27487 2008), establishes incentives for restoration projects or productive projects which involve planting and harvesting. In order to receive the incentive, scientific procedures for planting propagation material need to be available and used.

Misiones has enacted a series of provincial laws in order to promote restoration efforts of threatened forest and protect regional biodiversity. Provincial Law XVI n°53 (Ley Provincial XVI nº53 2010), utilizes protective forests and riparian corridors to aid in the recovery of stream and slope borders. The law establishes that restoration of border areas is mandatory, and that tree species may be used to improve restoration efforts based on a management plan previously approved by the relevant authorities. The Misiones Provincial Law of Renewable Energies XVI n°97 (Ley Provincial XVI nº97 2010), promotes the planting of species for biomass purposes. By means of the database of plant species, ideal plant material can be selected for restoration of a habitat. For example, plantation of potential native species for timber purposes can have benefits to forest restoration (Eibl et al. 2015a). Law XVI n°98 (Ley Provincial XVI nº 98 2010), of the Provincial Bank of Germplasm of Misiones, which aims to preserve the genetic base of native plants, facilitates the availability of seeds for conservation, research and productive use. Likewise, the Provincial Law XVI n°47 (Ley Provincial XVI nº 47 2010), regulates the conservation and sustainable use of biological diversity and its components, in accordance with the Global Strategy for the Conservation of Plant Species (GSPC 2020), of which Argentina is a signatory.

The implementation of these restoration efforts requires appropriate propagation material. In order to ensure plant material has the greatest benefit to a region while avoiding potential invasive or diseased species requires the application of specific certification protocols. These certifications ensure that restoration projects use propagation material that will maximize their contribution to the biodiversity of the region. In addition, successful restoration efforts require proper monitoring and verification in order for the desired functions and objectives of propagation material to be properly fulfilled.

The important restoration effort that Argentina is making by protecting forests and providing renewable energy fulfills its commitments made alongside a total of 195 countries at the United Nations Climate Change Conference in COP21 (2015). In COP25 (2019) Argentina was one of the Latin American countries that signed the "Declaration for Restoration," committing 30 million hectares within the region to be restored by 2030. This goal aims to bring back vital ecosystems and promote social and economic development of the region.

Forests are expected to make key contributions to reduce greenhouse gases (GHG) emissions and to ensure the long-term stability of global climate. Restoration and management of degraded forests can contribute to resilient landscapes by absorbing carbon and maintaining other ecosystem services, such as the provision of water for sustainable consumption and productive use. Conservation of forests and reforestation schemes contribute to achieving sustainable development goal (SDG) 15 on Life on Land by maintaining and increasing the proportion of the world's land area that is forested (IUCN 2020b). In fact, FAO's 2018 State of the World's Forests describes how forests and trees contribute to 28 targets relating to ten SDGs (FAO 2018; FAO 2020), showing the importance of degraded land restoration.

## 7.3 Strategies and Methods for Certification of Seed Sources of Native Species

The current strategies for conservation of native species for environmental and productive purposes in Argentina were tested in private reserves by technicians and professionals, and subsequently disseminated in workshops and participatory courses led by the School of Forestry of the National University of Misiones (FCF/UNaM). For each stage of the certification process, beginning by harvesting seeds and other specimens, there are a minimum of criteria to be considered before progressing to the next stage in order to ensure that species are properly conserved and successfully used in restoration practices.

Current conservation strategies include the establishment and upkeep of a registry of tree seed species, the storage of seeds of native species, and the certification of the characteristics of the seeds in their botanical, physical and physiological aspects as is typically done in a seed analysis laboratory. In addition, plants that are produced in nurseries receive the Certification of Nurseries of Native Species through specific protocols, which are then registered in the National Registry of Seed Trade and Control (RNCyFS) of the National Seed Institute (Instituto Nacional de Semillas, INASE). Certification protocols for trees follow the Forest Species Certification System, regulated by INASE Resolution n°256/99 (INASE 1999), which establishes a system of classes and categories for basic propagation material. The National Network of Germplasm Banks of Native Plant Species (RNBGEVN) contributes to the exchange of information and training on germplasm conservation of native species (De Viana et al. 2011).

### *7.3.1 Conservation Areas That Contain Propagation Material*

National parks and public nature reserves can serve as germplasm reservoirs for conservation or research. Private nature reserves can also provide germplasm for propagation and productive uses if the material is properly certified.

In the Province of Misiones, some private landowners participate in the country's private nature reserve network with their forest remnants (Schiaffino and Bertolini 2016). Most of the small traditional producers in the province maintain their own nature reserves because of their commitment to the environment, in response to legal requirements, and/or for the possibility of participating in certification systems for their products. It is important to generate resources for the owners who are currently conserving biodiversity to ensure the persistence of their private reserves and their willingness to protect them through time (Eibl 2006; Eibl and López 2017).

The existing national and provincial legal framework does not provide a basis for specifically certifying propagation material harvested from nature reserves, which could be a potential profitable resource as well as beneficial to the restoration process. Our project aims to provide specific actions to conserve remaining natural forests areas, to create new productive forests through natural succession, and to generate a productive process with native species that are part of restoration activities. Finally, our project attempts not only to provide specific conservation and restoration actions, but also to provide technical background for designing appropriate legal tools to certify areas that house these native seeds. These tools would establish a restoration process that is more efficient and easier, which would provide greater incentives for landowners to protect their land.

### *7.3.2 Seed Tree Network*

The Faculty of Forest Sciences who works on identification and registration of seed trees in areas of private reserves, developed their techniques in joint cooperation with private companies, using information that had been gathered in 20–30 years of projects where they had been participants. These projects were run or funded by organizations such as the Mellon Foundation (1990/1998), Wildlife Foundation (1998), Perez Companc (2002), Tajy Project of Tabaco Norte (2007/2010), Biofábrica Misiones SA (2010/2012), the Registry of Seed Areas in projects with yerba mate (*Ilex paraguariensis*), small farmers with the UCAR (Unidad de Cambio Rural), Ministry of Agriculture, and INTA (Instituto Nacional de Tecnología Agropecuaria) (2014/2017) (Eibl and López 2017; Niella et al. 2016, 2017). Using results of those projects, a list of potential native tree species was taken as a basis, with a total of 336 tree species represented in 67 botanical families that occupy the middle and upper stratum of the Misiones forest (Paranaense forest) (Gartland and Bohren 2008).

Based on the interest expressed by individual landowners, who were identified through local information and references, technical visits to each site were made. In each reserve area, species are selected from this list according to the preferences of the owners as well as the classification of chosen species as valuable, rare, vulnerable, and/or threatened species (IUCN 2020a). Specimens of greater size, with good health, and those whose tree form is the most representative for each of the species are given priority when choosing. These criteria are used for all species, with greater emphasis on species of ecological importance. If individual specimens are no longer found in desired conditions, they are registered in the state in which they appear, in order to guarantee natural regeneration of that species. When possible, each species is recorded at every site of their natural distribution in order to ensure accuracy in documentation, always considering the recommendations of Thomas et al. (2014) on genetic considerations in ecosystem restoration when using native tree species.

Seed trees are registered by their identity, geographical position (GPS point), health status, location on the site, and description of their surrounding environment. A data sheet is prepared for each individual, containing information on size and photographs. Isolated trees of interest are also considered for registration. A list of all selected individuals is prepared, using a code with acronyms for identifying the region, species, number of individuals and date (Eibl et al. 2001, 2002; Eibl and Báez 2004; Eibl and López 2017). This allows leaders of restoration efforts to have ample data available to them in order to make informed decisions on the proper propagation material to use to have the most effective impact on the intended area of conservation.

### 7.3.3 Harvest, Identification, and Conditioning of Fruits and Seeds

Fruit harvesting is done when there is a specific demand of propagation material. Desired seed plants are located based on a database of species already registered by the School of Forestry (FCF/UNaM). The time of end of maturation and beginning of dispersion is identified from the phenological observation of the individuals to be harvested. This is the stage in which fruits and seeds have the lowest dry weight and the greatest percentages of germination and vigor (Eibl et al. 2012). These location data and observations allow harvesters to easily find and collect fruit when needed.

In general, the species of dehiscent fruits which split at maturity releasing their contents must be harvested on the tree, while those of indehiscent fruits, such as legumes and berries, can be harvested from the soil, as well as on the tree when they are in the last stage of dispersion. Treetop harvesting is done by specially trained staff who climb up the trees. A net is previously placed on the ground to collect the fruits and seeds of both, fruits that fall naturally and those that have to be harvested (Gold et al. 2004).

The harvested fruits corresponding to each individual specimen represent the propagation material of that tree. These materials are labeled in the field and entered in the Seed Bank notebook, with a number that guarantees their traceability. This information includes owner name, GPS point, harvest date, tree number, species, and location. In addition to fruits and seeds of trees (Eibl 2013), collection of propagation material can also come from bushes, herbs, grasses, epiphytes, orchids, ferns, and others. After being harvested, propagation materials are classified and certified by origin, provenances and identity. Their identity is verified using the website darwin.edu.ar which is based on information from the Darwinion Botanical Institute (Instituto de Botánica Darwinion) (Flora del Conosur 2020).

After harvest, the seeds are conditioned by separating them from their fruits and cleaning them of impurities such as dust and woody sticks. The seeds are dried until they reach equilibrium with the environment and then they are prepared for storage and/or for nurseries. The seeds that are entered into active seed banks or germplasm banks are conditioned for their purpose using drying and/or ultra-drying techniques and are packed in suitable containers (to avoid excess moisture), in order to maintain their viability for a longer time (Eibl et al. 2012, 2013a).

### 7.3.4 Certification of Physical and Physiological Seed Quality

The germination to produce a normal seedling (percent germination, PG), the moisture of the seed at the time of harvest and during storage (H%), purity (whole seeds and/or portion of germinable seeds), and the weight of seeds (weight of 1000 seeds in grams) are determined in the laboratory, following recommendations from the International Seed Test Standards (ISTA 2019) and Ministério da Agricultura (2009). Germination tests are used to determine the best pre-germination treatments, while the germination conditions, times appropriate for each species, the minimum requirements for germination and purity percentages are established for each species following procedures described by Eibl et al. (2012).

### 7.3.5 Nursery and Field Planting of Native Species

For production of plants in the nursery, different technical specifications are required for each species in terms of pre-germination treatments, nursery times, substrates, containers and nutrients. Likewise, technical specifications are used in determining quality indicators for height and diameter at the collard or base (dac), slenderness (ratio of stem/dac), length, lignification of seedlings (highest dry weight) and the consistency of the roots with the substrate used for growing the seedlings (Eibl et al. 2013b).

Field planting of native species requires knowledge of the silvicultural system and certification requirements. The selection of the species to be planted depends on the owner's preferences and conditions of the site. Individual planting holes of at least 60–80 cm in depth promote the development of the trees, ensuring their support and improving their source of nutrition.

The natural vegetation of the site can protect the planted seedlings from direct sunlight, from high and low temperatures, and from frost, drought, wind and insects. If the natural vegetation is not present, pioneer or species tolerant to direct exposure should be planted first to ensure greatest survival and good initial growth. Prior evidence indicates that agroforestry systems (AFS) or trees planted in areas supporting early natural succession are likely to guarantee the survival of the seedlings (Montagnini et al. 2005, 2006; Barth et al. 2008; López et al. 2013; Eibl et al. 2015b), as well as to contribute to the recovery of degraded soils (Day et al. 2011; Eibl et al. 2019b).

For example, as part of our own research on restoration of degraded lands using native trees, we planted 11 native tree species in sites with extreme degradation due to intensive agricultural use and invasive grasses. The initial trees planted were heliophytes of rapid growth. In a second stage, we added valuable shade tolerant woody species, in order to recover the potential of the site to produce wood and fuelwood and fix atmospheric carbon (Montagnini et al. 2006; Eibl et al. 2019b). The evidence from this research showed that tree plantations can facilitate the establishment of natural regeneration, adding diversity of species to the site and allowing the generation of new multi-diverse forests by tending to the species of interest (Montagnini et al. 2005; Eibl et al. 2019c).

## 7.4 Advances in the Conservation and Certification of Seed Sources of Native Tree Species

### *7.4.1 Conservation Areas That Contain Propagation Material*

In a joint work with the National Seed Institute of Argentina (INASE), a new regulation for the Registration of Conservation Areas and the certification of propagation material was recently achieved. The Resolution N° 318/18 (INASE 2018) refers to the registration of the seed production area of native species and is being disseminated by technical personnel from the National Seed Institute and the Forestry School of the National University of Misiones (FCF-UNaM). In addition, through training and communication workshops, the Resolution is being distributed among interested provinces that must adhere to the proposal at the national level.

The Province of Misiones followed the national proposal with an agreement subscribed on September 24, 2019 (n°81), between INASE and MEyRNR (Ministry of Ecology and Renewable Natural Resources of Misiones). Simultaneously, an agreement was reached between the FCF-UNaM and MEyRNR for the implementation and the scientific and technical support of the proposal in the field. The identification and documentation of tree diversity that is contained in an area or private reserve that the owner agrees to protect are registered under a number assigned by the RNCyFS (Fig. 7.1) in order to keep accurate record of species counts and types present. Examples of lists of identified trees and seedlings in private reserves are shown in Tables 7.1 and 7.2.

The propagation material of all species (fruits, seeds, pollen, buds, branches, roots, others) that is harvested for use is certified. In designating this certification, several factors are considered, including the sustainability of the source. Certification allows for more informed decisions to be made regarding what propagation material will have the greatest beneficial impact to the land and not result in any invasive or diseased species. Therefore, a management plan designed by a qualified professional is critical and has to be appropriate to the area and approved by the MEyRNR of the Province of Misiones (Ley Provincial XVI n°105 2010).

The FCF-UNaM had previously worked in more than 30 natural reserves with a total area of 54 hectares of remnants of small and medium private forests, registering

**Fig. 7.1** Forest reserve with selected native tree species, in productive areas in a family farm in Andresito, Province of Misiones. (Photo: B. Eibl)

the trees and monitoring the biodiversity, providing valuable information prior to the creation of the new national regulation. A field demonstration of certification at the level of trees, shrubs, epiphytes, and herbs of the propagation material collected from the origin was carried out in these reserves. We also evaluated the opportunity to sell fruit, seeds and/or plant seedlings in the market, as a way to generate income to the owners to compensate them for their conservation efforts (Fig. 7.2).

## *7.4.2 Seed Tree Network*

In accordance with the existing seed tree network protocol, all the species living in each area are included in the registry, from the crown strata with epiphytes and ferns to the undergrowth with herbs and shrubs as well as mushrooms and mosses. Trees are registered in every landscape where they are present, whether they are part of forests, riverine strips or growing in fields with grasses (Table 7.3). For example, some hardwood species, such as *Anadenanthera colubrina*, *Astronium balansae*, *Handroanthus heptaphyllus,* and palms such as *Butia yatay*, and *Acrocomia aculeate* tend to predominate in open fields in the region.

**Table 7.1** Adult tree species in the forest reserve of the Stadler family farm, in General Alvear, Province of Misiones

| N° | Scientific name | Common name | Botanical family | Appearance/ frequency |
|---|---|---|---|---|
| 1 | ***Casearia sylvestris*** Sw. var. ***sylvestris*** | Burro caá | FLACOURTIACEAE | 1 |
| 2 | ***Cordia trichotoma***(Vell.) Arráb. exSteud. | Loro negro | BORAGINACEAE | 1 |
| 3 | ***Cupania vernalis*** Cambess. | Camboata Colorado | SAPINDACEAE | 1 |
| 4 | ***Matayba eleagnoides*** Radlk. | Camboata Blanco | SAPINDACEAE | 1 |
| 5 | ***Helietta apiculata*** Benth. | Canela de venado | RUTACEAE | 1 |
| 6 | ***Casearia decandra*** Jacq. | Guazatumba chica | FLACOURTIACEAE | 1 |
| 7 | ***Rauvolfia sellowii*** Müll. Arg. | Quina | APOCYNACEAE | 1 |
| 8 | ***Machaerium stipitatum*** (DC.) Vogel | Isapu'ymoroti | FABACEAE | 1 |
| 9 | ***Chrysophyllum gonocarpum*** (Mart. &Eichler) Engl. | Aguaí | SAPOTACEAE | 1 |
| 10 | ***Diatenopteryx sorbifolia*** Radlk. | María preta | SAPINDACEAE | 1 |
| 11 | ***Myrocarpus frondosus*** Allemão | Incienso | FABACEAE | 1 |
| 12 | ***Aralia warmingiana*** (Marchal) J. Wen | Sabuguero | ARALIACEAE | 1 |
| 13 | ***Zanthoxylum petiolare***A. St.-Hil. &Tul. | Fagara naranjillo | RUTACEAE | 1 |
| 14 | ***Chrysophyllum marginatum*** (Hook. &Arn.) Radlk. ssp. ***Marginatum*** | Vasuriña | SAPOTACEAE | 1 |
| 15 | ***Erythrina falcata***Benth. | Ceibo | FABACEAE | 1 |
| 16 | ***Syagrus romanzoffiana*** (Cham.) Glassman | Pindo | PALMAE | 1 |
| 17 | ***Lonchocarpus campestris*** Mart. exBenth. | Rabo ita | FABACEAE | 1 |
| 18 | ***Pisonia zapallo*** var. ***Zapallo*** Griseb. | Pisonia zapallo | NYCTAGINACEAE | 1 |
| 19 | ***Hovenia dulcis*** Thunb. | Hovenia | RHAMNACEAE | 1 |
| 20 | ***Picrasma crenata*** (Vell.) Engl. | Palo amargo | SIMAROUBACEAE | 1 |
| 21 | ***Ocotea puberula*** (Rich.) Nees | Laurel guaíca | LAURACEAE | 2 |
| 22 | ***Dendropanax cuneatus*** (DC.) Decne. &Planch. | Ombu ra | ARALIACEAE | 2 |
| 23 | ***Nectandra lanceolata*** Nees | Laurel amarillo | LAURACEAE | 2 |
| 24 | ***Myrsine balansae*** (Mez) Otegui | Pororoca | MYRSINACEAE | 2 |
| 25 | ***Sorocea bonplandii*** (Baill.) W.C. Burger, Lanj. &Wess. Boer | Ñandipa | MORACEAE | 2 |

(continued)

**Table 7.1** (continued)

| N° | Scientific name | Common name | Botanical family | Appearance/ frequency |
|---|---|---|---|---|
| 26 | ***Prunus brasiliensis*** (Cham. &Schltdl.) D. Dietr. | Persiguero | ROSACEAE | 2 |
| 27 | ***Jacaratia spinosa***(Aubl.) A. DC. | Yacaratia | CARICACEAE | 2 |
| 28 | ***Nectandra megapotamica*** (Spreng.) Mez | Laurel negro | LAURACEAE | 2 |
| 29 | ***Parapiptadenia rigida*** (Benth.) Brenan | Anchico Colorado | FABACEAE | 2 |
| 30 | ***Balfourodendron riedelianum*** (Engl.) Engl. | Guatambú Blanco | RUTACEAE | 3 |
| 31 | ***Cordia americana*** (L.) Gottschling& J.S. Mill. | Guayubira | BORAGINACEAE | 3 |
| 32 | ***Holocalyx balansae*** Micheli | Alecrín | FABACEAE | 3 |
| 33 | ***Apuleia leiocarpa*** (Vogel) J.F. Macbr. | Grapia | FABACEAE | 4 |
| 34 | ***Alchornea triplinervia*** (Spreng.) Müll. Arg. | Mora blanca grande | EUPHORBIACEAE | 4 |
| 35 | ***Maclura tinctoria*** ssp. ***Tinctoria*** (L.) Steud. | Mora amarilla | MORACEAE | 4 |
| 36 | ***Trichilia claussenii*** C. DC. | Catigua guazú | MELIACEAE | 4 |
| 37 | ***Cedrela fissilis*** Vell. | Cedro misionero | MELIACEAE | 5 |
| 38 | ***Aspidosperma austral*** Müll. Arg. | Guatambú Amarillo | APOCYNACEAE | 5 |
| 39 | ***Inga marginata***Willd. | Inga'i | FABACEAE | 6 |
| 40 | ***Cecropia pachystachya*** Trécul | Ambay | CECROPIACEAE | 7 |
| 41 | ***Cabralea canjerana*** (Vell.) Mart. | Cancharana | MELIACEAE | 12 |
| 42 | ***Ficus luschnathiana*** (Miq.) Miq. | Higuera de Monte | MORACEAE | 16 |

Source: Eibl et al. (2016a). Project: Productive experiences with multipurpose native species in environmental restoration areas and strategies for the conservation of forest remnants. Rural Change Unit (UDI) n°01/2013. Sustainable Forest Plantations (SFPC)

The Registry of Seed Trees currently contains more than 1000 individuals registered in different areas of the province since 1990, whose propagation materials are currently available to be harvested for a nursery supply. Endangered, vulnerable and rare species such as *Araucaria angustifolia*, *Aspidosperma polyneuron*, *Myrocarpus frondosus*, *Maclura tinctoria*, *Aralia warmingiana*, *Aspidosperma australe*, and others are the priority species in the records, even if the individuals are old or deteriorated, in order to prevent species extinction and preserve valuable genetic information (Fig. 7.3).

**Table 7.2** Regenerating tree seedlings in the forest reserve of Stadler family, in General Alvear, Province of Misiones

| N° | Scientific name | Common name | Botanical family | Appearance/ frequency |
|---|---|---|---|---|
| 1 | ***Casearia sylvestris*** Sw. var. ***sylvestris*** | Burro caá | FLACOURTIACEAE | 1 |
| 2 | ***Chrysophyllum gonocarpum*** (Mart. &Eichler) Engl. | Aguaí | SAPOTACEAE | 1 |
| 3 | ***Trichilia catigua*** A. Juss. | Catiguá Colorado | MELIACEAE | 1 |
| 4 | ***Matayba eleagnoides*** Radlk. | Camboata Blanco | SAPINDACEAE | 1 |
| 5 | ***Eugenia burkartiana*** (D. Legrand) D. Legrand | | MYRTACEAE | 1 |
| 6 | ***Jacaranda micrantha*** Cham. | Caroba blanca | BIGNONIACEAE | 1 |
| 7 | ***Ocotea puberula*** (Rich.) Nees | Laurel guaíca | LAURACEAE | 1 |
| 8 | ***Trichilia elegans***A. Juss. | Catigua Chico | MELIACEAE | 1 |
| 9 | ***Carica papaya L.*** | Mamon | CARICACEAE | 1 |
| 10 | ***Ilex brevicuspis*** Reissek | Caona | AQUIFOLIACEAE | 1 |
| 11 | ***Alchornea triplinervia*** (Spreng.) Müll. Arg. | Mora blanca grande | EUPHORBIACEAE | 1 |
| 12 | ***Diatenopteryx sorbifolia*** Radlk. | María preta | SAPINDACEAE | 1 |
| 13 | ***Ficus luschnathiana*** (Miq.) Miq. | Higuera de Monte | MORACEAE | 1 |
| 14 | ***Daphnopsis racemosa*** Griseb. | Daphnopsis | THYMELAEACEAE | 1 |
| 15 | ***Trichilia catigua*** A. Juss. | Catiguá Colorado | MELIACEAE | 1 |
| 16 | ***Citronella paniculata*** (Mart.) R.A. Howard | Pasto de anta | ICACINACEAE | 1 |
| 17 | ***Cecropia pachystachya*** Trécul | Ambay | CECROPIACEAE | 1 |
| 18 | ***Tabernaemontana catharinensis*** A. DC. | Horquetero | APOCYNACEAE | 1 |
| 19 | ***Solanum granulosum-leprosum*** Dunal | Fumo bravo | SOLANACEAE | 1 |
| 20 | ***Celtis iguanaea*** (Jacq.) Sarg. | Tala | CELTIDACEAE | 1 |
| 21 | ***Schefflera morototoni*** (Aubl.) Maguire, Steyerm. &Frodin | Cacheta | ARALIACEAE | 1 |
| 22 | ***Cordia americana*** (L.) Gottschling& J.S. Mill. | Guayubira | BORAGINACEAE | 2 |
| 23 | ***Guarea macrophylla*** ssp. ***Spicaeflora*** (A. Juss.) T.D. Penn. | Cedrillo | MELIACEAE | 2 |
| 24 | ***Hennecartia omphalandra*** J. Poiss. | Ñandipa'ra | MONIMIACEAE | 2 |
| 25 | ***Trema micrantha*** (L.) Blume | Palo polvorá | CELTIDACEAE | 2 |

(continued)

**Table 7.2** (continued)

| N° | Scientific name | Common name | Botanical family | Appearance/ frequency |
|---|---|---|---|---|
| 26 | ***Balfourodendron riedelianum*** (Engl.) Engl. | Guatambú Blanco | RUTACEAE | 2 |
| 27 | ***Nectandra lanceolata*** Nees | Laurel amarillo | LAURACEAE | 3 |
| 28 | ***Nectandra megapotamica*** (Spreng.) Mez | Laurel negro | LAURACEAE | 3 |
| 29 | ***Pilocarpus pennatifolius*** Lem. | Jaborandi | RUTACEAE | 3 |
| 30 | ***Cupania vernalis***Cambess. | Camboata Colorado | SAPINDACEAE | 3 |
| 31 | ***Maclura tinctoria*** ssp. ***Tinctoria*** (L.) Steud. | Mora amarilla | MORACEAE | 3 |
| 32 | ***Myrsine balansae*** (Mez) Otegui | Caápororocá | MYRSINACEAE | 4 |
| 33 | ***Trichilia claussenii*** C. DC. | Catigua guazú | MELIACEAE | 4 |
| 34 | ***Cedrela fissilis*** Vell. | Cedro misionero | MELIACEAE | 5 |
| 35 | ***Aspidosperma australe*** Müll. Arg. | Guatambú Amarillo | APOCYNACEAE | 5 |
| 36 | ***Cabralea canjerana*** (Vell.) Mart. | Cancharana | MELIACEAE | 8 |
| 37 | ***Inga marginata*** Willd. | Inga'i | FABACEAE | 8 |
| 38 | ***Sorocea bonplandii*** (Baill.) W.C. Burger, Lanj. &Wess. Boer | Ñandipa | MORACEAE | 23 |
| 39 | ***Actinostemon concolor*** (Spreng.) Müll. Arg. | Laranjeira | EUPHORBIACEAE | 31 |

Ref. Eibl et al. (2016a). Project: Productive experiences with multipurpose native species in environmental restoration areas and strategies for the conservation of forest remnants. Rural Change Unit (UDI) n°01/2013. Sustainable Forest Plantations (SFPC)

### 7.4.3 Harvest, Identification and Conditioning of Fruits and Seeds

As described in the methods, at the time of harvest, the fruits and seeds must be ripe, with the maximum dry weight and minimum humidity on the mother plant (a time that usually corresponds to the greatest germination and vigor). This moment coincides with the initial and maximum phase of fruit maturity and dispersion, as is the case for *Cabralea canjerana*, for example (Fig. 7.4).

Throughout the year, different species are found in their dispersal stage. When harvested, the containers are labeled with data on site, species, and date of harvest. The seeds of the recently collected ripe fruits have around 10–18% humidity for dry fruits (orthodox seeds) and over 40% for fresh fruits (recalcitrant seeds). Conditioning the fruits, which consists of separating the seeds to achieve maximum purity, is done manually or using sieves and fans to blow them into the seed containers.

**Fig. 7.2** Diversity monitoring guided by the owners in the natural reserve of the Stadler family farm in General Alvear, Province of Misiones. (Photo: B. Eibl)

### *7.4.4 Certification of the Physical and Physiological Seed Quality*

Based on our experiences following the protocols described, we have learned and refined the methodology in a number of ways. For instance, the storage unit or the nursery unit depends on the characteristic of the species. In general terms the unit is a seed; however, where the separation of the seed is difficult the unit is a fruit (i.e. *Balfourodendron riedelianum* and *Helietta apiculata*). The manual separation of the seeds allows purity levels to exceed 90% (Eibl et al. 2013a). For the purpose of certification, collected seed samples are expected to achieve a germination percentage higher than 70%. Exceptions are taken only in specific cases where germination is naturally low (under 50%), as is the case for *Bastardiopsis densiflora*, *Aralia warmingiana*, *Scheflera morototoni*, among others.

The germination rate is an expression of vigor and is linked to the storage potential of the specimen. As the germination rate increases, the chance of maintaining viability in storage also increases. For example, *Cedrela fissilis* seeds germinate faster and more evenly when the dry weight of the seeds is greater

**Table 7.3** Registered seed trees in the forest reserve of the Stadler family farm, in General Alvear, Province of Misiones

| Scientific name | Common name | Botanical family | Code |
|---|---|---|---|
| ***Alchornea triplinervia*** (Spreng.) Müll. Arg. | Mora blanca | MORACEAE | 13-ALCTRI-002-15 |
| ***Apuleia leiocarpa*** (Vogel) J.F. Macbr. | Grapia | FABACEAE | 13-APULEI-010-15 |
| ***Aspidosperma australe*** Müll. Arg. | Guatambú Amarillo | APOCYNACEAE | 13-ASPAUS-003-15 |
| ***Cabralea canjerana*** (Vell.) Mart. | Cancharana | MELIACEAE | 13-CABCAN-009-15 |
| ***Cabralea canjerana*** (Vell.) Mart. | Cancharana | MELIACEAE | 13-CABCAN-010-15 |
| ***Cedrela fissilis*** Vell. | Cedro Misionero | MELIACEAE | 13-CEDFIS-012-15 |
| ***Cordia trichotoma*** (Vell.) Arráb. ex Steud. | Peteribí | BORAGINACEAE | 13-CORTRI-022-15 |
| ***Cordia trichotoma*** (Vell.) Arráb. exSteud. | Peteribí | BORAGINACEAE | 13-CORTRI-023-15 |
| ***Holocalyx balansae*** Micheli | Alecrín | FABACEAE | 13-HOLBAL-008-15 |
| ***Maclura tinctoria*** (L.) Steud.ssp. ***tinctoria*** | Mora amarilla | MORACEAE | 13-MACTIN-006-15 |
| ***Maclura tinctoria*** (L.) Steud.ssp. ***tinctoria*** | Mora amarilla | MORACEAE | 13-MACTIN-007-15 |
| ***Rauvolfia sellowii*** Müll. Arg. | Quina de Monte | APOCYNACEAE | 13-RAUSEL-001-15 |
| ***Schefflera morototoni*** (Aubl.) Maguire, Steyerm. &Frodin | Cacheta | ARALIACEAE | 13-SCHMOR-004-15 |

Source: Niella F, Eibl B, Keller H, Maiocco D, Rocha P, Vega V (2017) Network of trees and seed areas for the conservation, domestication and rescue of native plant genetic resources in the Province of Misiones. SILVA N° 17/2013. Project Biodiversity Conservation Component in Forest Productive Landscapes. GEF 090118. Ministerio de Agricultura Ganaderia y Pesca (MAGyP)

(Ocampo et al. 2015). *Araucaria angustifolia* seeds also maintain their initial vigor in growth, two years after planted in the field (Eibl et al. 2017).

For seeds that are less likely to survive drying and freezing conditions during *ex-situ* conservation, known as recalcitrant seeds, the storage humidity should be generally maintained above 35%, and they should be preserved in moist sand under cold temperatures (Rodríguez et al. 2019). Such is the case of seeds of *Eugenia involucrata* of the Myrtaceae family, among others. The seeds of *Cabralea canjerana* are very recalcitrant. Even when stored in wet sand and a cold environment, they can only survive for 4 months (González et al. 2015).

**Fig. 7.3** Selected and registered tree of *Aspidosperma polyneuron* in productive areas in a family farm, in Andresito, Province of Misiones. (Photo: B. Eibl)

**Fig. 7.4** Ripe fruits and dispersion of *Cabralea canjerana* indicate the optimal moment for harvest. (Photo: B. Eibl)

Seeds of orthodox species should be dried until they reach equilibrium with the environment and subsequently placed in containers with silica-gel as a desiccator to reduce their humidity to 5% or less. They need to be stored in airtight containers and in a cold environment to maintain seeds long term in germplasm banks. In storage for conservation, the humidity must stay below 5% and germination must not fall below 70% (Eibl et al. 2013a). Ultra-dried seeds of *Aspidosperma polyneuron* were reported as viable after 5 years of storage (Otegui et al. 2016). This species' seeds were dehydrated to 5% and expressed 90% viability two (2) years after storage in dry and cold conditions (Eibl et al. 2019a).

The certification of the physical and physiological quality of the seeds is registered in a label that contains identity data, seed origin, purity, germination power, harvest date and weight, following procedures stated in resolution INASE n°42/2000 (INASE 2000) of seed marketing, and specifically for tree seeds, resolution INASE n°256/99.

## 7.4.5 Nurseries and Field Planting of Native Species

As part of our research, silvicultural knowledge is already available for more than forty native species of multiple use (firewood, wood, landscape, restoration, honey, food, medicines, others) (Eibl et al. 2015b, 2016a, 2019c). The nurseries distributed in different regions have received training and advice from personnel working at the Experimental Nursery of Native Species linked to the Seed Bank of the Forestry School of the University of Misiones in Eldorado, Misiones, Argentina. Seedlings of 30–40 cm in height of more than 40 different species, placed in rigid polypropylene containers of 100 to 230 $cm^3$, using composted pine bark substrates and slow-release fertilizer at a rate of 1.5 to 3 $kg/m^3$ were produced for field planting, attending to the development of proper root formation. The nursery time to achieve adequate quality plants is from 6 to 24 months, depending on the species.

Regarding the availability of propagation material, the selected trees present abundant fruiting every 3 years if weather conditions are adequate. Recalcitrant seeds and/or those that are difficult to store, as well as those of rare, vulnerable and/or valuable species such as *Handroanthus* sp., *Aspidosperma polyneuron*, *Myrocarpus frondosus*, *Araucaria angustifolia* and others, are not usually available for nurseries. Under those circumstances these species can be stored in nurseries in the form of living plants grown in rigid containers, for more than 3 years, as a Bank of Plants (Eibl et al. 2017).

Demonstration trials with native tree species in open plantations, enrichment planting in degraded forest, and species associated with agroforestry systems (AFS) form a network of demonstration productive experiences for the region that include more than 40 species of interest. Survival exceeding 70% and 200–300 plants achieved per ha are expected, given an initial implementation of 400 individual plants or more (Eibl et al. 2015a).

The technical certifications at 18 months after plantation allow the producer to access the benefits of forestation offered by National Law n°26,432. These payments can be approximately 500 USA $/ha, depending on whether the trees are planted for enrichment, restoration and/or in pure plantations with native species. According to National Law n°27,487, incentives also vary for medium (20–40 years) and/or long term (40–60 years) restoration projects, with time of tree harvest depending on the species and quality of the desired wood. In addition to the main goals of restoration practices towards conservation of species, these projects also provide ecological and productive benefits including mitigation of climate change and wood harvesting, even at the start of the project (Eibl et al. 2015b).

Results of our recent research attest to the positive impacts of some of these species on the restoration of degraded land. For example, in areas of degraded physical and chemical soil conditions, occupied by invasive pastures, 29 years after the intervention with an initial plantation of 11 native species, there were 62 different tree species represented by 28 families (Eibl et al. 2019b). The propagules that appeared on the site were mainly dispersed by birds, small animals, and wind (Dohman et al. 2019).

The higher growth and survival of species of economic interest, such as *Anadenanthera colubrina*, *Bastardiopsis densiflora, Nectandra lanceolata*, *Balfourodendron riedelianum, Peltophorum dubium* and *Machaerium stipitatum*, underscore their importance as tolerant to degraded soil conditions (Eibl et al. 2019c). These findings suggest the possibility of creating new productive islands of biodiversity from natural regeneration after an initial intervention with a plantation. This can be considered as an alternative in degraded ecosystems where no natural regeneration occurs. This also points to the need for creating an area of propagation material using the trees that grew successfully in the restored land.

Meetings were held at the Selva Misionera Botanical Garden (JBSM) as part of the Argentine Botanic Gardens Network (RAJB 2006), with local farmers, technicians, students, and staff from local institutions. The main purpose was to raise awareness about the environmental benefits of forests and knowledge of species and their multiple uses. These actions resulted in higher local commitment with environmental actions (Eibl et al. 2016a,b). Based on this initiative, the creation of botanical gardens in all the municipalities of the province is encouraged in order to display plants and facilitate the knowledge of species and their current and potential uses for society, in addition to their inherent role in ex-situ conservation (Fig. 7.5).

Several field experiences including native species have been established, with selected seed trees marked at their original site, which will annually provide part of the needed seeds. Several nurseries are already in production, thus ensuring the availability of quality plants for plantations that, in combination with several species of multiple uses, will allow future uses of wood and firewood, which may be marketed under the certification of sustainable management.

**Fig. 7.5** Traceability for certification, from seed origin and plants of *Araucaria angustifolia* to field plantations. (Photo: B. Eibl)

## 7.5 Conclusion

The certification of propagation material obtained from conservation areas, seed banks, nursery plants and tree plantations grown in the field follows a certification protocol for quality control that, under Resolution INASE n°318/18, allows native species to be included in productive systems with a purpose of sustainable management in productive plantations, restoration and/or conservation and to provide income to the owners.

This certification facilitates the care of the sources of seeds (propagation material), enable the traceability of the material used in the form of an eco-label, promotes dissemination and generates information for the work that currently includes productive uses (such as firewood, wood, medicinal, ornamental, honey, food, essences, landscape, restoration, others) for species from local plant diversity.

The establishment of biodiversity islands often necessitates restoration of degraded lands for which it is crucial to have the appropriate plant propagation material that can ensure their viability and diversity. The massive use of native species for multiple purposes is a contribution to the production, restoration, and conservation of their diversity.

## References

Armenteras D, Espelta JM, Rodríguez N, Retana J (2017) Deforestation dynamics and drivers in different forest types in Latin America: three decades of studies (1980–2010). Glob Environ Chang 46:139–147

Barth SR, Eibl BI, Montagnini F (2008) Adaptabilidad y crecimiento de especies nativas en áreas en recuperación del noroeste de la provincia de Misiones. 13as. Jornadas Técnicas Forestales y Ambientales. Facultad de Ciencias Forestales, Universidad Nacional de Misiones, y Estación Experimental Montecarlo del INTA, 5 al 7 de junio Eldorado, Misiones, Argentina. http://www.facfor.unam.edu.ar/la-facultad/secretarias/ciencia-tecnica-y-posgrado/jornadas-tecnicas-forestales-y-ambientales/

COP21 (2015) Convención Marco sobre el Cambio Climático. Acuerdo de las Partes15 de diciembre de 2015, Paris Francia. http://unfccc.int/resource/docs/2015/cop21/spa/l09s.pdf. Accessed 10/2016

COP25 (2019) Argentina entre los países latinoamericanos que se comprometen a restaurar 250 millones de hectáreas al 2050. https://www.argentinaforestal.com/2019/12/10/cop25-argentina-entre-los-paises-latinoamericanos-que-se-comprometen-a-restaurar-250-millones-de-hectareas-al-2050/. Accessed Jan 2020

Day S, Montagnini F, Eibl BI (2011) Effects of native trees in agroforestry systems on the soils and yerba mate in Misiones, Argentina. In: Montagnini F, Francesconi W, Rossi E (eds) Agroforestry as a tool for landscape restoration. Nova Science Publishers, New York, pp 99–112

De Viana ML, Morandini MN, Giamminola EM, Diaz RC (2011) Conservación ex situ, un banco de germoplasma de especies nativas. Lhavet 1(1):35–41

Di Bitetti MS, Placci G, Dietz LA (2003) Una Visión de Biodiversidad para la Ecorregión del Bosque Atlántico del Alto Paraná: Diseño de un Paisaje para la Conservación de la Biodiversidad y prioridades para las acciones de conservación. World Wildlife Fund, Washington DC. https://www.vidasilvestre.org.ar/?3000/Visin-de-Biodiversidad-de-la-Ecorregin-del-Bosque-Atlntico-del-Alto-Paran

Dohmann R, Suárez S, Eibl B, Dummel C, Robledo L, Montagnini F, López M (2019) Caracterización de grupos ecológicos y formas de dispersión en un área de restauración luego de 29 años de plantación de especies nativas en la Provincia de Misiones, Argentina. Actas de las XVIII Jornadas Técnicas Forestales y Ambientales. 17 al 19 de octubre, Eldorado, Misiones, pp 382–384. http://www.facfor.unam.edu.ar/la-facultad/secretarias/ciencia-tecnica-y-posgrado/jornadas-tecnicas-forestales-y-ambientales/

Eibl BI (2006) La estrategia global para la conservación de la biodiversidad al 2010. www.misionesonline.net edición n° 2510, del 21 de mayo del 2006

Eibl BI (2013) Biodiversidad. Semillas y Frutos de la Selva Misionera. Fundación Bosques Nativos Argentinos para la Biodiversidad. Año 3 N°6. 20 pp, www.bosques.org.ar

Eibl BI, Báez MN (2004) Áreas productoras de semillas de especies nativas de la Selva Misionera. Actas de las Jornadas de Mejoramiento Genético. Facultad de Ciencias Forestales. Instituto Nacional de Tecnología Agropecuaria. 14 y 15 de Julio de 2004. Centro de Convenciones. Posadas, Misiones, pp 22–28

Eibl BI, López MA (2017) Estrategias para la conservación y el uso productivo de especies nativas del Bosque Atlántico Interior (Selva Paranaense). Conferencia. Actas del IV Congreso Nacional de Ciencias Agrarias, Facultad de Ciencias Agrarias, Universidad Nacional de Asunción. 19 al 21 de abril, Campus San Lorenzo, Asunción, Paraguay, pp 135–141

Eibl BI, Paredes D, Gómez J (2001) Gigantes de Misiones. Comunicación Revista Forestal Yvyrareta 10:87–88

Eibl BI, Méndez RM, Di Stassi M, Bohren AV, Sosa G, Robledo F (2002) Selección de árboles semilleros en propiedades y reservas de la selva misionera. Trabajo n°60 en formato CD. Actas de las IX Jornadas Técnicas Forestales. 15 al 17 de Mayo de 2002. Facultad de Ciencias Forestales, Universidad Nacional de Misiones, Eldorado, Misiones

Eibl BI, González C, Otegui M, Dreyer N (2012) Protocolos tentativos para la propagación de 20 especies nativas de interés productivo de la selva misionera.15as Jornadas Técnicas Forestales y Ambientales. Facultad de Ciencias Forestales, UNaM – EEA Montecarlo, INTA. 7, 8 y 9 de Junio de 2012, Eldorado, Misiones, Argentina. http://www.facfor.unam.edu.ar/la-facultad/secretarias/ciencia-tecnica-y-posgrado/jornadas-tecnicas-forestales-y-ambientales/

Eibl BI, González C, Otegui M, Dreyer N, Mallozi G, Weinberger V (2013a) Conservación de semillas de especies nativas en bancos de germoplasma. Resumen Actas, p. 45. Trabajo completo n°13CA en CD. Jornadas Científico Tecnológicas 40 Aniversario UNAM, 15–17 mayo 2013, Posadas. Misiones

Eibl BI, González C, Dreyer N, Branco F, Bohren AV (2013b) Indicadores de calidad de plantas en vivero para el establecimiento de ensayos de producción de biomasa para energía con especies nativas. Resumen Actas, pp. 46. Trabajo completo n°14CA en CD. Jornadas Científico Tecnológicas 40 Aniversario UNAM.15 al 17 de mayo de 2013. Posadas. Misiones

Eibl BI, López MA, Bohren AV, Montagnini F, López LN, Barbaro L, Sforza O (2015a) Restauración de áreas mediante plantaciones de especies nativas con fines de biomasa leña y maderables en la Provincia de Misiones. IV Congreso Iberoamericano y del Caribe de Restauración Ecológica. Libro Actas M231, pp 131. SIACRE. 12–16 de Abril 2015, Buenos Aires, Argentina

Eibl BI, Montagnini F, López MA, Montechiesi R, Barth SR, Esterche E (2015b) *Ilex paraguariensis* A. St.-Hil., yerba mate orgánica bajo dosel de especies nativas maderables, una propuesta de producción sustentable. In: Montagnini F, Somarriba E, Murgueitio E, Fassola H, Eibl B (eds) Sistemas Agroforestales. Funciones productivas, socioeconómicas y ambientales. Serie Técnica Informe Técnico 402, CATIE, Turrialba, Costa Rica. Fundación CIPAV. Cali, Colombia, pp 158–177

Eibl BI, López MA, Barbaro L, Sforza O, López J, Branco F, López M, Stadler N, Méndez R, Steitzer N (2016a) Experiencias productivas con especies nativas multipropósito en áreas de restauración ambiental y estrategias para la conservación de remanentes de bosques. Actas XVII Jornadas Tecnicas Forestales y Ambientales. 17 al 19 de agosto 2018. Posadas, Misiones. http://www.facfor.unam.edu.ar/la-facultad/secretarias/ciencia-tecnica-y-posgrado/jornadas-tecnicas-forestales-y-ambientales/

Eibl BI, Rocha SP, González C, González CA, Silveira T, Krauczuk E, Suárez S, Hancherek FA (2016b) El Jardín Botánico Selva Misionera, una herramienta para la concientización ambiental. Actas de las XVII Jornadas Técnicas Forestales y Ambientales. 17 al 19 de octubre. Posadas, Misiones, http://www.facfor.unam.edu.ar/la-facultad/secretarias/ciencia-tecnica-y-posgrado/jornadas-tecnicas-forestales-y-ambientales/

Eibl BI, López MA, Stadler N, López M, González C (2017) Plantaciones productivas de *Araucaria angustifolia* (Bertol.) Kuntze utilizando progênies de dos orígenes de la Provincia de Misiones. Actas del IV Congreso Nacional de Ciencias Agrarias, Facultad de Ciencias Agrarias Universidad Nacional de Asunción. Formato CD. 19 al 21 de abril, Campus San Lorenzo, Asunción, Paraguay, pp 469–472

Eibl BI, González C, Lecanda Prado A, Ortega MN, Rodríguez GA, Kiefl JM (2019a) Germinación de *Aspidosperma polyneuron* Müll. Arg., luego del secado y el almacenamiento de las semillas. Actas de las XVIII Jornadas Técnicas Forestales y Ambientales. 17 al 19 de octubre, Eldorado, Misiones, pp 319–321. http://www.facfor.unam.edu.ar/la-facultad/secretarias/ciencia-tecnica-y-posgrado/jornadas-tecnicas-forestales-y-ambientales/

Eibl BI, López MA, Montagnini F, Suárez S, Dohman R, Dummel CJ, Robledo L (2019b) Diversidad de especies arbóreas luego de 29 años de la plantación de especies nativas en un área de restauración en la Provincia de Misiones, Argentina. Actas de las XVIII Jornadas Técnicas Forestales y Ambientales. 17 al 19 de octubre, Eldorado, Misiones, pp 307–309, http://www.facfor.unam.edu.ar/la-facultad/secretarias/ciencia-tecnica-y-posgrado/jornadas-tecnicas-forestales-y-ambientales/

Eibl BI, López MA, Montagnini F, Suárez S, Dohman R, Dummel C, Robledo L (2019c) Crecimiento de especies arbóreas luego de 29 años de la plantación de especies nativas en un área de restauración en la Provincia de Misiones, Argentina. Actas de las XVIII Jornadas Técnicas Forestales y Ambientales. 17 al 19 de octubre, Eldorado, Misiones, pp 310–312. http://www.facfor.unam.edu.ar/la-facultad/secretarias/ciencia-tecnica-y-posgrado/jornadas-tecnicas-forestales-y-ambientales/

FAO (2018) The state of the World´ s forests-forest pathways to sustainable development. http://www.fao.org/documents/card/en/c/I9535EN/

FAO (2020) Integrating the SDGs: showing contributions by forests and trees to the 2030 Agenda. Sustainable Forest Management (SFM) Toolbox. http://www.fao.org/sustainable-forest-management/toolbox/sfm-home/detail-events/en/c/1145494/

Flora del Conosur (2020) Instituto de Botánica Darwinion. http://www.darwin.edu.ar/Proyectos/FloraArgentina/BuscarEspecies.asp. Accessed on Jan 2020

Galindo-Leal C, Câmara IDG (2003) Atlantic forest hotspot status: an overview. In: Galindo-Leal C, IDG C (eds) Conservation International/Center for Applied Biodiversity Science. The Atlantic Forest of South America: biodiversity status, threats, and outlook. Island Press, Washington DC, pp 3–11

Gartland HM, Bohren AV (2008) Consideraciones sobre la biodiversidad forestal a nivel de especies arbóreas, en la selva paranaense de Misiones Argentina. Revista Forestal Yvyrareta 15:39–49

Gibbs HK, Ruesch AS, Achard F, Clayton MK, Holmgren P, Ramankutty N, Foley JA (2010) Tropical forests were the primary sources of new agricultural land in the 1980s and 1990s. Proc Natl Acad Sci U S A 107:16732–16737

Gold K, León-Lobos P, Way M (2004) Manual de recolección de semillas de plantas silvestres para conservación a largo plazo y restauración ecológica. Instituto de Investigaciones Agropecuarias, Centro Regional de Investigación Intihuasi, vol 110. Boletín INIA, La Serena, Chile, p 62

González C, Eibl BI, Dreyer NB (2015) Almacenamiento de semillas de *Cabralea canjerana* (Vell.) Mart. In: Actas versión Digital. Grupo forestales. Congreso Nacional de viveros cítricos, forestales y ornamentales. INASE, 4 al 6 de agosto de 2015, Posadas, Misiones

GSPC (2020). Global Strategy for Plant Conservation. https://www.plants2020.net

INASE (1999) Resolución n°256/99. Normas para la certificación, producción e importación de semillas de especies forestales. https://www.argentina.gob.ar/normativa/nacional/resolucion256-1999

INASE (2000) Resolución n°42/2000. Inscripción en el Registro Nacional de Comercio y Fiscalización de semillas y proceso de rotulación de semillas. https://www.argentina.gob.ar/normativa/nacional/resolucion42-2000

INASE (2018) Resolución n°318/18. Registro de Área Productora de Semillas de Especies Nativas (APSEN) https://www.argentina.gob.ar/normativa/nacional/resolucion318-2018

ISTA (2019) International Rules for Seed Testing. International Seed Testing Association (ISTA). Zürichstr. 50, CH-8303. Bassersdorf, Switzerland. 300 pp

IUCN (2020a) Red list of threatened species. https://www.iucn.org/resources/conservation-tools/iucn-red-list-threatened-pecies#RL_categories. Accessed Feb 2020

IUCN (2020b) Soluciones basada en naturaleza. UICN Sur. https://www.iucn.org/es/comisiones/commission-ecosystem-management/soluciones-basadas-en-la-naturaleza. Accessed Jan 2020

Keller HA (2007) *Solanum vaillantii* (Solanaceae), nueva cita para la flora Argentina. Bonplandia 16(3–4):275–278

Ley Nacional n°26331 (2007) Bosques Nativos. Presupuestos Mínimos de Protección Ambiental de los Bosques Nativos. Boletín Oficial Diciembre 26/2007. SAIJ, Sistema Argentino de Información Jurídica. Ministerio de Justicia, Buenos Aires, Argentina

Ley Nacional n°27487 (2008) Prorroga y Reforma de la Ley n°25080. Decreto Argentina, Boletín Oficial. 29/12/2008. Resolución134/2019-APN-SECAGYP. SAIJ Sistema Argentino de Información Jurídica. Ministerio de Justicia. Buenos Aires, Argentina

Ley Provincial de Energías Renovables XVI n°97 (ex Ley n°4439) (2010) Marco regulatorio de los recursos energéticos renovables. Provincia de Misiones. www.diputadosmisiones.gov.ar/digesto_juridico

Ley Provincial XVI N° 105 (ex ley n°3260) (2010) Provincia de Misiones el ordenamiento de los Bosques Nativos y los mecanismos a implementar en la para la Conservación, de los Bosques Nativos y el Régimen de Promoción de Manejo Sostenible. www.diputadosmisiones.gov.ar/digesto_juridico

Ley Provincial XVI n°29 (ex Ley n°2932) (2010) Áreas Naturales Protegidas. Provincia de Misiones. www.diputadosmisiones.gov.ar/digesto_juridico

Ley Provincial XVI N°47 (ex ley n°3337) (2010). Conservación y aprovechamiento sostenible de la diversidad biológica y sus componentes. www.diputadosmisiones.gov.ar/digesto_juridico

Ley Provincial XVI n°53 (ex Ley n° 3426) (2010) Bosques protectores de cursos de agua y pendientes. www.diputadosmisiones.gov.ar/digesto_juridico

Ley Provincial XVI n°60 (ex Ley n°3631) (2010) Área Integral de Conservación y Desarrollo Sustentable Corredor Verde. Provincia de Misiones. http://diputadosmisiones.gov.ar/digesto_juridico

Ley Provincial XVI N°98 (ex ley n°4464) (2010) Banco Provincial de Germoplasma de Misiones. http://diputadosmisiones.gov.ar/digesto_juridico/documentos

López MA, Eibl BI, Palavecino J (2013) Valoración de especies nativas en áreas de restauración ambiental, Provincia de Misiones, República Argentina. En: Educacao Ambiental: Premisa inafastavel ao desenvolvimento económico sustentavel. Editores Joao Almeida e Da Silva Almeida. Ed. Lumen Juris, Rio de Janeiro pp 371–391

Ministério da Agricultura (2009) Regras para análise de sementes. Ministério da Agricultura, Pecuária e Abastecimento, Secretaria de Defesa Agropecuária, Mapa/ACS, Brasília, Brasil. 399 pp

Mittermeier RA, Turner WR, Larsen FW, Brooks TM, Gascon C (2011) Global biodiversity conservation: the critical role of hotspots. In: Zachos FE, Habel JC (eds) Biodiversity hotspots. Distribution and protection of conservation priority areas. Springer, Berlin, Heidelberg, pp 3–22

Montagnini F, Cusack D, Petit B, Kanninen M (2005) Environmental services of native tree plantations and agroforestry systems in Central America. J Sustain For 21(1):51–67

Montagnini F, Eibl BI, Fernandez R (2006) Rehabilitation of degraded lands in Misiones, Argentina. Bois et Forests des Tropiques 288:51–65

Montagnini F, Levin B, Berg KE (2022) Introduction. Biodiversity Islands. Strategies for conservation in human-dominated environments. In: Montagnini F (ed) Biodiversity islands: strategies for conservation in human-dominated environments. Topics in Biodiversity and Conservation, Springer, Cham, pp 1–35

Niella F, Eibl B, Rocha P, Dummel C, Keller HA (2016) Estrategia para la conservación, domesticación y rescate de recursos fitogenéticos nativos a la selva paranaense. XVII Jornadas Técnicas Forestales y Ambientales, Posadas, Misiones. http://www.facfor.unam.edu.ar/modules/uploads/2017/03/Acta_de_Resumenes_XVII_Jotefa_2016.pdf

Niella F, Eibl B, Keller H, Maiocco D, Rocha P, Vega V (2017) Red de árboles y áreas semilleras para la conservación, domesticación y rescate de recursos fitogenéticos nativos de la Selva Paranaense. SILVA N°17/2013. Conservación de la Biodiversidad en Paisajes Productivos. GEF 090118. Ministerio de Agricultura, Ganadería y Pesca. https://www.magyp.gob.ar/sitio/areas/proyectos_forestales/biosilva/SILVA%2017%20Niella.pdf

Ocampo T, Eibl BI, González C (2015) Indicadores de calidad a partir de variables morfofisiológicas en frutos, semillas y plantas de *Cedrela fissillis* Vell. En: Actas versión Digital. Grupo forestales. Congreso Nacional de viveros cítricos, forestales y ornamentales. INASE, 4 al 6 de agosto de 2015, Posadas, Misiones

Otegui MB, Ojeda DS, Totaro ME, Eibl BI, Valdés MB, Mallozzi GY (2016) Comportamiento y Longevidad de *Aspidosperma polyneuron* Müll. Arg. frente al Ultrasecado. Seed conservation of palo-rosa. RECyT 18(25):15–19

RAJB (2006) Red Argentina de Jardines Botánicos. Boletín Año 2(3) https://inta.gob.ar/documentos/red-argentina-de-jardines-botanicos-rajb

Rodríguez G, González C, Eibl BI, González G, Grance L (2019) Viabilidad de semillas de *Plinia rivularis* (Cambess.) Rotman durante el almacenamiento. Actas de las XVIII Jornadas Técnicas Forestales y Ambientales, 17 al 19 de octubre, Eldorado, Misiones, pp 328–330. http://www.facfor.unam.edu.ar/la-facultad/secretarias/ciencia-tecnica-y-posgrado/jornadas-tecnicas-forestales-y-ambientales/

Schiaffino K, Bertolini MP (2016) Reservas Naturales Privadas, avances de un trabajo en Red. Conferencia. Actas de las XVII Jornadas Técnicas Forestales y Ambientales, 17–19 de agosto, Posadas, Misiones. http://www.reservasprivadas.org.ar. http://www.facfor.unam.edu.ar/la-facultad/secretarias/ciencia-tecnica-y-posgrado/jornadas-tecnicas-forestales-y-ambientales/

Thomas E, Jalonen R, Loo J, Boshier D, Gallo L, Cavers S, Bordacs S, Smith P, Bozzano M (2014) Genetic considerations in ecosystem restoration using native tree species. For Ecol Manage 333: 66–75

Ugarte-Guerra J, Alemán F, Nieto V, Prado L, Galvan O, Silva I, Ohashi S, García W (2010) Estudio de las cadenas de abastecimiento de germoplasma forestal en cinco países amazónicos. World Agroforestry Centre (ICRAF), Lima, Perú

Young CF (2003) Socieconomic causes of deforestation in the Atlantic forest of Brazil. In: Conservation International/Center for Applied Biodiversity Science (author). In: Galindo-Leal C, de Gusmao CI (eds) The Atlantic Forest of South America: biodiversity status, threats, and outlook. Island Press, Washintong/Covelo/London, pp 103–117

# Part III
# Biodiversity Islands Across the Globe: Case Studies

# Chapter 8
# Islands of Forests Among Savannas: Key Elements for Conservation and Production in the Paraguayan Humid Chaco

**Rafaela Laino, Karim Musalem, Luis Domingo Laino, Andrea Caballero-Gini, Diego Bueno-Villafñae, Lidia Aranda, Alberto Esquivel, Marcela Ferreira Riveros, Lía Romero Nardelli, Nicolás Cantero, and Rebeca Irala**

R. Laino (✉) · L. D. Laino · L. R. Nardelli
American Chaco Research Center, Benjamín Aceval, Paraguay
e-mail: info@chacoamericano.org; http://www.chacoamericano.org

K. Musalem
American Chaco Research Center, Benjamín Aceval, Paraguay

WWF Paraguay, Asunción, Paraguay
http://www.wwf.org.py

A. Caballero-Gini · D. Bueno-Villafñae
American Chaco Research Center, Benjamín Aceval, Paraguay

Instituto de Investigación Biológica del Paraguay, IIBP, Asunción, Paraguay

L. Aranda
American Chaco Research Center, Benjamín Aceval, Paraguay

Universidad Nacional de Asunción, San Lorenzo, Paraguay

A. Esquivel
WWF Paraguay, Asunción, Paraguay

Wildlife Paraguay, Luque, Paraguay

M. F. Riveros
American Chaco Research Center, Benjamín Aceval, Paraguay

Instituto de Investigación Biológica del Paraguay, IIBP, Asunción, Paraguay

Guyra Paraguay, Asunción, Paraguay
http://www.guyra.org.py

N. Cantero
Wildlife Paraguay, Luque, Paraguay

Universidad Nacional de Córdoba, Córdoba, Argentina

R. Irala
Wildlife Paraguay, Luque, Paraguay
http://www.wildlife.com.py



F. Montagnini (ed.), *Biodiversity Islands: Strategies for Conservation in Human-Dominated Environments*, Topics in Biodiversity and Conservation 20,
https://doi.org/10.1007/978-3-030-92234-4_8

**Abstract** This chapter provides information on islands of forests immersed in flood-prone savannah-dominated landscapes for biodiversity conservation and productive activities. Our study area in the Paraguayan Chaco is in a region dedicated to low intensity cattle rearing, with native grasslands as forage within a mosaic of natural plant formations. The region is highly diverse due to the variety of plant formations and the great amount of wild fauna. We present results of research projects conducted by the American Chaco Research Center, a Paraguayan research initiative. These results suggest that extensive or semi-extensive ecosystem-based livestock production could be an opportunity to maintain ecosystem services, including biodiversity, in the region via a production system that is highly compatible with the conservation of forest species. Adverse environmental conditions for production (such as floods from rain and periodic floods of the Paraguay River) underscore the importance of maintaining the regulating functions of natural ecosystems, which leads to an opportunity for the conservation of threatened biodiversity. The mosaic of forest islands, flood-prone palm savannas, and wetlands is fundamental for water quality and hydrological regulation as it reduces the intensity of the effects of flooding on neighboring areas, as well as for carbon sequestration. Inclusion of live fencing and the maintenance of continuous areas of intact or lightly disturbed vegetation can increase habitat connectivity, provide a barrier to manage fires, and serve as a refuge for animals when fires occur. These native ecosystems are at high risk of disappearing due to the intensification of agriculture and livestock production, as well as the expansion of urban areas, so strategies for their conservation, including tailor-made incentives are needed.

**Keywords** Biodiversity conservation · Ecosystem services · Floodable forest · Habitat connectivity · Livestock farming · Wildlife

## 8.1 Introduction

### *8.1.1 Forests as Islands Immersed in Flood-Prone Savannas*

This chapter describes the relevance of forest islands as part of a mosaic of savannas, palm groves and wetlands (an image of the study area is shown in Fig. 8.1). The presence of any of these formations is determined mainly by the geomorphology of its terrain. The landscape includes plains and depressions where water settles for different periods of time according to weather conditions (Mereles et al. 2020a). Wetlands occupy the lowest areas, generally with permanent waters where species linked to water develop with different life forms: floating, submerged (free or not) and rooted in the mud of the bottom. Some of the plant species recorded in the wetlands of the study area are *Cyperus giganteus, Typha domingensis* and *Sagittaria montevidensis*, with *Cyperus giganteus* being the most abundant and frequent (Macedo 2018). Palm groves are monotypical formations that can flood frequently.

**Fig. 8.1** Image of the study site in the Humid Chaco ecoregion. Dense forest island immersed in extensive savannas. (Photo: Gianfranco Mancusi)

The characteristic species is the native palm locally known as Karanday (*Copernicia alba*), accompanied by a rich herbaceous stratum, whose density varies according to the presence of water (Mereles and Rodas 2014; Mereles et al. 2013).

The landscape consists of dense forest islands immersed in savannas, palm groves and wetlands, resulting in different forest types: Dense Subtropical Forest, which occurs naturally on islands associated with palm groves in the floodplain of the Paraguay River (Pérez de Molas 2015); Dense Mesoxerophytic Forest, with Red Quebracho (*Schinopsis balansae*) being the dominant tree species (Mereles et al. 2020a); and Riparian Forest, which develops adjacent to the riverbeds forming strips 50–100 meters wide (Maturo et al. 2005; Peña-Chocarro et al. 2006). Main arboreal tree species found in the area are shown in Table 8.1.

This chapter is based on field studies that were conducted in rangelands located on the right bank of the Paraguay River (Fig. 8.2), where average rainfall is 1200 mm/year and the average temperature is 24 °C (Mereles et al. 2013). During some years, with the occurrence of the El Niño phenomenon, intense rainfall causes generalized flooding. In other years, drought is accentuated and prolonged (Ginzburg and Adámoli 2006; Junk et al. 2013). The study site itself has experienced both of these extremes (most recently in 2015 and 2019): over half of its surface area remained flooded during one part of the year and then suffered an extreme drought in the same year. The pulsing water-level in the rainy season and the pronounced dry and wet periods create an aquatic-terrestrial transition zone where important ecological processes occur (Mereles et al. 2020b). Most of the studies were conducted in nearby ranches, which are located within the Key Biodiversity Area 22 La Rafaela (Cartes and Clay 2009), and which have been identified as potential birding tour areas (Mamede et al. 2019) and as important corridors for connectivity of the Great American Chaco (Mereles et al. 2020b).

**Table 8.1** Forest types and tree species recorded in dense forest islands of the study area

| Forest type | Tree species | Source |
|---|---|---|
| **Dense subtropical forest** | *Peltophorum dubium, Enterolobium contortisiliquum, Ficus enormis, Ocotea diospyrifolia, Sapium haematospermum, Gleditzia amorphoides, Guazuma ulmifolia, Chloroleucon tenuiflorum, Handroanthus heptaphyllus, Syagrus romanzoffiana, Copernicia alba.* | El Raiss (2014) |
| **Dense mesoxerophytic forest** | *Schinopsis balansae, Rollinia emarginata, Aspidosperma quebracho blanco, Forsteronia sp., Tabernaemontana catharinensis, Syagrus romanzoffiana, Acrocomia aculeata, Copernicia alba, Tabebuia nodosa, Handroanthus heptaphyllus, Tabebuia aurea, Cordia americana, Carica papaya, Cecropia pachystachya, Celtis sp., Terminalia triflora, Sapium longifolium, Gleditsia amorphoides, Parapiptadenia rigida, Albizia niopoides, Enterolobium contortosiliquum, Prosopis affinis, Inga uraguensis, Peltophorum dubium, Copaifera langsdorfii, Pterogyne nitens, Vitex megapotamica, Ocotea diospyrifolia, Trichilia catigua, Trichilia pallida, Maclura tinctoria, Sorocea sprucei, Psidium guajava, Myrcianthes pungens, Genipa americana, Calycophyllum multiflorum, Zanthoxylum petiolare, Zanthoxylum riedelianum, Casearia sylvestris, Diplokeleba floribunda, Sapindus saponaria, Chrysophyllum marginatum, Guazuma ulmifolia, Seguieria paraguayensis, Ruprechtia laxiflora, Phyllostylon rhamnoides.* | Lubián (2014) |
| **Riparian forest** | *Ocotea diospyrifolia, Lonchocarpus sp., Terminalia triflora, Celtis sp., Peltophorum dubium, Nectandra angustifolia, Copernicia alba, Guazuma ulmifolia, Zanthoxylum petiolare, Machaonia spinosa, Xylosma venosa, Inga uraguensis, Vitex sp., Genipa americana, Pouteria glomerata, Albizia inundata, Chrysophyllum gonogocarpum, Zigia sp., Vitex megapotamica, Chrysophyllum marginatum, Enterolobium contortisiliquum, Sorocea sprucei, Myrsine sp.* | Macedo (2018) |

The chapter begins with a description of the study site explaining why these forests exist as islands immersed in flood-prone savannas. The next part focuses on ecosystem services such as carbon sequestration and water quality and describes the native fauna housed within the forest islands and surroundings. In the following sections, we present information about the role that these biodiversity islands play in functional connectivity and as fire-breaks in the savanna. Finally, we emphasize the benefits of forest islands to cattle ranching in native grasslands and the challenges for their conservation.

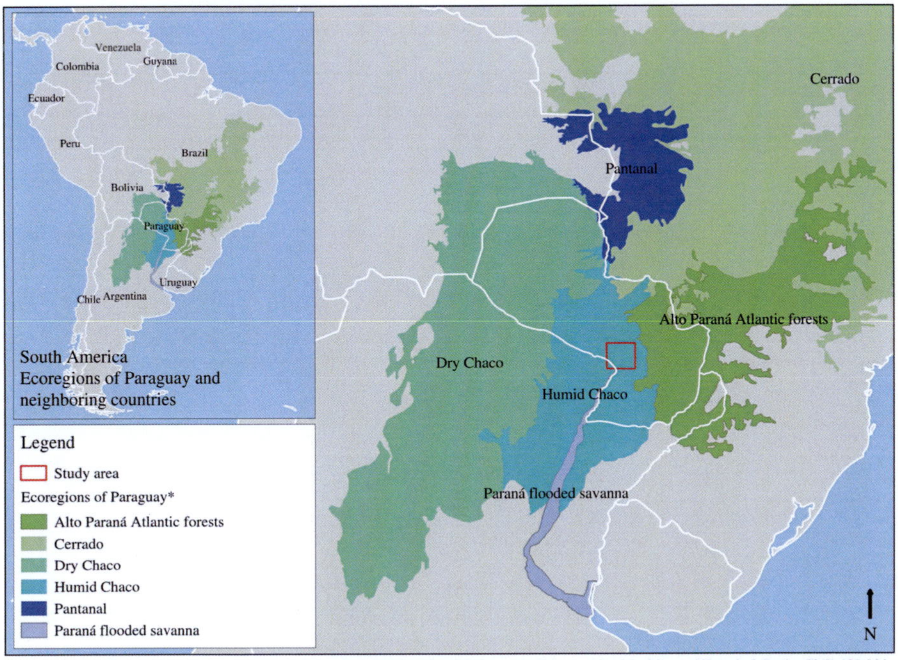


* Olson, D. M., Dinerstein, E. 2002. The Global 200: Priority ecoregions for global conservation. Annals of the Missouri Botanical Garden 89(2):199-224.

**Fig. 8.2** Study site in the Humid Chaco ecoregion. Landscape is dominated by naturally occurring savannas, palm groves, wetlands and dense forests

## 8.2 Ecosystem Services Evaluated in the Study Site

### *8.2.1 Carbon Sequestration and Water Quality*

Measurements of carbon storage were made in the dense forest islands, in the surrounding savannas and palm groves, and in the wetland soils. Lubián (2014) determined 254 tons of $CO_2$e/ha in the forest and 24.5 tons of $CO_2$e/ha in the palm groves and grasslands. Another evaluation focusing on the forest islands (Boródn 2015) calculated 175.8 tons $CO_2$e/ha (following IPCC 2005 report) and 291 tons $CO_2$e/ha (following Sato et al. 2014 equation) for the aboveground biomass, and 103.9 tons/ha for forest carbon soil. For wetland soil, Brun (2013) reported average values of 3.15 tons of $CO_2$/ha.

Wetlands are among the most productive primary ecosystems on the planet, depending on the hydrological regime, i.e., how often they are flooded and how long they remain flooded (Kandus et al. 2010). Water originates in the lowest areas, between savannas, marshlands and swamps with high productivity of the herbaceous stratum (Benzaquen et al. 2017). The herbaceous cover of the wetlands decreases water speed and facilitates the sedimentation and retention of suspended materials, thus improving water quality (Kandus et al. 2010). Wetlands of the study area were of 'good' qualitative conservation status according to the ECELS index[1] evaluated by Ferreira (2018) and Brun (2013). In addition, exploratory analysis of water quality (including pH, Electrical Conductivity, Suspended Solids, Nitrate, Total Phosphorus, Total Nitrogen, Chloride, Sulphate, Biochemical Oxygen Demand, Chemical Oxygen Demand, Coliforms and water temperature) found no relevant pollution in wetlands and rivers of the study area, except for specific sites where the flow decreased in the dry season (Chaparro 2014, PINV 15-143 2018).

In addition to maintaining water quality, the mosaic of forest islands, flood-prone palm savannas and wetlands is fundamental for hydrological regulation as it reduces the intensity of the effects of flooding on neighboring areas. Although these ecosystems do not prevent flooding, they reduce the river's flood peaks, retain excess runoff after rainfall and release it slowly afterwards, and encourage aquifer recharge (Kandus et al. 2010).

### *8.2.2 Wildlife Housed in the Forest Islands and Surroundings*

The study of amphibians, reptiles, mammals, and birds was carried out inside of the different forest islands (FI) and surrounding savannas (S), palm groves (PG) and wetlands (W) through different sampling methods.

Amphibian surveys were made using three different sampling techniques, as these vary in their effectiveness depending on intrinsic species characteristics (Ali et al. 2018). The three sampling methods used from August 2017 to June 2018 were visual encounter surveys (Crump and Scott 1994), pitfall traps with drift fences (Corn 1994) and PVC pipes of 40 mm of diameter as refuge for tree frog species (Boughton et al. 2000).

All together we registered 2449 individuals corresponding to 29 species included in the families: Bufonidae, Hylidae, Leptodactylidae, Phyllomedusidae, Microhylidae and Odontophrynidae, which represent 48% of all the species registered in the Humid Chaco ecoregion (Brusquetti and Lavilla 2006; Frost 2020). Some of these species are shown in Fig. 8.3. Regarding their conservation status, all species found are categorized as Least Concern (LC) at international (IUCN 2019) and national levels (Motte et al. 2019), except for the Rio Grande Dwarf Frog (*Physalaemus riograndensis*) (Motte et al. 2019). This species is considered to be in the data deficient category (DD) at the national level due to its scarce records (Table 8.2), which are mostly from the southern region of the country (Brusquetti and Lavilla 2006; Motte et al. 2019; Frost 2020). In addition, we obtained the first departmental record for the White spotted Humming frog (*Chiasmocleis albopunctata*),

[1] ECELS (Estado de Conservación de Ecosistemas Lénticos Someros Index) is a methodological tool used to determine the ecological status of wetlands, which was developed by Agencia Catalana del Agua (ACA 2004).

**Fig. 8.3** Amphibian species registered at the study area. A. Monkey frog (*Pithecopus azureus*). B. Polka-dot Tree Frog (*Boana punctata*). C. Male (on the top) and female of Dwarf tree-frog (*Dendropsophus nanus*). D. Uruguay Arlequin Frog (*Lysapsus limellum*). E. White lipped-tree frog (*Trachycephalus typhonius*). F. Two colored Oval frogs (*Elachistocleis bicolor*). G. Common

a rare species mainly due to its fossorial habits,[2] size and mimetic coloration. Only one individual was registered during the study time frame through the pitfall trap method. Nevertheless, the record is interesting since it extends the distribution range of the species 57 km northwest from the nearest locality at the Surubi'i Urbanization, Central Department (Aquino et al. 2004; Brusquetti and Netto 2008).

Through pitfall traps, visual encounter surveys and camera traps, we registered a total of 12 reptile species including snakes, lizards of the families Tupinambidae and Teiidae, and the Jacare caiman (*Caiman yacare*). Individuals of Jacare caiman were actively searched for at night, and we found them mostly isolated and using different water body types, including cattle ponds. When isolated sections of the wetlands were searched, we found juvenile aggregations along with adults, which indicates a preference in habitat use and reproductive success. Further research will improve our knowledge about the species habitat use, reproductive sites and its estimated density in the area, which harbors great potential for reproduction and conservation of Jacare caiman. Another great lizard found was the Black-and-white Tegu (*Salvator merianae*), which was recorded frequently with camera traps.

Fifteen mammal species were registered using camera traps that were placed in different types of environments in the area for three years. We obtained records of species categorized as vulnerable at the national level (APM and SEAM 2017), mainly as a consequence of fragmentation and habitat conversion associated with agricultural and livestock activities, illegal hunting, and road run overs, among others (Table 8.2). We found the Giant anteater (*Myrmecophaga tridactyla*) and even the Maned wolf (*Chrysocyon brachyurus*), which is typically an elusive species (Mujica 2014). Furthermore, we obtained records of Puma (*Puma concolor*) through camera traps. Although this species is categorized as least concern (APM and SEAM 2017), it still has conservation problems and its populations are decreasing (Nielsen et al. 2015). The Azara's night monkey (*Aotus azarae*), the Black and Gold howler monkey (*Alouatta caraya*) and the four-eyed gray opossum (*Philander opossum*) were also registered through night sightings. Presence of Jaguar (*Panthera onca*) was confirmed with camera traps on November 2020, after this chapter closed edition. Some of the registered mammal species are shown in Fig. 8.4.

The Humid Chaco harbors more than 430 species of birds, which represents ~60% of the avifauna of Paraguay (Del Castillo 2019). Specifically, in the Key Biodiversity Area 22 La Rafaela and nearby ranches, numerous census and surveys have been carried out by several ornithologists and researchers throughout the last decades. We compiled records obtained during the 2000–2020 period from available

**Fig. 8.3** (continued) Lesser Escuerzo (*Odontoprhynus americanus*). H. Klappenbachmmmmms Red-bellied Toad (*Melanophryniscus klappenbachi*) in amplexus. (Photographs: A - D, F: A. Caballero-Gini. E, G, H: M. Ferreira)

[2]Species that are adapted to digging and living in burrows.

**Table 8.2** Fauna species of national and global conservation concern, with details of habitat and breeding status

| Class | Common name | Scientific name | Habitat and ecology | Conservation status Paraguay | Conservation status IUCN | Breeding status |
|---|---|---|---|---|---|---|
| **Amphibia** | Rio Grande dwarf frog | *Physalaemus riograndensis* | S | DD | LC | – |
| **Birds** | Greater Rhea | *Rhea americana* | FI, S, PG | – | NT | BR |
| | Bare-faced curassow | *Crax fasciolata* | FI | Threatened | VU | BR |
| | Turquoise-fronted parrot | *Amazona aestiva* | FI, S, PG | – | NT | BR |
| | Bearded Tachuri | *Polystictus pectoralis* | S, PG | Threatened | NT | BR |
| | Sharp-tailed tyrant | *Culicivora caudacuta* | S | Endangered | VU | BR |
| | Dinellimmmmms Doradito | *Pseudocolopteryx dinelliana* | W | – | NT | AM |
| | Strange-tailed tyrant | *Alectrurus risora* | S | Endangered | VU | BR |
| | Dark-throated seedeater | *Sporophila ruficollis* | S | – | NT | BN, AM |
| | Rufous-rumped seedeater | *Sporophila hypochroma* | S | – | NT | BN, AM |
| **Mammalia** | Giant anteater | *Myrmecophaga tridactyla* | FI, W | Threatened | VU | – |
| | Maned wolf | *Chrysocyon brachyurus* | S, PG | Threatened | NT | – |
| | Neotropical otter | *Lontra longicaudis* | W | Least concern | NT | – |
| | Azarammmmms agouti | *Dasyprocta azarae* | FI | Least concern | DD | – |

**Habitat and ecology:** S, Savannas; FI, Forest islands; PG, Palm groves; W, Wetlands
**Conservation status Paraguay** (APM and SEAM 2017; MADES 2019; Motte et al. 2019): Threatened, "*Amenazada de extinción*"; Endangered, "*En peligro de extinción*"; DD, Data Deficient
**Conservation status IUCN** (IUCN 2019): LC, Least Concern; NT, Near Threatened; VU, Vulnerable; DD, Data Deficient
**Breeding status** (Guyra Paraguay 2004): BR, Breeding permanent resident; BN, Breeding resident but northern austral migrant; AM, Austral migrant

**Fig. 8.4** Mammals registered in the Humid Chaco (Paraguay) through camera traps and visual encounters. A. Giant anteater (*Myrmecophaga tridactyla*) carrying its cub. B. Gray Brocket (*Mazama gouazoubira*). C. Puma (*Puma concolor*). D. Lesser Capybara (*Hydrochoerus hydrochaeris)*. E. Collared peccary (*Pecari tajacu*). F. Neotropical otter (*Lontra longicaudis*). G. South American Coati (*Nasua nasua*). H. Gold howler monkey (*Alouatta caraya*). (Photographs: A - G: camera traps installed during 2017 to 2018. H: K. Musalem)

data in the eBird online platform (eBird 2020a, b), Guyra Paraguay's Biodiversity Database (www.guyra.org.py) and personal checklists, obtaining a total of 249 species included in 51 families, which represents 35% of the species of Paraguay and 58% of the species registered at the Humid Chaco ecoregion.

Habitat preferences of these species, based on our records in the area, have been identified for four main ecosystems: forest islands (FI), savannas (S), palm groves (PG) and wetlands (W). A total of 125 species (51% of the total species of the area) use FI as one of their habitats, and 82 species (33%) depend mainly on this habitat.

Nine bird species are globally threatened or near threatened (IUCN 2019), four of which are also nationally threatened or endangered (MADES 2019). Bare-faced curassow (*Crax fasciolata*), Sharp-tailed tyrant (*Culicivora caudacuta*) and Strange-tailed tyrant (*Alectrurus risora*) are categorized as Vulnerable (IUCN 2019), and the former depends strictly on forest island habitats in the area (Table 8.2). At the national level, the Bare-faced curassow and the Bearded tachuri (*Polystictus pectoralis*) are categorized as Threatened species, while the Sharp-tailed and Strange-tailed tyrant are Endangered species. Moreover, four species are endemic of the Chaco (Guyra Paraguay 2004): Chaco chachalaca (*Ortalis canicollis*), Cream-backed woodpecker (*Campephilus leucopogon*), Dinelli's doradito (*Pseudocolapterix dinelliana*) and Black-capped warbling-finch (*Microspingus melanoleucus*). Forty-one species are migratory. Six of those are austral migrants, found mainly during austral winter. Twenty-two are northern austral migrants, which breed in Paraguay but are less abundant or absent during the winter. Seven are southern austral migrants, breeding also in Paraguay, but increasing in numbers during the winter. Some bird species observed in the study area are shown in Fig. 8.5.

## 8.3 Role of Biodiversity Islands in Functional Connectivity at Local and Regional Levels

At a regional level, a recent study identified our study area as part of priority biological corridors relevant for connectivity of the Great American Chaco, an ecoregion that extends through Paraguay, Argentina and Bolivia (Mereles et al. 2020b). The study highlights the importance of biological corridors to maintain a long-term vision of biodiversity, which maintains the connection between key areas for conservation. Loss of continuous areas can lead to changes in the structure and function of the remaining fragments (Lindenmayer and Fischer 2006). One of the problems caused by habitat reduction and fragmentation is a contraction in population size, along with increases in inbreeding and the consequent decrease of genetic diversity (Shaffer 1990).

Although there is considerable uncertainty regarding how fast species respond to habitat loss, and how time-delayed responses vary in space (Semper-Pascual et al. 2018), we found evidence linked to population reduction sizes and fragmentation of

**Fig. 8.5** Birds registered in the Humid Chaco, Paraguay. A. Scarlet-headed Blackbird (*Amblyramphus holosericeus*). B. Strange-tailed Tyrant (*Alectrurus risora*). C. Crimson-crested Woodpecker (*Campephilus melanoleucos*). D. Black-collared Hawk (*Busarellus nigricollis*). E. Red-billed Scythebill (*Campylorhamphus trochilirostris*). F. Toco toucan (*Ramphastos toco*). G. Blue-crowned Trogon (*Trogon curucui*). (Photographs: A - B: N. Cantero. C - E, G: A. Esquivel. F: A. Merenciano)

habitat at Chaco and other regions (e.g., Gómez Fernández et al. 2016; Crooks et al. 2017; Zastavniouk et al. 2017; Pereyra et al. 2019). For example, according to Rodrigues et al. (2008), the main causes of decline of the Giant anteater populations are the reduction, deterioration and fragmentation of habitats. Habitat loss may lead to a decrease in population size and isolation among remnant populations. In this sense, Collevatti et al. (2007) warned that the population of *Myrmecophaga tridactyla* in Emas National Park (Brazil) has a low level of genetic diversity and a high level of inbreeding.

The Paraguayan Chaco is undoubtedly highly diverse due to the variety of plant formations and the great amount of wild fauna that remains (Mereles et al. 2020b). At a local level, the Humid Chaco ecoregion borders the Dry Chaco and the Pantanal to the North, and the Cerrado and the Alto Paraná Atlantic Forest to the Southeast (Rumbo 2010), therefore, the conservation of native ecosystems could benefit the fauna exchange with the other four ecoregions present in Paraguay. In both Humid Chaco and Pantanal ecoregions, the wooded formations of *Schinopsis balansae*, an emblematic tree species with a high content of natural tannins, are also found. These vegetation types constitute one of the most diverse ecotonal formations in the Great Chaco, where plant species converge from the Atlantic Forest, the Cerrado, the Amazon and the Dry Chaco, among others (Mereles et al. 2020b).

In the study site, El Raiss (2014) examined the functional connectivity of forest islands for the Black-and-gold Howler Monkey (*Alouatta caraya*) within the native silvopastoral system (i.e., the cow-calf system where animals graze on savannas and are interspersed with mixed native forests). The author's findings show an Equivalent Connected Area (ECA) of 79%, indicating a good connectivity condition of forest islands in the landscape. The author also points out that 10 to 15% removal of forest islets would drastically affect this connectivity due to this species' dependence on forest islets. Thus, the study concluded that the current production system is highly compatible with the conservation of this species.

In addition, the inclusion of live fencing, which is not a common practice in the region, can increase habitat connectivity. The maintenance of continuous areas of intact or lightly disturbed vegetation is a priority issue to consider in conservation policies.

## 8.4 Forest Islands as Fire-Breaks in the Savanna

Fires are a normal and frequent event that occur in flooded savannas around the world (Whelan 2006). For example, in South Africa, fire is considered a natural factor in the development and maintenance of the vegetation of the Kruger National Park (Govender et al. 2012). At the regional scale, Silveira et al. (1999) recommend a fire management program to minimize the danger of uncontrolled fires, using controlled burns on a rotational basis in different sections of the Emas National Park in Brazil. The program was also meant to improve the availability of food for herbivores and control the spread of alien grass species.

In addition to flooding, fire is a main agent of disturbance in the Chaco. According to Morello et al. (2009), floods put pressure on ecosystems with equal intensity as the fires. Fire is an ecological component in the vegetation distribution of the Humid Chaco, due to the high productivity of the herbaceous stratum during the wet season and the insufficient number of herbivores to assimilate all the production (Herrera et al. 2003). Fire consumes plant production that is not consumed by herbivores and other foraging species (such as ants) and re-opens space for growth. Extensive fires are provoked by natural causes such as electric storms or by anthropogenic burning during the management of pastures and forests (Ginzburg and Adámoli 2006; Benzaquen et al. 2017). In almost all of the world's natural savannas, the frequency of spontaneous fires is increased by human action and its effects can influence the existing balance in the natural vegetation (Whelan 2006).

Fire is used as a management tool in livestock production in savannas. After burning, the herbaceous stratum reaches a higher concentration of nitrogen and protein in the regrowth (Ginzburg and Adámoli 2006), and lignified grasses are eliminated, thereby stimulating the growth of native grasses, which are more palatable for cattle (López-Hernández and Hernández Valencia 2009). Inadequate fire management during pasture burning and the lack of regulation of the livestock load (such as intensity and frequency of grazing), lead to significant incidence of forest fires (Galindo et al. 2009). Inappropriate fire regimes threaten biodiversity conservation because high-intensity fires kill plants and animals and change the landscape for years, decades, or even centuries in some natural communities (Whelan 2006). Therefore, the conservation of forest islands in landscapes such as the Humid Chaco is of utmost importance as a barrier to manage fires that may get out of control. The islands of vegetation serve as a refuge for animals when fires occur (Silveira et al. 1999). In addition to the maintenance of forest islands immersed in savannas, controlled grazing is also proposed as a fire prevention measure (Ruiz-Mirazo et al. 2007).

## 8.5 Benefits of Forest Islands to Cattle Ranching in Native Grasslands

The farms located in the study area are usually dedicated to cattle raising, with a calving percentage[3] of 50% in cows and 80% in heifers, and weaned calves between six and eight months of age having a live weight of 130 to 150 kg. The stocking rate is 0.5 animal unit (of 400 kg) per ha, which is relatively low because it considers the possibility of flooding or drought, though it may increase during spring and summer in the crucial period of the birth of calves (Laino et al. 2017). Some ranches spare land purposely (as reserves, with no grazing animals) to avoid losing control of the

[3] Calving percentage (*porcentaje de parición*) is the percentage of females that give birth to calves from the total of cows/heifers serviced by bulls.

herd in difficult terrain or as a buffer area to avoid thefts, however no estimation is available of this practice.

Lubián (2014) determined the existence of a variety of native grasses that produce between 973 and 3612 kg of dry grass ha$^{-1}$ year$^{-1}$ in the study site, the most palatable being Clavel grass (*Hemarthria altissima*) and *Kapií-pytá* (*Andropogon lateralis*). Extensive cattle production is based mainly on three factors: the natural supply of grasses for cattle feed, the availability of water and its accumulation in savannas, and the presence of dense forest islands immersed in savannas, palm groves and wetlands in the study area. The forest islands play a role in livestock breeding since they provide shelter for animals during floods and in extreme conditions of heat (up to 50 °C) in summer and cold (−3 °C) weather in winter. In addition, management of forests, with occasional extractions of wood for corrals, cattle ponds,[4] fences and bridges, allows savings of up to 40% of infrastructure costs in a productive unit. These savings would not be possible if forests were converted to exotic pastures (Laino et al. 2017).

The combination of these three factors - natural grasses, water and forest islands - has allowed extensive livestock farming in the context of climate change adaptation. At the same time, frequent floods and droughts lead the farmer to seek economic stability at the producer level due to low and unstable profits (Laino et al. 2017). At this intersection of climate adaptation and the consequent search for economic viability, land use change happens through the clearing of forests for planting exotic pastures of high yield, which maximizes economic production, but constitutes a risk for conservation in the region.

## 8.6 Challenges for Conservation of Forests Islands in the Chaco Region of Paraguay

Ecosystem degradation in the Chaco is occurring at the regional level (WWF 2016). From 2012 to 2018, a total of 2,925,030 ha experienced land use change in the South American Great Chaco, according to data from the land use change monitoring carried out by the NGO Guyra Paraguay. The work demonstrates the gradual degree of modification of the Great Chaco ecoregion, which until recently comprised one of the largest natural areas in the world (Guyra Paraguay 2018). These studies are almost entirely based on monitoring of deforestation of dense forests, but no studies focus on savannas or open landscapes. In particular, the farms included in this chapter are at higher risk of deforestation because of their proximity to urban areas and also due to illegal extraction of trees by intruders during high floods when farms can be accessed by boat.

The South American Great Chaco is not only being affected by a very intense process of deforestation, but is also suffering a loss of natural grasslands, both on

[4]These are "tajamares" or ponds that have been dug out for rain water collection.

higher land and in wetlands, with a rate of disappearance even higher than that of the forests (Bucher 2016). In most tropical and subtropical biomes, conservation strategies are mainly focused on the preservation of forests, but savanna ecoregions and open habitats deserve conservation attention as well (Grau et al. 2015). Many South American countries have no specific wetland management programs. In areas with low population density and without agro-industrial activities, wetlands are less impacted (Junk et al. 2013). The complex interactions between biophysical and socioeconomic processes that drive the trends of Chaco natural grasslands represent a major scientific challenge to preserve this shrinking environment and its valuable biodiversity (Grau et al. 2015).

Although many variables can affect the impact of livestock practices on ecosystems, low intensity cattle ranching, with low densities of cattle on native pastures (savannas) and conserved forest islands, could be beneficial for the survival of the wild fauna that still remains. The presence of cattle consuming the high herbaceous productivity could be a key factor in avoiding extensive fires. This in turn also contributes to maintaining wetlands and associated riparian forests in the region, which serve as natural refuges for wildlife.

The high relative abundance of aquatic animal species increases the conservation value of the Paraguayan Humid Chaco, even though there are no endemic fauna species (Mereles et al. 2013). Despite its high biodiversity, the Humid Chaco does not have enough protected areas (Caballero-Gini et al. 2020). Thus, cattle ranchers of the study area play an important role in the conservation of species at the local level, since its livestock activities allow the coexistence of wild native and domesticated exotic species.

The vast majority of wild protected areas of the Paraguayan Chaco are located in the Dry Chaco, Continental Sand Dunes and Cerrado (Mereles et al. 2020b), therefore it is crucial to highlight the importance of conserving the ecosystems of the Humid Chaco as well. In addition, as described, the ecosystems of the Humid Chaco are also important in flood mitigation, aquifer recharge and water quality improvement (Benzaquen et al. 2017). Future economic development, combining production with biodiversity conservation in a sustainable way, may be possible in the region (Mereles et al. 2020b).

This chapter has described general patterns of the richness of species of flora and fauna of the region. The aim is to highlight the value of these areas for conservation, despite being intended predominantly for economic productive activities. However, it is not clear yet if there is a deliberate intention of farmers to conserve certain natural elements, or if it is simply an unintended consequence of their management (or lack thereof). While this production-based conservation model may be interpreted as a low income generating activity by some (or perhaps even inefficient in economic terms), it may alternatively be interpreted as an opportunity for conservation motivated by non-economic reasons such as cultural, family ties, research, or appreciation of nature, or as a combination of both. Further research is needed to understand drivers for intensification and also motivations to preserve natural ecosystems in the area. However, the key message is that the type of management discussed here allows for conservation of natural elements that more intensive economic activities may not allow. The chapter is not intended to present this production-conservation model as a substitute for the need to spare areas exclusively for biological conservation, but as a complement in the landscape.

The lack of a previous baseline of biodiversity present in the area before cattle ranching activities began (approximately 100 to 150 years ago), also limits our understanding of how cattle ranching activities have affected biodiversity in the past decades. Thus, this chapter provides only information of the current presence of fauna and flora under the existing management and makes no assumptions about the trends of the presence of the species. Comparisons through time, and especially with less intervened areas, are needed to understand the impact of the productive activity on biodiversity in the region over the long term.

## 8.7 Conclusion

Extensive or semi-extensive ecosystem-based livestock production could help to maintain ecosystem services in the region. The need to address adverse environmental conditions from the productive point of view, in this case floods from rain and periodic floods of the Paraguay River, could represent an opportunity for the conservation of threatened biodiversity. These native ecosystems are at high risk of disappearing due to the intensification of agriculture and livestock production, as well as the expansion of urban areas, and so strategies for their conservation, including tailored incentives, are needed.

Currently, an increase of higher yielding crops for the region are being proposed or piloted, and major roads and infrastructure are being planned or discussed locally for production of rice and other commodity crops. Dropping prices of meat in Paraguay and the poor recognition of these systems as a land sharing strategy, could in the future lead to other more profitable production systems, which could jeopardize relevant local ecosystem services such as wildlife, water quality, flood regulation, as well as global services such as carbon sequestration. Implementation of better management practices for cattle rearing, reducing taxes for sustainable farming, providing national and international incentives, increasing prices for meat produced under these environmentally friendly systems, and promoting ecotourism, among other strategies, are some of the possible solutions to promote conservation of the Humid Chaco forest islands, savannas and wetlands along with sustainable production.

**Acknowledgments** The American Chaco Research Center is a local initiative promoting biodiversity conservation, environmental research and sustainable cattle production in native grasslands of the Humid Chaco in Paraguay (www.chacoamericano.org). Institutionally, it is supported by the Manuel Gondra Foundation as well as the Estancia Playada associates.

This chapter is based on works carried out in the framework of several research projects: Project PINV 15-143 "Multidisciplinary contribution to the eco-hydrology of the wetlands of the Humid Chaco" and PINV 14-350 "State of the caiman populations in six localities of Paraguay" both financed by Consejo Nacional de Ciencia y Tecnología (CONACYT) of Paraguay, and the Project

"Initial evaluation of the richness of medium and large sized mammals using camera traps in a silvopastoral system in the American Chaco Research Center" carried out with the support of IDEA WILD. We also presented data obtained in field trips made by researchers and the results of theses of undergraduate and graduate students from national and foreign universities. We also thank Silvina Chaparro, Humberto Sánchez, Paloma Moreno, Lucas Cñaete and people in charge at Estancia Playada who helped during the fieldwork and for making our stay more enjoyable. We dedicate this chapter to the memory of Aldo Florentín, who worked with us in the field during the projects.

## References

Agencia Catalana del Agua (ACA) (2004) Caracterización, regionalización, elaboración de herramientas para la determinación del estado ecológico de los humedales de Catalunya. Catalunya, Espñaa. aca-web.gencat.cat/aca/documents/ca/directiva_marc/zones_humides.pdf. Accessed 2 Mar 2020

Ali W, Javid A, Bhukhari SM, Hussain A, Hussain SM, Rafique H (2018) Comparison of different trapping techniques used in herpetofaunal monitoring: a review. Punjab Univ J Zool 33(1): 57–68

Aquino L, Colli G, Reichle S, Silvano D, Scott N (2004) Chiasmocleis albopunctata. The IUCN red list of threatened species 2004, p e.T57748A11680057. https://doi.org/10.2305/IUCN.UK.2004.RLTS.T57748A11680057.en. Accessed 13 Jan 2020

Asociación Paraguaya de Mastozoología (APM), Secretaría del Ambiente (SEAM) (2017) Libro Rojo de los Mamíferos del Paraguay: especies amenazadas de extinción. Editorial CREATIO, Asunción, Paraguay, p 137

Benzaquen L, Blanco DE, Bo R, Kandus P, Lingua G, Minotti P, Quintana R (eds) (2017) Regiones de Humedales de la Argentina. Ministerio de Ambiente y Desarrollo Sustentable, Fundación Humedales/Wetlands International, Universidad Nacional de San Martín y Universidad de Buenos Aires, Buenos Aires

Boródn D (2015) Estimación del contenido de carbono en la cobertura forestal y en el suelo de un bosque mesoxerofítico, Chaco Paraguayo. Tesis de Ingeniería Forestal, Universidad Nacional de Asunción, Asunción, Paraguay

Boughton RG, Staiger J, Franz R (2000) Use of PVC pipe refugia as a sampling technique for hylid treefrogs. Am Midl Nat 144:168–177

Brun V (2013) Servicio ecosistémico de almacenamiento de carbono en el suelo del humedal del Río Negro, departamento de Presidente Hayes, Paraguay. Tesis de Ingeniería Ambiental, Universidad Nacional de Asunción, Asunción, Paraguay

Brusquetti F, Lavilla EO (2006) Lista comentada de los anfibios del Paraguay. Cuad Herpetol 20(2): 3–79

Brusquetti F, Netto F (2008) Geographic distribution: Chiasmocleis albopunctata. Herpetol Rev 39: 362

Bucher E (2016) El futuro incierto de los humedales del Chaco: el caso de los bñaados del río Dulce. Paraquaria Natural 4(2):11–17

Caballero-Gini A, Bueno-Villafñae D, Laino R, Musálem K (2020) Diversity of mammals and birds recorded with camera-traps in the Paraguayan humid Chaco. Bol Mus Nac Hist Nat Parag 24(1): 5–14

Cartes JL, Clay RP (2009) Paraguay. In: Devenish C, Díaz Fernández DF, Clay RP, Davidson I, Yépez Zabala I (eds) Important bird areas of the Americas – priority sites for biodiversity conservation. BirdLife international (BirdLife conservation series no. 16), Quito, Ecuador, pp 297–306

Chaparro P (2014) Análisis de la relación cobertura vegetal-calidad del agua en el humedal del Río Negro, Presidente Hayes, Paraguay. Tesis de Ingeniería Ambiental, Universidad Nacional de Asunción, Asunción, Paraguay
Collevatti R, Leite K, de Miranda G, Rodrigues F (2007) Evidence of high inbreeding in a population of the endangered giant anteater, Myrmecophaga tridactyla (Myrmecophagidae), from Emas National Park, Brazil. Genet Mol Biol 30(1):112–120
Corn P (1994) Straight-line drift fences and pitfall traps. In: Heyer WR, Donnelly MA, McDiarmid RW, Hayek LC, Foster MS (eds) Measuring and monitoring biological diversity. Standard methods for amphibians. Smithsonian Institution Press, Washington DC
Crooks K, Burdett C, Theobald D, King S, Moreno Di Marco C, Boitani L (2017) Quantification of habitat fragmentation reveals extinction risk in terrestrial mammals. PNAS 114(29):7635–7640. https://doi.org/10.1073/pnas.1705769114
Crump ML, Scott NJ (1994) Visual encounter surveys. In: Heyer WR, Donnelly MA, McDiarmid RW, Hayek LC, Foster MS (eds) Measuring and monitoring biological diversity. Standard methods for amphibians. Smithsonian Institution Press, Washington DC
Del Castillo H (2019) Segunda Actualización de la Lista Comentada de Aves del Paraguay (Guyra Paraguay 2004). Paraquaria Natural 7:8–18
eBird (2020a) EBird Basic Dataset. Cornell Lab of Ornithology, Ithaca, New York. https://ebird.org/paraguay/hotspot/L2762215?yr=all &m=&rank=mrec. Accessed 30 Jan 2020
eBird (2020b) EBird Basic Dataset. Cornell Lab of Ornithology, Ithaca, New York. https://ebird.org/paraguay/hotspot/L6126375?yr=all&m=&rank=mrec. Accessed 30 Jan 2020
El Raiss Z (2014) Análisis de la conectividad para el mono carayá (Alouatta caraya) en un sistema silvopastoril del chaco úhmedo sudamericano. Tesis para el Máster de Restauración de Ecosistemas, Universidad de Alcalá, Espñaa
Ferreira E (2018) Análisis del estado de conservación de tres humedales bajo distintos impactos ambientales antrópicos, Chaco Húmedo, Paraguay. Tesis de Ingeniería Ambiental, Universidad Nacional de Asunción, Asunción, Paraguay
Frost DR (2020) Amphibian Species of the World: an Online Reference. Version 6.0 (Date of access). Electronic Database accessible at http://research.amnh.org/herpetology/amphibia/index.html. American Museum of Natural History, New York, USA. https://doi.org/10.5531/db.vz.0001
Galindo I, Barrón J, Padilla JI (2009) Relación entre ganadería extensiva e incendios en zonas forestales del estado de Colima. Avances en Investigación Agropecuaria 13(3):17–33
Ginzburg R, Adámoli J (2006) Situación Ambiental en el Chaco Húmedo. In: Brown A, Martínez Ortiz U, Acerbi M, Corcuera J (eds) La Situación Ambiental Argentina 2005. Fundación Vida Silvestre Argentina, Buenos Aires, pp 103–113
Gómez Fernández A, Alcocer I, Matesanz S (2016) Does higher connectivity lead to higher genetic diversity? Effects of habitat fragmentation on genetic variation and population structure in a gypsophile. Conserv Genet 17:631–641
Govender N, Mutanga O, Ntsala D (2012) Veld fire reporting and mapping techniques in the Kruger National Park, South Africa, from 1941 to 2011. African J Range Forage Sci 29(2):63–73
Grau HR, Torres R, Gasparri NI, Blendinger PG, Marinaro S, Macchi L (2015) Natural grasslands in the Chaco. A neglected ecosystem under threat by agriculture expansion and forest-oriented conservation policies. J Arid Environ 123:40–46
Guyra Paraguay (2004) Lista comentada de las aves de Paraguay. Annotated checklist of the Birds of Paraguay. Guyra Paraguay, Asunción, Paraguay. 200 pp. www.guyra.org.py
Guyra Paraguay (2018) Monitoreo Mensual del Cambio de Uso y Cobertura de la Tierra, Incendios y Variación de la Cubierta de Aguas en el Gran Chaco Americano. Informe Técnico. Asunción, Paraguay. http://guyra.org.py/informe-deforestacion/
Herrera P, Torrella S, Adámoli J (2003) Los incendios forestales como modeladores del paisaje en la región chaqueña. In: Kunst C, Bravo S, Panigatti J (eds) Fuego en los ecosistemas argentinos. INTA, Santiago del Estero

IPCC (2005) Carbon dioxide capture and storage. Metz B, Davidson O, de Coninck H, Loos M, Meyer L (eds.) https://www.ipcc.ch/report/carbon-dioxide-capture-and-storage/
IUCN (2019) The IUCN Red List of Threatened Species. Version 2019–3. http://www.iucnredlist.org. Accessed 30 Jan 2020
Junk WJ, An S, Finlayson CM, Gopal B, Kveˇt J, Mitchell SA, Mitsch WJ, Robarts RD (2013) Current state of knowledge regarding the world's wetlands and their future under global climate change: a synthesis. Aquat Sci 75:151–167
Kandus P, Morandeira N, Schivo F (eds) (2010) Bienes y servicios ecosistémicos de los humedales del Delta del Paraná. Fundación Humedales/Wetlands International, Buenos Aires, Argentina
Laino L, Musálem K, Laino R (2017) Perspectivas para un desarrollo sustentable: un estudio de caso de producción ganadera en la región del Chaco paraguayo. Población y Desarrollo 45:95–106
Lindenmayer DB, Fischer J (2006) Habitat fragmentation and landscape change: An ecological and conservation synthesis. Island Press, Washington, DC
López-Hernández D, Hernández Valencia I (2009) Nutritional aspects in Trachypogon savannas as related to nitrogen and phosphorus cycling. In: Del Claro K, Oliveira P, Rico-Gray V (eds) Tropical biology and conservation management, volume X, savanna ecosystems. UNESCO/EOLSS, Oxford
Lubián A (2014) Valoración económica de sistemas de producción con inclusión de servicios ecosistémicos en el Bajo Chaco, departamento de Presidente Hayes, Paraguay. Tesis de Ingeniería Ambiental, Universidad Nacional de Asunción, Asunción, Paraguay
Macedo R (2018) Análisis de la estructura vegetal humedal-bosque ripario en el humedal del río Negro, Chaco Húmedo, Paraguay. Tesis de Ingeniería Ambiental, Universidad Nacional de Asunción. Asunción, Paraguay
MADES (2019) Resolución N° 254/19 "Por la cual se actualiza el listado de las especies protegidas de la vida silvestre de la clase aves". Ministerio del Ambiente y Desarrollo Sostenible. Asunción, Paraguay. 10 p
Mamede S, Benites M, Esquivel A, Clay R, de Lima MG, Rodrigues Alho CJ (2019) Turismo de observação de aves no Chaco: oportunidades e desafios ao Corredor Bioceânico, segmento Brasil/Paraguai. Interacções 20:159–177
Maturo HM, Oakley LJ, Prado DE (2005) Vegetación y posición fitogeográfica de la Reserva El Bagual. In: Di Giacomo AG, Krapovickas SF (eds) Historia natural y paisaje de la Reserva El Bagual, Formosa, Argentina. Asociación Ornitológica de La Plata, Buenos Aires. 592pp
Mereles F, Rodas O (2014) Assessment of rates of deforestation classes in the Paraguayan Chaco (great south American Chaco) with comments on the vulnerability of forests fragments to climate change. Clim Change 127:55–71
Mereles F, Cartes JL, Clay R, Cacciali P, Paradeda C, Rodas O, Yanosky A (2013) Análisis cualitativo para la definición de las ecorregiones de Paraguay occidental. Paraquaria Nat 1(2): 12–20
Mereles F, Céspedes G, De Egea-Elsam J, Spichiger R (2020a) Estudios fitosociológicos en el Gran Chaco: estructura, composición florística y variabilidad del bosque de Schinopsis balansae en el Chaco Húmedo Boreal, Paraguay. Bonplandia 29(1):39–55
Mereles MF, Cespedes G, Cartes JL, Goerzen R, De Egea-Elsam J, Rodriguez L, Yanosky A, Villalba L, Weiler A, Cacciali P (2020b) Biological corridors as a connectivity tool in the region of the great American Chaco: identification of biodiversity hotspots in the ecoregions of the Paraguayan Chaco. Res Ecology 2(1):27–36
Morello J, Rodríguez A, Silva M (2009) Clasificación de ambientes en áreas protegidas de las ecorregiones del Chaco Húmedo y Chaco Seco. In: Morello J, Rodriguez A (eds) El Chaco sin bosques: la pampa o el desierto del futuro. Orientación Gráfica Editora, Buenos Aires, pp 53–91
Motte M, Zaracho V, Caballero-Gini A, Ferreira-Riveros M, Romero Nardelli L, Coronel-Bejarano-D, Netto F, Carosini A, Rojas V, Bueno D, Cabral H, Martínez N (2019) Estado de conservación y lista roja de los anfibios del Paraguay. Bol Mus Nac Hist Nat Parag 23(1):1–62

Mujica N (2014) Habitat factors affecting occupancy and detection of mammals in the Paraguayan Chaco. Master's Thesis, Southern Illinois University, Carbondale, Illinois, USA
Nielsen C, Thompson D, Kelly M, Lopez-Gonzalez CA (2015) Puma concolor (errata version published in 2016). The IUCN Red List of Threatened Species 2015: e.T18868A97216466. https://doi.org/10.2305/IUCN.UK.2015-4.RLTS.T18868A50663436.en. Accessed 8 Jan 2020
Peña-Chocarro M, De Egea J, Vera M, Maturo H, Knapp S (2006) Guía de árboles y arbustos del Chaco úhmedo. Guyra Paraguay, Asunción, Paraguay, 291pp
Pereyra M, Pol RG, Galetto L (2019) Ant community patterns in highly fragmented Chaco forests of Central Argentina. Austral Ecol 44(4):668–679
Pérez de Molas L (2015) Manual de familias y géneros de árboles del Paraguay. FAO, San Lorenzo, p 216
PINV 15-143 (2018) Proyecto Aporte multidisciplinario a la eco-hidrología de los humedales del Chaco Húmedo. Centro de Investigación del Chaco Americano y Consejo Nacional de Ciencia y Tecnología (CONACYT). del Paraguay, Asunción, Paraguay
Rodrigues FHG, Medri IM, De Miranda GHB, Camilo-Alves C, Mourao G (2008) Anteater behavior and ecology. In: Vizcaíno SF, Loughry WJ (eds) The biology of the Xenarthra. Florida University Press, Gainesville, pp 257–268
Ruiz-Mirazo J, Robles Cruz AB, Jiménez Piano R, Martínez Moya JL, López Quintanilla J, González Rebollar JL (2007) La prevención de incendios forestales mediante pastoreo controlado: el estado del arte en Andalucía. Sesión 3, Wildfire, Sevilla, Espñaa
Rumbo M (2010) Análisis biogeográfico de los mamíferos de Paraguay. Bol Mus Nac Hist Parag 16(1):18–29
Sato JH, Figueiredo CC, Marchão RL, Madari BE, Benedito LEC, Busato JG, Souza DM (2014) Methods of soil organic carbon determination in Brazilian savannah soils. Sci Agric 71(4): 302–308. https://doi.org/10.1590/0103-9016-2013-0306
Semper-Pascual A, Macchi L, Sabatini F, Decarre J, Baumann M, Blendinger P, Gomez-Valencia B, Mastrangelo M, Kuemmerle T (2018) Mapping extinction debt highlights conservation opportunities for birds and mammals in the south American Chaco. J Appl Ecol 55(3):1–12
Shaffer M (1990) Population Viability Analysis. Conserv Biol 4:39–40
Silveira L, Rodrigues FHG, de Almeida Jácomo AT, Diniz Filho JAF (1999) Impact of wildfires on the megafauna of Emas National Park. Central Brazil Oryx 33(2):108–114
Whelan RJ (2006) The ecology of fire - developments since 1995 and outstanding questions, proceeding of bushfire 2006 - life in a fire-prone environment: translating science into practice conference. Griffith University, Brisbane
World Wildlife Fund (WWF) (2016) Análisis social, económico y ambiental de la producción de soja y carne en Paraguay. WWF-Paraguay, Asunción, Paraguay, Asunción, Paraguay
Zastavniouk C, Weir LK, Fraser D (2017) The evolutionary consequences of habitat fragmentation: body morphology and coloration differentiation among brook trout populations of varying size. Ecol Evol 7(17):6850–6862

# Chapter 9
# Biodiversity Islands and Dominant Species in Agricultural Landscapes of the South Western Amazon, Perú

**Zoyla Mirella Clavo Peralta, Jorge Washinton Vela Alvarado, and Carlos Mariano Alvez-Valles**

**Abstract** The Ucayali region, in the Peruvian Amazon, is characterized by forests that host a mega diversity of species. These forests have been affected by land use changes that were in some cases supported by public policies, such as in the 1960s, that encouraged the inhabitants of the Sierra (highlands) and Costa (coastal) regions of Peru with incentives to populate the Amazon. The majority of settlers in Ucayali were initially from the Sierras and had livestock rearing backgrounds. They became established along the road that connects the Ucayali region with the city of Tingo María in the Huánuco region. Most of them established their pastures by felling nearly all the trees, though fragments of different sizes of primary and secondary forest of different regeneration ages were conserved and used for various subsistence activities. Many farmers value the trees and use live fences as limits of their properties or their paddocks. Over the years, and due to the boom in agro-industrial crops, most farmers reduced their grazing fields to plant oil, palm, and cocoa. The objective of this chapter is to describe these islands of biodiversity and the dominant species in the agricultural landscapes of this region. We describe the characteristics of the Peruvian Amazon with an emphasis on the Ucayali region, its predominant land uses, and the deforestation and degradation therein; the characteristics of the biodiversity islands and their uses; the dominant flora species and their conservation value and adaptation ranges; biological connectivity; and conservation strategies for designing public policies. Effective design of land use or conservation programs in accordance with current needs will only succeed with knowledge and dissemination of the current state of our Amazon.

---

Z. M. Clavo Peralta (✉) · C. M. Alvez-Valles
Facultad de Medicina Veterinaria, Instituto Veterinario de Investigaciones Tropicales y Altura (IVITA), Universidad Nacional Mayor de San Marcos (UNMSM), Ucayali, Peru
e-mail: zclavop@unmsm.edu.pe; calvezv@unmsm.edu.pe

J. W. Vela Alvarado
Facultad de Ciencias Agropecuarias, Centro de Investigación de Fronteras Amazónicas (CIFA), Universidad Nacional de Ucayali (UNU), Ucayali, Peru

 
F. Montagnini (ed.), *Biodiversity Islands: Strategies for Conservation in Human-Dominated Environments*, Topics in Biodiversity and Conservation 20,
https://doi.org/10.1007/978-3-030-92234-4_9

**Keywords** Conservation and development · Deforestation · Flora · Rainforest · Ucayali region

## 9.1 Introduction

The Peruvian Amazon is a megadiverse region that provides important ecosystem services, such as carbon storage and biodiversity. The region holds diverse flora and fauna, as well as high cultural diversity. It also supplies natural resources such as hydrocarbons, and minerals, and its lands are used for the execution of development projects in agriculture and livestock. For many political administrations, it represents the largest area of expansion of the agricultural frontier.

At present times, legal and illegal productive and extractive anthropic activities directly and indirectly cause deforestation and habitat loss. This directly contributes to the present danger of a loss of balance in forest ecosystems and biodiversity. At this crucial time, as researchers from two academic institutions invited by the author of this book, we have reviewed and systematized the rich existing secondary information, in addition to experiences from our own research on the forest resources of the Peruvian Amazon in the Ucayali region, focusing on biodiversity and its conservation potential through current strategies such as the framing tool of biodiversity islands.

Forest fragmentation and habitat loss are the result of environmental expansion and degradation due to human activities (e. g. deforestation), all leading to a gradual deterioration of the quality of remaining ecosystems (Morera et al. 2007). This also results in changes in the forest microclimate (higher temperatures, lower humidity), and in the dynamics of biological communities (increased predation and decreased productivity) (Bierregaard et al. 1992). Likewise, the decrease in the size of forest remnants and geographic isolation of natural ecosystems reduces the potential for dispersion and colonization (Bennett 1998), restricting the natural movement of species, reducing the possibility of genetic exchange, and consequently decreasing biodiversity (Bennett 1998; Kattan 2002). The risk of extinction, especially for species that depend solely on certain habitats to survive, also increases as a consequence of forest fragmentation and habitat loss (McIntyre 1995).

The problem with decreasing the size and quality of habitat for species is that it leads to a reduction of their populations (García 2002). Plant populations and the populations of organisms with which they interact (pollinators, dispersers, herbivores) are reduced. Therefore, the species richness of a forest remnant depends on the suitability of the remnant as a habitat for several species (Begon et al. 1996). The larger the remnant, the greater the possibility of keeping interior habitat or the greater the probability of existence of a variety of habitats, which leads to a greater possibility of species survival (Begon et al. 1996). The smaller the population, the greater its probability of extinction. Consequently, when the population has been greatly reduced, genetic deterioration is a big threat (Kattan 2002).

Considering the current high rates of deforestation and fragmentation due to the increase in anthropic pressure on natural cover, which on a large scale can radically alter the physical environment and climate, there is a need to implement conservation strategies in altered and fragmented environments. One of these strategies is the establishment of connectivity between fragments of isolated or poorly connected forests through the use of ecological corridors, which can facilitate the structural and functional connectivity of biotic elements (Colorado Zuluaga et al. 2017). This strategy should not only focus on the diversity attributes of biological communities, but also on the maintenance of their natural dynamics, including the conservation of their habitats and ecological processes at different spatial and temporal scales (i.e. ecosystem functionality) that they require for their sustainability and biodiversity conservation (Nott and Pimm 1997; Armenteras and Vargas 2016). Habitat connectivity facilitates the dispersion and migration of species (their entry and exit flow) through the landscape to meet basic habitat requirements (Bergoeing 1998).

In the southwestern Amazon, specifically the Ucayali Department of the Peruvian Amazon, several immigrants from the coast and highlands ("Sierra") of Peru settled over time. These settlers, many of who have experience with livestock, were encouraged by public policies over time with the goal of populating and utilizing the Amazonian lands. For this reason, cattle raising was the main activity of these families for their income. The cutting and burning of forests for pasture cultivation resulted in the loss of large areas of vegetation cover. Nowadays, it is notable to see these large areas without forests, which later were transformed into monocultures such as oil palm and cocoa. Following this activity, an agricultural landscape can be observed where large areas of pastures are dominant and the few areas with vegetation cover are small patches of different sizes and areas of forests, called "biodiversity islands." These islands play a fundamental role in the conservation of biodiversity because they act as shelters for species of flora and fauna. In addition, these forest fragments promote connectivity while maintaining ecological integrity and increasing landscape diversification.

The content of this chapter is based on secondary information on the Peruvian Amazon region, as well as on research carried out by the authors over many years of work on the subject as university professors as well as in research institutions. Researchers working at the Veterinary Institute for Tropic and Altitude Research (IVITA), Pucallpa, and of the Faculty of Veterinary Medicine of the Universidad Nacional Mayor de San Marcos, have been conducting research since the 1980s on the species of trees that predominate in the pastures as products of natural regeneration, on tree species that are suited for agricultural production, on the characterization of plant succession after pastures or agricultural crops, on tree species resistant to fire, on species with adequate vegetative reproduction suited for live fences ("cercos vivos"), and on the use of the various species by the rural population. Since the 1990s, faculty from the National University of Ucayali, Faculty of Agricultural Sciences (UNU) and the Experimental Station of the National Institute of Agrarian Research (INIA), Pucallpa, have generated knowledge on animal production systems, the management of tropical pastures, silvopastoral systems from natural regeneration of tree species, and tree planting.

The chapter has four parts: it begins with the characterization of the Ucayali region of the Peruvian Amazon and the classification of its vegetation; the second part describes land use, conservation, deforestation and forest fragmentation, and drivers of deforestation; the third part describes the dominant flora species, their conservation value and key species; and the fourth section characterizes the ranges of adaptation and biological connectivity. The information presented can be used to guide strategies for conserving the biodiversity of the region.

## 9.2 Characterization of the Peruvian Amazon and the Ucayali Region

The Peruvian Amazon is located at the eastern Andean flank and covers an area of 78.5 million ha, of which 23.11% correspond to the so-called High Forest which is located between the Andean mountains at 500–2000 m above sea level, and 76.89% correspond to the Low Forest located at less than 500 m above sea level (Gazzo 1982). The Ucayali region is located in the southwestern Amazon and in the central and eastern part of the Peruvian territory, between the following coordinates: by the North 7 0 20′ 00″ S Latitude and 74 0 32′ 05″ West Longitude; by the East, 9 0 25′ 09″ of S and 70 0 29′ 46″ W; by the South, 11 0 27′ 35″ S and 72 0 34′ 55″ W; and from the West, to 08 0 40′ 19″ S and 75 0 58′ 08″ W (Vivanco 2004) (Fig. 9.1).

The average temperature is higher than 24 °C in the tropical area and around 22 °C in the sub-tropical parts. The precipitation varies according to the life zone: in the subtropical area, the precipitation exceeds 3000 mm per year, while in the tropical humid area it approaches 2000 mm per year and in the tropical dry area it approaches 1200 mm per year.

The total area of the Ucayali region is 102,410.55 km$^2$, corresponding to 13.15% of the total area of the Peruvian Amazon and to 7.96% of Peru, being the second largest region of the country, after the Loreto region (INEI, cited by Vivanco 2004). The population of the Ucayali region according to the 2017 census was 496,459 inhabitants, of which approximately 75% live in urban areas, with a population density of 5 inhabitants/km$^2$ (INEI 2018).

Three morphologically different altitudinal floors are found in the Ucayali region: Ceja de Selva ("Forest Brow"), Selva Alta ("High Forest"), and Selva Baja ("Low Forest"). The Ceja de Selva begins at the headwaters of the Sepa, Unini, and Catsingari rivers in the Atayala province at 1000 m above sea level, and in the headwaters of the Aguaytía and Yurac rivers in the Padre Abad province, which reaches as high as 3000 m above sea level. The Selva Alta is found in valley bottoms of high altitude and narrow width, with terraces staggered in four levels, the lowest having agricultural aptitude. Finally, the Selva Baja occupies the largest area in the region and has acidic soils with low content of organic matter and phosphorous, low base saturation, and high levels of aluminum (Table 9.1) (MINAGRI 2012).

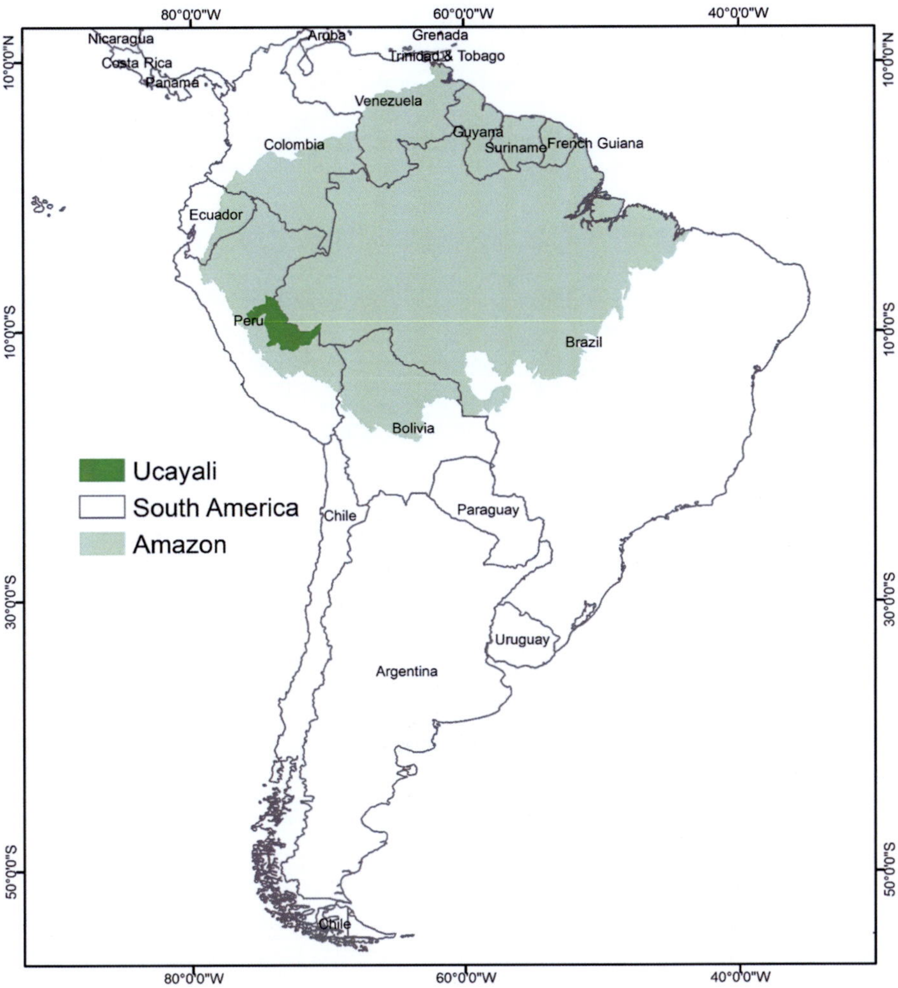


**Fig. 9.1** Geographical location of the Ucayali region in the southwestern Amazon. (Source: Sonaira Silva. Federal University of Acre. Floresta Campus. Cruzeiro Do Sul)

**Table 9.1** Morphological characteristics of the forests of the Ucayali region

| Altitudinal floors | Area ($km^2$) | Altitude (masl) | Characteristics |
|---|---|---|---|
| Ceja de Selva ("forest brow") | 1135.00 | 1000–3000 | It begins in the provinces of Atalaya and Padre Abad |
| Selva Alta ("high forest") | 12,948.00 | 500–1000 | Neshuya districts, Curimaná, Campo Verde |
| Selva Baja (low forest") | 88,434.00 | 0–500 | Callería districts, Masisea |

Source: Adapted from MINAN (2020), GeoBosques Program

The Ucayali region is crossed from South to North by the Ucayali River, which constitutes the middle course of the Amazon River. The Ucayali river is born at the junction of the Tambo and Urubamba rivers and culminates when its waters pour into the Marañón River. From that confluence it begins to be called Amazon River, the largest river in Peru, which is approximately 3000 km long with a width that varies from 400 to 2000 m.

The Ucayali river has a winding riverbed that forms many meanders along its way. Some of them become "cochas" or "tipishcas," large areas which are very rich in fish, and some of them have also become touristic zones.

The river level rises between the months of September and October and reaches its maximum in February. The water levels start to recede in May, reaching their minimum level in July and August; according to bathymetric studies, there is a difference of 12 m between the highest and lowest water levels, each cycle lasting an average of 6 months. This cycle generates a distinct type of coastal tropical or subtropical moist broadleaf forest area called "restinga" in the river shores, as well as beaches and mud areas that form a complex of banks, in which large amounts of silt and fine sediments accumulate. These sediments originate from materials that are dragged along the rivers of the slopes of the Eastern Andes (Kalliola and Puhakka 1993). The sediments originate soils of high fertility, used for planting yellow corn, plantain, grain legumes, rice, cassava, peanuts, and soybeans. The agriculture of these crops plays an important role in food security of the riverine population as well as a high percentage of the urban population.

### *9.2.1 Vegetation Classification of the Peruvian Amazon*

At first look, Amazonian vegetation gives an impression of homogeneity, but this cannot be farther from reality. The large number of species makes it difficult to identify the different existing vegetation units. Different geological and geomorphological formations create habitats with different drainage conditions and very diverse soils, which cause differences in the structure and floristic composition of the vegetation.

In tropical America, numerous studies have been carried out to classify the vegetation, for example, those of Holdridge (1967), Encarnación (1985), Malleux (1971, 1982), all cited by Tuomisto (1993). According to the Holdridge system, the life zones of the Ucayali region are classified mostly as Tropical Dry Forest and Tropical Humid Forest. The Tropical Dry Forest begins in the southern Contamana district, follows the Ucayali river upstream, and ends at the confluence of the Urubamba and Tambo rivers. The Tropical Humid Forest begins at the border with the Loreto region and follows the border with Brazil to the province of Purús in the Ucayali region. The wettest natural life zone in Peru is located between Tingo María and Pucallpa, where the sub-tropical rainforest and the very humid tropical forests are located (Figs. 9.2 and 9.3) (Cochrane and Sánchez 1982).

**Fig. 9.2** Tropical dry forest, Abujao River Basin, Pucallpa, Perú. (Photo: Jorge Vela, National University of Ucayali. Pucallpa, Perú)

Malleux (1971, 1982) and Encarnación (1985), both cited by Tuomisto (1993) have defined different types of vegetation for the Peruvian Amazon. Malleux based his definitions on visible characteristics as seen in photographs, especially the topography of the land and the texture of the vegetation cover, while Encarnación based his on deep knowledge of the vegetation of the Loreto region, and used the same vernacular nomenclature used by the inhabitants of the region. Both authors classify the vegetation as: (a) Swamp forest, such as "aguajales"; (b) Temporarily flood forests, such as riparian forests and restinga forests; and (c) mainland forest, such as terrace forests, hill forests, secondary and degraded forests, and protection forest (Figs. 9.4 and 9.5).

**Fig. 9.3** Tropical humid forest, Yurac river basin, border of Pucallpa and Tingo María, Padre Abad Province, Ucayali region. (Photo: Mirella Clavo, IVITA, Pucallpa, Peru)

## 9.3 Land Use, Conservation, Deforestation, and Fragmentation in the Ucayali Region, Southwestern Amazon

### *9.3.1 Land Use*

Land use capacity is an important variable given by the soil and its territorial distribution, posing real challenges in the exploitation of this resource. The Peruvian Amazon has an approximate extension of 78,308,801 ha, of which 69,380,729 (92.7%) are of mature forests (primary forest and secondary forests older than 12 years). The Ucayali region has aptitude and vocation for forestry, and a great richness of natural resources, with high availability of water and diversity of flora and fauna. Of the 10,515,536 ha of territorial extension of the Ucayali region, 88.41% corresponds to forests, and the rest to agriculture, grasslands, wetlands and rural settlements as shown in Table 9.2.

**Fig. 9.4** Swamp forest "aguajal". Abujao river basin, Coronel Portillo Province, Ucayali region. (Photo: Jorge Vela, National University of Ucayali, Pucallpa, Peru)

### *9.3.2 Conservation of Natural Resources*

Peru is one of the ten megadiverse countries in the world, housing much of the biological diversity of the planet. It has 28 climates, 84 out of the total 104 life zones, and eight biogeographic provinces. It has three major river basins that contain 12,201 lakes and ponds, 1007 rivers, and 3044 glaciers. Much of this natural wealth is conserved in the country's 77 Natural Protected Areas, 17 Regional Conservation Areas and 108 Private Conservation Areas, which in total conserve 22,584,586.19 ha (MINAM 2020). Most of the protected area belongs to the Amazon region, specifically to the Ucayali region; however, it represents only a small fraction of its territory.

The Ucayali region harbors a wealth of natural resources that combines well with the cultural diversity of its people, wealth that must be valued in order to achieve a harmonious balance between humans and nature to support the life of this planet. Although the Ucayali region is one of the most biodiverse in the Peruvian Amazon, only 12% of its territory is protected by different conservation categories. It only has three National Parks, two Communal Reserves and one Regional Reserve, covering a total of 1,261,864.32 ha. Furthermore, these areas, despite being protected by the government, are undergoing a deforestation process, with 18,335 ha deforested so far (MINAM 2020).

**Fig. 9.5** Forest remnants in cattle ranches, in areas of the Pucallpa to Tingo María road. (Photo: Jorge Vela, National University of Ucayali, Pucallpa, Peru)

**Table 9.2** Land use in the Ucayali region

| Greater use capacity | Aptitude or vocation | Area (ha) | % |
|---|---|---|---|
| Forest land | Forest | 9,296,795 | 88.41 |
| | Flooded areas in forests | 95,601 | 0.91 |
| Agricultural land | Agriculture | 297,378 | 2.83 |
| Prairies | Secondary vegetation | 165,650 | 1.58 |
| | Pastures/grasslands | 320,210 | 3.05 |
| | Hydromorphic Savannah | 83 | 0.00 |
| Wetlands | Water body | 285,354 | 2.71 |
| | Flooded areas in non-forest sites | 36,503 | 0.35 |
| Settlement | Artificial areas | 17,731 | 0.17 |
| | Mining areas | 221 | 0.001 |
| Other lands | Bare land | 10 | 0.001 |
| **Total** | | **10,515,536** | **100.00** |

Source: Adapted from MINAN (2020) GeoBosques program

In Peru, Private Conservation Areas are the conservation category with the highest number of protected areas, and yet, none of them are in the Ucayali region (MINAM 2020). In addition to conservation areas promoted by the government, the government of Peru also grants land to native communities, currently covering an area of 1793.2 $km^2$ and representing 17.06% of the territory of the Ucayali region (MINAM 2020).

### 9.3.3 *Deforestation and Forest Fragmentation in the Amazon*

Currently, the balance of biological diversity worldwide is threatened. The crisis is generated by anthropic activities of extractive and productive nature, such as illegal and indiscriminate logging, the extraction of non-timber forest products, exploitation of hydrocarbons, mining, hunting, illegal trade in native species, drug trafficking, migratory agriculture, and extensive cattle ranching among others. These activities have resulted in the disappearance of invaluable habitats and ecosystems. Likewise, ethnodiversity, whose security and permanence has been threatened since ancient times, is also currently in crisis because of the intrusion of entities or ideologies that undermine ancestral customs, issues of territoriality, and a lack of resources. These cause the decline of native human groups and the progressive excision or cultural disappearance of these populations due to the loss of traditional knowledge.

Peru, as part of the Amazon basin, is no stranger to this deforestation process that progressively removes a high percentage of its forests year after year and consequently its biodiversity. MINAM (2020) has monitored deforestation in the Ucayali region, calculating that from 2001 to 2018, 384,474 ha have been lost. In 2013, 36,793 ha were lost, representing the year of highest deforestation. In the following years, that number has gradually decreased. For example, in 2018, deforestation decreased to 25,991 ha (Fig. 9.6).

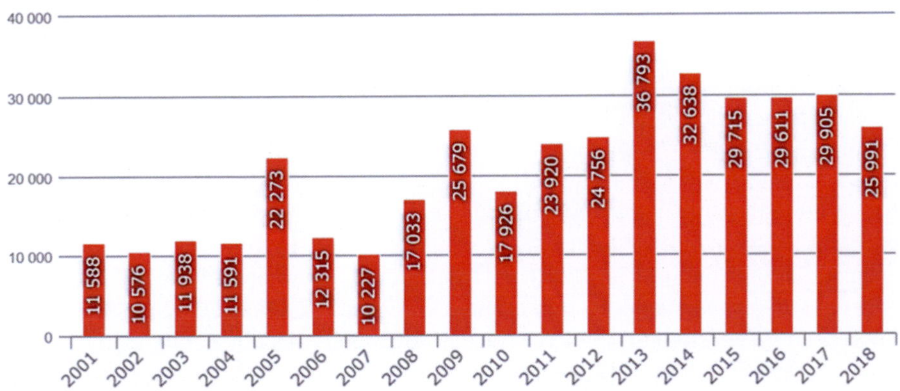


**Fig. 9.6** Forest loss in the Ucayali region 2001–2018 (hectares). (Source: Adapted from MINAN, Programa GeoBosques 2020)

### 9.3.4 Drivers of Deforestation and Fragmentation

The occupation of the Amazonian territory, and consequently, the impacts of deforestation have been occurring for many years, which continues to be both a cause for concern and a subject of studies by many researchers and scientists. Toledo and Serrao (1982) were of the opinion that the wisest decision to conserve the Amazon would be to completely preserve it without modifying the ecosystem. They also however raised the question of how long it would be possible to prevent human occupation of this region. At that point in time, it was already too late. The Amazon was already being invaded by people driven by socioeconomic pressures, as well as demographic and regional integration policies in countries such as Colombia, Ecuador, Peru, Bolivia and Brazil.

Rodríguez (2003) indicated that the Amazon, in the context of several development models of the Peruvian economy, had only fulfilled roles as a provider of natural resources. At the time, this market had a national and international heyday, increasing demand for products like rubber cases, rosewood, *Lonchocarpus nicou* ("barbasco"), wood, oil, and *Theobroma cacao* ("cocoa"). As promoted by the governments of the day, infrastructure changed through the construction of access roads and development projects for easier access to these profitable resources. Thus, the Amazon became a sort of "escape valve" to mitigate the poverty of the Andean populations. It was also subject to border colonization projects, implemented as a security measure due to its status as a border region with countries such as Colombia, Ecuador, Bolivia and Brazil.

According to Dourojeanni (1987), Robiglio et al. (2015), and Porro et al. (2015), in the 1980s, agriculture and livestock activities were the main direct causes of deforestation in Peru. Urban development, building of communication infrastructure, mining and oil exploitation were also other direct causes of deforestation (Dourojeanni 1987). This author also stated that logging was not a direct cause of deforestation at the time. While it is true that those activities were not major direct causes of deforestation at the time due to their relatively small area, they were the spearhead for the entry of large projects for agricultural and livestock development, which take advantage of the facilities of the communication infrastructure and internal roads and highways generated by logging. These large agriculture and livestock projects included the sowing of large areas of oil palm, cocoa, coffee, plantain, "camu camu" (*Myrciaria dubia*), and cattle raising for meat and milk. Furthermore, the planting of illicit crops such as coca, which has been an important economic engine in the Ucayali region since the 1980s, is not mentioned in official statistics. The province of Padre Abad in Ucayali has the highest levels of coca expansion in the country (Salisbury and Fagan 2011; Porro et al. 2015).

Deforestation and fragmentation of the forest of the Peruvian Amazon continues rapidly today. The Monitoring of the Andean Amazon Project (MAAP), an initiative of Amazon conservation (ACCA), uses satellite technology to monitor deforestation in real time on the megadiverse Andean Amazon (Amazon basin of Peru, Colombia, Ecuador and Bolivia) (ACCA 2020). In the last 17 years (2001–2017), approximately 2.1 million ha of Peruvian forests have been lost, representing 50% of the total Andean Amazon forests (ACCA 2020). Some of the main drivers of deforestation are: (a) agriculture, which includes crops such as oil palm, cocoa and other crops of small and large scale; (b) livestock, (c) gold mining, (d) forest roads and (e) highways; with agriculture and livestock the drivers that most threaten the entire Amazon region.

Most of the deforested areas (74%) are of small scale, with an average of less than 5 ha, which makes forest fragmentation very marked. Satellite imagery reveals many small and large patches, as well as isolated forest fragments, known as biodiversity islands (Finer and Mamani 2018).

ACCA (2020), MINAN (2020), and AIDER (2020) report that the greatest amount of forest loss in the Ucayali region occurs in the province of Coronel Portillo, in the Campo Verde and Nueva Requena districts, with 138,111 ha, representing 4% of its area; and in the province of Padre Abad, in the districts of Curimaná, Neshuya, Von Humboldt, Irazola and Padre Abad, with 162,581 ha, representing 25% of its area. For Robiglio et al. (2015), the loss of forest is influenced by the opening of land communication routes, expansion of the agricultural frontier, and development of permanent crop plantations motivated by legal financial institutional incentives, as well as the existence of perverse incentives that promote the shift from forest use to crops.

Such is the case of the processes for granting proof of possession and titling promoted by the National Commission for Development and Life without Drugs (DEVIDA) that seek to eradicate the cultivation of coca. DEVIDA promotes uses in a variety of agricultural activities, without necessarily considering or valuing the importance of maintaining forests to ensure the quality of life and food security of the population, as well as the important ecosystem services they provide (Robiglio et al. 2015).

In addition, the study carried out by Porro et al. (2015) in the provinces of Coronel Portillo and Padre Abad with samples of mestizo populations and indigenous peoples with diverse ethnic composition and different environments shows that, on average, almost 40% of annual income comes from forests and environmental products (including fish), followed by agriculture (25%), wages (17%), livestock activity and animal products (11%), demonstrating the importance of multiple uses of forest products for a large part of the local population. Public policies for sustainable use could therefore be expected to promote a balance between forest conservation and use.

## 9.4 Dominant Flora Species, Conservation Value, Key Species

### *9.4.1 Characteristics of the Secondary Forests of the Peruvian Amazon*

Several studies on the floristic composition of vegetation succession after the use of the primary forest were initiated in the 1980s. Secondary forests known as "purmas", or fallows, are a consequence of the type of shifting agriculture practiced in the region. After cutting and burning the forest, maize and rice are grown in the first year and cassava or plantains in the second year. Then, the farmer makes the decision to either plant grass or to leave the area abandoned to start the successional process. They usually wait for a minimum of 7 years, a period considered to be enough for the soil to regain its fertility, and then they can begin the cycle again. The period of time that the land is left to recover its fertility is currently shorter than in the previous generations, when farmers waited for periods of 10–15 years before starting a new agricultural cycle (Toledo and Serrao 1982; Tuomisto 1993; Clavo et al. 2006).

Observations and studies indicate that within the flora of these secondary or fallow forests there are species with diverse use potential (wood, firewood, fruits, fibers, medicinal, among others), including fast-growing species such as *Cecropia sp, Ochroma piramidale* and *Guazuma* sp. These fast-growing species are relatively easy to manage because they do not require good soil quality and can withstand variations in environmental conditions (Sabogal et al. 2001; Porro et al. 2015). The more traditional use of the land, including shifting cultivation in large areas of land and allowing fields to remain fallow for at least 10 years, gave rise to extensive areas of abandoned fields with a population density of two or three people per square kilometer. This migratory cultivation system was sustainable and self-sufficient and also allowed for the conservation of biodiversity (Chávez 1991).

In recent years, the population has increased to five inhabitants per square kilometer and new crop initiatives have been started (INEI 2018). New inhabitants with large capitals have established monocultures of crops such as oil palm and rice on large areas of fallows or "purmas" at different stages of succession. There are also grasslands with scattered trees and wetlands. Forest fragmentation has increased, especially along the road that connects the department of Ucayali with Lima.

According to Colan, cited by Sabogal et al. (2001), in the Peruvian Amazon region, secondary vegetation is distributed in small patchy areas within the agricultural units, usually located near the residual forests that have some seed trees of commercial species. Small producers dedicated to the production of crops or livestock generally establish at least 2 ha of oil palm, taking advantage of the monetary incentives provided by the government, and leave, according to the extension of land that they own, small areas of primary or secondary forest of about a hectare or more. Farmers also leave vegetation around water sources, which they use as watering holes for cattle, indirectly conserving riparian vegetation and water. These constitute the so-called islands of biodiversity that not only conserve the plant and animal species but also the ecological processes and the landscape.

The presence of scattered trees in the pastures or fields also contributes to the conservation of biodiversity and to biological connectivity. For example, a total of 87 tree species were identified in studies carried out in the department of Ucayali in cultivated pastures of 40 farmers (Clavo and Fernández-Baca 1999; Riesco et al. 1995). The most frequent and economically valuable species that can be considered as key species were *Cordia ucayalensis*, *Ochroma pyramidale, Handroanthus serratifolius, Terminalia* spp., *Trema micranta and Trichilia* sp. These key species are fast growing, have high regeneration capacity with good ability for regrowth and good seed production, and some are fire resistant. In recent evaluations, we found an increase in individuals per tree species that were growing in the boundaries of the farms or as live fences of the pastures, providing biological connectivity and acting as biological corridors (Vela et al. 2019).

### *9.4.2 Vegetation Studies in Different Stages of Succession*

IVITA researchers, as part of a project on Sustainable Amazon Systems (SAS), characterized the composition of secondary forest vegetation in different successional stages following abandoned agriculture or pastures (Riesco et al. 1995). A total of 16 tree and shrub species were found in a 2-year old secondary forest developed after the abandonment of agricultural use. In this forest, 98% of the shrub and arboreal stratum was made up of individuals over 2 m high and less than 5 cm in diameter at breast height (DBH). Among these, the species "llausaquiro" (*Heliocarpus popayenensis*) was the most frequent, followed by the "sachahuaca negra" (*Baccharis floribunda*). In this successional stage, the proliferation of individuals is large, due to the availability of light and the large number of seeds produced, but few individuals manage to establish and develop due to competition (Table 9.3). We must emphasize that "sachahuaca negra" is always present in the first successional stages of the vegetation in the region, along with other species of the Asteraceae family, constituting one of the botanical families that is characteristic of the vegetation recovery of the Peruvian Amazon forests (Riesco et al. 1995; Clavo and Fernández-Baca 1999).

In the 5-year old secondary forests of agricultural origin, 34 shrub and tree species were identified, with the most frequent being "black ocuera" (*Vernonia baccharoides*) with a frequency of 12.36%, followed by "black shimbillo" and "alcanfor moena" (Table 9.4). In this successional stage, it is also common to find trees of *Piper hispidum, Jacaranda copaia, Cecropia membranacea* and *Cordia ucayalensis*. The abundance and density of herbaceous vegetation begins to decrease at this stage (Table 9.4) (Riesco et al. 1995; Clavo and Fernández-Baca 1999).

In the 10-year old secondary forests of agricultural origin, 61 shrubs/trees and 8 liana species were identified. The most frequent tree species was the "white tahuari" (*Handroanthus* sp.), followed by "black cordoncillo" (*Piper laevigatum*) and "shimbillo bean" (*Inga pundata*). At this stage, the amount of herbaceous species decreases, the number of trees increases, and liana species such as "cat's

**Table 9.3** Frequency and density of tree species by diameter classes in 2-year old secondary forest of agricultural origin

| No. | Scientific name | Common name | Diametric classes | | | Total | |
|---|---|---|---|---|---|---|---|
| | | | >2 to <5 cm | 5 to 10 cm | >10 cm | Number of trees/ha | Frequency (%) |
| 1 | *Heliocarpus popayanensis* H.B.K | Llausaquiro | 10,133.00 | 979.78 | 40.0 | 11,152.78 | 47.16 |
| 2 | *Ochroma pyramidale* Urbam. | Topa negra | 1333.33 | 162.96 | 24.0 | 1520.2.09 | 6.43 |
| 3 | *Solanum grandiflorum* R&P. | Shucahuito | 2133.33 | 133.33 | 0 | 2266.66 | 9.58 |
| 4 | *Cordia ucayalensis* Jonhst | Añallu caspi | 133.33 | 29.63 | 0 | 162.96 | 0.69 |
| 5 | *Inga thiboudiana* D.C | Shimbillo negro | 533.33 | 14.81 | 0 | 548.14 | 2.32 |
| 6 | *Baccharis floribunda* H.B.K | Sachahuaca negra | 3733.33 | 0 | 0 | 3733.33 | 15.79 |
| 7 | *Cecropia membranacea* Trecul | Cetico blanco | 533.33 | 0 | 0 | 533.33 | 2.26 |
| 8 | *Vernonia baccharoides* R. et P. | Ocuera negra | 1333.33 | 0 | 0 | 1333.33 | 5.64 |
| 9 | *Hura crepitans* L. | Catahua | 133.33 | 0 | 0 | 133.33 | 0.56 |
| 10 | *Hymatanthus Sucuuba* (Spruce) Wood | Bellaco caspi | 400.0 | 0 | 0 | 400.0 | 1.69 |
| 11 | *Trema micrantha* (L.) Blume | Atadijo | 666.67 | 0 | 0 | 666.67 | 2.82 |
| 12 | *Acacia* sp | Pashaquilla colorada | 400.0 | 0 | 0 | 400.0 | 1.69 |
| 13 | *Vitex* sp | Cormiñon colorado | 266.67 | 0 | 0 | 266.67 | 1.13 |
| 14 | *Banara Guianensis* Aubl. | Varilla blanca | 133.33 | 0 | 0 | 133.33 | 0.56 |
| 15 | *Bahuinia* sp. | Pashaquilla pata de vaca | 133.33 | 0 | 0 | 133.33 | 0.56 |
| 16 | *Lisianthus alatus* (Aubl) | Tabaquillo | 266.62 | 0 | 0 | 266.62 | 1.13 |
| **Total** | | | **22,266.2** | **1320.5** | **64.0** | **23,650.7** | **100.0** |

Source: Riesco et al. (1995)

**Table 9.4** Frequency and density of tree species by diameter classes in 5-year old secondary forest of agricultural origin

| No. | Scientific name | Common name | Diametric classes >2 to <5 cm | 5 to 10 cm | >10 cm | Total Number of trees/ha | Frequency (%) |
|---|---|---|---|---|---|---|---|
| 1 | *Apeiba tibourbou* Aubl. | Maquisapa ñacha negra | 0 | 14.81 | 4.0 | 18.82 | 0.21 |
| 2 | *Casearia* sp | Sanipanga | 0 | 0 | 4–0 | 4.0 | 0.04 |
| 3 | *Cecropia membranacea* Trecul | Cetico blanco | 533.33 | 162.96 | 12.0 | 708.3 | 7.81 |
| 4 | *Cecropia engleriana* Snethlage | Cetico shiari | 133.33 | 103.70 | 24.0 | 261.04 | 2.88 |
| 5 | *Cordia ucayalensis* Jonhst | Añallu caspi | 0 | 148.14 | 4.0 | 152.15 | 1.68 |
| 6 | *Dictyoloma peruviana* Plach. | Huamanzamana negra | 0 | 29.63 | 4.0 | 33.63 | 0.37 |
| 7 | *Rollinia insignis* R.E. Frus | Anonilla blanca | 133.33 | 0 | 4.0 | 137.33 | 1.51 |
| 8 | *Inga Thiboudiana* D.C | Shimbillo negro | 666.66 | 385.18 | 48.0 | 1099.85 | 12.13 |
| 9 | *Jacaranda copaia* (Aubl.) D.Don | Huamanzamana blanca | 0 | 29.63 | 16.0 | 45.63 | 0.5 |
| 10 | *Miconia* sp | Rifari | 0 | 0 | 4.0 | 4.0 | 0.04 |
| 11 | *Leonia glycicarpa* R. & P. | Tamara negra | 0 | 0 | 4.0 | 4.0 | 0.04 |
| 12 | *Rollinia ulei* Diels | Sacha anona, anonilla | 0 | 44.44 | 8.0 | 52.44 | 0.58 |
| 13 | *Sclerolobium* sp | Ucshaquiro blanco | 133.33 | 133.33 | 32.0 | 298.67 | 3.29 |
| 14 | *Solanum grandiflorum* R&P. | Shucahuito | 0 | 0 | 4.0 | 4.0 | 0.04 |
| 15 | *Banara* sp | Varilla | 0 | 14.81 | 0 | 14.82 | 0.16 |
| 16 | *Alchornea triplinervia* | Uchumullaca | 0 | 14.81 | 0 | 14.82 | 0.16 |
| 17 | *Banara guianensis* Aubl. | Varilla blanca | 266.66 | 44.44 | 0 | 311.11 | 3.43 |
| 18 | *Vismia baccifera* sub sp. *ferruginea* (Kunth)Ewam | Pichirina colorada | 0 | 14.81 | 0 | 14.82 | 0.16 |
| 19 | *Trattinickia* sp | Tanque negro | 533.33 | 29.63 | 0 | 562.96 | 6.21 |
| 20 | *Cordia nodosa* Lam. | Pucaruro caspi negro | 0 | 14.81 | 0 | 14.82 | 0.16 |
| 21 | *Hymatanthus sucunba* (Spruce) Woodson | Bellaco caspi | 0 | 14.815 | 0 | 14.82 | 0.16 |
| 22 | *Inga* att. *Alba* (Swartz) Willo | Shimbillo blanco | 666.66 | 14.81 | 0 | 681.48 | 7.51 |

(continued)

**Table 9.4** (continued)

| No. | Scientific name | Common name | Diametric classes | | | Total | |
|---|---|---|---|---|---|---|---|
| | | | >2 to <5 cm | 5 to 10 cm | >10 cm | Number of trees/ha | Frequency (%) |
| 23 | *Micropholis venulosa* Mart. & Eichl | Tanque blanco | 0 | 14.81 | 0 | 14.82 | 0.16 |
| 24 | *Theobroma subincanum* Mart. | Cacahuillo | 0 | 14.81 | 0 | 14.82 | 0.16 |
| 25 | *Vernonia baccharoides* R. et P. | Ocuera negra | 913.33 | 207.40 | 0 | 1120.74 | 12.36 |
| 26 | *Banara arguta* Briq. | Varilla negra | 266.66 | 0 | 0 | 266.67 | 2.94 |
| 27 | *Baccharis floribunda* H.B.K | Sachahuaca negra | 533.33 | 0 | 0 | 533.33 | 5.88 |
| 28 | *Bactris* sp | Ñejilla | 533.33 | 0 | 0 | 533.33 | 5.88 |
| 29 | *Naucleopsis glabra* Sprice ex Beill | Tamamurillo | 133.33 | 0 | 0 | 133.33 | 1.47 |
| 30 | *Bahuinia* sp. | Pashaquilla pata de va | 266.66 | 0 | 0 | 266.67 | 2.94 |
| 31 | *Capirona decorticans* Spruce | Capirona negra de altura | 266.66 | 0 | 0 | 266.67 | 2.94 |
| 32 | *Ocotea killipii* A.C. Smith | Alcanfor moena | 800 | 0 | 0 | 800.00 | 8.82 |
| 33 | *Piper hispidum* SW | Cordoncillo | 533.33 | 0 | 0 | 533.33 | 5.88 |
| 34 | *Cassia ruiziana* Benth. | Mataru amarillo | 133.33 | 0 | 0 | 133.33 | 1.47 |
| **Total** | | | **7446.6** | **1451.8** | **172.0** | **9070.5** | **100** |

Source: Riesco et al. (1995)

claw" (*Uncaria guianensis*) and "clavo huasca" (*Tinanthus palyanthus*), both of medicinal and commercial value, appear more frequently (Table 9.5) (Riesco et al. 1995; Clavo and Fernández-Baca 1999).

When comparing forests of the same successional ages but of different origin, differences in composition of species can be found in the early stages of secondary succession. Secondary forests that originated from abandoned grassland have a greater amount of herbaceous species than those from abandoned agricultural use, and they are all similar (Riesco et al. 1995; Clavo and Fernández-Baca 1999).

In a study of the composition of seed banks of secondary forests (fallow) of different successional stages and different origin, López et al. (2005) found no significant differences in seed diversity, with a similar value of the Shannon index. Fallows of agricultural origin, in different stages, had a smaller value than those of pasture origin, probably due to the greater quantity of herbaceous species. The Shannon indices were as follows: 1-year fallow of agricultural origin 2.58, 1-year of pasture origin 3.09, fallow of pasture origin of 5 years 2.96 and fallow of agricultural origin of 5 years 2.50. Herbaceous species that are more abundant in fallows of pasture origin may contribute to the higher values of Shannon index diversity. The species of greatest commercial value such as cedar, mahogany, kapok and hardwoods such as "shihuahuaco" (*Dipteryx* sp.) were not recorded in any successional stage probably due to their seed characteristics, which are more affected by fire and lack of seed dispersers (Riesco et al. 1995; Clavo and Fernández-Baca 1999).

### 9.4.3 Use of Species from Secondary Forests and Islands of Biodiversity

The secondary forests of the Peruvian Amazon are generally underutilized. Lack of knowledge of the species is a constraint on their use and affects farm productivity. Farmers in the region prefer species that provide them with products to increase their income throughout the year and are diversifying their farms with this objective. The secondary forest is part of the farm as a component for the future, waiting for the recovery of soils and as a provider of some resources. However, secondary forests constitute part of the biodiversity islands of the region, so it is essential to improve knowledge of their species composition, and to quantify the products that they can provide so that they can be managed properly to enhance their biodiversity conservation (Riesco et al. 1995; Clavo and Fernández-Baca 1999).

Studies conducted in Pucallpa, department of Ucayali, show that the settlers (producers) know the name of desirable and undesirable plants, but they have less knowledge of their classification than other native peoples elsewhere in Peru or in other parts of the world. Farmers in Pucallpa do not use plants as indicators of soil fertility, they know few uses, such as building materials and edible fruits, and they have very little knowledge on medicinal plants. They do, however, have good

**Table 9.5** Frequency and density of tree species by diameter classes in 10-year old secondary forest of agricultural origin

| | | | Diametric classes | | | Total | |
|---|---|---|---|---|---|---|---|
| No. | Scientific name | Common name | >2 to <5 cm | 5 to 10 cm | >10 cm | Number of trees/ha | Frequency (%) |
| 1 | *Cassia spectabilis* D.C. | Retama negra | 0 | 0 | 20.0 | 20.0 | 0.32 |
| 2 | *Cecropia membranacea* Trecul | Cetico blanco | 0 | 0 | 36.0 | 36.0 | 0.58 |
| 3 | *Trichilia rubra* C.D,C. | Tushmo blanco | 0 | 14.81 | 28.0 | 42.81 | 0.69 |
| 4 | *Inga edulis* Mart. | Guabilla | 0 | 29.63 | 32.0 | 61.63 | 1.0 |
| 5 | SD | Chupo huayo | 133.33 | 0 | 0 | 133.33 | 2.16 |
| 6 | *Ceiba samauma* (Mart.) K. Schum. | Huimba negra | 0 | 0 | 12.0 | 12.0 | 0.19 |
| 7 | *Cassia lucens* Vagel | Retama colorada | 0 | 59.25 | 48.0 | 107.25 | 1.74 |
| 8 | *Heliocarpus popayanensis* H.B.K. | Llausaquiro | 0 | 14.81 | 8.0 | 22.81 | 0.37 |
| 9 | *Guatteria phanerocampta* Diels | Carahuasca negra | 0 | 44.44 | 8.0 | 52.44 | 0.85 |
| 10 | *Vismia cayennensis* (Jack) Pers. | Pichirina blanca | 0 | 0 | 12.0 | 12.0 | 0.19 |
| 11 | *Inga brachyrhachis* Harms | Charapa shimbillo | 0 | 59.25 | 16.0 | 75.25 | 1.22 |
| 12 | *Terminalia oblonga* (R&P) Steud. | Yacushapana blanca | 0 | 0 | 8.0 | 8.0 | 0.13 |
| 13 | *Guazuma crinita* C. Mart | Bolaina blanca | 0 | 0 | 28.0 | 28.0 | 0.45 |
| 14 | *Cochlospermum orinocense* (Kunth) Steud. | Topa negra | 0 | 0 | 16.0 | 16.0 | 0.26 |
| 15 | *Dictyoloma peruviana* Plach. | Huamanzamana negra | 0 | 14.81 | 4.0 | 18.81 | 0.31 |
| 16 | *Cordia ucayalensis* Jonhst | Añallu caspi | 0 | 0 | 12.0 | 12.0 | 0.19 |
| 17 | *Eschweilera coriácea* (DC)S.A.Mori | Machimango blanco | 0 | 0 | 4.0 | 4.0 | 0.06 |
| 18 | *Bactris gassipaes* Kunth | Pijuayo | 0 | 0 | 4.0 | 4.0 | 0.06 |
| 19 | *Schizolobium amazonicum* Huber ex Ducke | Pashaco colorado | 0 | 0 | 32.0 | 32.0 | 0.52 |
| 20 | *Talisia* cf. *cerasina* (Benth.) Radlk. | Caspi ubos | 0 | 0 | 4.0 | 4.0 | 0.06 |
| 21 | *Guazuma ulmifolia* Lam. | Bolaina negra | 0 | 0 | 8.0 | 8.0 | 0.13 |
| 22 | *Vitex klugii* Mold | Cormiñon | 0 | 0 | 12.0 | 12.0 | 0.19 |
| 23 | *Croton matourensis* Aubl | Aucatadijo | 0 | 0 | 4.0 | 4.0 | 0.06 |
| 24 | *Lonchocarpus spiciflorus* Mart. ex Benth | Maria buena | 0 | 0 | 4.0 | 4.0 | 0.06 |

| | | | | | | | |
|---|---|---|---|---|---|---|---|
| 25 | *Piptadenia pteroclada* Benth | Pashaco blanco | 0 | 0 | 4.0 | 4.0 | 0.06 |
| 26 | *Hura crepitans* L. | Catahua | 0 | 0 | 4.0 | 4.0 | 0.06 |
| 27 | SD | Desconocido cortezaun | 133.33 | 0 | 0 | 133.33 | 2.16 |
| 28 | *Acacia loretensis* J.F. Macbr. | Pashaco negro con espina | 0 | 14.81 | 8.0 | 22.81 | 0.37 |
| 29 | *Uncaria guianensis* (Aubl.) Gmel. | Uña de gato | 933.33 | 0 | 0 | 933.33 | 15.14 |
| 30 | *Trema micrantha* (L.) Blume | Atadijo | 0 | 0 | 12.0 | 12.0 | 0.19 |
| 31 | *Inga calantha* Ducke | Rosca pacaya | 0 | 14.81 | 4.0 | 18.81 | 0.31 |
| 32 | *Rollinia insignis* R.E. Frus | Anonilla | 0 | 0 | 4.0 | 4.0 | 0.06 |
| 33 | *Sapium marmieri* Huber | Caucho masha | 0 | 0 | 8.0 | 8–0 | 0.13 |
| 34 | *Astrocaryum* sp | Huicungo | 0 | 0 | 4–0 | 4.0 | 0.06 |
| 35 | *Cordia nodosa* Lam. | Pucaruro caspi negro | 0 | 14.81 | 0 | 14.81 | 0.24 |
| 36 | *Hymatanthus sucuuba* (Spruc.) Wood | Bellaco caspi | 0 | 14.81 | 0 | 14.82 | 0.24 |
| 37 | *Erythroxylum* cf. k Penth | Charichuela blanca | 0 | 44.44 | 0 | 44.44 | 0.72 |
| 38 | *Eschweilera* sp | Machimango colorado | 0 | 14.81 | 0 | 14.81 | 0.24 |
| 39 | *Ichnosiphon* cf. *rotundifolius* | Trompetero sacha | 573.33 | 0 | 0 | 573.33 | 9.3 |
| 40 | *Siparuna guianensis* Aubl. | Pishi huayo | 0 | 29.63 | 0 | 29.63 | 0.48 |
| 41 | *Terminalia amazonia* (J.F. Gmel) Exel | Yacushapana negra | 0 | 14.81 | 0 | 14.81 | 0.24 |
| 42 | *Aegiphila* sp | Sacha tol blanco | 133.33 | 59.25 | 0 | 192.59 | 3.12 |
| 43 | *Banara guianensis* Aubl. | Varilla blanca | 0 | 14.81 | 0 | 14.81 | 0.24 |
| 44 | *Cassia macrophilla* Kunth. | Maturu negro | 0 | 14.81 | 0 | 14.81 | 0.24 |
| 45 | *Tovomita umbellata* Benth. | Chullachaqui caspi rojo | 0 | 14.81 | 0 | 14.81 | 0.24 |
| 46 | *Inga pundata* Willd. | Poroto shimbillo | 133.33 | 14.81 | 0 | 148.14 | 2.4 |
| 47 | *Lacmella arborescens* (Muell. Arg.) Monach. | Sachavaca huayo | 0 | 14.81 | 0 | 14.81 | 0.24 |
| 48 | *Tococa guianensis* Aubl. | Pucaruro sacha | 0 | 14.81 | 0 | 14.81 | 0.24 |
| 49 | *Morisonia oblongifolia* Britt. | Tamara blanca | 0 | 14.81 | 0 | 14.81 | 0.24 |
| 50 | *Adenocalymna impressum* (Rusby) Sandw. | Huangana huasca blanca | 0 | 14.81 | 0 | 14.81 | 0.24 |
| 51 | *Naucleopsis glabra* Sprice ex Beill | Tamamurillo | 133.33 | 0 | 0 | 133.33 | 2.16 |

(continued)

**Table 9.5** (continued)

| No. | Scientific name | Common name | Diametric classes >2 to <5 cm | 5 to 10 cm | >10 cm | Total Number of trees/ha | Frequency (%) |
|---|---|---|---|---|---|---|---|
| 52 | *Casearia aculeata* Jacq. | Limon casha | 0 | 14.81 | 0 | 14.81 | 0.24 |
| 53 | *Perebea guianensis* Aubl. | Pama negra | 133.33 | 14.81 | 0 | 148.15 | 2.4 |
| 54 | *Handroanthus* sp | Tahuari blanco | 266.66 | 14.81 | 0 | 281.48 | 4.56 |
| 55 | *Garcinia gardneriana* (Planch & Triana) Zappi | Charichuela negra | 0 | 14.81 | 0 | 14.81 | 0.24 |
| 56 | *Ruizodendron ovale* (R&P) R.E.Fr. | Paujil ruro | 0 | 14.81 | 0 | 14.81 | 0.24 |
| 57 | *Tinanthus polyanthus* | Clavo huasca | 666.66 | 0 | 0 | 666.66 | 10.81 |
| 58 | *Doliocarpus* dentatus (Aubl)Standl | Paujil chaqui | 133.33 | 0 | 0 | 133.33 | 2.16 |
| 59 | *Ocotea killipii* A.C. Smith | Alcanfor moena | 133.33 | 0 | 0 | 133.33 | 2.16 |
| 60 | *Simarouba amara* Aubl. | Marupa | 133.33 | 0 | 0 | 133.33 | 2.16 |
| 61 | *Solanum grandiflorum* R&P. | Shucahuito | 133.33 | 0 | 0 | 133.33 | 2.16 |
| 62 | *Calicobolus sericeus* (HBX) House. | Camote huasca | 266.67 | 0 | 0 | 266.66 | 4.32 |
| 63 | *Banara arguta* Briq. | Varilla negra | 133.33 | 0 | 0 | 133.33 | 2.16 |
| 64 | *Costus* sp | Sacha huiro blanco | 133.33 | 0 | 0 | 133.33 | 2.16 |
| 65 | *Piper laevigatum* H.B.K | Cordoncillo negro | 266.67 | 0 | 0 | 266.66 | 4.32 |
| 66 | *Trattinickia aspera* (Standl) Swcrtz | Sacha uvilla | 133.33 | 0 | 0 | 133.33 | 2.16 |
| 67 | *Couepia* sp | Parinari | 133.33 | 0 | 0 | 133.33 | 2.16 |
| 68 | *Dioscorea decorticans* Presl. | Soga alambre | 133.33 | 0 | 0 | 133.33 | 2.16 |
| 69 | *Trichilia* sp | Requia blanca | 133.33 | 0 | 0 | 133.33 | 2.16 |
| **Total** | | | **5106.6** | **651.8** | **408** | **6166.5** | **100** |

Source: Riesco et al. (1995). SD: unidentified

knowledge of timber species that can be commercialized (Fujisaka et al. 2000). It is important to note that most of the population in the productive area of Pucallpa are migrants from different regions of Peru, and they try to use and manage the local species in the same way as they did in their region of origin. Some farmers are not even interested in knowing the uses of the plants they find in their plots. Thus, it is important to disseminate the various uses of the species so that farmers may value them better and so that they manage properly those species that are part of the islands of biodiversity (Riesco et al. 1995; Clavo and Fernández-Baca 1999).

As previously mentioned, the islands of biodiversity in the area have different sizes and shapes: some are fragments of secondary forests, others have originated from primary forests, and many are riparian forests whose species contribute to hydrological regulation through processes such as water storage and decreasing the impact of floods. The riparian plant species give stability to the water channels and provide refuge for the fauna, as well as climate regulation. Given the high value of the ecosystem services that they provide for human societies, these biodiversity islands need strategies to become areas of conservation with recognized value by the government and with incentive mechanisms that ensure their permanence over time (Riesco et al. 1995; Clavo and Fernández-Baca 1999).

These islands of biodiversity also house medicinal and food plants whose use must be promoted so that the population can value them. In this manner, the biodiversity islands can become Areas of Conservation Value, taking into consideration the definitions of High Conservation Value Resource[1] Network and the criteria used by the Forest Stewardship council (FSC) to establish Conservation Value Areas.

## 9.5 Landscape Fragmentation and Biological Connectivity

The fragmentation processes cause a decrease in the vegetation covers, leaving the original vegetation of a given area as small fragments isolated from each other. Increasing the number of fragments, reducing their area, and increasing the distance between them, are limitations for ecological processes such as seed dispersal, colonization, migration and interaction between species (Bennett 1998). Therefore, landscape fragmentation is a process that severely affects biodiversity (Bennett 1998; Kattan 2002; Morera et al. 2007), and as habitat loss increases, connectivity decreases and the edge effect increases (Fig. 9.7).

The Ucayali region presents alteration in its landscapes due to the different land uses carried out there, mainly agricultural activities with crops of oil palm, coca, banana, pineapple and cacao; and cultivation of pastures for livestock. Taking into

[1] High conservation value of natural resources in an area is established according to the criteria and principles of responsible forest management. Sustainable forest Deforestation hotspots in the Peruvian Amazon (FSC 2020).

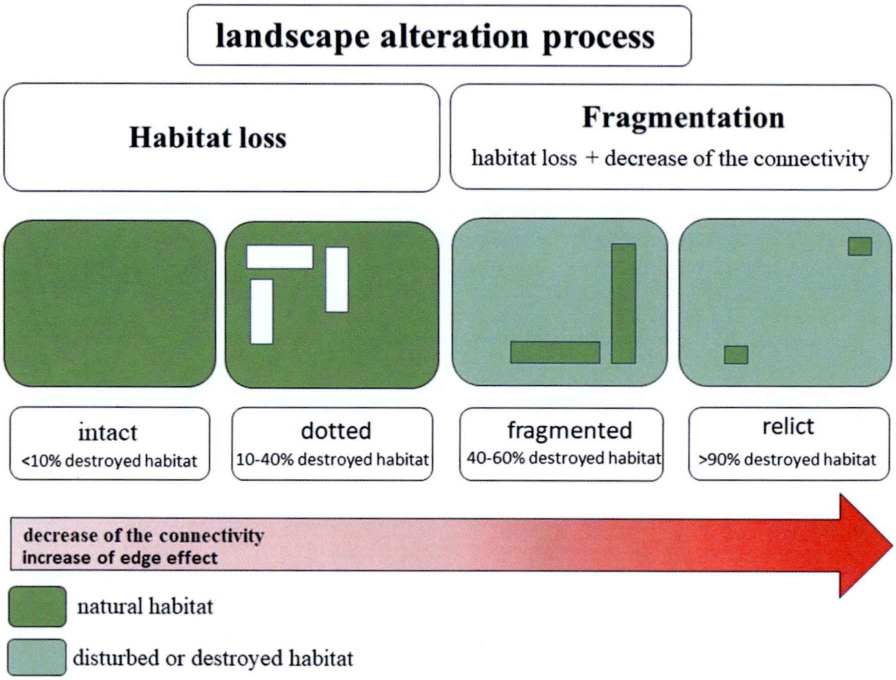


**Fig. 9.7** Landscape alteration process in relation to habitat loss and fragmentation. (Source: Modified from Hobbs and Wilson 1998)

account the existence of altered and fragmented landscapes, it is necessary to implement conservation strategies in these types of landscapes. Ecological corridors are a strategy that facilitates biological connectivity between isolated or poorly connected forest fragments (Colorado Zuluaga et al. 2017). This connectivity facilitates the dispersal and migration of species across the landscape to meet basic habitat requirements (Bergoeing 1998). For this reason, connectivity between forests in a fragmented landscape is important for the conservation and maintenance of their integrity and biodiversity. Within these fragmented forests in the Ucayali region, there are native species of various uses such as timber, medicinal, fruit or other, which places greater emphasis on their conservation.

One of the alternatives that is being promoted as a biological corridor in the Ucayali region is the implementation of agroforestry systems (AFS), mainly with silvopastoral systems (SPS) as land use alternatives in agricultural landscapes. These systems can serve to recover degraded areas, mitigate climate change by reducing deforestation pressure in forests, increase biodiversity, and reestablish ecological interactions through planting trees and perennial species. Despite the fact that this research is still in progress, some areas have already been established with silvopastoral systems, including one in the research areas of the IVITA station – Pucallpa (Fig. 9.8). Positive results are being obtained regarding the increase in

**Fig. 9.8** Left. Silvopastoral system with *Simarouba amara* ("marupa") plantations. Right. Live fences with *Erythrina* sp. ("Amasisa"). Both established at the IVITA – Pucallpa station. (Photos: Carlos Mariano Alvez-Valles and Zoyla Mirella Clavo Peralta, IVITA, Pucallpa, Peru)

biodiversity, mainly with records of bird and amphibian species (Vela et al. ongoing research project). Additionally, Redondo-Brenes and Montagnini (2010) mention that agrosilvopastoral systems present a greater ecological complexity than that of monocultures and, in turn, are found within land uses that harbor intermediate values of bird species richness (143) within the Paso de la Danta Biological Corridor, south of Costa Rica.

Live fences ("cercas vivas") are also part of the AFS that provide land cover and can serve as agroecological strategies to promote landscape connectivity (Schelhas 2007; Francesconi and Montagnini 2015). Such is the study by Gabriel and Pizo (2005) in southwestern Brazil, which documents the frequent use of live fences by birds during their movement and vocalizations, whether for rest or foraging. In the Ucayali region, these systems are widely used in agricultural landscapes, where planted trees and/or shrubs delimit the divisions of the field and the boundaries between farms (Fig. 9.8), but further studies are still lacking in this part of the Amazon confirming this system as an ecological corridor.

Riparian forests, apart from functioning as habitats, also function as ecological corridors. The interaction between vegetation and river dynamism causes a heterogeneous mosaic of very productive habitats, due to the combination of fertility and water availability of their soils (Martín et al. 2013). The most relevant factors that affect the viability and functionality of riparian forest as biological corridors are the existence of forest cover and the continuous or temporary presence of water. This connective function is not only restricted to the circulation of living beings, since these types of forest also act as conduits for runoff, sediment, organic matter and other material (Martín et al. 2013). Therefore, the restoration of riparian forests promotes their role as ecological corridors, facilitating displacement and providing extensive resources as food and refuge for wildlife in fragmented landscapes (Estrada and Coates-Estrada 2001; Gurrutxaga 2004).

Finally, Francesconi and Montagnini (2015) mention that agroforestry systems (AFS) increase the vegetation cover of the landscape and therefore, can act as ecological strategies to restore functional connectivity in fragmented landscapes, but still need further studies to confirm this function. Likewise, it is necessary to consider and prioritize conservation at the landscape level in areas where the native habitat is almost always fragmented (Vandermeer et al. 2008). In promoting conservation strategies of plant diversity in the fragments, their use by local human populations should not be ignored, since in most cases forest fragments positively favor the local economy (Brokaw 2002).

## 9.6 Conclusions

The southwestern Amazon is a very complex region, rich in life formations, megadiverse and of great importance for the conservation and maintenance of the ecological balance of the planet. There is wide and varied information on the ecological characterization, types of vegetation, and ecosystem services of the southwestern Amazon. The research carried out in the southwestern Amazon dates back a long time, with a strong initial onset that has since apparently decreased in its intensity, when compared with studies of the Amazon forest conducted in other countries such as Brazil, Colombia, and Ecuador. The approaches were also different, initially using aerial photography, field work and local knowledge, and then varying as new technologies became available. Currently, with the advancement of satellite technology, remote sensing, specialized software, and drones, among others, deforestation and its causes can be detected in real time. It is very important to have early warning of land use changes so that the competent authorities can take immediate action.

In the 1980s, due to the expansion of agriculture and livestock in the southwestern Amazon, large areas of forest were lost. After growing one or two crops, these lands were abandoned to recover their soil fertility. As a result of this anthropic activity, large tracts of "purmas", or fallows, appeared, many of them forming fragments of forests, patches or islands of biodiversity. The majority of unbridled areas are on average less than 5 ha.

The diverse species of plants present in the islands of biodiversity in the southwestern Amazon fulfill different functions that contribute to ecological processes, with many key species of large size that ensure their own continuity as well as the conservation of several other species of flora and fauna. These areas of conservation value are often used by the local population and communities, who know their various uses that facilitate their daily lives, and maintain this knowledge from generation to generation. However, the more recent settlers who may not be aware of these varied uses could benefit from greater dissemination of knowledge around these species to enhance their conservation value. Fragments of forests or relics, known as islands of biodiversity, can be an alternative for the conservation of biodiversity and all its ecological processes in areas close to cities or agricultural landscapes. With the adequate management of the species they contain, the biological connectivity necessary for the survival of the species can be achieved.

## References

AIDER (2020) Promoviendo y concertando el desarrollo rural bajo en emisiones en la Región Ucayali bajo el enfoque de producción, protección e inclusión (GCF-TF/PNUD Ucayali). Análisis integrado de causas y mecanismos causales de deforestación y cambio de uso. Informe Final de Análisis. Equipo técnico de Apoyo Regional Ucayali. Pucallpa, Perú, 71p

Armenteras D, Vargas O (2016) Patrones del paisaje y escenarios de restauración: acercando escalas. Acta Biol Colomb 21((1) Supl):S229–S239. https://doi.org/10.15446/abc.v21n1sup.50848

Begon M, Harper JL, Townsend CR (1996) Ecology individuals, populations and communities. Blackwell Science, London, p 1068

Bennett A (1998) Enlazando el paisaje: el papel de los corredores y la conectividad en la conservación de la vida silvestre. IUCN, Cambridge, 276pp

Bergoeing J (1998) Geomorfología de Costa Rica. Instituto Geográfico Nacional, San José, 408pp

Bierregaard RO, Lovejoy TE, Kapos V, dos Santos AA, Hutchings RW (1992) The biological dynamics of the tropical rainforest fragments: a prospective comparison of fragments and continuous forest. Bioscience 42:859–866

Brokaw N (2002) La persistencia de las especies vegetales en los fragmentos de bosque neotropical. Source: Kattan GH, Fragmentación: patrones y mecanismos de extinción de especies. In: Guariguata MR, Katan GH (eds) Ecología y Conservación de Bosques Neotropicales. EULAC – GTZ, 1st edn. Ediciones LUR, Cartago, pp 564–565

Chávez F (1991) Composición florística de los bosques secundarios en tres etapas sucesionales en la zona de Pucallpa. Tesis para optar el grado de Bachiller en Ciencias Biológicas. Universidad Nacional San Antonio Abad del Cusco, Perú

Clavo M, Fernández-Baca J (1999) Regeneración natural de especies arbóreas para el establecimiento de sistemas silvopastoriles. Revista de Investigaciones Veterinarias del Perú 10(1):71–81

Clavo Z, Roncal S, Ricse A, Sabogal C (2006) Composición florística post-quema en áreas degradadas por la agricultura en la Región Ucayali, Amazonia peruana. Revista Forestal del Perú 29(1–2):78–91

Cochrane T, Sánchez P (1982) Recursos de tierras, suelos y su manejo en la región Amazónica: Informe acerca del estado de conocimientos. In: Hecht SB (ed) Amazonía, Investigación sobre agricultura y uso de tierras. Centro Internacional de Agricultura Tropical, Cali, pp 141–218

Colorado Zuluaga GJ, Vásquez Muñoz JL, Mazo Zuluaga IN (2017) Modelo de conectividad ecológica de fragmentos de bosque andino en Santa Elena (Medellín, Colombia). Acta Biol Colomb 22(3):379–393

Conservación Amazónica–ACCA (2020) Hotspots y tendencias de deforestación en la Amazonía. Monitoring of the Andean Amazon Project. https://maaproject.org/2020/sintesis-2019/. Consulted 18 May 2020

Dourojeanni M (1987) Aprovechamiento del barbecho forestal en áreas de Agricultura migratoria en la Amazonía peruana. Revista Forestal del Perú CEDINFOR14(2):1–33

Estrada A, Coates-Estrada R (2001) Bat species richness in live fences and in corridors of residual rain forest vegetation at Los Tuxtlas, Mexico. Ecography 24:94–102

Finer M, Mamani N (2018) Deforestation hotspots in the Peruvian Amazon. MAAP (Mapping the Andean Amazon Project). Síntesis #3. https://maaproject.org/2018/sintesis3/. Consulted 26 Jan 2020

Forest Stewardship Council (2020) Principios y criterios del FSC para el manejo forestal responsable. FSC-STD-01-001 V5-2 E. Secretariado Internacional Charles-de-Gaulle-Str.5. 53113 Bonn, Germany, 32p. https://fsc.org/es/quienes-somos. Consultado el 17 May 2020

Francesconi W, Montagnini F (2015) Los SAFs como estrategias para favorecer la conectividad funcional del paisaje fragmentado. In: Montagnini F, Somarriba E, Murgueitio E, Fassola H, Eibl B (eds) Sistemas Agroforestales. Funciones Productivas, Socioeconómicas y Ambientales. Serie técnica. Informe técnico 402. CATIE, Turrialba. Editorial CIPAV, Cali, 454p

Fujisaka S, Escobar G, Veneklaas E (2000) Weedy fields and forests: interactions between land use and the composition of plant communities in the Peruvian Amazon. Agric Ecos Environ 78: 175–186

Gabriel VA, Pizo MA (2005) The use of fencerows by birds in a fragmented landscape of the Atlantic rainforest. Natureza e Conservacão 3:79–89(portugués), 182–192(inglés)

García R (2002) Biología de la conservación: conceptos y prácticas. INBio, Santo Domingo de Heredia, 166pp

Gazzo J (1982) Políticas y planes de desarrollo para la región amazónica del Perú. Amazonía, Investigación sobre agricultura y uso de tierras. In: Hecht SB (ed) Centro Internacional de Agricultura Tropical, Cali, pp 87–108

Gurrutxaga M (2004) Conectividad Ecológica del Territorio y Conservación de la Biodiversidad. Nuevas Perspectivas en Ecología del Paisaje y Ordenación del Territorio. Departamento de Agricultura y Pesca, Gobierno Vasco

Hobbs RJ, Wilson AM (1998) Corridors: theory, practice and achievement of conservation objetives. In: Dover W, Bunce RGH (eds) Key concepts in landscape ecology. International Association for Landscape Ecology, Preston, pp 265–279

Instituto Nacional de Estadística e Informática (INEI) (2018) Perú, Perfil Sociodemográfico. Censos Nacionales 2017: XII de Población, VII de Vivienda y III de Comunidades Indígenas. Lima, Perú, 641p

Kalliola R, Puhakka M (1993) Geografía de la Selva Baja Peruana. In: Kalliola R, Puhakka M, Danjoy W (eds) Amazonía Peruana. Vegetación húmeda tropical en el llano Sub Andino. Publicado por Proyecto Amazonía, Universidad de Turko y la Oficina Nacional de Evaluación de Recursos Naturales, Lima, pp 9–21

Kattan G (2002) Fragmentación de patrones y mecanismos de extinción de especies. In: Guariguata MR, Katan GH (eds) Ecología y Conservación de Bosques Neotropicales. EULAC – GTZ, 1st edn. Ediciones LUR, Cartago, pp 561–590

López E, Baldoceda R, Clavo ZM, Veneklaas E (2005) Estudio del banco de semillas presentes en suelos de barbechos originados por efecto de agricultura y pastura. Revista Forestal de Ucayali 2(2):1–15

Martín J, Fernández L, Urios G (2013) Los Bosques isla en Andalucía. Consejería de Medio Ambiente y Ordenación del Territorio, Junta de Andalucía, Sevilla, 192pp

McIntyre NE (1995) Effects of forest patch size on avian diversity. Landsc Ecol 10(2):85–99

Ministerio de Agricultura (MINAGRI) (2012) Plan Estratégico Regional del sector agrario de Ucayali. Gobierno Regional de Ucayali. Dirección Regional Agraria de Ucayali. Periodo: 2008–2012. Pucallpa, Perú

Ministerio del Ambiente (MINAN). (2020). Programa GeoBosques. Plataforma de monitoreo de cambios sobre la cobertura de los bosques. Bosque y pérdida de bosque en la región Ucayali. Lima. Perú. http://geobosques.minam.gob.pe/geobosque/view/perdida.php. Accesed on line 28 Jan 2020

Ministerio del Ambiente (MINAM) (2020) Servicio Nacional de Áreas Naturales Protegidas (SERNANP). Sistema de Áreas Naturales Protegidas del Perú. Áreas naturales protegidas de administración nacional con categoría definitiva. Lima Perú. https://www.sernanp.gob.pe/documents/10181/165150/Listado+ANP+20.01.2020.pdf/5e1a7af5-ab65-455c-9094-f86bf9b994d4. Consulted 28 Jan 2020

Morera C, Pintó J, Romero M (2007) Paisaje, procesos de fragmentación y redes ecológicas: aproximación conceptual. In: Chassot O, Morera C (eds) Corredores Biológicos: acercamiento conceptual y experiencias en América. Centro Científico Tropical/Universidad Nacional de Costa Rica, San José, 128pp

Nott MP, Pimm SL (1997) The evaluation of biodiversity as a target for conservation. In: Pickett STA, Ostfeld RS, Shachak M, Likens GE (eds) The ecological basis of conservation. Springer, Boston

Porro R, López A, Vela J (2015) Forest use and agriculture in Ucayali, Peru: livelihood strategies, poverty and wealth in an Amazon frontier. Forest Policy Econ 51:47–56

Redondo-Brenes A, Montagnini F (2010) Contribution of homegardens, agrosilvopastoral systems, and other human-dominated land-use types to the avian diversity of a biological corridor in Costa Rica. Adv Environ Res 2(2):111–148

Riesco A, Ara M, De La Torre M (1995) Informe final Proyecto Sistemas Amazónicos Sostenibles. SAS. Financiado por CIID Canadá, Ejecutado por IVITA-Pucallpa

Robiglio V, Reyes M, Castro E (2015) Diagnóstico de los productores familiares en la Amazonia Peruana. ICRAF, Oficina Regional para América Latina, Lima, Perú. Por encargo de GGGI&DIE

Rodríguez F (2003) Perú: Ordenamiento Territorial con base en la Zonificación Ecológica Económica en la Amazonia. Programa de Ordenamiento Ambiental. Instituto de Investigaciones de la Amazonía Peruana. Iquitos, Perú

Sabogal C, Colán V, Galván O, Current D, Dominguez G, Yepes F (2001) Aspectos metodológicos para determinar opciones de manejo de bosques secundarios (purmas) en la Amazonia peruana, con especial referencia a la zona de Pucallpa. Revista Forestal del Perú 24(1–2):85–109

Salisbury DS, Fagan C (2011) Coca and conservation: cultivation, eradication, and trafficking in the Amazon borderlands. GeoJournal 11:1–20

Toledo J, Serrao A (1982) Producción de pastos y ganado en la Amazonía. In: Hecht SB (ed) Amazonía, Investigación sobre agricultura y uso de tierras. Centro Internacional de Agricultura Tropical, Cali, pp 295–324

Tuomisto H (1993) Clasificación de vegetación en la Selva Baja Peruana. In: Kalliola R, Puhakka M, Danjoy W (eds) Amazonía Peruana. Vegetación húmeda tropical en el llano Sub Andino. Publicado por Proyecto Amazonía, Universidad de Turko y la Oficina Nacional de Evaluación de Recursos Naturales, Lima, pp 103–112

Vandermeer J, Perfecto I, Philpott S, Chappell MJ (2008) Reenfocando la conservación en el paisaje: la importancia de la matriz. In: Harvey CA, Sáenz JC (eds) Evaluación y conservación de biodiversidad en paisajes fragmentados de Mesoamérica. INBio, Santo Domingo de Heredia, pp 75–104

Vela J, Clavo M, Alvez-Valles C, Mori J, Panduro N, Ramírez N, Bautista J, Tavera I, Fernández V (2019) Construcción de un modelo de ganadería sostenible con pequeños productores de leche como una opción de mitigación y adaptación al calentamiento global. Amazonía Sur Occidental, Pucallpa, Perú. Informe de Avance 2019. Facultad de Ciencias Agropecuarias, Vicerrectorado de investigación, Universidad Nacional de Ucayali, Pucallpa, Perú, 40p

Vivanco L (2004) Gran enciclopedia de la región Ucayali. Amazonía Peruana: Identidad Regional Pucallpa. Perú, 303p

# Chapter 10
# The Monteverde Cloud Forest: Evolution of a Biodiversity Island in Costa Rica

**Quint Newcomer, Fabricio Camacho Céspedes, and Lindsay Stallcup**

**Abstract** Monteverde, Costa Rica represents an example of dynamic processes shaping an ever-changing, landscape-scale biodiversity island. Monteverde is internationally renowned for biodiversity conservation efforts initiated by non-governmental organizations and private citizens that led to the creation of the Monteverde Reserve Complex—a network of reserves spanning the region. Located in the Tilarán Mountain Range, an area of high endemism in the Central American isthmus, Monteverde's reserves provide habitat for over half of the species found in the entire country of Costa Rica, including 55 species of birds, mammals, amphibians and reptiles with some degree of threatened status on the IUCN Red List. One characteristic that makes Monteverde unusual is the number of research scientists that have settled in the area and studied the region over multiple decades. Some of these scientists helped secure international funding to purchase land for the Monteverde Cloud Forest Biological Preserve and the Children's Eternal Rainforest and participated in the creation of local non-governmental organizations to promote conservation, education, and sustainable community development. Recognizing that Monteverde's biodiversity island requires habitat connectivity across a larger landscape to support seasonal migratory species, Monteverde's organizations established the Bellbird Biological Corridor. The impacts of changing climate conditions—in particular, the increase of daily minimum temperatures and the increase in number of consecutive dry day periods—are being observed in Monteverde's cloud forests and further threaten the conservation of habitat and species. Holistic policies and programs spanning tourism, agriculture, transportation, energy, and environmental sectors are needed for continued conservation successes.

Q. Newcomer (✉)
University of Georgia, Warnell School of Forestry and Natural Resources, Athens, GA, USA

F. Camacho Céspedes
Tropical Science Center, Monteverde, Costa Rica

L. Stallcup
Monteverde Conservation League, Monteverde, Costa Rica



F. Montagnini (ed.), *Biodiversity Islands: Strategies for Conservation in Human-Dominated Environments*, Topics in Biodiversity and Conservation 20,
https://doi.org/10.1007/978-3-030-92234-4_10

**Keywords** Biodiversity conservation · Biological corridors · Endangered species · Grassroots conservation · Private land conservation · Tropical Forest conservation

## 10.1 Introduction

The arc of the Central American isthmus spans approximately 1,835 km (1,140 mi) from the southeast in Panama to the northwest in Guatemala. At its narrowest point, the isthmus is only 50 km (30 mi) wide. The region serves as a land bridge between North and South America, a critical path and melting pot for species from both hemispheres that colonized the isthmus during the past three million years. Coastal plains on both Atlantic and Pacific slopes are connected by a mountainous spine. Yet the mountains also form a barrier of separation between Atlantic and Pacific slopes, which have distinct ecological conditions shaped by both geology and climate patterns. As a result, in a relatively small area, Central America has great biodiversity with many micro-climates. In Costa Rica alone, Holdridge (1967) described 13 life zones.

Monteverde, Costa Rica is a small town (population less than 1,000) at the end of a meandering mountain road. The town of Monteverde lends its name to the surrounding region, which includes the population center of Santa Elena, and several other communities including San Luis, Cerro Plano, Cañitas, and La Lindora. Monteverde also lends its name, particularly within the international tourism industry, to the region's natural environment—the Monteverde Cloud Forest.

In this chapter, we describe the broader region of Monteverde, including both the town and the region's cloud forests, as a case study example of the processes shaping a dynamic, large-scale biodiversity island.[1] Monteverde is internationally renowned for privately-owned biodiversity conservation efforts that led to the creation of the greater Monteverde Reserve Complex—a network of reserves spanning the region. Its exceptional natural environment is complemented by a distinctive socio-cultural environment which was fundamental to the establishment of the network of reserves and more recently the expansion of the biological corridor extending outward from Monteverde through the region (see Sect. 3.3). Monteverde's outstanding conservation successes are tempered by many challenges. Based on our personal experience as administrators of local organizations and active participants in community-based conservation initiatives during the past three decades, it the authors' perspective that the ability of Monteverde's citizenry—including governmental offices,

[1] For background and general description of the region's ecology and conservation history, we rely heavily on the seminal work of Nalini Nadkarni and Nathaniel Wheelwright, *Monteverde: Ecology and Conservation of a Tropical Cloud Forest*, first published in 2000 with updated chapters published in 2014. We encourage readers interested in a more in-depth understanding of Monteverde to reference this outstanding resource. Our reflections on conservation initiatives in Monteverde, including the Bellbird Biological Corridor, the Children's Eternal Rainforest, Enlace Verde, and local reforestation projects are largely derived from our personal involvement in these initiatives and organizations.

non-governmental organizations, private-sector businesses, and engaged individual community members—to self-reflect, adapt to local issues and to external, global-scale threats, and continue to lead biodiversity conservation efforts makes for an interesting case study.

Nestled high in the Tilarán Mountain Range in northwestern Costa Rica, the Monteverde region is home to the Monteverde Reserve Complex, a 27,500-hectare block of protected forest comprised of three privately owned reserves (the Monteverde Cloud Forest Biological Preserve, Children's Eternal Rainforest, and Bosqueterno) and the state-owned Santa Elena Cloud Forest Reserve (see Fig. 10.1). The Reserve Complex is the centerpiece of a larger protected area, which we will refer to as the Monteverde Arenal Bioregion (MAB), that includes the state-owned Arenal Volcano National Park and Alberto Manuel Brenes Biological Reserve (see Fig. 10.2) and other smaller reserves. The MAB is connected to other protected areas via biological corridors, including the Bellbird Biological Corridor, which aims to connect Monteverde with the mangrove forests on Costa Rica's Pacific gulf coast.

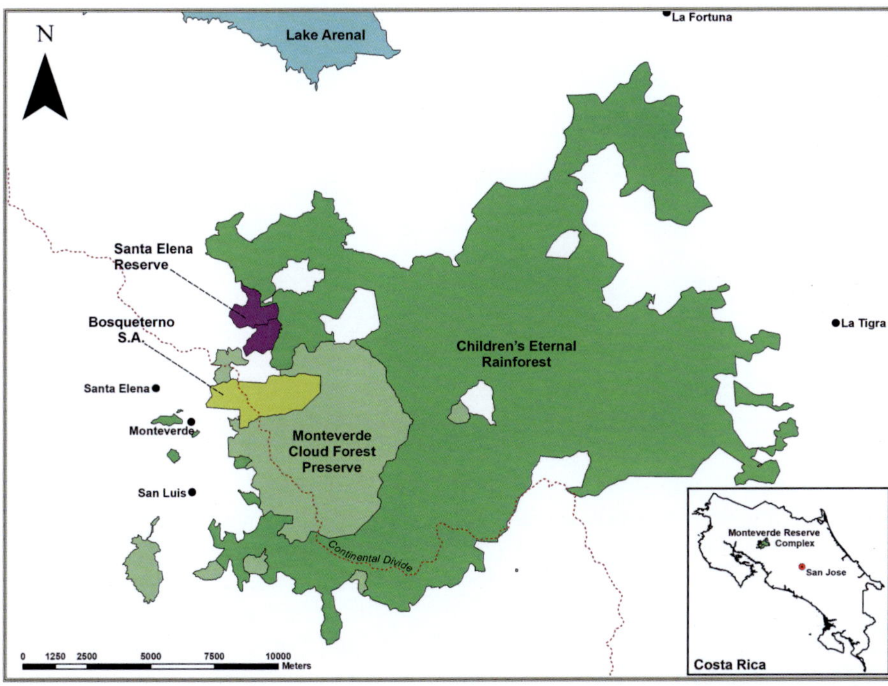


**Fig. 10.1** The Monteverde Reserve Complex is comprised of three privately-owned reserves—the Children's Eternal Rainforest, owned by the Monteverde Conservation League; the Monteverde Cloud Forest Biological Preserve, owned by the Tropical Science Center; and Bosqueterno S.A., a corporation whose land assets are managed by the Tropical Science Center and which is owned by Quaker families who settled in Monteverde in the 1950s. Map Source: Yuber Rodríguez, Monteverde Conservation League 2020

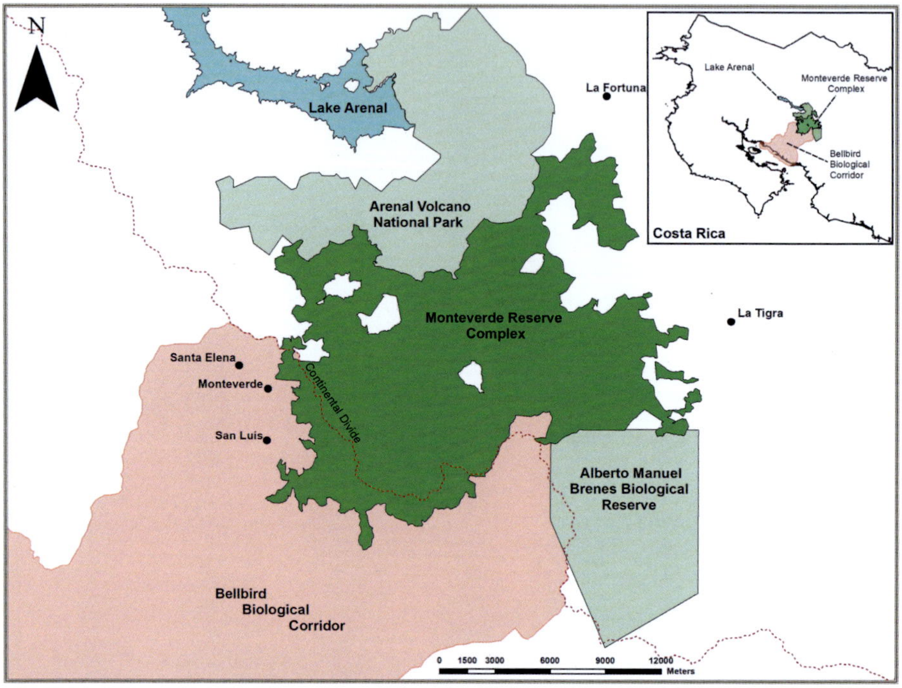


**Fig. 10.2** The Monteverde Arenal Bioregion surrounding the Monteverde Reserve Complex includes two large government-owned protected areas—Arenal Volcano National Park and the Alberto Manuel Brenes Biological Reserve, several smaller reserves and protected areas, and the Bellbird Biological Corridor which connects Monteverde with the Gulf of Nicoya. Map Source: Yuber Rodríguez, Monteverde Conservation League 2020

The properties within the Monteverde Reserve Complex span seven Holdridge life zones (Haber 2000a; Holdridge 1967; Tosi et al. 1969). The elevational gradient of the Monteverde Reserve Complex ranges from 700 meters above sea level (masl) on the Pacific slope up to the Continental Divide—1,850 masl at its highest point—and down the Atlantic slope to 466 masl. The species on the Pacific slope have greater affinity with the tropical dry forest ecosystem, while organisms on the Atlantic slope tend to be more adapted to the humid conditions of the tropical rain forest. The species found at higher elevations are also tolerant to cool, humid conditions and are adapted to resist strong easterly trade winds, heaviest between December – March with gusts up to 80 kmph (Tropical Science Center unpublished data). Much of the Monteverde region falls within the tropical montane cloud forest vegetation type (see Fig. 10.3), which, according to Nadkarni and Wheelwright (2000: 9) is "one of the world's most threatened ecosystems."

As traditionally understood, islands undergo geological processes of formation—the eruption of volcanoes creating igneous landforms, and the collision and subduction/abduction of continental plates lifting up land masses. While geological forces created the underlying conditions for the establishment of the cloud forest and the

**Fig. 10.3** A landscape view of the Monteverde Cloud Forest Preserve and the Children's Eternal Rainforest from the Continental Divide looking down the Atlantic slope. The Monteverde Reserve Complex represents the largest privately-owned protected area in Central America. Image Source: Fabricio Camacho Céspedes

evolution of the ecological niches that established along the Continental Divide in the Tilarán Mountain range in Costa Rica, the formation of Monteverde's biodiversity island was driven by strong socio-political, economic, cultural components over decades, rather than millennia. Expansionist development policies encouraged human population incursion into forested regions. Frontiersmen's axes felled seemingly inexhaustible tropical forests, fires "cleaned" the land, and cattle farms expanded across the landscape. As Harvey and Haber (1999) describe, this process resulted in the creation of small, sometimes single remnant tree biodiversity islands throughout the Monteverde region.

Agricultural expansion in the surrounding landscape completed the transformation of the remaining forests in the Monteverde region into a biodiversity island. Forested highland hillsides were replaced by coffee plantations and dairy cattle operations. Rivers flowing to the Caribbean were dammed and diverted to the Pacific, providing hydropower and supplying irrigation to the dry northwestern province of Guanacaste. The arid tropical dry forest and drained coastal wetlands were replaced with extensive fields of cattle, rice, sugar cane, and eventually pineapple. To the north of the Monteverde region, extensive tourism development modified the landscape near the Arenal Volcano; ornamental plants, pineapple, and family farms dominated areas east of the Monteverde Reserve Complex.

Yet in the midst of this landscape transformation, there were also concerted efforts to protect the region's natural resources. In the 1950s, Quaker settlers in Monteverde recognized the need to maintain forested lands that protected highland springs, the community's primary water source. This area, known as Bosqueterno, would eventually become the nucleus of the Monteverde Cloud Forest Biological Preserve, which was founded in 1972. In 1977, the Costa Rican government declared protected areas in the region (the Arenal Forest Reserve, later to become the Zona Protectora Arenal Monteverde, or ZPAM). The government was unable to purchase or effectively manage the lands it had declared as a reserve, however, and private landowners continued agricultural expansion into their farms in this area. Private conservation initiatives, led by the Tropical Science Center and Monteverde Conservation League, sprang up in an effort to purchase primary rainforest before it was felled (see Sect. 3.1). In the 1980s and 1990s, Costa Rica's international debt would be swapped for investments into the expansion of the Monteverde biodiversity island, eventually—in combination with funds raised by youth around the world—creating the Children's Eternal Rainforest.

The focus has shifted over time from protecting water to protecting specific endangered species, then to ecosystem conservation (Burlingame 2000). Most recently, the Monteverde community is engaged in an evolving initiative integrating farm-scale to watershed-scale ecosystem management and sustainable economic development with the goal of climate change resiliency for the region's biodiversity and the growing human population (Brenes et al. 2019). Simultaneously, research scientists and conservation organizations are working to develop a more unified approach to share data, collaborate across disciplines, and perhaps develop new methods and approaches for understanding and disseminating information about the region's ecology (Allen and Hoekstra 1992; Zamsow et al. 2018).

The following sections highlight natural history characteristics as well as some of the organizations and initiatives that make Monteverde's biodiversity island unique and of critical importance as a large node in Costa Rica's national landscape matrix of conservation. We then describe some of the challenges for ongoing biodiversity conservation, many of which are not unique to the Monteverde region. We remind the reader that the complete story of biodiversity conservation in Monteverde, Costa Rica is far more complex than we are able to present in this snapshot.

## 10.2 Species Accounts

> *"Biodiversity in the broadest sense is a characteristic feature of Monteverde"*. Wheelwright 2000:420.

The diversity of micro-climates (or life zones) in the Monteverde region positively impacts the area's overall biodiversity. The Atlantic slope of Monteverde has the greatest bird diversity in Costa Rica—40% of Central America's mammal species have been identified in Monteverde. The area is home to one-third of Costa Rica's plant species, including as many families of epiphytes as the entire country of Mexico. Ten percent of Monteverde's plants are endemic to Costa Rica's Tilarán mountain range (Wheelwright 2000).In this section, we offer a brief glimpse into Monteverde's natural history. For a much more complete and detailed description refer to Nadkarni and Wheelwright (2000, 2014).

### *10.2.1 Epiphytes, Orchids*

Characteristic to Monteverde's vegetation are its diverse epiphyte communities which represent the most diverse plant group in Monteverde with more than 800 species including 471 species of monocots, 230 species of dicots, and 177 species of ferns (Haber 2000a, b). These communities not only add to the overall biodiversity in Monteverde, but also play a significant role in cloud forest ecosystem dynamics. For example, many of these plants decompose in the forest canopy, producing an abundant accumulation of canopy soil that is rich in organic matter and which contributes to nutrient cycling and conservation (Nadkarni 1981; Nadkarni et al. 2002). The canopy epiphyte root-humus mat exhibit high potential water storage capacity (Köhler et al. 2007). However, the actual water availability in this canopy-mat is conditioned to rainfall and evaporation patterns (Köhler et al. 2007), which is one of the main contributing factors that suggest that this may be one of the most vulnerable communities to the effects of climate change (Clark et al. 2014).

Orchids represent the most diverse epiphyte family in Monteverde with more than 450 species (Haber 2000a, b), of which more than 30 species are new to science (Atwood and Dressler 1995). This richness makes Monteverde one of the most, if not the most, orchid-diverse locations on the planet (Haber 2000a, b). The most important aspects that add up to this diversity are the abundance of life zones, pollinators and the intensity of studies in the area (Atwood and Dressler 1995). Orchid diversity in Monteverde represents about one third of the estimated number of orchid species in Costa Rica (Atwood and Dressler 1995). Although it is one of the most prominent families in terms of species diversity, many species known as miniature orchids are inconspicuous and difficult to recognize among other epiphytes by the untrained eye.

### *10.2.2 Vascular Plants*

Within the seven Holdridge life zones (Holdridge 1967) included in the Monteverde Flora Project study area, there are three forest types—Pacific slope seasonal forest, cloud forest, and Atlantic slope rain forest (Haber 2000a). The Monteverde Flora Project was an initiative of the Missouri Botanical Garden and the Manual to the Plants of Costa Rica Project. Through the Monteverde Flora Project, the flora of the

Monteverde area, from the Continental Divide down both Atlantic and Pacific slopes to 700 masl, was collected and identified (Haber 1991, 2000a, b).

In addition to the Monteverde Flora Project, multiple studies since the mid-1970s have contributed to the identification and classification of Monteverde's flora (e.g., Dyer 1979; Hartshorn 1983). By 2000, the list of vascular plants for Monteverde included 3,021 species with 755 species of trees (Haber 2000a, b). This represents about one-third of all vascular plant species in Costa Rica (Haber 2000a, b). About 10% of Monteverde's flora species are endemic to the Tilarán Mountains (Haber 2000a, b), and over 10% of the 216 of the common or characteristic tree species of the life zones of Monteverde face some degree of threat as per the IUCN Red List (see Table 10.1).

The diversity of wild avocados—66 species of Lauraceae representing eight genera (Haber 2000a, b)—in Monteverde is notably greater than in adjacent areas. Lauraceae play a key role as a food source for many frugivorous birds, including Three-wattled Bellbirds (*Procnias tricarunculata*) and Resplendent Quetzals (*Pharomacros moccino*) (see Fig. 10.4), which synchronize their migration routes to follow the fruiting patters of these and other tree species (Powell and Bjork 1994). The lipids contained in the wild avocados serve as one of the main sources of energy to support the reproduction of these iconic birds.

### 10.2.3 Avifauna

Many early scientific publications (e.g., Snow 1977; Wheelwright 1983) focused on Monteverde's spectacular avifauna, capturing the attention of the international birding community. Natural history films (e.g., BBC's *Forest in the Clouds*), books (e.g., National Geographic's *Mountain Worlds*), and magazine articles (e.g., *International Wildlife's* "Is This the Garden of Eden?") based on the region's impressive biodiversity also brought international attention to the area (Burlingame 2000).

To date, more than 400 of Costa Rica's 850 avian species have been reported in the area (Fogden 2000), including a number of IUCN Red Listed species such as the Great Curassow (*Crax rubra*), Keel-billed Motmot (*Electron carinatum*), Blue-and-gold Tanager (*Bangsia arcaei*), the above-mentioned Three-wattled Bellbird and Resplendent Quetzal, and the Bare-necked Umbrellabird (*Cephalopterus glabricollis*) (see Table 10.1). Avian diversity is amplified by the combination of resident and migratory communities, which include latitudinal and altitudinal migrants. Among the latter, the most economically important species is the Resplendent Quetzal, due to its importance to Monteverde's ecotourism industry.

**Table 10.1** Species found in the Monteverde area which are listed as Near Threatened, Vulnerable, Endangered, Critically Endangered, Extinct in the Wild, or Extinct on the IUCN Red List

| Species | Common name | IUCN Red List Status | Citation |
|---|---|---|---|
| **Plants**[a] | | | |
| *Agonandra macrocarpa* | | Vulnerable | Nelson (1998a) |
| *Bombacopsis quinata* | | Vulnerable | Sandiford (1998) |
| *Capparis discolor* | | Near Threatened | Mitré (1998a) |
| *Chrysophyllum hirsutum* | | Near Threatened | World Conservation Monitoring Centre (1998a) |
| *Costus nitidus* | | Endangered | Skinner (2014) |
| *Dichapetalum costarricense* | | Vulnerable | World Conservation Monitoring Centre (1998b) |
| *Elaeagia uxpanapensis* | | Endangered | World Conservation Monitoring Centre (1998c) |
| *Eugenia salamensis* | | Endangered | Nelson (1998b) |
| *Ficus lateriflora* | | Critically Endangered | Page (1998) |
| *Ilex costaricensis* | | Vulnerable | Mitré (1998b) |
| *Magnolia poasana* | | Near Threatened | Khela (2014) |
| *Ocotea monteverdensis* | Quizzará Blanco | Critically endangered | Joslin et al. (2018) |
| *Ocotea viridiflora* | | Vulnerable | World Conservation Monitoring Centre (1998d) |
| *Oreomunnea pterocarpa* | | Endangered | Americas Regional Workshop (1998) |
| *Persea schiedeana* | Coyo Avocado | Endangered | Wegier et al. (2017) |
| *Pouteria austin-smithii* | | Vulnerable | World Conservation Monitoring Centre (1998e) |
| *Pouteria congestifolia* | | Vulnerable | World Conservation Monitoring Centre (1998f) |
| *Pouteria fossicola* | | Vulnerable | World Conservation Monitoring Centre (1998g) |
| *Sideroxylon capiri* | | Near Threatened | World Conservation Monitoring Centre (1998h) |

(continued)

**Table 10.1** (continued)

| Species | Common name | IUCN Red List Status | Citation |
|---|---|---|---|
| *Sideroxylon persimile* | | Near Threatened | World Conservation Monitoring Centre (1998i) |
| *Terminalia bucidoides* | | Endangered | Nelson (1998c) |
| *Ticodendron incognitum* | Duranzo de Ardilla | Near Threatened | Rivers et al. (2019) |
| *Zinowiewia costaricensis* | | Near Threatened | Mitré (1998c) |
| ***Amphibians***[2] | | | |
| *Agalychnis annae* | Blue-sided Treefrog | Endangered | IUCN SSC Amphibian Specialist Group and NatureServe (2014) |
| *Agalychnis lemur* | Lemur Leaf Frog | Critically Endangered | Solís et al. (2008d) |
| *Atelopus varius* | Variable Harlequin Frog | Critically Endangered | Pounds et al. (2010) |
| *Bolitoglossa subpalmata* | La Palma Salamander | Endangered | Pounds et al. (2008c) |
| *Craugastor andi* | Atlantic Robber Frog | Critically Endangered | Pounds et al. (2008a) |
| *Craugastor angelicus* | Angel Robber Frog | Critically Endangered | Pounds et al. (2008b) |
| *Craugastor podiciferus* | Cerro Utyum Robber Frog | Near Threatened | Solís et al. (2010a) |
| *Duellmanohyla uranochroa* | Costa Rica Brook Frog | Endangered | NatureServe and IUCN SSC Amphibian Specialist Group (2013) |
| *Ecnomiohyla fimbrimembra* | Heredia Treefrog | Endangered | Solís et al. (2008a) |
| *Ecnomiohyla miliaria* | Cope's Brown Treefrog | Vulnerable | Solís et al. (2010b) |
| *Incilius periglenes* | Golden Toad | Extinct | Savage et al. (2008) |
| *Isthmohyla angustilineata* | Narrow-lined Treefrog | Critically Endangered | Solís et al. (2008c) |
| *Isthmohyla rivularis* | American Cinchona Plantation Treefrog | Critically Endangered | Solís et al. (2010c) |
| *Isthmohyla tica* | Starrett's Treefrog | Critically Endangered | IUCN SSC Amphibian Specialist Group and NatureServe (2013a) |

(continued)

**Table 10.1** (continued)

| Species | Common name | IUCN Red List Status | Citation |
|---|---|---|---|
| *Isthmohyla zeteki* | Zetek's Treefrog | Near Threatened | Solís et al. (2008b) |
| *Lithobates vibicarius* | Rancho Redondo Frog | Vulnerable | IUCN SSC Amphibian Specialist Group and NatureServe (2013b) |
| *Nototriton gamezi* | Monteverde Moss Salamander | Vulnerable | Pounds et al. (2008f) |
| *Nototriton picadoi* | La Estrella Salamander | Near Threatened | Bolaños et al. (2008b) |
| *Oedipina poelzi* | Quarry Worm Salamander | Endangered | Bolaños et al. (2008a) |
| *Oedipina uniformis* | Cienega Colorado Worm Salamander | Near Threatened | Wake et al. (2008) |
| *Pristimantis altae* | Mountain Robber Frog | Near Threatened | Pounds et al. (2008d) |
| *Pristimantis caryophyllaceus* | La Loma Robber Frog | Near Threatened | Pounds et al. (2008e) |
| **Reptiles**[b] | | | |
| *Celestus hylaius* | Rain Forest Caiman Lizard | Near Threatened | Solórzano et al. (2013) |
| *Trimetopon simile* | Dunn's Tropical Ground Snake | Endangered | Porras et al. (2013) |
| ***Insects (Odonata)***[c] | | | |
| *Libellula mariae* | | Near Threatened | Paulson (2009b) |
| *Palaemnema baltodanoi* | | Endangered | Paulson and von Ellenrieder (2006a) |
| *Perigomphus pallidistylus* | | Vulnerable | Paulson and von Ellenrieder (2006b) |
| **Birds**[d] | | | |
| *Ara ambiguus* | Great Green Macaw | Endangered (once abundant in Atlantic Slope lower elevations, but has not been recently sighted) | BirdLife International (2016a) |
| *Antrostomus carolinensis* | Chuck-will's Widow | Near Threatened | BirdLife International (2018a) |
| *Bangsia arcaei* | Blue-and-gold Tanager | Near Threatened | BirdLife International (2018b) |
| *Buteogallus solitarius* | Black Solitary Eagle | Near Threatened | BirdLife International (2016b) |

(continued)

**Table 10.1** (continued)

| Species | Common name | IUCN Red List Status | Citation |
|---|---|---|---|
| *Cephalopterus glabricollis* | Bare-necked Umbrellabird | Endangered | Birdlife International (2016c) |
| *Contopus cooperi* | Olive-sided Flycatcher | Near Threatened | BirdLife International (2017a) |
| *Crax rubra* | Great Curassow | Vulnerable | BirdLife International (2016d) |
| *Electron carinatum* | Keel-billed Motmot | Vulnerable | BirdLife International (2016e) |
| *Grallaricula flavirostris* | Ochre-breasted Antpitta | Near Threatened | BirdLife International (2017b) |
| *Hylocichla mustelina* | Wood Thrush | Near Threatened | BirdLife International (2017c) |
| *Morphnus guianensis* | Crested Eagle | Near Threatened | BirdLife International (2017d) |
| *Patagioenas subvinacea* | Ruddy Pigeon | Vulnerable | BirdLife International (2016f) |
| *Pharomachrus mocinno* | Resplendent Quetzal | Near Threatened | Birdlife International (2016g) |
| *Procnias tricarunculatus* | Three-wattled Bellbird | Vulnerable | Birdlife International (2016h) |
| *Sclerurus albigularis* | Gray-throated Leaftosser | Near Threatened | BirdLife International (2016i) |
| *Setophaga cerulea* | Cerulean Warbler | Near Threatened | BirdLife International (2019a) |
| *Spizaetus ornatus* | Ornate Hawk-Eagle | Near Threatened | BirdLife International (2016j) |
| *Sturnella magna* | Eastern Meadowlark | Near Threatened | BirdLife International (2019b) |
| *Tinamus major* | Great Tinamou | Near Threatened | BirdLife International (2017e) |
| *Touit costaricensis* | Red-fronted Parrotlet | Vulnerable | BirdLife International (2018c) |
| *Vermivora chrysoptera* | Golden-winged Warbler | Near Threatened | BirdLife International (2018d) |
| ***Mammals***[5] | | | |
| *Ateles geoffroyi* | Black-handed Spider Monkey | Endangered | Cuarón et al. (2008) |
| *Ectophylla alba* | Caribbean White Bat | Near Threatened | Rodriguez and Pineda (2015) |
| *Leopardus tigrinus* | Northern Tiger Cat | Vulnerable | Payan and de Oliveira (2016) |
| *Leopardus wiedii* | Margay | Near Threatened | de Oliveira et al. (2015) |

(continued)

**Table 10.1** (continued)

| Species | Common name | IUCN Red List Status | Citation |
|---|---|---|---|
| *Myrmecophaga tridactyla* | Giant Anteater | Vulnerable (once present in Monteverde, but not recently sighted) | Miranda et al. (2014) |
| *Panthera onca* | Jaguar | Near Threatened | Quigley et al. (2017) |
| *Sylvilagus brasiliensis* | Tapeti | Endangered | Ruedas and Smith (2019) |
| *Tapirus bairdii* | Baird's Tapir | Endangered | García et al. (2016) |
| *Tayassu pecari* | White-lipped Peccary | Vulnerable (once present in Monteverde, but not recently sighted) | Keuroghlian et al. (2013) |
| *Vampyrum spectrum* | Spectral Bat | Near Threatened | Solari (2018) |

[a]Based on database search of IUCN Red List (iucnredlist.org) using the 216 species listed in Table 3.2. Common or characteristic tree species of the life zones in the Monteverde area in the chapter Plants and Vegetation (Haber 2000a, b) in Nadkarni and Wheelwright (2000). An exhaustive list of the status of all flora in the Monteverde area is beyond the scope of this chapter. Furthermore, many of the 3021 species of vascular plants known to the Monteverde area have not been evaluated for their IUCN Red List endangered status. Therefore, a representative sample of the most common or characteristic tree species (216 of the 755 known tree species in Monteverde) was deemed appropriate for the purposes of this chapter

[b]Based on database search of IUCN Red List (iucnredlist.org) using the complete species list of Amphibians and Reptiles of Monteverde provided in Pounds and Fogden (2000). This was also cross-referenced with the most recent unpublished Amphibians and Reptiles species lists for the Monteverde Cloud Forest Biological Preserve compiled by the Tropical Science Center (2019 unpublished data) to ensure any new species added to the lists of observed species in the Monteverde area were included

[c]An exhaustive study of all insect orders in the Monteverde area has yet to be completed. We have chosen to use the order Odonata as an example subset of insects representing the Monteverde area. Based on database search of IUCN Red List (iucnredlist.org) using the complete species list ($n = 102$) of Odonata collected at 700 m or higher on both Pacific and Atlantic slopes within the Monteverde area provided by Haber (personal communication, 2020 unpublished data). An additional species of Odonata, *Epigomphus subsimilis*, found in the lower Pacific slopes (355 m) is IUCN Red Listed as Endangered (Paulson 2009a, b). 51 of the 102 species identified by Haber in the Monteverde area have not been evaluated for inclusion in the IUCN Red List Database, and five of those 51 species were not included in the Catalogue of Life database (http://www.catalogueoflife.org/col/search/all). One Odonata species was included in the IUCN Red List database, however its status could not be determined due to deficient data

[d]Based on database search of IUCN Red List (iucnredlist.org) using the complete species list of Birds of the Monteverde Area provided in Fogden (2000). This was also cross-referenced with the most recent unpublished Bird species list for the Monteverde Cloud Forest Biological Preserve compiled by the Tropical Science Center (2019 unpublished data) to ensure any new species added to the lists of observed species in the Monteverde area were included

[e]Based on database search of IUCN Red List (iucnredlist.org) using the complete species list of Mammals of Monteverde provided in Timm and LaVal (2000). This was also cross-referenced with the most recent unpublished Mammals species lists for the Monteverde Cloud Forest Biological Preserve compiled by the Tropical Science Center (2019 unpublished data) to ensure any new species added to the lists of observed species in the Monteverde area were included

**Fig. 10.4** The Resplendent Quetzal (*Pharomacros moccino*) and the Three-wattled Bellbird (*Procnias tricarunculata*) represent two of Monteverde's iconic, IUCN Red Listed bird species. Both species have attracted scientific researchers to the region and have played an important role in attracting bird watchers and nature-based tourism in general to Monteverde. (Image Sources: Alvaro Cubero (Resplendent Quetzal) and Orlando Calvo (Three-wattled Bellbird))

Even this flagship species has not escaped environmental threats; the Tropical Science Center[2] estimates a current population of 55 individual quetzals (male, female and juveniles) in the Monteverde Cloud Forest Biological Preserve (TSC unpublished data) in contrast to the 50 reproductive couples estimated by Wheelwright (1983). Hamilton et al. (2003) report habitat loss and fragmentation on the Pacific Slope as the main causes of population declines of Three-wattled Bellbirds in Monteverde. This species declined from 135 individuals in 1997, to just 90 in 2002. Current monitoring efforts by the Tropical Science Center in the Monteverde region revealed that the population of Bellbirds may be increasing; 102 individuals were identified in 2018 and 120 in 2019 (TSC unpublished data).

Over one-third of Costa Rica's hummingbird species have been identified in Monteverde (Feinsinger 1977). It is believed that cooler air temperatures at higher elevations may result in a lower abundance of pollinating insects, thus providing conditions that support a higher diversity of hummingbirds due to lower resource competition. Several species of vascular plants, including *Justicia sp.* (Acanthacea)

[2]The Tropical Science Center (TSC) is Costa Rica's oldest environmental non-profit, non-governmental organization. TSC was established in 1962 by research scientists Leslie Holdridge, Joseph Tosi, and Robert Hunter together with several local businessmen. See Sect. 3.1.1 for further description of TSC's role in the Monteverde Reserve Complex.

and *Satyria sp.* (Ericaceae), have coevolved pollination systems to capitalize on diverse hummingbird phenotypes (Deliso 2008).

### 10.2.4 Insects

There has been no comprehensive study of insects in Monteverde, by far the most diverse and abundant group of organisms in the region. The region is home to 102 species of damselflies and dragonflies (W Haber personal communication). Monteverde's butterflies have been studied by Haber (1993) and Stevenson (W Haber and R Stevenson, unpublished data), who indicate a total of 658 species in the Monteverde zone (Stevenson and Haber 2000, W Haber personal communication). More than half of the area's butterfly species are altitudinal migrants, moving up and down the Pacific and Caribbean slopes with the wet and dry seasons (Stevenson and Haber 2000). Habitat availability is critically important, and changing climate conditions are driving up-slope migration of Pacific lowland butterfly species—in addition to lowland birds and mammals—into the Monteverde area (J Porter personal communication; Wheelwright 2000; Timm and LaVal 2000).

Ant faunas of cloud forests are strongly specialized for cloud forest habitat and sharply differentiated from the adjacent lowlands. The ant fauna of Monteverde has been extensively studied by Longino, who estimated that about 70 species are cloud forest specialists (J Longino personal communication). About a third of these are widespread, known from other cloud forest sites in Central America. Most of the rest are restricted to the cloud forests of Costa Rica and adjacent parts of Panama. A few are known only from Monteverde, but it is difficult to know if this is true local endemism or a result of under-sampling elsewhere. Recent DNA studies are showing that some of the species shared among different Costa Rican mountain ranges actually have deep genetic divergences and have been isolated from each other for millions of years (J Longino personal communication).

### 10.2.5 Larger Mammals

Three species of monkeys—Mantled Howler Monkeys (*Alouatta palliata)*, White-faced Capuchins (*Cebus capucinus*), and the IUCN Red List endangered Black-handed Spider Monkeys (*Ateles geoffroyi*); agoutis (*Dasyprocta punctate*), pacas (*Cuniculus paca*), and prehensile-tailed porcupines (*Coendou mexicanus*) among other rodents; coatis (*Nasua narica*), peccaries (*Tayassu pecari* and *Pecari tajacu*), sloths (*Bradypus variegatus* and *Choloepus hoffmanni*), kinkajous (*Potos flavus*), olingos (*Bassaricyon gabbii*), and opossums (*Didelphis marsupialis* and *Marmosa mexicana*) may be observed during visits to one of Monteverde's reserves. The area is home to all six of Costa Rica's wild felines (*Herpailurus yagouaroundi, Leopardus pardalis, Leopardus tigrinus, Leopardus wiedii, Panthera onca, Puma concolor*). Camera trap research shows regular and widespread presence of pumas and ocelots, and occasional presence of jaguars (Zamzow et al. 2018). Baird's Tapir (*Tapirus bairdii*), Costa Rica's largest land mammal and IUCN Red Listed as endangered (see Table 10.1), is also present in lower abundance. As is the case for other taxonomic groups in the Monteverde region, comprehensive monitoring programs to estimate populations and conservation status of mammals have not been established in Monteverde.

By night, insectivorous, nectar-feeding, and frugivorous bats parallel the daytime roles of birds—pollinating, dispersing seeds, and preying on invertebrates in Monteverde's forests (Timm and LaVal 1998; Muchhala 2003). These ecosystem services are essential in maintaining the functional integrity of the cloud forest ecosystem. Montverde's bat diversity includes 58 species, which represents nearly half of the total bat diversity in Costa Rica (Timm and LaVal 2000; Wainwright 2007). Migration outside of Monteverde's protected areas on the Pacific Slope for foraging and pollination make bats especially important in the process of natural habitat restoration, as they help disperse fruits and seeds of pioneer species such as Piperaceae and Solanaceae. More research on bat migration is needed, however, in order to more fully understand their contribution to habitat restoration, which may in turn promote greater bat conservation (Caughlin et al. 2007).

### *10.2.6 Amphibians*

Historically, Monteverde was home to 60 species of amphibians, including 2 caecilians, 5 salamanders, and 53 anurans, many of which are part of a "distinctive upland assemblage rich in endemic species" (Pounds and Fogden 2000). As is the case globally, Monteverde's amphibians have experienced dramatic and alarming declines over the past four decades due to habitat loss as well as the amphibian chytrid fungus, *Batrachochytrium dendrobatidis (Bd)* (James et al. 2015; Whitfield et al. 2017; Whitfield et al. 2016). In 1990, researchers found only 25 of the 50 anuran species expected to be present in the Monteverde area; notably absent was the iconic and endemic Golden Toad (*Incilius periglenes*) (Pounds and Fogden 2000).

However, Monteverde's extensive protected areas apparently function as a refuge and foster recovery for some species. For example, Forrer's Leopard Frog (*Lithobates forreri*) was later found (Pounds and Fogden 2000), the Green-eyed Frog (*L. vibicarius*) (see Fig. 10.5) and Starrett's tree frog (*Isthmohyla tica)* also reappeared in remote areas of the Children's Eternal Rainforest, and other species have since been rediscovered as well (Garcia-Rodriguez et al. 2012; Whitfield et al. 2017; M. Wainwright personal communication). In the case of *L. vibicarius*, research indicates that these remnant populations can survive and persist even with Bd infection (Whitfield et al. 2017).

**Fig. 10.5** The Green-eyed Frog (*Lithobates vibicarius*) was thought to be extinct but later reappeared in the Children's Eternal Rainforest. (Image Source: Mark Wainwright)

### *10.2.7 Soils and Microbial Communities*

The taxonomy and ecology of soil fauna are largely missing from biodiversity inventories in Monteverde, although two studies have examined the biology of cloud forest canopy soils (Nadkarni et al. 2002; Rains et al. 2003). This could be the next frontier of biological research in the area. The implementation of new DNA sequencing technologies available in Costa Rica provides a significant opportunity to study the composition of microorganisms in cloud forest soil ecosystems.

### *10.2.8 Endemism, Climate Change, and Species Decline*

The Monteverde region exhibits high endemism rates in comparison to areas at lower elevations. For example, Haber (2000a, b) reports approximately 10% of Monteverde's plant species to be endemic to the Tilarán mountain range. A comprehensive study of endemic species, invasive species, and the vulnerability of endemic species to invasive species and changes in climate is needed for the region.

In the 50 years between 1940 and 1990, deforestation and agricultural expansion were the primary forces creating the island-effect around Monteverde; however, the primary current threat to Monteverde's biodiversity is climate change. Strikingly, the number of dry days and the number of dry days in runs—five or more consecutive days with no precipitation, which have an impact on mist frequency in the cloud forest—have shown a steadily increasing trend over time (Pounds et al. 2006). Through the 1970s, there were less than 10 dry days in runs per year, with some years having no such runs (Pounds et al. 2006); 108 total dry days were recorded in 2019, with most of the zero precipitation days during January–April (Monteverde Institute, unpublished data). The shift of environmental variables regulating ecosystem function may produce a rapid transformation in life zone distribution and species composition across the region. Should this change happen more quickly than ecosystems and species are able to adapt, species declines—and extinctions—will likely result.

New disease vectors invading the Monteverde region may become a leading cause of species declines (Pounds et al. 2006). This is perhaps most vividly highlighted by the example of amphibian species declines and extinctions at the hands of the chytrid fungus *B. dendrobatidis*, mentioned above. This phenomenon may be exacerbated by changing climate conditions.

## 10.3 Biodiversity Conservation in Monteverde

> *"Within Costa Rica, the Monteverde Zone is atypical, with its multi-cultural population, the large number of people within high education al levels, sharp awareness of conservation and sustainable development, relative prosperity based on dairy farming and ecotourism, and ability to create grassroots organizations to deal with local issues".* (Burlingame 2000: 374)

Given the Monteverde region's biodiversity, as highlighted in Sect. 2, and in light of past and current threats to the region's forests, biodiversity conservation initiatives have played a central role in maintaining the region's biodiversity and in the emergence of Monteverde as a world-renowned nature-based tourism destination. The Monteverde Reserve Complex—the amalgamation of the various protected areas described in the Introduction (see Sect. 1, Fig. 10.1)—is the accomplishment of private citizens who established organizational structures and carried out fundraising to acquire and maintain large swaths of land in private ownership for conservation purposes. Initiatives from two Costa Rican NGOs, the Tropical Science Center (TSC, owner of the Monteverde Cloud Forest Biological Preserve) and the Monteverde Conservation League (MCL, owner of the Children's Eternal Rainforest) and their international fundraising counterparts grew out of the recognition that the Costa Rican government did not have the funds to effectively purchase or steward this area.While requiring Herculean effort to successfully carry out fundraising for land purchase, Monteverde had a compelling story about immediate threats to species such as the charismatic Resplendent Quetzal and the endemic Golden Toad. Furthermore, the concept of a Children's Eternal Rainforest—a privately-owned protected area established by many small contributions from children all over the world—began with Swedish elementary school students, whose fundraising efforts subsequently spread around the globe thanks to several international NGOs that were formed to collect funds raised by the students. The international fundraising efforts were able to raise hundreds of thousands of dollars from tens of thousands of small donations. In doing so, they brought further attention to and grew the mystique of Monteverde as a special place on the planet where biodiversity protection is paramount.

One of the great challenges of large-scale land conservation, whether public or private, is the cost of management. Purchase is a one-time event; maintenance and stewardship are forever. Neither the TSC nor the MCL was able to raise sufficient funding to establish long-term endowment funds to support the staff, materials, and other costs related to forest protection and management. They depend in part on revenue-generating use of the reserves to support annual budgets, which requires ongoing investment in infrastructure, and larger annual budgets for operations and maintenance. Both organizations have experienced periods of institutional growth to ramp up operations and generate more revenue, followed by periods of contraction and re-organization as revenue cycles ebbed.

How these organizations managed reserve borders and encroachments has been an important factor in the conservation successes of the Monteverde Reserve Complex. The MCL and TSC realized early on that effective conservation was better achieved through non-confrontational means, and, in contrast to government park rangers, their park guards do not carry weapons (Burlingame 2000). When park guards come across poachers and others encroaching on reserve properties, the guards take the approach of dialog and explain conservation goals. Many of the early forest guards were former hunters and farmers themselves, and they understood the mindset of those living around the reserves. While this approach did not always work in terms of stopping encroachment, it did succeed in building trust and respect between reserve personnel and neighbors. Anecdotal conversations with park guards reveal that while poaching continues to be a problem in the private reserves, there is much less hunting today than in the past; some hunters also simply move to neighboring properties (including state owned reserves) in order to respect the boundaries of the Monteverde Cloud Forest Biological Preserve and Children's Eternal Rainforest (H Chacón personal communication). Some neighboring landowners have seen the economic opportunities associated with growing ecotourism, particularly from birdwatching (e.g., the 83 ha Curi Cancha Reserve), and have adopted conservation measures, leaving remnant forest trees on their properties and allowing pastures to return to forest.

## *10.3.1 The Monteverde Reserve Complex*

### 10.3.1.1 Tropical Science Center, the Monteverde Cloud Forest Biological Preserve, and Bosqueterno, S.A.

The Monteverde Cloud Forest Biological Preserve (MCFBP), owned and operated by the Tropical Science Center, was founded in 1972. Through strategic land purchases, the MCFBP has grown from its initial 328 ha to 4,125 ha in 2019 (Burlingame 2000; C Hernandez personal communication).

Biologists who came to Monteverde to study the cloud forest were instrumental in fundraising and land purchase efforts, and helped generate broad interest about conservation of this unique ecosystem. In particular, George Powell played a crucial role in raising awareness—and funds—on an international scale in support of conservation of the Golden Toad, Resplendent Quetzal, Bare-necked Umbrellabird, and wild feline species. Powell was also the person who first approached the TSC, Costa Rica's first nonprofit conservation organization, about taking ownership of the newly acquired lands (Burlingame 2000).

As part of an effort to expand the MCFBP, in 1974 the TSC secured a 90-year lease on 554 ha of land belonging to Bosqueterno S.A., a Costa Rican corporation established by the Quaker community.[3] When Quakers originally settled in Monteverde in the 1950s, they set aside this land—high on the mountain, shrouded in dense cloud forest, and unsuitable for farming—in order to protect the community's principal water source (Burlingame 2000). A small portion of the Bosqueterno property forms part of the main trail system open to visitors in the MCFBP.

Visitation at the MCFBP has grown from 471 visitors in 1973–1974 (Burlingame 2000) to more than 100,000 visitors in 2019 (Y Méndez personal communication). As the Monteverde area's most visited reserve, the MCFBP has established a maximum limit of 250 persons at any given time on its 13 km of trails; the maximum limit for daily visitation is 450 people. Of the MCFBP's 4,125 ha, slightly more than 80 ha (or about 2% of the total reserve area) are open to the public (http://www.cct.or.cr/contenido/our-protected-areas).

The MCFBP's environmental education program, founded in 1992, brings local students to the cloud forest for talks, workshops, and hikes. Environmental educators

[3]The 554 ha owned by Bosqueterno, S. A. and managed by the Tropical Science Center are not included in the 4,215 ha owned by the Tropical Science Center.

also visit local classrooms and participate in environmental education activities in conjunction with the Monteverde Environmental Education Commission. The MCFBP's research and monitoring program is important for carrying out censuses of birds, amphibians, mammals, and other groups, as well as monitoring Resplendent Quetzals and other species of special interest. Most of the MCFBP's operations and programs—including environmental education, research and monitoring, forest protection, and maintenance—are funded via tourist visitation to the reserve, interest on investments, and participation in the Costa Rican government's Payments for Environmental Services (PES) program.[4]

Also located in the Monteverde area, the TSC's 250 ha San Luis Biological Reserve protects a critical tract of habitat connecting the cloud forest to the lower dry forests on the Pacific slope within the Bellbird Biological Corridor. This property is not open for tourism and is primarily used for conservation and research.

#### 10.3.1.2 Monteverde Conservation League and the Children's Eternal Rainforest

One of the first objectives of the Monteverde Conservation League, a Costa Rican nonprofit organization founded in 1986 by a group of Monteverde residents, was to raise funds for the purchase and preservation of forest that would otherwise be lost or severely degraded by agricultural expansion. The land purchased would soon become known as the Children's Eternal Rainforest (CER), today Costa Rica's largest privately-owned reserve. The 22,600-hectare forested expanse traverses seven geopolitical districts and three provinces, bridges an elevational range of >1,200 m, and spans the Continental Divide that separates Atlantic and Pacific watersheds. As the centerpiece of the Monteverde Reserve Complex, the CER is a vital nexus for natural habitats and populations.

The CER (known locally as "Bosque Eterno de los Niños") was purchased and protected thanks to donations from children, adults, and organizations in more than 40 countries around the globe. In addition to its key role in biodiversity conservation, the CER also benefits Costa Ricans through the conservation of five major watersheds that provide a continuous supply of clean water for human consumption, agriculture, and hydroelectric production; opportunities in the ecotourism sector; and via innovative outreach services such as facilitating participation by neighboring landowners in national PES programs. The MCL was also instrumental in early reforestation efforts on private farms in the Monteverde region (see Bosques en Fincas, Section 3.3.1, below), and has maintained an active environmental education program since 1986. In this way, the reach of the CER extends beyond the borders of

[4]Mechanisms for the Costa Rican Payment for Environmental Services (PES) program and the National Forestry Financing Fund (FONAFIFO) were established in Costa Rica's Forestry Law 7575 in 1996 (Asamblea Legislativa de la República de Costa Rica 1996). FONAFIFO is the administrative entity for the national program. See Sect. 4 for further discussion of PES as a tool for conservation in Monteverde.

the Monteverde Reserve Complex, creating the human and natural connections needed to ensure the welfare of human and nonhuman communities in the future.

As is the case for many conservation organizations, the quest for economic stability over time has been one of the MCL's main challenges. Currently, about half of the MCL's total gross income comes from PES, including participation in Costa Rica's national program with FONAFIFO, and contracts with two private hydroelectric companies in recognition of the ecosystem services (abundant, clean water throughout the year) provided by thousands of hectares of protected forest upstream of their dams. Unfortunately, the cooperative spirit that led to the original signing of the private PES contracts did not last, and the MCL has had to fight to defend both agreements—including one that is still in the appeals process in the Costa Rican court system.

The MCL also receives important income from visitation at its field stations and trails. Though most of the CER is not open to the public, there are four visitation centers, including the Bajo del Tigre Reserve, which receives the most visitation (approximately 9,000 visitors in 2019). Two field stations, San Gerardo and Pocosol, offer rustic lodging, meals and trails to student groups and ecotourism visitors. Finca Steller, on the eastern border of the CER, is home to the MCL's environmental education program and native tree nursery, and also has a small trail system.

Donations and grants continue to provide crucial funding as well, although the increase in environmental crises on a global scale—NOAA (the United States Department of Commerce's National Oceanic and Atmospheric Administration; (2020)) reports that droughts, flooding, freezing, severe storms, tropical cyclones, wildfires, and winter storms caused $531.7 billion in damage between 2015 and 2019[5]—combined with changes in US tax law increasing the threshold for deductions for charitable giving (many of the CER's donors are US-based), have brought new challenges to the nonprofit financing landscape.

#### 10.3.1.3 Santa Elena Cloud Forest Reserve

The Santa Elena Reserve (SER) is another example of a unique idea for biodiversity conservation piloted in Monteverde. The Santa Elena Technical-Professional High School (CTPSE, Colegio Técnico Profesional de Santa Elena)—the public high school which serves students from communities throughout the Monteverde region—provides technical skill training in areas relevant to the local economy. As tourism's influence grew to a substantial portion of the local economy, preparing local youth for careers in the tourism sector became a priority. The CTPSE had

[5]These NOAA (2020) figures do not account for other global environmental crises, including habitat loss due to large-scale Amazonian fires, the current Australian wildfires, world-wide coral reef decline, impacts of increasing plastic contamination in the world's oceans, melting glaciers and ice caps, and localized disruptions due to coastal zone flooding. Dollar values are based on losses which would not have been incurred had the event not taken place and include both insured and non-insured losses.

previously signed a long-term lease with the Costa Rican government for a 310-ha farm; however, using this farm for agricultural purposes did not prove successful (Burlingame 2000). In 1992, the Santa Elena Reserve was established as a training ground for ecotourism and gave students the opportunity to learn about park management and natural history guiding. The SER borders MCL's CER (see Fig. 10.1) and includes about 80% primary forest on the upper Atlantic slope (Burlingame 2000). In 2019, the SER received 51,164 visitors (Y Arias personal communication). Given its more remote location from the towns of Monteverde and Santa Elena, the SER receives fewer visitors than the MCFBP; however, this has made the SER an attractive alternative for birdwatchers and tourists who prefer less crowded environments. For researchers and ecology students, the SER also provides a good site for comparative research and study with respect to other reserves in the area.

#### 10.3.1.4 Alberto Manuel Brenes Biological Reserve, Arenal Volcano National Park, and Other Surrounding Protected Areas

The Alberto Manuel Brenes Biological Reserve (RBAMB, Reserva Biológica Alberto Manuel Brenes), located contiguous to the CER and the north-eastern edge of the Bellbird Biological Corridor, spans 7,800 ha and is managed by the University of Costa Rica. Although very remote in terms of access via poorly-maintained roads from the nearest town, the RBAMB includes basic lab and dormitory space to support field research and field study courses.

The Arenal Volcano National Park (PNVA, Parque Nacional Volcán Arenal) spans 29,692 ha and includes its namesake, the Arenal Volcano, which was one of the hemisphere's most active volcanoes from the late 1960s through the first decade of the 2000s. Established in 1994, PNVA is an important buffer region on the Atlantic slope of the Monteverde Reserve Complex, extending forest connectivity to the north and east through lower elevations and protecting critical habitat and connectivity for migratory birds and large mammals requiring expansive ranges.

The Zona Protectora Arenal Monteverde[6] (Arenal-Monteverde Protected Zone) includes the MCFBP, CER, SER, Bosqueterno, RBAMB, PNVA, and a number of private farms, as well as a sizeable property to the north of Arenal Volcano National Park belonging to ICE, Costa Rica's electric utility (see Fig. 10.2). These are primarily located on the region's Atlantic slope. All of these additional protected areas have helped to expand the region's biodiversity island across the landscape.

[6]In Costa Rica, Zona Protectora is one of seven management designations for "protected wildlife areas." These also include National Parks, Forest Reserves, Biological Reserves, National Wildlife Refuges, Wetlands, and National Monuments. These are defined in Chapter VII, Article 32 of the Ley Orgánica del Ambiente, No. 7554, of October 4, 1995.

### 10.3.2 Sustainable Agriculture and Agroecology

Agricultural production in the earlier years of Monteverde's settlement was largely focused on subsistence because of the region's remote location and poor road access to external markets. Partly due to lack of access and partly because of personal values, many local farmers did not use agrochemicals, thereby contributing to more sustainable small-scale production (Griffith et al. 2000); nonetheless, agricultural expansion was a leading cause of deforestation in the Monteverde region through the 1970s.

Monteverde's Quaker settlers developed the nucleus of a large dairy production region, the "Monteverde milkshed," supplying the Productores de Monteverde S.A. dairy which, at its peak, grew into a large-scale producer of cheeses distributed throughout Central America (Griffith et al. 2000; Burlingame 2000). Due to low prices for milk, some of the pastures previously dedicated to dairy and mixed dairy-beef production have been abandoned and have returned to secondary growth forests (Stuckey et al. 2014).

Agriculture continues to be an important component of the local economy, but it does not currently contribute to continued deforestation. Instead, agriculture has moved toward a more sustainable model, particularly at higher elevations (Stuckey et al. 2014). Lack of support from extension agents is an ongoing challenge for local producers who have adopted organic and other sustainable farming methodologies (O Salazar personal communication).

While coffee has long played an important role in the local agricultural sector, the focus of Monteverde's coffee sector has shifted toward organic/sustainable production and is marketed primarily toward tourism (Stuckey et al. 2014). This includes direct sales to tourists as well as agroecology tours of sustainable farms, which diversifies the income stream for the farmers and their families. Integrated sustainably-produced coffee and agrotourism projects represent a rapidly-growing segment of Monteverde's agricultural economy.

A recent study by the TSC found a growing local market for sustainably produced agricultural products, but insufficient organization to connect producers to consumers and promote this emerging market for locally and sustainably produced foods (Tropical Science Center 2018). Consequently, TSC has initiated a new program, Encadenamientos Productivos (or Agricultural Products Network), to link local farmers with restaurants and other consumers, particularly within the Bellbird Biological Corridor. Some local businesses have independently developed similar initiatives. For example, the University of Georgia Costa Rica Campus (UGA CR) implemented the policy of purchasing the majority of food from within a 400 km radius of the campus, with an emphasis on on-site organic production and purchasing from local farmers in the Monteverde area and in the nearby lowlands who also followed sustainable agricultural practices. The Hotel Belmar and several other Monteverde-based hotels and restaurants have established on-site food production and local food purchasing components to their operations. Weekly farmer's markets in Santa Elena and Guacimal offer outlets for connecting local farmers with local consumers.

### 10.3.3 Biological Corridors

*"The main focus of conservation biology at Monteverde since 2000 has been on the role of landscape features in preserving biodiversity, particularly connectedness between habitats at different spatial scales".* (Wheelwright 2014: 2).

This section describes some of the ways in which Monteverde has integrated landscape-level conservation into the surrounding communities, functionally extending the limits of the Monteverde biodiversity island beyond the physical limits of the core protected areas.

#### 10.3.3.1 Bosques en Fincas

In the mid-1990s, the MCL administered the program Bosques en Fincas (or Forests in Farms), which encouraged local farmers to maintain existing forest fragments and connect them by planting windbreaks. Dairy farmers realized the effectiveness of windbreaks at maintaining pasture productivity during the extremely windy season. Many of these windbreaks are now 20–30 years old. The Bosques en Fincas program set the stage for expanding forest connectivity across the privately-owned landscape surrounding the Monteverde Reserve Complex and spawned research suggesting the critical importance of remnant trees for biodiversity conservation (Guindon 1996; Harvey and Haber 1999). Brownson et al. (2019) describe Bosques en Fincas as an example of a successful local PES program, in which farmers were engaged by the MCL's outreach team and given the species of trees they were interested in planting on their farms. This program triggered a cultural shift toward reforestation in the region (K Brownson personal communication), which set the stage for other reforestation programs focusing on native species reforestation (see Sect. 3.3.4, below).

#### 10.3.3.2 Enlace Verde

In the mid-1990s, the Monteverde Institute promoted a corridor program, Enlace Verde, aimed at protecting existing forests on individual farms within the town of Monteverde using conservation easements. This program intended to connect the Children's Eternal Rainforest with the Bajo del Tigre Reserve, both owned by the MCL. At the time, the San José, Costa Rica-based Environmental and Natural Resource Law Center (CEDARENA, Centro de Derecho Ambiental y de los Recursos Naturales) was introducing the concept of conservation easements within Costa Rica and partnered with the Monteverde Institute to support the Enlace Verde program. Three easements were signed in 1998, including unique reciprocal easement contracts. Despite multiple charrette exercises and individual meetings with landowners, including preparation of drafts of easement contracts and accompanying maps, the overall program did not expand as hoped across some fifty properties located in six neighborhood clusters (Scrimshaw et al. 2000:382).

#### 10.3.3.3 Reforestation Initiatives

Following in the footsteps of the MCL's Bosques en Fincas program, the Fundación Conservacionista Costarricense (FCC, or Costa Rican Conservation Foundation) was established in 2002 in response to the observed need to expand Pacific slope habitat for the Three-wattled Bellbird and other migratory species in this range. The FCC's reforestation program was started in 1998 as the "Bellbird Project" with the support of local organizations and funds provided by the British Embassy in San José, Costa Rica (D Hamilton personal communication). Following several years of project implementation, the formal non-profit foundation was established.

The FCC's tree nursery produces native forest species that provided free of charge to local landowners. The FCC has studied multiple areas where trees have been planted to understand and develop best practices that yield the greatest benefit-to-cost related to minimizing mortality rates and maximizing growth rates during establishment (D Hamilton personal communication). The FCC does not use formal agreements with recipient landowners to maintain the trees. To date, FCC has planted over 250,000 trees of more than 143 species and 42 families (D Hamilton personal communication).

The former University of Georgia Costa Rica Campus (UGA CR)[7] reforestation program was established in 2008 to offset University of Georgia students' emissions related to study abroad in Costa Rica, and ran for 11 years until 2019. In 2010 the FCC partnered with UGA CR, providing funding to expand the tree nursery. UGA CR's native species nursery primarily focused on species native to the San Luis Valley, where the campus was located. Planting was done both on-campus and on neighboring farms in San Luis, with saplings given to local landowners. UGA CR asked participant landowners to sign agreements which noted the numbers and species of saplings provided and stated that landowners agreed to maintain the saplings through establishment, however there was no enforcement. Beginning in 2011, UGA CR student groups and interns monitored planting sites to determine mortality and measure growth rates.

Between 2008 and 2016, the UGA CR program planted more than 35,000 trees representing over 90 species from the local forests (UGA CR unpublished data). For carbon offset estimates, mortality was estimated at 25%; however, at several sites, monitoring revealed high mortality due to cattle encroachment on one property and development of a housing lot on another. Other sites with effective protection for saplings during establishment had less than 5% mortality (UGA CR unpublished data).

Reforestation projects with non-binding agreements for landowner care of the trees planted are only as effective as the quality of the management provided until tree establishment. In their study of formal and informal payment for ecosystem services programs (see Sects. 3.4 and 4 for further discussion), Brownson et al.

---

[7]The University of Georgia sold this property in 2019. Long-term research and reforestation programs were not maintained by the current owner.

(2019) found that the more informal local reforestation programs in the Monteverde region promoted greater tree species diversity and more effectively engaged with less economically prosperous landowners than the formal national ecosystem services program.

#### 10.3.3.4 Corredor Biológico Pájaro Campana (Bellbird Biological Corridor)

At the same time the Monteverde Institute was working on Enlace Verde, the TSC proposed the concept of a biological corridor following the Guacimal River watershed from Monteverde down the Pacific slope to the mangrove forests along the Gulf of Nicoya. As a step toward implementing this project, TSC acquired a 240 ha (593 acre) farm in the lower San Luis Valley, securing the largest remaining Pacific slope forest patch in the upper elevation of this watershed (Burlingame 2000). Further expansion of the corridor was not immediately put into place; yet the idea for a corridor initiative, like the seed of an *Ocotea monteverdensis* tree buried in the forest floor, was firmly planted and waiting for the right conditions to germinate.

The corridor initiative was revived in 2006, and in 2007 a group of local conservation and education organizations established the Local Advisory Committee for the Corredor Biológico Pájaro Campana (CBPC, or Bellbird Biological Corridor), including representatives from TSC, MCL, UGA CR, FCC, and the Monteverde Institute. The CBPC has been successful in large part due to common interests, shared vision, cooperative efforts, and commitment of time and funding by the organizations involved. The Local Advisory Committee successfully addressed differences in opinion, listened to and incorporated concerns from a wide body of stakeholder groups, and implemented the initiative in a professional manner.

The CBPC spans 88,738 hectares within three adjacent watersheds which flow from the Monteverde cloud forest to the Gulf of Nicoya—the Aranjuez, Guacimal, and Lagarto River watersheds (see Fig. 10.2). Eleven of Costa Rica's 13 Holdridge Life Zones (Holdridge 1967) are represented within the corridor, and approximately 50% of Costa Rica's terrestrial vertebrate species are found within this region (Welch et al. 2011).

The CBPC was recognized by the Costa Rican government as part of the formal national network of biological corridors connecting Costa Rica's larger parks and reserves. The CBPC's Local Advisory Committee secured funding from the UNDP Small Grants Program to develop a strategic plan for the corridor (Welch et al. 2011). This strategic plan defined key areas of common interest, and sub-committees then developed sector-specific work plans from these broadly defined goals (J Welch personal communication). Over the course of the next decade, the Local Advisory Committee was able to secure additional grant funding from the UNDP Small Grants Program to develop promotional and educational materials, hold workshops in communities throughout the corridor region, and establish multiple community-based sub-committees throughout the three watersheds. The Local Advisory Committee was sensitive to the appearance of conservation initiatives being driven by the organizations—both literally and figuratively—"at the top of the mountain" (and in the communities which received the most direct economic benefits from ecotourism), and worked to build broad common interest in the concept of maintaining and expanding habitat connectivity. Local concerns were incorporated into the strategic plan and, subsequently, continued UNDP Small Grant Program funding helped to establish sub-regional committees to generate locally driven initiatives (J Welch personal communication).

A growing number of PhD dissertations, master's theses, and undergraduate thesis projects have been completed related to the Bellbird Corridor (e.g., Brownson 2019; Camacho Céspedes 2019; Padgett-Vasquez 2019; O'Halloran et al. 2018; Powlen 2018; Allen 2016; Chinchilla Ramos 2015; Piedrahíta López 2013). The University of Georgia and Lynchburg College carried out a long-term water quality monitoring program in the region for 8 years, documenting the macroinvertebrate communities in the three watersheds (T Shahady personal communication).

The CBPC also is the focus of current land purchase priorities for the Monteverde Conservation League, whose goal is to increase ecological connectivity between the Children's Eternal Rainforest and forested areas at lower elevations on the Pacific slope. The MCL has purchased more than 100 hectares in the CBPC since 2014 and continues to raise funds for land purchase and conservation in the area.

### *10.3.4 Payment for Environmental Services (PES)*

Costa Rica's national Payment for Environmental Services (PES, or Pago por Servicios Ambientales) program administered by the National Forestry Financing Fund (FONAFIFO, Fondo Nacional de Financiamiento Forestal) has been extensively studied by academics and practitioners alike. The early years of Costa Rica's PES program coincided with a noticeable trend of afforestation across the country, following decades where Costa Rica experienced some of the highest deforestation rates in the world (FONAFIFO 2001). The successes of PES in helping to protect privately-owned forest land in Costa Rica is highlighted by figures such as over 10% of the country having been protected via PES forest conservation contracts (Ringhofer et al. 2013).

Within the CBPC, 11 Holdridge life zones were represented by PES contracts issued between 2008 and 2012, with most of these PES contracts located in areas not represented by the Monteverde Reserve Complex (Padgett-Vasquez 2019). During this time period, 51 properties throughout the CPBC region were inscribed by FONAFIFO in the PES program (Padgett-Vasquez 2019). These properties ranged in size from 11 to 300 hectares (Padgett-Vasquez 2019). The majority of the 51 properties were located along or in very close proximity to rivers and streams, which suggests that the protection of these forested areas through PES contributes to maintaining the functional integrity of the region's ecological systems (Padgett-Vasquez 2019).

### 10.3.5 The Monteverde-Arenal Bioregion Initiative (MABI)

One of the characteristics that makes Monteverde unique is the number of research scientists that have settled in the area and continued to study the region over multiple decades. The Monteverde-Arenal Bioregion Initiative (MABI) was initiated in 2014 by several of Monteverde's long-term research scientists and educators to coordinate and expand research, teaching, and outreach collaborations in support of biodiversity conservation. Given the ongoing observations and growing collective understanding about changing climate conditions in the Monteverde region and the likely impacts for biodiversity conservation (e.g., Pounds et al. 1999; Pounds et al. 2006; Nadkarni and Solano 2002; Eaton et al. 2012), MABI was created to develop shared database resources and coordinate regular conferences which serve to share information, define research priorities for the region, develop coordinated research proposals, and support each other's research, teaching, and outreach. To date, funding limitations have limited new region-wide collective research efforts, however the multiple MABI conferences held during 2014–2016 served to share information and build new relationships among the many scientists and educators working in the region (N Nadkarni personal communication).

### 10.3.6 The Monteverde Commission for Resilience to Climate Change (CORCLIMA)

The same capacity for community mobilization and organization and deeply-rooted environmental values that led to the creation of the Monteverde Reserve Complex and more recently the Bellbird Biological Corridor has also spawned a new initiative, The Monteverde Commission for Resilience to Climate Change (CORCLIMA). Promoting Costa Rica's National Strategy for Climate Change, local residents and institutions have joined forces and are coordinating across sectors and scales with the goal of leading the Monteverde Zone to become carbon neutral and develop climate resilience strategies for the region (Brenes et al. 2019, corclima.org). While the carbon balance benefits are global in scale, the adjustments Monteverde seeks to achieve in meeting carbon neutral goals and implementing climate change resilience will enhance the quality of life for local residents, help further protect habitat to support the region's biodiversity, and promote Monteverde's reputation as an authentic, sustainable tourism destination. CORCLIMA's actions serve as a demonstration for other communities seeking to implement similar plans.

## 10.4 Challenges for the Future

How to sustain the local economy, improve quality of life for a growing local population, and continue to protect the region's natural capital while adapting to changing climate conditions is the biggest overarching challenge facing Monteverde today. Should the projected changes in climate impact the cloud forest habitat to the extent that it further threatens the populations of quetzals, bellbirds and other iconic species which draw many tourists to the region, what impact will that have on tourism and thus the local economy? Long-term biodiversity conservation is tied to economic stability, and economic downturn can lead to increased pressure on forests and forest resources. How can Monteverde diversify the local economy and also find alternative funding mechanisms for biodiversity conservation to balance fluctuations in tourism?

Addressing the challenges requires both top-down and continued grass-roots leadership. Coordinated strategic visioning with an alignment of national policies integrating environmental conservation, agricultural production, and continued development for sustainable tourism, sustainable manufacturing, and service sectors are also needed. Costa Rica's national law recognizing biological corridors as regions where such integrated conservation and development takes place provides a broad legal framework and national administrative structure to support local initiatives. Yet as Padgett-Vasquez (2019) notes, one of the challenges in landscape-level conservation is to establish functional wildlife corridors within these broadly declared conservation regions spanning privately-owned property. The accomplishments of the Bellbird Biological Corridor reflect how local leadership capacity and coordination is critical to implement such broad national conservation plans. The challenge of funding staff and administrative overhead to support the ongoing coordination of the CBPC reflects the common economic struggle faced by grass-roots conservation initiatives around the world. Volunteer boards comprised of dedicated representatives of local NGOs can only carry conservation initiatives so far, particularly in small communities.

Planning and zoning is of critical importance as populations grow and small rural towns like Santa Elena undergo rapid urbanization. Of the 82 cantons in Costa Rica, 40 have developed zoning plans, 21 have comprehensive environmental planning regulations, and only four have completed studies of hydrological vulnerability and established guidelines regarding water resource protection (INVU 2019). While the springs emanating from the protected reserves supply clean water for the region, wastewater management and contamination of local waterways continues to threaten public health as well as Monteverde's aquatic diversity outside of the protected areas.

A growing body of literature through the first two decades of the twenty-first Century describes the shortcomings of public land conservation programs to provide sufficient habitat to address long-term biodiversity conservation concerns (Norton 2000; Chacón 2005; Mayer and Tikka 2006; Pasquini et al. 2011; Kamal et al. 2015). In particular, there is growing recognition that connectivity across a wider landscape is critical to provide sufficient habitat to support migratory species through an increasingly more densely human-dominated landscape, and to support long-term ecological resiliency and both floral and faunal adaptation to changing global climate conditions (Zamzow et al. 2018). Yet for privately-owned projects and conservation initiatives led by non-governmental organizations to be successful, there must be supportive public policy in place to encourage and support conservation practices (Pasquini et al. 2011). Private nature reserves and those owned by non-governmental organizations have some advantages compared to public protected areas—management flexibility, ability to make decisions more rapidly, and broader economic opportunities (Pasquini et al. 2011). Indeed, in the case of Monteverde, initial government designation of the region as a Zona Protectora was ineffective and led to the local establishment of NGOs to fill the conservation gap. Private land conservation—including NGO-owned conservation reserves—cannot exist in a vacuum, however, and requires lock-step coordination with the public sector and the long-term public sector commitment to establish and maintain such policy. Comprehensive conservation policy supporting private land conservation must address ecological, economic, and social aspects, including public education required to embed the value and importance of biodiversity conservation as core to the human condition. From the local to global-scale, there must be coordination and commitments from the public sector reflecting that biodiversity conservation is a priority beyond partisan politics.

For private land conservation programs to successfully meet conservation goals, there is a need for thoughtfully designed, contextually-appropriate incentive programs to encourage private land conservation that address both ecosystem conservation goals as well as economic productivity which is considered a right of private land ownership (Mayer and Tikka 2006). The ability of farmers to adopt more sustainable agricultural practices is largely influenced by broader socio-economic trends, national-level rural development policies, and institutional capacity of agencies, such as the Ministry of Agriculture, to provide technical support for organic and other forms of sustainable agriculture (Wegner 2016). Conversations with coffee farmers in Monteverde reveal the complete absence of Ministry of Agriculture extension support for farmers practicing organic agriculture (O Salazar personal communication), despite the growing international demand for certified organic products.

As a tool in the biodiversity conservation toolbox, payment for ecosystem services has been employed in Costa Rica since the mid-1990s. While studies of the earlier years of the PES program found that PES was effective at increasing forest cover within biological corridors in Costa Rica (Morse et al. 2009; Newcomer 2007), most evidence in Costa Rica, however, points toward PES not being particularly helpful in changing landowner decision-making, and there is mixed evidence at best about whether PES payments broadly contribute to avoided deforestation (Allen 2018, K Allen personal communication). In the case of private landowners in the Monteverde area, PES subsidizes land abandonment and/or protection of forests which are not necessarily in imminent threat (K Allen personal communication) and, in general across Costa Rica, has favored larger landowners (Zbinden and Lee 2005;

Newcomer 2007). Furthermore, individuals have expressed a lack of trust with the government (Newcomer 2007; Allen and Padget-Vasquez 2017), in part driven by restrictions placed on properties in PES during and following the termination of PES contracts (S Padgett-Vasquez personal communication). Thus, local PES programs, such as the reforestation programs described in Sect. 3.3.3, are effective ways to build trust relationships with smaller local landowners, increase tree species diversity across the landscape, and provision ecosystem services (Brownson et al. 2019).

Despite these noted issues, however, the national PES program is an important mechanism that provides critical funding for ongoing habitat protection and biodiversity conservation of large protected areas owned by NGOs such as MCL. Fundraising for the purchase of land for conservation purposes is challenging; fundraising for ongoing maintenance and protection of these properties is far more difficult. While Costa Rica's PES program promotes biodiversity conservation as one of the four primary ecosystem services recognized in all PES contracts, carbon sequestration and hydrological services are the primary services connecting buyers (e.g., those purchasing gasoline, those using water in manufacturing) to sellers (i.e., forest land owners). In the case of NGOs, including MCL, which own large tracts of land that serve as regional biodiversity islands, the service provision of biodiversity conservation is a high value-added contribution to the bundle of ecosystem services. Given FONAFIFO's ongoing struggles to fund the national PES program via gas taxes, water fees, and subsidies from international agencies (including the World Bank and the Global Environment Fund), a more coordinated, holistic national policy strategy might incorporate a small fee for the provision of biodiversity services as a mechanism to help support local NGOs' land conservation initiatives, such as the Monteverde Cloud Forest Preserve and the Children's Eternal Rainforest.

Efforts to integrate natural capital accounting into the national accounting structure can help Costa Rica better value and manage its natural resources. Providing such an overarching framework allows private industry, non-governmental organizations (NGOs), and individuals to then pursue innovative approaches toward achieving a more sustainable economic and ecological future for Costa Rica. Such a holistic approach toward national development policy would provide the framework for sustaining and expanding biodiversity islands such as Monteverde.

## 10.5 Conclusion

Monteverde reflects a unique conservation situation where the national government of Costa Rica recognized the importance of the region for biodiversity conservation but was not able to dedicate resources to establish a public protected area. International research scientists partnered with existing Costa Rican NGOs and established new local NGOs, filling the federal conservation gap by securing funds to purchase and manage large swaths of forest across the region. At the heart of the region's privately-held protected lands are 554 hectares of forest that Quaker settlers set aside to protect their community's water source some 20 years prior to the establishment of larger protected areas.

Many of the international scientists studying the area's ecology decided to make Monteverde their home and/or have spent significant amounts of time in the region over decades. They helped to establish local NGOs that not only protect land and study the region's ecology, but also help to share this knowledge with the local community and teach local people about what they are studying. Over time, they have created a vast body of knowledge about Monteverde's biodiversity, some of which we have highlighted in Sect. 2 as representative of the Monteverde region. In turn, many local residents have subsequently committed their professional careers to support conservation initiatives, working as naturalist guides, park guards, research technicians, administrators, and educators. Together, this blended community established local schools that incorporate experiential learning and environmental stewardship into their curriculum, spreading conservation values among the younger generations. As described, the NGOs translate knowledge into outreach programs supporting local farmers and broader community interests, all the while maintaining a core mission of habitat protection. The local NGOs capitalized on the increased popularity of Costa Rica as an ecotourism destination and the region's charismatic species and were able to move more nimbly than many of the public parks, setting up infrastructure, administration, and support services to promote nature-based tourism. This grew the local economy and increased general prosperity in the region. As described in previous sections, the diversity of different types of initiatives—university partnerships, sustainable tourism, international fundraising campaigns, purchasing policies for sustainable products, carbon offsets, farmer and landholder outreach, technical school partnerships, etc.—have formed a patchwork quilt of biodiversity conservation that has endured for nearly 50 years. As some pieces of this quilt fray, the Monteverde community has remained dedicated to the long-term biodiversity conservation goals and has developed new initiatives to fill the gaps. From the initial successes of establishing the MVCFBP, the CER, and the SER, the scale of conservation initiatives has expanded from bounded protected areas focused on Monteverde's cloud forests to landscape-level protection spanning both Atlantic and Pacific slopes.

Changing climate conditions now make Monteverde an interesting place to study the impacts of these changes on the cloud forest ecosystem. How this biodiversity island will evolve, how biological corridors will support species migration, and how climate impacts and projected resource constraints will affect the region's socioeconomic conditions and how that, in turn, impacts biodiversity conservation remains to be seen. If there is a big-picture takeaway message from the case study of Monteverde, it is that biodiversity conservation—through a portfolio of private land conservation initiatives, public parks and reserves, and public-private partnerships—is a continuously evolving process made ever more challenging by changing climate conditions, yet made possible and successful by highly engaged, coordinated, cooperative private-sector, public-sector, and NGO-led initiatives.

**Acknowledgements** We are grateful to our many colleagues in Monteverde for the knowledge and insights they have shared with us through the years living and working in Monteverde. In particular, thanks to Orlando Calvo, Alvaro Cubero, Bill Haber, Deb Hamilton, Russ Kumai, Jack Longino, Yoryineth Méndez, Nalini Nadkarni, Alan Pounds, Yúber Rodríguez, Katy VanDusen, Mark Wainwright and Justin Welch for their contributions to this chapter.

## References

Allen K (2016) Conservation planning in a global Era: the role of market-based mechanisms in promoting sustainable development in Costa Rica. PhD dissertation, The University of Georgia, Athens, GA

Allen K (2018) Why exchange values are not environmental values: explaining the problem with neoliberal conservation. Conserv Soc 16(3):243–256

Allen TFH, Hoekstra TW (1992) Toward a unified ecology. Columbia University Press, New York

Allen KE, Padgett-Vasquez S (2017) Forest cover, development, and sustainability in Costa Rica: can one policy fit all? Land Use Policy 67:212–221. https://doi.org/10.1016/j.landusepol.2017.05.008

Americas Regional Workshop (Conservation & Sustainable Management of Trees, Costa Rica, November 1996) (1998) Oreomunnea pterocarpan. The IUCN Red List of Threatened Species 1998: e.T30663A9570512. https://doi.org/10.2305/IUCN.UK.1998.RLTS.T30663A9570512.en. Downloaded on 24 January 2020

Asamblea Legislativa de la República de Costa Rica (1996) No 7575 Ley Forestal. Alcance No 21 a La Gaceta No 72, La Uruca, San José, Costa Rica, 16 April 1996, 8 pp

Atwood JT, Dressler RL (1995) Another new species of *Stellilabium* from Monteverde, Costa Rica. Selbyana 16:239–241

BirdLife International (2016a) Ara ambiguous. The IUCN Red List of Threatened Species 2016: e.T22685553A93079606. https://doi.org/10.2305/IUCN.UK.2016-3.RLTS.T22685553A93079606.en. Downloaded on 25 January 2020

BirdLife International (2016b) Buteogallus solitarius. The IUCN Red List of Threatened Species 2016: e.T22695849A93530532. https://doi.org/10.2305/IUCN.UK.2016-3.RLTS.T22695849A93530532.en. Downloaded on 25 January 2020

BirdLife International (2016c) Cephalopterus glabricollis. The IUCN Red List of Threatened Species 2016: e.T22700933A93805184. https://doi.org/10.2305/IUCN.UK.2016-3.RLTS.T22700933A93805184.en. Downloaded on 19 January 2020

BirdLife International (2016d) Crax rubra. The IUCN Red List of Threatened Species 2016: e.T22678521A92776389. https://doi.org/10.2305/IUCN.UK.2016-3.RLTS.T22678521A92776389.en. Downloaded on 25 January 2020

BirdLife International (2016e) Electron carinatum. The IUCN Red List of Threatened Species 2016: e.T22682989A92971734. https://doi.org/10.2305/IUCN.UK.2016-3.RLTS.T22682989A92971734.en. Downloaded on 25 January 2020

BirdLife International (2016f) Patagioenas subvinacea. The IUCN Red List of Threatened Species 2016: e.T22690328A93269904. https://doi.org/10.2305/IUCN.UK.2016-3.RLTS.T22690328A93269904.en. Downloaded on 27 January 2020

BirdLife International (2016g) Pharomachrus mocinno. The IUCN Red List of Threatened Species 2016: e.T22682727A92958465. https://doi.org/10.2305/IUCN.UK.2016-3.RLTS.T22682727A92958465.en. Downloaded on 19 January 2020

BirdLife International (2016h) Procnias tricarunculatus. The IUCN Red List of Threatened Species 2016: e.T22700946A93806341. https://doi.org/10.2305/IUCN.UK.2016-3.RLTS.T22700946A93806341.en. Downloaded on 19 January 2020

BirdLife International (2016i) Sclerurus albigularis. The IUCN Red List of Threatened Species 2016: e.T22702968A93897178. https://doi.org/10.2305/IUCN.UK.2016-3.RLTS.T22702968A93897178.en. Downloaded on 25 January 2020

BirdLife International (2016j) Spizaetus ornatus. The IUCN Red List of Threatened Species 2016: e.T22696197A93548774. https://doi.org/10.2305/IUCN.UK.2016-3.RLTS.T22696197A93548774.en. Downloaded on 25 January 2020

BirdLife International (2017a) Contopus cooperi (amended version of 2016 assessment). The IUCN Red List of Threatened Species 2017: e.T22699787A110734937. https://doi.org/10.2305/IUCN.UK.2017-1.RLTS.T22699787A110734937.en. Downloaded on 25 January 2020

BirdLife International (2017b) Grallaricula flavirostris (amended version of 2016 assessment). The IUCN Red List of Threatened Species 2017: e.T22703362A110974550. https://doi.org/10.2305/IUCN.UK.2017-1.RLTS.T22703362A110974550.en. Downloaded on 25 January 2020

BirdLife International (2017c) Hylocichla mustelina (amended version of 2016 assessment). The IUCN Red List of Threatened Species 2017: e.T22708670A111170926. https://doi.org/10.2305/IUCN.UK.2017-1.RLTS.T22708670A111170926.en. Downloaded on 25 January 2020

BirdLife International (2017d) Morphnus guianensis. The IUCN Red List of Threatened Species 2017: e.T22695991A118209977. https://doi.org/10.2305/IUCN.UK.2017-3.RLTS.T22695991A118209977.en. Downloaded on 27 January 2020

BirdLife International (2017e) Tinamus major (amended version of 2016 assessment). The IUCN Red List of Threatened Species 2017: e.T22678148A110915916. https://doi.org/10.2305/IUCN.UK.2017-1.RLTS.T22678148A110915916.en. Downloaded on 25 January 2020

BirdLife International (2018a) Antrostomus carolinensis. The IUCN Red List of Threatened Species 2018: e.T22689778A131764040. https://doi.org/10.2305/IUCN.UK.2018-2.RLTS.T22689778A131764040.en. Downloaded on 27 January 2020

BirdLife International (2018b) Bangsia arcaei. The IUCN Red List of Threatened Species 2018: e.T22722558A132016322. https://doi.org/10.2305/IUCN.UK.2018-2.RLTS.T22722558A132016322.en. Downloaded on 25 January 2020

BirdLife International (2018c) Touit costaricensis. The IUCN Red List of Threatened Species 2018: e.T22686017A131517274. https://doi.org/10.2305/IUCN.UK.2018-2.RLTS.T22686017A131517274.en. Downloaded on 25 January 2020

BirdLife International (2018d) Vermivora chrysoptera. The IUCN Red List of Threatened Species 2018: e.T22721618A132145282. https://doi.org/10.2305/IUCN.UK.2018-2.RLTS.T22721618A132145282.en. Downloaded on 25 January 2020

BirdLife International (2019a) Setophaga cerulea. The IUCN Red List of Threatened Species 2019: e.T22721740A153691320. https://doi.org/10.2305/IUCN.UK.2019-3.RLTS.T22721740A153691320.en. Downloaded on 25 January 2020

BirdLife International (2019b) Sturnella magna (amended version of 2018 assessment). The IUCN Red List of Threatened Species 2019: e.T22735434A155622113. https://doi.org/10.2305/IUCN.UK.2018-2.RLTS.T22735434A155622113.en. Downloaded on 25 January 2020

Bolaños F, Chaves G, Wake D, Savage J (2008a) Oedipina poelzi. The IUCN Red List of Threatened Species 2008: e.T59321A11916146. https://doi.org/10.2305/IUCN.UK.2008.RLTS.T59321A11916146.en. Downloaded on 19 January 2020

Bolaños F, Wake D, Savage J (2008b) Nototriton picadoi. The IUCN Red List of Threatened Species 2008: e.T59299A11902609. https://doi.org/10.2305/IUCN.UK.2004.RLTS.T59299A11902609.en. Downloaded on 20 January 2020

Brenes J, VanDusen K, Welch J (2019) Pathways toward Climate Change Resilience in Monteverde, Costa Rica. Report prepared for the Monteverde Community Fund, Monteverde, Costa Rica. http://corclima.org/en/wp-content/uploads/pdf/Pathways-Climate-Resilience-Monteverde.pdf

Brownson K (2019) From incentives to impact: an evaluation of the socioecological impacts of payments for ecosystem services. PhD dissertation, The University of Georgia, Athens, GA

Brownson K, Guinessey E, Carranza M, Esquivel M, Hesselbach H, Madrid Ramirez L, Villa L (2019) Community-Based Payments for Ecosystem Services (CB-PES): implications of community involvement for program outcomes. Ecosyst Serv 39:Article 100974. https://doi.org/10.1016/j.ecoser.2019.100974

Burlingame L (2000) Conservation in the Monteverde zone: contributions of conservation organizations. In: Nadkarni NM, Wheelwright NT (eds) Monteverde: ecology and conservation of a tropical cloud Forest. Oxford University Press, New York, pp 351–388

Camacho Céspedes F (2019) Biofertilización orgánica de almácigos de café (*Coffea arabica* L.) con compost producido a partir de residuos biomásicos locales, microorganismos de montaña y lodos digeridos de biodigestor en la región de Monteverde, Costa Rica. PhD dissertation, Universidad Estatal a Distancia, Programa DOCINADE, San José, Costa Rica

Caughlin TT, Ganesh T, Lowman MD (2007) COS 159-8: the role of frugivorous birds and giant fruit bats as agents for reforestation in South India. Ecological Society of America/Society of Restoration Ecology Joint Meeting, August 5–August 10, 2007, San Jose, California. https://eco.confex.com/eco/2007/techprogram/P5436.HTM

Chacón CM (2005) Fostering conservation of key priority sites and rural development in Central America: the role of private protected areas. Parks Prot Areas Program Priv Protected Areas 15(2):39–47

Chinchilla Ramos R (2015) Conservación y manejo integral a través del análisis del uso de la tierra y la fragmentación boscosa en el Corredor Biológico Pájaro Campana, Pacífico Central, Puntarenas. Thesis for Licenciatura, University of Costa Rica, San José, Costa Rica

Clark KL, Lawton RO, Butler PR (2014) The physical environment – update 2014. In: Nadkarni NM, Wheelwright NT (eds) Monteverde: ecology and conservation of a tropical cloud forest – 2014 updated chapters. Oxford University Press, New York

Cuarón AD, Morales A, Shedden A, Rodriguez-Luna E, de Grammont PC, Cortés-Ortiz L (2008) Ateles geoffroyi. The IUCN Red List of Threatened Species 2008: e.T2279A9387270. https://doi.org/10.2305/IUCN.UK.2008.RLTS.T2279A9387270.en. Downloaded on 19 January 2020

de Oliveira T, Paviolo A, Schipper J, Bianchi R, Payan E, Carvajal SV (2015) Leopardus wiedii. The IUCN Red List of Threatened Species 2015: e.T11511A50654216. https://doi.org/10.2305/IUCN.UK.2015-4.RLTS.T11511A50654216.en. Downloaded on 19 January 2020

Deliso E (2008) Climate Change and the Hummingbirds of the Monteverde Cloud Forest, Costa Rica. Centro Científico Tropical 1 ed. San José, Costa Rica 42 pp:22 X 28 cms

Dyer V (1979) List of plants collected. Unpublished report, Tropical Science Center, San José, Costa Rica

Eaton WD, Roed M, Chassot O, Barry D (2012) Differences in soil moisture, nutrients and the microbial community between forests on the upper Pacific and Caribbean slopes at Monteverde, Cordillera de Tilaran: implications for responses to climate change. Trop Ecol 53(2):235–240

Feinsinger P (1977) Notes on the hummingbirds of Monteverde, Cordillera de Tilarán, Costa Rica. Wilson Bull 89(1):159–164

Fogden MP (2000) Appendix 9: birds of the Monteverde area. In: Nadkarni NM, Wheelwright NT (eds) Monteverde: ecology and conservation of a tropical cloud Forest. Oxford University Press, New York, pp 541–552

FONAFIFO (2001) Reactivación de la Reforestación Comercial en Costa Rica. Proyecto REFORESTA. Propuesta de Asistencia Técnica y Financiera. MINAE-FONAFIFO. San José, Costa Rica

García M, Jordan C, O'Farril G, Poot C, Meyer N, Estrada N, Leonardo R, Naranjo E, Simons Á, Herrera A, Urgilés C, Schank C, Boshoff L, Ruiz-Galeano M (2016) Tapirus bairdii. The IUCN Red List of Threatened Species 2016: e.T21471A45173340. https://doi.org/10.2305/IUCN.UK.2016-1.RLTS.T21471A45173340.en. Downloaded on 19 January 2020

García-Rodríguez A, Chaves G, Benavides-Varela C, Puschendorf R (2012) Where are the survivors? Tracking relictual populations of endangered frogs in Costa Rica. Divers Distrib 18:204–212. https://doi.org/10.1111/j.1472-4642.2011.00862.x

Griffith K, Peck DC, Stuckey J (2000) Agriculture in Monteverde: moving toward sustainability. In: Nadkarni NM, Wheelwright NT (eds) Monteverde: ecology and conservation of a tropical cloud Forest. Oxford University Press, New York, pp 389–417

Guindon CF (1996) The importance of Forest fragments to the maintenance of regional biodiversity in Costa Rica. In: Schelhas J, Greenberg RS (eds) Forest patches in tropical landscapes. Island Press, Washington, DC, pp 168–186

Haber WA (1991) Lista provisional de las plantas de Monteverde, Costa Rica. Brenesia 34:63–120

Haber WA (1993) Seasonal migration of monarchs and other butterflies in Costa Rica. In: Malcolm SB, Zalucki M (eds) Biology and conservation of the monarch butterfly. Los Angeles County Museum of Natural History, Los Angeles, pp 201–207

Haber WA (2000a) Plants and vegetation. In: Nadkarni NM, Wheelwright NT (eds) Monteverde: ecology and conservation of a tropical cloud forest. Oxford University Press, New York, pp 39–94

Haber WA (2000b) Appendix 1: vascular plants of Monteverde. In: Nadkarni NM, Wheelwright NT (eds) Monteverde: ecology and conservation of a tropical cloud forest. Oxford University Press, New York, pp 457–518

Hamilton D, Molina V, Powell G (2003) El Estatus del Pájaro Campana (*Procnias tricarunculata*): Un Ave en Peligro de Extinción. Zeledonia 7:15–24

Hartshorn GS (1983) Plants. In: Janzen DH (ed) Costa Rican natural history. University of Chicago Press, Chicago, pp 118–157

Harvey CA, Haber WA (1999) Remnant trees and the conservation of biodiversity in Costa Rican pastures. Agrofor Syst 44:37–68. https://doi.org/10.1023/A:1006122211692

Holdridge LR (1967) Life zone ecology, Rev. edn. Tropical Science Center, San Jose

Instituto Nacional de Vivienda y Urbanismo (INVU) (2019) Planes Reguladores. https://www.invu.go.cr/planes-reguladores

IUCN SSC Amphibian Specialist Group and NatureServe (2013a) Isthmohyla tica. The IUCN Red List of Threatened Species 2013: e.T55675A3031693. https://doi.org/10.2305/IUCN.UK.2013-1.RLTS.T55675A3031693.en. Downloaded on 19 January 2020

IUCN SSC Amphibian Specialist Group and NatureServe (2013b) Lithobates vibicarius. The IUCN Red List of Threatened Species 2013: e.T58746A3072875. https://doi.org/10.2305/IUCN.UK.2013-1.RLTS.T58746A3072875.en. Downloaded on 19 January 2020

IUCN SSC Amphibian Specialist Group and NatureServe (2014) Agalychnis annae. The IUCN Red List of Threatened Species 2014: e.T55288A3027705. https://doi.org/10.2305/IUCN.UK.2014-3.RLTS.T55288A3027705.en. Downloaded on 20 January 2020

James TY, Toledo LF, Rödder D, da Silva LD, Belasen A, Betancourt-Román C, Jenkinson TS, Soto-Azat Z, Lambertini C, Longo AV, Ruggeri J, Collins JP, Burrowes PA, Lips KR, Zamudio KR, Longcore JE (2015) Disentangling host, pathogen, and environmental determinants of a recently emerged wildlife disease: lessons from the first 15 years of amphibian chytridiomycosis research. Ecol Evol 5:4079–4097

Joslin JD, Haber WA, Hamilton D (2018) Ocotea monteverdensis. The IUCN Red List of Threatened Species 2018: e.T48724260A117762662. https://doi.org/10.2305/IUCN.UK.2018-1.RLTS.T48724260A117762662.en. Downloaded on 19 January 2020

Kamal S, Grodzińska-Jurczak M, Brown G (2015) Conservation on private land: a review of global strategies with a proposed classification system. J Environ Plan Manag 58(4):576–597. https://doi.org/10.1080/09640568.2013.875463

Keuroghlian A, Desbiez A, Reyna-Hurtado R, Altrichter M, Beck H, Taber A, Fragoso JMV (2013) Tayassu pecari. The IUCN Red List of Threatened Species 2013: e.T41778A44051115. https://doi.org/10.2305/IUCN.UK.2013-1.RLTS.T41778A44051115.en. Downloaded on 19 January 2020

Khela S (2014) Magnolia poasana. The IUCN Red List of Threatened Species 2014: e.T193992A2293588. https://doi.org/10.2305/IUCN.UK.2014-1.RLTS.T193992A2293588.en. Downloaded on 24 January 2020

Köhler L, Tobón C, Frumau KA, Bruijnzeel LS (2007) Biomass and water storage dynamics of epiphytes in old-growth and secondary montane cloud forest stands in Costa Rica. Plant Ecol 193(2):171–184

Mayer AL, Tikka PM (2006) Biodiversity conservation incentive programs for privately owned forests. Environ Sci Pol 9(7–8):614–625. https://doi.org/10.1016/j.envsci.2006.07.004

Miranda F, Bertassoni A, Abba AM (2014) Myrmecophaga tridactyla. The IUCN Red List of Threatened Species 2014: e.T14224A47441961. https://doi.org/10.2305/IUCN.UK.2014-1.RLTS.T14224A47441961.en. Downloaded on 19 January 2020

Mitré M (1998a) Capparis discolor. The IUCN Red List of Threatened Species 1998: e.T37802A10077032. https://doi.org/10.2305/IUCN.UK.1998.RLTS.T37802A10077032.en. Downloaded on 24 January 2020

Mitré M (1998b) Ilex costaricensis. The IUCN Red List of Threatened Species 1998: e.T30657A9570425. https://doi.org/10.2305/IUCN.UK.1998.RLTS.T30657A9570425.en. Downloaded on 24 January 2020

Mitré M (1998c) Zinowiewia costaricensis. The IUCN Red List of Threatened Species 1998: e.T30585A9558303. https://doi.org/10.2305/IUCN.UK.1998.RLTS.T30585A9558303.en. Downloaded on 24 January 2020

Morse WC, Schedlbauer JL, Sesnie SE, Finegan B, Harvey CA, Hollenhorst SJ, Kavanagh KL, Stoian D, Wulfhorst JD (2009) Consequences of environmental service payments for Forest retention and recruitment in a Costa Rican biological corridor. Ecol Soc 14(1):23

Muchhala N (2003) Exploring the boundary between pollination syndromes: bats and hummingbirds as pollinators of *Burmeistera cyclostigmata* and *B. tenuiflora* (Campanulaceae). Ocecologia 134:373–380

Nadkarni NM (1981) Canopy roots: convergent evolution in rainforest nutrient cycles. Science 214(4524):1023–1024

Nadkarni NM, Solano R (2002) Potential effects of climate change on canopy communities in a tropical cloud forest: an experimental approach. Oceologia 131:580–586

Nadkarni NM, Schaeferz D, Matelson TJ, Solano R (2002) Comparison of arboreal and terrestrial soil characteristics in a lower montane forest, Monteverde, Costa Rica. Pedobiologia 46:24–33

Nadkarni NM, Wheelwright NT (eds) (2000) Monteverde: Ecology and conservation of a tropical cloud forest. Oxford University Press

Nadkarni NM, Wheelwright NT (eds) (revised 2014) Monteverde: Ecology and conservation of a tropical cloud forest. Oxford University Press

NatureServe and IUCN SSC Amphibian Specialist Group (2013) Duellmanohyla uranochroa. The IUCN Red List of Threatened Species 2013: e.T55314A3028798. https://doi.org/10.2305/IUCN.UK.2013-1.RLTS.T55314A3028798.en. Downloaded on 19 January 2020

Nelson C (1998a) Agonandra macrocarpa. The IUCN Red List of Threatened Species 1998: e.T36132A9984004. https://doi.org/10.2305/IUCN.UK.1998.RLTS.T36132A9984004.en. Downloaded on 24 January 2020

Nelson C (1998b) Eugenia salamensis. The IUCN Red List of Threatened Species 1998: e.T36135A9984195. https://doi.org/10.2305/IUCN.UK.1998.RLTS.T36135A9984195.en. Downloaded on 24 January 2020

Nelson C (1998c) Terminalia bucidoides. The IUCN Red List of Threatened Species 1998: e.T32502A9710189. https://doi.org/10.2305/IUCN.UK.1998.RLTS.T32502A9710189.en. Downloaded on 24 January 2020

Newcomer Q (2007) Innovations in Private Land Conservation: An Integrated Evaluation of Payment for Environmental Services in the Path of the Tapir Biological Corridor in Costa Rica. PhD dissertation Yale University School of Forestry and Environmental Studies, New Haven, CT

NOAA National Centers for Environmental Information (NCEI) (2020) U.S. Billion-Dollar weather and climate disasters. https://www.ncdc.noaa.gov/billions/

Norton DA (2000) Conservation biology and private land: shifting the focus. Conserv Biol 14(5): 1221–1223

O'Halloran DT, Mowatt E, Leclerc ED, Bhatia M (2018) Designing a marketing plan for rural Tourism in Costa Rica. Worcester Polytechnic Institute Interactive Qualifying Projects. Retrieved from https://digitalcommons.wpi.edu/iqp-all/3024

Padgett-Vasquez S (2019) Forest connectivity from a Bird's eye view: linking Cloud Forests to Mangroves in the Bellbird Biological Corridor of Costa Rica. PhD dissertation, The University of Georgia, Athens, GA

Page W (1998) Ficus lateriflora. The IUCN Red List of Threatened Species 1998: e. T30556A9552527. https://doi.org/10.2305/IUCN.UK.1998.RLTS.T30556A9552527.en. Downloaded on 24 January 2020

Pasquini L, Fitzsimons J, Cowell S, Brandon K, Wescott G (2011) The establishment of large private nature reserves by conservation NGOs: key factors for successful implementation. Oryx 45(3):373–380. https://doi.org/10.1017/S0030605310000876

Paulson DR (2009a) Epigomphus subsimilis. The IUCN Red List of Threatened Species 2009: e. T7836A12854505. https://doi.org/10.2305/IUCN.UK.2009-2.RLTS.T7836A12854505.en. Downloaded on 19 January 2020

Paulson DR (2009b) Libellula mariae. The IUCN Red List of Threatened Species 2009: e. T165077A5960306. https://doi.org/10.2305/IUCN.UK.2009-2.RLTS.T165077A5960306.en. Downloaded on 30 January 2020

Paulson D, von Ellenrieder N (2006a) Palaemnema baltodanoi. The IUCN Red List of Threatened Species 2006: e.T60261A12317735. https://doi.org/10.2305/IUCN.UK.2006.RLTS. T60261A12317735.en. Downloaded on 30 January 2020

Paulson D, von Ellenrieder N (2006b) Perigomphus pallidistylus. The IUCN Red List of Threatened Species 2006: e.T29603A9503838. https://doi.org/10.2305/IUCN.UK.2006.RLTS. T29603A9503838.en. Downloaded on 30 January 2020

Payan E, de Oliveira T (2016) Leopardus tigrinus. The IUCN Red List of Threatened Species 2016: e.T54012637A50653881. https://doi.org/10.2305/IUCN.UK.2016-2.RLTS. T54012637A50653881.en. Downloaded on 19 January 2020

Piedrahíta López C (2013) Lineamientos Técnicos para la Planificación de la Adaptación al Cambio Climático en Áreas Funcionales para la Conservación de la Biodiversidad. MSc thesis Centro Agronómico Tropical de Investigación y Enseñanza (CATIE), Turrialba, Costa Rica

Porras LW, Chaves G, Solórzano A (2013) Trimetopon simile. The IUCN Red List of Threatened Species 2013: e.T203605A2769074. https://doi.org/10.2305/IUCN.UK.2013-2.RLTS. T203605A2769074.en. Downloaded on 20 January 2020

Pounds JA, Fogden MP (2000) Appendix 8: amphibians and reptiles of Monteverde. In: Nadkarni NM, Wheelwright NT (eds) Monteverde: ecology and conservation of a tropical cloud forest. Oxford University Press, New York, pp 537–538

Pounds J, Fogden M, Campbell J (1999) Biological response to climate change on a tropical mountain. Nature 398:611–615. https://doi.org/10.1038/19297

Pounds JA, Bustamante M, Coloma L, Consuegra JA, Fogden MPL, Foster PN, La Marca E, Masters KL, Merino-Viteri A, Puschendorf R, Ron SR, Sánchez-Azofeifa GA, Still CJ, Young B (2006) Widespread amphibian extinctions from epidemic disease driven by global warming. Nature 439:161–167. https://doi.org/10.1038/nature04246

Pounds J, Bolaños F, Chaves G (2008a) Craugastor andi. The IUCN Red List of Threatened Species 2008: e.T56414A11473154. https://doi.org/10.2305/IUCN.UK.2008.RLTS. T56414A11473154.en. Downloaded on 20 January 2020

Pounds J, Bolaños F, Chaves G (2008b) Craugastor angelicus. The IUCN Red List of Threatened Species 2008: e.T56418A11473990. https://doi.org/10.2305/IUCN.UK.2008.RLTS. T56418A11473990.en. Downloaded on 20 January 2020

Pounds J, Bolaños F, Chaves G, Wake D, Savage J (2008c) Bolitoglossa subpalmata. The IUCN Red List of Threatened Species 2008: e.T59212A11897656. https://doi.org/10.2305/IUCN.UK. 2008.RLTS.T59212A11897656.en. Downloaded on 20 January 2020

Pounds J, Bolaños F, Solís F, Ibáñez R, Chaves G, Savage J, Jaramillo C, Fuenmayor Q (2008d) Pristimantis altae. The IUCN Red List of Threatened Species 2008: e.T56406A11470632. https://doi.org/10.2305/IUCN.UK.2008.RLTS.T56406A11470632.en. Downloaded on 19 January 2020

Pounds J, Bolaños F, Solís F, Ibáñez R, Chaves G, Savage J, Jaramillo C, Fuenmayor Q, Castro F, Grant T, Lynch J, Lips K (2008e) Pristimantis caryophyllaceus. The IUCN Red List of Threatened Species 2008: e.T56497A11475825. https://doi.org/10.2305/IUCN.UK.2008.RLTS.T56497A11475825.en. Downloaded on 20 January 2020

Pounds J, Bolaños F, Wake D, Savage J (2008f) Nototriton gamezi. The IUCN Red List of Threatened Species 2008: e.T59294A11900971. https://doi.org/10.2305/IUCN.UK.2008.RLTS.T59294A11900971.en. Downloaded on 19 January 2020

Pounds J, Puschendorf R, Bolaños F, Chaves G, Crump M, Solís F, Ibáñez R, Savage J, Jaramillo C, Fuenmayor Q, Lips K (2010) Atelopus varius. The IUCN Red List of Threatened Species 2010: e.T54560A11167883. https://doi.org/10.2305/IUCN.UK.2010-2.RLTS.T54560A11167883.en. Downloaded on 19 January 2020

Powell G, Bjork R (1994) Implications of altitudinal migration for conservation strategies to protect tropical biodiversity: a case study of the Resplendent Quetzal *Pharomacrus mocinno* at Monteverde, Costa Rica. Bird Conserv Int 4(2–3):161–174. https://doi.org/10.1017/S0959270900002744

Powlen K (2018) The motivations and barriers of landowner participation in reforestation in the bellbird biological corridor in Costa Rica. MSc Thesis Colorado State University, Fort Collins, CO

Quigley H, Foster R, Petracca L, Payan E, Salom R, Harmsen B (2017) Panthera once (errata version published in 2018). The IUCN Red List of Threatened Species 2017: e.T15953A123791436. https://doi.org/10.2305/IUCN.UK.2017-3.RLTS.T15953A50658693.en. Downloaded on 19 January 2020

Rains KC, Nadkarni NM, Bledsoe CS (2003) Epiphytic and terrestrial mycorrhizas in a lower montane costa Rican cloud forest. Mycorrhiza 13:257–264

Ringhofer L, Singh SJ, Smetschka B (2013) Climate change mitigation in Latin America: a mapping of current policies, plans and programs. Institute for social ecology working paper 143

Rivers MC, Barstow M, Fuentes ACD (2019) Ticodendron incognitum. The IUCN Red List of Threatened Species 2019: e.T37468A128258819. https://doi.org/10.2305/IUCN.UK.2019-3.RLTS.T37468A128258819.en. Downloaded on 24 January 2020

Rodriguez B, Pineda W (2015) Ectophylla alba. The IUCN Red List of Threatened Species 2015: e.T7030A22027138. https://doi.org/10.2305/IUCN.UK.2015-4.RLTS.T7030A22027138.en. Downloaded on 19 January 2020

Ruedas L, Smith AT (2019) Sylvilagus brasiliensis. The IUCN Red List of Threatened Species 2019: e.T87491102A45191186. https://doi.org/10.2305/IUCN.UK.2019-2.RLTS.T87491102A45191186.en. Downloaded on 19 January 2020

Sandiford M (1998) Bombacopsis quinata. The IUCN Red List of Threatened Species 1998: e.T32905A9738742. https://doi.org/10.2305/IUCN.UK.1998.RLTS.T32905A9738742.en. Downloaded on 24 January 2020

Savage J, Pounds J, Bolaños F (2008) Incilius periglenes. The IUCN Red List of Threatened Species 2008: e.T3172A9654595. https://doi.org/10.2305/IUCN.UK.2008.RLTS.T3172A9654595.en. Downloaded on 19 January 2020

Scrimshaw N, Gibbons W, Burlingame LJ (2000) Conservation Easements in Monteverde: the Enlace Verde project. In: Burlingame L Conservation in the Monteverde Zone in: Nadkarni NM and Wheelwright NT (eds) Monteverde: ecology and conservation of a tropical cloud forest. Oxford University Press, New York, pp. 381–382

Skinner D (2014) Costus nitidus. The IUCN Red List of Threatened Species 2014: e.T56347808A56353108. https://doi.org/10.2305/IUCN.UK.2014-1.RLTS.T56347808A56353108.en. Downloaded on 19 January 2020

Snow BK (1977) Territory behavior and courtship of the male three-wattled bellbird. Auk 94:623–645

Solari S (2018) Vampyrum spectrum. The IUCN Red List of Threatened Species 2018: e.T22843A22059426. https://doi.org/10.2305/IUCN.UK.2018-2.RLTS.T22843A22059426.en. Downloaded on 19 January 2020

Solís F, Ibáñez R, Chaves G, Savage J, Jaramillo C, Fuenmayor Q (2008a) Ecnomiohyla fimbrimembra. The IUCN Red List of Threatened Species 2008: e.T55482A11304367. https://doi.org/10.2305/IUCN.UK.2008.RLTS.T55482A11304367.en. Downloaded on 20 January 2020

Solís F, Ibáñez R, Chaves G, Savage J, Jaramillo C, Fuenmayor Q, Bolaños F, Lips K(2008b) Isthmohyla zeteki. The IUCN Red List of Threatened Species 2008: e.T55701A11351517. https://doi.org/10.2305/IUCN.UK.2008.RLTS.T55701A11351517.en. Downloaded on 27 January 2020

Solís F, Ibáñez R, Pounds J, Bolaños F, Chaves G (2008c) Isthmohyla angustilineata. The IUCN Red List of Threatened Species 2008: e.T55390A11290945. https://doi.org/10.2305/IUCN.UK.2008.RLTS.T55390A11290945.en. Downloaded on 27 January 2020

Solís F, Ibáñez R, Savage J, Jaramillo C, Fuenmayor Q, Kubicki B, Pounds J, Chaves G, Jungfer K, Lips K (2008d) Agalychnis lemur. The IUCN Red List of Threatened Species 2008: e.T55855A11381418. https://doi.org/10.2305/IUCN.UK.2008.RLTS.T55855A11381418.en. Downloaded on 20 January 2020

Solís F, Ibáñez R, Chaves G, Savage J, Jaramillo C, Fuenmayor Q, Bolaños F (2010a) Craugastor podiciferus. The IUCN Red List of Threatened Species 2010: e.T56870A11547021. https://doi.org/10.2305/IUCN.UK.2010-2.RLTS.T56870A11547021.en. Downloaded on 20 January 2020

Solís F, Ibáñez R, Chaves G, Savage J, Jaramillo C, Fuenmayor Q, Kubicki B, Bolaños F (2010b) Ecnomiohyla miliaria. The IUCN Red List of Threatened Species 2010: e.T55561A11319134. https://doi.org/10.2305/IUCN.UK.2010-2.RLTS.T55561A11319134.en. Downloaded on 20 January 2020

Solís F, Ibáñez R, Pounds JA, Bolaños F, Chaves G, and Lips K (2010c) Isthmohyla rivularis. The IUCN Red List of Threatened Species 2010: e.T55627A11343375. https://doi.org/10.2305/IUCN.UK.2010-2.RLTS.T55627A11343375.en. Downloaded on 20 January 2020

Solórzano A, Acosta Chaves V, Chaves G (2013) Celestus hylaius. The IUCN Red List of Threatened Species 2013: e.T203032A2758756. https://doi.org/10.2305/IUCN.UK.2013-2.RLTS.T203032A2758756.en. Downloaded on 20 January 2020

Stevenson R, Haber WA (2000) Section 4.5.2: Migration of butterflies through Monteverde. In: Hanson P (ed) Insects and spiders. In: Nadkarni NM, Wheelwright NT (eds) Monteverde: ecology and conservation of a tropical cloud Forest. Oxford University Press, New York, pp 118–119

Stuckey J, Camacho Céspedes F, Vargas Leitón G, Stuckey SA, Vargas Leitón JL (2014) Agriculture in Monteverde, moving toward sustainability—update 2014. In: In Nadkarni NM, Wheelwright NT (eds) Monteverde: ecology and conservation of a tropical cloud forest – 2014 updated chapters. Oxford University Press, New York

Timm RM, LaVal RK (1998) A field key to the bats of Costa Rica, Occasional publication series 22: 1–30. Center of Latin American Studies, The University of Kansas, Lawrence

Timm RM, LaVal RK (2000) Appendix 10: Mammals of Monteverde. In: Nadkarni NM, Wheelwright NT (eds) Monteverde: ecology and conservation of a tropical cloud Forest. Oxford University Press, New York, pp 553–557

Tosi JA, Holdridge LR, Centro Científico Tropical (1969) República de Costa Rica, mapa ecológico San José. Centro Científico Tropical, Costa Rica. http://nla.gov.au/nla.obj-2568086976. Web. 18 December 2021

Tropical Science Center (2018) Promoción de Encadenamientos Productivos Sostenibles en el Corredor Biológico Pájaro Campana. Programa de Corredores Biológicos, Gerencia de Desarrollo, Centro Científico Tropical, San José, Costa Rica

Wainwright M (2007) The mammals of Costa Rica: a natural history and field guide, No. 599 W142m. Cornell University Press, Ithaca

Wake D, Savage J, Bolaños F (2008) Oedipina uniformis. The IUCN Red List of Threatened Species 2008: e.T59327A11918239. https://doi.org/10.2305/IUCN.UK.2008.RLTS.T59327A11918239.en. Downloaded on 20 January 2020

Wegier A, Lorea Hernández F, Contreras A, Mastretta-Yanes A (2017) Persea schiedeana. The IUCN Red List of Threatened Species 2017: e.T34402A61528670. https://doi.org/10.2305/IUCN.UK.2017-3.RLTS.T34402A61528670.en. Downloaded on 24 January 2020

Wegner GI (2016) Payments for ecosystem services (PES): a flexible, participatory, and integrated approach for improved conservation and equity outcomes. Environ Dev Sustain 18:617. https://doi.org/10.1007/s10668-015-9673-7

Welch J et al (2011) Plan Estratégico del Corredor Biológico Pájaro Campana (CBPC) 2011–2016. Monteverde, Costa Rica

Wheelwright NT (1983) Fruits and the ecology of resplendent quetzals. Auk 100:286–301

Wheelwright N (2000) Conservation biology. In: Nadkarni NM, Wheelwright NT (eds) Monteverde: ecology and conservation of a tropical cloud forest. Oxford University Press, New York, pp 419–456

Wheelwright N (2014) Conservation biology – update 2014. In: Nadkarni NM, Wheelwright NT (eds) Monteverde: ecology and conservation of a tropical cloud Forest – 2014 updated chapters. Oxford University Press, New York

Whitfield SM, Lips KR, Donnelly MA (2016) Amphibian decline and conservation in Central America. Copeia 104(2):351–379

Whitfield SM, Alvarado G, Abarca J, Zumbado H, Zuñiga I, Wainwright M, Kerby J (2017) Differential patterns of *Batrachochytrium dendrobatidis* infection in relict amphibian populations following severe disease-associated declines. Dis Aquat Org 126:33–41. https://doi.org/10.3354/dao03154

World Conservation Monitoring Centre (1998a) Chrysophyllum hirsutum. The IUCN Red List of Threatened Species 1998: e.T37729A10074123. https://doi.org/10.2305/IUCN.UK.1998.RLTS.T37729A10074123.en. Downloaded on 24 January 2020

World Conservation Monitoring Centre (1998b) Dichapetalum costaricense. The IUCN Red List of Threatened Species 1998: e.T36492A9999314. https://doi.org/10.2305/IUCN.UK.1998.RLTS.T36492A9999314.en. Downloaded on 19 January 2020

World Conservation Monitoring Centre (1998c) Elaeagia uxpanapensis. The IUCN Red List of Threatened Species 1998: e.T31855A9665319. https://doi.org/10.2305/IUCN.UK.1998.RLTS.T31855A9665319.en. Downloaded on 24 January 2020

World Conservation Monitoring Centre (1998d) Ocotea viridiflora. The IUCN Red List of Threatened Species 1998: e.T30597A9560236. https://doi.org/10.2305/IUCN.UK.1998.RLTS.T30597A9560236.en. Downloaded on 24 January 2020

World Conservation Monitoring Centre (1998e) Pouteria austin-smithii. The IUCN Red List of Threatened Species 1998: e.T34409A9864731. https://doi.org/10.2305/IUCN.UK.1998.RLTS.T34409A9864731.en. Downloaded on 24 January 2020

World Conservation Monitoring Centre (1998f) Pouteria congestifolia. The IUCN Red List of Threatened Species 1998: e.T32686A9717700. https://doi.org/10.2305/IUCN.UK.1998.RLTS.T32686A9717700.en. Downloaded on 24 January 2020

World Conservation Monitoring Centre (1998g) Pouteria fossicola. The IUCN Red List of Threatened Species 1998: e.T37716A10072704. https://doi.org/10.2305/IUCN.UK.1998.RLTS.T37716A10072704.en. Downloaded on 24 January 2020

World Conservation Monitoring Centre (1998h) Sideroxylon capiri subsp. capiri. The IUCN Red List of Threatened Species 1998: e.T36041A9977046. https://doi.org/10.2305/IUCN.UK.1998.RLTS.T36041A9977046.en. Downloaded on 24 January 2020

World Conservation Monitoring Centre (1998i) Sideroxylon persimile subsp. subsessiliflorum. The IUCN Red List of Threatened Species 1998: e.T36057A9971292. https://doi.org/10.2305/IUCN.UK.1998.RLTS.T36057A9971292.en. Downloaded on 24 January 2020

Zamzow BK, Nieman SJ, Davis CN, Garro Cruz M, Monroe A, Stallcup L, Moran MD (2018) Status of large terrestrial vertebrates in the Monteverde-Arenal bioregion, northwestern Costa Rica. Tropical Conservation Science. https://doi.org/10.1177/1940082918809617

Zbinden S, Lee DR (2005) Paying for environmental services: an analysis of participation in Costa Rica's PSA program. World Dev 33:255–272

# Chapter 11
# A Highly Productive Biodiversity Island Within a Monoculture Landscape: El Hatico Nature Reserve (Valle del Cauca, Colombia)

**Zoraida Calle D, Carlos Hernán Molina C, Carlos Hernando Molina D, Enrique José Molina D, Juan José Molina E, Bernardo Murgueitio C, Amalia Murgueitio C, and Enrique Murgueitio R**

**Abstract** This chapter describes the landscape-scale, national and regional influence of a rural property that forms a biodiversity island within a monoculture landscape. Managed continuously by nine generations of the Molina family, El Hatico Nature Reserve embodies a set of values grounded in a deep connection to the land. Between 1960 and 1990, the fertile flatlands of the Cauca River valley lost almost all dry forest remnants, wetlands, traditional annual crops, and agroforestry systems, adopting a uniform method of sugarcane production that modified stream banks, eliminated the small-scale topographic heterogeneity, and integrated periodic burning and herbicide applications as part of the management protocols. Meanwhile, El Hatico gained tree cover, enhanced its soil quality, conserved its forest fragments, transformed its conventional pastures into biodiverse silvopastoral systems and transitioned to agroecological sugarcane production. El Hatico's long tradition of agricultural and livestock research and detailed production records helped develop

Coauthor Carlos Hernán Molina passed away during the final stage of editing this book. None of the innovations described in this chapter would exist were it not for his leadership, moral courage and commitment to generational exchange and agroecology. With admirable clarity and wisdom, he guided his family through decades of dramatic land use change and stood up for a diversified agricultural production against the ravaging advance of monoculture. Thanks to Carlos Hernán, El Hatico persists as a biodiversity island.

Zoraida Calle D (✉) · Bernardo Murgueitio C · Amalia Murgueitio C · Enrique Murgueitio R
Centro para la Investigación en Sistemas Sostenibles de Producción Agropecuaria (CIPAV), Cali, Colombia
e-mail: zoraida@fun.cipav.org.co

Carlos Hernando Molina D · Enrique José Molina D · Juan José Molina E
Centro para la Investigación en Sistemas Sostenibles de Producción Agropecuaria (CIPAV), Cali, Colombia

El Hatico Nature Reserve, El Cerrito, Colombia



F. Montagnini (ed.), *Biodiversity Islands: Strategies for Conservation in Human-Dominated Environments*, Topics in Biodiversity and Conservation 20,
https://doi.org/10.1007/978-3-030-92234-4_11

the highly efficient intensive silvopastoral systems in which cattle graze on nitrogen-fixing fodder shrubs interspersed with grasses and under the shade of native trees. These silvopastures have inspired thousands of technical assistants, extension workers and farmers in Colombia and other Latin American countries to undertake the transformation of conventional cattle ranching systems. Simultaneously, El Hatico developed organic sugarcane by applying the principles of agroecology to produce sugar while storing carbon, enhancing the soil biota and making an efficient use of water. El Hatico's forest fragment is surrounded by a wildlife-friendly silvopastoral matrix that is permeable to the movements of birds and arthropods. This property's unique combination of land uses provides a model for the integration of agroecology, agroforestry and ecological restoration.

**Keywords** Agroecology · Ecological restoration · Functional biodiversity · Intensive silvopastures · Silvopastoral systems · Sugarcane

## 11.1 Introduction

Can a single rural property make a difference for biodiversity and sustainability? If so, on which scales can that occur? This chapter explores the spheres of influence of a family estate that stands out as a biodiversity island within an intensive monoculture landscape. Seen from the air, El Hatico Nature Reserve, located in the fertile flatlands of the geographic Cauca river valley in Colombia, is the only woodland in a landscape dominated by sugarcane plantations. On closer inspection, what appears to be a secondary forest is in fact an old growth dry forest remnant, partially surrounded by silvopastures with a high density and diversity of trees. On one edge, this forest fragment borders a sugarcane plantation that has been managed agroecologically for two decades, produces certified organic sugar, and has lines of native palms and trees between rows of sugarcane plants.

The first section of this chapter provides a brief historical context of land use change in the geographic Cauca river valley and the lower Amaime river basin, where El Hatico is located. The following sections describe two land-use changes that were occurring at El Hatico while the surrounding landscape was following the opposite trends: the transition from conventional cattle ranching to silvopastoral systems and the adoption of agroecological practices in sugarcane production. Then, we summarize the results of research projects done at El Hatico on the spatial distribution of ants, parasitic wasps, spiders and birds, and the functional biodiversity in silvopastures and sugarcane. In the final section, we explore El Hatico's influence at larger spatial scales through research, training and inspiring farmers, extension workers, and decision makers. We also discuss El Hatico's impacts on land use policy related to livestock, sustainable agriculture, and private conservation initiatives.

This chapter focuses on why El Hatico is an outstanding leader in tropical sustainable agriculture, with broad impacts from local to global scales. The farm provides critical habitat for biodiversity within its agricultural matrix, modeling the alignment of increased productivity with enhanced ecological functioning.

## 11.2 Historical Context: Land Use Change in the Geographic Cauca River Valley

The geographic Cauca river valley comprises 421,000 hectares of flatlands in the high Cauca river basin, surrounded by the central and western Andean ranges (Cordillera Central and Cordillera Occidental) in the Colombian departments of Valle del Cauca, Cauca and Risaralda. Between 1957 and 1986, 66% of the forest cover in the upper Cauca river basin was transformed into agricultural land (CVC 1990). Today this landscape retains only 1.76% of its original forest, represented by scattered fragments with a mean area of 6 ha (Arcila et al. 2012). Wetlands once occupied almost one quarter of the geographic Cauca river valley; the few remaining wetlands are besieged by urban and agricultural expansion (Rivera et al. 2007).

As a result of fragmentation, the extreme isolation of forest remnants and the opportunity cost of the land, the agricultural landscape of the geographic Cauca River valley has very limited opportunities for conservation based on protected areas and restored biological corridors alone. Close to 75% of the remaining forest fragments are separated from their nearest neighbor by distances of 500 m or more (Arcila et al. 2012). Approximately 50% of these fragments lack a real forest interior area when assuming a 50 m edge effect. The fact that the isolated forest patches are surrounded by hostile matrices that barely promote the movement of organisms heralds an uncertain future for the regional biodiversity. Small populations of animals and plants in isolated remnants tend to have a high probability of local extinction, related to low genetic variability combined with demographic and stochastic effects (Eibl et al. 2022; Niella et al. 2022). All conservation or restoration initiatives in the geographic Cauca river valley should involve redesigning and managing the agricultural matrix to enhance the movement of plants and animals in a landscape dominated by sugarcane (Calle et al. 2012, 2013).

El Hatico Nature Reserve is located at the lower Amaime river basin, close to the site where the Amaime river joins the larger Cauca river (162 km south-west of Pinzacuá farm; see Fig. 12.1 in Montes-Londoño et al. 2022). The lower Amaime basin retained an extensive forest cover until the 1950s. Between 1950 and 1970, agriculture expanded in this area with public and private investment at the expense of forests, wetlands and even cattle ranching (Murgueitio 2019). Large areas were planted with cotton, millet, corn, soy, beans, rice and sugarcane. Landscape transformation in the 1960s was driven mainly by the economic blockade of Cuba and the increasing demand for sugar in the United States. The growing dominance of intensive sugarcane production has continued until the present.

**Fig. 11.1** Sugarcane, silvopastures and forest at El Hatico Nature Reserve. (Photo: Juan Diego Vanegas)

Murgueitio (2019) analyzed the changes in land cover that took place between 1986 and 2018 in the lower Amaime river basin. Land covered by perennial crops (mainly sugarcane) increased from 62.2 to 74%, while annual crops declined from 15.3 to 2.2%. Areas occupied by infrastructure increased by 1.9%, areas covered by natural vegetation increased by 0.8%, and pastures declined by 1.6%.

El Hatico Nature Reserve occupies 0.64% of the lower Amaime river basin but conserves 4% of its forested areas (Murgueitio 2019). This 288-ha property combines silvopastoral systems (140 ha), sugarcane (100 ha), forest (14 ha), *Guadua angustifolia* forest (26 ha), restored biological corridors (2 ha) and mixed fruit trees (5 ha); Figs. 11.1 and 11.2). El Hatico is located at 1000 m of altitude and has average temperature of 24 °C, average annual rainfall of 750 mm and 75% relative humidity (Molina-Castro et al. 2012). Due to its location at the center of the geographic Cauca river valley, evapotranspiration at El Hatico (1600 mm year$^{-1}$) far exceeds rainfall, creating a significant moisture deficit. The Cauca river valley is considered a dry tropical forest according to the Holdridge life zone system, although most trees retain their foliage even during the driest periods. This mild deciduousness is explained by the bimodal distribution of rainfall and the superficial water table.

In 1942, land use in El Hatico showed a pattern similar to other rural properties in the area, with one forest fragment, giant bamboo (*Guadua angustifolia*) stands, pastures with a low density of trees, and sugarcane plantations. However, an aerial photograph from 1986 shows a radically different landscape context; El Hatico's forest had become the only remaining fragment in the area, and the surrounding matrix had been transformed into a simplified grid of sugarcane plantations with few strips of giant bamboo forest. Satellite images from 2007 and 2018 show another

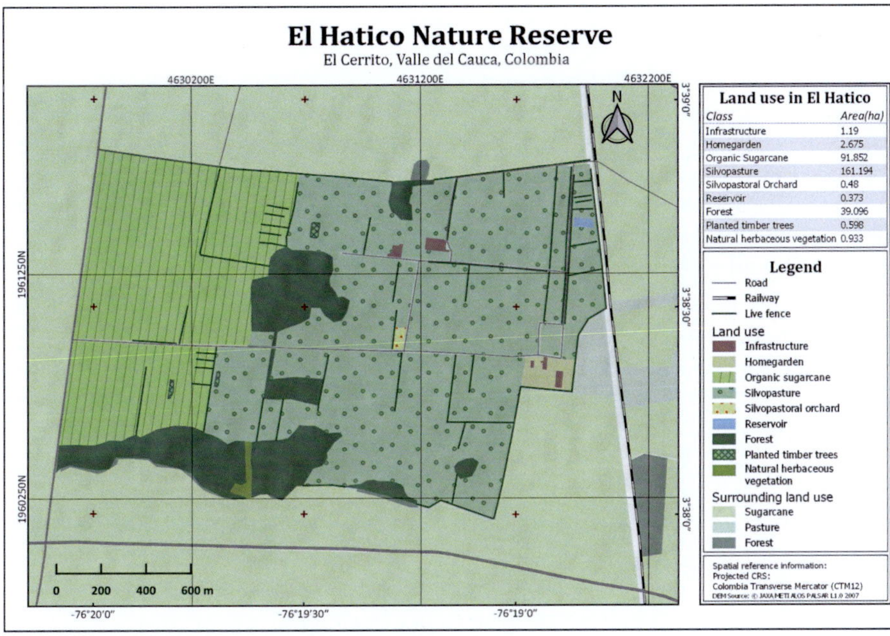


**Fig. 11.2** Land use distribution at El Hatico Nature Reserve

reality, in which tree density and canopy cover had increased in El Hatico's silvopastures, creating an open woodland physiognomy in which the edges of the forest fragment had become blurred (Molina-Castro et al. 2012). In two decades, vegetation structure was enhanced and the once simplified agricultural matrix became suitable habitat for diverse groups of organisms.

## 11.3 From Conventional Cattle Ranching to Silvopastoral Systems

El Hatico has a long tradition of cattle farming. For several generations, livestock production focused on beef, horses, mules and bovine breeding stock. Dairy farming played a marginal role until the development of local road infrastructure in the 1950s.

The foundations of El Hatico's silvopastoral systems go back to the first decades of the twentieth century. The first Colombian publication on the role of trees in cattle pastures, written by a member of the Molina family, mentions the failure of the initial trials with temperate legumes planted as feed for cattle in the tropics and highlights the nutritional value of fruits from native leguminous trees such as *Samanea saman* and *Enterolobium cyclocarpum* (Molina-Garcés 1938).

Trees of a high-yield ecotype of *Prosopis juliflora* were planted in paddocks of El Hatico in the 1940s to provide nutritious fruits for the cattle. These trees offer a high energy supplement in the form of sweet pods during the driest months of the year, when grass biomass drops to its lowest level. In the 1980s, *Gliricidia sepium* trees were planted at high density to provide cut-and-carry fodder; this *Gliricidia* fodder bank was used for more than two decades. However, the manual harvest of forage is labor-intensive and most tree and shrub species with highly nutritional forage do not tolerate cattle browsing. This motivated El Hatico's team to integrate leguminous shrubs that could be browsed directly in the paddocks. An empirical observation of cattle browsing *Leucaena leucocephala* trees led to the initial idea of planting this species in an intensive silvopastoral system (Murgueitio et al. 2015).

### *11.3.1 Intensive Silvopastoral Systems*

El Hatico pioneered the adoption of silvopastoral systems in the 1970s and the development of intensive silvopastoral systems (ISS) in the 1990s. ISS are characterized by the high-density cultivation of fodder shrubs (5000–80,000 plants $ha^{-1}$) interspersed with improved tropical grasses, legumes, trees and palms (Murgueitio et al. 2015; Chará et al. 2017; Santos-Gally and Boege 2022).

Today, El Hatico's ISS form a complex and wildlife friendly agricultural matrix that combines grass species such as *Cynodon plectostachyus*, *Megathyrsus maximus*, *Cynodon dactylon* and *Paspalum notatum*; a high density of the nitrogen-fixing tree *Leucaena leucocephala* (up to 30,000 plants $ha^{-1}$ managed as fodder shrubs); 30–50 medium sized planted and regenerating trees $ha^{-1}$; a lower density of large shade and timber trees and palms (Murgueitio et al. 2011; Calle et al. 2013; Chará et al. 2015). Paddocks are separated by 40 km of live fences of trees such as *Gliricidia sepium* (some of which are more than 100 years old), broad leaf mahogany (*Swietenia macrophylla*), *Guazuma ulmifolia, Maclura tinctoria,* the large bromeliad *Bromelia plumieri* and mixed live fences formed by naturally regenerating trees, shrubs and herbs (Molina-Castro et al. 2012). Table 11.1 presents the most common tree and palm species found in silvopastures; Figure 11.3 shows a live fence of broad-leaf mahogany.

The grazing method applied in El Hatico's ISS combines high animal loads (50 Lucerna dairy cows with an average weight of 450 kg in 4000 $m^2$ paddocks) with brief rotations (2 days), followed by long periods of recovery (45 days). Therefore, throughout the year, each individual paddock is grazed intensively for a total of 16 days and recovers during the remaining 349 days.

El Hatico has access to two sources of irrigation: a concession agreement providing a fixed volume of water from the Amaime river and groundwater that is pumped from a deep well (100 m) at a high energetic cost. Both sources were used for several decades to irrigate pastures. The gradual adoption of complex silvopastoral systems allowed El Hatico to increase its per-hectare productivity from 7436 l of milk in 1996 to 18,299 l in 2004 (El Hatico, unpublished data). After reaching that historical peak yield, the owners decided to suspend irrigation,

**Table 11.1** Common trees and palms in silvopastures and live fences at El Hatico

| Scientific name | Local name | Family | Origin | Uses |
|---|---|---|---|---|
| *Attalea butyracea* (Mutis ex L. f.) Wess. Boer | Palma corozo de puerco | Arecaceae | P, NR | FW |
| *Roystonea regia* (Kunth) O.F. Cook | Palma real | Arecaceae | P, NR | A |
| *Syagrus sancona* H. Karst. | Palma zancona | Arecaceae | P | A, FW |
| *Bromelia plumieri* (E. Morren) L.B. Sm. | Piñuela | Bromeliaceae | P | LF, FW |
| *Enterolobium cyclocarpum* (Jacq.) Griseb. | Orejero | Fabaceae | NR | T, FC |
| *Gliricidia sepium* (Jacq.) Kunth ex Walp. | Matarratón | Fabaceae | P | LF, T |
| *Prosopis juliflora* (Sw.) DC. | Mezquite, Algarrobo | Fabaceae | P, NR | T, FC |
| *Samanea saman* (Jacq.) Merr. | Samán | Fabaceae | NR | T, FC |
| *Senna spectabilis* (DC.) H.S. Irwin & Barneby | Vainillo | Fabaceae | NR | FC |
| *Ceiba pentandra* (L.) Gaertn. | Ceiba | Malvaceae | NR | A |
| *Guazuma ulmifolia* Lam. | Guácimo | Malvaceae | NR | FC, LF |
| *Guarea guidonia* (L.) Sleumer | Cedro macho | Meliaceae | NR | T, FW |
| *Swietenia macrophylla* King | Caoba | Meliaceae | P | T, LF |
| *Cedrela odorata* L. | Cedro rosado | Meliaceae | P, NR | T |
| *Maclura tinctoria* (L.) D. Don ex Steud. | Dinde | Moraceae | P, NR | LF, T, FW |
| *Zanthoxylum rhoifolium* lam. | Tachuelo | Rutaceae | NR | T, FW |

Origin: Planted (P), Natural Regeneration (NR); Uses: Aesthetics (A), Fruits for cattle (FC), Fruits for wildlife (FW), Live fence (LF), Timber (T)

making their milk production depend on the water stored in the soil, known as "green water". Since then, ISS became 100% rainfed and milk yield stabilized around 15,000 l $ha^{-1}$ $year^{-1}$ (Molina-Castro et al. 2012).

Organic dairy products from El Hatico (and the closely related Lucerna farm[1]) have been a logical consequence of the superior quality and safety of this agroecological milk, rather than being an explicit goal of the adoption of silvopastoral systems. By replacing its conventional monocultures of African star grass (*Cynodon plectostachyus*) with ISS, El Hatico cut down fixed costs, added value to the milk through organic certification and increased the profitability of dairy farming. The cost of mineral salt was reduced by 42% because ISS provide abundant minerals and the animals limit their salt intake. The costs of irrigation and fertilizer were completely eliminated. Without ISS, the current per hectare profit would be US$ -27 $ha^{-1}$ $month^{-1}$ for a milk price of US $0.35. With ISS, the monthly per hectare profit is $206 for a market price of US $0.47 $l^{-1}$ for certified organic milk (El Hatico, unpublished records).

[1]Lucerna is another rural property dedicated to organic milk and sugarcane production, located in Bugalagrande, Valle del Cauca.

**Fig. 11.3** Live fence of broad-leaf mahogany (*Swietenia macrophylla*). (Photo: Carlos Pineda)

### *11.3.2 Soil Recovery in ISS*

Several studies on soil conditions and other environmental factors that influence crop and animal productivity have been conducted at El Hatico. For example, Vallejo et al. (2010) studied the microbiological, physical and chemical properties of the soil in a chronosequence of silvopastoral systems (3–6, 8–10 and 12–15 years old), and compared them to a conventional pasture outside of El Hatico that had been grazed intensively for 35 years and illustrates the baseline condition for the farm's silvopastures. In the conventional pasture, lower microbial responses (hydrolytic enzyme activities) from bacteria involved in decomposition and nutrient mineralization were explained by the high bulk density and penetration resistance that create a less favorable environment for root exploration of nutrients and water (Vallejo et al. 2010). In contrast, lower penetration resistance values in silvopastoral soils indicate improved soil aggregation and greater pore space, both of which enhance microbial habitats and activity.

The oldest silvopastures at El Hatico (12–15 years) showed the highest microbial biomass and enzyme activity levels when normalized for carbon or clay contents (Vallejo et al. 2010). The microbial responses summarized in Table 11.2 indicate that silvopastures are improving soils; however, the results of this research suggest that it takes at least 8 years (though likely >12 years) to fully appreciate this effect under the conditions of El Hatico (Vallejo et al. 2010).

The diverse and multi-layered plant community (as shown in Fig. 11.4) is one of the factors that explains greater microbial responses in ISS. With the adoption of

**Table 11.2** Total carbon (per clay unit), bulk density, soil penetration resistance and enzyme activity in silvopastures, conventional pasture and forest at El Hatico Nature Reserve

| Land Use | Age (years) | C clay$^{-1}$ (x100%) | BD (g cm$^{-3}$) | SPR (MPa) | Relative enzyme activity per unit of C: (B-glucosidase, alkaline phosphatase and urease) | | |
|---|---|---|---|---|---|---|---|
| SS12 | 12–15 | $10.0^{a}$ | $1.39^{b}$ | $3.30^{b}$ | *** | *** | *** |
| SS8 | 8–10 | $8.5^{a}$ | $1.40^{b}$ | $2.47^{b}$ | ** | ** | ** |
| SS3 | 3–6 | $8.6^{a}$ | $1.44^{b}$ | $2.85^{b}$ | ** | * | ** |
| CP | >30 | $7.9^{a}$ | $1.52^{a}$ | $3.98^{a}$ | ** | * | ** |
| F | >100 | $8.3^{a}$ | $1.21^{c}$ | $1.49^{c}$ | *** | * | *** |

Source: Vallejo et al. (2010, 2012)
*C clay$^{-1}$* Total carbon per clay unit basis, *BD* Bulk density, *SPR* Soil penetration resistance. Asterisks indicate the relative magnitude of enzyme activity levels; land uses sharing the same number of asterisks are not significantly different at $P < 0.05$

silvopastures, fodder biomass production at El Hatico increased from 23 to 39 Mg ha$^{-1}$ year$^{-1}$ (Mahecha 2003). Higher biomass production and fodder yield in these systems imply a higher return of carbon to the soil. Animal manure and urine enhance soil microbial activity by providing readily usable substrates (Clegg et al. 2006). The root systems of multi-layered silvopastures likely have a higher biomass that fully explores the soil profile. Increased above and belowground carbon inputs seem to drive the microbial responses in older silvopastures. The results summarized in Table 11.2 suggest that silvopastoral systems promote agroecosystem sustainability by improving the ability of soils to perform decomposition and nutrient mineralization, as reflected by hydrolytic enzyme activities (Vallejo et al. 2010).

The enhanced soil moisture and water use efficiency of ISS is probably related to the recovery of soil organic matter (Vallejo et al. 2010, 2012). The baseline condition for soil organic matter (SOM) in El Hatico's paddocks was 2.9% in 1994, while the reference condition for the forest is 4.3% (Arias 1994). The chronosequence of ISS illustrates two contrasting situations: 3.4 to 3.7% of SOM outside tree crowns and 4.4 to 4.9% under trees (Vallejo et al. 2012). These results show that the soil under ISS as the one shown in Fig. 11.3 can store significant amounts of carbon and that trees form high-carbon "islands" within the paddocks.

An additional benefit of El Hatico's ISS is the reduction of greenhouse gas emissions. Methane is a potent greenhouse gas with a global warming potential equivalent to 25 times that of carbon dioxide ($CO_2$). In cattle ranching, this gas results mostly from the activity of anaerobic microorganisms of the Archaea domain, that hydrolyze proteins, starches and cell wall components in the rumen. Animals with high fiber and dry matter intakes (DMI) release more methane. However, diet composition affects enteric methane production; when forages have high digestibility, less methane is produced per unit of DMI. Also, the presence of condensed tannins in leucaena is known to reduce methane production by inhibiting the growth

**Fig. 11.4** Intensive silvopasture with *Leucaena leucocephala* and native trees. (Photo: Zoraida Calle)

of cellulolytic and proteolytic bacteria. Methane released by the cattle per unit of degraded dry matter is 30% lower in El Hatico's ISS with leucaena than in conventional star grass monocultures (Molina et al. 2015, 2016 and references therein).

## 11.4 Agroecological Sugarcane

For several decades, sugarcane in the geographic Cauca river valley has been planted in intensive monocultures that use chemical fertilizers, herbicides and insecticides. Pre-harvest herbicide is applied to induce stress and increase the concentration of saccharose in the sugarcane. Pre- and post-harvest burning are done to facilitate the manual harvest and to eliminate crop residues, respectively. Conventional practices result in significant water pollution, soil compaction, greenhouse gas emissions and an increasing vulnerability of sugarcane to pests. Currently, the sector is transitioning from manual to mechanical harvest with heavy machinery in the region; this makes burning unnecessary but will bring new challenges related to soil physical degradation.

El Hatico joined the sugarcane industry in 1960 and implemented pre- and post-harvest burning in 1972. In 1994, they realized that 50% of the soil organic matter had been lost after two decades of conventional sugarcane cultivation (Arias 1994).

Since 1993 they had been monitoring parameters such as per hectare monthly sugarcane biomass production with the Colombian Sugarcane Research Center (Cenicaña, www.cenicana.org).

The transition to organic sugarcane production included the adoption of agroecological practices such as minimum tillage, mulching crop residues instead of burning them, and integrating nitrogen-fixing legumes (*Vigna unguiculata* and *Crotalaria juncea*) between cane rows to provide green manure. Herbicides were replaced with African hair sheep (breed of *Ovis aries* that does not grow wool) that consume all grasses and weeds, complemented with selective manual weed control. These practices have promoted functional biodiversity and water economy (Sadeghian and Madriñan 2000; Hincapié et al. 2019).

Stem borers (*Diatraea* spp., Lepidoptera: Crambidae) are considered the main pests of sugarcane in the Americas and cause severe economic losses (Vargas et al. 2015; Solis and Metz 2016). Conventional biological control involves releasing parasitoids such as *Lydella minense, Billaea claripalpis, Trichograma demanun* and *Cotesia flavipes*, which have low survival rates in conventional sugarcane plantations, where they lack vital resources such as pollen and nectar. Natural control of *Diatraea* spp. by ants such as *Solenopsis achinid, Wasmannia auropunctata, Ectatomma ruidum, Pheidole* spp., *Solenopsis* spp., *Camponotus* sp., *Nylanderia* spp., and *Pachycondyla ferruginea* has been observed at El Hatico and other organic farms (Gómez and Vargas 2014; Rivera et al. 2019). The Tachinid fly *Genea jaynesi* is a native parasitoid that controls this pest in the Cauca river valley. Its mass-rearing has not been successful in the laboratory, but it thrives in native vegetation strips, where it feeds on flowers of common weeds (Vargas et al. 2006, 2015; Cenicaña 2017; Rivera et al. 2019).

In 1996, El Hatico obtained its organic certification for sugarcane and livestock production. Shortly after, they started to integrate lines of native trees such as the endangered *Caesalpinia ebano* and palms (*Sabal mauritiiformis* and *Syagrus sancona*) within the sugarcane plots as shown in Fig. 11.5. Palm and tree lines were established between the sugar cane lines, spaced 36 m apart, which is equivalent to 24 lines of sugarcane, since the distance between sugarcane lines is 1.5 m. Planting distance between palms was 2 m to facilitate the manual harvest of the palm leaves for thatching, and there were 5 m between trees. Additional *Syagrus sancona* palms are being planted between each pair of trees.

### *11.4.1 Yields of Agroecological Sugarcane at El Hatico*

In the geographic Cauca river valley, the yield of sugarcane per kilogram of nitrogen applied to the crop has declined steadily in conventional sugarcane, from 1.03 ton in 1985 to 0.58 ton in 2015 (Cenicaña, unpublished results). This waning response to chemical nitrogen fertilization has triggered a soil degradation alert in the region. In contrast to conventional sugarcane, the yield of El Hatico's agroecological sugarcane varied between 1.3 and 1.87 ton per kilogram of nitrogen (from organic

**Fig. 11.5** Line of *Sabal mauritiiformis* palms in agroecological sugarcane plantation at El Hatico. (Photo: Carlos Pineda)

fertilizer and green manure) between 2002 and 2018 (El Hatico, unpublished yield records). Conventional producers currently use 180–220 kg of synthetic N $ha^{-1}$ $year^{-1}$, compared to 80 kg $ha^{-1}$ $year^{-1}$ from poultry manure used at El Hatico. Crop residues, green manure crops and free-living N-fixers such as soil bacteria provide approximately 80 kg of additional nitrogen at El Hatico (Cenicaña, unpublished data); however, this nitrogen is more difficult to quantify.

Between 2001 and 2018, El Hatico's agroecological sugarcane consistently outperformed its conventional counterpart (average yield: 9.99 vs. 8.58 tons of sugarcane $ha^{1}$ $month^{-1}$, respectively). Even more important than the higher yield is the fact that agroecological sugarcane behaves as a long-lived perennial crop. Several plots at El Hatico have been harvested 18 times between 2001 and 2020 without a decline in yield (El Hatico, unpublished yield records); in contrast, the conventional counterparts are being replaced after only five harvests due to decreasing productivity (Cenicaña 2001–2018, annual reports). This frequent replanting of sugarcane has high financial and environmental costs.

### *11.4.2 Environmental Impacts of Conventional and Agroecological Sugarcane*

In conventional sugarcane, herbicides are used periodically to eliminate grasses and other weeds in plots and alleys between plots (which occupy 10% of the land in plantations). El Hatico solved the problem of weed control by integrating African hair sheep. This makes sense, both financially and environmentally. Sheep use marginal areas that produce a high biomass of grasses, legumes and other species. Grazing replaces costly and unsustainable management practices such as the application of herbicides and mechanic weed control in sugarcane alleys and plantations. In addition to making marginal areas productive, sheep reduce the cost of weed control by 35% (Molina et al. 2013, 2014). Each hectare planted with sugarcane supports two to three adult hair sheep, transforming weeds into high-quality meat and manure that can be used as organic fertilizer (Molina et al. 2013, 2014). Some manure remains in the crop and another part is collected in enclosures (750–1000 g are collected daily from each 30–40 kg animal) (Fig. 11.6).

Agroecological management of sugarcane enhances soil biological activity (Sadeghian and Madriñan 2000). Data summarized in Table 11.3 show that populations of beneficial soil fungi and bacteria differ significantly between conventional and agroecological sugarcane plantations at different depths in the soil profile (Manrique et al. 2006).

Pardo (2009) found dramatic differences in macroinvertebrate biomass related to the type of management (agroecological at El Hatico vs. conventional in a nearby property) and between the moist and dry seasons. Macroinvertebrate biomass was 15.3 times higher in agroecological than in conventional sugarcane during the moist season (26.44 vs. 1.73 g per 0.75 $m^3$ of soil), and 59% higher during the dry season (6.56 vs. 4.11 g per 0.75 $m^3$ of soil). These differences were largely driven by variations in the biomass of earthworms and millipedes during the moist season, and earthworms during the dry season (Pardo 2009; Pardo et al. 2017).

Since 1994 soil organic matter (SOM) has increased from 2% to 4% at El Hatico's agroecological sugarcane, which means that it has almost recovered the forest reference value (4.2%). With the increase in SOM, the use of irrigation water has declined from 10,000 to 6000 $m^3$ $ha^{-1}$ $cycle^{-1}$ at El Hatico (a cycle at El Hatico is 12 months). This water economy has huge economic and environmental implications. The doubling of SOM in less than three decades highlights the climate change mitigation potential of agroecological sugarcane. If the whole region transitions to sustainable practices, sugarcane producers will be able to sequester large amounts of carbon, save precious water resources and cut down production costs while increasing yields significantly. Scaling-up agroecological practices in sugarcane would enhance ecosystem services and revitalize the regional economy.

**Fig. 11.6** African hair sheep under a line of endangered *Caesalpinia ebano* trees. (Photo: Zoraida Calle)

## 11.5 Biodiversity Studies at El Hatico: Arthropods and Birds

### *11.5.1 Ants*

Ants are considered good indicators of biodiversity, disturbance, ecological succession and ecosystem rehabilitation (Dominguez-Haydar and Armbrecht 2011). Armbrecht and Chacón (1997, 1999) studied ant species richness and diversity in

**Table 11.3** Populations of beneficial fungi and soil bacteria at different depths in the soil profile in Hatico's agroecological sugarcane plantations and reference conventional plantations

| **Beneficial fungi**[a] | **Conventional** | **Agroecological** |
|---|---|---|
| 0–5 cm | 37 | 88 |
| 5–10 cm | 30 | 60 |
| 10–20 cm | 23 | 37 |
| 20–40 cm | 10 | 30 |
| >40 cm | 7 | 25 |
| **Beneficial bacteria**[b] | **Conventional** | **Agroecological** |
| 0–5 cm | 400 | 580 |
| 5–10 cm | 280 | 500 |
| 10–20 cm | 250 | 480 |
| 20–40 cm | 155 | 400 |
| >40 cm | 50 | 100 |

Source: Manrique et al. (2006)
[a]Colony forming units $g^{-1} \times 10^4$
[b]Colony forming units $g^{-1} \times 10^6$

the seven largest forest fragments of the Cauca river valley and their surrounding agricultural matrixes, including El Hatico as one of their study sites. They found a total of 137 morphospecies, grouped into 37 genera and 6 subfamilies; 90% of the ant species were captured in forest fragments and 54% in their surrounding matrixes. They also found a significant correlation between ant species richness in the forest fragments and their matrixes (Armbrecht and Chacón 1999).

In these studies, the highest species richness and diversity values were found at El Hatico: a total of 81 ant species (66 in the forest and 35 in the surrounding silvopastoral matrix). El Hatico also had the highest number of exclusive ant species (ants that were not found in any other study site) in the agricultural matrix (6). Interestingly, the diversity index for El Hatico's silvopastoral systems ($H' = 3.07$) was higher than the diversity indexes of four of the seven forest fragments ($H'$ between 1.96 and 2.87) (Armbrecht and Chacón 1999). The studied agricultural matrixes were on average 7.7 °C warmer than the forest fragments; however, El Hatico was only 3.7 °C warmer. El Hatico's silvopastures formed the coolest matrix, with an average temperature of 30.1 °C (Armbrecht 1995).

In another study conducted in dry forest fragments in this same region Armbrecht et al. (2001) found that the assemblages of ant species in dry forest fragments were not simple subsets of the regional pool of species, but rather, that each fragment preserved an assembly with unique elements. The loss of any one of these small forest patches would cause the disappearance of a fraction of the regional ant diversity. This study highlights the importance of every remaining forest fragment for the long-term viability of the populations of several ant species (Armbrecht et al. 2001).

### 11.5.2 Spiders

Spiders (Aranae order) are considered as appropriate models for studies of community structure, composition and dynamics because they are diverse and abundant in terrestrial ecosystems, and their communities are affected by habitat, land use, vegetation structure and plant species composition (Pearce and Venier 2006). In tropical ecosystems, structurally complex vegetation tends to support diverse spider assemblages (Baldissera et al. 2012).

Delgado et al. (2014) studied the species composition and diversity of spiders at El Hatico in 1-ha plots located in a silvopastoral system, a forest fragment and agroecological sugarcane. They collected 3635 adult spiders, belonging to 156 morphospecies and 30 families; these species represent approximately 75% of El Hatico's total estimated spider fauna (Delgado et al. 2014). The silvopastoral system plot had the highest number of spider species (74), followed by the forest (71 species), and the agroecological sugarcane plot (46 species). Average similarity between land uses in the composition of spider assemblages was 45.4%; this suggests a high beta diversity or species turnover between land uses at El Hatico. The forest fragment and the silvopastoral system shared 54.4% of spider species (Delgado et al. 2014).

In the study conducted at El Hatico, the silvopastoral system was the most diverse habitat type, while the forest fragment was the least diverse one, given that half of the collected spiders belonged to a single species (*Leucauge* sp.). El Hatico's spider fauna found in this research included eight different guilds; this ecological diversity shows that different land uses provide a variety of microhabitats and resources for spiders with different hunting strategies. A high diversity of spiders suggests the presence of diverse prey species at El Hatico.

### 11.5.3 Parasitoid Wasps

Parasitoid wasps form a large group of hymenopterans that lay their eggs on (or inside) the bodies of other arthropods, causing the slow death of their hosts. Different parasitoid groups specialize in hosts from different insect orders. López et al. (2013) studied the diversity of parasitoid wasps in four silvopasture plots and the forest fragment at El Hatico. They collected 1376 parasitoids belonging to 7 super-families, 18 families and 42 morphospecies.

Vegetation structure and plant species composition in different land uses at El Hatico were clearly related to the abundance of parasitoids. These small wasps were more abundant in the silvopastures (319–364 individuals captured in traps throughout the study) than in the forest fragment (11 captured individuals). Species richness was also higher in silvopastures (22–26 species) compared to the forest fragment (11 species). However, the diversity index ($H'$) was higher at the forest ($H' = 2$) compared to the silvopastures ($H'$ values from 1 to 1.22), as a result of the higher equitability index (0.572 in the forest and 0.281–0.397 in the silvopastures).

The abundance of parasitoid wasps was correlated with plant species richness. The diversity of parasitoids in silvopastures is related to shade and the diversity of complementary food resources (nectar and pollen), including common weeds such as *Lantana camara* (Verbenaceae), *Parthenium hysterophorus* (Asteraceae) and *Sida acuta* (Malvaceae); parasitoids require nectar and pollen apart from insect prey.

The parasitioid wasp fauna identified at El Hatico includes species that have been released in the farm since the 1960s as biological control agents for pest species such *Diatraea saccharalis* (Pyralidae) in sugarcane and *Spodoptera* sp. (Noctuidae) in corn and sorghum (López et al. 2013). Apparently, these commercial biological control agents have established populations at El Hatico's silvopastures, which probably enhances biological pest control.

### *11.5.4 Birds*

El Hatico's bird fauna is outstanding. In a one-year study based on periodic bird censuses, Cárdenas (1998) observed a total of 135 species, belonging to 39 families and 17 orders. Two thirds of the birds (89 species) used agroecosystem habitats during the censuses. Tyrannidae was the richest bird family with 19 species (14%), followed by Fringillidae (11 species, 8%) and Thraupidae (8 species, 6%); 30 families were represented by 4 species or less. Waterfowl richness and the presence of species belonging to specialized taxonomic groups were also noteworthy. A total of 51 bird species showed evidence of reproductive activity. The bird list compiled by Cárdenas (1998) includes nine species that had not been reported previously in the Valle del Cauca and that expand the latitudinal distribution described by Hilty and Brown (1986) for the birds of Colombia.

This study found a considerable bird species richness in silvopastoral systems: 57 species in a system with fruit trees, 46 in an ISS with *Leucaena leucocephala* shrubs and 43 in a star grass silvopasture. The forest and the agroecological sugarcane had 33 bird species each. The lowest bird species richness was observed in the bamboo forest with 29 species and in a conventional sugarcane (outside of El Hatico) with 19 species (Cárdenas 1998).

Diversity indices (H′) were consistent with those of species richness: 3.21 for the silvopasture with fruit trees, 3.07 for ISS with *L. leucocephala*, 2.98 for the silvopasture with star grass, 2.86 for the forest, 2.73 for the bamboo forest, 2.43 for the agroecological sugarcane and 1.53 for the conventional sugarcane (Cárdenas 1998). Bird diversity and richness in the agroecological sugarcane are likely related to the proximity of the forest fragment and the bamboo forest, as well as the presence of live fences and tree-lined alleys that enhance structural and floristic diversity.

In a more recent but shorter study of El Hatico's bird fauna, Hurtado-G et al. (2016) observed a total of 109 bird species belonging to 37 families and 16 orders. They observed the highest species richness in the agroecological sugarcane system (57), followed by the fruit orchards (50), forest and silvopastoral systems (43 each). Conventional sugarcane plantations outside of El Hatico had 40 bird species (Hurtado-G et al. 2016).

Of the bird species observed by Hurtado-G et al. (2016), 9.2% were migratory. The forest and agroecological sugarcane had the highest number of migratory birds (4 species each), followed by the silvopastoral system (3 species), conventional sugarcane (2) and fruit trees (1).

Migratory birds are not restricted to pristine areas in the tropics; instead, they use agroecosystems during their winter residence. Trees in crops or paddocks favor the arrival and permanence of Nearctic-Neotropical migratory species by providing perches, shelter, foraging substrates and corridors (Rice and Greenberg 2004).

Hurtado-G et al. (2016) also studied the diet of migratory bird species at El Hatico. Food fragments obtained from fecal samples of migratory birds included mostly arthropods (99%) from the orders Coleoptera (64%), Hymenoptera (18%), Araneae (9%), Hemiptera (5%), Diptera (1.6%), Lepidoptera (1.6%), Acari (0.4%) and Psocoptera (0.4%). Most migratory bird species fed on similar items, although in different proportions related to their foraging habits. Beetles were important components of the diet of most migratory birds, varying from 28% for *Catharus ustulatus* to 72% for *Setophaga petechia*. Consumption of Hemiptera and spiders was common among all species except for *Hirundo rustica*; this bird consumed a high proportion of Hymenoptera (40%). Hurtado's results confirm that migratory birds that visit El Hatico each year are mostly insectivorous.

## 11.6 Functional Biodiversity: Natural Enemies of Pests

The term functional biodiversity includes the value and range of species and organismal traits that influence ecosystem functioning (Tilman 2001). Here we use the term in the narrower agroecological sense, referring to species with positive effects on agroecosystems such as natural enemies of pests.

Many organisms that behave as pests in conventional cattle ranching and agriculture are controlled naturally by different species at El Hatico without external inputs or energy. Some examples include:

- Birds such as the cattle egret (*Bubulcus ibis*), the yellow-headed caracara (*Milvago chimachima*) and the smooth-billed ani (*Crotophaga ani*) contribute to the integrated management of ticks *(Rhipicephalus microplus)* and other ectoparasites of cattle.
- The ant *Ectatomma ruidum,* the entomopathogenic fungus *Nomuraea rileyi* and minute wasps *(Trichogramma* sp.*)* that are endoparasitoids of insect eggs control periodic outbreaks of the lepidopteran *Azeta versicolor*, which completely defoliates *Gliricidia sepium* trees (Gómez et al. 2002).
- The cattle egret, together with spiders and entomopathogenic fungi, control two important pests of sugarcane, the fall armyworm (*Spodoptera frugiperda*) and the small mocis moth (*Mocis lapites*), rendering insecticides unnecessary in agroecological sugarcane (CH Molina, personal observations).

- *Bacillus thuringiensis*, *Trichogramma* sp. and paper wasps (*Polistes erythrocephalus*, Vespidae) controlled an outbreak of a defoliator worm (Noctuidae) that reduced the available biomass of *Leucaena leucocephala* by 30% in El Hatico's ISS during the El Niño drought in 2008–2010 (Montoya et al. 2010).
- Lacewings (Chrysopidae) and at least 10 parasitoid wasps from 8 different families control the microlepidopteran *Eccopsis galapagana*, which defoliates *Prosopis juliflora* trees in El Hatico's ISS (Reyes et al. 2012).
- The smooth-billed ani (*Crotophaga ani*), a common large bird in the cuckoo family, controls outbreaks of *Calligo illioneus* butterflies, which lay their eggs on the leaves of sugarcane (Gómez and Lastra 1998).
- Dung beetles such as *Dichotomius belus, Ontophagus marginicollis* and *Coprophanaeus jasius* disrupt the biological cycle of flies by burying manure, thus controlling populations of the stable fly *Stomoxys calcitrans* and the horn fly *Haematobia irritans,* both of which suck blood from cattle (Giraldo et al. 2018b).
- Ants such as *Ectatomma ruidum, Crematogaster* sp., *Nylanderia fulva, Azteca* sp. and *Dolichoderus* sp., wasps (*Polistes erytrociphalus*), together with spiders and birds such as *Theristicus caudatus, Vanellus chilensis, Crotophaga ani, Milvago chimachima* and *Bubulcus ibis* are predators of leafcutter ants (considered the most problematic herbivores in the Neotropics) at El Hatico. Arthropods and birds that control winged leafcutter ants prevent the proliferation of new colonies (Castaño et al. 2019).

Diverse birds, arthropods (including parasitoids) and fungi play important productive roles at El Hatico as they help avoid the economic and environmental costs of pest control in animal and plant-based systems. As seen, predation by birds, together with entomopathogenic fungi and management practices facilitate the production of organic milk by helping avoid the use of toxic substances for tick control (Giraldo and Uribe 2007). It is remarkable that judicious management and associated biodiversity control pests and weeds in agroecological sugarcane, thus making the use of toxic agrochemicals unnecessary. Hopefully, farmers in the region and beyond will try to emulate these examples, as discussed in the following sections.

## 11.7 El Hatico's Influence on Society at Different Scales

El Hatico has strong a commitment to the generation of knowledge on sustainable agriculture and livestock production. Technical knowledge is shared openly with research centers, networks, universities, farmers and decision makers; a total of 13,732 people from 815 institutions and 59 countries visited El Hatico between 2005 and 2016 (El Hatico, unpublished records). This section describes the scales of influence of El Hatico, demonstrating how a single rural property can make a difference for biodiversity and sustainability at a regional scale and beyond.

### 11.7.1 Promoting Sustainable Cattle Ranching

The local or landscape-scale impacts of sustainable ranching practices developed at El Hatico have been modest because cattle grazing is no longer an important economic activity in the geographic Cauca river valley. However, El Hatico has become a national and international reference for sustainable cattle ranching. Field visits of ranchers, extensionists and decision makers have been instrumental to overcome skepticism about tree-based ranching practices (Calle et al. 2012, 2013). El Hatico played a key role in convincing the Colombian government and the national cattle ranchers association (FEDEGAN) about the strategic value of silvopastoral practices. Extensionists and other members of the technical team of the Colombian Sustainable Cattle Ranching Project[2] were trained at El Hatico and later influenced 4200 farms (95,000 ha) in five regions of Colombia (Giraldo et al. 2018a). In 2011, El Hatico received FEDEGAN's national sustainable livestock award.

Ten Colombian universities have integrated field visits and scientific knowledge developed at El Hatico into their academic programs of agronomy, environmental studies, animal science, veterinary medicine and biology. Different research groups have studied the productivity and ecosystem services of silvopastoral systems at El Hatico (Chará et al. 2015).

Through a strategic alliance with the Celema dairy plant in Manizales, El Hatico and Lucerna farm developed Colombia's first organic ultra-pasteurized or extended shelf-life milk and a variety of certified organic dairy products that have penetrated national markets.

### 11.7.2 The Expansion of Agroecological Sugarcane

Agroecological sugarcane developed by El Hatico's team has impacted land use change at the local and landscape scales, and has inspired a large number of international producers. The agroecological protocol developed for the farm's 100 ha of sugarcane has influenced the regional sugar sector by guiding producers in the adoption of sustainable practices. More than 20,000 ha of sugarcane that supply eight sugar mills have shifted to organic production in the Valle del Cauca and more than 20% of the sugarcane in the lower Amaime river basin (surrounding

[2]The Colombian Sustainable Cattle Ranching Project (CSCRP) was designed by an alliance between the Global Environment Fund (GEF), the UK government, FEDEGAN, The Nature Conservancy (TNC), CIPAV and Fondo Acción, under the supervision of The World Bank. It took place from 2010 to 2020 in five ecoregions where cattle ranching exists close to protected areas, and aimed to overcome the main barriers to the adoption of sustainable practices.

El Hatico), is now organic (Murgueitio 2019). Agroecological and biodiversity-friendly farming practices have entered the vocabulary of the regional sugar industry and have influenced Cenicaña's research agenda (Torres 2006; Hincapié et al. 2019).

El Hatico and its closely related farm Lucerna inspired the adoption of more efficient practices for brown sugar production in Mexico. Sugar mills from El Salvador, Brazil and Argentina began the transition to organic sugarcane after their teams visited El Hatico. The Better Sugarcane Initiative of Bonsucro (global sugarcane platform, www.bonsucro.com) integrates technical knowledge developed at El Hatico. Organic sugar from El Hatico and Lucerna is exported to Europe, Asia and North America.

### 11.7.3 Rural Sustainability Values and Generational Exchange

El Hatico is a founding member of Resnatur,[3] the Colombian Network of Private Nature Reserves, which currently joins 168 farm-scale conservation initiatives that preserve 57,296 ha of natural ecosystems in the five regions of the country. El Hatico influenced Resnatur's vision of biodiversity conservation integrated to sustainable agriculture and livestock production. Currently, the Colombian System of Protected Areas (SINAP, https://www.parquesnacionales.gov.co/portal/es/sistema-nacional-de-areas-protegidas-sinap/) recognizes this conservation strategy and its philosophy of sustainable production.

El Hatico's vision of integrated livestock production, agriculture and forestry, rooted in the principles of agroecology and ecological restoration, has reached national and regional audiences through ELTI,[4] CIPAV[5] and SOCLA's[6] field courses on Agroecology and Restoration. Case studies of El Hatico are available for global audiences through ELTI's Online Program. Since 1986, training events offered by CIPAV and El Hatico have inspired thousands of alumni from 40 countries to undertake sustainable livestock production and agriculture projects.

---

[3] www.resnatur.org.co

[4] Environmental Leadership & Training Initiative at Yale University's School of the Environment, a program dedicated to capacity building for forest and landscape restoration in the tropics.

[5] Center for Research on Sustainable Agricultural Production Systems (www.cipav.org.co), a Colombian organization dedicated to research, training and outreach on sustainable agriculture and livestock production, ecosystem services, water-based systems and ecological restoration.

[6] The Latin American Scientific Society for Agroecology (www.soclaglobal.com)

## 11.8 Influencing Policy

Knowledge on sustainable cattle ranching generated at El Hatico has influenced projects developed by The World Bank, FAO, IDB, The Nature Conservancy and CIAT. Silvopastoral systems were included in Colombia's public policy and the current government's development plan (DNP 2019). In addition, tree-based sustainable ranching is part of Colombia's strategy to meet the 20% reduction goal for greenhouse gas emissions committed by the country for 2030 (DNP 2019).

In 2014, El Hatico hosted the meeting of the institutional members of the Global Agenda for Sustainable Livestock.[7] This was a unique opportunity for livestock specialists from around the world to closely analyze El Hatico's silvopastoral systems. One effect of that meeting was the creation of the Global Network on Silvopastoral Systems that brings together 107 members from 50 institutions and 29 countries. This multi-stakeholder partnership works to strengthen and scale-up silvopastoral systems worldwide, by generating, exchanging and disseminating knowledge, documenting public policy and facilitating dialogue to address the challenges related to the sustainable development goals (https://globalsilvopastoralnetwork.org/). A case study of silvopastoral systems developed by FAO, CIPAV and Agri Benchmark (Germany) analyzes natural resource use efficiency, land productivity, economic performance and environmental benefits of ten silvopastoral production models in Colombia (including El Hatico's ISS), Mexico and Argentina (Chará et al. 2019). This document provides policy recommendations for promoting and scaling-up silvopastoral systems in Latin America and other regions.

## 11.9 Conclusions

El Hatico has played a pivotal role in spreading two ideas that are key for implementing sustainable farming systems. First, farm-scale biodiversity conservation should go beyond the protection of natural ecosystem remnants to include productive areas. Second, the redesign of farming systems based on the principles of agroecology can result in higher yields, better quality of agricultural products, lower production costs and multiple ecosystem services, without known trade-offs. Such changes can take place when landowners have a strong intergenerational

[7]The Global Agenda for Sustainable Livestock, established in 2011, is a multi-stakeholder partnership with the aim of fostering and guiding the sustainable development of the global livestock sector in alignment with the SDG framework of the UN Agenda 2030. It provides a platform to address comprehensively the sector's multiple challenges towards sustainable development by facilitating global dialogue and encouraging local practice and policy change, focusing on innovation, capacity building, incentive systems and enabling environments (www.livestockdialogue.com)

commitment to the land, motivation to innovate and improve their farming systems, and a desire to share knowledge with peers, scientists and decision makers.

Biodiversity conservation in fertile productive landscapes, such as the geographic Cauca river valley, should include the agricultural matrix with actions that enhance habitat for native species and facilitate wildlife movements. Even small forest fragments such as the one protected at El Hatico can contribute to regional conservation if surrounded by wildlife-friendly agroecosystems.

Synergies between sustainable livestock production, agroecology and ecological restoration can allow rural properties to increase their profitability and productivity, while protecting forests and integrating native trees into cattle grazing areas and agriculture. Innovative and efficient farming systems such as those found at El Hatico can inspire positive change at local, regional and global scales by motivating producers to enhance thousands of hectares. Mahatma Gandhi wrote that *"a small body of determined spirits fired by an unquenchable faith in their mission can alter the course of history"*. El Hatico inspires people to believe that the sum of individual decisions in the right direction has the power to transform whole regions.

## References

Arcila AM, Valderrama C, Chacón P (2012) Estado de fragmentación del bosque seco en la cuenca alta del río Cauca, Colombia. Biota Colombiana 13(2):86–101

Arias J (1994) Evaluación exploratoria del comportamiento de la materia orgánica por prácticas de quema en molisoles del Valle del Cauca. Tesis de Posgrado en Suelos y Aguas. Facultad de Ciencias Agropecuarias, Universidad Nacional de Colombia, Palmira

Armbrecht I (1995) Comparación de la mirmecofauna en fragmentos boscosos del Valle Geográfico del río Cauca, Colombia. Bol Mus Ent Univ Valle 3(2):1–14

Armbrecht I, Chacón P (1997) Composición y diversidad de hormigas en bosques secos relictuales y sus alrededores en el Valle del Cauca, Colombia. Revista Colombiana de Entomología 23(1–2):45–50

Armbrecht I, Chacón P (1999) Rareza y diversidad de hormigas en fragmentos de bosque seco colombianos y sus matrices. Biotropica 31(4):646–653

Armbrecht I, Tischer I, Chacón P (2001) Nested subsets and partition patterns in ant assemblages (Hymenoptera, Formicidae) of Colombian dry forest fragments. Pan Pac Entomol 77(3): 196–209

Baldissera R, Rodriguez E, Hartz S (2012) Metacommunity composition of web spiders in a fragmented neotropical forest: relative importance of environmental and special effects. PLoS One 7(10):1–9

Calle Z, Murgueitio E, Chará J (2012) Integrating forestry, sustainable cattle-ranching and landscape restoration. Unasylva 239(63):31–40

Calle Z, Murgueitio E, Chará E, Molina CH, Zuluaga AF, Calle A (2013) A strategy for scaling-up intensive silvopastoral systems in Colombia. J Sustain For 32(7):677–693. https://doi.org/10.1080/10549811.2013.817338

Cárdenas G (1998) Comparación de la composición y estructura de la avifauna en diferentes sistemas de producción. BS Thesis, Universidad del Valle, Cali

Castaño K, Giraldo C, Rodríguez J, Giraldo NV, Calle Z, Reyes K, Molina CH, Molina EJ, Molina JJ, Armbrecht I, Montoya J (2019) Aspectos ecológicos, enemigos naturales y métodos de control de la hormiga arriera *Atta cephalotes* (Hymenoptera: Myrmicinae) en la Reserva Natural

El Hatico. In: Castaño K, Chará J, Giraldo C, Calle Z (eds) Manejo integrado de insectos herbívoros en sistemas ganaderos sostenibles. CIPAV, Cali, pp 150–179

Cenicaña (2001–2018) Informes anuales. Cenicaña, Cali. https://www.cenicana.org/publicaciones-de-cenicana/

Cenicaña (2017) Manual de reconocimiento de arvenses en el cultivo de la caña de azúcar. Cenicaña. 168 p https://www.cenicana.org/pdf_privado/documentos_no_seriados/manual_arvenses/manual_reconocimiento_arvenses.pdf

Chará J, Camargo JC, Calle Z, Bueno L, Murgueitio E, Arias L, Dossman M, Molina EJ (2015) Servicios ambientales de sistemas silvopastoriles intensivos: mejoramiento del suelo y restauración ecológica. In: Montagnini F, Somarriba E, Murgueitio E, Fassola H, Eibl B (eds) Sistemas agroforestales: funciones productivas, socioeconómicas y ambientales, Serie Técnica, Informe Técnico 402. CATIE/Editorial CIPAV, Turrialba/Cali, 454 p

Chará JD, Rivera J, Barahona R, Murgueitio E, Deblitz C, Reyes E, Martins Mauricio R, Molina JJ, Flores M, Zuluaga AF (2017) Intensive silvopastoral systems: economics and contribution to climate change mitigation and public policies. In: Montagnini F (ed) Integrating landscapes: agroforestry for biodiversity conservation and food sovereignty, advances in agroforestry 12. Springer, Cham, pp 395–416

Chará J, Reyes E, Peri P, Otte J, Arce E, Schneider F (2019) Silvopastoral systems and their contribution to improved resource use and sustainable development goals: evidence from Latin America. FAO, CIPAV and Agri Benchmark. Cali, 58 p. License CC BY-NC-SA 3.0 IGO. ISBN: 978-958-9386-84-2 http://www.cipav.org.co/pdf/SPS_Report_ISBN_FAO.pdf

Clegg CD, Attard E, Degrange V, Klumpp K, Richaume A, Soussana JF, Le Roux X (2006) Impact of cattle grazing and inorganic fertiliser additions to managed grasslands on the microbial community composition of soils. Soil Biol Biochem 31:73–82

Corporación Autónoma Regional del Valle del Cauca (CVC) (1990) Comparación de cobertura de bosques y humedales entre 1957 y 1986 con delimitación de las comunidades naturales críticas en el valle geográfico del río Cauca. CVC Report 90–7. Cali, 84 p

Delgado J, Armbrech I, Flórez E, Molina CH (2014) Arañas (Arachnida: Araneae) asociadas a cuatro tipos de manejo del hábitat en la Reserva Natural El Hatico (Colombia). Revista Ibérica de Aracnología 25:59–69. http://gia.sea-socios.com/PDF/25/059-069RIA25ArachnidaColombia2.pdf

Departamento Nacional de Planeación (2019) Plan Nacional de Desarrollo 2018–2022: Pacto por Colombia, pacto por la equidad. Bogotá, Colombia, 477 p

Dominguez-Haydar Y, Armbrecht I (2011) Response of ants and their seed removal in rehabilitation areas and forests at El Cerrejón coal mine in Colombia. Restor Ecol 19:178–184. https://doi.org/10.1111/j.1526-100X.2010.00735.x

Eibl B, Montagnini F, López MA, Lavignolle P, Cortes J, De La Vega M, López LN, Dummel C, Küppers G (2022) Conservation, connectivity and registration of seed areas in remnants of natural reserves in the province of Misiones, Argentina. In: Montagnini F (ed) Biodiversity islands: strategies for conservation in human- dominated environments. Topics in Biodiversity and Conservation, Springer, Cham, pp 157–181

Giraldo C, Uribe F (2007) Manejo integrado de garrapatas en sistemas sostenibles de producción ganadera. Red Electrónica de Garrapatas y Enfermedades Transmitidas por Garrapatas en América Latina y El Caribe. FAO, Rome

Giraldo C, Chará J, Uribe F, Gómez JC, Gómez M, Calle Z, Valencia LM, Modesto M, Murgueitio E (2018a) Ganadería Colombiana Sostenible: entre la productividad y la conservación de la biodiversidad. In: Halffter G, Cruz M, Huerta C (eds) Ganadería sustentable en el Golfo de México. Instituto de Ecología, A.C., Xalapa, pp 31–61

Giraldo C, Montoya S, Escobar F (2018b) Escarabajos del estiércol en paisajes ganaderos de Colombia. CIPAV, Cali

Gómez LA, Lastra LA (1998) *Caligo illioneus*: el gusano cabrito. Carta Trimestral Cenicaña 20(1): 8–13. https://www.cenicana.org/pdf_privado/carta_trimestral/ct1998/ct1_98/ct1_98_p8-13.pdf

Gómez LA, Vargas G (2014) Los barrenadores de la caña de azúcar, *Diatraea* spp., en el valle del río Cauca: investigación participativa con énfasis en control biológico. Documento de Trabajo 734, Cenicaña, Cali

Gómez ME, Rodríguez L, Murgueitio E, Ríos CI, Rosales M, Molina CH, Molina CH, Molina EJ, Molina JP (2002) Árboles y arbustos utilizados en alimentación animal como fuente proteica. Editorial CIPAV, Cali

Hilty SL, Brown WL (1986) A guide to the birds of Colombia. Princeton University Press, Princeton

Hincapié E, Sánchez J, Carbonell JA (2019) IoT network applied to agriculture: monitoring stations for irrigation management in soils cultivated with sugarcane. In: Corrales J, Angelov P, Iglesias J (eds) Advances in information for communication technologies for adapting agriculture to climate change II, AACC 2018. Advances in intelligent systems and computing, vol 893. Springer, pp 249–259. https://link.springer.com/chapter/10.1007%2F978-3-030-04447-3_17

Hurtado-G A, Cruz-B L, Molina-D EJ (2016) Dieta de aves migratorias en un sistema agroecológico del Valle del Cauca, Colombia. Bol Cient Mus Hist Nat U de Caldas 20(2): 151–163. https://doi.org/10.17151/bccm.2016.20.2.11

López L, Armbrecht I, Montoya-Lerma J, Molina EJ (2013) Diversidad de avispas parasitoides en un sistema silvopastoril orgánico de producción ganadera de Colombia. Revista Avances en Investigación Agropecuaria 17(1):65–78. http://ww.ucol.mx/revaia/portal/pdf/2013/enero/3.pdf

Mahecha L (2003) Importancia de los sistemas silvopastoriles y principales limitantes para su implementación en la ganadería colombiana. Rev Col Cienc Pec 16:11–18

Manrique GM, Guzmán ML, Victoria JI (2006) Flora microbiana en suelos cultivados con caña de azúcar y otros sistemas de exploración. In: Torres JS (ed) Manejo del cultivo en condiciones de caña verde, Serie Técnica 35. Cenicaña, pp 54–63. https://www.cenicana.org/pdf_privado/serie_tecnica/st_35/st_35.pdf

Molina EJ, Molina CH, Molina JJ, Suárez JF, Uribe P, Valencia LM, Murgueitio E, Uribe F (2013) Ovinos y caña de azúcar. Revista DeCarne 7:55–59

Molina EJ, Molina CH, Molina JJ, Suárez JF, Uribe P, Valencia LM, Murgueitio E, Uribe F (2014) Integración con beneficios. Revista DeCarne 8:48–51

Molina IC, Donney G, Montoya S, Rivera JE, Villegas G, Chará J, Barahona R (2015) La inclusión de *Leucaena leucocephala* reduce la producción de metano de terneras Lucerna alimentadas con *Cynodon plectostachyus y Megathyrsus maximus*. Livestock Research for Rural Development 27, Article 96. Retrieved June 12, 2020, http://www.Irrd.org/Irrd27/5/moli27096.html

Molina IC, Angarita EA, Mayorga OL, Chará J, Barahona R (2016) Effect of *Leucaena leucocephala* on methane production of Lucerna heifers fed a diet based on *Cynodon plectostachyus*. Livest Sci 185:24–29

Molina-Castro CH, Molina-D CH, Molina-D EJ, Molina-E JJ (2012) Manejo agroecológico de caña de azúcar y sistemas silvopastoriles intensivos. Revista Tecnicaña 29:28–35. https://issuu.com/revistatecnicana/docs/tec_no29_2012/37

Molina-Garcés C (1938). Árboles para sombrío y forraje. Revista de la Academia Colombiana de las Ciencias Exactas, Físicas y Naturales 2(6): 273–278

Montes-Londoño I, Calle A, Montes-Calderon A, Montes-Botero O (2022) Hacienda Pinzacuá: an island of regenerative agriculture in the piedmont of the Colombian Central Andes. In: Montagnini F (ed) Biodiversity islands: strategies for conservation in human-dominated environments. Topics in Biodiversity and Conservation, Springer, Cham, pp 305–355

Montoya J, Calle Z, Murgueitio E (2010) Biodiversidad funcional como reguladora de insectos plaga en los sistemas silvopastoriles durante la sequía del fenómeno del niño 2009–2010. In Memorias del VI Congreso Latonoamericano de Agroforestería para la Producción Pecuaria Sostenible, Panama, Sept 28–30, 2010. CATIE/CIPAV. Turrialba/Costa Rica, pp 29–30

Murgueitio B (2019) Análisis del cambio de la cobertura del suelo en la zona baja de la cuenca del río Amaime, periodo 1986–2018. Colegio La Arboleda, Cali, 138 p

Murgueitio E, Calle Z, Uribe F, Calle A, Solorio B (2011) Native trees and shrubs for the productive rehabilitation of tropical cattle ranching lands. For Ecol Manag 261:1654–1663

Murgueitio E, Xóchitl M, Calle Z, Chará J, Barahona R, Molina CH, Uribe F (2015) Productividad en sistemas silvopastoriles intensivos en América Latina. In: Montagnini F, Somarriba E, Murgueitio E, Fassola H, Eibl B (eds) Sistemas agroforestales: funciones productivas, socioeconómicas y ambientales, Serie Técnica, Informe Técnico 402. CATIE/Editorial CIPAV, Turrialba/Cali, pp 59–101. 454 p

Niella F, Rocha P, Tuzinkievicz A, Buchwies R, Bulman C, Thalmayr P, González J, Montagnini F (2022) Native multipurpose species in the Yaboti Biosphere Reserve, Misiones-Argentina. In: Montagnini F (ed) Biodiversity islands: strategies for conservation in human-dominated environments. Topics in Biodiversity and Conservation, Springer, Cham, pp 461–483

Pardo LC (2009) Macroinvertebrados edafícolas en sistemas agrícolas del municipio de El Cerrito (Valle del Cauca), con énfasis en la comunidad de escarabajos Melolonthidae (Coleoptera: Scarabaeoidea). Tesis de doctorado. Universidad del Valle, Cali

Pardo LC, Sevilla F, Amézquita E (2017) Macroinvertebrados edáficos en cultivos de caña de azúcar, ganadería silvopastoril y un relicto forestal en el Valle del Río Cauca (Colombia). Boletín de la Sociedad Entomológica Aragonesa 60:301–312

Pearce JL, Venier LA (2006) The use of ground beetles (Coleoptera: Carabidae) and spiders (Araneae) as bioindicators of sustainable forest management: a review. Ecol Indic 6:780–793

Reyes K, Montoya J, Giraldo C, Brown JW (2012) Description of the early stages of *Eccopsis galapagana* Razowski & Landry (Tortricidae), a defoliator of *Prosopis juliflora* (SW.) DC. (Fabaceae) in Colombia. J Lepidopt Soc 66(3):156–164

Rice RA, Greenberg R (2004) Silvopastoral systems: ecological and socioeconomic benefits and migratory bird conservation. In: Schroth G, Fonseca GAB, Harvey CA, Gascon C, Vasconcelos A (eds) Agroforestry and biodiversity conservation in tropical landscapes. Island Press, Washington, DC, pp 453–472

Rivera C, Naranjo L, Duque AM (2007) De María a un mar de caña: Imaginarios de naturaleza en la transformación del paisaje vallecaucano entre 1950 y 1970. Universidad Autónoma de Occidente, Cali

Rivera LF, Escobar F, Philpott SM, Armbrecht I (2019) The role of natural vegetation strips in sugarcane monocultures: ant and bird functional diversity responses. Agric Ecosyst Environ 284:106603. https://doi.org/10.1016/j.agee.2019.106603

Sadeghian KH, Madriñan R (2000) La actividad microbiana-CO2 en suelos cultivados en caña de azucar con y sin quema. Universidad Nacional de Colombia (Colombia)

Santos-Gally R, Boege K (2022) Biodiversity islands: the role of native tree islands within silvopastoral systems in a neotropical region. In: Montagnini F (ed) Biodiversity islands: strategies for conservation in human-dominated environments. Topics in Biodiversity and Conservation, Springer, Cham, pp 117–138

Solis MA, Metz MA (2016) An illustrated guide to the identification of the known species of *Diatraea* Guilding (Lepidoptera, Crambidae, Crambinae) based on genitalia. ZooKeys 565:73–121

Tilman D (2001) Functional diversity. In: Levin SA (ed) Encyclopedia of biodiversity, vol 3. Academic Press, NewYork, pp 109–120

Torres JS (ed) (2006) Manejo del cultivo en condiciones de caña verde. Cenicaña, Serie Técnica 35 https://www.cenicana.org/pdf_privado/serie_tecnica/st_35/st_35.pdf

Vallejo V, Roldán F, Dick RP (2010) Soil enzymatic activities and microbial biomass in an integrated agroforestry chronosequence compared to monoculture and a native forest of Colombia. Biol Fertil Soils 46(6):577–587

Vallejo V, Arbelia Z, Teránb W, Lorenzc N, Dickc RP, Roldan F (2012) Effect of land management and Prosopis juliflora (Sw.) DC trees on soil microbial community and enzymatic activities in intensive silvopastoral systems of Colombia. Agric Ecosyst Environ 150:139–148

Vargas GA, Obando V, Gómez LA (2006) *Jaynesleskia jaynesi*: otra alternativa para el manejo de Diatrea spp. Carta Trimestral Cenicaña (Colombia) 28(2):3–5

Vargas G, Gómez LA, Michaud JP (2015) Sugarcane stem borers of the Colombian Cauca River valley: pest status, biology and control. Fla Entomol 98:728–735

# Chapter 12
# Hacienda Pinzacuá: An Example of Regenerative Agriculture Amidst a Transformed Landscape in the Colombian Andes

**Irene Montes-Londoño, Alicia Calle, Olimpo Montes, and Arturo Montes**

**Abstract** The central Andes of Colombia is a region of high biodiversity that has been intensely transformed, mostly by unsustainable agricultural practices, and especially by extensive cattle grazing. Whereas cattle farms are often considered ecological deserts, cattle production can be approached in a different way: by integrating more trees into the pastures, introducing better animal management practices, and restoring protective forests, so that productivity, biodiversity, and the flow of ecosystem services can be positively impacted. In this case study, we describe the agroecological transformation of Hacienda Pinzacuá, a 45-hectare farm that has become an island of regenerative agriculture amidst a highly fragmented landscape. We explain how the farm's land use history led to the severe degradation of its once fertile soils; how key land management decisions were made and gradually implemented through trial and error; and the significant land cover changes that occurred over 20 years of transformation. We also provide data on how these changes have impacted productivity, biodiversity, and ecosystem services to illustrate how conservation and production can work synergistically to transform the land and the people. Today Pinzacuá stands out as an island of vegetation in an otherwise treeless landscape and has become a high-quality matrix that serves as habitat or refuge for a variety of taxa striving to persist in this fragmented landscape. Finally, we reflect on the challenges faced along the process, and the prospects for maintaining Pinzacuá as both an island of biodiversity and an example for other farmers seeking more resilient productive alternatives.

I. Montes-Londoño (✉) · O. Montes · A. Montes
Hacienda Pinzacuá, Alcalá, Valle del Cauca, Colombia

Miami, FL, USA

A. Calle
Environmental Leadership & Training Initiative, Yale School of the Environment, New Haven, CT, USA



F. Montagnini (ed.), *Biodiversity Islands: Strategies for Conservation in Human-Dominated Environments*, Topics in Biodiversity and Conservation 20,
https://doi.org/10.1007/978-3-030-92234-4_12

**Keywords** Agroforestry · Biodiversity conservation · Degraded landscapes · Ecosystem services · Landscape restoration · Silvopastoral systems

## 12.1 Introduction

The Andean region of Colombia is part of the Northern Andes biodiversity hotspot, an area known for its ecosystem diversity and high levels of endemism, and its significance for global biodiversity (Mittermeier et al. 2011). The region is home to 34% of the country's population—almost 17 million people—and generates 32% of the national Gross Domestic Product (GDP) (Delgado and Pérez 2018). Human activities such as agriculture, cattle ranching, urbanization, infrastructure development, and species introductions have rapidly transformed this landscape, creating enormous pressure on the remaining natural ecosystems, their biodiversity, and the vital services they provide (Kattan and Alvarez-López 1996). Today, only 18–25% of the original forest cover remains, mostly as fragmented remnants embedded in a matrix of agricultural land uses (Etter 1998). As a result, Colombia's Andean ecosystems are among the most threatened in the world (Mittermeier et al. 2011), and the human populations who inhabit them face increasing risks from the impacts of land degradation.

Across Latin America, this pattern of ecosystem transformation has led to a widespread biodiversity crisis that is further compounded by climate change. A variety of nature-based solutions—actions to protect, restore and sustainably manage both natural and transformed landscapes—have been identified that can contribute to preserve biodiversity, combat climate change, and improve human well-being (IUCN 2020). For example, improving the management of agricultural lands to achieve a balance between a sustainably managed matrix, protected ecosystem remnants, and high connectivity is critical to support biodiversity, sustain the flow of ecosystem services, and enhance climate change resilience (Chazdon et al. 2009; Perfecto and Vandermeer 2010; Vílchez et al. 2013; Griscom et al. 2017; Kremen and Merenlender 2018).

Increasing tree cover through practices such as adding live fences, planting, establishing or retaining trees in pastures, and using complex agroforestry and silvopastoral systems can enhance the conservation value of agricultural landscapes without compromising production (Perfecto and Vandermeer 2010; Harvey et al. 2011; Tscharntke et al. 2012; Mendenhall et al. 2014; Prevedello et al. 2018). Although lack of access to technical assistance and financial incentives still limit the widespread adoption of many such practices (Calle et al. 2013; Calle 2020), a growing number of landowners are realizing the need to re-evaluate their conventional practices and test alternative methods to work more closely with nature.

This chapter presents a case study examining the transformation of Pinzacuá, a farm located in the central Andean region of Colombia and whose landowners, having witnessed the degradation of their once fertile lands, decided to change course. We describe the efforts made over 20 years to restore productivity in this 45-hectare farm by increasing tree cover and recovering soil health. We begin by

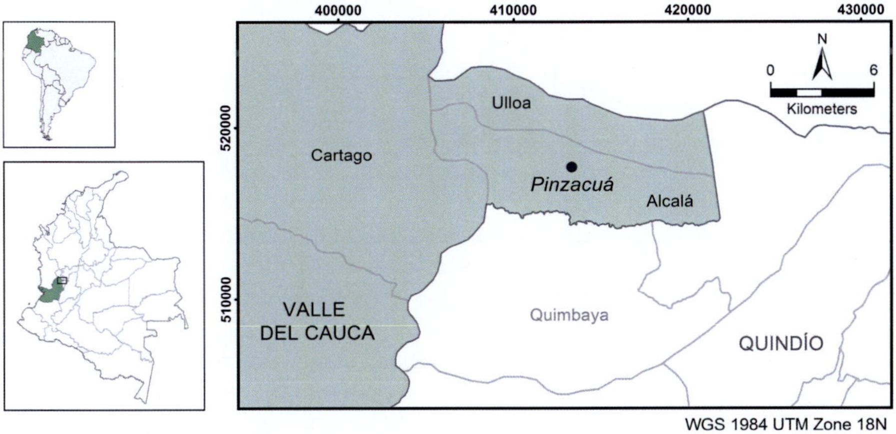


**Fig. 12.1** Location of Pinzacuá in the La Vieja River watershed in Colombia's Andean region. Map: Alicia Calle

explaining the landscape context and land use history, then describe how gradual changes were implemented, and the impacts of the process on biodiversity, farm productivity, and ecosystem services. Finally, we reflect on the future of the farm and the challenges and opportunities for others who may decide to tackle such an effort.

## 12.2 Landscape and Social Context

Hacienda Pinzacuá is located on the western slope of Colombia's Central Cordillera, in the La Vieja river central watershed which marks the limits between the departments of Quindío and Valle del Cauca (Fig. 12.1). The farm is adjacent to Alcalá, a town of 20,000 people located at 1290 m.a.s.l. Annual precipitation averages 1900 mm distributed bimodally from March to May and September to November; average annual temperature is 23.4 °C and relative humidity is 85% (Comisión Conjunta 2008). The area's original vegetation is typical of the transition between premontane and low montane humid forest (Espinal 1977). However, only 11% of the original forest cover remains, mostly in bamboo forests and unconnected fragments <5 hectares (Camargo and Cardona 2005). Soils are Inceptisols and Andisols of quaternary alluvial and volcanic origins respectively, which are young and deep (20–30 m), ranging from sandy to clayey loam in texture, free of rock and acidic (pH 4.5–5.5), and highly productive (Montes Londoño et al. 2017).

While the region has a strong agricultural tradition, the type of agriculture practiced has changed over the past century in response to external drivers. The original forests were first replaced by traditional coffee agroforests, and eventually converted to single-crop coffee plantations of the more productive sun-loving varieties. The collapse of coffee prices in the 1980s and 90s resulted in further conversion, mostly to highly fertilized pastures for cattle ranching (Calle and Piedrahíta 2007), disregarding the fact that the fragile soils and steep slopes are unsuitable for grazing (Sadeghian et al. 2001). Today, the area is an agricultural mosaic of cash crops (e.g., coffee, plantain, citrus) and cattle pastures, which occupy 33% of the land (DANE 2014). Pinzacuá is surrounded by undulating hills covered by a mix of croplands with some isolated trees, monocultures of African grasses with sparse tree cover, some live fences, small fragments of disturbed secondary forests and riparian areas, and small bamboo forest patches.

Historically, this region enjoyed a strong economy and a high quality of life for two main reasons. First, it is located at the center of the so-called "golden triangle" formed by Bogotá, Medellín and Cali, Colombia's main cities. Second, for most of the twentieth century, this was the country's prime coffee-growing region, and the majority of its nearly four million inhabitants were dedicated to this activity. High productivity and a strong international coffee market resulted in an economic boom that allowed for the development of excellent infrastructure and services, and the highest quality of life in the country (Toro Zuluaga 2005; Mejía Cubillos 2013). The crash of the coffee markets in the 1990s led to a general economic contraction and the emergence of new problems: unemployment, loss of income, unequal access to essential public services, and increased levels of drug-related violence, all compounded by the arrival of people displaced by the armed conflict that was taking place elsewhere in the country (Toro Zuluaga 2005). Ultimately, the human development index (HDI) stagnated for almost a decade and its gap with respect to the national average widened (UNDP 2004).

The decline of coffee production gave way to new sectors such as telecommunications, services, tourism, construction and commerce that replaced agriculture. Urban centers now concentrate most economic activities and the majority of the population, while less than 15% of the people remain in rural areas (Mejía Cubillos 2013). Moreover, the farmer population is aging; more than 40% are between 40 and 54 years old (DANE 2014). The remaining farms are mostly privately owned, small and medium properties between 5 and 200 hectares (UPRA 2019a, b, c). Smaller farms are usually family-owned and subsistence-oriented, with a mix of cash crops and livestock. Meanwhile, larger farms are largely monoculture-based commercial operations. As farming declines, the price of land in this densely populated region has skyrocketed: the price of an average hectare of cropland in 2014 was $4000 (Becerra et al. 2017a, b, c), but near urban or touristic areas it can be as high as $40,000 (López Murillo 2015). High prices are forcing a land ownership transformation in this landscape where only large developers or consolidated farm owners can afford the land, and the latter tend to stick to the conventional high-input-high-output monoculture model.

## 12.3 Transforming Pinzacuá: An Incremental Process

### 12.3.1 History of a Transformation

Pinzacuá is a 45-hectare family-owned and operated business dedicated to raising high-quality female Brangus cattle for breeding and beef production. The farm was acquired in 1985, and its land use history since then can be divided in two distinct periods: before and after 1997. A summary of the main events in the farm's history is provided in Table 12.1.

Before 1997, the farm followed the prevailing high-input, treeless pasture model dictated by the dominant agricultural intensification trend of the 1960s, 70s and 80s. Under this model, a succession of different crops was planted: first coffee, then tobacco, and finally pasture. Despite the high yields, the farm was never profitable because the prices of these commodities could not compensate for the excessive costs of the imported agrochemicals on which production depended. Moreover, as the soils were gradually depleted, higher doses of fertilizers were required every year simply to maintain production. After a decade of struggling to keep the farm afloat, the need to reduce input dependency and keep production costs under control became evident.

In 1997 the decision was made to stop chemical fertilization altogether. This not only required cutting animal load by half, but it also revealed the severity of soil degradation. Finding a strategy to restore soil fertility without applying external inputs thus became the priority. Experience and self-reflection about traditional production systems eventually led to a question: if the soils of coffee systems shaded with nitrogen-fixing *Inga* trees were so rich that they required little fertilization and no pesticides, could the same principles apply to pastures? So, defying the widespread idea that trees and pastures do not mix, in 2002 a 2-hectare trial plot was established with *Inga* spp. trees planted at a density of 100 trees ha$^{-1}$.

The first 2 years of this trial were difficult: establishing trees in active pastures turned out to be more challenging and expensive than anticipated. Damage by cattle, leaf-cutter ants and humans all contributed to the high mortality rate of the planted seedlings. The idea that trees do not belong in pastures was deeply engrained, and farm workers were reluctant to protect and care for the seedlings which resulted in the need to replant constantly. Protecting the young trees from cattle browsing and trampling was also problematic and required creative solutions (Fig. 12.2). Trees were initially protected with individual bamboo corrals, but eventually a better method was developed by planting lines of trees protected with electrical fence. Fellow cattle ranchers mocked these efforts arguing that intentionally increasing tree cover in the paddocks was nonsense as it would reduce pasture productivity.

In 2002, as the landowner began to question the introduction of *Inga* trees in the pastures, news about the launching of a sustainable ranching project in the region reached him. The Regional Integrated Silvopastoral Ecosystem Management (RISEM) project was providing technical assistance and payments for ecosystem services to promote the adoption of silvopastoral systems, previously unknown in

**Table 12.1** Main events in Pinzacuá's land-use history

| Year | Event |
|---|---|
| 1985 | Pinzacuá is acquired. The farm is a traditional agroforestry plantation with coffee growing under diverse assemblages of shade trees, mainly *Inga*. |
| 1986 (Inspired by the Green Revolution, the National Coffee Federation promoted intensification efforts during these years, paying farmers to eliminate tree cover and replace traditional shade coffee for more productive sun-loving varieties. Tree cover was eliminated in 50% of this territory as a result of both coffee intensification campaigns and a push to replace coffee with pastures in the 1990s (Rodríguez et al. 2004). These changes led to landscape-level degradation and fragmentation of the remaining natural habitats, and in some cases triggered local or regional extinctions (Kattan et al. 2004). Although yields did increase, approximately half of the increased production was lost due to a labor shortage) | Shade coffee is replaced with sun-coffee managed with intensive application of synthetic fertilizers and pesticides, and shade trees are completely eliminated. |
| 1993 | Sun coffee is eliminated due to low market prices, and replaced by a monoculture of the African grass *Cynodon plectostachium* without trees, and a monocrop plantation of tobacco. |
| 1994 | Tobacco is eliminated and the entire farm is converted to pasture. Pasture management based on intensive application of fertilizers and amendments (1 ton of urea plus 2 tons of dolomite lime $ha^{-1}$ $year^{-1}$) continues in order to sustain a high animal load (10 heads $ha^{-1}$). |
| 1998 | Use of agrochemicals becomes economically unviable and is completely eliminated. As a result, animal loads are reduced and soil degradation becomes evident. Shifting away from grass monoculture to a regenerative system becomes the priority. |
| 2000 | A trial plot with *Inga* trees is established with the hope of restoring soil fertility. |
| 2002 | Using a farm planning approach, plots are classified according to their potential and limitations. Streams are isolated and riparian areas reforested with native trees and bamboo (*Guadua angustifolia*). |
| 2015 | Fodder hedgerows are established within some paddocks. |
| 2016 | A small agroforestry plot with shade-loving arabica coffee is established. |
| 2018 | A forage bank to feed goats is established. |

**Fig. 12.2** Methods to plant and protect trees in established active pastures. Individual bamboo and shade cloth corral method (left), and 2 m-wide strips protected with electrical fence (right), an innovation devised in Pinzacuá and later replicated in other farms. Photos: Álvaro Zapata

the region (Pagiola and Ríos 2013). Pinzacuá, along with other 74 farms, joined the project and started working with assistance from the Center for Research on Sustainable Agriculture Production Systems (CIPAV, CIPAV.org.co), whose technical staff advocated increasing tree cover throughout the farm and encouraged the *Inga* trial. Although the project's economic incentive was capped at US $6500 per farm and only covered approximately 30% of the implementation costs incurred, the motivation and support provided by CIPAV were instrumental for Pinzacuá to persevere in its tree-planting efforts.

From that point on, tree planting became the centerpiece of Pinzacuá's transformation. A land use plan was devised to utilize each plot according to its potential. Trials planting a variety of trees in different spacing arrangements were expanded throughout the farm to identify the optimal combinations. Paddocks were divided with live fences and cattle were rotated in short periods. All streams and water courses were fenced off, and riparian buffer areas were reforested with native tree species and *Guadua angustifolia,* a native giant bamboo. The steepest areas of the farm, which were unsuitable for cattle, were planted with mixed native species. More recently, an agroforestry system with arabica coffee was established and other non-timber forest products such as vanilla and pepper were tested. Fodder hedgerows were planted in some paddocks and a forage bank to feed goats was established.

The changes implemented in Pinzacuá described in the following sections were inspired by a variety of sources. As a former coffee grower, the landowner drew initial ideas for tree planting densities and pruning regimes from technical manuals for shade coffee published by Colombia's Coffee Grower's Federation (federaciondecafeteros.org). Drawing on decades of experience developing silvopastoral systems and management, CIPAV staff offered guidance on tree and fodder species and pasture management, and helped adjust the systems to Pinzacuá's conditions. Visits to demonstration farms, especially to the flagship silvopastoral farm El Hatico Nature Reserve (Calle et al. 2022), enabled farmer-to-farmer learning and inspired many ideas for small scale trials. Ultimately, the landowner's own curiosity and innovative spirit, observation skills, and proclivity to old-fashioned empirical learning were critical for the refinement of techniques that worked for Pinzacuá, some of which went on to become standard silvopastoral practice (Calle 2008).

### *12.3.2 Trees and Living Soils, the Pillars of Change*

Ever since coffee was introduced in Colombia, trees have played a key role in traditional coffee agroforestry systems. Pinzacuá's transformation and its current land use practices are rooted in this traditional knowledge and in the realization that the mechanisms that sustained high natural productivity in shaded coffee systems are largely based on the benefits provided by trees. Today, the entire farm is planted with trees and all land uses—silvopastures, riparian areas, and the homegarden—have some form of tree cover (Fig. 12.3).

The impact of trees on productivity is not direct, but rather mediated by their effect on different components of the production system, from the livestock to the soils. The strength and quality of high-performing soils rest on the interactions between a multitude of beneficial macro and microorganisms that constitute the soil food web. Earthworms, arthropods, nematodes, fungi, protozoa and bacteria consume energy-rich materials (e.g., leaves, roots, root exudates) all of which are directly or indirectly sourced from plants, especially trees, via litterfall and roots. Without plentiful and diverse plant-derived organic inputs, the soil food web cannot thrive (Moebius-Clune et al. 2016). For this reason, and despite the inherent competition between trees and pastures, the main objective of all actions implemented in Pinzacuá since 1997 has been to adequately feed and provide the best conditions for the life belowground. Plants, and especially trees, have been the farm's best allies in restoring soils and making them alive, dynamic, and productive.

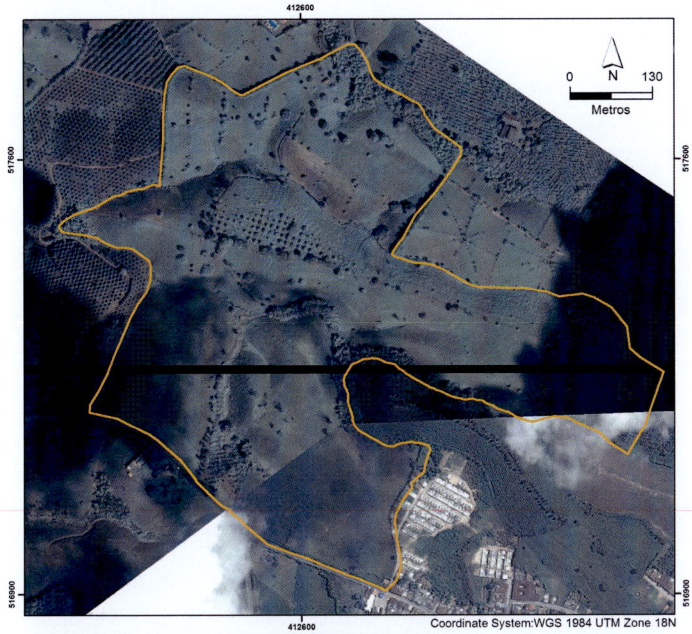


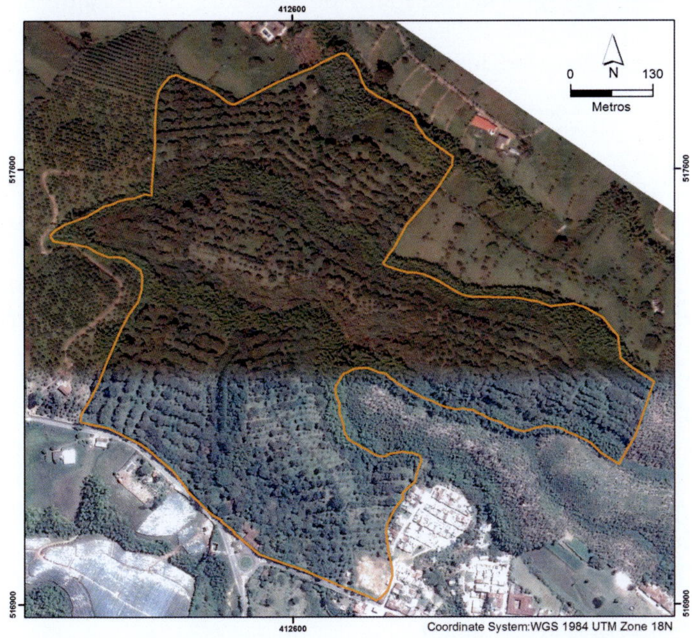


**Fig. 12.3** Aerial views of Pinzacuá in 2003 (top) and 2016 (bottom). Main land use in 2003: treeless pasture; in 2016: silvopastoral systems, riparian forests, and homegarden. Land uses in neighboring farms can also be observed. Photos: RISEM Project

### *12.3.3 Land Use Planning*

Today, the level and type of tree cover (i.e., tree density and arrangement) are intentionally managed throughout the farm. As a general rule, soil suitability (e.g., fertility, depth, slope, vulnerability to erosion and compaction) determines the type of land use, which in turn determines tree cover and tree species (Table 12.2). For example, steep hillsides, which are vulnerable to erosion and compaction, are not used for cattle but instead are planted with trees at a high density using a mixed native species forestry model described below. By contrast, the flat areas atop hills, which are more fertile and less prone to erosion, are reserved for cattle production and planted with trees at a lower density to ensure that pastures receive sufficient sunlight.

### *12.3.4 Tree Species Selection*

Pinzacuá's planned tree diversity includes more than 60 different species that have been intentionally incorporated into the different land uses, including some of conservation concern. With the exception of the homegarden, native species are preferred over exotics because they are better adapted to the local climatic and edaphic conditions and provide food and shelter for the organisms above and below ground.

Trees for the silvopastoral systems were selected for specific characteristics including: (i) nitrogen fixation or high production of rapidly decomposing litter (e.g., *Inga edulis*); (ii) hardiness to withstand herbivory, both from cattle and leaf cutter ants (e.g., *Maclura tinctoria*); (iii) rapid growth and vigor to outcompete grasses (e.g., *Gliricidia sepium*); (iv) seeds, fruits and forage production to supplement cattle and horse nutrition (e.g., *Senna spectabilis*); and (v) timber production for farm use or local markets (e.g., *Anacardium excelsum*). Meanwhile, riparian area reforestation favored two fast-growing native species: *Guadua angustifolia*, a giant bamboo with high local economic and cultural value; and *Colubrina* sp., a tree with good timber properties.

The homegarden includes a mix of native and exotic species of cash crops (e.g., *Vanilla planifolia*), medicinal plants (e.g., *Sechium edule*), and ornamentals (e.g., *Anthurium* sp.), as well as timber trees (e.g., *Swietenia macrophylla*), fruit trees (e.g., *Garcinia madruno*), multipurpose trees (e.g., *Inga edulis*) and palms (e.g., *Aiphanes caryotifolia*).

For the forestry plantations, dinde (*Maclura tinctoria* (L.) D. Don *ex* Steud) was the main species selected because of its timber quality, high survivorship relative to other native species, and ability to provide fruit and habitat for birds and other wildlife (Martins and Setz 2000; Chízmar-Fernández 2009; Suárez et al. 2012; Montes-Londoño et al. 2017). Also known as old fustic or Argentine osage orange, dinde is valued for heavy construction, flooring, furniture, turnery, fence posts and

**Table 12.2** Current land uses and their extension, relief, and species planted

| Land use and area | Relief type | Tree density (trees $ha^{-1}$) | Plant species |
|---|---|---|---|
| Silvopastoral system<br>15 ha | Hilltops | 100–150 | *Albizia guachapele, Anacardium excelsum, Anadenanthera peregrina, Brachiaria* sp., *Cassia grandis, Cynodon plectostachyus, Enterolobium cyclocarpum, Gliricidia sepium, Hymenaea courbaril, Inga edulis, Leucaena leucocephala, Pennisetum* sp., *Psidium guajava, Samanea saman, Senna spectabilis, Syagrus sancona* |
| Mixed planted forest<br>15 ha | Hillsides | Initial density: 1100; | |
| Final density: 250. | *Anacardium excelsum, Aniba perutilis, Cedrela odorata, Lafoensia acuminata, Maclura tinctoria, Magnolia hernandezii, Ocotea helicterifolia, Swietenia macrophylla, Vachellia macracantha* | | |
| Riparian areas<br>5 ha | Depressions | >900 | *Anacardium excelsum, Guadua angustifolia* |
| Shaded coffee<br>1 ha | Hillside | 500 shade trees, 2000 coffee shrubs | *Albizia carbonaria, Anacardium excelsum, Caesalpinia ebano* (threatened in Colombia), *Citrus* sp., *Inga edulis, Jacaranda mimosifolia, Pourouma cecropiifolia, Pouteria caimito, Quararibea cordata, Swietenia macrophylla* |
| Natural regeneration and enriched secondary forest<br>1 ha | Hillside/ depression | 500–900 | *Cedrela odorata, Colubrina* sp., *Erythina fusca, Erythrina poeppigiana, Juglans neotropica, Magnolia hernandezzi, Montanoa quadrangularis* |
| Homegarden<br>0.24 ha | Hilltop | 500 | *Acca sellowiana, Aiphanes caryotifolia, Anthurium* sp., *Arachis pintoi, Bactris gasipaes, Bismarckia nobilis, Brownea ariza, Bougainvillea* |

(continued)

**Table 12.2** (continued)

| Land use and area | Relief type | Tree density (trees ha$^{-1}$) | Plant species |
|---|---|---|---|
| | | | sp., *Cassia grandis, Ceiba pentandra, Citrus* sp., *Codiaeum variegatum, Crescentia cujete, Diospyros* sp., *Eugenia stipitata, Euphorbia pulcherrima, Garcinia madruno, Garcinia mangostana, Inga edulis, Juglans neotropica, Myrciaria cauliflora, Passiflora edulis, Passiflora ligularis, Persea americana, Piper nigrum, Sechium edule, Selenicereus megalanthus, Senna spectabilis, Swietenia macrophylla, Tamarindus indica, Vanilla planifolia* |
| Forage/protein bank 0.5 ha | Hillside | 3000–4000 | *Alocasia macrorrhiza, Boehmeria nívea, Guazuma ulmifolia, Pennisetum* sp., *Tithonia diversifolia, Trichanthera gigantea.* |

railroad crossties, and specialty wood items, and is culturally and economically important throughout Latin America. Dinde is the source of fustic, a yellow pigment used as dye for khaki and other color textiles (Rangel 1949; Roig 1974). The leaves, sap and wood have been used in traditional medicine, and have potential for extraction of a non-toxic broad-spectrum antioxidant (Cioffi et al. 2003). Once common in La Vieja river watershed, dinde populations have been decimated; today it is rarely found in the forests or in the markets, although it is considered a species of least concern (LC) for conservation (Rivers et al. 2017).

## 12.3.5 Tree Density and Shade Management

### 12.3.5.1 Silvopastures

The main factor affecting pasture productivity in silvopastures is shade, which in turn depends on the level of canopy cover and the characteristics of the tree canopy. Thus, identifying the level of shade that maximizes pasture productivity is key to successful silvopasture management. In Pinzacuá, informal trials manipulating shade with different planting densities and pruning intensities have shown that the pasture of choice, star grass (*Cynodon plestostachys*), performs best with <50% shade.

**Fig. 12.4** Final tree spacing arrangement in the silvopastoral system. Trees are planted in an east-west direction to allow maximum light penetration to the understory. Photo: Fernando Uribe, 2018

Three different spatial arrangements identified through empirical observation were tested to determine the optimal tree density to both restore the soils and allow enough light into the understory: (i) a 10 × 10 m grid (100 trees ha$^{-1}$); (ii) rows planted every 20 m with 5 m between trees (100 trees ha$^{-1}$); and (iii) rows planted every 15 m with 5 m between trees (133 trees ha$^{-1}$). The latter design has worked best as the initial planting arrangement, and is then gradually reduced to a final density of approximately 100 trees ha$^{-1}$ to avoid canopy closure (Montes Londoño unpublished data) (Fig. 12.4).

The optimal level of shade at this tree density is maintained by pruning three fourths of the tree crown every 2 years. Branches growing in east-west direction are retained while those growing north-south and directly shading the pasture corridors are removed (Fig. 12.5). While pruning is labor intensive, the residue is used to make artisanal charcoal and sold to cover the labor costs and to get some additional revenue (Fig. 12.5). Furthermore, the residue from the charcoal heaps is used as a high-quality biochar amendment to improve soil fertility throughout the farm.

#### 12.3.5.2 Forestry Plantations

The forestry plantations were established with a mix of native timber trees planted in a 3 × 3 m spacing arrangement. The grid pattern spreads trees out to minimize competition, and the high initial density prevents early branching and promotes self-thinning, reducing the need for frequent pruning. Because dinde is good at self-thinning, plantations require less intensive management than silvopastures.

**Fig. 12.5** Pruning of *Inga* trees (left) and artisanal charcoal production from the pruning material (right). Photos: Álvaro Zapata

Thinning is performed according to a specific guideline developed for dinde (Montes-Londoño et al. 2017).[1]

## 12.4 Impact on Land Cover and Biodiversity

### *12.4.1 Land Cover Change*

Although Pinzacuá's transformation has been slow and gradual, 20 years into the process the accumulated changes in land use and tree cover are substantial. A spatio-temporal analysis based on high resolution satellite images revealed that between 2003 and 2016 land uses with high tree cover (i.e., forests, silvopastures and live fences) increased by 74.9% at the expense of uses with low tree cover (i.e., treeless pastures and croplands) (Table 12.3, Fig. 12.6) (Alicia Calle 2018, unpublished data). Forests included both reforestation and forestry plantations with a closed canopy and natural regeneration, and silvopastures were all areas of pasture with intermediate or high tree density. The most important changes during this period occurred in forests, which increased by 14.4% (6.3 ha); pastures with tree cover, which increased by 62.9% (27.6 ha); and pastures without trees, which initially

[1]Given the lack of information on growth and silvicultural management for dinde, research was conducted on 12 farms of the region to develop preliminary spacing and thinning guidelines (see Montes et al. 2017). Dinde performed similar to other trees used for reforestation in the humid American tropics, with a Mean Annual Increment in Diameter at Breast Height (MAIDBH) of 2.56 cm $year^{-1}$, a growth rate which is higher than it has been reported for sites in Honduras and Cuba (Cordero and Boshier 2003) but lies on the lower end of fast-growing native and exotic species in Central America. This growth rate stands overall within the range that has been reported for other native species in tropical humid regions elsewhere in Latin America (Piotto et al. 2004; Wishnie et al. 2007; van Breugel et al. 2011).

**Table 12.3** Land cover change in Pinzacuá and the surrounding landscape from 2003 to 2016 (Alicia Calle, 2018, unpublished data)

| | Pinzacuá | | | | Surrounding landscape |
|---|---|---|---|---|---|
| | 2003 Area (ha) | 2016 Area (ha) | Change (ha) | % Change | % Change |
| Forests | 2.85 | 9.18 | 6.33 | 14.44 | 2.82 |
| Silvopastures | 1.75 | 29.31 | 27.56 | 62.86 | 0.80 |
| Live fences | 1.23 | 0.17 | −1.06 | −2.43 | 0.62 |
| Pasture no tree cover | 34.74 | 2.82 | −31.92 | −72.81 | −15.85 |
| Cropland (no trees) | 1.21 | 0.05 | −1.16 | −2.64 | 10.92 |
| Other/No information | 2.06 | 2.32 | 0.25 | 0.58 | 1.38 |
| Total | 43.84 | 43.84 | | | |

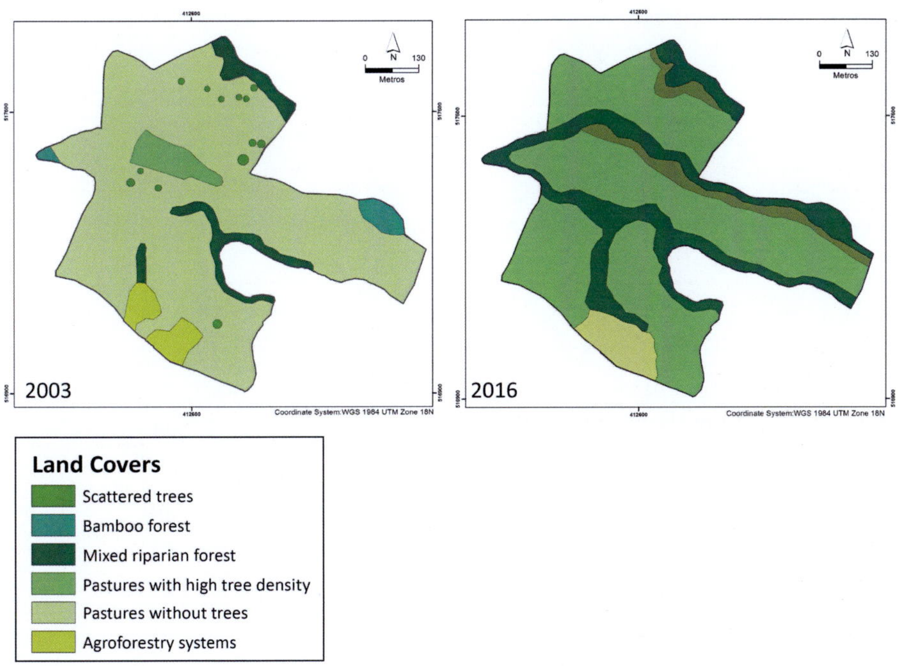


**Fig. 12.6** Land cover in Pinzacuá, 2003 and 2016. Maps: Alicia Calle

comprised most of the farm but decreased by 72.8% (31.9 ha) (Calle unpublished data). Given the small size of the property, the decision to release areas for forest restoration and convert all treeless pastures to silvopastures underscores the landowners' appreciation for both the direct and indirect benefits of trees.

The magnitude of changes implemented in Pinzacuá, however, is perhaps best appreciated in the context of the surrounding landscape. During the same 13-year period, land cover change in a total of 557.4 hectares covered by farms surrounding Pinzacuá followed a different trend. Forest cover increased modestly (2.8%) and treeless pastures decreased by 15.9%, mostly to accommodate cropland expansion (10.9%) with only marginal improvements in pasture tree cover (1.4%) (Calle 2020) (Table 12.3, Fig. 12.6). The neighboring farms, which were not part of the RISEM project, had no access to payments for ecosystem services or technical assistance to improve their pasture areas. Instead, they appear to have planted cash crops in response to short-term spikes in market prices. Although spillover of silvopastoral systems onto neighboring farms did happen in other project farms, Pinzacuá's neighbors were not interested. Location adjacent to a large town and a main road connecting to markets may partly explain why other farmers opted for high-value cash crops.

## 12.4.2 *Biodiversity*

Since the transformation began, Pinzacuá has been the site of numerous biodiversity studies. While the lack of baseline assessments does not allow for direct comparisons, the minimal vegetation cover and severe land degradation evident in the 2003 satellite image (Fig. 12.6) suggest low levels of biodiversity at the baseline. The studies and species documented on the farm are summarized in Table 12.4.

### 12.4.2.1 Trees

Pinzacuá's transformation has relied heavily on planting trees and managing woody natural regeneration on degraded pastures for different purposes. Riparian forests were established to improve and protect the water supply; forestry plantations were established in the steepest parts of the farm that were not apt for grazing; and different silvopastoral arrangements were implemented to reduce cattle heat stress and improve productivity. As a result, overall tree diversity has increased dramatically over time. The farm also propagates several tree species and shares seedlings with others interested in improving on-farm tree cover.

To date, studies of tree diversity on the farm have identified 45 different tree species as adults or seedlings across the different land uses (Table 12.4). Thirty-six woody species have been identified across the different forest fragments, 17 as adult trees and 28 as regenerating seedlings. Eight tree species were also recorded in the mixed plantation (Calle and Méndez 2017, unpublished data, Giraldo et al. 2019), and 16 additional tree species have been recorded in the silvopastures.

A survey of woody vegetation structure and composition in 20 recovering forests protected during the RISEM project included two young riparian forests in Pinzacuá. The study found five species in the canopy and 25 species regenerating in the farm's

**Table 12.4** Biodiversity studies and species reported in Pinzacuá

| Woody plants | | Author and methods |
|---|---|---|
| 11 species<br>219 individuals | *Cedrela odorata, Cestrum* sp., *Cordia alliodora, Gliricidia sepium, Inga edulis, Leucaena leucocephala, Maclura tinctoria, Psidium guajava, Senna spectabilis, Vachellia macracantha, Vernonanthura patens*<br>*Anacardium excelsum, Erythrina poeppigiana, Syagrus sancona* (recorded outside the plots) | 1000 $m^2$ transects (Calle and Méndez 2017, unpublished data) |
| 16 species<br>64 individuals | *Anacardium excelsum, Bauhinia picta, Cecropia angustifolia, Cedrela odorata, Cordia alliodora, Cupania americana, Erythrina fusca, Erythrina poeppigiana, Gliricidia sepium, Juglans neotropica, Montanoa quadrangularis, Nectandra turbacensis, Psidium guajava, Samanea saman, Solanum aphyodendron* | 1000 $m^2$ transects (Giraldo et al. 2019, unpublished data) |
| Woody regeneration | | |
| 28 species<br>834 individuals | *Anacardium excelsum, Bauhinia picta, Cecropia angustifolia, Celtis schippii, Cinnamomum triplinerve, Citrus sinensis, Cordia alliodora, Cupania americana, Erythrina poeppigiana, Eugenia sp, Ficus insipida, Ficus tonduzii, Hymenaea courbaril, Inga edulis, Maclura tinctoria, Myrcia sp, Nectandra lineata, Nectandra turbacencis, Ocotea macropoda, Oroepanax cecropifolium, Persea americana, Sapium laurifolium, Senna spectablis, Sorocea trophoides, Tetrochidium rubrinervium, Trichilia pallida, Zanthoxylum rhoifolium* | 1000 $m^2$ transects (Calle and Méndez 2017, unpublished data) |
| Dung beetles | | |
| 11 species<br>97 individuals | *Canthidium* sp., *Delthochilum* sp., *Dichotumius alyattes, Dichotomius* sp., *Eurysternuspp; —mexicanus, Eurysternus plebejus, Ontherus azteca, Onthophagus acuminatus, Onthophagus nasutus, Onthophagus* sp., *Oxysternon conspicillatum* | Pitfall traps (Giraldo et al. 2019, unpublished data) |
| Birds | | |
| 34 species<br>104 individuals | *Amazilia tzacatl, Amazona ochrocephala, Anthracothorax nigricollis, Ara severus, Bubulcus ibis, Coccyzus pumilus, Coereba flaveola, Columbina talpacoti, Dendroplex picus, Icterus nigrogularis, Lepidocolaptes souleyetii, Melanerpes formicivorus, Melanerpes rubricapillus, Myiarchus apicalis*[a], *Myiodynastes maculatus,* | Point counts (Giraldo et al. 2019, unpublished data) |

(continued)

**Table 12.4** (continued)

| Woody plants | | Author and methods |
|---|---|---|
| | *Myiopagis viridicata, Picumnus granadensis*[a], *Pionus chalcopterus, Pionus menstruus, Piranga rubra*[b], *Pitangus sulphuratus, Polioptila plúmbea, Pyrocephalus rubinus*[b], *Setophaga petechia*[b], *Sicalis flaveola, Synallaxis azarae, Thamnophilus multistriatus, Thraupis episcopus, Thraupis palmarum, Tiaris olivaceus, Todirostrum cinereum, Troglodytes aedon, Tyrannus melancholicus, Zimmerius chrysops* | |
| 89 species 28 families 1069 individuals | *Bubulcus ibis, Camptostoma obsolete, Columbina talpacoti, Columbina minuta, Dendroica petechia, Egretta thula, Eucometis penicillate, Leptopogon superciliaris, Myiodynastes maculatus, Phaeothlypis fulvicauda, Pachyramphus rufus, Pygochelidon cyanoleuca, Pyrocephalus rubinus, Sporophila schistacea, Thraupis episcopus, Tyrannulus elatus, Tyrannus melancholicus, Vireo olivaceus, Xiphorhynchus susurrans,* among others | Point counts (Sánchez and Camargo 2015) |
| Butterflies | | |
| 16 species | *Actinote ozomene, Anartia amathea, Anthanassa Drusilla, Caligo eurilochus, Ceratinia tutia, Cissia pompilia, Dryas iulia, Eurema daira, Heliconius charitonia, Heliconius erato, Papilio thoas, Pteronymia artena, Pteronymia picta, Siproeta stelenes, Timolus echion, Urbanus proteus* | Visual encounter surveying (VES) (Montes-Londoño 2019, unpublished data) |
| Mammals | | |
| 8 species | *Sciurus granatensis, Didelphis marsupialis, Dasyprocta spp., Dasypus novemcinctus, Cerdocyon thous, Puma yagouaroundi, Cuniculus paca, Leopardus pardalis* | Observations by managers and workers |
| Reptiles | | |
| 6 species | *Chironius monticola, Bothriechis schlegelii, Micrurus mipartitus, Oxybelis fulgidus, Oxyrhopus petola, Spilotes pullatus* | Observations by managers and workers |

[a]Endemic species
[b]Migratory species

bamboo forest, and another 11 species in the canopy and 11 species regenerating in the mixed planted forest (Calle and Méndez 2017, unpublished data) (Table 12.4). Pinzacuá's bamboo forest had the highest abundance of woody regeneration with 785 seedlings, including many species that are not currently represented in its forest canopy. The study suggests that these riparian sites that were retired from cattle production over a decade ago are slowly recovering, and the total number of woody species will likely increase over time (Calle and Holl 2019).

#### 12.4.2.2 Birds

Birds play a number of important roles in both natural and agricultural ecosystems: they contribute to natural pest control by consuming pest insects and organisms; they pollinate flowers helping with fruit production; they disperse seeds facilitating natural regeneration; and they add aesthetic and cultural value to landscapes (Giraldo et al. 2019).

Two different bird studies have been conducted in Pinzacuá. Sánchez and Camargo (2015) examined the relationships between bird diversity and landscape structure in four farms with different land uses (i.e., conventional treeless pastures, silvopastoral systems and agroforestry systems) connected to bamboo forests. The highest bird diversity and abundance were recorded in Pinzacuá, where 1069 individuals belonging to 89 species and 28 families, including 3 migratory and 3 endemic species (Table 12.4), were sighted in the silvopastoral systems connected to bamboo forests. The structural and functional attributes of these systems create connectivity to the forest edges, effectively providing food and habitat resources needed for the permanence of bird species, both local and migratory.

The abundance of birds in the Tyrannidae family (180 species) suggests that silvopastoral systems offer optimal conditions for insectivorous species. These birds are known to reduce insect pest outbreaks and herbivory in multi-strata cocoa and coffee agroforestry systems (De Beenhouwer et al. 2013; Maas et al. 2013; Peters and Greenberg 2013; Poch and Simonetti 2013). Although the same could be true in silvopastoral systems, where the main pests are ticks and flies, the role of insectivorous birds in controlling pests in these systems has not been studied.

Another study comparing bird diversity across three different land uses in Pinzacuá (i.e., forest, silvopastoral systems, and fodder hedges) recorded 104 individuals belonging to 34 species (Giraldo et al. 2019) (Table 12.4). Most species were open area generalists but at least four were forest dwellers; three species were migratory and three others were endemic. Silvopastoral systems with scattered trees had the highest bird diversity, followed closely by the forest; they also hosted a variety of common generalists that have been displaced by the elimination of coffee shade and overall tree cover across the region.

The presence of birds with a wide range of habitat preferences in Pinzacuá suggests that the complex and perennial habitat diversity that was intentionally created in the farm facilitates bird movement in fragmented agroecosystems and can provide important habitat for farmland species and migratory birds (Philpott and Bichier 2012; Vílchez et al. 2013).

#### 12.4.2.3 Dung Beetles

Dung beetle communities are shaped by both dung availability and vegetation structure. Because beetle assemblages can be severely modified by farming intensification practices and the elimination of tree cover, dung beetles are useful indicators of land use change and pasture health (Davis et al. 2004; Giraldo et al. 2011). In grazing landscapes, dung beetles mediate important ecosystem functions: they rapidly remove dung piles from the pasture surface by incorporating them into the soil and improving its fertility; they reduce soil compaction and improve soil structure; and they interrupt the life cycles of flies and parasites that lay their eggs on dung piles and affect cattle health (Giraldo et al. 2019).

A study comparing dung beetle diversity across three land uses in Pinzacuá found 97 individuals of 11 species, of which 24 individuals of five species were in the pastures with scattered trees (silvopastures); 20 individuals of seven species were in the fodder hedges; and 53 individuals of eight species were in the planted forest (Table 12.4). Species diversity was twice as high in the forest than in the silvopastures, and similar in silvopastures and fodder hedges (Giraldo et al. 2019). Structural complexity, high canopy cover, and the presence of a leaf litter layer that retains moisture explain the higher abundance and richness in the forests, as well as the diverse assemblage in the pastures with high tree density.

#### 12.4.2.4 Other Species

No formal studies of mammals or reptiles have been conducted to date on Pinzacuá. However, squirrels, opossums, agoutis, and armadillos are common on the farm, crab eating fox (*Cerdocyon thous*) and jaguarundi (*Puma yagouaroundi*) have been occasionally seen, and lowland pacas (*Cuniculus paca*) and ocelots (*Leopardus pardalis*) are likely present as they have been spotted in neighboring farms. In addition, three species of snakes including the false coral (*Oxyrhopus petolarius*), the green vine snake (*Oxybelis fulgidus*), and the yellow rat snake (*Spilotes pullatus*) are also on site.

Overall, Pinzacuá's efforts to increase vegetation cover and complexity by planting and protecting a variety of tree species have not only increased tree diversity, but they have provided adequate resources for other wildlife species, thereby transforming this previously degraded farm into an island of biodiversity amidst an otherwise impoverished landscape. Furthermore, Pinzacuá has become a site for biodiversity research and has contributed to our understanding of how the incorporation of trees can aid in the recovery of degraded farmlands.

**Table 12.5** Productivity indicators for two intensive cattle ranching systems in Colombia

| Productivity indicators | High input treeless pastures, Cimitarra, Santander[a] | Silvopastoral system, Pinzacuá |
|---|---|---|
| Stocking rate (animal units[b] $ha^{-1}$) | 7 | 5.5 |
| Meat production (gr $animal^{-1}$ $day^{-1}$) | 600 | 600 |
| Conception rate (excluding first time pregnancies) | 80% | 95% |
| Veterinary care costs (vet, pest control, vitamins, etc.) | 4% of total input cost | 0.5% of total input cost |
| Calf mortality rate | 5% | 3% |

[a]Personal communication with Antonio José Uribe, CIPAV
[b]1 animal unit = 450 Kg

## 12.4.3 Impact on Productivity

In terms of productivity, the silvopastoral systems implemented over the past 20 years have allowed the farm to support cattle stocking rates similar to those seen in high-input cattle grazing systems, while also eliminating the use of chemical fertilizers completely. During this period fodder biomass production in the pastures has increased from two tons $ha^{-1}$ in 1997 when fertilization ceased, to 11 tons $ha^{-1}$ in the silvopastoral systems. Compared to high-input grazing systems, Pinzacuá's calf mortality rate is 2% lower and the birth rate is 10% higher, with a calving interval of 13 months (Table 12.5) (Reyes et al. 2017). In addition, the shade provided by trees has reduced the risk of heat stress, creating a more comfortable environment in which animals stay hydrated and maintain their normal feeding behavior (Broom et al. 2013). Shade also provides habitat for predators that control ticks, flies, and other common cattle pests, thereby reducing the incidence of disease. These improvements in livestock's' bodily condition reduce costs in veterinary care and medications. Overall, the direct and indirect benefits provided by trees are reflected in better animal welfare and higher productivity.

## 12.4.4 Impact on Ecosystem Services

### 12.4.4.1 Water Quality and Quantity in Restored Streams

The re-establishment of riparian corridors in the farm's previously unprotected steams has also impacted water quality. The upstream town of Alcalá discharges raw sewage from approximately 1400 people into the main stream that flows through Pinzacuá at a rate of 10.6 L $s^{-1}$. An evaluation of the stream's self-purification capacity was conducted in 2010, approximately 10 years after 5 to 10-meter-wide strips were fenced off and reforested with guadua (*Guadua angustifolia*) on both streambanks. Analyses of three composite water samples taken at 400-m intervals showed the following removal efficiencies for five pollutants along the 800-m stretch: 52% for nitrogen, 49% for phosphorus, 88% for chemical oxygen demand (COD), 92% for total coliforms and 97% for fecal coliforms (Montes Londoño and Rodríguez Susa 2011). Compared to similar studies, the stream appears to retain its self-purification capacity thanks in part to resemblance of the restored forests to the basin's historical guadua forest cover. Despite the lack of baseline data, workers and landowners report substantial changes in water quality parameters such as color, odor, turbidity and the presence of foam. Thus, riparian restoration appears to have had a positive effect on the water, underscoring the links between riparian area protection, land management practices, and water quality.

#### 12.4.4.2 Climate Change Mitigation

Whereas greenhouse gas (GHG) emissions, carbon stocks and carbon sequestration have not been directly measured at Pinzacuá, numerous studies from tropical and sub-tropical Latin America highlight the potential of tree-based grazing systems for climate mitigation. Silvopastoral systems can contribute in two important ways: via carbon sequestration in both the woody biomass and the soils, and via an improved animal diet that reduces emissions from enteric fermentation (Ibrahim et al. 2007; Arias Giraldo et al. 2009; Mesa Arboleda 2009; Amézquita et al. 2010; Naranjo et al. 2012; Montagnini et al. 2013; Aynekulu et al. 2019).

Silvopastoral systems store more carbon than grass-only systems because of their higher amount of above and belowground biomass. A study conducted in the RISEM project farms estimated a total carbon storage of 153 Mg C in 1 ha of silvopasture consisting of improved grass planted with 83 native trees (Arias Giraldo et al. 2009). Silvopastures also perform better in their total GHG balance, as shown in a study comparing four different types of pasture systems in Colombia. The two conventional systems—degraded treeless pastures and improved treeless pastures—were net GHG sources, emitting 3.2 and 3.3 tons $CO_2e$ $ha^{-1}$ $year^{-1}$ respectively. Meanwhile, the two silvopastoral systems—one combining improved grasses with *Leucaena leucocephala* (60,000 shrubs $ha^{-1}$) and a similar one that also had mixed species of timber trees—were net sinks, respectively removing 8.8 and 26.6 ton $CO_2e$ $ha^{-1}$ $year^{-1}$ (Naranjo et al. 2012).

Sequestration potential varies with factors such as soil type, tree species, stand age, and management (Nair et al. 2010). However, these studies suggest that given the farm's baseline condition of mostly degraded soils and treeless pastures, and the significant increase in tree cover over the past two decades, Pinzacuá has become a provider of climate mitigation services.

## 12.5 Discussion and Conclusions

### *12.5.1 Pinzacuá as an Island of Biodiversity*

Twenty years ago, Pinzacuá began a gradual transformation of its land use and management practices with the goal of improving productivity and ensuring economic viability. The results are evident.

Increasing farm productivity was the key first step that allowed the farm to reduce the production area to accommodate more areas for conservation. Pasture monocultures were replaced with live fences, silvopastures, riparian buffers, and connectivity corridors, transforming the farm into a mosaic of vegetation structures associated to different land uses. Today, the added diversity and complexity provides refuge for birds, bats, and other mammals; native flowering trees offer resources for wildlife and a variety of beneficial insects; canopy cover creates favorable conditions for ground-dwelling ants and dung beetles that require moisture and tree litter to move through the pasture; trees support the web of underground biodiversity that underpins soil fertility; and riparian corridors and high tree cover provide connectivity throughout the farm. In short, Pinzacuá has become a high-quality matrix that serves as permanent habitat or temporary refuge for a variety of taxa that strive to persist in this highly fragmented landscape. As seen in the satellite images, Pinzacuá stands out, literally, as an island of vegetation in an otherwise treeless landscape (Calle 2020).

Beyond the physical transformation, Pinzacuá's process is the expression of a deep cultural change that began at a personal level and eventually became a family project. However, Pinzacuá alone is not enough, and its impact will be limited unless other farms make similar management decisions and intentional efforts to integrate more trees and change harmful practices.

### *12.5.2 Challenges and Setbacks*

While transitioning from input-based conventional systems to agroecological-based systems can entail a productivity trade-off, silvopastoral systems may be a notable exception. The extensive cattle production model currently used across Latin America is known for its low productivity (0.6 animal units $ha^{-1}$ (FAO 2006) and high environmental impact. By contrast, the higher productivity of silvopastoral systems is rooted precisely in their environmental benefits, and both cannot be decoupled (Murgueitio et al. 2011; Chará et al. 2019). The question then is, if the systems is so good why are more cattle ranchers not adopting it? The answer is simple: silvopastoral systems are knowledge and management intensive, and they require a significant up-front investment that takes time to recover. Whereas Pinzacuá's owner was able to keep the farm afloat during the transition period by drawing from another business, many farmers are not in a position to do the same. Unfortunately, the structures needed to overcome these entry barriers are currently not in place in most countries. In recent years, however, the RISEM and other projects have shown that economic incentives and technical assistance can effectively address these barriers and deliver lasting impacts (Calle 2020; Pagiola et al. 2020).

Pinzacuá's transformation clearly illustrates the potential to increase productivity while restoring and conserving the natural capital, a process known as ecological intensification (Gaba et al. 2014). However, this transformation requires a radical shift in mindset, from one of controlling nature to one of learning from nature, as well as perseverance in facing challenges. Below we describe some of the failed trials and unexpected setbacks faced at Pinzacuá, as well as some of the emerging challenges.

One of the earliest failures was the implementation of an intensive system with *Leucaena leucocephala*, a nutritious nitrogen-fixing fodder tree/shrub (Murgueitio et al. 2011). Although expensive to implement, the system's high productivity can quickly offset the investment. However, Pinzacuá's soils proved to be too acidic for *Leucaena*, and the time and resources invested were lost. A number of failures were related to tree-planting. For example, different versions of corrals built with barb wire or bamboo and efforts to spray cattle dung and other substances to deter herbivores failed to keep leaf-cutter ants and cattle at bay. Eventually better protection methods were developed through trial and error. The high mortality of healthy-looking tree seedlings, also frustrating early on, was later attributed to roots coiling in the small bags used by commercial nurseries. Careful sourcing of plant material and direct seeding techniques eventually helped overcome the problem. Pinzacuá even faced challenges with the propagation of *Gliricidia sepium* live fences, which is commonly done using plant stakes. After strong winds repeatedly unrooted the new fences, the farm modified its propagation approach planting *Gliricidia* directly from seed to obtain stronger roots.

Besides tree-planting, other setbacks are related to human factors. After transitioning to agroecological methods, the use of chemical inputs –even fertilizers to facilitate tree growth– has become a difficult choice to make. Although this has led to creative workarounds like the use of charcoal residue as biochar for the soils, developing and perfecting new methods has been time consuming. Perhaps one of the most recurring and persisting frustrations is related to training and retaining of farm workers. Silvopastoral systems represent a radical departure from conventional ranching and requires a completely different mindset. Training a new worker to do things differently requires an investment of time and resources, and losing a trained worker represents a huge setback. To alleviate labor, Pinzacuá has also experimented with agri-voluntourism but so far volunteers have been ill-prepared for the difficult field tasks, and often just interested in how the experience looks on their resume and on social media.

Over the past 20 years, Pinzacuá has faced a number of setbacks and learned many important lessons. Moving forward, the main challenges appear to be less related to increasing or maintaining productivity, and more to the numerous local socio-economic factors that threaten the farm's long-term sustainability. Today, the most significant production cost in the farm is labor. Compared to conventional extensive production, silvopastoral systems require more labor for activities such as herd rotation and tree maintenance. Production costs can be diluted either by increasing productivity per unit area or by increasing the size of the herd (Holmann et al. 2003). Although higher productivity in silvopastoral systems allows for higher stocking rates and lower costs, labor costs are barely covered at the current herd size and herd expansion is limited by the size of the farm.

Pinzacuá's proximity to an urban center raises other cost-related concerns. As land prices increase, so does the opportunity cost of using the property for conservation-friendly farming, and the pressure to plant cash crops or sell to developers increases. Proximity to town also exacerbates the labor shortage as younger generations increasingly reject farming in favor of urban employment (Calle 2020). Here, as elsewhere in the world, rural life has been downgraded and youth are concerned with the lack of future and opportunities in the agricultural sector.

The potential to diversify production and increase farm revenue through the integration of high-value timber and non-timber products is often highlighted as one of the benefits of silvopastoral systems (Somarriba 1997; Pezo and Ibrahim 1998). However, Pinzacuá has faced a different reality. While sales of bamboo posts alone can be up to ten times more profitable than cattle breeding, the excessive paperwork and bureaucracy required for harvesting and transporting planted timber make it difficult to realize these diversified revenue streams. The lack of developed value chains and local markets for both planted native hardwoods and non-timber forest products currently limits access to these other sources of revenue that could potentially complement income from livestock. Thus, although ongoing trials with vanilla, pepper, and dinde show promise, their future will depend on the development of local markets.

Being an island of biodiversity in a degraded landscape is certainly reason for pride, but at times it also feels like an uphill battle. The farm has suffered the spillover effects from unsustainable practices used in the surroundings. For example, in only 2 years the apiary, which provides additional revenue, has lost 30% of the beehives due to agrochemical drift from neighboring farms. In addition, the use of unconventional methods elicits mockery by other farmers, increasing the sense of isolation in an already lonely occupation.

Finally, Pinzacuá faces challenges related to monitoring of both productive and ecological indicators. Basic farm data are essential for sound decision making, especially in agroecological systems where management—not inputs—determines productivity, and introducing timely changes is essential. Simple tasks such as recordkeeping, measuring pasture capacity or tree growth, testing for diet deficiencies, and other forms of data collection remains challenging as the farm lacks the technological and human capacity to systematically compile and analyze this information.

Likewise, monitoring ecological data in a simple and cost-effective way poses a challenge moving forward. The studies conducted to date, mostly through collaborations with individual researchers, have sparked the landowners' interest in understanding the underlying ecological processes. Implementing simple protocols to consistently monitor key ecological indicators in the long-term would therefore be desirable. For example, understanding soil dynamics and the impact of organic management on sensitive groups such as butterflies and amphibians, is of particular interest. Thus, the challenge will be to build long-term partnerships with universities or research institutions in order to develop a structured long-term research plan.

### *12.5.3 Opportunities*

Despite these challenges, the farm continues to seek new opportunities to remain competitive while respecting its commitment to biodiversity-friendly production. For example, Pinzacuá has already positioned itself as a demonstration farm and hosts 400+ visitors every year, including local and international groups of school and university students, farmers, researchers, as well as NGO and government workers. Aside from providing additional income, these visits foster opportunities to exchange knowledge and ideas, strengthen networks, and contribute to scale-up sustainable land use practices across the region (Calle et al. 2013).

Pinzacuá is located within a larger region known as the Coffee Cultural Landscape of Colombia, an UNESCO World Heritage site and the country's second most popular tourist destination with approximately 100,000 visitors every year (La Patria 2019). Agritourism is already one of the main sources of income across the region, and the farm could potentially combine nature-based tourism with biodiversity-friendly farming. Furthermore, ecotourism and proximity to urban centers are potential new markets for organic produce, which the farm already grows, and direct marketing of specialty crops such as mushrooms or underutilized species such as natural dyes or medicinal plants. But this will require efforts to identify short value chains supplied by many small producers, and to market the services.

Finally, agri-voluntourism and scientific tourism could potentially offer an alternative to facilitate the collection and analysis of both productivity and ecological data. A careful selection process to identify volunteers with the right skills set will be required to realize this opportunity.

## 12.6 Conclusion

Pinzacuá's efforts to increase vegetation cover and complexity by planting and protecting a variety of trees have not only improved the farms' productivity but have also increased its biodiversity. By providing adequate resources for wildlife species, this previously degraded farm has been transformed into an island of biodiversity amidst an otherwise impoverished landscape. Furthermore, Pinzacuá has become a farm that serves as an example for others interested in restorative agriculture, and a site for biodiversity research that has contributed to our understanding of how agroforestry and sustainable management can aid in the recovery of degraded farmlands. Today, Pinzacuá stands as living proof that by mimicking natural ecosystems, the ecological interactions and synergies among the components of the agroecosystem can be re-established, and land degradation can be reversed to restore soil productivity, support biodiversity, and recover important ecosystem services. The process has been long and not without setbacks, but the results have led to an undeniable transformation of both the land and its people. As new challenges continue to emerge, Pinzacuá will draw on the lessons learned to continue applying agroecological principles, and perhaps more importantly, to remain open to experimenting with new ideas.

## References

Amézquita MC, Murgueitio E, Ibrahim RB (2010) Carbon sequestration in pasture and silvopastoral systems compared with native forests in ecosystems of tropical America. In: Abberton M, Conant R, Batello C (eds) Grassland carbon sequestration: management, policy and economics: proceedings of the workshop on the role of grassland carbon sequestration in the mitigation of climate change. Food and Agriculture Organization of the United Nations (FAO), Rome

Arias Giraldo LM, Camargo García JC, Cardona Trujillo H (2009) Carbono orgánico edáfico en rodales de guadua, Guadua angustifolia Kunth., Poaceae y en pasturas arborizadas en la zona cafetera de Colombia. In: Murgueitio E, Cuartas C, Naranjo J (eds) Ganadería del futuro: Investigación para el desarrollo. Cali, Fundación CIPAV, pp 246–261

Aynekulu E, Suber M, Zomer R, Mboi D, Arango J, Rosenstock TS (2019) Mitigation benefits from expansion of trees on rangeland: an analytical proof of concept for Colombia. CCAFS working paper no. 295. CGIAR Research Program on Climate Change, Agriculture and Food Security (CCAFS). Wageningen. Available online at: www.ccafs.cgiar.org

Becerra I, Becerra M, Montes N (2017a) Dinámica de mercado: departamento de Quindío. Unidad de Planificación Rural Agropecuaria, UPRA, Bogotá

Becerra I, Becerra M, Montes N (2017b) Dinámica de mercado: departamento de Caldas. Unidad de Planificación Rural Agropecuaria, UPRA, Bogotá

Becerra I, Becerra M, Montes N (2017c) Dinámica de mercado: departamento de Risaralda. Unidad de Planificación Rural Agropecuaria, UPRA, Bogotá

Broom DM, Galindo FM, Murgueitio E (2013) Sustainable, efficient livestock production with high biodiversity and good welfare for animals. Proc R Soc Biol Sci 280:2013–2025

Calle A (2008) What makes an early adopter? Transforming landscapes one farmer at a time. Trop Resour 27:7–14

Calle A (2020) Can short-term payments for ecosystem services deliver long-term tree cover change? Ecosyst Serv 42(C). https://doi.org/10.1016/j.ecoser.2020.101084

Calle A, Holl KD (2019) Riparian forest recovery following a decade of cattle exclusion in the Colombian Andes. For Ecol Manage 52(August):10.1016/j.foreco.2019.117563

Calle Z, Méndez LE (2017) Assessment of the regenerating woody vegetation in Pinzacua. Unpublished data

Calle Z, Piedrahita L (2007) ¿Cómo diseñar estrategias para el manejo de plantas de interés para la conservación en paisajes ganaderos? Agroforestería en las Américas 45:117–122

Calle Z, Murgueitio E, Chará J, Molina CH, Zuluaga AF, Calle A (2013) A strategy for scaling-up intensive silvopastoral systems in Colombia. J Sustain For 32(7):677–693

Calle Z, Molina C CH, Molina D CH, Molina D EJ, Molina E JJ, Murgueitio C B, Murgueitio C A, Murgueitio R E (2022) A highly productive biodiversity island within a monoculture landscape: El Hatico nature reserve (Valle del Cauca, Colombia). In: Montagnini F (ed) Biodiversity islands: strategies for conservation in human-dominated environments. Topics in Biodiversity and Conservation, Springer, Cham, pp 279–304

Camargo JC, Cardona GA (2005) Análisis de fragmentos de bosque y guaduales. In: Consultancy report – regional integrated silvopastoral approaches to ecosystem management project. The World Bank, GEF, CIPAV, p 38

Chará J, Reyes E, Peri P, Otte J, Arce E, Schneider F (2019) Silvopastoral systems and their contribution to improved resource use and sustainable development goals: Evidence from Latin America. FAO, CIPAV and Agri Benchmark, Cali. Retrieved from http://www.livestockdialogue.org/fileadmin/templates/res_livestock/docs/2018_Ulaanbataar/Silvopastoral_Systems_and_their_contribution_to_improved_resource_use_and_SDG.pdf

Chazdon RL, Harvey C, Komar O, Griffith DM, Ferguson BG, Martínez-Ramos M, Morales H, Nigh R, Soto-Pinto L, van Bruegel M, Philpott S (2009) Beyond reserves: a research agenda for conserving biodiversity in human-modified tropical landscapes. Biotropica 4:142–153

Chízmar Fernández C (2009) Plantas comestibles de Centroamérica. Instituto Nacional de Biodiversidad – INBio. Santo Domingo de Heredia, Costa Rica

Cioffi G, Morales-Escobar L, Braca A, De Tommasi N (2003) Antioxidant chalcone glycosides and flavanones from Maclura (Chlorophora) tinctoria. J Nat Prod 66:1061–1064

Comisión Conjunta (2008) Plan de Ordenación y Manejo Cuenca Hidrográfica (POMCH) del Rio La Vieja. Corporación Autonóma Regional del Quindío, Corporación Autonóma Regional de Risaralda, Corporación Autonóma Regional del Valle del Cauca, Cali

Cordero J, Boshier DH (2003) Descripción de Especies: Maclura tinctoria. In: Cordero J, Boshier DH (eds) Arboles de Centroamérica: un manual para extensionistas. Oxford Forestry Institute (OFI, Oxford University, Oxford, UK) and Centro Agronómico Tropical de Investigación y Enseñanza (CATIE), Turrialba, pp 690–697

DANE (2014) Tercer Censo Nacional Agropecuario 2014 Tercer CNA. Departamento Administrativo Nacional de Estadística, Gobierno Nacional, Bogotá DC

Davis ALV, Scholtz CH, Dooley PW, Bham N, Kryger U (2004) Scarabaeine dung beetles as indicators of biodiversity, habitat transformation and pest control chemicals in agro-ecosystems. S Afr J Sci 100:415–424

De Beenhouwer M, Aerts R, Honnay O (2013) A global meta-analysis of the biodiversity and ecosystem service benefits of coffee and cacao agroforestry. Agric Ecosyst Environ 175:1–7

Delgado M, Pérez C (2018) Proyecciones de actividad económica regional 2017-2021. Fedesarrollo, Bogotá

Espinal S (1977) Zonas de Vida y Formaciones Vegetales de Colombia, vol. XIII, no. 11. Instituto Geográfico Agustín Codazzi–IGAC, Bogotá

Etter A (1998) General ecosystem Map of Colombia (1:1,500,000). Instituto Alexander von Humboldt, Bogotá

FAO (2006) Informe Pecuario. Subdirección de Políticas y Apoyo en Materia de Publicación Electrónica. FAO, Rome

Gaba S, Bretagnolle F, Rigaud T, Philippot L (2014) Managing biotic interactions for ecological intensification of agroecosystems. Front Ecol Evol 2. https://doi.org/10.3389/fevo.2014.00029

Giraldo C, Escobar F, Chará JD, Calle Z (2011) The adoption of silvopastoral systems promotes the recovery of ecological processes regulated by dung beetles in the Colombian Andes. Insect Conserv Divers 4:115–122. https://doi.org/10.1111/j.1752-4598.2010.00112.x

Giraldo C, Hernández M, Giraldo AM, Calle A, Mendivil J, Quevedo C, Velásquez A, Perdomo A, Castaño K, Giraldo NV, Chará J (2019) Resultados de monitoreo de biodiversidad – Finca de Referencia Pinzacuá. Proyecto Ganadería Colombiana Sostenible. CIPAV, Cali

Griscom BW, Adams J, Ellis PW, Houghton RA, Lomax G, Miteva DA, Schlesinger WH, Shoch D, Siikamäki JV et al (2017) Natural climate solutions. Proc Natl Acad Sci 114(44):11645–11650. https://doi.org/10.1073/pnas.1710465114

Harvey CA, Villanueva C, Esquivel H, Gómez R, Ibrahim M, López M, Martinez J, Muñoz D, Restrepo C, Saénz JC, Villacís J, Sinclair FL (2011) Conservation value of dispersed tree cover threatened by pasture management. For Ecol Manag 261(10):1664–1674. https://doi.org/10.1016/j.foreco.2010.11.004

Holmann F, Rivas L, Carulla J, Rivera B, Giraldo LA, Guzman S, Martinez M, Medina A, Farrow A (2003) Evolution of milk production systems in tropical latin america and its interrelationship with markets: an analysis of the Colombian case (online). Livest Res Rural Dev 15-(9) Available at: http://www.lrrd.org/lrrd15/9/holm159.htm

Ibrahim M, Chacón M, Cuartas C, Naranjo J, Ponce G, Vega P, Casasola F, Rojas J (2007) Carbon storage in soil and biomass in land use systems of ranchlands of Colombia, Costa Rica and Nicaragua. Agroforestería en las Américas 45:27–36

IUCN Commission on Ecosystem Management (2020) Nature based solutions. Retrieved from https://www.iucn.org/commissions/commission-ecosystem-management/our-work/nature-based-solutions on May 24, 2020

Kattan GH, Alvarez-López H (1996) Preservation and management of biodiversity in fragmented landscapes in the Colombian Andes. In: Schelhas J, Greenberg R (eds) Forest patches in tropical landscapes. Island Press, Washington, D. C

Kattan GH, Franco P, Rojas V (2004) Biological diversification in a complex region: a spatial analysis of faunistic diversity and biogeography of the Andes of Colombia. J Biogeogr 31: 1829–1839

Kremen C, Merenlender AM (2018) Landscapes that work for biodiversity and people. Science 362(6412):eaau6020

La Patria (2019) Las cuentas del turismo en el Eje Cafetero. Monday, April 15, 2019. Retrieved from https://www.lapatria.com/economia/las-cuentas-del-turismo-en-el-eje-cafetero-435307 on February 25, 2020

López Murillo E (2015) El valor del suelo. Columna de opinión La Cronica del Quindío. Retrieved on June 1, 2020 from https://www.cronicadelquindio.com/noticia-noticia_opinion-titulo-el-valor-del-suelo-cronica-del-quindio-op-11685

Maas B, Clough Y, Tscharntke T (2013) Bats and birds increase crop yield in tropical agroforestry landscapes. Ecol Lett 16:1480–1487

Martins MM, Setz EZF (2000) Diet of buffy tufted-eared marmosets (Callithrix aurita) in a forest fragment in southeastern Brazil. Int J Primatol 21:467–476

Mejía Cubillos J (2013) Perfil económico del Eje Cafetero. Análisis con miras a la competitividad territorial. Adaptación del informe final de la consultoría "Competitividad territorial de la Ecorregión Eje Cafetero. Análisis desde el perfil económico del Eje Cafetero como aporte a la línea de base de la Agenda de Desarrollo Sostenible" realizada para Universidad Tecnológica de Pereira. MPRA Paper No. 43873. Retrieved on June 1, 2020 from https://mpra.ub.uni-muenchen.de/43873/

Mendenhall C, Karp D, Meyer C, Hadly E, Daily G (2014) Predicting biodiversity change and averting collapse in agricultural landscapes. Nature 509:213–217. https://doi.org/10.1038/nature13139

Mesa Arboleda HF (2009) Balance de gases de efecto invernadero en un modelo de producción de ganadería doble propósito con alternativas silvopastoriles en Yaracuy, Venezuela. MS thesis. CATIE, Turrialba, 225 p

Mittermeier RA, Turner WR, Larsen FW, Brooks TM, Gascon C (2011) Global biodiversity conservation: the critical role of hotspots. In: Zachos FE, Habel JC (eds) Biodiversity hotspots. Springer, London, pp 3–22

Moebius-Clune BN, Moebius-Clune DJ, Gugino BK, Idowu OJ, Schindelbeck RR, Ristow AJ, van Es HM, Thies JE, Shayler HA, McBride MB, Kurtz KSM, Wolfe DW, Abawi GS (2016) Comprehensive assessment of soil health – the Cornell framework, Edition 3.2. Cornell University, Ithaca

Montagnini F, Ibrahim M, Murgueitio E (2013) Silvopastoral systems and climate change mitigation in Latin America. Bois et forêts des Tropiques 316:3–16

Montes-Londoño (2019) Monitoring of species richness of butterfly in Pinzacua through Visual Encounterying surveying. Unpublished data

Montes Londoño I, Rodríguez Susa MS (2011) Evaluación de la capacidad de autodepuración de una quebrada con bosque de guadua ribereño en la cuenca del Río La Vieja. Undergraduate thesis. Universidad de Los Andes, Bogotá

Montes-Londoño I, Montagnini F, Ashton MS (2017) Allometric relationships and reforestation guidelines for Maclura tinctoria, an important multi-purpose timber tree of Latin America. New For. https://doi.org/10.1007/s11056-017-9617-1

Murgueitio E, Calle Z, Uribe F, Calle A, Solorio B (2011) Native trees and shrubs for the productive rehabilitation of tropical cattle ranching lands. For Ecol Manag 261(10):1654–1663

Nair PKR, Nair VD, Kumar BM, Showalter JM (2010) Carbon sequestration in agroforestry systems. Adv Agron 108:237–307

Naranjo JF, Cuartas CA, Murgueitio E, Chará J, Barahona R (2012) Balance de gases de efecto invernadero en sistemas silvopastoriles intensivos con Leucaena leucocephala en Colombia. Livestock research for rural development, 24, Article #150. Retrieved May 31, 2020 from http://www.lrrd.org/lrrd24/8/nara24150.htm

Pagiola S, Ríos AR (2013) Evaluation of the impact of payments for environmental services on land use change in Quindío, Colombia. PES learning papers. World Bank, Washington DC

Pagiola S, Honey-Rosés J, Freire-González J (2020) Assessing the permanence of land-use change induced by payments for environmental services: evidence from Nicaragua. Trop Conserv Sci 13:1940082920922676

Perfecto I, Vandermeer J (2010) The agricultural matrix as alternative to the land sparing/agriculture intensification model. PNAS 107:5786–5791

Peters VE, Greenberg R (2013) Fruit supplementation affects birds but not arthropod predation by birds in Costa Rican agroforestry systems. Biotropica 45:102–110

Pezo D, Ibrahim M (1998) Sistemas Silvopastoriles. Materiales de Enseñanza no. 44/CATIE CATIE-GTZ Turrialba, 276 pp

Philpott SM, Bichier P (2012) Effects of shade tree removal on birds in coffee agroecosystems in Chiapas, Mexico. Agric Ecosyst Environ 149:171–180

Piotto D, Montagnini F, Kanninen M, Ugalde L, Viquez E (2004) Forest plantations in Costa Rica and Nicaragua: performance of species and preferences of farmers. J Sustain For 18(4):59–77

Poch TJ, Simonetti JA (2013) Ecosystem services in human-dominated landscapes: insectivory in agroforestry systems. Agrofor Syst 87:871–879

Prevedello JA, Almeida-Gomes M, Lindenmayer DB (2018) The importance of scattered trees for biodiversity conservation: a global meta-analysis. J Appl Ecol 55:205–214

Rangel A (1949) Maderas industriales de Colombia Caribb For 10(3):161–162

Reyes E, Bellagamba A, Molina JJ, Izquierdo L, Deblitz C, Chará J, Mitchell L, Romanowicz B, Gómez M, Murgueitio E (2017) Measuring sustainability on cattle ranches. Working paper on Silvopastoral Systems. Agribenchmark, CIPAV, Fedegan, World Animal Protection. Retrieved from http://www.cipav.org.co/pdf/ReportSPS6-Colombiancasestudies.pdf

Rivers MC, Barstow M, Mark J (2017) Maclura tinctoria. The IUCN red list of threatened species 2017: e.T61886731A61886745. Retrieved from https://doi.org/10.2305/IUCN.UK.2017-3.RLTS.T61886731A61886745.en. Feb 29, 2020

Rodríguez N, Armenteras D, Morales M, Romero M (2004) Ecosistemas de los Andes Colombianos. Instituto de Investigación de Recursos Biológicos Alexander von Humboldt, Bogotá

Roig JT (1974) Plantas medicinales, aromáticas o venenosas de Cuba. Ed Ciencia y Técnica, Instituto del Libro. La Habana, Cuba

Sadeghian S, Murgueitio E, Mejía C, Rivera JM (2001) Ordenamiento ambiental y reglamentación del uso y manejo del suelo en la zona cafetera. En: Suelos del eje cafetero. Universidad Tecnológica de Pereira, GTZ, Fondo Editorial del Departamento de Risaralda, Pereira, pp 96–108

Sánchez Gómez EL, Camargo García JC (2015) Diversidad de avifauna en paisajes rurales de la cuenca del río La Vieja, Eje Cafetero de Colombia. Recursos Naturales y Ambiente:83–89

Somarriba E (1997) Pastoreo bajo plantaciones forestales. Agroforestería en las Américas 4:26–28

Suárez A, Williams-Linera G, Trejo C, Valdez-Hernández JI, Cetina-Alcala VM, Vibrans H (2012) Local knowledge helps select species for forest restoration in a tropical dry forest of central Veracruz, Mexico. Agrofor Syst 85:35–55

Toro Zuluaga G (2005) Eje Cafetero colombiano: compleja historia de caficultura, violencia y desplazamiento. Documento preparado para el Congreso 2004 de la Asociación de Estudios Latinoamericanos (LASA), Las Vegas, Nevada, octubre 7-9, 2004. Revista de Ciencias Humanas UTP 35: 127–149

Tscharntke T, Clough Y, Wanger TC, Jackson L, Motzke I, Perfecto I, Vandermeer J, Whitbread A (2012) Global food security, biodiversity conservation and the future of agricultural intensification. Biol Conserv 151:53–59

Unidad de Planificación Rural Agropecuaria, UPRA (2019a) Gestión de información agropecuaria y planificación del desarrollo agropecuario: Qundío. Retrieved on June 1, 2020 from https://drive.google.com/file/d/1Wnmc5v8CGRi0q0-Hz65bAqIV_4yjZL0V/view

Unidad de Planificación Rural Agropecuaria, UPRA (2019b) Gestión de información agropecuaria y planificación del desarrollo agropecuario: Caldas. Retrieved on June 1, 2020 from https://drive.google.com/file/d/11013KahlDDAN628zQrY6lR0OcQBvQaD9/view

Unidad de Planificación Rural Agropecuaria, UPRA (2019c) Gestión de información agropecuaria y planificación del desarrollo agropecuario: Risaralda. Retrieved on June 1, 2020 from https://drive.google.com/file/d/1cgTexaHDDyXOrl8QhhLeH4bjx3aackm4/view

United Nations Development Programme (UNDP) (2004) Informe Regional de Desarrollo Humano. PNUD, Manizáles

van Breugel M, Hall JS, Craven DJ, Gregoire TG, Park A, Dent DH, Wishnie MH, Mariscal E, Deago J, Ibarra D, Cedeño N, Ashton MS (2011) Early growth and survival of 49 tropical tree species across four sites differing in soil fertility and rainfall. For Ecol Manag 261:1580–1589

Vílchez S, Harvey C, Sáenz JC, Casanoves F, Carvajal JP, Villalobos JG, Hernandez B, Medina A, Montero J, Merlo D, Sinclair FL (2013) Consistency in bird use of tree cover across tropical agricultural landscapes. Ecol Appl 24:158–168

Wishnie MH, Dent DH, Mariscal E, Deago J, Cedeño N, Ibarra D, Condit R, Ashton MS (2007) Performance of 24 tropical tree species in relation to reforestation strategies in Panama. For Ecol Manag 243:39–49

# Chapter 13
# Islands of Trees in Long-Fragmented Landscapes in Great Britain

**Keith J. Kirby**

**Abstract** The British countryside contains scattered 'islands' of ancient woodland and groups of ancient trees in 'seas' of managed farmland and plantations of non-native trees. However, do they function like islands for the plants and insects particularly associated with these patches? Is the absence of species from newer islands because of dispersal limitations, or because their habitat requirements are not met in newly formed sites?

Past fragmentation may mean remnant patches hold more species than can be sustained in the longer-term. However, if the fragmentation of the original wildwood happened hundreds, even thousands of years ago, the woodland species that have survived and are valued today, of necessity must be able to live in relatively small patches with a predominance of edge conditions. Some can also be found associated with individual trees and hedges scattered through the intervening farmland.

The island analogy is useful but has limitations in guiding future conservation practice. The extent and pattern of patches as perceived by researchers, i.e. what is mapped as woodland or wood-pasture, may be very different to the actual patch size used by the species. In managed landscapes there is the possibility of creating or restoring habitats in the places that appear to offer the best opportunities for allowing species movement through a landscape. Colonisation credits should be as important a focus as extinction debts. This might involve translocating ground flora plants to new woods and deliberately damaging mature trees to expedite the decay processes normally associated with ancient trees.

**Keywords** Ancient woodland · Parkland · Saproxylic beetles · Vascular plants · Veteran trees · Wood-pasture · Woodland management

---

K. J. Kirby (✉)
Department of Plant Sciences, University of Oxford, Oxford, UK
e-mail: keith.kirby@bnc.oxon.org

 
F. Montagnini (ed.), *Biodiversity Islands: Strategies for Conservation in Human-Dominated Environments*, Topics in Biodiversity and Conservation 20,
https://doi.org/10.1007/978-3-030-92234-4_13

## 13.1 Introduction

During the 1980s, a new law, The Wildlife and Countryside Act, was introduced in Great Britain. This law gave much stronger protection to the series of sites identified by the state nature conservation agencies as important for biodiversity (Sheail 1998). These areas (examples of woodland, heathland, meadow, wetland, etc.) covered about 7% of the land surface. However, despite this and subsequent stronger legislation to protect such sites, many species continued to decline at a national scale (Defra 2017). It was recognised that conservation needed to work across the countryside as a whole.

The principles and approaches of landscape ecology, e.g. Forman and Godron (1986) gained ground amongst the conservation movement in Great Britain during the 1980s. Ideas based around island biogeography theory seemed particularly appropriate to the British conditions. Frequently there are small patches of habitat with a high biodiversity value (the "islands") distributed amongst large areas of low biodiversity habitat (the "seas") consisting of arable crops, highly-improved grassland or monocultural plantations of introduced trees managed for production of food and fibre (Peterken 2002) (Fig. 13.1). The enduring influence of landscape ecological concepts can be seen in Fig. 13.2 based on an influential report on the state of

**Fig. 13.1** A view of Devon, south-west Britain, showing 'islands' of woodland in an agricultural 'sea'. (Photo: Keith Kirby)

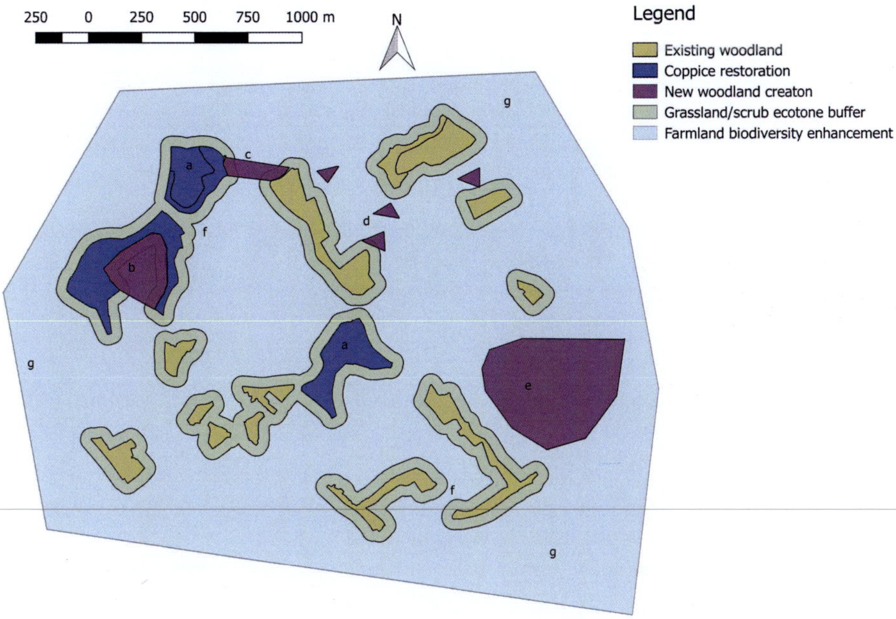


**Fig. 13.2** Enhancing ecological networks based on *Making Space for Nature.* (Lawton et al. 2010) Approaches needed include: (**a**) improving the quality of habitat patches; (**b**) making existing sites bigger; enhancing connectivity through (**c**) a new woodland corridor or (**d**) stepping stone patches; (**e**) creating new woods; reducing pressures on sites through (**f**) establishing grass-scrub buffers and ecotones, or (**g**) more general biodiversity enhancement in the surrounding farmland.

England's protected sites produced for the Department of Environment, Food and Rural Affairs (Defra) (Lawton et al. 2010).

This chapter explores the degree to which patches of ancient woodland and wood-pasture behave as 'island' habitats in the cultural landscapes of Great Britain. Peterken (1977, 1993) and Rackham (1976, 2003) summarised the conservation importance of ancient woodland for ecological and cultural reasons. The significance of veteran and ancient trees, especially oak in Britain – often occurring in more open conditions such as parkland – for a range of different species groups such as fungi, lichens and saproxylic invertebrates has been highlighted by Rose (1976, 1993), Harding and Rose (1986) and contributions in Farjon (2017).

My focus is on two species groups: vascular plants that have a high affinity with ancient woodland; and saproxylic invertebrates such as the stag beetle, *Sinodendron cylindricum,* that are strongly associated with a continuity of ancient trees such as are often found in wood-pastures (Table 13.1). However, the real world is more variable in space and time than the conditions and assumptions that underpinned some landscape models.

**Table 13.1** Some terms used in this paper to describe the tree and woodland resource in Britain

| |
|---|
| 'Wood', 'woodland' and 'forest' are used generally to describe tree-covered lands. |
| 'Forest' (capital F) is also used in a more specialised sense where it refers to land subject to Forest law in the medieval period; the land to which it is applied might or might not be covered by trees. |
| 'Wood-pasture' refers to landscapes grazed by domestic stock or deer with an open tree cover. |
| 'Wildwood' is used for the pre-Neolithic landscape (prior to c.7000 years BP) when tree-cover was probably at its most extensive in Britain. |
| 'Ancient woods' (or ancient forests) are those where there has been continuous woodland cover since 1600 AD in England and Wales and 1750 AD in Scotland. |
| 'Coppicing' refers to the practice of repeatedly cutting trees close to ground level, at intervals of between five and thirty years, then allowing regrowth of multiple stems from the stump to provide the next harvest. 'Pollarding' is a similar process, but the cut is made at 2-3 m above the ground so that the regrowth is out of the reach of browsing animals. |
| 'Ancient trees' are those old for their species with features such as cavities or a hollow trunk, bark loss over sections of the trunk and a large quantity of dead wood in the canopy. The broader term 'veteran trees' includes younger individuals that have developed similar characteristics, perhaps due to adverse growing conditions or injury. |
| 'Hedges/hedgerows' are lines of (usually thorny) shrubs managed to provide boundaries between fields. Large mature trees are often scattered along the hedge. |
| 'Saproxylic species' are those that live on or in dead and decaying wood, with deadwood beetles being one of the most studied groups. |

## 13.2 An Ancient Island Pattern

During the Holocene, much of Britain came to be tree-covered below about 300 m altitude, reaching its maximum extent about 7000 years BP (Kirby and Watkins 2015). Thereafter, tree cover dropped as farming developed and land was cleared for crops and for livestock grazing. By 1000 years BP there may have been less than 20% cover; clearance continued and c.100 years BP woodland cover was about 4–5%. There had been some large-scale planting of new forests in the nineteenth century and this accelerated during the twentieth century such that woodland cover is currently about 13% (Forestry Commission 2019). However, patches of woodland composed of native trees, mainly broadleaved species, tend to be small, less than 50 ha.

Early studies confirmed the expectation that larger patches of native trees do tend to contain more species than smaller patches for butterflies, birds, and vascular plants (Moore and Hooper 1975, Shreeve and Mason 1980, Usher et al. 1992). The history of the patches is also important. More of the specialised woodland vascular plant species occur in patches that have existed for several hundred years at least (ancient woods, see Table 13.1 and Peterken (1977)). The distribution of ancient woods across Britain has been mapped by the Nature Conservancy Council and its successors, as illustrated in Fig. 13.3 (Goldberg et al. 2011, Goldberg 2015).

Ancient woods do contain generalist plant species such as brambles (*Rubus fruticosus*) that are widespread through the rest of the landscape. However ancient woods are distinctive in being richer than more recently created woodland of a

**Fig. 13.3** The distribution of ancient and recent woodland in part of eastern England. The intervening land is arable and improved grassland. In the last 150 years the development of the recent woodland means that the ancient woods are now *less* isolated than they were. (Based on data from the Ancient Woodland Inventory (Spencer and Kirby 1992))

similar extent in a suite of vascular plants that are generally more shade-tolerant, stress-tolerant and poor-colonisers (often apparently reliant on ants for dispersal, e.g. *Anemone nemorosa*, *Mercurialis perennis, Primula vulgaris*) compared to the non-woodland flora (Peterken 1974, 2000, Kirby et al. 2012, Kimberley et al. 2013). Peterken and Game (1984) showed that for these vascular plants new woods that had developed adjacent to ancient woodland were richer than more remote new woods, suggesting isolation as an important factor. Similar findings have been reported from elsewhere in north-western Europe, e.g. Brunet and Von Oheimb (1998).

Britain also has many areas of open ancient wood-pasture, with large-crowned, open-grown trees, generally old parks and Royal Forests, grazed by deer or livestock (Kirby and Perry 2014). These may or may not be mapped as woodlands depending on the extent of canopy cover (Fig. 13.4), but they include many of the sites of highest value for epiphytic lichens (e.g *Lobaria pulmonaria, L. amplissima, Thelotrema lepadinum*) and for saproxylic invertebrates (e.g. *Ampedus cinnabarinus, Gnorimus variabilis, Limoniscus violaceus*) (Harding and Rose 1986). Both specialist lichens and invertebrates appear to be very specific in their requirements for dead wood conditions and have poor rates of dispersal (Rose 1993, Scheidegger and Werth 2009, Irmler et al. 2010). For these species, the extent of the site (hectares) is less important than the number of suitable habitat trees, which may be scattered or clustered across a larger or smaller grazed area. There is still a broad relationship between site size and richness because large sites tend to have more

**Fig. 13.4** (**a**) Glenamara Park, Cumbria: an upland wood-pasture showing areas of scattered and clustered veteran trees; (**b**) Moccas Park, Herefordshire: a lowland wood-pasture with scattered trees. (Photos: Keith Kirby)

trees and a greater variety within the tree population in terms of species and structural forms, but the area effect is an indirect one.

## 13.3 Real Islands Versus Tree Patches

Real islands are surrounded by a matrix habitat, the sea or another water body, that cannot support the terrestrial island species. By contrast, the matrix habitats (farmland, plantations) around ancient woods and parklands often can support some woodland or wood-pasture species albeit generally at low population levels. Specialist woodland flora do occur in and spread along hedges between fields; ancient trees, with associated decaying wood fauna, may similarly occur in fields and along boundaries (Fig. 13.5).

During the last two centuries, particularly since the end of World War II, there have been losses of many of these microhabitats in the matrix through agricultural intensification and increased fragmentation of patches by major roads and railway lines, e.g. Peterken and Allison (1989). This has increased the effective isolation of the woodland and wood-pasture patches. However, there have also been substantial increases in woodland and tree cover as a consequence of changing forestry policies

**Fig. 13.5** A farmed landscape near Oxford, with abundant hedges and hedgerow trees that might act as habitat for woodland species. (Photo: Keith Kirby)

during the twentieth century that should eventually increase the available habitat extent and reduce effective isolation between woods. An increase in new woodland is a poor substitute for the loss of long-established or ancient woods, but new farm woods can within a century start to harbour woodland specialist vascular plants (Usher et al. 1992, Harmer et al. 2001).

Saproxylic species may benefit from the aging of oak trees established in the nineteenth Century along hedges and in small woods, or from those left to grow on as timber trees in woods that were once cut-over regularly as coppice, but in most cases were largely neglected in the late twentieth century. These trees are or will soon be moving into the age and size classes where decaying wood starts to form hollows in the main trunk.

## 13.4 Recent Versus Long-Fragmented Systems

Fragmentation models often start with a large more-or-less continuous block of habitat – say a forest – which is then split up into small isolated blocks leading to the loss of species with large area requirements; a loss of 'interior' species and an increase in 'edge' species (the latter generally seen as less valuable); and an ongoing loss of other species for a while – the extinction debt – because the smaller patches created by fragmentation initially contain more species than can be sustained in the reduced area. How well does this model fit with current-day ancient woods and wood-pastures?

In Britain there is debate as to what was the nature of the 'continuous habitat' – the wildwood of the pre-Neolithic era – from which our ancient woodland and wood-pasture have been derived. If much of the country were closed forest (Peterken 1996), perhaps similar to that in Bialowieza National Park, Poland (Falinski 1986), then the initial clearance and fragmentation effects would have led to the sorts of major changes in species richness suggested by models and studies of recent fragmentation of tropical forests. However, if the wildwood was a more open mixture of blocks of tree cover, scrub and open grassland, such as proposed by Vera (2000), then the contrast between the wildwood and modern landscapes (particularly those prior to the mid-twentieth century agricultural intensification) would be much less and we might suppose that the changes in species abundances would be correspondingly lower. This debate has not been fully resolved.

The initial clearance and fragmentation happened centuries or millenia ago. Any species that required large blocks of continuous woodland would be unlikely to have survived into the historic period. We have lost most of the larger wild mammals (*Alces alces, Ursus arctos, Bos primigenius, Lynx lynx, Canis lupus, Sus scrofa, Castor fiber*) at least in part from reductions in their potential habitat area (Yalden 1999), but our ancient woods have then had 500–1000 years to adapt to their absence. Even some very small woods may hold diverse and valued assemblages of ancient woodland plants. Furthermore, these may now be threatened by the recent

**Table 13.2** Ellenberg's Light Scores as modified for British conditions by Hill et al. (2004) for British woodland specialist vascular plants (Kirby et al. 2012)

| Ellenberg's score | Meaning (Ellenberg 1988) | Number of woodland specialist species with that score |
|---|---|---|
| 1 | Plants in deep shade, may be less than 1%, seldom more than 30% relative light. | 0 |
| 2 | Between 1 and 3 | 2 |
| 3 | Shade plants, mostly less than 5% relative light, but also in lighter places. | 111 |
| 4 | Between 3 and 5 | 39 |
| 5 | Plants of half shade, rarely in full shade but generally more than 10% relative light. | 48 |
| 6 | Between 5 and 7. | 36 |
| 7 | Plants generally in well-lit places but also occur in partial shade. | 23 |
| 8 | Light loving plants, rarely found where there is less than 40% relative light. | 7 |
| 9 | Plants in full light, found only in full sun; rarely in less than 50% relative light. | 0 |

spread of large herbivores (mainly deer, both native and introduced) and wild boar (escaped from farms), e.g. Kirby (2001), Sims et al. (2014).

Our native broadleaved woods are small. Riutta et al. (2014) calculated that even if woods less than 2 ha were excluded, about 45% of the total woodland area of England was within 60 m of an open edge. Many of the plants that we seek to conserve are 'edge' species, found in partially open conditions, rather than deep shade interior species (Table 13.2).

Similarly, many of the best places for saproxylic beetle assemblages are where the oaks are growing in relatively open, sunny conditions – edge type environments, with a different suite of species to that found where ancient trees occur in dense stands. This is one argument in favour of the Frans Vera's model of the wildwood. In his model transitions from open to closed cover, i.e. 'edge' could have been widespread as part of the natural state.

High levels of edge conditions may not be therefore a negative feature of British ancient woods and wood-pastures, as they have been a characteristic that has existed for such a long time. However, there are new threats associated with edges that have developed in the last century. The most significant of these are the potential for increased eutrophication at edges (from emissions from roads, farmland etc); pesticide drift in from adjacent farmland; and increased drying-out of the edge zone under climate change (Pitcairn et al. 2002, Gove et al. 2007, Riutta et al. 2012).

When large areas of forest are broken-up, leaving smaller patches, these patches may initially contain more species than can be sustained in the long-term, so that they then gradually lose species. This phenomenon is known as 'extinction debt.' Over much of Britain, however, ancient woods and patches of ancient trees have been relatively stable in size and isolated in the landscape for centuries. Their species richness is likely to be in equilibrium with their current area. Peterken and Game (1984) did not find any evidence for an extinction debt in woods that had been recently reduced in area compared to ancient woods whose extent had long been stable, although Vellend et al. (2006) did find such an effect in ancient woods studied in Belgium. Even if, at the species level there is spread to new sites, this may include only part of the genetic variation that is present in the source population (Scheidegger and Werth 2009), such that there could still be a genetic debt.

Of greater significance for British conditions may be how to make the most of the 'colonisation credit' that is building up. Over much of Britain the extent of broadleaved woodland has been increasing over the last century, but not all the species that this larger area can support have yet colonised it. Spread of specialist woodland plants to new sites does occur but can be slow even when they are directly adjacent to an ancient woodland source (Peterken and Game 1984, Rackham 2003). Should we therefore deliberately sow such species into new woods e.g. Francis and Morton (2001), Worrell and Francis (2003)? Conservation managers have also been experimenting with whether the processes of wood decay in mature trees can be speeded up through management (Lonsdale 2013). The trees are deliberately damaged by cutting into the bark in order to bring forward the time when they might be suitable for colonisation by the saproxylic species characteristic of ancient trees.

There are periodic discussions as to whether it is better to have a single large patch as a reserve or several smaller patches (SLOSS debate (Tjørve 2010)). In long-fragmented landscapes where the species assemblages are in equilibrium with patch area some of the arguments for preferring the single large reserve are less relevant. Collections of small woods often contain more species than a single site of the same total area because they are, for example, spread across different soil types or are in different ownerships and hence have different past management and woodland structures. Recent research results suggest that small patches can have greater importance for biodiversity conservation than previously anticipated (Montagnini et al. 2022). Results from the Lacandona rainforest of Mexico, where the effect of forest patch size on species density of different taxonomic groups was examined, add to the increasing evidence that, on a per-sample area basis, small patches are valuable for conservation of forest-specialist species, and are not the near-exclusive habitat of generalist species (Arroyo-Rodríguez et al. 2022).

## 13.5 Natural Disturbances Versus Regular Human Intervention

Pickett and Thompson (1978) discuss the idea of the minimum dynamic area in nature reserve design. A reserve needs to be large enough that it can always include patches of recent disturbance, whatever form that disturbance takes. Many woodland plants including some ancient woodland species are mainly found in the temporary

**Fig. 13.6** Recently cut areas often show increased flowering of many woodland specialist plants, the yellow of *Lamiastrum galeobdolon* contrasting with the blue of *Hyacinthoides non-scripta*. (Photo: Keith Kirby)

open gaps (Fig. 13.6), so ideally woods should be big enough that there are always some trees falling and creating gaps.

This concept can be difficult to apply with natural woodland disturbance patterns because gap creating events are generally unpredictable in size and frequency. Where woods are managed and the disturbance is from felling the trees, then the minimum dynamic area can be specified more precisely. For example, many ancient woods in England were managed by the coppice system in the past where stands were cut on rotations of between about 5 and 30 years. Often key species are found mainly in the first few years after cutting (Buckley and Mills 2015a, b) (Fig. 13.6). If we assume that a minimum gap extent of 1 ha is required for the key species and that gaps remain suitable as habitat for 2 years, then an area of 1 ha must be felled at least every 2 years. If the trees are managed on a coppice rotation of 20 years that implies a minimum woodland area of 10 ha; but if the trees are to be grown for 100 years as high forest, 50 ha of woodland would be needed. If the woodland extent is fixed (which is usually the case) the procedure can be used to calculate the frequency of gap creation cutting needed to sustain the gap species.

In the example given above, the rotation length in a 20-ha wood should not be more than 40 years. Most ancient woods are less than 20 ha (Spencer and Kirby 1992) so the shift from coppice management (5–30 year rotations) to high forest

(60–120 year rotations) over the last 80 years has generally been accompanied by reductions in the more light-demanding elements of the vascular plant flora because the frequency of gap creation has been reduced, e.g. Kirby et al. (2005).

An additional consideration is how close a patch that is becoming suitable needs to be to a source patch for it to be useful. The butterfly *Melitaea athalia* in south-east England was commonly found in recently cut coppiced areas where one of its larval food plants, *Melampyrum pratense*, grew and flowered in abundance for a few years before declining as the canopy reformed. However, while adults readily colonised new patches within 300 m of a source patch, they rarely did so when the patches were 600 m or more apart (Warren 1987).

The same approach can be taken for any other groups of species that are found largely in one stage in the tree/stand life cycle. For saproxylic species, the minimum dynamic area is that that can contain sufficient trees across all generations to ensure there are some that have developed sufficient maturity to replace the current ancient trees when they finally die (Kirby 2015). For example, if the aim is to sustain two ancient trees (>400 years old) per hectare and on average half the trees in any age cohort die, then this might mean 32 trees per hectare (<100 years old), 16 trees (101–200 years old), 8 trees (201–300 years old); and 4 trees (301–400 years old). Most wood-pasture sites in Britain do not contain such a balanced population structure. The potential younger age classes of trees as do exist in the countryside may be too far from the current ancient trees for reliable colonisation to be assumed, although rare long-distance colonisation events cannot be ruled out, partly because they can be very difficult to detect (Jonsell et al. 2003).

## 13.6 Cost-Dispersal Distances and Developing Networks

Arable fields, i.e. those lands that are under use for crops, or hold vegetation of low stature that allows for ploughing, are likely to be less favourable for the spread to new sites of both woodland specialist plants and saproxylic beetles than, for example, fields of tall grass with scattered scrub. Measures of how 'joined-up' the landscape is, that is, the likelihood of species being able to move from one site to another, should take into account the nature of the intervening landscape as well as the absolute distance between patches (Adriaensen et al. 2003). This is then used to measure the effective l distance between two woodland patches to give a better idea of the degree of connectivity within a landscape. Such analyses might suggest whether species populations in a group of woods were likely to be linked through exchange of individuals and genes; or they might be used in locating new woodland to provide the most useful stepping stones or linkages for species dispersal through the landscape (Catchpole 2006, Latham et al. 2013).

The results from such modelling can inform the various initiatives seeking to establish networks of sites across the country (Crick et al. 2020). For example, which of the current habitat patches may be most critical in allowing species to spread, where are there significant gaps or bottlenecks that hinder dispersal? Practical conservation action can be targeted to key locations, for example to improve the condition of a group of old trees that are an important stepping stone between two important wood-pastures, to create new woodlands and hedges to provide potential links between ancient woods, or to encourage the growth of new trees along existing hedges where currently the trees are widely spaced.

## 13.7 Conclusions

Landscape ecology has proved a useful framework for looking at how individual patches of woodland or wood-pasture relate to the ecological processes taking place across the surrounding landscape. Among the lessons that have emerged from their application to our long-fragmented cultural landscapes are the following.

The extent and pattern of patches as perceived by researchers, i.e. what is mapped as woodland or wood-pasture, may be very different to the actual patch size used by the species. The effective patch size may be smaller because a species uses only part of the woodland cycle, e.g. the vascular plants found mainly in the short-lived gap phase, or the beetles found only in the trees over 300 years-old in a parkland; or larger because a species can also use part of the adjacent matrix. Patch boundaries, their extent and pattern in the landscape are species specific.

The processes of species colonisation of new woodland are not simply the reverse of those that apply when a continuous piece of habitat is broken up: if a piece of ancient woodland is converted to an arable field, the specialist woodland plant species may be lost overnight. If an arable field is converted to woodland, it may take centuries for some of those specialist species to colonise that area. A 400-year-old tree can be felled in a day, but it will take many years before a replacement young tree has grown sufficiently to build up the equivalent levels of decaying wood habitat (Watts et al. 2020).

In old cultural landscapes, most of the land has been managed in the past and species occurrence may be closely linked to how habitats such as ancient woods and wood-pastures have been treated. Some species may survive in smaller patches than might be expected under more natural conditions because controlled disturbances may allow a smaller minimum dynamic area.

In managed landscapes there is the possibility of creating or restoring habitats in the places that appear to offer the best opportunities for allowing species movement through a landscape. In planning restoration strategies, colonisation credits should be as important a focus as extinction debts. This might involve translocating ground flora plants to new woods and deliberately damaging mature trees to expedite the decay processes normally associated with ancient trees.

It is difficult to detect and allow for rare long-distance dispersal, but even so we can increase the likelihood of its success simply by building up source populations in existing woods and wood-pastures through improved habitat management; and increasing the extent of new woodland and wood-pasture patches in the target area.

## References


Adriaensen F, Chardon JP, De Blust G, Swinnen E, Villalba S, Gulinck H, Matthysen E (2003) The application of 'least-cost' modelling as a functional landscape model. Landsc Urban Plan 64: 233–247

Arroyo-Rodríguez V, Arasa-Gisbert R, Arce-Peña N, Cervantes-López MJ, Cudney-Valenzuela SJ, Galán-Acedo C, Hernández-Ruedas MA, San-José M, Fahrig L (2022) The importance of small rainforest patches for biodiversity conservation: a multi-taxonomic assessment in the Lacandona region, Mexico. In: Montagnini F (ed) Biodiversity islands: strategies for conservation in human-dominated environments. Topics in Biodiversity and Conservation, Springer, Cham, pp 41–60

Brunet J, Von Oheimb G (1998) Migration of vascular plants to secondary woodlands in southern Sweden. J Ecol 86:429–438

Buckley G, Mills J (2015a) Coppice silviculture: from the mesolithic to 21st century. In: Kirby KJ, Watkins C (eds) Europe's changing woods and forests: from wildwood to managed landscapes. CABI, Wallingford, pp 77–92

Buckley G, Mills J (2015b) The flora and fauna of coppice woods: winners and losers of active management or neglect. In: Kirby KJ, Watkins C (eds) Europe's changing woods and forests: from wildwood to managed landscapes. CABI, Wallingford, pp 129–139

Catchpole RDJ (2006) Planning for biodiversity – opportunity mapping and habitat networks in practice: a technical guide. English Nature Research Reports Number 687, Northminster House, Peterborough PE1 1UA

Crick H, Crosher I, Mainstone C, Taylor S, Wharton A, Langford P, Larwood J, Lusardi J, Appleton D, Brotherton P, Macgregor N (2020) Nature networks evidence handbook, Natural England Research Report 81. Natural England, Peterborough

Defra, Department for Environment, Food and Rural Affairs (2017) Wild bird populations in the UK 1970 to 2016. Defra, London

Ellenberg H (1988) The vegetation ecology of Central Europe. Cambridge University Press, Cambridge

Falinski JB (1986) Vegetation dynamics in temperate lowland primeval forests. W. Junk, Dordrecht

Farjon A (2017) Ancient oaks in the English landscape. Royal Botanic Gardens, Kew, Surrey. Forestry Commission (2019) Forestry Statistics 2019. Forestry Commission, Edinburgh

Forman RTT, Godron M (1986) Landscape ecology. Wiley, New York

Francis JL, Morton A (2001) Enhancement of amenity woodland field layers in Milton Keynes. British Wildlife 12:244–251

Goldberg EA (2015) The UK's ancient woodland inventory and its use. In: Kirby KJ, Watkins C (eds) Europe's changing woods and forests: from wildwood to managed landscapes. CABI, Wallingford, pp 326–336

Goldberg EA, Peterken GF, Kirby KJ (2011) Origin and evolution of the ancient woodland inventory. British Wildlife 23:90–96

Gove B, Power SA, Buckley GP, Ghazoul J (2007) Effects of herbicide spray drift and fertilizer overspread on selected species of woodland ground flora: comparison between -term and long-term impact assessments and field surveys. J Appl Ecol 44:374–384

Harding PT, Rose F (1986) Pasture-woodlands in Lowland Britain. Institute of Terrestrial Ecology Abbots Ripton, Huntingdon

Harmer R, Peterken GF, Kerr G, Poulton P (2001) Vegetation changes during 100 years of development of two secondary woodlands on abandoned arable land. Biol Conserv 101:291–304

Hill MO, Preston CD, Roy DB (2004) PLANTATT: attributes of British and Irish plants. Biological Records Centre (NERC), Cambridge

Irmler U, Arp H, Nötzold R (2010) Species richness of saproxylic beetles in woodlands is affected by dispersion ability of species, age and stand size. J Insect Conserv 14:227–235

Jonsell M, Schroeder M, Larsson T (2003) The saproxylic beetle *Bolitophagus reticulatus*: its frequency in managed forests, attraction to volatiles and flight period. Ecography 26:421–428
Kimberley A, Blackburn GA, Whyatt JD, Kirby KJ, Smart SM (2013) Identifying the trait syndromes of conservation indicator species: how distinct are British ancient woodland indicator plants from other woodland species? Appl Veg Sci 16:667–675
Kirby KJ (2001) The impact of deer on the ground flora of British broadleaved woodland. Forestry 74:219–229
Kirby KJ (2015) What might a sustainable population of trees in wood-pasture sites look like? Hacquetia 14:43–52
Kirby KJ, Perry SC (2014) Institutional arrangements of wood-pasture management: past and present (UK). In: Hartel T, Pleininger T (eds) European wood-pastures in transition: a social-ecological approach. Earthscan/Routledge, Abingdon, pp 254–270
Kirby KJ, Watkins C (eds) (2015) Europe's changing woods and forests: from wildwood to managed landscapes. CABI, Wallingford
Kirby KJ, Smart SM, Black HJ, Bunce RGH, Corney PM, Smithers RJ (2005) Long-term ecological changes in British woodland (1971-2001), English Nature Research Report 653. English Nature, Peterborough
Kirby KJ, Pyatt DG, Rodwell JS (2012) Characterization of the woodland flora and woodland communities in Britain using Ellenberg values and functional analysis. In: Rotherham ID, Jones M, Handley C (eds) Working and walking in the footsteps of ghosts: volume 1 the wooded landscape. Wildtrack Publishing, Sheffield
Latham J, Sherry J, Rothwell J (2013) Ecological connectivity and biodiversity prioritisation in the terrestrial environment of Wales. CCW Staff Science Report No. 13/3/3, Terrestrial Ecosystem Group, Countryside Council for Wales, Bangor
Lawton JH, Brotherton PNM, Brown VK, Elphick C, Fitter AH, Forshaw J, Haddow RW, Hilborne S, Leafe RN, Mace GM, Southgate MP, Sutherland WJ, Tew TE, Varley J, Wynne GR (2010) Making space for nature: a review of England's wildlife sites and ecological network. Defra, London
Lonsdale D (2013) Ancient and other veteran trees:further guidance on management. The Tree Council, London
Montagnini F, Levin B, Berg KE (2022) Introduction. Biodiversity Islands: Strategies for conservation in human-dominated environments. In: Montagnini F (ed) Biodiversity islands: strategies for conservation in human-dominated environments. Topics in Biodiversity and Conservation, Springer, Cham, pp 1–35
Moore NW, Hooper MD (1975) On the number of bird species in British woods. Biol Conserv 8: 239–250
Peterken GF (1974) A method for assessing woodland flora for conservation using indicator species. Biol Conserv 6:239–245
Peterken GF (1977) Habitat conservation priorities in British and European woodlands. Biol Conserv 11:223–236
Peterken GF (1993) Woodland conservation and management, 2nd edn. Chapman and Hall, London
Peterken GF (1996) Natural woodland: ecology and conservation in northern temperate regions. Cambridge University Press, Cambridge
Peterken GF (2000) Identifying ancient woodland using vascular plant indicators. British Wildlife 11:153–158
Peterken GF (2002) Reversing the habitat fragmentation of British woodlands. WWF-UK, Godlaming
Peterken GF, Allison H (1989) Woods, trees and hedges: a review of changes in the British countryside. Nature Conservancy Council, Peterborough
Peterken GF, Game M (1984) Historical factors affecting the number and distribution of vascular plant species in the woodlands of Central Lincolnshire. J Ecol 72:155–182

Pickett STA, Thompson JN (1978) Patch dynamics and the design of nature reserves. Biol Conserv 13:27–37

Pitcairn C, Skiba U, Sutton M, Fowler D, Munro R, Kennedy V (2002) Defining the spatial impacts of poultry farm ammonia emissions on species composition of adjacent woodland groundflora using Ellenberg Nitrogen Index, nitrous oxide and nitric oxide emissions and foliar nitrogen as marker variables. Environ Pollution 119:9–21

Rackham O (1976) Trees and woodland in the British landscape. Dent, London

Rackham O (2003) Ancient woodland: its history, vegetation and uses in England (revised ed.). Castlepoint Press, Dalbeattie

Riutta T, Slade EM, Bebber DP, Taylor ME, Malhi Y, Riordan P, Macdonald DW, Morecroft MD (2012) Experimental evidence for the interacting effects of forest edge, moisture and soil macrofauna on leaf litter decomposition. Soil Biol Bioche 49:124–131

Riutta T, Slade EM, Morecroft MD, Bebber DP, Malhi Y (2014) Living on the edge: quantifying the structure of a fragmented forest landscape in England. Landsc Ecol 29:949–961

Rose F (1976) Lichenological indicators of age and environmental continuity in woodlands. In: Brown DH, Hawksworth DL, Bailey RH (eds) Lichenology: progress and problems. Academic, London

Rose F (1993) Ancient British woodlands and their epiphytes. British Wildlife 5:83–93

Scheidegger C, Werth S (2009) Conservation strategies for lichens: insights from population biology. Fungal Biol Rev 23:55–66

Sheail J (1998) Nature conservation in Britain - the formative years. The Stationery Office, London

Shreeve T, Mason C (1980) The number of butterfly species in woodlands. Oecologia 45:414–418

Sims NK, John EA, Stewart AJ (2014) Short-term response and recovery of bluebells (*Hyacinthoides non-scripta*) after rooting by wild boar (*Sus scrofa*). Plant Ecol 215:1409–1416

Spencer JW, Kirby KJ (1992) An inventory of ancient woodland for England and Wales. Biol Conserv 62:77–93

Tjørve E (2010) How to resolve the SLOSS debate: lessons from species-diversity models. J Theoretical Biol 264:604–612

Usher MB, Brown AC, Bedford SE (1992) Plant species richness in farm woodlands. Forestry 65: 1–13

Vellend M, Verheyen K, Jacquemyn H, Kolb A, Van Calster H, Peterken GF, Hermy M (2006) Extinction debt of forest plants persists for more than a century following habitat fragmentation. Ecology 87:542–548

Vera FWM (2000) Grazing ecology and forest history. CABI, Wallingford

Warren M (1987) The ecology and conservation of the heath fritillary butterfly, Mellicta athalia. III. Population dynamics and the effect of habitat management. J Appl Ecol:499–513

Watts K, Whytock RC, Park KJ, Fuentes-Montemayor E, Macgregor NA, Duffield S, Mcgowan PJ (2020) Ecological time lags and the journey towards conservation success. Nature Ecol Evol 4: 304–311

Worrell R, Francis J (2003) Establishing woodland flora in broadleaf and conifer woodlands. Scott For 57:203–210

Yalden D (1999) The history of British mammals. Poyser, London

# Chapter 14
# Natural Landscape of the Pampa Region in Santa Fe Province, Argentina: Environmental Resilience and Opportunity for Changing the Agri-Food Paradigm

**Libertario Hugo González, Germán Neffen, Victoria Benedetto, Marta Sánchez Miñarro, Andrea García, Ricardo Biasatti, Pablo Rimoldi, Cristian Alesio, and Daniel Paiz**

**Abstract** The Argentinean humid pampas were once a large grassland biome. Starting when the European colonizers arrived five centuries ago, and lasting until present, the original grassland ecosystems suffered an extreme transformation and have now almost completely disappeared. Today this region is under pressure from an "agriculture without farmers" dependent on external inputs which are increasingly scarce. This has serious social, environmental and economic implications. The

---

The human being disconnected from nature, conceived it as an object of knowledge, then as an object of domination and finally as a mere commodity. [...] With the advance of modernity, a paradigm of the individual was consolidated whose relationship with the "others" is one of domination and exploitation, a concept of the individual as an uprooted being, with weakened and fragmented collective ties (Gauna Zotter and Rey 2017)

---

L. H. González (✉) · V. Benedetto · A. García
Experiment Station (EEA) INTA Oliveros, Santa Fe, Argentina
e-mail: gonzalez.libertario@inta.gob.ar

G. Neffen
Agricultural Producer, Pergamino, Buenos Aires, Argentina

M. Sánchez Miñarro
Economist and Director of the School of Social Entrepreneurships, Municipality of Rosario, Santa Fe, Argentina

R. Biasatti
Faculty of Veterinary Sciences, National University of Rosario, Santa Fe, Argentina

P. Rimoldi
Chair of General Zoology, Faculty of Agricultural Sciences, National University of Rosario, Santa Fe, Argentina

C. Alesio · D. Paiz
Chair of Biology and Ecology, Faculty of Veterinary Sciences, National University of Rosario, Santa Fe, Argentina

 
F. Montagnini (ed.), *Biodiversity Islands: Strategies for Conservation in Human-Dominated Environments*, Topics in Biodiversity and Conservation 20,
https://doi.org/10.1007/978-3-030-92234-4_14

landscape matrix has been increasingly alternating its seasonal composition between chemical fallows and monocultures of wheat, corn, and soybean. If allowed to continue, this artificialization of ecosystems, dependent on fossil fuels, chemical inputs, and biotechnology, represents a serious threat to present and future social and environmental wellbeing. We have identified surviving land areas that persisted in the increasingly anthropic landscape of this region, constituting patches of biodiversity referred to in this chapter as "islands of resilience." In this chapter, we propose a multidisciplinary approach which utilizes the existing resilience islands in rural areas and in bordering urban zones. We analyze a case study located in the city of Las Rosas (Santa Fe province, Argentina) and the land's current and potential ecological wellbeing. We examined several socioeconomic and agroecological strategies associated with the production of local healthy food supplied to simplified marketing frameworks. These strategies are developed with production-consumption, community, and cooperative practices in mind. In this case study, we are guided by principles of cooperation, mutual support and solidarity, which build the foundation for an eco-social and resilient transition to more widespread biodiversity friendly agricultural landscapes.

**Keywords** Agri-food system · Agroforestry systems · Biodiversity Islands · Common goods · Landscape ecology · Production models · Sustainability

## 14.1 Introduction

The term "islands of resilience" arises from empirical observation of the recurrent vegetation that stands out in industrial agriculture's landscape matrix. The tree, shrub and herbaceous composition of these islands is unexpectedly dynamic, making the islands a symbol of resistance to a dominant agricultural production model that transformed natural ecosystems. These islands or patches have varied forms, including circular, ovoid and irregular. They contrast with the extensive degraded checkerboard plains of the humid pampas of southern Santa Fe, Argentina (Fig. 14.1).

In addition to disrupting human-dominated land, resilience islands also reflect the complex structure and functionality of ecosystems that maintain stability, despite external disturbances. Satellite images of these resilience islands can allow observers to develop their own reasoning for the persistence of the patches. But it is in situ, when travelling along provincial and national routes, when one can appreciate their presence and observe the different stages of plant succession in which these relics are found. This chapter provides a detailed description of the history of the landscape and the current state of ecological wellbeing of the islands in order to offer a logical explanation for their persistence.

In this chapter we describe the change over time of the natural landscape matrix of the Argentinean humid pampas. This land transformation is depicted through the key historical events of the southern region of the Santa Fe province, from the original

**Fig. 14.1** Island of resilience inserted in the matrix of the dominant landscape. (Photo: Pablo Olive)

**Fig. 14.2** Ecosystems recovery capacity in the Pampean region of Santa Fe, Argentina. (Photos: Ricardo Biasatti) (see text for explanation)

predominance of the pastures to the implementation of an extensive system of monocultures. This chapter utilizes research and information on the flora and fauna that are crucial to aiding in the regeneration of resilience islands. In addition, the chapter integrates agroecological proposals based on the experience of the agroecology module of the National Institute of Agricultural Technology (INTA, Oliveros Experiment Station) (Benedetto et al. 2017). These publications and proposals demonstrate the capacity of ecosystems to recover their structural and functional parameters after land degradation due to external pressures (Biasatti et al. 2013, 2016) (Fig. 14.2).

The images in Fig. 14.2 correspond to the physiognomy of the landscape that is the subject of this work. The first image represents the matrix of the landscape that predominates in the region, in which the agro-productive activity has entirely modified the environment. The second image shows a representative example of an island of biodiversity. The third image shows the role of biological corridors (a stream in the area) contributing to landscape connectivity. For our case study, this capacity for recovery or resilience is a starting point for developing a new model of the pampean rurality, including agricultural production areas with greater biodiversity and landscape connectivity.

## 14.2 Passages from the History of the Santa Fe Pampa Region

### *14.2.1 The New Configuration of the Landscape Matrix*

Our analysis of landscape transformation in the Pampa region included climate change as an influencer of structural change in the landscape matrix, and humans in their role as decision makers on whether to work with nature or oppose it. In the study region, indigenous people used to work with nature. Evidence of mega-fauna grazing activities, dated at circa (ca.) 12,200–10,000 years, proves that the inhabitants of the Pampean zone were hunters (Politis et al. 2014). Their assimilation to the territory and movement in search of food was part of the pre-Hispanic culture before the Spanish conquest of the region.

The year 1492 was a turning point in the history of this hemisphere of the world. America did not appear on any map prior, so the Spanish and Portuguese, as the first European occupants, named the continent and initiated a territorial organization similar to that already existing in Spain and Portugal (Mignolo 2007). The European arrival in the Americas was a crucial event that led to a history of European imperial expansion and developed the existence of a European lifestyle that became prominent in Argentine society. The Americas emerged in the European consciousness as a great expanse of land to be conquered with a people to be evangelized and exploited (Mignolo 2007).

The start of imperialistic expansion into this region led to the first signs of the fragmentation of its pristine landscape. This event initiated the fragmentation seen throughout the history of the Pampa, where the colonizers began to transform the unaltered South American landscape as they had similarly done in their own land in Europe. This fragmentation was due in part to the introduction of new animal species such as cattle and horses by the Spanish upon their arrival, which significantly altered the ecosystems of the pampean grassland biome (Brailovsky 2006).

### 14.2.2 Territory Configuration or Landscape Fragmentation?

In two and a half centuries of colonial hegemony, the independence movement, which started in 1810 in Argentina, grew significantly due to an economic power imbalance between the locals and the colonizers (Donghi 2014). At the beginning of the nineteenth century, a new configuration of the economy and politics was structured around Buenos Aires, Argentina, in the context of an economic junction favorable to the trade of cattle hides. The litoral region (riverine coastal area) which had been exhausted as a supplier of leathers for overseas markets, was replaced by the Buenos Aires markets, which profited as the cattle industry rapidly expanded, providing an opportunity for territorial expansion (Donghi 2000).

As the wealth and power of Buenos Aires increased, its political decision makers planned and executed military campaigns of extermination and subjugation of the native inhabitants, with the aim of expanding the border that was under their control. The result of these campaigns was an immense increase of available land, which the Argentine government publicly offered in "emphyteusis". This system consists of a contract for the use of the land in exchange for taxing the production. This implies the temporary cession of the land, but use of this land ultimately became perpetual, requiring very low tax payments (Donghi 2000).

However, these newly acquired lands were only granted to landowners of the lands of old colonization, political and military leaders of the provinces, and holders of urban mercantile wealth. None of the lands' original inhabitants were given parcels. These elite groups formed the social nucleus of the landowning class (Donghi 2000). In this way, the system of latifundia (mega-landowners) was consolidated, resulting in the fragmentation of the landscape matrix, as proprietors implemented barriers such as fences to contain the cattle and to delimit private property.

### 14.2.3 Benefit or Penalty? Geopolitics of Sacrifice

At the end of the nineteenth century, the Argentine government, led by President Nicolás Avellaneda, continued its military campaigns, taking the lands of the aborigines to be exploited. However, as the Argentine government expanded its control over land in the region, the available territory to be conquered diminished. Finally, on October 5, 1878, Law 947 was passed in order to obtain funds for the "Conquest of the Desert" to further expand the territory under Argentine control (Del Corro 2003). Avellaneda affirmed that the elimination of Native American populations in the frontiers was meant to allow Argentine people to "populate the desert." He defined settling as the replacement of the original inhabitants who were not linked to the national and international markets by others who were. Avellaneda further explained: "to govern is to populate, and to populate is to change nomads for sedentary salaried workers to develop a new livestock production model" (Brailovsky and Foguelman 2009).

Under the impetus of a sustained British demand for foods and raw materials, the leading sectors oriented the use of natural resources to meet global market demands (Brailovsky and Foguelman 2009). Thus, between 1860 and 1930 Argentina entered the global market as a producer of meat, wool and cereals. Argentina's growing involvement in international markets led the country to be considered "the granary of the world" (Brailovsky and Foguelman 2009). The same president, two years before Law 947 was passed, passed law N° 817, which regulated immigration and colonization, helping to provide the region with a labor force and further grow Argentina's economy (Sili and Soumoulou 2011).

As a consequence, at the beginning of the twentieth century, the Argentine agricultural sector had been separated into two distinct facets. One was the ranchers who owned the land, most of them dedicated to cattle raising. The other was small and medium-sized farmers who produced cereals and lived in the colonies, either using their own land or renting out plots in the large ranches (Sili and Soumoulou 2011).

## 14.2.4 *History of Agriculture in the Santa Fe Pampa*

The history of agriculture in the Santa Fe province of Argentina is strongly tied to the history of the work of immigrant families and their descendants (Cloquell et al. 2007). These actors played a leading role in the transformation of the landscape, which is evident in their productive activities and in the increase in population over time. Over almost 40 years, the proportion of immigrants grew from 10% of the total population in 1858 to 42% in 1895 (Cloquell et al. 2007).

While older production methods did contribute to the fragmentation of the landscape matrix, they were not as harmful as subsequent production methods utilized at the end of the 20th and beginning of the twenty-first century. In contrast, the first settlers' belief toward nature was that they not only "lived in a territory," but they "lived from the territory" they inhabited. This ideology explains why these original landowners did not decimate their habitat through their productive actions because of their care for the environment.

### 14.2.4.1 Argentina as "The Granary of the World"

The expansion of Argentina's borders, the introduction of machinery, and massive immigration caused exportation of cereals and flax to increase, allowing Argentina to prosper greatly and become known as "the granary of the world" (Volkind 2009). Unlike the United States or Canada, Argentina's agricultural policy did not encourage small and medium independent producers, but rather aimed at incorporating the land into the national agricultural economy by creating incentives for owners to put it into production (Bidaseca and Lapegna 2006). This resulted in an asymmetric relationship between the producer, who leases land, and the owner, who signs short-term leases and has the flexibility to frequently increase rental fees (Cloquell et al. 2007). However, producers held the belief that they were lessees only temporarily, as one must first lease the property before eventually owning it. Their goal was to transition from being a part of the labor force to being owner-producers, thereby transitioning their immigrant status (Cloquell et al. 2007).

It is important to note that some contemporaries of the time, such as the geographer and naturalist Kropotkin, in his 1892 book "The Conquest of Bread," warned of what was happening in that region of the world. He explained that as the productivity of land increases, the workers' profits do not increase to the same degree as the profits of the landowners and the government. This inequity results from an increase in taxes towards the farmers as their production rises (Kropotkin 1977).

This era could have been a time in Argentina's history for equitable redistribution of land among farmers, in consonance with the social struggles that were taking place, in which the access, use, and occupation of the land for production and habitation were disputed. However, in the framework of the international economic crisis of 1930, a mass migration of farmers occurred from the countryside to the city due to their financial fragility and dependence on international prices, resulting in the concentration of land into the hands of few (Cloquell et al. 2007).

Towards 1940, a new government policy enabled productive units under lease to be bought by the tenant producer, either by benefits of credit policies, or by tacit agreements due to the situation (Cloquell et al. 2007). Despite the struggle for control of this land, the Pampean landscape was still relatively untamed up to this point in time. However, the land was subsequently quickly transformed and fragmentated. The islands of resilience, which are the subject of this chapter, are surviving relics of this process, and they are exemplified by our case study, the farm located in Las Rosas, home to the Fontana family, in the south of the Santa Fe province.

These trends of transformation and fragmentation continued to the end of the 1950s, when the national political and economic situation changed abruptly. This societal change was brought about by an international context that, concluding the post-war reconstruction stage, sought to reduce the role of the government in the national economy and to promote the liberation of economic relations (Cloquell et al. 2007). Along with this change, new strategic scientific and technological advances were promoted. Agricultural production was given strong incentives to fulfill its traditional role as a provider of foreign exchange to the national economy. These incentives launched a new phase of capital accumulation that furthered Argentina's role in the international market (Cloquell et al. 2007).

#### 14.2.4.2 The "Modernization" of the Agricultural Sector

Around 1960, the region experienced increases in production and productivity based on the application of the industrial-based technological model. This began the process of "modernization," which led the agriculture sector to maximize the process of capital accumulation, with different impacts on the social actors of the agriculture industry (Cloquell et al. 2007).

In our case study island of resilience in Las Rosas, the recollections of a family member, Angel Fontana, serve to illustrate this transition. As Angel explains, in his farm they used to share equipment with neighbors, using herbicides and planting non-hybrid seeds. Later on, due to the challenges of harvesting crops in the 1970s, in addition to the impacts of weeds, insects, and other pests, they incorporated dairy farming to have a more regular income. Over time, they began to rely more on additional contract workers and rental equipment to face increasing challenges.

As Angel relates, "In those days, in some farms they worked the land with horses, but not in ours where there was already a small tractor, which we shared with other family members so that they could work on their own fields. In the early days, some seed was selected at the time of harvest for replanting, but little by little the commercial hybrid seed gained ground. Herbicides were there as long as I can remember, but the fertilizers arrived much later. Being almost a teenager, we started with my father a small dairy farm to get a monthly income that made it possible for us not to wait until the agricultural harvest. Milk delivery was not a problem, since the creameries sent the truck to the house. This experience was from 1970 to 1980, when I saw different crops grow, potatoes, watermelons, melons and fruit trees. Over time it was impossible to fight off weeds, insects and other pests. In the sixties, someone used to pass by with a vehicle to get eggs, milk, and chickens and offered in exchange sugar, and oil in smaller quantities because we used pork fat (the product of periodic meatings). Then I saw the incorporation of wheat, corn and soybean planting. The soil was broken and planting was implemented with the use of zero-tillage or direct sowing. Through the years everything changed. From having our own tools we moved on to the work done by a contractor and rental that continues to this day" (Angel Fontana, personal communication).

This first-person experience provides evidence of the gradual incorporation of new agricultural technologies, collectively known as "the green revolution". The integration of these technologies was promoted by government agencies, facilitating the advancement of private companies into the region. This integration of new inputs and technologies, which was more prevalent in crops than in cattle raising, began in the second half of the 1950s, grew in the 1970s and 1980s, and was accentuated in the 1990s (Pizarro 2003).

With the start of the green revolution, new agricultural practices began to be utilized across the landscape matrix in order to increase profits. Among these new practices, the increasingly common use of monocultures and limited productive diversification led to a destructive occupation of the land. In fact, other qualities of the land, such as its social or cultural benefits, were disregarded as productive value became the dominant productive focus. In this mindset for the use of the territory, land was considered socially disposable, simply designating it as "sacrifice areas" for the sake of progress (Svampa and Viale 2014).

Agribusiness, and especially transgenic soybeans (*Glycine max*), is currently the core of Argentina' s extractive matrix. Currently, the country is among the four main world producers of transgenic soybeans with almost 24 million hectares cultivated. Since the end of the 1990s, when the use of transgenic soy was approved, the expansion of agribusiness was tied to the use of genetically-modified organisms (GMOs), which led to a global restructuring of the traditional agrarian ways (Svampa and Viale 2020).

This new agricultural model spread not only in the Pampas region, but also in marginal areas, such as in the northern and coastal regions of Argentina. Today it occupies 23 to 33 million hectares of land, of which 90% is dedicated to soybeans. The great concentration of transgenics and agrochemicals across this soybean-dominated land presents dangers to the country due to prolonged exposure to these chemicals (Svampa and Viale 2020). The massive scale of this exposure within the territory and on the bodies of people constitutes a true model of "bad-development" similar to the Chernobyl case, which illustrates the dangers and their disregard by governments and economic actors involved (Svampa and Viale 2020).

### 14.2.5 Resilience of the Santa Fe Pampean Ecosystem

Throughout the history of the Pampa plains of the Santa Fe province, there have been significant changes to the landscape, the results of which will still play out over years to come. The era in which human actions on their environment started to dominate the landscape led to the designation of the time of the Anthropocene (Herrero 2017). During this era, the Pampean grassland biome was transformed into an agroecosystem in which the economic-financial goals were prioritized. This prioritization disregarded the multiple dimensions that must be addressed for a productive system that is sustainable over time and that allows humans to monitor their actions on the environment and its natural commons such as water, soil, air, and biodiversity.

Our case study shows that the establishment of resilient systems in the Pampean grasslands, with the aim of restoring biotic communities congruent with the original landscape physiognomy, has already been observed in the region. While we were performing our work for this chapter, some restrictions on land use were established. This allowed for the development of plant secondary succession that in relatively short periods of time (2 to 4 years) has shown the system's capacity to recover spontaneous biodiversity, increase soil coverage, and allow a successive occupation of the space with progressive substitution of species (Biasatti et al. 2013, 2016).

Other observations of these resilience islands have revealed their potential for developing new specific ecological niches and a variety of new habitats or micro-habitats according to local soil, hydrological and even climatic variations (Biasatti et al. 2013, 2016). This evidence of the land's persistence encourages us to assess the viability of this land for the development of agroecologically-based production practices. These new practices would help strengthen food sovereignty and security, improving human welfare and benefitting other forms of life in the context of the dominant agricultural matrix. In addition, this proposal could be extrapolated to other regions of the country and to other parts of the world which are withstanding similar ecological conditions.

The next sections of the chapter offer a characterization of the islands of resilience, focusing on their flora and fauna, and arguing for their integration into the landscape matrix. We propose to apply agroecological principles to crop management around the islands, and we relate experiences of local markets using agroecological production. The path we choose, among the different paths that intersect in human development, can undoubtedly determine the future of our species for centuries to come (Bookchin 2019).

## 14.3 The Islands of Resilience and their Context in South Santa Fe

### *14.3.1 Recovering Areas with Real Potential for Hosting Biodiversity*

In this section, islands of resilience are defined from the perspective of landscape ecology as "patches", as they are areas with greater biodiversity than the surrounding anthropic matrix, which is dominated by an extensive homogenized and simplified territory associated with the agro-industrial production model or system of soy cultivation (Forman and Grodon 1986, Biasatti et al. 2007). These patches can take many forms, such as the one described in this chapter, as well as marginal lands, floodable lands, and lands with limitations for agriculture.

Our case study provides evidence of the depopulation of rural areas due to the development of soybean production methods over the last decades. As previously explained, in the Argentinean Humid Pampas, farming originally consisted of mixed production, with the producer's family living in the countryside and producing a variety of goods and foods. The arrival of soybean cultivation introduced a new model of production with less reliance on people, which led to the migration of the rural settler and the abandonment of their homes and surroundings. This new model of agricultural production, which heavily relied on the soybean, replaced diverse cropping systems and became dominant.

These abandoned settlements were largely left untouched, which allowed for the development of plant secondary succession, sheltered by the infrastructure that had been left behind and its generally wooded surroundings. The abandoned areas contained exotic species such as fruit trees and others that would provide shade, shelter from the winds and some wood products, such as firewood for cooking and heating.

Not all of these abandoned relics have been preserved. However, it is still common to observe these "patches" in the pampas landscape of monoculture, where nature reveals its biotic potential. This shows the capacity of resilience of the Pampean grasslands to recolonize these relics through successive occupation of space with spontaneous species, starting with pioneers and then moving into stages of greater forest maturity.

The province of Santa Fe has several examples that show the resilience of the Pampean grasslands. Government strategies have created several biological corridors, rescuing relics of the original ecosystems. These have quickly shown their effectiveness for restoration, not only of plant species, but also of animals, including birds and mammals of medium and large size. These species also can take advantage of corridors to move around the landscape and establish themselves in recovered spaces.

The conservation model that utilizes these strategies is based on the configuration of a "reticulated system for the conservation of biological diversity" through the implementation of biological corridors around lotic water bodies (rivers and streams) or communication routes, roads, and railways (Cracco and Guerrero 2004, Biasatti et al. 2013). The implementation of this proposal was carried out by preventing human intervention in the areas of study and through the monitoring and management of spontaneous secondary succession. In none of the case studies was secondary succession induced by the planting of species or the transfer of animals to increase biological richness. Instead, succession was self-generated by the favorable conditions created by protective barriers. Corridors also aided in ecological succession by connecting previously isolated relics which developed into existing patches, or "islands of resilience."

Due to the large-scale agricultural use of the humid Pampa, human intervention has led to a significant transformation, including the creation of "resiliency islands" in an agricultural landscape. Human domination in this region has reached levels that exceed 95% of the available land used for productive purposes or for urban areas. The large scale of human occupancy has transformed the landscape into a new anthropic and homogeneous matrix. Contrary to other regions, the strategy necessary to conserve biodiversity and restore land in the humid pampas requires the fragmentation of the anthropized matrix. This phenomenon was studied and developed under the concept of "inverse fragmentation," within the framework of a strategy aimed at establishing mechanisms compatible with the joint practices of production and conservation (Biasatti et al. 2019).

"Resiliency islands" are key elements of this strategy because they help recover spaces with real potential to host biodiversity. Many animal species typical of the Pampean grasslands, such as the Pampean fox (*Lycalopex gymnocercus*), the wild cat (*Leopardus geoffroyi*), or the coipo (*Myocastor coypus*) are threatened because they do not cross cultivated environments. However, the presence of natural vegetation corridors allows these species to reach these biodiversity patches, aiding in

**Fig. 14.3** The island of resilience studied and its immediate environment, highlighting the contrast in the configuration of the landscape of the region. (Google image)

species longevity (Rimoldi and Chimento 2018, Biasatti et al. 2019). Plant species can also thrive in these areas, moving from one habitat patch to another as they are carried by wind or animals that move seeds or propagules (eggs, larvae, vegetatively reproducing plant parts) through corridors. This dispersal mode has the additional advantage of enriching the corridors. Biological corridors facilitate the movement of native species even in landscapes that are severely hostile due to their high degree of homogenization and habitat simplification generated by the expansion of the agricultural frontier.

Since the aesthetic or ethical appeal for protection alone may not suffice, the conservation of "resiliency islands" requires the application of sustainable productive practices. In addition to providing potential productive benefits, these islands contribute to economic development by maintaining species and their genetics, aiding in species diversification and promoting variability in an otherwise uniform landscape (Biasatti et al. 2016, Biasatti et al. 2017).

The previously mentioned model of a "reticulated system for the conservation of biological diversity" constitutes a strategy that relies on islands of resilience integrated into a network of connectivity devices designed as biological corridors. This relationship between patches and corridors creates a system capable of strengthening both the dispersion and the colonization of spontaneous species in a highly anthropized productive region that invests material and energy in maintaining broad spaces with minimal biodiversity (i.e. crops). "Resiliency islands," by their definition, are isolated from other biodiversity, as can be seen in Fig. 14.3. However,

**Fig. 14.4** The island of resilience and its potential links to strengthen a system that increases connectivity, allows the transfer of spontaneous species, and promotes biodiversity and coexistence with alternative production systems. (Satellite image, Google Earth font)

these patches can be part of a large, interconnected web of islands that utilize corridors (Fig. 14.4). This network constitutes a viable and effective alternative for establishing a new system of production that allows for the conservation of regional biodiversity. From the spatial unit of the island of resilience that is the subject of this chapter (see Fig. 14.1) it is possible to extend the scope by zooming out to a regional scale in which the sum of other islands linked by biological corridors constitutes a viable and effective alternative for establishing new models of production and conservation, as can be seen in Fig. 14.3.

### *14.3.2 Case Study in the Center-South of the Province of Santa Fe*

Our case study is an island of resilience that covers an area of approximately one hectare, located in the south-central province of Santa Fe (32°31′06″S 61°33′26″W), near the town of Las Rosas and a few meters from National Route No. 178 (Fig. 14.5). With regards to purely anthropic structures, there is a mill and

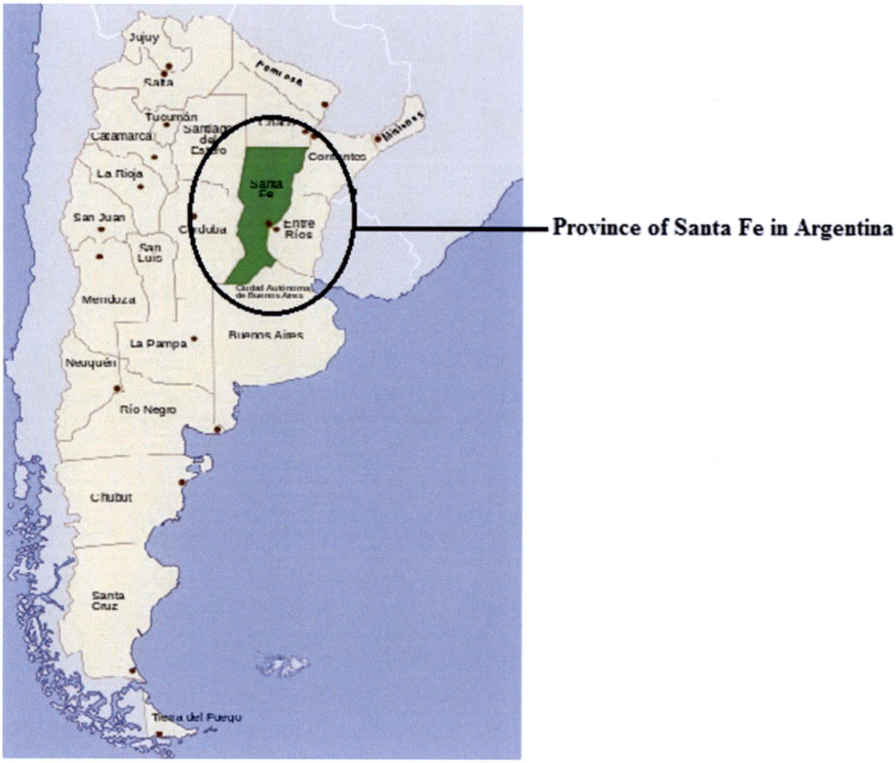


**Fig. 14.5** Province of Santa Fe in Argentina. (Google Image)

a small "tapera" (abandoned rural housing), which can potentially be used as a refuge by the local fauna.

This area of study corresponds to the "humid pampas" or "pampas grasslands" biome which is in a transition to an "espinal" (thorny scrub) biome. The region has a temperate climate of marked seasonality, average temperatures between 17–18 °C, an annual rainfall close to 1000 mm, and soils with a low slope and a well-developed A horizon. The land in this area has good quality and productive potential, which has led to its transformation for agricultural purposes.

In the studied island of resilience, the dominant tree species is the paper mulberry (*Broussonetia papyrifera*). Other less abundant species include eucalyptus (*Eucalyptus* sp.), paradise tree (*Melia azedarach*), evergreen (*Ligustrun* sp.), three-thorned acacia (*Gleditsia triacanthos*), olive tree (*Olea europaea*), peach tree (*Prunus persica*), fig tree (*Ficus carica*), *Cereus* sp., ceibo (*Erythrina crista-galli*), ombú (*Phytolacca dioica*), and palo borracho (*Chorisia* sp.). At one edge of the island there is a small group (about 20m$^2$) of canes (*Arundo donax*). Young individuals of cina cina (*Parkinsonia aculeata*) and carob tree (*Prosopis* sp.) were planted to enrich the patch with native species and fruit trees like apple trees (*Malus domestica*) for their food production (Table 14.1). Below the tree canopy and in areas without tree cover, communities of herbaceous species cover the soil.

**Table 14.1** Plant species recorded in island of resiliency in Las Rosas, Santa Fe, Argentina

| Family | Scientific name | Origin/ Observations |
|---|---|---|
| Amaranthaceae | *Amaranthus hybridus* L. *ssp. Hybridus* | Native/ Direct observation |
| Apiaceae | *Bowlesia incana* Ruiz &Pav. | Native/ Direct observation |
| Asteraceae | *Ambrosia cumanensis* Kunth | Exotic/ Direct observation |
| | *Artemisia absinthium* L. | Exotic/ Direct observation |
| | *Carduus acanthoides* L. | Exotic/ Direct observation |
| | *Carduus nutans* L. | Exotic/ Direct observation |
| | *Conyza bonariensis* (L.) Cronquist *var. bonariensis* | Native/ Direct observation |
| | *Cynara scolymus* L. | Exotic/ Direct observation |
| | *Helianthus tuberosus* L. | Exotic/ Direct observation |
| | *Sonchus oleraceus* L. | Exotic/ Direct observation |
| Boraginaceae | *Borago officinalis* L. | Exotic/ Direct observation |
| Brassicaceae | *Raphanus sativus* L. | Exotic/ Direct observation |
| Cactaceae | *Cereus argentinensis* Britton & Rose | Native/ Direct observation |
| Commelinaceae | *Commelina erecta* L. | Native/ Direct observation |
| Cannaceae | *Canna indica* L. | Native/ Direct observation |
| Convolvulaceae | *Ipomoea grandifolia* (Dammer) O´Donell | Native/ Direct observation |
| Euphorbiaceae | *Euphorbia peplus* L. | Exotic/ Direct observation |
| Fabaceae | *Parkinsonia aculeata* L. | Native/ Direct observation |
| | *Erythrina crista-galli* L. *var. crista-galli* | Native/ Direct observation |
| | *Prosopis alba* Griseb. *var. alba* | Native/ Direct observation |
| | *Gleditsia triacanthos* L. | Exotic/ Direct observation |
| Malvaceae | *Sida rhombifolia* L. | Native/ Direct observation |
| | *Ceiba speciosa* (A. St.-Hil.) Ravenna | Native/ Direct observation |

(continued)

**Table 14.1** (continued)

| Family | Scientific name | Origin/ Observations |
|---|---|---|
| Meliaceae | *Melia azedarach* L. | Exotic/ Direct observation |
| Moraceae | *Broussonetia papyrifera* (L.) Vent. | Exotic/ Direct observation |
| | *Morus alba* L. | Exotic/ Direct observation |
| | *Ficus carica* L. | Exotic/ Direct observation |
| Oleaceae | *Olea europaea* L | Exotic/ Direct observation |
| Phytolaccaceae | *Phytolacca dioica* L. | Native/ Direct observation |
| Poaceae | *Bromus catharticus* Vahl *var. Catharticus* | Native/ Direct observation |
| | *Sorghum halepense* (L.) Pers. *var. Halepensé* | Exotic/ Direct observation |
| | *Cynodon dactylon* (L.) Pers. *var. Dactylon* | Exotic/ Direct observation |
| | *Nassella neesiana* (Trin. &Rupr.) Barkworth | Native/ Direct observation |
| | *Arundo donax* L. | Exotic/ Direct observation |
| Rosaceae | *Malus domestica* Borkh. | Exotic/ Direct observation |
| | *Prunus persica* (L.) Batsch *var. Pérsica* | Exotic/ Direct observation |
| Rhamnaceae | *Hovenia dulcis* Thunb. | Exotic/ Direct observation |
| Solanaceae | *Salpichroa origanifolia* Baill. | Native/ Direct observation |
| | *Cestrum parqui* L'Hér. | Native/ Direct observation |
| Urticaceae | *Parietaria debilis* G. Forst. | Exotic/ Direct observation |
| | *Urtica urens* L. | Exotic/ Direct observation |
| Verbenaceae | *Aloysia citrodora* Palau | Native/ Direct observation |
| | *Aloysia polystachya* (Griseb.) Moldenke | Native/ Direct observation |
| Vitaceae | *Vitis vinifera* L. | Exotic/ Direct observation |

Biological corridors in this area that help increase connectivity are relatively simple yet successful interventions to promote the growth of spontaneous communities while also creating small refuge areas. These corridors often use plot edges or rural roads to create the naturally interconnected network described above without affecting the surrounding agricultural fields (Fig. 14.2; Biasatti et al. 2013).

The location of the island of study is within a larger system of similar islands connected through potential biological corridors, which allows for each patch to remain connected. This network provides an opportunity to study the relationship between landscape ecology, agroforestry systems, and agroecology models with scientific and epistemological support. The study of the intersection of these disciplinary fields provides valuable information that can be integrated into conservation strategies. This research adds to the already rich and diverse knowledge that those disciplinary fields supply, which serves to provide recommendations for combined systems that integrate trees, shrubs, and wild plants with crops and domestic animals. This intersection lays the foundation for local food security and food sovereignty (Montagnini et al. 2015).

## 14.4 The Role of Islands of Resilience in Wildlife Conservation

Loss of animal species diversity is a critical problem in both community ecology and conservation biology (Moreno et al. 2011). Its study has gained greater relevance globally in recent years as the issue has worsened due to human activities (Moreno et al. 2011).

As previously explained, the agricultural production model in this region of Argentina has had severe effects on biodiversity. The model of maximizing agricultural profits has led to a significant decrease in wellbeing of the natural environment and a change in the structure and functioning of the local ecosystem. Anthropic actions have had a clear effect on species' populations, wildlife in particular, affecting the abundance and range of species (Dirzo et al. 2014). Wildlife species have different levels of sensitivity to human alterations, depending on their space requirements, feeding needs, and reaction to anthropization (e.g., Fox and Fox 2000, Smith et al. 2000, Poiani et al. 2001, Abba et al. 2007).

The effect of anthropization on fauna is a multidimensional phenomenon, the study of which requires a multidisciplinary approach to best understand the dynamics and structure of species and predict their maintenance in modified environments. One important contributor to the perseverance of species in a rapidly anthropizing region is the presence of resiliency islands. These relics have the potential to serve as ecological sanctuaries and thus provide the minimum resources necessary to meet the biological needs of wildlife.

**Table 14.2** Animal species (mammals) recorded in island of resiliency in Las Rosas, Santa Fe, Argentina

| Common name | Scientific name | Origin/ Observations |
|---|---|---|
| Comadreja overa | *Didelphisal biventris* | Native/ Indirect records |
| Peludo | *Chaetophractus villosus* | Native/ Indirect records |
| Zorro | *Lycalopex gimnocercus* | Native/ Indirect records |
| Gato montés | *Leopardus geoffroyi* | Native/ Indirect records |
| Zorrino | *Conepatus chinga* | Native/ Indirect records |
| Hurón menor | *Galictis cuja* | Native/ Indirect records |
| Ratón del pastizal | *Akodon azarae* | Native/ Indirect records |
| Colilargo chico | *Oligoryzomys flavescens* | Native/ Indirect records |
| Laucha manchada | *Calomys laucha* | Native/ Indirect records |
| Laucha bimaculada | *Calomys musculinus* | Native/ Indirect records |
| Cuis | *Cavia aperea* | Native/ Indirect records |

### *14.4.1 Measuring Biodiversity in Resiliency Islands*

In order to measure the diversity of terrestrial vertebrates, indirect records such as footprints or caves/burrows left by the transit of some mammals were utilized. The methodology used was the delineation and examination of linear transects in search of signs of wildlife activity. These observations were made at an average speed of 500 m/hour, which allowed the observer to carefully examine the area of study. The field work was carried out between late winter and early spring.

Due to the presence of caves/burrows, we can say that several of the observed mammals can be considered "engineers" of the ecosystems because they influence the availability of resources for other species through changes in the state of the biotic or abiotic materials used to build their shelters (Jones et al. 1994). Although the concept of "ecological engineers" has been refuted because of its possible application to all organisms, it is useful as a reference for studying biotic interactions caused by changes to a habitat (Jones and Gutiérrez 2007). Excavating mammals, for example, modify their environment by creating entrances and conduits in the ground that cause changes related to the soil formation processes and increase the availability of shelter for other vertebrate species (Jones et al. 2006). Particularly in highly anthropized ecosystems, caves or burrows represent a valuable resource as a refuge to avoid predators and mitigate worsening environmental conditions. It is common to observe the reuse of a cave over time by individuals of the same or different species (Roldan and Saurthier 2016). The species that occupy these caves can save time and energy intended for building their own shelter or protecting themselves. For example, foxes take refuge mainly in caves and holes of different types and sizes that may belong to hairy armadillos (*Chaetophractus villosus*) or skunks (*Conepatus chinga*).

Utilizing these shelters as well as other footprints and signs, observations were made of a total of 11 species of mammals belonging to 8 families (Didelphidae, Dasypodidae, Canidae, Felidae, Mephitidae, Mustelidae, Cricetidae and Caviidae) and 4 orders (Didelphimorphia, Cingulata, Carnivora and Rodentia). These observed mammals include the following: 1 marsupial species, 1 xenarthran (an exclusively American placental mammal in a clade that include anteaters, armadillos, and sloths), 4 carnivores, and 5 rodents (Table 14.2). For amphibians and reptiles, no samples were taken during the period of observation, but tracks have shown the presence of the overo lizard (*Tupinambis merianae*).

Regarding birds, the records obtained allowed us to determine the presence of 39 species, which included 9 orders and 18 families (Table 14.3). This field work was carried out between late winter and early spring and focused on pre-established transects. Each transect had a fixed-radius of 10 meters and was 20 meters high, separated from each other by a distance of 25 meters. Intervals for counting specimens were 15 minutes long. Counts were made in the morning, starting 20 minutes after sunrise and up to no more than 4 hours after sunrise. The results allowed us to determine that the largest number of individuals observed belong to the order Passeriformes, which contributed 22 species, followed by the order Falconiformes and Columbiformes, with 4 species.

Bird nests also played an important role in the evaluation of species richness. The construction of nests for reproduction, hibernation and/or rest is a very common activity in birds. Bird nests are a large determining factor in the reproductive success of the nesters because of the role nests play in sexual selection, protection against potential predators, and maintenance of suitable temperature and pH level, which prevents the proliferation of pathogens. Due to the importance of nests in bird survival, the presence of nests in the study area led us to conclude that the existing resources satisfy the previously mentioned needs, which are nearly nonexistent in the surrounding matrix.

The existence of these small green patches in a dominant matrix that is highly disturbed by anthropic activity provides an opportunity for the recovery of favorable environmental conditions for the conservation of biodiversity. However, in order to enhance conservation, it is fundamental to establish the basis for future monitoring of the species involved in this study. Tracking species interactions within the ecosystem is crucial in evaluating the importance of the resiliency island in animals' feeding and breeding habits and therefore assessing the ecological importance of the islands in the region.

**Table 14.3** Animal species (Birds) recorded in island of resiliency in Las Rosas, Santa Fe, Argentina

| Common name | Scientfic name | Origin/ Observations |
|---|---|---|
| Paloma Doméstica | *Columba livia* | Exotic/ Direct Observation |
| Paloma Manchada | *Patagioenas maculosa* | Native/Directly Observed |
| Paloma Picazuro | *Patagioenas picazuro* | Native/Directly Observed |
| Torcaza Común | *Zenaida auriculata* | Native/Directly Observed |
| Pirincho | *Guira guira* | Native/Directly Observed |
| Chiflón | *Syrigmasi bilatrix* | Native/Directly Observed |
| Picaflor Verde | *Chlorostilbon lucidus* | Native/Directly Observed |
| Tero común | *Vanellus chilensis* | Native/Directly Observed |
| Milano Blanco | *Elanus leucurus* | Native/Directly Observed |
| Taguató Común | *Rupornis magnirostris* | Native/Directly Observed |
| Carpintero real Común | *Colaptes melanolaimus* | Native/Directly Observed |
| Carpintero Campestre | *Colaptes campestroides* | Native/Directly Observed |
| Carancho | *Caracara plancus* | Native/Directly Observed |
| Chimango | *Phalcoboenus chimango* | Native/Directly Observed |
| Halconcito Colorado | *Falco sparverius* | Native/Directly Observed |
| Halcón Plomizo | *Falco femoralis* | Native/Directly Observed |
| Cotorra | *Myiopsitta monachus* | Native/Directly Observed |
| Hornero | *Furnarius rufus* | Native/Directly Observed |
| Piojito Gris | *Serpophaga nigricans* | Native/Directly Observed |
| Benteveo Común | *Pitangus sulphuratus* | Native/Directly Observed |
| Suiriri Real | *Tyrannus melancholicus* | Native/Directly Observed |
| Churrinche | *Pyrocephalus rubinus* | Native/Directly Observed |
| Monjita Coronada | *Xolmis coronatus* | Native/Directly Observed |
| Monjita Cabeza Negra | *Microspingus melanoleucus* | Native/Directly Observed |
| Tijereta | *Tyrannus savana* | Native/Directly Observed |
| Golondrina Barranquera | *Pygochelidon cyanoleuca* | Native/Directly Observed |
| Tacuarita Azul | *Polioptila dumicola* | Native/Directly Observed |
| Ratona Común | *Troglodytes aedon* | Native/Directly Observed |
| Calandria Grande | *Mimus saturninus* | Native/Directly Observed |
| Calandria Real | *Mimus triurus* | Native/Directly Observed |
| Gorrión Común | *Passer domesticus* | Exotic/ Direct observation |
| Cachilo Ceja Amarilla | *Ammodramus humeralis* | Native/Directly Observed |
| Chingolo | *Zonotrichia capensis* | Native/Directly Observed |
| Tordo Pico Corto | *Molothrus rufoaxillaris* | Native/Directly Observed |
| Tordo Renegrido | *Molothrus bonariensis* | Native/Directly Observed |
| Tordo Músico | *Agelaioides badius* | Native/Directly Observed |
| Corbatita Común | *Sporophila caerulescens* | Native/Directly Observed |
| Jilguero Dorado | *Sicalis flaveola* | Native/Directly Observed |
| Misto | *Sicalis luteola* | Native/Directly Observed |

## 14.5 Proposals for Promoting Alternative Agro-Food Systems

### *14.5.1 An Agroecological Approach for Rethinking Food Production from Islands of Resilience*

Landscape ecology conceptualizes landscapes as complex, adaptive systems composed of ecological and social processes interacting at different levels (Zacagnini 2014). There are many other fields of study that consider the potential interaction between ecological systems and agricultural practices. Agroecology is one such field that focuses on agricultural systems and the role ecological processes can and should have in productive models (Levin 2022). Due to agroecology's focus on production and conservation, it is possible to address the socioeconomic, cultural, ecological, and ethical complexity of food, fiber, and energy production, as well as processing systems, by providing a set of principles for planning local food systems (Wezel et al. 2009). Therefore, agroecology serves as an alternative to current production models that would transform the way we produce and consume food into a more sustainable practice (Nyéléni Boletin 2013, Levin 2022).

Permaculture, regenerative agriculture, agroforestry, biodynamics, and natural agriculture are models with a strong ecological component in their conceptualizations. These practices integrate agricultural needs of local communities with sustainable farming practices to better protect local ecosystems. Therefore, these various models of sustainable agriculture can be classified as agroecological strategies. With regards to the pampean grasslands, it is appropriate, and potentially beneficial, to use these agricultural practices that provide diversity and require engagement from producers, as opposed to current mechanized agricultural practices. In order to protect biodiversity, it is necessary to promote productive processes that incorporate landscape ecology and agroecology practices to maintain production while benefitting the local ecosystems.

Developing a new productive system that utilizes the islands of resilience and local ecology requires an understanding of the biological interactions in the ecosystems. For example, the design must consider effects on the soil, nutrient recycling, biodiversity, and the optimal use of energy. Permaculture practices provide efficient energy use by locating and designing landscape elements according to their capacity of use. Areas that are visited daily or with greater frequency are located near homes or other central components of a landscape, while less-visited sites are located farther away, thereby minimizing energy spent in transit and protecting more productive areas from unintentional impacts. This design is related to the multifunctionality of the ecosystem and the complementary relationships between its components. Another key design point is the promotion of secondary ecological succession to reestablish areas of greater biodiversity, whether spontaneous or induced, in accordance with the local soil characteristics (Fig. 14.6).

**Fig. 14.6** This diagram represents a reticulated system to contribute to the conservation of biological diversity, in order to integrate it with the productive systems. (Satellite image intervened by architect Mercedes Machado)

With regards to the reticulated system for biodiversity conservation, spaces between corridors can be illustrated in diagrams linking multifunctional components of systems. These designs can integrate permaculture design components, laid out in concentric rings or strips that relate to the various production systems (Biassatti et al. 2013, 2017). These productive systems can include vegetables, legumes, cereals, fruit trees, pastures with animal integration or agroforestry systems.

Moving towards sustainable food production through agroecological principles requires several simultaneous transitions at different scales and in different contexts, such as social, biological, economic, cultural, institutional and/or political (Tittonell 2019). Appropriate agroecological practices differ according to initial site conditions due to differing needs and associated conservation strategies (Marasas 2012). Through careful design and planning, we can use these "islands," that could be considered "abandoned," as central elements of ecosystem conservation practices, achieving desired goals for the ecosystem. Many of these transformations are already occurring as young people and adults are organizing in pursuit of a healthier environment, a better quality of life for their children, and a harmonious relationship with nature.

### 14.5.2 Promotion of Productive Practices in the Islands of Resilience

Management of islands of resilience should promote biodiversity with emphasis on soil improvement, using the famous "living soil" approach developed by agronomist Ana Primavesi in "Ecological Soil Management" (Primavesi and Molina 1984). Agroecological approaches are described elsewhere in this book, such as the chapter "Regenerative Agricultural Systems as Biodiversity Islands" (Levin 2022).

Here, we propose land management tactics that can sustainably benefit production outside the "islands," improving soil health and biodiversity management. This will serve as a guide that can be useful for planning sustainable productive systems.

#### 14.5.2.1 Soil Management

Soil is a living system, and its characteristics are properties that emerge from the networks of interacting parts, not explainable by any single component (Restrepo et al. 2000). The physical, chemical and biological properties of a soil are interdependent and generate favorable bio-structural and nutrient conditions, as well as biological interactions that support the health of plants and the food chain.

Many agro-industrial production systems are unable to store stable carbon in the soil due to bare soil, the lack of sufficient photosynthetic capacity of planted crops, and/or the use of synthetic fertilizers or other chemicals which inhibit plant-soil microorganism associations. The soil must be treated as a living organism and covered with actively growing plants which provide root exudates and organic substances that are critical to the cycling of nutrients and surface organic matter, processes which are necessary for maintaining healthy soil and therefore increasing the wellbeing of the ecosystem as a whole (Jones 2008).

Successful soil management balances the inputs and outputs of organic matter by keeping the soil covered as well as maintaining living roots within the soil to promote the normal functions of living soil without the use of chemical inputs (Benedetto et al. 2019). Rotating crops that add nutrients and improve soil functioning also add structure.

The incorporation of cover crops (CC), plants covering the soil for the purpose of its protection and improvement rather than to be harvested, is one of the most commonly adopted practices of soil conservation. CC contribute significant amounts of organic matter to the soil, increasing the soil's microbial life. In addition, CC can reduce soil and nutrient losses due to surface runoff (Capurro 2018). However, the soil may still be threatened by mechanical disruption from the harvest of products or by water use by plants.

An optimal CC mixture depends on the sequence in which crops are to be planted, and on the landowner's objectives. CC compositions can be modified to address specific crop needs or other soil constraints. For example, if the previous crop is a legume, the CC could be a mixture of grasses. If the soil is compacted, we can add species with deep tap roots that make channels and break up the dense soil layers. Several species that can help loosen soil include wild turnip (*Brassica rapa* L.), fodder cabbage (*Brassica napus* x *B. oleracea* cv Interval), cow's tongue (*Rumex crispus*), and chicory (*Cichorium intybus*). In some cases, a light grazing of grass species can be integrated, contributing manure, biological pest control, and also aiding in weed control.

#### 14.5.2.2 Cropping Systems

Several techniques can improve cropping systems, including intercropping, strip cropping, and multistrata cropping. Each has its own benefits to production and conservation. In addition to improving productive capabilities of crops, implementing a sequence of diverse crops generates favorable environments for insect populations, increasing biological interactions among different components of biodiversity, which further promotes key ecological functions and processes.

#### 14.5.2.3 Promotion of Biodiversity

The promotion of biodiversity contributes to the health of the agro-ecosystem. The different strategies for protecting the ecosystem contribute to preserving and increasing predatory and parasitic insects, conserving healthy levels of pest species, including birds, as well as micro and mesofauna. The presence of multistrata borders, hedges and patches of plants in landscape design establishes refuge zones for insects, as well as barriers to threatening anthropized areas.

Biodiversity corridors, integrating species from the Umbelliferae, Brassicaceae (Cruciferaceae), Asteraceae (Compositaceae) and Fabaceae (Leguminosae) families can be established to attract pollinators and other beneficial insects, further increasing biodiversity.

### *14.5.3 Key Factors for Achieving Ecosystem Recovery*

Land tenure is a key factor to consider in the design of productive agro-ecosystems. With guaranteed tenure, the producer is able to sustainably plan for long-term ecological recovery, utilizing long-lived, regionally appropriate species. Public policies linked to the promotion, support, strengthening and financing of agroecological food production are also necessary. Producers decide to make changes to their production strategy depending on many factors such as their presence on the farm, personal observations, and farmers' connection to changes in the local environment. They also often work towards their goals of becoming independent of external inputs, demonstrating opinions on appropriate uses of agriculture technologies, which often contrast with predominant trends.

Understanding nature and ecological principles should guide us in the design and management of agroecosystems, discarding a priori designed formulas or one-dimensional responses. The complexity of the relationships and interactions within the environment and between species, including humans, should be the foundation for planning agroecological models of food production. This growing emphasis on relationships in nature is reflected in the many and varied experiences that are described in the following section.

## 14.6 Integrating Ecology and Economy in the Region

In most societies around the world, the dominant economic system has relied on a similar method of anthropic transformation of natural landscapes, which has increased significantly in scale over the last three centuries. This reliance on anthropization has had serious negative impacts on the environment, which has perverted the original objectives of the economy as an agent of transforming nature to satisfy human needs. This system becomes further distorted by the fact that human society is a subordinate system to the environment, which provides all of its shelter and resources. In fact, the biased and erroneous hierarchization of the economy over the environment compromises the sustainability of current and future societies. As the Romanian economist Georgescu-Rogen states, "the only characteristic that differentiates humanity from all other species... is that we are the only species that in its evolution has violated biological limits" (Georgescu-Roegen 1977).

Therefore, it is crucial to incorporate bioeconomic practices into the current economic systems, integrating economic activities into natural ecological contexts (Passet 1996). The bioeconomic practices have been developed by "ecological economists" such as José Manuel Naredo (2015) or Joan Martínez Alier (2005). The original definition of the Greek term "economy" ("oikos" = home or household, "nemein" = management and dispensation) refers to the practice of managing resources, materials, and natural goods for the support and welfare of the family home (Aristotle, Politics, Book I, Chap. III) (Aristóteles 2005). This definition illustrates the fundamentals of the economy in which humans and society depend on nature and its resources for survival and social reproduction. Humans are therefore responsible for designing an "instituted process of interaction between them and their natural environment, in order to provide themselves with material means to satisfy their needs," (Polanyi 1976). Therefore, the "primary objective of economic activity is the conservation of the human species" (Georgescu-Roegen 1971). However, the human species will fail to be conserved if the natural environment, in which human life originates, develops and endures, is destroyed.

Historically, several communities used a balanced logic of exchange with the natural environment that included principles such as reciprocity, redistribution and exchange. Today, many economies consider these practices, which were harmoniously and wisely linked to the surrounding environment, primitive (Polanyi 2011).

Over time, the center of the economy in most societies around the world has shifted to the market and capital, which are both increasingly financialized and speculative, further neglecting the importance of the environment.

However, in many countries around the world, and in Argentina in particular, despite threats to the environment posed by profit-focused economies, family agroecosystems that produce healthy food continue to exist. They allow us to reconnect with the original empirical and substantive sense of the economy. Present mainly in increasingly environmentally threatened rural and peri-urban areas, they become small resilient strongholds in areas where an economy linked to the extractive model dominates, which seriously compromises the health of the environment, people and local economies.

These local economic systems linked to family agroecosystems are centered on people and sustainable productive work. They link production, transformation, distribution and consumption in order to meet the needs of community members while still considering the wellbeing of the ecosystem. In this sense, they involve "forms of production and reproduction of their material conditions of existence, based on a balanced metabolism with nature" (Toledo and González de Molina 2007). In these family agroecosystems, the household unit is integrated to the production process, forming a nucleus of social management of the agroecosystem (Petersen et al. 2017). The creation of this nucleus is the foundation for implementing diversified economies, integrating diverse and complementary activities linked to the possibilities and limitations of the territory and maintaining a reduced economic and productive scale, since the primary purpose of the system is the continuation of the family economic unit.

Our proposal is to recover and promote these economic and agroecological experiences with local communities, mainly in the humid pampa region of Santa Fe, Argentina. This would require the use of the interstitial spaces provided by the "islands of resilience" and the peri-urban ranges or strips of the cities. The proposal also urges for the economy to be restructured to incorporate ecologically-mindful practices, and particularly, agroecology.

Locally based agro-food systems are also a key important aspect of this new proposed economic system, and further demonstrate the need to redesign local economic systems for the purposes of increasing healthy food production for the community. By shifting control of the economy and of the various components of the agroecosystem to actors of the local community, agro-food systems would be given more consideration because of the numerous benefits they provide to community members. This will further aid in the proposal to integrate "resiliency islands" and the larger ecosystem into the economy in order to increase wellbeing of both the community and local biodiversity.

### 14.6.1 The Potential of Local Food Systems

The current global food system is unsustainable from an economic, social and environmental perspective. It is based on the concentration of the food supply into the hands of a small number of actors who decide what we eat, where we acquire that food, and what prices we pay. They also have the ability to introduce increasingly unhealthy, non-nutritional, and processed food, which may travel long distances, unleashing serious social and environmental impacts. These are the "mileage foods" described by the organization Friends of the Earth (www.foe.org) in their 2012 article called "Mileage foods: $CO_2$ emissions from food imports into Spain". In addition to challenges with food supply and demand, the global trend of urbanization, a large-scale societal transition from periods of progressive expansion and growth to the unsustainable mega-cities, threatens food supply, generates waste, and promotes high levels of energy and materials use. Therefore, it is vital for economies to be centered more regionally, relocating food production and consumption to the local level, within the framework of a true eco-social transition.

In many areas in the Santa Fe Pampa, the local demand for food could be supplied by the diversified production of healthy and fresh food coming from the region itself. This goal could only be achieved if the conditions were enabled for the local economy to be adequately strengthened and enhanced. However, in most rural areas of the region, there is currently a dependence on food from distant places, which negatively affects regional food security. Local communities are losing autonomy and capacity to influence their own food production and consumption models as they rely more on global food production and a food culture that is increasingly homogeneous, and less healthy and diverse.

Therefore, we need to recover and energize fair, healthy, sustainable and sovereign food production systems. This transformation must occur at the local level, revitalizing towns, villages and rural areas, identifying, characterizing, and enhancing local "islands of resilience," while supporting local, agroecology-based food systems.

### 14.6.2 Examples of Local Food Strategies in the Province of Santa Fe, Argentina

In economic systems focused on food production, potential exists for varied and creative initiatives that link primary production, product transformation or value addition, distribution, marketing and consumption. For instance, in the Province of Santa Fe, a digital platform is currently being used which was implemented in Rosario, called EcoAlimentate, the purpose of which is "connecting producers and consumers." The platform maps existing farms in Argentina that sell directly to consumers, providing their location, reviews and contact information (https://ecoalimentate.org.ar/). This increases transparency and connection between local people and producers.

One such farm is "El Hornerito," an agroecological farm that has both educational and productive goals and is managed cooperatively by a group of farmers in the town of Totoras, in the south-central part of the province of Santa Fe (www.facebook.com.granjaelhornerito). This project carries out primary production (milk, eggs, garden vegetables, chickens, corn, etc.) and value addition of products (cheese, jams, etc.) with an agroecological approach. In addition, the farm sells its products directly to the local community through unique marketing strategies, such as the sale of pre-ordered baskets of seasonal vegetables.

Another example located in the southcentral region of the province is "El Manso" farm, a business operated by a family of five (Granja El Manso). This farm is a small-scale operation, raising chickens and laying hens outdoors, fed with a balanced diet of chickpeas, corn and ground peas. The operation also utilizes organic compost, created from pruning waste and other organic residues from the farm and surrounding areas. In addition, the family's income is diversified and supplemented by vegetables grown in its garden, consumed on site. Some of these products are commercialized in the farm's store "Las 3 Ecologías" in the center of the city of Rosario (1551 Julio Cortázar street).

The "Common Land" project (http://www.suelocomun.com/), which originated in 2012 in the town of Lucio V. López, 44 km northwest of Rosario and 132 km from Santa Fe, the provincial capital, produces a variety of foods without agrochemicals. It is a self-management and cooperative project developed to produce, add value to, and market healthy foods (vegetables, seedlings, honey, etc.) grown with an agroecological approach. In addition to the original products, some agroecological producers involved with this project in the "green belt" of Rosario (or peri-urban belt in "agroecological transition") also have diversified their commercial offerings. This project enables customers in Rosario to place orders in advance, which are then delivered on a weekly basis. This allows producers to gradually engage local consumers, as well as better plan and organize their production and offerings of healthy food. Producers are also able to better engage with their consumers and tell the stories of their sustainable methods of production and their benefits to local ecologies and economies. The project recently has grown to have a fixed sales location, its own concessionary station in the food area of the promenade known as "Mercado del Patio" (https://www.mercadodelpatio.gob.ar/), a large store strategically located in front of the bus terminal in Rosario.

Popular, traditional food fairs (markets), which originated in the northeast region of Argentina, can also be utilized to promote a local foods movement. In these rural fairs, local producers can directly connect with their consumers, generating bonds of trust and transparency in the construction of prices and traceability of their food. In urban areas of the region (such as the cities of Rosario or Santa Fe), there are also "consumer groups," in which neighbors or consumers are organized, driving the supply of local agroecological producers who are typically linked through periodic weekly or biweekly commitments of healthy consumer baskets.

Other more complex forms of engaging production and consumption locally are consumer cooperatives or producer and consumer cooperatives (called "prosumers"). In prosumer cooperatives some of the consumers are also local producers themselves and are committed members of the social organization that constructs the system supply, distribution logistics and demand. There are also some worker cooperatives that operate as distributors and marketers of agroecological products, such as Pronoar (https://pronoar.wordpress.com/), or In lak' ech Almacen Natural (In lak' ech Natural Store, 1967 Brown Street, in Rosario).

It is increasingly imperative that governments play an active role in promoting sustainable food policies. Municipalities must engage with groups of producers and consumers, aid in the direct transportation and purchase of products from local producers for school lunchrooms or vulnerable social groups. In addition, local authorities should generate institutional and regulatory mechanisms that guarantee the safety of foods and improve their marketing. They should support agroecological production which guarantees socio-environmental health. An example of this is found in the Municipality of Bellavista, province of Corrientes, which has passed a law recognizing the "agroecological label," derived from the participatory guarantee system – or SPG – carried out by the organized network of public and private actors and the local community itself.

Other examples of creative and innovative mechanisms for supporting local foods systems can be found in Europe. In Spain, for example, the government allows for the establishment of "hubs" that concentrate the logistics and distribution of healthy food, eliminating unnecessary and/or speculative intermediaries and building fair prices. In Catalonia (Spain), the design and implementation of agroecological food banks offers a healthy food supply as an alternative to conventional "food banks," which involve large distributors of industrialized food with low nutritional value.

Finally, "cooperative supermarkets," in several places in the United States and Europe, also offer a valuable alternative to traditional food markets. They are self-managing and involve participation of consumers, with their focus being the provision of ecological food to urban consumers. The institutional precedent is the "Park Slope Food Coop," in Brooklyn (https://www.foodcoop.com/), which presents a model that has been exported and adapted to countries like France or Spain (cooperative supermarket "La Louve" in Paris and "La Osa," in Madrid, https://cooplalouve.fr/ and https://laosa.coop/, respectively). These coops are alternatives to traditional capitalist supermarkets, helping organic producers and promoting healthy food consumption.

### 14.6.3 *Towards Regionalized Food Systems*

All these projects show socio-productive and economic strategies that facilitate the production and consumption of healthy and agroecological foods (primary and/or with added value) at the local level. Furthermore, these are locally based food systems in which there are few intermediaries or speculative actors, directly (or nearly directly) linking producers and consumers, allowing for the building of trust and transparency between both.

Current global food supply chains sustain their dynamics with a few concentrated agro-industrial actors (often large globalized economic groups) and multiple intermediaries, inserting high quantities of processed, expensive foods with low nutritional value into markets. In contrast, the models proposed in this section are based on regional food systems with an agroecological approach, rooted in community. Food systems benefit the local and regional economies by boosting socio-productive conditions, improving the social structure, and recovering local food culture. They also provide multiple environmental and social benefits to the region and local communities.

## 14.7 Conclusions

This chapter condenses the history of the Pampean grassland biome, where a series of significant political, economic and social events led to the fragmentation of the landscape matrix, homogenizing and simplifying the natural ecosystems. As in other regions of the planet, in this part of the world, the domination and subjugation of nature prevailed historically, considered as a repository of resources at the service of an economy based on accumulation, growth and unlimited consumption. In contrast, in this chapter we propose a strategy of balanced integration between human communities and ecosystems, in order to embark on another path that will gradually reverse the degradation of natural resources and make it possible to sustain life in a broad sense for present and future generations.

The identification of "islands of resilience" in the landscape matrix surrounding industrial agriculture represents an interesting discovery of enclaves where flora and fauna are able to retreat, protect themselves and reproduce. For example, in the case study that we discussed, 23 families of tree, shrub and herbaceous plants, 11 species of mammals and 39 species of birds were observed, as well as the presence of burrows and nests. All of these are indicators of ecological success.

Moreover, islands of resilience can contribute to strengthening the economy and endogenous development of local communities. Thus, the sustainable management of local agroecosystems leads to the development of territorialized food systems that link the production of healthy food with local consumption. The "islands of resilience" and the peri-urban fringes of our cities are potential interconnected "nodes" of production, transformation, organization and distribution of healthy food to the surrounding communities. Our research showed a wide range of creative strategies based on local economic circuits. These strategies include practices and actors such as prosumers, consumer cooperatives, local food fairs or small local family-owned farms. In all of them, producers, consumers, local government, social organizations and other relevant local actors link, agree and collaborate to build regional, endogenous and healthy food systems, while forging new socioeconomic and community links.

We cannot ignore the fact that human beings today are faced with extraordinary problems that are radically different from anything they have faced before in human history. They must ask themselves whether organized human society can survive in a recognizable form. And the answer cannot be delayed (Chomsky 2020). No matter how much this society is disguised as green or how many speeches there are about the need for an ecological perspective: the way society actually functions cannot be transformed unless it undergoes a deep structural transformation, namely by replacing competition with cooperation, and the pursuit of economic profit with relationships based on mutual solidarity and concern (Bookchin 2019).

## References


Abba AM, Cassini MH, Vizcaíno SF (2007) Effects of land use on the distribution of three species of armadillos (Mammalia, Dasypodidae) in the pampas, Argentina. J Mammal 88:502–507

Aristóteles (2005) Política, Libro I, Capítulo III. Ediciones Istmo (Colección Fundamentos; Serie Clásicos del pensamiento político), Madrid

Benedetto V, Gamundi JC, Timoni R, Cardozo F, Longo AD, Guglielmone P, Aradas Díaz ME (2017) Experimental module of agroecological transition. Magazine to Improve Production N ° 56. INTA Oliveros, Santa Fe, Argentina

Benedetto V, Aradas Díaz M E, Cardozo FV, Longo AD, Timoni RE (2019) Experience of extensive Agroecological production with integral and multidimensional approach in INTA EEA Oliveros, Santa Fe. Book of abstracts, 1st Argentine Congress of Agroecology. National University of Cuyo. Mendoza, Argentina. ISBN 978-987-575-210-8

Biasatti NR, Bedetti F, Colomar G, Di Nucci D, Marc LB, Romano MC, Spiaggi EP (2007) El impacto de la actividad agropecuaria sobre la conservación de la biodiversidad en la pampa húmeda. Rev. UNR/AMBIENTAL N°7, Comité Universitario de Política Ambiental, SECyT (UNR), Rosario, Argentina, 203p. ISBN 950-817-116-2

Biasatti NR, Avogradini F, Rapalino M (2013) Gridded system for biodiversity conservation in the province of Santa Fe, Argentina. Mag Argentine Assoc Landscape Ecol 4(2):181–189

Biasatti NR, Rozzatti JC, Fandiño B, Pautaso A, Mosso, E, Marteleur G, Algañaraz N, Giraudo A, Chiarulli C, Romano M, Ramirez Llorens P, Vallejos L (2016) Las eco-regiones: su conservación y las áreas naturales protegidas de la provincia de Santa Fe. Secretaría de Medio Ambiente. Imprenta UNL, Santa Fe. Primera Edición, 244 p. ISBN 978-987-23776-2-5

Biasatti NR, Marc L, Rimoldi P, Spiaggi E (2017) Los valores ambientales de ecosistemas relictuales en la pampa húmeda: caso de "El Espinillo" y el tramo medio del río Carcarañá en Santa Fe". Centro de Estudios Ambientales (CEAV). Facultad de Ciencias Veterinarias. Universidad Nacional de Rosario (UNR). UNR Editora, Rosario, 62p. ISBN 978-987-702-262-9

Biasatti NR, Rimoldi P, Cabrera L (2019) Desafíos de la conservación biológica en el espinal y la pampa húmeda santafesina, los mamíferos del sur de Santa Fe. Ministerio de Medio Ambiente. Provincia de Santa Fe, Santa Fe, Argentina.106p. ISBN978-987-45488-3-2

Bidaseca K, Lapegna P (2006) El grito de Alcorta revisitado: cultura y sentimientos en la acción colectiva. Anuario 2006_09 Nov06.qxp 09/11/2006 06:04 am, p 309

Bookchin M (2019) La próxima revolución. Las asambleas populares y la promesa de la democracia directa. Virus Editorial, Barcelona ISBN 978-84-92559-96-1

Brailovsky A (2006) Ecological history of Ibero-America: from the Mayas to Don Quixote. 1st ed, Editorial Capital Intelectual, Buenos Aires, 240 p ISBN 987-1181-45-0

Brailovsky A, Foguelman D (eds) (2009) Green memory. Ecological History of Argentina. 7th Edition. ISBN 987-1138-30-X. Ed Sudamericana, Buenos Aires, Argentina

Capurro JE (2018) Efectos de un cultivo de cobertura sobre propiedades edáficas y uso del agua en ambientes con erosión hídrica y monocultivo de soja, en el sur de la provincia de Santa Fe. Tesis de Magister en Manejo y Conservación de Recursos Naturales. Facultad de Ciencias Agrarias. Universidad Nacional de Rosario, Rosario, Argentina

Chomsky N (2020) Climate crisis and the global green new Deal: the political economy of saving the planet. Ed Capital Intelectual, Buenos Aires, 188 p, ISBN 978-987-614-607-4

Cloquell S, Albanesi R, Propersi P, Preda G, De Nicola M (2007) Rural families. The end of a story in the beginning of a new agriculture. Editorial Homo Sapiens, Rosario, 200 p. ISBN 978-950-808-524-5

Cracco M E, Guerrero S (2004) Aplicación del enfoque ecosistémico a la gestión de corredores en América del Sur. Memorias taller regional. 3 al 5 de junio, Quito, Ecuador

Del Corro F (2003) Apropiación de la tierra a los aborígenes y genocidios en el Rio de la Plata. Rebanadas de Realidad, Buenos Aires 11/08/03

Dirzo R, Young S, Galetti M, Ceballos G, Isaac N, Collen B (2014) Defaunation in the Anthropocene. Science 345:401. https://doi.org/10.1126/science.1251817. www.sciencemag.org/content/345/6195/401/suppl/DC1

Donghi TH (2000) From the Independence revolution to the Rosista confederation (Argentine History, VOL.III). Ed Paidos Iberoamerica, Buenos Aires, Argentina, ISBN 978-950-127-723-4

Donghi TH (2014) Revolution and war: building a leading elite in Argentina creole. Siglo Veintiuno Publishing House, Mexico, 480 p, ISBN 978-987-629-379-2

Forman RT, Grodon M (1986) Landscape ecology. Wiley, New York

Fox B, Fox M (2000) Factors determining mammal species richness on habitat islands and isolates: habitat diversity, disturbance, species interactions and guild assembly rules. Glob Ecol Biogeogr 9:19–37

Friends of the Earth (2012) Food miles: CO2 emissions from food imports into Spain (www.foe.com)

Gauna Zotter J, Rey M (2017) Tecnología y Sociedad. Ed EDUCA, Buenos Aires. Available in: http://bibliotecadigital.uca.edu.ar/repositorio/revistas/svampa-maldesarrollo-argentina-extractivismo.pdf

Georgescu-Roegen N (1971) The entropy law and the economic process. Princeton University Press, Princeton

Georgescu-Roegen N (1977) Bioeconomics. Princeton University Press, Princeton

Herrero A (2017) Navigating through the turbulent times of the Anthropocene. Political ecology. International debate notebooks, June 2017. ISSN: 1130-6378. Eds: Martínez Alier J, Puig Ventosa I, Monjo A. Ed Fundació ENT / Icaria editorial, Barcelona

Jones C (2008) The liquid carbon pathway. Australia. Extract from Australian Farm Journal Ed. 338, www.amazingcarbon.com

Jones CG, Gutiérrez JL (2007) On the purpose, meaning, and usage of the ecosystem engineering concept. In: Ciddington K, Byers JE, Wilson WG, Hastings A (eds) Ecosystem engineers: plants to protists. Academic, Cambridge, MA, pp 3–24

Jones CG, Lawton JH, Shachak M (1994) Organisms as ecosystem engineers. Oikos 69:373–386

Jones CG, Gutierrez JL, Groffman PM, Shachak M (2006) Linking ecosystem engineers to soil processes: a framework using the Jenny state factor equation. Eur J Soil Biol 42:39–53

Kropotkin P (1977) The conquest of bread. Jucar Editions, Gijón. ISBN 84-334-1555-7

Levin B (2022) Regenerative agricultural systems as biodiversity islands. In: Montagnini F (ed) Biodiversity islands: strategies for conservation in human-dominated environments, Series: Topics in Biodiversity and Conservation. Springer, Cham, pp 61–88

Marasas M (2012) El camino de la transición agroecológica. Ediciones INTA, Ciudad Autónoma de Buenos Aires, Argentina. ISBN 978-987-679-104-5 1

Martínez Alier J (2005) The environmentalism of the poor. Environmental conflicts and languages of valuation, 3rd ed. Antrazyt, Icaria Editorial 207, Barcelona ISBN: 978-84-7426-743-3
Mignolo W D (ed) (2007) The idea of Latin America. Editorial Gedisa, S.A., Buenos Aires. ISBN: 978-84-9784-094-1
Mollison B (1994) Introduction to permaculture. Tagari Publications, Stanley, Tasmania, Australia. ISBN 0 908228 09
Montagnini F, Somarriba E, Murgueitio E, Fassola H, Eibl B (Eds) (2015) Sistemas Agroforestales. Funciones productivas, socioeconómicas y ambientales. Serie Técnica Informe Técnico 402, CATIE, Turrialba, Costa Rica. Fundación CIPAV. Cali, Colombia, 454p
Moreno C, Barragán F, Pineda E, Pavón N (2011) Reanálisis de la diversidad alfa: alternativas para interpretar y comparar información sobre comunidades ecológicas. Revista Mexicana de Biodiversidad 82:1249–1261
Naredo JM (2015) Economic roots of ecological and social deterioration. Ed Siglo XXI, Madrid
Nyéléni Boletin (2013) Número 13, Marzo 2013, www.nyeleni.org - info@nyeleni.org
Passet R (1996) Principios de bioeconomía. Fundación Argentaria. Visor distribuciones S.A., Colección Economia y Naturaleza, Madrid-Bilbao
Petersen P, Marçal da Silveira L, Bianconi Fernandes G, Gomes de Almeida S (2017) Método de Análise Econômico-Ecológica de Agroecossistemas. AS-PTA, Rio de Janeiro, 246 p
Pizarro JB (2003) La evolución de la producción agropecuaria pampeana en la segunda mitad del siglo XX. Revista Interdisciplinaria de Estudios Agrarios, numero 18, 1er semestre 2003, Universidad de Buenos Aires, Buenos Aires
Poiani K, Merrill M, Chapman K (2001) Identifying conservation-priority areas in a fragmented Minnesota landscape based on the umbrella species concept and selection of large patches of natural vegetation. Conserv Biol 15:513–522
Polanyi K (1976) El sistema económico como proceso institucionalizado. In: Godelier M, Antropología y economía. Anagrama, Barcelona, pp 155–178
Polanyi K (2011) The great transformation. The political and economic origins of our time. Ed. Economic Culture Fund, Mexico
Politis G, Gutierrez MA, Scabuzzo C (2014) Chapter 7: The vertebrates of the site Arroyo Seco 2. Palaeo-environmental implications in current status of the investigations in the archaeological site Arroyo Seco 2, Tres Arroyos County, 1st ed. National University of the Center of the Province of Buenos Aires. Faculty of Social Sciences, Tandil, 514 p, ISBN 978-950-658-338-5
Primavesi A, Molina JS (1984) Ecological soil management: agriculture in tropical regions. Ed El Ateneo, Buenos Aires, 5th edition, 499p
Restrepo JM, Ángel SD, Prager M (2000) Agroecology. Centro para el Desarrollo Agropecuario y Forestal, Inc (CEDAF), Santo Domingo, Dominican Republic. ISBN 99934-8-002-9
Rimoldi PG, Chimento NR (2018) Diversidad de mamíferos nativos medianos y grandes en la cuenca del rio Carcarañá, provincia de Santa Fe (Argentina). Rev Mus Argentino Cienc Nat 20(2):333–341
Roldan AV, Sauthier UD (2016) Madrigueras de *Dolichotis patagonum* como recurso para otros vertebrados en Península Valdés. Mastozoología Neotropical 23(2):515–520
Sili M, Soumoulou L (2011) The land issue in Argentina. Conflicts and dynamics of use, tenure and concentration. Ministry of Agriculture, livestock and fisheries. Presidency of the Nation, Buenos Aires
Smith W, Solow A, Chu C (2000) An index of the contribution of spatial community structure to the species accumulation curve. Ecology 81:3233–3236
Svampa M, Viale E (2014) Bad development. The Argentina of extractivism and dispossession. Katz Publishers, Buenos Aires. ISBN Argentina: 978-987-1566-92-1
Svampa M, Viale E (2020) The ecological collapse is here. XXI Century Publishers, Buenos Aires, 296p. ISBN 978-987-801-027-4
Tittonell P (2019) Agroecological transitions: multiples scales, levels and challenges. Mendoza, Argentina. Rev. FCA UNCUYO 51 (1): 231-246. ISSN (online) 1853 8665

Toledo VM, González de Molina M (2007). El metabolismo social: las relaciones entre la sociedad y la naturaleza. In: Garrido F, González de Molina M (Ed.). El paradigma ecológico en las ciencias sociales. Icaria Editorial. Barcelona, España. ISBN 978-84-7426-756-3, p. 85–112

Volkind P (2009) El acuerdo de 1920 entre la Federación Agraria Argentina y la Federación Obrera Regional Argentina (IX Congreso): alcances y limites en el marco de la conflictividad agraria de la época. Revista Interdisciplinaria de Estudios Agrarios n° 31-2do semestre de 2009. Buenos Aires, Argentina

Wezel A, Bellon S, Doré T, Francis C, Vallod D, David C (2009) Agroecology as a science, a movement and a practice. A review. Agronomy Sustain Dev 29(4):503–515. https://doi.org/10.1051/agro/2009004

Zaccagnini ME (2014) Manual of good practices for the conservation of soil and biodiversity and its ecosystem services. UNDP, Secretary of the Environment and Sustainable Development of the Nation, INTA, Buenos Aires. ISBN 978-987-1560-55-4

# Chapter 15
# Residential Garden Design for Urban Biodiversity Conservation: Experience from Panama City, Panama

**Helen R. C. Negret, Rafael Negret, and Irene Montes-Londoño**

**Abstract** Urban expansion encroaches on natural areas causing habitat and species loss. However, cities can offer ecological spaces that harbor high proportions of regional and local species. In addition to public urban green spaces, private residential gardens are important for biodiversity conservation particularly if spatially arranged to maximize habitat-patch sizes and minimize isolation from remnants of native habitat in the city. Urban growth is projected to increase considerably, including in biodiversity hotspots, many of which are in developing tropical countries. In urban areas of these countries, residential "ornamental" gardening is not as widespread as in temperate developed countries where a multimillion-dollar industry supports garden design and maintenance. This case study discusses residential garden design frameworks for tropical biodiversity conservation that, if adopted at scale, could channel private finance to conservation in urban areas. It documents the establishment and management of a residential ornamental garden designed to protect native fauna and flora in an urban landscape in Panama City, Panama. It describes the design elements and records the positive impact on biodiversity over 15 years in a 1700 $m^2$ property. Grass areas were reduced by 80%, and 64% of the property was planted, increasing vascular plant species from 10 to at least 180 and birds from 9 to 157 species. Management approaches, and challenges of increasing habitat alongside human wellbeing benefits from the garden, are presented. Recommendations and required attitude changes are outlined for garden practitioners, urban planners and policymakers to replicate the design elements of this biodiversity garden island in Panama City, and beyond.

H. R. C. Negret (✉)
Retired/former environmental advisor, United Nations Development Program, Panama City, Panama, Republic of Panama

R. Negret
Retired/former environmental advisor, Interamerican Development Bank, Panama City, Panama, Republic of Panama

I. Montes-Londoño
Hacienda Pinzacuá, Alcalá, Valle del Cauca, Colombia



F. Montagnini (ed.), *Biodiversity Islands: Strategies for Conservation in Human-Dominated Environments*, Topics in Biodiversity and Conservation 20,
https://doi.org/10.1007/978-3-030-92234-4_15

**Keywords** Attitude changes · Biodiversity gardens · Garden practitioners · Tropical gardens · Urban green spaces · Urban policy and planning

## 15.1 Introduction

### *15.1.1 Urbanization and Biodiversity*

Urban expansion is a major driver of habitat conversion. It transforms habitats critical for biodiversity, decreasing and fragmenting natural patches by altering their size, shape, and their connectivity in the landscape (Ricketts 2001; Alberti 2005; McDonald et al. 2008). Consequently, urbanization poses one of the largest threats to global biodiversity (Luck et al. 2009; Seto et al. 2012; McDonald et al. 2013). Growth models project the global total urban area to triple by 2030, with respect to 2000 levels. Across the world this expansion will be $4 \pm 0.8$ times in biodiversity hotspots,[1] with local and global ecological consequences (Seto et al. 2012; Güneralp and Seto 2013; McDonald et al. 2013). In absolute area, South America is expected to have the largest increase of urban land in biodiversity hotspots with an increase of more than $100{,}000 \pm 25{,}000$ km$^2$, corresponding to nearly a $3.5 \pm 0.5$-fold increase. The largest proportional increase, of about $14 \pm$ three-fold, is forecasted to be in Mid-Latitudinal Africa (Güneralp and Seto 2013). Biodiversity loss is inevitable with this urban expansion; however, its extent will depend on urban and landscape planning approaches in public and private land.

### *15.1.2 Biodiversity in Cities*

Urbanization fragments natural ecosystems and alters ecosystem processes, but also creates social and economic opportunities to build unique ecological spaces that can conserve native species (Muller et al. 2013). Studies have shown that globally declining taxa can attain high densities in urban habitats (McFrederick and LeBuhn 2006; Osborne et al. 2008). In this context, many taxonomic groups have 50% or more of the regional or even national species assemblage found in cities (Sec CBD 2012; McDonald et al. 2013). Two-thirds of plant species occurring in urban areas tend to be native to the region of each city; the proportion of native bird species being considerably higher than that of other taxa (94%) (McDonald et al. 2013).

[1] Biodiversity hotspots have high endemism, are threatened by human inhabitation and have been designated as priority areas for conservation (Myers et al. 2000). Of the 36 biodiversity hotspots identified globally, many are in the tropics and all contain urban areas. Panama City is within the Mesoamerica Corridor Biodiversity Hotspot.

Thus, cities can play important roles in the conservation of biodiversity. Within cities, urban green spaces (UGS) can be critical for native biodiversity conservation, providing habitat that supports threatened and specialist species, and which warrant conservation (Aronson et al. 2014; Ives et al. 2016; Lepczyk et al. 2017). UGS comprise all-natural, semi-natural, and artificial ecological systems within and around a city (Cilliers et al. 2013). They include a range of habitat types from public parks with well conserved remnant patches of native vegetation, urban wastelands, private gardens/yards, to engineered green infrastructure such as green roofs (Aronson et al. 2017). These UGS vary in their conservation value, depending on the biodiversity present, but all could represent opportunities to contribute to conservation if designed and managed appropriately.

### *15.1.3 Urban Residential Gardens and Biodiversity Conservation*

In this case study, a distinction is made between an urban residential garden and a homegarden. The latter is the agroforestry practice of planting a mixed patch of livelihood-oriented perennial and annual species within a clearly bounded area near the homes in rural or urban landscapes. The role of homegardens in biodiversity conservation is important in agricultural landscapes (Kumar and Nair 2006; Montagnini 2006; Galluzzi et al. 2010). Agricultural production in these homegardens in tropical cities constitutes a substantial element of food production, while also playing a role in urban biodiversity conservation (González-García and Gómez Sal 2008).

In contrast, residential gardens, especially in urban settings, focus primarily on the provision of other benefits such as aesthetic, therapeutic, and sensory pleasures. These gardens, referred to by a variety of names such as residential, domestic, ornamental and private, are the focus of this article. The term residential garden is used hereafter in this article.

In some cities, residential gardens make up a large part of the urban area and its UGS. In the UK, residential gardens typically represent 27% of urban land and up to 47% of the UGS area, e.g. in Leicester (Goddard et al. 2010). In developing countries, residential gardens are also an important resource for ornamental and conservation purposes, as well as for food. In Latin America, for example, household "patios" make up 85% of overall UGS area in Nicaragua (González García and Gómez-Sal 2008) and 45% of UGS in Asunción, Paraguay, which constitutes 23% of the city's land area (Yanosky et al. 2014).

Though private green spaces have traditionally been less studied than public ones due to access difficulties (Goddard et al. 2010), some literature has discussed their contribution to species diversity in urban areas. In the UK for example, 22.7 million households (87% of homes) have gardens (Thompson and Head 2014), harboring an estimated 28.7 million trees, at least 4.7 million nest boxes and up to 3.5 million ponds (Davies et al. 2009). A study of 61 residential gardens in the city of Sheffield, UK, found 4000 species of invertebrates, 80 species of lichens, and more than 1000 species of plants (Gaston et al. 2007).

Habitat patch size, quality, pattern and connectedness in the landscape are important for the survival of several taxa, so, to achieve long term conservation from residential gardens, land use planning, policy and coordination of actors is needed across an entire urban space (Evans et al. 2009; Goddard et al. 2010). However, urbanization rates are highest in those regions of the world where there are generally under-resourced urban governance arrangements (Muller et al. 2013) and where financial constraints and livelihood options may continue to place priority on food-production focused gardening. A multimillion-dollar industry has arisen to support the design and maintenance of residential gardens particularly in the northern hemisphere.[2] Although figures on financial investment by private citizens are not easily available for all countries, much of urban expansion will occur in developing Middle-Income Countries, such as all but one of the countries in Latin America. Mobilizing more gardeners to adopt biodiversity conservation garden design and management in these countries could harness a largely untapped resource for achieving local and national conservation targets and provide opportunities for private citizens to become stewards of biodiversity and ecosystem services within city boundaries. For this to occur, several challenges need to be overcome including knowledge gaps on biodiversity garden design and individuals' attitudes to gardening and biodiversity conservation.

### 15.1.4 Garden Design for Biodiversity Conservation

There is increasing data in the literature on the role of residential gardens, and in general UGS, in biodiversity conservation (Goddard et al. 2010; Aronson et al. 2017); however, this focuses mainly on temperate countries in the northern hemisphere and with a high proportion in the United States (Muller et al. 2013). This is not the case in the developing world, particularly in the tropics where there is a scarcity of empirical, systematically obtained data on biodiversity conservation in urban residential gardens (Gaston et al. 2005; Jaganmohan et al. 2012). Nonetheless, it can be expected that urban residential gardens in the tropics play a role in biodiversity conservation as well. Further studies can provide guidance for scaling up this practice. Our case study reports empirical observations from our residential garden, providing information that could be useful to scientists and practitioners in understanding and promoting these ecosystems in tropical urban settings.

[2] In the UK alone, households spent around £7.5 billion (or US$9.21 billion) on garden goods in 2017, equivalent to 1% of household spending (Oxford Economics 2017). In the USA, in 2018 77% of American gardeners reported spending a record US$47.8 billion on lawn and garden retail sales, with an average household expenditure of US$503 (NGS 2018).

Studies in temperate countries show that form and management of gardens are essential variables influencing the benefits they provide, including biodiversity (Cameron et al. 2012). Historically, the legacy of green spaces intended to deliver aesthetic and recreational benefits has been the simplification of habitats (Davies et al. 2019). For example, pruning and removing trees, shrubs, and leaf litter, dramatically simplifies the garden structure (Aronson et al. 2017). Gardeners often clear away dead wood for potential safety and aesthetic reasons, yet this negatively affects species that rely on coarse woody debris (e.g., woodpeckers; Kane et al. 2015). Furthermore, very few green spaces are designed and managed to deliver synergistic biodiversity conservation and wellbeing and health benefits (Davis et al. 2019). Guidelines for amateur gardeners could support a broad range of biodiversity conservation, in balance with the aesthetic and wellbeing values of the garden.

Garden design is a complete package that defines a vision and provides guidance and a road map for developing a garden over time.[3] Classical garden design books for residential gardens follow a variety of schools, or styles, largely developed in temperate climates[4] and provide practical, off-the-peg options that may not necessarily include the conservation of natural assemblages across different taxa. Garden books for the tropics (e.g. Warren 1991; Wijaya 1999; 2007; Neal 2012) often focus on showcasing renowned gardens or listing plants that can be used in tropical settings, and may not always propose specific garden design elements and guidance for the establishment of new gardens specifically for biodiversity conservation. Hence there is a need for a residential garden design framework for the tropics that addresses conservation of the broad range of biodiversity, from wildlife to native plants within urban settings. This chapter is a step towards filling these gaps.

The case study presented in this chapter describes the establishment and management of an urban residential garden for biodiversity conservation. It reports the observations of biodiversity increases over 15 years and provides design guidelines with biodiversity conservation elements alongside those from traditional garden design. In this sense, it is one of the first documented long-term experiences of converting a barren grassland residential property into a garden rich in native species in a tropical developing country where there are still only incipient urban biodiversity conservation actions. By describing the main elements of this garden design framework and how it could be replicated, the case study also provides an example of how private initiatives can contribute to biodiversity conservation strategies in tropical developing countries.

[3] Garden design includes three elements (Newbury 2000): (i) Definition of function, style and feel of the garden (ii) Costs of initial construction (borders), garden features (paths, ponds), plantings (trees, bushes, annuals), and maintenance; (iii) Physical nature of the garden plot (temperature, rainfall, soil, slopes, drainage, aspect)

[4] Garden design schools include *formal* with axial symmetry and small numbers plant species; *Japanese* with simple styles and intellectual and philosophical elements; *sensory* with sounds and colors to stimulate people with disabilities; *entertainment* with features for outdoor living; and *wildlife* with food and shelter for animals (Newbury 2000).

## 15.2 Garden Location, Design and Establishment

### *15.2.1 Location of Garden and Surrounding Environment*

The residential garden is situated in Clayton, in the Ancon District of Panama City. It is located near two national parks in an area that was an American army base called Fort Clayton until its return to the Panamanian Government at the start of 2000. The area was designated, through Law 21,1997, as a "garden city" which advocated the pursuit of knowledge and housed the "City of Knowledge" (Gaceta Official 1997; Conniff and Bigler 2019). The army base headquarters, and most of the nearby military houses, were built on steep slopes refilled by heavy clay and stones from the Panama Canal excavation. The houses had no formally established residential gardens and were joined by communal open spaces of grass dotted with high Royal Palm trees.[5]

Two years after the transfer of the base to the government of Panama, in 2002, the army houses were sold, and an increasing number of families moved from the city center to Clayton. Since 2000, the population of Ancon has jumped from 11,700 to an estimated 79,301 in 2020. The annual growth rate from 2000 to 2010 was 10.3%, compared to just 0.34% in Panama City Center (National Institute of Statistics and Census, https://www.inec.gob.pa/). Since 2002, a series of new housing development projects with small gardens were started, often with buildings encroaching into forest fragments. Active hunting of wildlife is practiced by construction workers, many of whom immigrated to the city from the interior of the country where hunting still occurs in some rural areas. Nonetheless, there is still significant tropical forest habitat in and around Clayton and other large blocks in the nearby Metropolitan and Camino de Cruces National Parks.

Panama City is known for its warm climatic conditions with annual average temperatures of around 27 °C and a 24 (minimum) to 32 °C (maximum) range. Relative humidity is high (80 to 100%) with a total precipitation of 1930 mm, most of which falls over 8 months (May to December) with a monthly average of over 200 mm. The remaining 4 months (January to April) are very dry (less than 20 mm monthly precipitation) with strong winds and high evapotranspiration (WMO 2019). The garden faces east-southeast.

### *15.2.2 Garden Design Approach*

When we bought the property in 2005, the house, and two neighboring homes, were set in an expanse of grass covering 9000m$^2$, located about 60 m away from a patch of

[5]The Royal Palm scientific name is *Roystonea regia,* native to Mexico and Central America. From here on all plants and animals will referred to with the common name. Scientific names are shown in Table 15.1

**Table 15.1** Scientific names of species cited in text

| Cited Animals | | | Cited Plants | | |
|---|---|---|---|---|---|
| Taxa | | | Type | Scientific name | English common names |
| **Birds** | *Ortalis cinereiceps* | Gray-headed Chachalaca | **Bushes** | *Sanchezia especiosa* | Sanchezia |
| | *Buteo nitidus* | Gray Hawk | | *Brunfelsia paiciflora* | Yesterday, Today & Tomorrow |
| | *Gampsonyx swainsonii* | Pearl Kite | | *Megaskepasma erythrochlamys* | Brazilian red cloak |
| | *Piaya cayana* | Cuckoo Squirrel | | *Duranta erecta* | Pigeon berry |
| | *Amazilia tzacatl* | Rufous-tailed Hummingbird | | *Stacytarpheta jamiacensis* | Snakeweed |
| | *Trogon violaceus* | Violaceous Trogon | | *Ixora coccinea* and varieties | Jungle geranium |
| | *Trogon massena* | Slaty tailed Trogon | | *Bougainvillea spp.* | Bougainvillea |
| | *Momotus momota* | Blue-crowned Motmot | | *Rusellia equisetiformis* | Firecracker |
| | *Thamnophilus doiatus* | Barred Ant shrike | **Flowers** | *Allamanda cathartica* | Golden Trumpet |
| | *Myozetetes granadensis* | Social Flycatcher | | *Lantana camara* | Lantana |
| | *Zimmerius vilissimus* | Paltry Tyrannulet | | *Hibiscus spp* | Rose Mallow |
| | *Vireo flavoviridis* | Yellow-green Vireo | | *Ruella simplex* | Mexican petunia |
| | *Contopus sordidulus* | Western Wood Pewee | | *Cuphea hyssopifolia* | Mexican heather |
| | *Euphonia luteicapilla* | Yellow-crowned Euphonia | | *Alpinia spp* | Red, Pink and Shell Ginger |
| | *Troglodytes aedon* | House Wren | | *Heliconia spp* | Parrot´s Beak |
| | *Dendroica petechia* | Yellow Warbler | **Foliage Tall, medium and ground cover** | *Spathiphyllum willis* | Peace Lily |
| | *Thraupis episcopus* | Blue-gray Tanager | | *Dieffenbachia amoena* | Dumbcane |
| | *Thraupis palmarum* | Palm Tanager | | *Tradescantia spp* | Wandering Jew |
| | *Piranga rubra* | Summer Tanager | | *Calathea spp.* | Prayer plants |
| | *Piranga olivacea* | Scarlet Tanager | | *Philodendron spp.* | Climbing and non-climbing |

(continued)

**Table 15.1** (continued)

| Cited Animals | | | Cited Plants | | |
|---|---|---|---|---|---|
| Taxa | | | Type | Scientific name | English common names |
| | *Ramphocelus dimiditus* | Crimson-backed Tanager | | *Arachis pintoi* | Pinto peanut |
| | *Turdus grayi* | Clay-colored Thrush | | *Maranta leuconeura* | Prayer plant |
| | *Seiurus noveboracensis* | Northern Water Thrush | **Palms** | *Roystonea regia* | Royal Palm |
| | *Quiscalis mexicanus* | Grackle | | *Bactris gasipaes* | Peach fruit Palm |
| | *Tityra semifasciata* | Masked tityra (Cotinga) | | *Elaeis oleifera* | American Oil Palm |
| | *Icterus galbula* | Baltimore Oriole | **Trees** | *Calliandra surinamensis* | Pink powder puff Tree |
| | *Icterus spurius* | Orchard Oriole | | *Caesalpinia pulcherrima* | Peacock flower Tree |
| | *Basiliscus basiliscus* | Basilisk Lizard | | *Tabebuia rosea* | Pink trumpet Tree |
| **Reptiles** | *Iguana iguana* | Green Iguana | **Fruit trees** | *Theobroma cacao* | Cacao |
| | *Bothrops asper* | Fer-de-lance | | *Annona muricata* | Soursop |
| | *Micrurus nigrocinctus* | Central American Coral | | *Myrciaria cauliflora* | Jabuticaba |
| **Mammals** | *Sciurus variegatoides* | Variegated Squirrel | | *Carica papaya* | Papaya |
| | *Dasyprocta punctata* | Agoutis | | *Psidium guajava* | Guava |
| | *Cuniculus paca* | Paca | | *Mangifera indica* | Mango |
| | *Saguinus geoffroyi* | Geoffroy's tamarin | **Vines** | *Thunbergia sp* | Kings mantle |
| | *Tamandua mexicana* | Northern anteater | | *Petrea sp* | Sandpaper vine |

forest connected to the Camino de Cruces National Park (approx. 4000 ha), a kilometer away. This park, in turn, is adjacent to the Soberanía National Park (22,104 ha). Our property was an island in a sea of grass with no physical connection to the forest and its biodiversity. As far as we observed, there were only a few mammals (such as agoutis and coatis), surprisingly few birds and insects, and no flowers. As professional ecologists, life-long ornithologists, and keen gardeners, we decided to take on the challenge of designing and establishing a diverse garden reflecting and protecting the rich biodiversity of the surrounding area.

A key long-term goal was to increase connectivity with the native forest patch, providing a wildlife haven within the garden and a safe transit across the garden to a second smaller patch of forest behind two other houses nearby. We developed a framework to guide a garden design that would contribute to biodiversity conservation whilst fulfilling other important aesthetic, recreational, therapeutic, and spiritual attributes of gardens. This framework draws from our professional work in biodiversity conservation and includes elements from different gardening schools. It adopts a gardening approach that replicates processes that are known to increase connectivity in natural habitats such as hedgerows and large canopy trees. It uses aspects of the English garden such as paths with curves to entice the walker to discover new species and spaces at each bend. Japanese elements include a Zen-style pool with wide underwater steps to sit on and contemplate the forest. Elements from modernistic tropical designs include those of the Brazilian landscaper Roberto Burle Marx, which uses tropical native plants, especially bromeliads and heliconias, and landscaping to create structured elements that capture tropical exuberance.

The process aimed to create a structured design in the short and medium term, so that the garden did not look like a "work-in-progress" but rather, had a specific and satisfying feel and character in each phase. Thus, from the start we defined the long-term vision and identified five main elements of design to repeat and expand over the next 15 years and beyond as we gradually achieved our goal. These elements are hedgerows, large flowerbeds/islands with mixes of species native to Panama or other neotropical countries, trees, paths, and bodies of water. Figure 15.1 demonstrates the integration of these five elements into the landscape design of our garden. Together, these elements provide various benefits for wildlife and people (Table 15.2).

### *15.2.3 Garden Design Framework*

Windbreaks and hedgerows have a role in tropical forest regeneration particularly if they are connected to forest (Harvey 2000; Chacon and Harvey 2006; Leon and Harvey 2006). Applying this same principle to our residential urban garden, we first started work at one edge of three properties that bordered the forest patch. These properties, collectively covering 9000 $m^2$, were largely grassland. Starting at the edge of the forest, and using native forest trees species, we planted a hedgerow that separated these properties from others nearby and from a main road. It has been left to develop naturally, and now has 20 m high trees. We then planted narrower hedgerows along the borders of our property and the adjacent properties, again starting at the edge of the forest. We opted for creating hedgerows, not hedges, to create greater complexity (species mix and dimensional structure) and optimize biodiversity conservation (Daniels and Kirkpatrick 2006; Smith et al. 2006a, b; Gonzalez-Garcia et al. 2009). Normally used in rural areas, a hedgerow is a row of usually untended, closely planted bushes or trees of different species. A hedge, by contrast, is a thicket of bushes, usually of one species and trimmed regularly to form a fence between two portions of land or parts of the garden. If designed carefully both provide food, shelter, and nesting sites for wildlife, and act as a corridor for their safe passage between different habitats (Peoples Trust for Endangered Species 2019; Woodland Trust 2019). As in other ecosystems, ecological corridors in the urban context can play an effective role for the distribution of organisms with low dispersal capabilities and should be included in planning strategies (Vergnes et al. 2013).

The next step was to plant large flowerbeds or islands of different types of flowers and shrubs to attract native fauna, provide diverse habitats and nesting sites, and facilitate movement across the garden and between the hedgerows. To create these islands, native or naturalized species were used that attract a wide variety of pollinators including insects, hummingbirds, and bats. Mixes of hardy flowering

**Fig. 15.1** Garden Habitat changes over fifteen years
(1) Growth of Hedgerows and Flowerbed-Islands in the Garden. *Top image* First row, left to right: Garden in 2006, 2012; Second row, left to right: Garden in 2015, 2020
(2) The Five Elements of Garden Design Framework. *Bottom image*. Left column, top to bottom. Mixed Flower Islands, Trees (Pink powder puff and Peacock flower tree), Water: fountain in bromeliad area. Right column, top to bottom: Hedgerow, Path with bench and resting area at the end

**Fig. 15.1** (continued)

bushes were favored as these withstand tropical downpours and require low maintenance. As the aim was to optimize biodiversity conservation, we used many different species. However, in keeping with ornamental gardening advice, in some flowerbeds, we clumped several plants of one species together for creating more visual impact. For the same reason, we also created smaller flowerbeds planted entirely with only one species.

A third element of the framework was to plant trees, especially those with larger canopies and edible fruits. These were planted strategically to improve connectivity between and within the larger flowerbed islands and hedgerows and to increase structural complexity. Also, they were planted between existing trees and the forest so as provide canopy level connectivity with the forest. The selection of the locations and structure of the tree species was made to optimize different conservation strategies and to provide fruit for wildlife and people.

**Table 15.2** The Five Elements of the Garden Design Framework and their Benefits

| Framework Element | Area established | Main Ecological Function | Human wellbeing | Species Examples (Scientific names in Table 15.1) |
|---|---|---|---|---|
| **Hedgerow** | $480m^2$ outside garden<br>480 $m^2$ in 4 garden border hedgerows; each with ~ area of $120m^2$. Individual hedgerows have 58 plant spp.;<br>30 spp.,18 spp. and 22 spp.(average is 32 spp.). | Structurally complex corridors connecting different habitats<br>Shelter from predators<br>Food & nesting sites<br>Carbon sink | Sound and windbreaks<br>Privacy<br>Visual diversity<br>Scientific knowledge<br>Multidimensional complexity and sensorial development | Trees<br>e.g. American oil palm; pink trumpet tree<br>Large bushes<br>e.g. *Sanchezia* sp. (flowers attract hummingbirds);<br>Yesterday, today and tomorrow;<br>Brazilian red cloak<br>Vines<br>e.g. Sandpaper vine; Kings mantle<br>Medium height<br>Heliconias, Dumbcane, Gingers<br>Lower foliage<br>e.g. Philodendron; Calatheas |
| **Flowerbed/ islands** | 4 large islands ($>12m^2$) with mixed species. Total area $219m^2$; ~ individual area $55m^2$; with ~12spp. each | Safe transit across garden between hedgerows<br>Nesting habitats and food<br>Shelter from strong wind and heavy rain | Visual beauty from plant colors and mixes and variety of pollinator species including hummingbirds and bees<br>Creativity (shapes and mixes)<br>Olfactory diversity<br>Alignment with different gardening schools | Flowering bushes: e.g. Pigeon berry, large Ixora, Hibiscus, Snakeweed, Bougainvillea<br>Flowers: Peace lilies, small Heliconias, Mexican heather, Mexican petunias |
| | 6 smaller islands ($<12m^2$) totaling 55 $m^2$; ~ area $8.5m^2$; ~ 4spp. | | | Small shrubs: Golden trumpet, Lantana, Firecracker<br>Flowers: Mexican heather<br>Ground cover: *Tradescantia* spp., small prayer plants |
| | 18 m long flowerbeds with fewer species, total 291 $m^2$; ~$16m^2$; ~2spp. / flowerbed | | | *Lantana* spp.<br>*Heliconia* spp.<br>Mexican petunias, dwarf jungle geraniums |
| | | | | Golden trumpet<br>Hibiscus |

(continued)

**Table 15.2** (continued)

| Framework Element | Area established | Main Ecological Function | Human wellbeing | Species Examples (Scientific names in Table 15.1) |
|---|---|---|---|---|
| | 4 small flowerbeds 56m$^2$; ~2.65m$^2$; ~3spp. | | | |
| **Trees in garden including in hedgerows** | 14 existing at start.<br>64 new trees planted from 31spp. (including 21 fruit trees from 9 spp. and 3 self-planting Royal Palms)<br>5 stand-alone bushes | Canopy level connectivity with forest<br>Increased habitats, nesting sites and food<br>Carbon sink<br>Shade | Sensory stimulus movement, light and shade, visual beauty<br>Fruit trees | Flowering trees: Pink powder puff, peacock flower<br>Fruit trees: Jabuticaba, papaya, soursop, breadfruit, guava, mango, Banana, coconut, cacao<br>Palms: Peach fruit palm, American oil palm |
| **Paths and resting places** | 144m$^2$ paths<br>6 seating areas / decks<br>100m$^2$ Burle Marx bromeliad garden with 27 bromeliad species | Paths provide transit across garden for small mammals e.g. agoutis<br>Seating areas/ decks provide feeding spaces | Access to, and discovery of, garden diversity and spaces<br>Spaces for relaxation, meditation, observation | Paths of granite stones used in building and cement<br>Resting places of cobblestones with rockeries of large rocks unearthed from flowerbeds and pool excavations<br>Decks from ceramic and coralline tiles<br>Painted benches, aluminum chairs and tables |
| **Water** | Garden fountain<br>Swimming pool | Drinking and cooling water for birds and small mammals<br>Hunting ground for insectivorous animals; attracts migratory birds; increases humidity in dry season | Sensory benefits, therapeutic sounds, visual interest, and facilitates wildlife sighting<br>Pool provides exercise reducing stress and relaxation, reflection and meditation areas | Garden fountain: 3-tiered trickle-down fountain attracting tanagers (blue-grey, crimson-backed), and small mammals e.g. agoutis<br>Pool: 12x 4 meters headed with raised waterfall, submerged steps facing pool for relaxation and meditation.<br>Attracts green iguana; basilisk lizard (Jesus Christ lizard); social flycatcher birds |

The fourth element of the framework was garden features, including pathways and resting places. An important element of a garden is to have easy and safe access to its different corners, so we gradually installed paths made from porous material such as stones, pebbles and bricks for drainage. Since more effort will likely be put into the management and conservation of a garden where enjoyment and wellbeing can be derived, we created resting places with benches or tables along the paths. Over time, as new trees grew, shade became too dense for grass, so we laid pebbles dotted with shade-loving species. In the largest of these areas, we created a bromeliad garden using unearthed rocks and driftwood collected from beaches. This provided visual diversity and new habitats for many animals, especially invertebrates.

The fifth element of the framework is water. Annual rainfall in Panama is high, but four months are very dry, hot, and windy, creating harsh conditions for many animals and plants. We first installed a small garden fountain in the pebbled area. It provides drinking and bathing water for different species and a soft background murmuring that contributes to human wellbeing. We later added a swimming pool to a steep, exposed, grassy slope that was hard to mow. The pool is headed by a waterfall that provides additional soothing sounds. It is surrounded by multiple plant species that attract many animals including native bees and hummingbirds and has provided a new hunting ground for others such as lizards and fly-catchers. We have even seen iguanas drink from the pool during the dry season (Fig. 15.4).

### *15.2.4 Observations on Biodiversity Changes in the Garden Over 15 Years*

Over 15 years, the property has been transformed from a sea of grass to a garden with at least 180 vascular plant species from 120 genera and 48 families (varieties and cultivars are not included). There has been an 80% reduction of grass-areas and, using mainly those native to the Neotropics, the planting of 64% of the total area as hedgerows, mixed and single plant species flowerbeds, and 64 trees (Figs. 15.2). The different garden-design elements have provided biodiversity and human wellbeing benefits (Tables 15.2 and 15.3).

In 2005, the only wildlife we observed in the garden were a few agoutis, occasionally groups of coatis and grass seed-eating birds. Over the years we have seen an increase in wildlife species in the garden, except for coatis which have diminished considerably. In mammals, new species include the paca, the northern anteater, Geoffroy's tamarin, the nine-banded armadillo, and variegated squirrels. We have observed that these mammals move to and from the larger forest patch along the corridors formed by hedgerows. The number of agoutis in the hedgerows and large flowerbeds also have increased considerably. Between 10 and 20 agoutis can be seen feeding in the garden in the early mornings. We have seen an increase in the litter size of agoutis that reproduce in the garden. The litter of agouti is usually 1 or 2 offspring (Smythe 1978). For the first 5 years, this was the case in the female

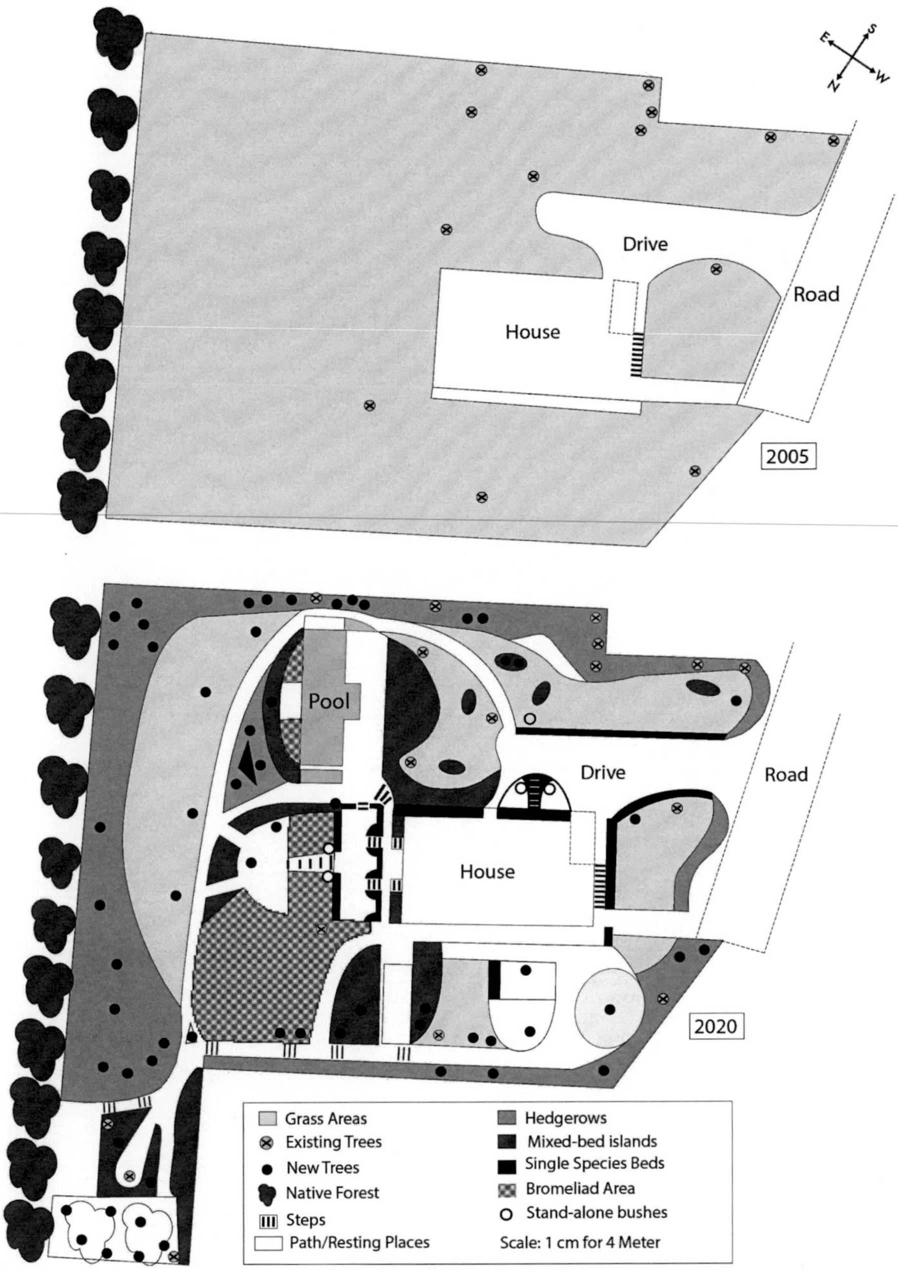


**Fig. 15.2** Diagrammatic Illustration of the Increase in Area of the Different Garden Design Elements from 2005 to 2020

**Table 15.3** Examples of Successful Plant-types Used in the Garden Design Framework. (All native to the Americas except for those marked * which are Asian in origin, now naturalized throughout the tropics)

| Plant Type | Examples and Ecological Functions and Human wellbeing |
|---|---|
| **Flowering bushes and trees** | Pigeon berry: (Neotropical) plants of different sizes with yellow berries, with white, pink or blue flowers. Dwarf plants good for low borders and standard size for shaping into hedges. Attracts bees as pollinators in gardens and vital for 70% of crops that feed 90% of global population. |
| | Sanchezia: Plants native to South America; flowers year-round with colored leaves, bracts and large orange or yellow tube-shaped flower that attract many pollinators, especially hummingbirds. Grows well in dappled shade in the garden as medium size bushes. Good for hedgerows |
| | Yesterday, today and tomorrow: With blooms lasting 3 days starting violet-purple as a signal to pollinators that the flower is full of nectar and pollen; then fading to pale lavender-blue and then white that are no longer attractive, increased pollination efficiency. |
| **Colorful hardy bracts withstanding the heavy rains** | Heliconia: e.g. Parrot's beak flowers, false birds paradise, with 195 known species most native to tropical America, provide color, aesthetic shapes and lush foliage year-round; propagate very quickly, attract hummingbirds and are low maintenance. Some rhizomes are medicinal. |
| | Bougainvillea: (native to South America) visual beauty from large colorful bracts (pink, magenta, purple, red, orange, white) surrounding 3 simple waxy white or pale yellow flowers; thorny stems provide predator protection for some nesting birds. Leaves are medicinal. |
| | Red and pink ginger*: Provides long-lasting color and yearlong lush foliage affording protection for wildlife transit. Once mature, ginger produces offshoots from the flower bracts that enable easy propagation. Although Asian in origin now naturalized throughout the tropics. |
| **Hardy flowers withstanding heavy rains** | Jungle geranium (Ixora*): Colorful, tropical shrub with numerous named cultivars differing in flower color (white, yellow, pink, orange) and plant size (dwarf through king), easy to maintain, good for borders, foundation, large flowerbeds or standalone shrubs of small tree. Attracts numerous native bee species. Native to tropical and sub-tropical areas throughout the world (although world center of diversity is Tropical Asia). |
| | Mexican heather: Small hardy flowers, attracts numerous native bee species. |
| **Delicate flowers but with copious, year-round flowering** | Lantana: Small, tubular blooms with abundant nectar attract many pollinators particularly butterflies and colorful Euglossa bees. |
| | Snakeweed: Year-round flowering, attracts pollinators for native bees, especially colorful Euglossine bees, hummingbirds and butterflies; leaves are toxic to insects so maintains shape well, can be used for medicinal tea. Requires frequent pruning and can be well shaped to different heights. |

(continued)

**Table 15.3** (continued)

| Plant Type | Examples and Ecological Functions and Human wellbeing |
|---|---|
| | Firecracker: Colorful year-round flowers (red or yellow) that attract pollinators including hummingbirds. Tolerates full sun to partial shade, grows quickly and has dense foliage good for overhanging walls. Provides visual diversity due to weeping form and thin leaves. |
| **Bromeliads** | Provide visual diversity, are easy to maintain and useful in tree trunks or shady steep spaces, excellent for stony resting spaces for meditation. *Guzmania* spp. and *Nidularium* have colored long lasting flowers; *Tislandia* and *Vriesa* have smaller but brightly colored flowers. Other genera such as *Neoregelia,* and *Crypanthus* have small flowers but have colorful leaves year-round. Some such as pineapple have edible fruits. Complex architecture and role in water reservoirs make bromeliads biodiversity amplifiers, providing habitats for frogs, insects and spiders. 3499 of the 3500 species are from the Neotropics (Baench and Baench 1994). |
| **Foliage** | Ground cover e.g. pinto peanut and small Maranta prayer plants, provide year emerald color. *Tradescantia* spp. provides purple and pink colors. |
| | Medium height foliage Larger Prayer plants, including the *Calathea* genus, provide a variety of shapes and colors of leaves and flourish in shady conditions |
| | Taller foliage Dumbcane is hardy and shade-loving with large, oblong, and cream or light-yellow leaves with deep green spots and stripes. Can reach up to 6 feet; Peace lilies: Large green leaves year-round and white cup shaped flowers. |
| | Climbing and ground: Philodendrum, glossy leaves, fast growing. |
| **Fruit trees** | Peach palm (Pisba), soursop, mangos, guava, and jabuticaba provide feeding sources for a variety of animals and food for humans if people manage to pick them before they are eaten by wildlife. |
| **High volume flowering tree** | Pink and red powder puff trees provide food for many species of insects, birds, butterflies and moths. Flowers come out in the early morning and attract daytime pollinators, and again early evening for nocturnal species. The flower-laden, umbrella-shape provides shade for garden tables.<br>Peacock flower tree: Seed pods provide food for birds such as parrots and squirrels. |

agoutis we saw in the garden with their offspring. At least 3 pairs now breed in the garden in burrows under the pool and along one hedgerow. In two of these pairs, the litter size has increased to three, which is normally only found in food rich and safe environments or captivity (Smythe 1978; Aliaga–Rossel et al. 2008). Also, we have observed an increase in litter size in the Geoffroy's tamarin that started coming near the house in 2014 once the new trees provided some canopy connectivity with the forest 50 m away. In 2015 the pair had one offspring, in 2017 two offspring and this year (2020) a further two (Fig. 15.3 (3)).

**Fig. 15.3** Examples of Wildlife Conservation Effectiveness
(1) Examples of Bird Species Observed in the Garden as Habitat Diversity Increased. *Top left image*. Left column, top to bottom: Resident birds: Blue-gray tanager and Social Flycatcher. Right column, top to bottom: Migrant birds: Baltimore Oriole, Summer Tanager, Yellow Warbler
(2) Forest-dwelling Bird Species Observed in Garden as Habitat Diversity Increased. *Top right image*. Top left to right, Blue Crowned Motmot, Barred Ant Strike. Bottom left to right: Slaty tailed Trogon, Violaceous Trogon
(3) Observed Increased Reproductive Success (increased litter size). *Bottom left image*. Three Young Agoutis Feeding (top) and Geoffroy's Tamarin with Young (bottom)
(4) Examples of Increased Species of Bee Pollinators. *Bottom right image*. Right column top to bottom, Euglossine Bee spp. on Hibiscus and Snakeweed flowers. Left column, Native Bees on Peacock Flower Tree, Jungle Geranium and Periwinkle flowers.

The insect biota has also increased. In 2005, the main insects in the garden were mosquitos, ants, fireflies and a few butterflies. Now countless species of insects can be seen flying between the canopy of trees and understories in the hedgerows and large bushes in flowerbeds. Many of these insects are pollinators, such as 10 species of native bees (Fig. 15.3 (4)), at least 7 wasp species, and a range of butterflies including the blue morpho. Butterflies in the genus *Morpho* are noticeable components of neotropical forests and fly at either canopy or understory height with characteristic gliding and flapping flight patterns (DeVries et al. 2010). We now frequently see this forest species at midday flapping along the hedgerows and across the garden along flowerbed islands.

The most dramatic and well registered increase in wildlife has been in bird species. Bird watching is a constant activity for us and has included, on average, one-hour observation periods from the same spots and times, twice a day, 5 days a week over the 15 years. In 2005, some nine species frequented the garden. These were mainly grass seed eaters. We have now recorded 157 bird species in the garden from 43 families and 101 genera. Based on our observation and knowledge of birds, along with bird literature (Angehr 2010; Audubon 2019), we have categorized these birds into three main groups. The first group is native bird species that permanently feed in the garden, often also nesting in it. Examples of this group are Tanagers (crimson-backed, blue-and-grey, and palm), Euphonias, House Wren, Paltry Tyrannulet, Social Flycatcher and Rufous-tailed Hummingbird.

The second group is native bird species that frequent the garden temporarily for feeding or at different times in the year, but do not nest in it. These include Cotingas, Cuckoo Squirrel, Great Kiskadee, Streaked Flycatcher, Tropical Kingbird, Gray-headed Chachalaca and birds of prey such as the Gray Hawk and Pearl Kite. It also includes bird species that normally dwell and hunt in the forest and are now seen in the garden. An example of these are the Violaceous and Slaty tailed Trogons. They usually live in the forest, feeding on insects and fruits, but after seven years, started coming into our garden where they are now seen often. Another forest species that now comes into our garden is the Blue-crowned Motmot. A third is the Barred Ant shrike that feeds on insects in tangled and dense mid-story vegetation but that has recently started visiting the garden (Figs. 15.3 (1) and (2)).

The third group includes migratory birds that are known to stop over in Panama (Angehr 2010). We have seen migratory birds in the garden in September through November coming from the northern hemisphere autumn to the southern summer, and in April and May on the return journey. Of the 31 migratory species we have identified during our bird observation, some use the garden as temporary stop overs, staying for a few days or weeks as they fly south (e.g. Western-wood Pee-wee, some Warblers, Baltimore and Orchard Orioles). Other migratory species stay more time in the garden (e.g. the Summer Tanager) until flying back north, and in a few cases nest here (e.g. the Yellow-green Vireo). We do not know if other individuals of these species stop over before flying on to more southern sites. Our bird watching observations have recorded migrant bird diversity increasing for the first 8 years, but in the last 7 years there has been a reduction in some migratory birds such as the Baltimore and Orchard Oriole, several warbler species, the Scarlet Tanager and

Northern Water Thrush. This may reflect the loss of avifauna in North America and the steep decline in migrating birds from the use of pesticides and habitat loss there and along their routes (Eng et al. 2019; Rosenberg et al. 2019).

Observations of increased reproductive success and number of insects and pollinator species, such as native bees and hummingbirds, suggest the effectiveness of the garden as a biodiversity island for wildlife. Insect biota in tropical rain forests is distributed between the forest canopy and understory that form a vertical, structurally complex continuum with differing temperature, humidity, light levels and plant life (DeVries et al. 2010). Structural complexity and vascular plant composition in gardens also have been shown to be correlated to native biodiversity, especially invertebrate richness (Smith et al. 2006a, b). This appears to be the case in our structurally complex hedgerows and large flowerbeds that now house a plethora of insects. Their growth also roughly correlates to the observed increase in forest mammals in the garden. Increased vegetation and connectivity to the forest most likely has facilitated the movement of these mammals through the sheltered areas rich in food. We also suggest that the increase in litter size of agoutis and tamarins may be attributed to the increase of suitable habitat, fruits and seeds in the garden. This is also the case with birds. The increase in bird species and sightings of forest birds in the garden became more frequent as hedgerows and flowerbeds grew and structural complexity increased, suggesting their effectiveness in habitat diversification and connectivity with the forest.

## 15.3 Challenges

### *15.3.1 Challenges in Establishing our Garden*

There have been several challenges in creating diverse habitat and increasing biodiversity. One has been overcoming the physical limitations of the garden lot and the climate in Panama. Soil enrichment has been a main concern as soil in the garden was poor and had high clay content. Much of the soil for the first flowerbeds was brought in from nurseries. Since then, we have used humus-rich soil formed from decomposed garden plant material left for several seasons under the forest cover.

Whilst temperatures permit all year growing, water shortage in the dry season is acute. Animals such as agoutis dig up roots or gnaw trunks for water thus requiring the netting of some plants. Although the amount of foliage is greatly reduced in the dry season, many plants can be kept alive by watering daily. Many species flower in the dry season, but are not all able to withstand the extremely heavy rains of the wet season. Placing the most delicate plants in sheltered environments or in beds under the house eaves, enables flowering across seasons. In open spaces, we have planted small bushes with hardy flowers, or plants with colorful bracts that can withstand the rain and provide year-round color. Also, in the dry season, winds are high and blow palm leaves down, causing damage to plants and potentially to humans.

New palms have been located only on hedgerows. The native Royal Palms are self-seeding and fast growing. Although the agoutis help control their spread by eating large numbers of their seeds, several have grown in the outer edges of the forest with the danger of displacing other species. We now remove younger seedlings and reforest the forest edge with other native species. For this, we collect seeds from forest species.

Overcoming these challenges required time and knowledge, often obtained from trial and error. These challenges were exacerbated by a lack of well-stocked nurseries. In general, non-native species are preferred over native species in urban designs because of their ornamental qualities rather than their ecological function (Quigley 2011), which leads to a homogenization of plants available in nurseries worldwide (Ignatieva 2011). In 2005 there was little variety of native plants available in nurseries in Panama, and many were from other regions. Nursery staff were unable to provide guidance on suitable native plants for different light, soil and temperature conditions. Books on tropical plants from nearby countries that described their characteristics, growth, and management were available (e.g., Gentry 1996; Zuchowski and Forsyth 2007). While these are important as field guides for naturalists, they are not user-friendly gardening books with information on mixes of species to be used in urban residential gardens.

City infrastructure and human activity can change the local light, temperatures and hydrologic systems (Güneralp and Seto 2013) and native plantings often fail because species are not adapted to urban environments (Quigley 2011). Even within the garden, these conditions have changed over time, particularly with the growth of large canopy trees changing the light and to some extent temperature conditions of original planting sites, thus requiring relocation of plants. Our experiences have provided several recommendations of plant-types that can be used in different garden-design elements (Table 15.3).

Associated with lack of knowledge is the challenge of using native versus naturalized plant species to maintain the balance between optimizing biodiversity and enhancing more traditional aesthetic values. Native plant species grow especially well as they are suited to both the weather conditions of the area and have been in co-evolution with native insects and animals. However, many of the most visually attractive and more well-known garden plants come from other regions and have been the focus of horticulturalist cultivars and gardening guidance.

Some of these species are so widely used in gardening that they have become naturalized in many parts of the world, adjusting to the local environment and sustaining populations without direct intervention from humans. While some can help in the process of ecological restoration, some exotic plants can however become invasive, with adverse effects on the local ecosystem (Parrota et al. 1997; Senbeta et al. 2002; Carpenter et al. 2004; McNamara et al. 2006; Berens et al. 2008; Lamb and Madsen 2012). As much as possible, we use only species native to Panama or other Neotropical countries. The exception is some species that are naturalized throughout the tropics and have been established in Panama for many years such as the *Alpinia* (ginger) and *Ixora* (jungle geranium) varieties and cultivars. Both provide high visual impact and appear to offer shelter and food for native fauna. We have observed wildlife moving freely through ginger plants near the forest and ixoras attract a wide range of native bees.

Another challenge that will affect the long-term sustainability of our garden biodiversity and the replication of our design framework at scale is the attitude and practices of neighboring homeowners. The first hedgerows were established in cooperation with keen gardening neighbors that shared costs and helped with planting. The gardens of the current surrounding homeowners are less well-tended and their owners may not have biodiversity conservation in mind. Many are regularly sprayed with pesticides to keep insects and vermin at bay and to decrease the spread of tropical diseases such as dengue. Pesticide use is detrimental to biodiversity. Recent work indicates that the population dynamics of pollinator communities are less resilient to chemical control of plants and insects in fragmented urban habitats (Muratet and Fontaine 2015). Few detailed studies have examined impacts of the amount and type of chemical applications in urban areas, and more should be done to promote the use of alternatives to chemical control. The use of pesticides and herbicides in public areas should be managed with the conservation of adjacent residential gardens in mind.

In pursuit of our biodiversity conservation goal, we have kept the use of pesticides and fungicides to a minimum. The proximity to the forest with large nests of leaf cutter ants has made this a challenge. When large numbers of ants have come into the garden, stripping entire trees and bushes, we have used limited amounts of site-specific insecticide to reduce populations. The control of grasshoppers, caterpillars and green iguanas has also been a challenge. As they contribute to the overall biodiversity of the garden, we have not used manufactured pesticides to control their populations. The use of natural repellents (garlic, tobacco and pepper), manual removal, and replenishment of plants has let us maintain populations of these herbivores and enjoy flowering plants and butterflies. However, we have had to change our attitude, relinquishing the vision of pristine, shiny whole leaves depicted in traditional gardening books, and accept the intrinsic beauty of insect-eaten leaves as part of the attraction of our biodiverse garden (Fig. 15.4).

### *15.3.2 Challenges in Scaling-up Biodiversity Gardens*

Individual residential gardens can contribute to biodiversity conservation in cities, particularly when garden design and maintenance frameworks are developed accordingly. Their contribution will depend in part on the number of gardens within a city that adopt a biodiversity garden design approach. There are few examples of this available in the tropics, but this case study has provided a number of lessons that could be used for scaling up. Nonetheless, many knowledge gaps and questions remain and could be addressed with input from the scientific community. For example, we recorded a significant increase in bird species in the garden over 15 years and hypothesize that this is related to increased habitat that provides food and shelter. However, an important element of sampling is to determine what

**Fig. 15.4** Sharing the Garden Space with Wildlife
Top left image: Green Iguana drinking from pool and eating garden plants.
Top Right Image: Caterpillar eating Golden Trumpet leaves
Bottom image. Insect eaten leaves (part of the intrinsic beauty of a biodiversity garden)

constitutes a count: a single individual of a species or a viable population (Muller et al. 2013). For this, more systematic and uniform monitoring of populations is required.

It is also important to define potential wildlife carrying capacities within the garden space. In this case study, our own observations seem to indicate that bird population increases have plateaued and several native species such as the Clay-colored Thrush, Palm tanagers, Blue and Gray tanager, Gray-headed Chachalaca and Crimson-backed Tanagers are increasing. The effect these have on displacing other native species remains to be seen. The recent absence of the Tody Flycatcher from the garden could be a sign of such displacement.

Similarly, research and community collaboration are needed to identify methods to control the spread of more aggressive birds in cities. In Clayton, for example, we have observed a growing population of the Great-tailed grackle (Talingos) preying on young iguanas and other lizards and on eggs in the nests of native birds. They were absent from the area until 4 years ago, but can now be seen in new housing development areas during construction, and in nearby gardens where we have observed them eating pet food left outside. Panama City has several scientific institutions that could provide this type of research, including its national universities, branches of international ones such as Florida State and McGill, and the Smithsonian Tropical Research Institute.

Garden-size is another factor that will affect the impact on biodiversity. The relatively small spaces of some gardens may not guarantee enough foraging space or breeding habitats for long-term survival of some wildlife. Small, stand-alone gardens may be able to support small populations of some species, but increased isolation tends to reduce population and gene-flow among patches with the risk of deterioration of gene pools (McDonald et al. 2013). In this case study we had direct intervention over $1700m^2$ and an indirect impact over $9000m^2$ via our hedgerows connecting these areas to forest patches adjacent to a large national park. Impacts on biodiversity could therefore be expected to be considerable. However, residential gardens in Panama City are rarely this size, although several new housing developments have gardens and big apartment complexes are creating large ornamental gardens in residential common spaces.

The location of gardens relative to each other and across the city also will affect their contribution to biodiversity. Improving connectivity between gardens and other UGS within city limits will increase the positive contribution from small garden sizes. For this, the role of residential gardens should be recognized in policy frameworks and biodiversity strategies to facilitate incentives and focus priority action in specific urban settings (Rudd et al. 2002). Their role should also be integrated into urban land use planning and monitored. In this case, our garden is in an area designated by law as a "garden city" with land-use planning and norms that favor the role of gardens in conservation. Although the law, for example, requires garden fences to leave gaps for wildlife transit, this is largely ignored by homeowners, and enforcement is missing.

Networks of like-minded gardeners that adopt similar approaches to managing their properties can also enhance benefits over larger scales. The importance of networks of gardens for conserving avian diversity is well-documented (White et al. 2005; Palomino and Carrascal 2006). Similarly, networks for planting native vegetation in gardens adjacent to creeks have been advocated for enhancing habitat connectivity in riparian corridors in New South Wales, Australia (Parker et al. 2008). In this case study, we were able to plant hedgerows along bordering properties with likeminded gardeners, but this needs to be extended across larger areas. Homeowner associations can provide a vehicle for setting up networks (Lerman et al. 2012) and could be promoted in new housing developments here. Similarly, NGOs have a role, for instance, the USA National Audubon Society offers the *Healthy Yard Pledge* for gardeners to commit to specific management principles.

The Panama Audubon Society could support networks for monitoring of bird species and populations in gardens, including migratory species using these as stop-over sites.

In addition to garden networks, research has shown that all UGS could interact synergistically to support biodiversity when clustered together (Colding 2007), and that the presence of adjacent gardens can increase the species richness of urban parks (Chamberlain et al. 2004). Our garden is situated near the visitor center of one of the most visited parks in the city, the Metropolitan National Park. Our garden's location next to the park could serve to showcase how biodiversity friendly gardens adjacent to parks could contribute to the conservation of their biodiversity. The visitor center is also close to two large plant nurseries. Nursery staff could be trained on native plant species and on the conditions for growing them in residential gardens here. Ideally, they should avoid stocking exotic plants that have the potential to become invasive and focus on species native to the country or region. They could also be stocked with user-friendly guidelines, in local languages, on garden design and species that contribute to biodiversity conservation. Such guidelines still need to be developed.

For these upscaling approaches to gain traction, undoubtedly the most important challenge is that of the choices and attitudes of individuals. Neighbors and city planners' attitudes and behavior will determine whether or not our garden will remain as a biodiversity island, and whether other property owners adopt similar approaches. The choices people make in terms of what they plant, and how they structure and manage their garden (behavior), will influence biodiversity outcomes. Landscape aesthetics and lifestyle factors often take precedence over ecological concerns (Larson et al. 2009; Kiesling and Manning 2010). For behavioral change to occur, a first step is to overcome knowledge gaps. Attitudes determine behaviors, but actual behavior change is the final step in a transition from increasing knowledge, to attitude change, to the intention to change behavior, and finally to actual behavior change itself (Orams 1999).

This case study provides an example of what can be achieved when choices are made that favor biodiversity conservation. In Panama, and in many other tropical developing countries, closing knowledge gaps needs to be complemented by strengthening awareness around biodiversity and residential gardening. Multimedia awareness programs are needed on the role gardens play in biodiversity conservation and in the provision of ecosystem services on which the city depends (Gómez-Baggethun et al. 2013). This includes biodiversity conservation as a natural solution in the fight against climate change and in controlling its effects, such as flooding, in urban areas. Recent research has presented ways of engaging citizens in exploring biodiversity and ecosystem services in residential gardens (Beumer and Martens 2015), and has shown that learning-focused initiatives can enhance biodiversity on private property (Diduck et al. 2020). For this and for self-guided learning, a wider range of supportive materials are needed. This includes developing user-friendly gardening design books in Spanish.

Awareness building and behavioral change are a long-term endeavor, but there are some favorable conditions for this in Panama. Plants grow very quickly here, and change occurs in a relatively short timescale, thus avoiding one of the challenges faced in biodiversity-focused attitude-change initiatives in the UK (van Heezik et al. 2012). Ultimately, however, behavioral change will only come about once individuals experience first-hand the many benefits that gardens and gardening provide to human wellbeing, including the aesthetic, therapeutic, and spiritual values of a garden (Lin et al. 2017; Raymond et al. 2019). In addition to biodiversity benefits, the design elements we developed in our garden have provided benefits important in our wellbeing. These include food provision (fruit trees) and exercise (gardening and swimming), as well as benefits from observing and listening to the more subtle voices of the garden. There is increasing recognition of positive associations between species diversity and psychological and physical well-being (Barton and Pretty 2010; Aerts et al. 2018); of beneficial aspects of the biodiversity and spiritual well-being nexus (Irvine et al. 2019); and of the correlation between biodiversity and spiritual and aesthetic experiences (De Lacy and Shackleton (2017). In our experience, spiritual and aesthetic benefits from the garden have risen alongside the increase in habitats and biodiversity. These include *mindfulness* benefits such as meditation and relaxation; *sensory benefits* such as light and shadow, tactile, olfactory, and auditive senses; and *scientific benefits* such as knowledge, observation and garden science. All of these augment our fulfillment from the garden and sustain our commitment to its continued management.

## 15.4 Conclusions

The case study presents the changes in biodiversity over 15 years in a residential garden established using a design framework to enhance biodiversity conservation. The design framework included creating hedgerows and large flowerbeds of native and naturalized species managed with a minimal amount of pesticide. New habitat patches were established across and along the garden, creating connectivity between these and the forest, reducing fragmentation and providing corridors for wildlife. Thus, as well as increasing biodiversity in the garden, this design framework has likely reduced the edge effect on forest species, potentially contributing to the long-term survival of the native forest remnant.

Our observations and supporting literature have led us to hypothesize that the increase in wildlife that we have observed is a consequence of the creation of a structurally complex and plant species-rich garden that has been successful in supporting high levels of biodiversity. But this hypothesis needs to be tested. Replication and complementary studies are needed. The garden is in a privileged location near a native forest patch. It is not universally replicable, but there are large intact forest patches in Panama City that could be favorable to replicating the experience, especially in Clayton and in new urbanizations such as Panama Pacific, Sendero de Cruces, Portanorte, and Green City.

Whilst direct replication may be restricted to areas closer to native forest, smaller garden spaces further away are important for creating new habitats and conserving many native species. However, to optimize the contribution of all gardens to conservation, independent of their size, specific biodiversity considerations should be included in the garden design framework. These have been outlined in the garden design presented in this case study. Recommendations for plants that can be used in the different elements of the garden design have been provided.

To achieve uptake of this approach, and take it to scale, recommendations have been made in the previous section that could be applied in Panama City. Thus far, scaling up has been documented in developed countries where gardening cultures and urban planning are strong. Ironically, this does not often correlate with the richest biodiversity under threat and where urban expansion is likely to be highest. Promoting residential gardens as urban islands of biodiversity in biodiversity hotspots in the tropics could bring multiple biodiversity benefits in threatened areas, along with improved physical and mental health to large parts of urban populations.

**Acknowledgements** We thank our parents for instilling in us the love of nature and gardens and their importance in human wellbeing. We would also like to thank Alexander Coles from the Florida State University in Panama, who first inspired us to share the design elements of our garden with his students. Thanks also to all others that have showed interest in the garden, not least of which is our gardener, Poto, who has helped us for 15 years with the most strenuous of our tasks, and those that have visited our Instagram account @voicesofthegarden.

## References

Aerts R, Honnay O, Van Nieuwenhuyse A (2018) Biodiversity and human health: mechanisms and evidene of the positive health effects of diversity in nature and green spaces. Br Med Bull 127: 5–22

Alberti M (2005) The effects of urban patterns on ecosystem functions. Int Reg Sci Rev 28:168–192

Aliaga-Rossel E, Kays RW, Fragoso JMV (2008) Home-range use by the Central American agouti (*Dasyprocta punctata*) on Barro Colorado Island, Panama. J Trop Ecol 24:367–374

Angehr GR (2010) The birds of Panama. A field guide. Comstock Publishing Associates, Cornell University Press, Ithaca

Aronson MFJ, La Sorte FA, Nilon CH (2014) A global analysis of the impacts of urbanization on bird and plant diversity reveals anthropogenic drivers. P Roy Soc B 281:20133330

Aronson MFJ, Lepczyk CA, Evans K (2017) Biodiversity in the city: key challenges for urban space management. Front Ecol Environ 15(4):189–196

Audubon (2019). http://www.audubon.org/bird/at_home/. Accessed July 2019

Baench U, Baench U (1994) Blooming Bromeliads. Tropical Beauty Publishers, Nassau

Barton J, Pretty J (2010) What is the best dose of nature and green exercise for improving mental health? A multi-study analysis. Environ Sci Technol 44:3947–3955

Berens DG, Farwig N, Schaab G, Böhning-Gaese K (2008) Exotic guavas are foci of Forest regeneration in Kenyan farmland. Biotropica 40:104–112

Beumer C, Martens P (2015) Biodiversity in my (back)yard: towards a framework for citizen engagement in exploring biodiversity and ecosystem services in residential gardens. Sustain Sci 10(1):87–100

Cameron RWF, Blanusa T, Taylor JE, Salisbury A, Halstead AJ, Henricot B, Thompson K (2012) The domestic garden – its contribution to urban green infrastructure. Urban For Urban Greening 11(2):129–137

Carpenter FL, Nichols JD, Sandi E (2004) Early growth of native and exotic trees planted on degraded tropical pasture. For Ecol Manag 196:367–378

Chacon M, Harvey CA (2006) Live fences and landscape connectivity in a neotropical agricultural landscape. Agrofor Syst 68:15–26

Chamberlain DE, Cannon AR, Toms MP (2004) Associations of garden birds with gradients in garden habitat and local habitat. Ecography 27:589–600

Cilliers S, Cilliers J, Lubbe R, Siebert S (2013) Ecosystem services of urban green spaces in African countries – perspectives and challenges. Urban Ecosyst 16:681–702

Colding J (2007) Ecological land-use complementation for building resilience in urban ecosystems. Landsc Urban Plan 81:46–55

Conniff ML, Bigler GE (2019) Modern Panama: from occupation to crossroads of the Americas. Cambridge University Press, Cambridge

Daniels GD, Kirkpatrick JB (2006) Does variation in garden characteristics influence the conservation of birds in suburbia? Biol Conserv 133:326–335

Davies Z, Fuller R, Loram A, Irvine K, Sims V, Gaston K (2009) A national scale inventory of resource provision for biodiversity within domestic gardens. Biol Conserv 142:761–771

Davies ZG, Dallimer M, Fisher JC, Fuller RA (2019) Biodiversity and health: implications for conservation. In: Marselle M, Stadler J, Korn H, Irvine K, Bonn A (eds) Biodiversity and health in the face of climate change. Springer, Cham, pp 283–294

De Lacy P, Shackleton C (2017) Aesthetic and spiritual ecosystem services provided by urban sacred sites. Sustainability 9(9):1–14

DeVries PJ, Penz CM, Hill RI (2010) Vertical distribution, flight behavior and evolution of wing morphology in Morpho butterflies. J Animal Ecol 79:1077–1085

Diduck AP, Raymond CM, Rodela R, Moquin R, Boerchers M (2020) Pathways of learning about biodiversity and sustainability in private urban gardens. J Environ Plan Manage 63(6): 1056–1076

Eng ML, Stutchbury BJM, Christy A, Morrissey CA (2019) A neonicotinoid insecticide reduces fueling and delays migration in songbirds. Science 365(6458):1177–1180

Evans KL, Newson SE, Gaston KJ (2009) Habitat influences on urban avian assemblages. Ibis 151: 19–39

Galluzzi G, Eyzaguirre P, Negri V (2010) Home gardens: neglected hotspots of agro-biodiversity and cultural diversity. Biodivers Conserv 19:3635–3654

Gaston KJ, Smith RM, Thompson K, Warren PH (2005) Urban domestic gardens (II): experimental tests of methods for increasing biodiversity. Biodivers Conserv 14(2):395–413

Gaston KJ, Cush P, Ferguson S, Frost P, Gaston S, Knight S, Loram A, Smith RM, Thompson K, Warren PH (2007) Improving the contribution of urban gardens for wildlife: some guiding propositions. British Wildlife 18(3):171–177

Gazeta Official NO 23323 Republic of Panama (1997). https://www.gacetaoficial.gob.pa /gacetas/ 233231997.pdf Accessed January 2020

Gentry AH (1996) A field guide to the families and genera of Woody plants of North west South America. Conservation International, Washington DC

Goddard M, Dougill A, Benton T (2010) Scaling up from gardens: biodiversity conservation in urban environments. Trends Ecol Evol 25:90–98

Gómez-Baggethun E, Gren A, Barton DN, Langemeyer J, McPhearson T, O'Farrell P, Andersson E, Hamstead Z, Kremer P (2013) Urban ecosystem services. In Elmqvist et al. (eds) Urbanization, biodiversity and ecosystem services: challenges and opportunities: a global assessment. Springer, Dordrecht, pp. 201–277

González-García A, Gómez Sal A (2008) Private urban greenspaces or "patios" as a key element in the urban ecology of tropical Central America. Hum Eco 36:291–300

Gonzalez-Garcia A, Bellure J, Gomez Sal A (2009) The role of urban greenspaces in fauna conservation: the case of the iguana *Ctenosaura similis* in the ¨patios¨ of Leon city, Nicaragua. Biodivers Conserv 18:1909–1020

Güneralp B, Seto KC (2013) Futures of global urban expansion: uncertainties and implications for biodiversity conservation. Environm Res Letters 8:014025

Harvey CA (2000) Colonization of agricultural windbreaks by forest trees: effects of connectivity and remnant trees. Ecol Appl 10(6):1762–1773

Ignatieva M (2011) Plant material for urban landscapes in the era of globalization: roots, challenges, and innovative solutions. In: Richter M, Weiland U (eds) Applied urban ecology: a global framework. Blackwell Publishing, Hoboken, pp 139–161

Irvine KN, Hoesly D, Bell-Williams R, Warber SL (2019) Biodiversity and spiritual well-being. In: Marselle M, Stadler J, Korn H, Irvine K, Bonn A (eds) Biodiversity and health in the face of climate change. Springer, Cham, pp 213–250

Ives CD, Lentini PE, Threlfall CG, Ikin K, Shanahan DF, Garrard GE, Bekessy SA, Fuller RA, Mumaw L, Rayner L, Rowe R, Valentine LE, Kendal D (2016) Cities are hotspots for threatened species. Global Ecol Geog 25:117–126

Jaganmohan M, Vailshery LS, Gopal D, Nagendra H (2012) Plant diversity and distribution in urban domestic gardens and apartments in Bangalore. India Urban Ecosyst 15:911–925

Kane, Warren PS, Lerman SB (2015) A broad scale analysis of tree risk, mitigation and potential habitat for cavity-nesting birds. Urban For Urban Green 14(4):1137–1146

Kiesling FM, Manning CM (2010) How green is your thumb? Environmental gardening identity and ecological gardening practices. J Environ Psychol 30:315–327

Kumar BM, Nair PKR (eds) (2006) Tropical homegardens: a time-tested example of sustainable agroforestry. Springer, Dordrecht

Lamb D, Madsen SJ (2012) What is Forest landscape restoration? In: Stanturf J, Lamb D, Madsen P (eds) Forest landscape restoration: integrating natural and social sciences. World Forests XV, Springer, Dordrecht, pp 3–24

Larson KL, Casagrande D, Harlan SL, Yabiku ST (2009) Residents' yard choices and rationales in a desert city: social priorities, ecological impacts, and decision tradeoffs. Environ Manag 44:921–937

Leon MC, Harvey CA (2006) Live fences and landscape connectivity in a neotropical agricultural landscape. Agrofor Syst 68(1):15

Lepczyk A, Aronson MFJ, Evans KL (2017) Biodiversity in the city: fundamental questions for understanding the ecology of urban green spaces for biodiversity conservation. BioSci 67:799–807

Lerman SB, Turner VK, Bang C (2012) Homeowner associations as a vehicle for promoting native urban biodiversity. Ecol Soc 17(4):45

Lin BB, Gaston KJ, Fuller RA, Wu D, Bush R, Shanahan DF (2017) How green is your garden? Urban form and socio-demographic factors influence yard vegetation, visitation, and ecosystem service benefits. Landsc Urban Plan 157:239–246

Luck GW, Smallbone LT, O'Brien R (2009) Socioeconomics and vegetation change in urban ecosystems: patterns in space and time. Ecosystems 12:604–620

McDonald RI, Kareiva P, Forman R (2008) The implications of urban growth for global protected areas and biodiversity conservation. Biol Conserv 141:1695–1703

McDonald RI, Marcotullio PJ, Güneralp B (2013) Urbanization and global trends in biodiversity and ecosystem services. In: Elmqvist T et al (eds) Urbanization, biodiversity and ecosystem services: challenges and opportunities. Springer, Dordrecht, pp 60–81

McFrederick QS, LeBuhn G (2006) Are urban parks refuges for bumble bees *Bombus* spp. (Hymenoptera: Apidae)? Biol Conserv 129:372–382

McNamara S, Viet Tinh D, Erskine PD, Lamb D, Yates D, Brown S (2006) Rehabilitating degraded forest land in Central Vietnam with mixed native species plantings. For Ecol Manag 233:358–365

Montagnini F (2006) Homegardens of Mesoamerica: biodiversity, food security and nutrient management. In: Kumar BM, Nair PKR (eds) Tropical Homegardens: a time-tested example of sustainable agroforestry. Springer, Dordrecht, pp 61–84

Muller N, Ignatieva M, Nilon CH, Werner P, Zipperer WC (2013) Patterns and trends in urban biodiversity and landscape design. In: Elmqvist T et al (eds) Urbanization, biodiversity and ecosystem services: challenges and opportunities. Springer, Dordrecht, pp 149–200

Muratet A, Fontaine B (2015) Contrasting impacts of pesticides on butterflies and bumblebees in private gardens in France. Biol Conserv 182:148–154

Myers N, Mittermeier RA, Mittermeier CG, da Fonseca GAB, Kent J (2000) Biodiversity hotspots for conservation priorities. Nature 403(6772):853–858

National Gardener Survey (2018) A comprehensive study of consumer gardening practices, trends and products sales. GardenReseachcom https://gardenresearch.com/national-gardening-survey-2018-edition/. Accessed January 2020

Neal N (2012) Gardener's Guide to Tropical Plants (Gardener's Guides). Cool Springs Press, Minneapolis

Newbury T (2000) The ultimate garden designer. Seven Dials Cassell & Co., London

Orams MB (1999) The effectiveness of environmental education: can we turn tourists into 'greenies'? Prog Tour Hosp Res 3:295–306

Osborne JL, Martin AP, Shortall CR, Todd AD, Goulson D, Knight ME, Hale RJ, Sanderson RA (2008) Quantifying and comparing bumblebee nest densities in gardens and countryside habitats. J Appl Ecol 45:784–792

Oxford Economics (2017) The economic impact of ornamental horticulture in the UK pp 50 www.oxfordeconomics.com/recent-releases/3b5ce883-cc72-4cf9-910e-be267fe93f46 Accessed January 2020

Palomino D, Carrascal LM (2006) Urban influence on birds at regional scale. A case study with the avifauna of northern Madrid Province. Landsc Urban Plan 77:276–290

Parker K, Head L, Chisholm L, Feneley N (2008) A conceptual model of ecological connectivity in Shellharbour local government area, New South Wales, Australia. Landsc Urban Plan 86:47–59

Parrotta JA, Turnbull JW, Jones N (1997) Catalyzing native forest regeneration on degraded tropical lands. For Ecol Manag 99:1–7

Peoples Trust for Endangered Species (2019). https://ptes.org/get-informed/publications/guidance-leaflets/hedgerow-guide-web-version/. Accessed February 2019

Quigley MF (2011) Potemkin gardens: biodiversity in small designed landscapes. In: Niemelä J (ed) Urban ecology: patterns, processes, and applications. Oxford University Press, New York

Raymond CM, Diduck AP, Buijs A, Boerchers M, Moquin R (2019) Exploring the co-benefits (and costs) of home gardening for biodiversity conservation. Int J Justice Sustain 24(3):258–273

Ricketts TH (2001) The matrix matters: effective isolation in fragmented landscapes. Am Nat 158(1):87–99

Rosenberg KV, Adriaan M, Dokter AM, Blancher PJ, Sauer JR, Smith AC, Smith PA, Stanton JC, Panjabi A, Helft L, Parr M, Marra PP (2019) Decline of the North American Avifauna. Science 366(6461):120–124

Rudd H, Vala J, Schaefer V (2002) Importance of backyard habitat in a comprehensive biodiversity strategy. A connectivity analysis of urban green spaces. Restor Ecol 10:368–375

Secretariat of the Convention on Biological Diversity (2012) Cities and biodiversity outlook. Montreal. www.cbd.int

Senbeta F, Beck E, Lüttge U (2002) Exotic trees as nurse-trees for the regeneration of natural tropical forests. Trees 16:245–249

Seto KC, Güneralp B, Hutyra LR (2012) Global forecasts of urban expansion to 2030 and direct impacts on biodiversity and carbon pools. Proc Nat Acad Sci 109(40):16083–16088

Smith R, Gaston K, Warren P, Thompson K (2006a) Urban domestic gardens (VI): environmental correlates of invertebrate richness. Biodivers Conserv 15:2415–2438

Smith R, Thompson K, Hodgson JG, Warren PH (2006b) Urban domestic gardens (IX): Composition and richness of the vascular plant flora, and implications for native biodiversity. Biodiv Conserv 129:312–332

Smythe N (1978) The natural history of the Central American agouti (Dasyprocta punctata). In: Smithsonian contributions to zoology 257. Smithsonian Institution Press, Washington DC

Thompson K, Head S (2014) Gardens as a resource for wildlife. Wildlife Gardening Forum Nov 2019. http://www.wlgf.org. Accessed November 2019
van Heezik YM, Dickinson KJM, Freeman C (2012) Closing the gap: communicating to change gardening practices in support of native biodiversity in urban private gardens. Ecol Soc 17(1):34
Vergnes A, Kerbiriou C, Clergeau P (2013) Ecological corridors also operate in an urban matrix. A test case with garden shrews. Urban Ecosyst 16:511–525
Warren W (1991) Tropical garden plants. Thames and Hudson, London
White J, Antos M, Fitzsimons J, Palmer G (2005) Non uniform bird assemblages in urban environments: the influence of streetscape vegetation. Landsc Urban Plan 71:123–135
Wijaya M (1999) Tropical garden design. Archipelago Press and Wijaya Words, Singapore
Wijaya M (2007) Modern tropical garden design. Thames and Hudson, London
Woodland Trust (2019). https://www.woodlandtrust.org.uk/media/1800/wood-wise-hedgerows-and-hedgerow-trees.pdf Accessed February 2019
World Meteorological Organization (2019) World weather information service – Panama City. https://worldweather.wmo.int/en/city.html?cityId=1221 Accessed December 2019
Yanosky A, Ferreira A, Jent A, Brítez Sánchez FJ, Cabral H, Del Castillo H, Benítez Stanley MA, Velilla M, Ruiz Diaz M, Mereles Haidar NM, Rodas Insfran OI, Cacciali Sosa P, Clay RP, Granada López YE (2014) Libro Verde de Asunción: Justificación técnica y biológica para la declaración de Asunción como Capital Verde de Iberoamérica de 2015. Asociación Guyra Paraguay
Zuchowski W, Forsyth T (2007) Tropical plants of Costa Rica: a guide to native and exotic Flora. Zona Tropical Publications, Cornell University Press, Ithaca

# Chapter 16
# Biodiversity Islands at the World's Southernmost City: Plant, Bird and Insect Conservation in Urban Forests and Peatlands of Ushuaia, Argentina

**Rosina Soler, Julieta Benítez, Francisco Sola, and María Vanessa Lencinas**

**Abstract** Ushuaia, a coastal city located in Patagonia, Argentina, amid unique natural habitats such as sub-antarctic *Nothofagus* forests and *Sphagnum* peatlands, preserves green areas of natural ecosystems within the urban matrix. Forests and peatlands have persisted as natural islands of different sizes, conservation status and anthropic disturbances. These urban green areas are underappreciated and administratively abandoned, and several threats could lead to their irreversible transformation (e.g., replacement by new neighborhoods, urban waste deposits). However, natural green urban areas can serve as biodiversity islands that make substantial contributions to conserve local biota and also to house new species assemblages (different from those in pristine areas), bringing native forest and peatland species closer to the local residents and visitors. These green urban islands can also help to preserve traditional cultural values, enhance education and generate local touristic attractions. In this chapter, we assessed the assemblages of plants and birds in urban native forests, and plants and insects in urban peatlands of Ushuaia city considering two levels of urbanization surrounding these green areas, and analyzed their variations compared with assemblages in nearby similar non-urban (low level of disturbance) ecosystems. Despite some differences in richness, abundance, biodiversity indices, and species composition of vascular plants, birds and insects, natural urban and peri-urban forests and peatlands still conserve several species and characteristics similar to those of unmodified ecosystems. However, the presence of new species, mainly plants, introduced into these urban patches has modified the original communities, establishing new assemblages that may contribute positively to urban biodiversity islands.

---

R. Soler (✉) · J. Benítez · M. V. Lencinas
Centro Austral de Investigaciones Científicas (CONICET), Ushuaia, Tierra del Fuego, Argentina
e-mail: rosina.soler@conicet.gov.ar

F. Sola
Instituto de Ciencias Polares, Ambiente y Recursos Naturales (ICPA), Universidad Nacional de Tierra del Fuego, Ushuaia, Tierra del Fuego, Argentina

 
F. Montagnini (ed.), *Biodiversity Islands: Strategies for Conservation in Human-Dominated Environments*, Topics in Biodiversity and Conservation 20,
https://doi.org/10.1007/978-3-030-92234-4_16

**Keywords** Endemic species · Green areas · *Nothofagus* · Patagonia · Refuge habitats · *Sphagnum* · Tierra del Fuego

## 16.1 Introduction

Cities located in mountainous regions and coastal areas are one of the most attractive destinations for tourism. These areas are still perceived as natural treasures, offering striking landscapes and a wide variety of highly attractive plant and animal species. In Patagonia, one of the main attractions are the pristine landscapes surrounding the cities and the possibility to be in close contact with wildlife through the broad spectrum of existing opportunities for outdoor activities. In addition, natural areas within the urban matrix (i.e., parks or wild areas containing original vegetation) are of great value to residents and visitors, as urban green areas provide a number of ecosystem services to the community. These ecosystem services include regulating (e.g., noise reduction, modulation of temperature, removal of air pollution, protection of water quality), supporting (e.g., increased biodiversity, habitat, soil formation and storage and cycling of nutrients), cultural (recreation, enhancement of property value, community cohesion, source of knowledge), and provisioning (e.g., food, water, fuel) services (Millennium Ecosystem Assessment 2005; Gómez-Baggethun and Barton 2013). To benefit from ecological services provided by urban green areas it is necessary to incorporate these overlooked areas in the urban planning of our cities. The places where people live and work need to be designed so that they can serve to promote contact with the natural world (Miller 2005). Even those areas whose original biological composition has been profoundly modified are still important for people. The benefits provided by green urban areas, even those that are somewhat altered, are invaluable (Thomas and Geller 2013).

### 16.1.1 *Biodiversity in Urban Green Areas*

From the perspective of island biogeography (MacArthur and Wilson 1967), urban green areas can be considered as a type of ecological island more or less isolated from the other green habitats by the surrounding urban landscape (McGregor-Fors et al. 2011). Urban green areas, even with some levels of disturbance, are considered to function as biodiversity islands providing potential habitat refuge for different organisms within the urban matrix (Matthies et al. 2017; Montagnini et al. 2022). They are also important in maintaining key ecosystem services (e.g., water purification and regulation) which are necessary for human well-being and public health in cities (Gómez-Baggethun and Barton 2013). Biodiversity is essential for maintaining ecosystem function (Maestre et al. 2012; Reich et al. 2012) therefore it is important to maintain urban green areas as undisturbed as possible.

Plants play a significant role in animal diversity because they constitute the first trophic level, i.e. they are the primary producers of energy that is used by other members of the food chain. Poorly managed urban green areas have lower biodiversity compared to undisturbed ones (Burghardt et al. 2010; Tallamy et al. 2010). Similarly, poorly managed landscapes can promote the introduction of non-native plants (e.g., Rozzi et al. 2003; Marzluff et al. 2008; Ceplová et al. 2017; Lencinas et al. 2017) which can have negative effects on native plant diversity and microhabitat conditions (Ceplová et al. 2017). Non-native plants sustain significantly fewer insects (e.g., caterpillars) than native-plants (Tallamy et al. 2010). Insects play a significant role in supporting further biodiversity because they are eaten by other animals (e.g., birds) and represent the main energy source in critical stages of the life cycle of many species (e.g., many birds feed their hatchlings with insects). Furthermore, insects are involved in soil formation and organic matter decomposition, which affect both the diversity and structure of the vegetation (Hartley and Jones 2008).

The integration of green urban areas for biodiversity conservation in city planning is among the most difficult problems to resolve by public and local policies (Murphy 1999; Nilon et al. 2017). Even the effort to keep the native biota intact in urban green areas may represent an unlikely goal to achieve for local administrations. This is the case of young and developing cities which suffer from intense and regular disturbances (e.g., constant building, new streets) which significantly alter local biota in green areas (Lososová et al. 2011). In contrast, green areas in less urbanized habitats (e.g., those that suffer irregular and less strong disturbances) support rich species diversity and native biota (Zerbe et al. 2003; Ceplová et al. 2017).

In order to identify conservation priorities and/or prevent future environmental problems in cities it is necessary to understand how urbanization influences different biological groups. It is important to assess the potential role of biodiversity islands to conserve local biota and ultimately avoid its contamination, transformation and total loss inside an urban matrix.

### *16.1.2 Case Study: Ushuaia City and its Urban Green Areas*

Ushuaia, a coastal city founded in 1884, is one of the southernmost urban settlements in the world, and the most populated city (56,900 inhabitants, INDEC 2010) at such latitude. It is the political and administrative capital of the Argentinian province of Tierra del Fuego, Antarctica and Islands of the South Atlantic, located in the biogeographic region of the Andean-Patagonian forest (Fig. 16.1.)

Ushuaia's urban development has had a short and hurried history since its foundation, marked by a penal colony established in 1896 (closed down in 1947) that started to bring people to the city, and by the Industrial Promotion Law (1972) that stimulated regional economic development. During 1972–1990, the city underwent a population boom, which tripled its number of residents and expanded the urban area into the surrounding natural forests to settle new neighborhoods. After the 1990s, the unexpected increase of the population strongly promoted building-up areas (houses, schools, etc.) and installing public services, which reduced the

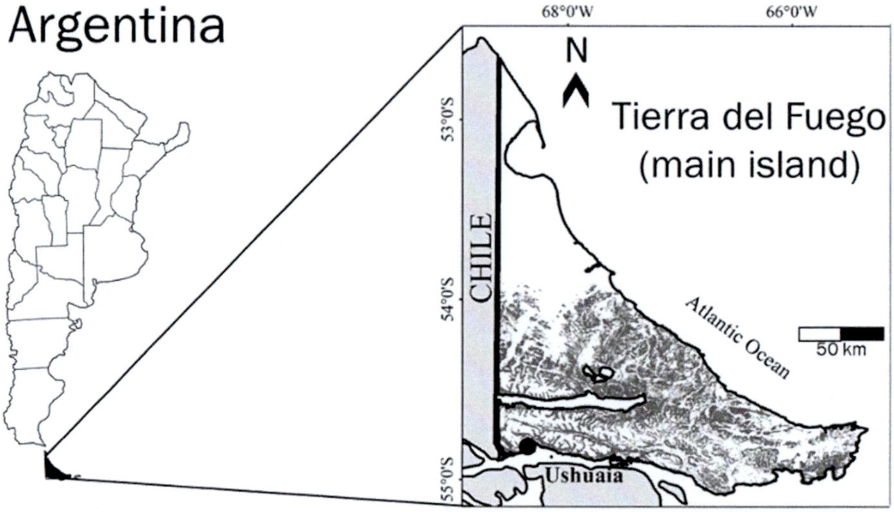


**Fig. 16.1** Map of Tierra del Fuego, Argentina, showing Ushuaia city, which is located on the coastal Nothofagus forests of the Beagle Channel

**Fig. 16.2** Inside view from urban *Nothofagus* forests (**a**) and *Sphagnum* peatlands (**b**) in Ushuaia city, Tierra del Fuego, Argentina. (Photos: Julieta Benítez and Francisco Sola)

forested areas around the unplanned urban layout, a situation that continues in present days. Therefore, existing forest patches inside the city are currently remnants of old or secondary native forests (~50 years old) originated from old cuts or fires (Fig. 16.2). The urbanization growth also gradually surrounded and isolated the existing peatlands within the city (Fig. 16.2). Small peatlands (<0.3 ha) were filled with soil, rubble and rocks, and therefore irreversibly transformed to building areas. But the larger ones, deemed non-suitable for building were left as "brown areas" immersed in the urbanization, exposed to degrading uses (e.g., substrate extraction, artificial drainage, urban waste accumulation, refuge of abandoned domestic animals, human trespassing). A similar situation is currently found in some patches of urban forests, which are also used to install gardens or squares, or exceptionally preserved areas (e.g., the Parque Yatana Urban Reserve, located in the center of Ushuaia).

Due to the unplanned nature of the city's sprawl, recreational and protected areas are also unplanned and scarce, resulting in the misuse and/or overuse of pre-existent natural patches inside and outside the city. This is the consequence of a cultural and social necessity, where the lack of accessible leisure outlets is compensated with available natural resources, resulting in its unregulated use and eventual deterioration. There is still the possibility to restore these sites as natural biodiversity refuges while still making them available to the public under sustainable conditions. Although Ushuaia city is surrounded by natural ecosystems, the natural green areas in the most densely urbanized area barely reach 10% of the surface. While these forests and peatlands inside the urban area were not created consciously to increase the proportion of green areas available to residents, they can provide the areas of green spaces that are necessary for a healthier and more sustainable city.

Unfortunately, these urban natural green areas are underappreciated and administratively abandoned. Moreover, we have a limited understanding of how green spaces are used for refuge, foraging and reproduction by different animal species. Urban green areas, especially if they are patches of pre-existing native forests and peatlands, retain part of the local diversity. In this chapter we aim to determine the role of natural urban green areas as biodiversity islands within Ushuaia city as a case study from Patagonia, Argentina. We present results of studies on species assemblages (species richness, relative abundance, diversity, evenness and similarity indices) of understory vascular plants and birds in forests, and vascular plants and insects in peatlands of Ushuaia city. We also compare species composition of patches inside the urban matrix (two levels of urbanization surrounding these green areas) with similar and near undisturbed patches outside the urban matrix (non-urban patches). We propose these natural urban islands could potentially serve as places to promote conservation of local biodiversity, bringing native forest and peatland species closer to people. This can contribute to biodiversity conservation in a broad sense, which could also help to preserve traditional cultural values, improve education and enhance touristic attractions.

## 16.2 Methods

### *16.2.1 Study Area*

Ushuaia city is located in the north bank of the Beagle Channel, Argentina (Figs. 16.1 and 16.3) in the southern end of the South American Cone (54°48′S, 68°19′W). Mountains and glaciers surround the city, and the natural vegetation is primarily comprised of forests that reach the coast. These native forests were the main source of firewood and timber since the city foundation. Most of them are deciduous forests dominated by *Nothofagus pumilio* (lenga, Nothofagaceae), which develop mainly in mountain slopes and valleys. Mixed deciduous-evergreen forests

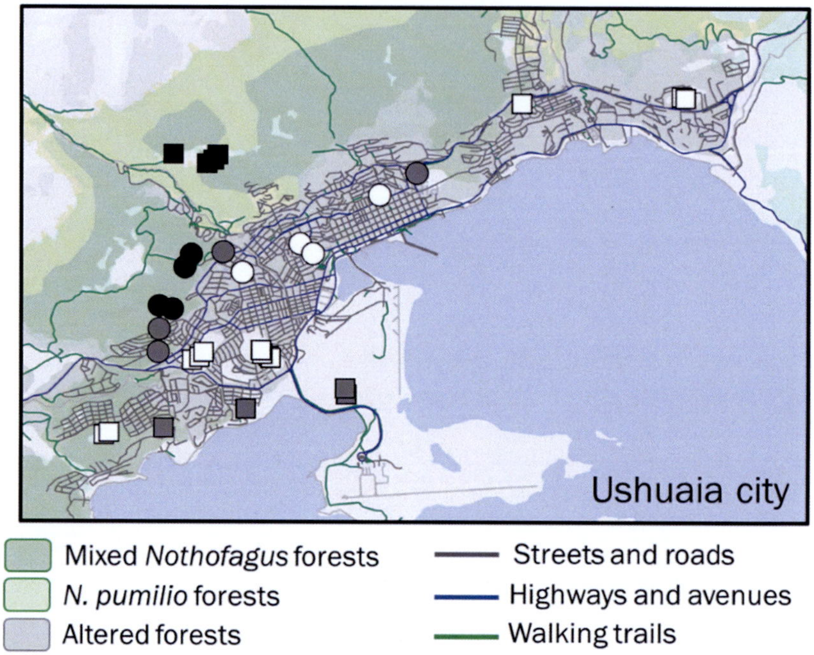


**Fig. 16.3** Location of the study area in Ushuaia city, Tiera del Fuego, Argentina. Circles = forest patches, squares = peatlands. The color indicates the urban categories studied: urban (white), peri-urban (gray) and non-urban (black)

of lenga and *N. betuloides* (guindo) also occur on the coast and mid slopes and include other woody species such as *Drymis winterii* (canelo), *Embothrium coccineum* (notro) and *Maytenus magellanica* (leña dura) as small trees in the understory (Moore 1983). On the other hand, *N. antarctica* (ñire) is found surrounding peatbogs and inhabiting mixed coastal forests. The vascular plants in the understory usually include few shrubs (e.g., *Berberis* and *Ribes*) that produce berries, and abundant forbs, graminoids, fungi, mosses, liverworts and ferns. In the whole landscape, forest masses are intermingled with peatlands and grasslands of different size and composition according to elevation above sea level. Peatlands formed by accumulation of *Sphagnum* spp. mosses during geological times support other species of vascular plants like Cyperaceae (e.g., *Carex* spp.), *Astelia* sp., and *Drosera* sp. (Moore 1983). The climate in Ushuaia city is sub-antarctic (Tuhkanen 1992) with short, cold summers (7 °C average temperature) and long winters (−1 °C average temperature) with regular snow and frost. Mean annual rainfall is 611 mm near the coast and more than 1000 mm in the mountains.

For our study, we classified forests and peatlands in Ushuaia city according to the current degree of surrounding urbanization (i.e., their respective locations in the urban matrix): (1) Urban: patches inside the urban matrix, 100% surrounded by buildings, streets or waste disposal areas; (2) Peri-urban: peripheral patches but still

**Table 16.1** Summarized description (patch level) of native forests and peatlands studied in the urban landscape of Ushuaia, Tierra del Fuego (Argentina)

| | | Urban | Peri-urban | Non-urban |
|---|---|---|---|---|
| General location | | **Surrounded by urbanization** | **Partially surrounded by urbanization** | **Not surrounded by urbanization** |
| Area | **Forests** | 9–3 ha | 10–9 ha | > 10 ha |
| | **Peatlands** | | 0.3–12 ha | |
| Main anthropic use and frequency | | Recreation, educative, passage, frequent | Recreation, sports, sporadic | Trekking, very sporadic |
| Main contamination | | Household and recreational waste | Recreational waste | Without contamination |
| Domestic species and frequency of visit | | Dogs, cats, horses, very frequent | Dogs and horses, frequent | Almost none, very sporadic |

inside the urban matrix, and partially surrounded (approximately 50%) by the urban matrix; and (3) Non-urban: patches outside the urban matrix (i.e., without direct contact with urbanization but still close to the city), with low-impact recreational use (Fig. 16.3). The degree of urbanization is also determined by the main anthropic uses of such natural habitats, the use frequency or intensity, and consequently, the potential pollution and habitat degradation (Table 16.1).

### *16.2.2 Sampling Design: Surveys of Plant, Bird and Insect Species*

For each forest type, we selected four forest patches (replicates) located at the same elevation range (38 to 170 m above sea level (m.a.s.l.). These were at least 150 m apart from each other and from other types of structures, as well as at least 50 m from the edge of the same forest type, in order to avoid the influence of other environments. In the case of peatlands, we selected 12 urban, 4 peri-urban and 4 non-urban locations (elevation range 20 to 200 m.a.s.l.) separated by at least 150 m from each other.

Vegetation surveys were carried out in each patch (forests and peatlands) during 2016–2017, using a circular plot of approximately 30 m radius centered in each sampling site. In each plot, we identified vascular plants (Dicotyledonae, Monocotyledonae and Pteridophytae) at the species level (Moore 1983; Correa 1969–1998). We estimated ground cover of vascular plants using a modification of the relevé method of Braun-Blanquet proposed by Pauchard et al. (2000). The relevé method involves describing recognizable units in the vegetation of a region by the characterization of the vegetation in a single representative standard plot—a relevé—within each unit. The relevés from many units are then analyzed to develop descriptions and classifications of the vegetation in the study region and have been increasingly used in other kinds of vegetation studies as a practical, relatively fast means of collecting information on vegetation (Minnesota Department of Natural Resources 2013). Ground cover was estimated for each species separately and then values were added to obtain family and total vascular plant cover. The complementary covers to reach 100% ground cover (data not shown) were: bryophyte (mosses and liverworts), woody debris (> 3 cm diameter), and bare soil including litter. Species richness was calculated as the total number of vascular plant species identified in each replica of treatments. Voucher specimens were deposited in the Herbarium of Tierra del Fuego at the Centro Austral de Investigaciones Científicas (Austral Center of Scientific Research, CADIC-CONICET) in Ushuaia, Argentina.

Bird surveys in forest patches were conducted during the 2016–2017 breeding season in a 4-h period at sunrise, always under similar climatic conditions (i.e., avoiding rainy or foggy days). We identified bird assemblages by using the point-count method with unlimited distances, which had been used widely in previous studies in Tierra del Fuego (see Deferrari et al. 2001; Lencinas et al. 2005, 2014, 2019). We selected one observation point in each forest patch ($n = 12$), with two observers during each survey conducting direct sightings (with binoculars) during 13 min of effective observation period (leaving 2 min prior to recording to allow for birds' habituation to the observers). We registered the birds to the species level according to the classification proposed by Remsen et al. (2019). From these surveys, we calculated bird species richness and relative frequency (number of individuals). Forests patches were revisited three times per month for bird observations during the complete breeding season (from October to February). Species richness represent the total number of species recorded during that period, and the relative abundance was calculated as the sum of the total number of individuals observed for the same species during visits.

Epigean insects were collected in peatlands during January–February 2016. Sampling was done using pitfall traps (plastic containers 12 cm in diameter and 14 cm height) buried and filled to 1/3 of their volume with soapy water (300 ml) to trap and kill arthropods which fell in. We used pitfall traps in sets of five in each plot, and contents of the five traps in a plot were pooled and used as a single sample in order to calculate total capture per sample (Lencinas et al. 2014). Traps were arranged placing one at the center and the remaining four at 5 m from the first, and at 90 degrees from each other, and were left open at ground level for one week before being collected. We obtained 20 samples from 100 pitfall traps. All samples were taken to the laboratory for specimen identification and quantification. Identifications were performed under a binocular dissecting microscope to genus or species level when possible (Roig-Juñent et al. 2002; Marvaldi and Lanteri 2005; Posadas 2012). Due to a lack of complete taxonomic data on Patagonian beetles, some specimens could not be determined to the species level. We employed the recognizable taxonomic unit or morphospecies concept (Oliver and Beattie 1993; Gerlach et al. 2013) when the former could not be determined (hereafter, "species"). This has been demonstrated to be a good tool for insect diversity studies in Patagonian ecosystems, such as *Nothofagus* forests (Lencinas et al. 2014; Sola et al. 2016; Cárcamo et al. 2019).

### 16.2.3 Statistical Analyses

Data collected were used to determine species richness, relative abundance (ground cover for plants, and number of individuals for birds and insects), Shannon-Wiener diversity (H′) and Pielou evenness (J') indices of plants and birds in each forest, and plants and insects in each peatland types (Pielou 1975). We used one-way ANOVAs after validation of the statistical assumptions of homoscedasticity (homogeneity) and normality, to evaluate species richness, relative abundance, Shannon-Wiener diversity and Pielou evenness indices of vascular plants, birds and insects, considering the category of forests and peatlands (urban, peri-urban and non-urban) as the main factor.

Also, differences in assemblage composition of plants, birds and insects along the urban categories were examined by the non-Metrical Multidimensional Scaling (NMDS) ordination method using a manual methodology, with a Bray-Curtis distance and 250 iterations. A Monte Carlo test was used to evaluate stress in randomized data; probability was presented for each axis.

Finally, we calculated the abundance-based Chao-Sørensen similarity index (Chao et al. 2005) to determine differences among categories of urbanization, both for forests and peatlands, analyzing each taxonomic group separately. This index, varying between zero and one, considers explicitly the relative abundance of both common and rare species, and estimates the extent of shared species taking into account unseen shared species, based on the number of observed rare (singletons and doubletons, as well as unique and duplicate), shared species between two sites. The use of this index is recommended when study sites could be under-sampled and contain only a substantial fraction of the rare species (Chao et al. 2005). We used the software EstimateS (Colwell 2009) for each pairwise evaluation and compared averages and standard deviation for each urbanization category.

## 16.3 Results and Discussion

Plants, birds and insects have been widely used as bio-indicators of human impact on natural ecosystems (Lencinas et al. 2005, 2014, 2017; Gerlach et al. 2013; Rozzi and Jiménez 2014; Contador et al. 2015) and urban biodiversity islands (Noreika et al. 2015; Matthies et al. 2017). These three groups, due to high latitude and extreme climatic conditions of Southern Patagonia, sustain the greatest biological diversity in the natural ecosystems, with many endemic species and guilds (Roig-Juñent et al. 2002; Díaz 2012; Rozzi and Jiménez 2014; Cárcamo et al. 2019). Plants are among the more characteristic and recognized organisms for people, both in forests and peatlands, while birds are the most conspicuous groups in forests, and arthropods are more visible and easily distinguishable in peatlands (open areas).

### 16.3.1 Vascular Plants

In Ushuaia city, vascular plants were affected differently in forest patches and peatlands. Species richness did not strongly differ along the urbanization categories, but the relative abundance of vegetation and species diversity indexes were strongly modified in urban and peri-urban peatlands (Table 16.2). Urbanization significantly increased the relative abundance of vascular plants, due to the increase of some plant species favored by microclimatic modifications (e.g., light, soil moisture). According to our data, forbs and graminoids such as *Gunnera magellanica*, *Acaena magellanica*, and *Carex* sp. doubled their cover, while other species (e.g., the native fern *Blechnum penna-marina* and the exotic herbs *Achilea millefolium*, *Rumex crispus*, *Taraxacum officinale*, *Poa annua*), which were absent in peri-urban and non-urban peatlands, entered in the urban ones. Soil drainage, building and protection against extreme temperatures and winds could favor herbaceous development on peatlands (Pinceloup et al. 2020).

**Table 16.2** Species richness, relative abundance, Shannon-Wiener diversity (H′) and Pielou evenness (J') indices of vascular plants, birds and insects occurring in urban, peri-urban and non-urban forests and peatlands of Ushuaia city

| | Type | Richness | Abundance | H′ | J' |
|---|---|---|---|---|---|
| Forest plants | Urban | 16.0 | 67.2 | 4.2 | 0.82 |
| | Peri-urban | 19.5 | 48.2 | 2.4 | 0.83 |
| | Non-urban | 15.5 | 37.5 | 2.2 | 0.86 |
| | *F (p)* | *0.40 (0.679)* | *0.80 (0.480)* | *0.15 (0.859)* | *0.15 (0.865)* |
| Forest birds | Urban | 9.8 | 93.7 b | 1.9 | 0.82 |
| | Peri-urban | 10.0 | 62.5 a | 2.0 | 0.88 |
| | Non-urban | 7.8 | 39.8 a | 1.8 | 0.87 |
| | *F (p)* | *1.54 (0.266)* | *13.98 (0.002)* | *0.98 (0.411)* | *0.71 (0.518)* |
| Peatland plants | Urban | 10.7 | 89.6 b | 1.4 a | 0.60 a |
| | Peri-urban | 8.0 | 86.7 b | 1.1 a | 0.53 a |
| | Non-urban | 13.5 | 66.2 a | 2.1 b | 0.81 b |
| | *F (p)* | *1.59 (0.232)* | *14.5 (<0.001)* | *3.81 (0.0502)* | *4.53 (0.048)* |
| Peatland insects | Urban | 5.6 | 22.4 a | 1.17 | 0.68 |
| | Peri-urban | 8.3 | 44.3 ab | 1.38 | 0.65 |
| | Non-urban | 8.8 | 92.2 b | 1.29 | 0.60 |
| | *F (p)* | *1.75 (0.206)* | *9.34 (0.002)* | *0.18 (0.838)* | *0.13 (0.883)* |

F (p) Fisher statistic with probability in parentheses. Different letters show differences by Tukey test at $p < 0.05$. Values followed by different letters in each column and for each factor are significantly different according to the Tukey test at a $p \leq 0.05$

Relative abundance shows: vascular plants= total ground cover (%), birds = average of total observations along five months during the breeding season (n° individuals), and insects = average of adult individuals collected in pitfall traps for 15 days in the summer season (n° individuals)

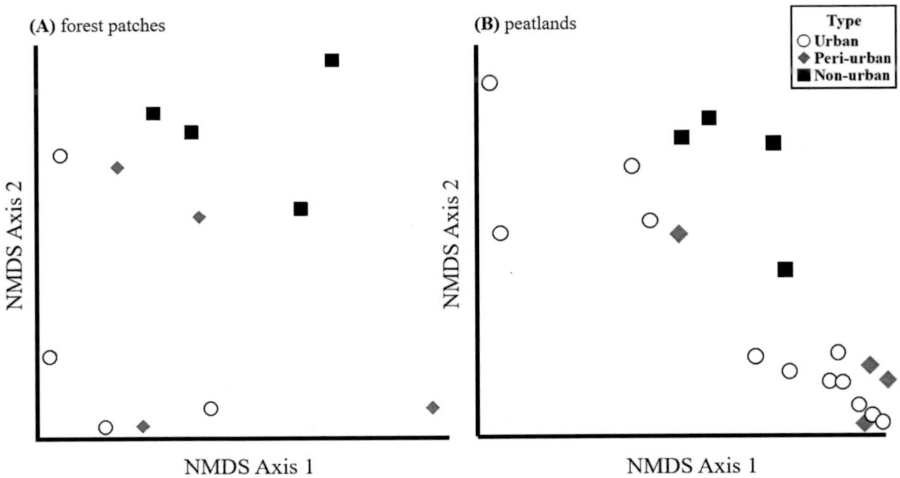


**Fig. 16.4** Results of analysis of differences in assemblage composition of plants examined by the Non-Metrical Multidimensional Scaling (NMDS) ordination method, based on understory plants composition in different urban categories of forest patches (**a**) and peatlands (**b**)

On the other hand, urban and peri-urban forests showed changes in their community composition in some patches (Fig. 16.4a), by the introduction of exotic cultivated shrubs (*Ribes idaeus* and *Ribes rubrum*) and potentially invasive forbs (*Ranunculus acris* and *Hypochaeris radicata*), which could be intentionally planted by people or escaped from home gardens, orchards and parks. Urbanization also differently influenced the Shannon diversity and Pielou evenness indices for vascular plants in forests patches and peatlands. While both indices were quite similar in forest patches along the urban categories, Shannon diversity and Pielou evenness indices of urban peatlands decreased by half compared to those in non-urban peatlands. Plant composition in urban peatlands was similar to vegetation in peri-urban areas but differed from non-urban peatlands located far from urbanization (Fig. 16.4b).

### *16.3.2 Birds*

With respect to birds, urban assemblages in Ushuaia's forests are conformed by bird species commonly observed in *Nothofagus* forests (Deferrari et al. 2001; Lencinas et al. 2005), which correspond to thirteen families of Passeriformes, and one family of five orders (Charadriiforme, Falconiforme, Pelecaniforme, Piciforme, Psittaciforme (Benítez et al. 2020). Some species detected in these urban forests are endemic to Patagonia (i.e. *Aphrastura spinicauda*, *Phrygilus patagonicus*, *Pygarrichas albogularis*, *Enicognathus ferrugineus* and *Campephillus magellanicus*) (Deferrari et al. 2001). Insectivore and omnivore birds are the most important groups, being more than a half of all resident species (Benítez et al. 2020).

In our study, species richness remained similar in peri-urban and non-urban patches (Table 16.2). However, bird abundance significantly increased in urban forests, with slight changes in peri-urban and non-urban forest patches. This was explained by the high abundance of generalist bird species and/or the occurrence of non-specialist forest birds in altered urban patches (pers. obs.). We observed that urban forests offered different food sources (e.g., waste, fruits/seeds) and nesting alternatives (e.g., open places, canopy openings) compared to native forests, which coincides with other studies (Marzluff and Rodewald 2008). This favored the occurrence of bird species from open habitats such as *Vanellus chilensis* (Marín 2014).

The availability of extra food and greater hunting possibilities favored the occurrence of scavenging raptors such the *Milvago chimango*, as was also reported by Biondi et al. (2005). Moreover, urbanization in Ushuaia increased the frequency of exotic bird species (e.g., *Passer domesticus*). Contrary to this, native birds such as woodpeckers and hole-nesters birds (*C. magellanicus*, *P. albogularis*, *Troglodytes aedon* and *A. spinicauda*) showed an increasing trend in richness and abundance from the urban to the periphery environments. These charismatic species prefer forests with dense vegetation (Vergara and Schlatter 2006), but they can be seen while feeding in urban habitats (pers. obs.). Analyzing the whole species assemblages of urban forests (Fig. 16.5a), great differences in biodiversity along the urbanization categories were possible to distinguish. Plant and bird communities in urban forests differed from the communities of non-urban forests (Figs. 16.4a and 16.5a).

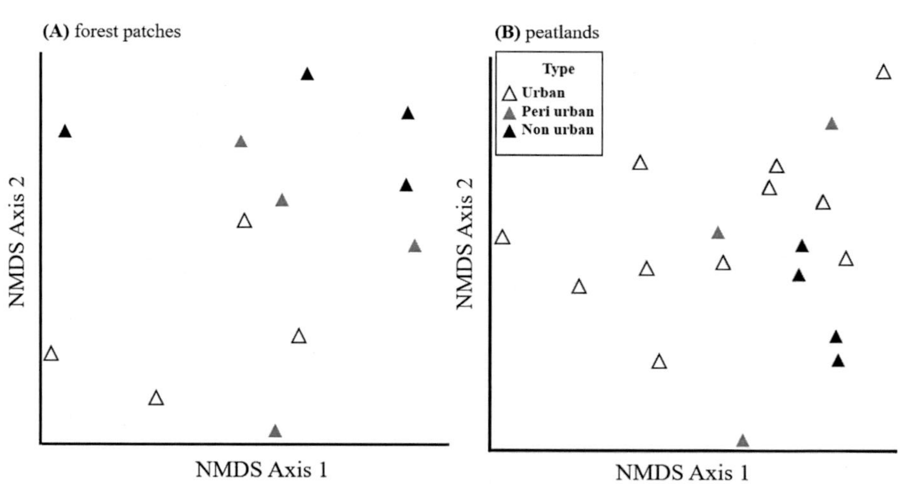


**Fig. 16.5** Results of analysis of differences in bird abundance in different urban forests (**a**), and insect abundance in different urban peatlands (**b**), examined by the Non-Metrical Multidimensional Scaling (NMDS) ordination method

### 16.3.3 Invertebrates

Peatland invertebrate communities are almost unknown, and even more the effects that anthropic disturbances have on their diversity. Recent surveys in urban and non-urban peatlands of Ushuaia city (Sola et al. 2018) indicate the presence of a wide variety of invertebrates: snails (Gastropoda), spiders (Araneae), and harvestmen (Opiliones), mites (Acari), woodlice (Isopoda), and many orders of insects. Due to their diversity, abundance, easy sampling, and rapid response to environmental changes, insects are useful bio-indicators of habitat degradation (Bolger et al. 2000; Gerlach et al. 2013; Moreno et al. 2013), and have been widely used to study the effects of habitat modification even in south Patagonia (Sola et al. 2016; Lencinas et al. 2014, 2017, 2019; Cárcamo et al. 2019). In particular, the study of ground dwelling insects has important functional implications because of their ecological roles, as they include predators, scavengers, and herbivores.

According to our results, the main insect orders present in all categories included beetles (Coleoptera), moths (Lepidoptera), true bugs (Heteroptera), springtails (Collembola), leafhoppers (Auchenorrhyncha), and wasps (Hymenoptera). Insect richness in peatlands (including Coleoptera, Lepidoptera, Heteroptera, Auchenorrhyncha, Collembola and Hymenoptera) remained similar throughout the urbanization categories (Table 16.2). However, insect abundance significantly decreased in urban peatlands compared with those in non-urban; while peri-urban peatlands showed intermediate values. The most affected groups were Coleoptera, Auchenorrhyncha and Collembola, which reduced by half the number of individuals within these orders. All three of these groups are weaker fliers (in the case of beetle species sampled), making them more susceptible to local impacts. Furthermore, Auchenorrhyncha are herbivorous and although the ecology of local species is unknown, they could be susceptible to changes observed in local flora at each site. Collembolans, as other detritivores, can be associated with available organic matter; a fall in their abundance could be indicative of reduced productivity, and may also result in slowed decomposition rates and nutrient cycling, essential in these cold, high latitude landscapes. Finally, beetles sampled fell under the categories of scavengers or predators (e.g., Curculionidae, Carabidae) indicating a higher trophic position. These can ultimately serve as indicators of the health of plants and prey in the ecosystem, both in diversity and abundance. Although richness values recorded in our study did not significantly differ among the three urbanization levels, they hide changes in assemblages (Fig. 16.5b). Nevertheless, the proximity between points entails the occurrence of some species both in urban and non-urban sites.

The Chao-Sørensen similarity index varied among taxonomic groups and urban categories (Fig. 16.6), showing different tendencies among urbanization levels. Plants showed less similarity for all urban categories in forests (0.57 in average) than in peatlands (0.73 on average), with the lower values in urban patches. Greatest similarities in plants occurred in peri-urban forests and in non-urban peatlands, where we also found the lowest variation. In contrast, birds showed great similarity values (0.85 in average) in all urban categories, with slightly higher values in urban

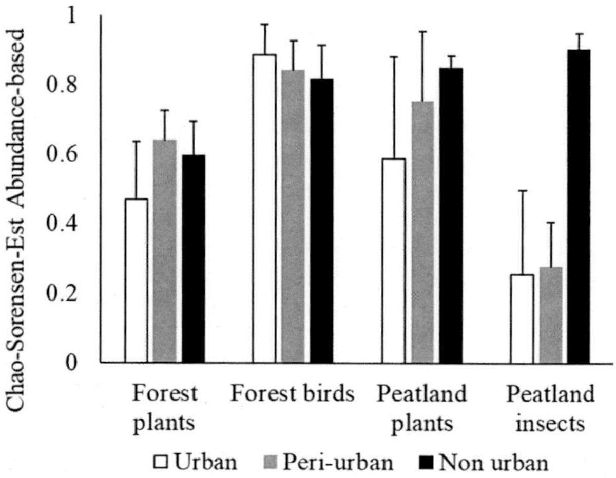


**Fig. 16.6** Abundance-based Chao-Sorensen similarity index (mean ± SE) for forest plants and birds, and peatland plants and insects among the urban categories (urban, peri-urban, and non-urban)

than in peri-urban than in non-urban forests. On the other hand, insects showed high similarity only in non-urban peatlands (0.90), with very low values in urban and peri-urban patches (0.26 in average), highlighting the greater sensitivity of this group compared with plants and birds.

### *16.3.4 Main Factors Affecting Biodiversity in Urban Forests and Peatlands in Ushuaia City*

Despite some differences in richness, abundance, biodiversity indices, and composition for vascular plants, birds and insects, urban and peri-urban forests and peatlands still conserve several species and characteristics similar to non-urban ecosystems. However, the occurrence of new species, mainly plants, introduced into these urban ecosystems, modified the original communities into new assemblages, which offer another food supply to the fauna (i.e., birds and insects) (Marzluff and Rodewald 2008). The probability of greater dispersion of these newly introduced exotic plants in the non-urban ecosystem could depend on the mobility of dispersing agents, and on the connectivity and distance between urban forests and peatland patches to non-urban ecosystems. Previous studies affirm that urban green areas are ecological islands isolated from others in and around the city (Ceplová et al. 2017), but this depends on the size of cities and green areas. In addition, the relative importance of local- versus larger-scale variables in influencing urban biodiversity remains under discussion and will vary with the scale dependencies of different taxa (Goddard et al. 2010). The influence of local factors, especially patch size and quality, for urban forests and peatlands in Ushuaia city has not been analyzed, either for urban planning or conservation strategies.

The preservation of forests and peatlands as biodiversity islands within the urban matrix generates benefits for the environment and also for citizens living there, promoting ecological and socio-cultural functions (Rozzi et al. 2012) that would be completely lost if such natural islands were not preserved. Unfortunately, the link between the identity of inhabitants and care of regional habitats in Ushuaia is weak (Orzanco 1999; Kizman 2012) and much of the ancestral knowledge about natural ecosystems and their species has been lost. Moreover, the inadequate planning of public services (e.g., streets, sewage) to attend to the growth rate of the city, and bulky waste management are considered the main conflicts between urbanization and sustainability in Ushuaia city (Orzanco 1999). All these factors negatively influence the structure and functioning of biodiversity islands in the city (Fig. 16.7).

Because urban development is guided by human values, there is a need to inform people about the relevance of natural ecosystems supporting multiple services that increase human well-being and encourage caring for the environment. Ushuaia is a unique city in Argentina where peatlands are part of the urban space. However, their value as part of the natural heritage has not yet been recognized by society, and peatlands are considered obstacles to urban development rather than areas of natural, recreational and/or educational interest. However, peatlands were recently declared Environmental Reserves for their conservation and protection as Historical and Cultural Heritage by the Municipality of Ushuaia (Environment Office of Ushuaia 2018).

Although in Ushuaia urban forests and peatlands are connected to the natural undisturbed forests and peatlands occurring in the surrounding landscape, especially by birds and also by some insects (e.g., wasps) and some plants, this is generally not common knowledge for local people (Environment Office of Ushuaia 2018). This generates a twofold problem: on one hand, urban green areas do not receive the attention they rightfully deserve and are frequently modified irreversibly; and on the other, natural landscapes on the periphery are in danger from freely introduced

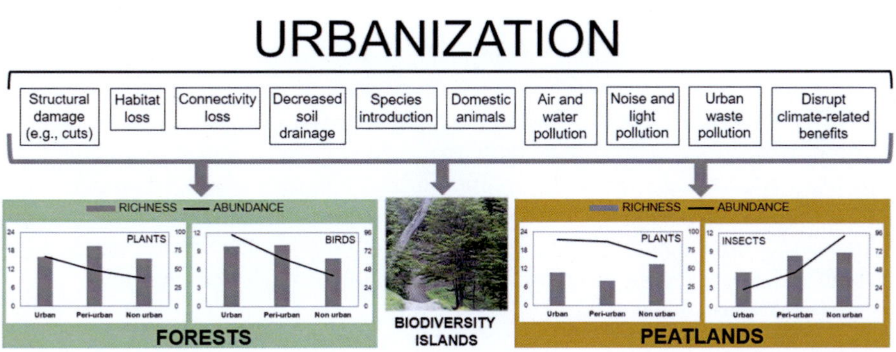


**Fig. 16.7** Conceptual framework to understand the main threats (top panels) affecting biodiversity (species richness and abundance) in urban forests and peatlands in Ushuaia city. Biodiversity islands serve as refuge for some native species that manage to survive despite structural changes and isolation from other natural environments, although they are also prone to invasion by introduced species (e.g., exotic plants) dispersed from homegardens, parks, etc

exotics, either by negligence (weeds, insects) or intent (ornamental plants, pets). We recommend controlling and avoiding the installation of exotic plants, especially trees and bushes with fruits eaten by birds, in these urban green areas, to limit their spread and/or invasion to undisturbed environments.

## 16.4 Conclusions and Recommendations

Our results confirm that the urban forests and peatlands of Ushuaia are valuable reservoirs of biodiversity, even under anthropogenic influence. We recommend preserving existing natural green areas for their role as biodiversity islands in the urban landscape, due to their function as a refuge for native plants, birds and insects. In order to increase biodiversity and improve the connectivity of biodiversity islands within Ushuaia city we consider necessary the preservation and/or delimitation of new natural green areas (e.g., within new neighborhoods). Moreover, we highlight their value compared with parks and squares since these are non-natural, and mostly contain exotic vegetation (e.g., *Salix* sp., *Populus* sp., *Pseudotsuga* sp.).

According to our results, the urban peatlands would be more susceptible to deterioration and loss of species typical of these habitats than the urban forests. Therefore, it is important to implement urgent conservation plans to avoid the loss of these remaining natural habitats (indeed, some of the studied green areas have nowadays already been transformed and urbanized). Moving forward, the level and magnitude of the urbanization impact (i.e., distance to the urban center, transformation degree in the surroundings) on biodiversity deserve more analyses.

Urban islands must contribute to strengthening the link between society and nature, while still offering refuge, food and habitat for plant and animal species. These ecosystems also play an important role in maintaining the minimum environmental (e.g., carbon sinks, water purification) and socio-cultural functions (e.g., recreation, sense of belonging).

We also recommend including environmental education programs to facilitate the understanding of the importance of these biodiversity islands in the city, spread knowledge about native species (especially those inconspicuous or less charismatic), and promote the sustainable use of these environments. For local governments and planners, it is essential to plan the retention of natural green spaces when planning new neighborhoods, buildings, schools and parks. Moreover, conserving, designing, and managing urban green spaces requires balancing human perceptions, needs, and use with ecological requirements for preserving and enhancing biodiversity.

## References

Benítez J, Pizarro JC, Blazina AP, Lencinas MV (2020) Response of bird communities to native forest urbanization in one of the southernmost city of the world. Urban Urban Gree 126887. https://doi.org/10.1016/j.ufug.2020.126887

Biondi LM, Bó MS, Favero M (2005) Dieta del chimango (*Milvago chimango*) durante el periodo reproductivo en el sudeste de la provincia de Buenos Aires, Argentina. Ornitol Neotrop 16:31–42

Bolger DT, Suarez AV, Crooks KR, Morrison SA, Case TJ (2000) Arthropods in urban habitat fragments in southern California: area, age and edge effects. Ecol Appl 10:1230–1248

Burghardt KT, Philips CR, Tallamy DW, Shropshire KJ (2010) Non-native plants reduce abundance, richness, and host specialization in Lepidoptera communities. Ecosphere 1(5):11. https://doi.org/10.1890/ES10-00032.1

Cárcamo JR, Contador T, Gañán M, Convey P, Kennedy J, Rozzi R (2019) Altitudinal gradients in Magellanic sub-Antarctic lagoons: the effect of elevation on freshwater macroinvertebrate diversity and distribution. Peerj 7:e7128–e7128. https://doi.org/10.7717/peerj.7128

Ceplová N, Kalusová V, Lsosová Z (2017) Effects of settlement size, urban heat island and habitat type on urban plant biodiversity. Landsc Urban Plan 159:15–22

Chao A, Chazdon RL, Colwell RK, Shen TJ (2005) A new statistical approach for assessing similarity of species composition with incidence and abundance data. Ecol Lett 8:148–159

Colwell RK (2009) EstimateS: statistical estimation of species richness and shared species from samples. http://purl.oclc.org/estimates

Contador T, Kennedy JH, Rozzi R, Villarroel JO (2015) Sharp altitudinal gradients in Magellanic sub-Antarctic streams: patterns along a fluvial system in the Cape Horn biosphere reserve. Polar Biol 38:1853–1866

Correa MN (1969–1998) Flora Patagónica. Colección Científica INTA Tomo 8. Partes II, III, IVb, V, VI y VII. Buenos Aires

Deferrari G, Camilión C, Martínez Pastur GM, Peri PL (2001) Changes in *Nothofagus pumilio* forest biodiversity during the forest management cycle. 2: birds. Biodivers Conserv 10:2093–2108

Díaz S (2012) Biología y conservación de la cachaña (*Enicognathus ferrugineus*) en Argentina. El Hornero 27:17–25

Environment Office of Ushuaia (2018) Plan de manejo de la Reserva Natural Urbana Bahía Encerrada, Municipalidad de Ushuaia, Tierra del Fuego, 98 p

Gerlach J, Samways M, Pryke J (2013) Terrestrial invertebrates as bioindicators: and overview of available taxonomic groups. J Insect Conserv 17:831–850

Goddard MA, Dougill AJ, Benton TG (2010) Scaling up from gardens: biodiversity conservation in urban environments. Trends Ecol Evol 25:90–98

Gómez-Baggethun E, Barton DN (2013) Classifying and valuing ecosystem services for urban planning. Ecol Econ 86:235–245

Hartley SE, Jones TH (2008) Insect herbivores, nutrient cycling and plant productivity. In: Weisser WW, Siemann E (eds) Insects and ecosystem function. Springer, Berlin, pp 27–52

INDEC (2010) Censo Nacional De Población, Hogares y Viviendas 2010. Documento Metodológico Preliminar. Available at: http://censo2010.indec.gov.ar/metodologia.asp. Accessed 1 Apr 2020

Kizman S (2012) Diagnóstico participativo: desarrollo de actividades sustentables. Parque Nacional Tierra del Fuego. APN, Tierra del Fuego. www.argentina.gob.ar.tierradelfuego

Lencinas MV, Martínez Pastur G, Medina M, Busso C (2005) Richness and density of birds in timber *Nothofagus pumilio* forests and their unproductive associated environments. Biodivers Conserv 14:2299–2320

Lencinas MV, Martínez Pastur G, Gallo E, Cellini JM (2014) Decreasing negative impacts of harvesting over insect communities using variable retention in southern Patagonian forests. J Insect Conserv 18:479–495

Lencinas MV, Sola FJ, Martínez Pastur G (2017) Variable retention effects on vascular plants and beetles along a regional gradient in *Nothofagus pumilio* forests. For Ecol Manag 406:251–265

Lencinas MV, Sola FJ, Cellini JM, Peri PL, Martínez Pastur GJ (2019) Land sharing in South Patagonia: conservation of above-ground beetle diversity in forests and non-forest ecosystems. Sci Total Environ 690:132–139

Lososová Z, Horsák M, Chytry M, Cejka M, Danihelka T, Fajmon J et al (2011) Diversity of Central European urban biota: effects of human-made habitat types on plants and land snails. J Biogeogr 38:1152–1163

MacArthur RH, Wilson EO (1967) The theory of island biogeography. Princeton University Press, Princeton

Maestre FT, Castillo-Monroy AP, Bowker MA, Ochoa-Hueso R (2012) Species richness effects on ecosystem multifunctionality depend on evenness, composition and spatial pattern. J Ecol 100: 317–330

Marín M (2014) Distribución, fenología reproductiva, e historia natural del queltehue (*Vanellus chilensis*) en la zona central de Chile. Bol Mus Nac Hist Nat 63:119–126

Marvaldi AE, Lanteri AA (2005) Key to higher taxa of South American weevils based on adult characters (Coleoptera, Curculionoidea). Rev Chil Hist Nat 78:65–87

Marzluff J, Rodewald A (2008) Conserving biodiversity in urbanizing areas: nontraditional views from a bird's perspective. Cities Environ 1(2):6–27

Marzluff JM, Bradley G, Shulenberger E, Endlicher W, Alberti M, Bradley G, Ryan C, ZumBrunnen C, Simon U (eds) (2008) Urban ecology. An international perspective on the interaction between humans and nature. Springer, Berlin

Matthies SA, Rüter S, Schaarschmidt F, Prasse R (2017) Determinants of species richness within and across taxonomic groups in urban green spaces. Urban Ecosyst 20:897–909

McGregor-Fors I, Morales-Pérez L, Schondube JE (2011) Does size really matter? Species–area relationships in human settlements. Divers Distrib 17:112–121

Millennium Ecosystem Assessment (2005) Ecosystems and human well-being. Island Press, Washington

Miller JR (2005) Biodiversity conservation and the extinction of experience. Trends Ecol Evol 20: 430–434

Minnesota Department of Natural Resources (2013) A handbook for collecting vegetation plot data in Minnesota: The relevé method. 2nd ed. Minnesota Bio-logical Survey, Minnesota Natural Heritage and Nongame Research Program, and Ecological Land Classification Program. Biological Report 92. St. Paul, Minnesota Department of Natural Resources

Montagnini F, Levin B, Berg KE (2022) Introduction. Biodiversity Islands. Strategies for conservation in human-dominated environments. In: Montagnini F (ed) Biodiversity islands: strategies for conservation in human-dominated environments. Topics in Biodiversity and Conservation, Springer, Cham, pp 1–35

Moore DM (1983) Flora of Tierra del Fuego. Anthony Nelson

Moreno ML, Fernández MG, Molina SI, Valladares G (2013) The role of small woodland remnants on ground dwelling insect conservation in Chaco Serrano, Central Argentina. J Insect Sci 13:40. https://doi.org/10.1673/031.013.4001

Murphy D (1999) Challenges to biological diversity in urban areas. In: Wilson EO (ed) Biodiversity. National Academy Press, Washington, pp 71–76

Nilon CH, Aronson MFJ, Cilliers SS, Dobbs C, Frazee LJ, Goddard MA, O'Neill KM, Roberts D, Stander EK, Werner P, Winter M, Yocom KP (2017) Planning for the future of urban biodiversity: a global review of city-scale initiatives. Bioscience 67:332–342

Noreika N, Pajunen T, Kotze DJ (2015) Urban mires as hotspots of epigaeic arthropod diversity. Biodivers Conserv 24(12):2991–3007

Oliver I, Beattie AJ (1993) A possible method for the rapid assessment of biodiversity. Conserv Biol 7(3):562–568

Orzanco MG (1999) Problemas ambientales detectados por la población de Ushuaia (Tierra del Fuego, Argentina). Invest Geogr 40:85–98

Pauchard A, Ugarte E, Millán J (2000) A multiscale method for assessing vegetation baseline of environmental impact assessment (EIA) in protected areas of Chile. In: McCool SF, Cole DN, Borrie WT, O'Loughlin J (eds) Wilderness science in a time of change conference. USDA Forest Service, Rocky Mountain Research Station (United States), pp 111–116

Pielou EC (1975) Ecological diversity. Wiley, New York

Pinceloup N, Poulin M, Brice M-H, Pellerin S (2020) Vegetation changes in temperate ombrotrophic peatlands over a 35 years period. PLoS One 15(2):e0229146. https://doi.org/10.1371/journal.pone.0229146

Posadas P (2012) Species composition and geographic distribution of Fuegian Curculionidae (Coleoptera: Curculionoidea). Zootaxa 3303:1–36

Reich PB, Tilman D, Isbell F, Mueller K, Hobbie S, Flynn D (2012) Impacts of biodiversity loss escalate through time as redundancy fades. Science 336(6081):589–592

Remsen JV, Areta JI, Cadena CD, Claramunt S, Jaramillo A, Pacheco JF, Robbins MB, Stiles FG, Stotz DF, Zimmer KJ (2019) A classification of the bird species of South America. American Ornithologists' Union. http://www.museum.lsu.edu/~Remsen/SACCBaseline.htm

Roig-Juñent S, Crisci JV, Posadas P, Lagos S (2002) Áreas de distribución y endemismo en zonas continentales. In: Costa C, Vanin SA, Lobo JA, Melic A (eds) Red Iberoamericana de Biogeografía y Entomología Sistemática (PrIBES). Sociedad Entomológica Aragonesa, Zaragoza, pp 247–266

Rozzi R, Jiménez JE (2014) Magellanic sub-antarctic ornithology: first decade of long-term bird studies at the Omora Ethnobotanical Park, Cape Horn biosphere reserve, Chile. University of North Texas Press, Dallas

Rozzi R, Massardo F, Silander J, Dollenz O, Connolly B, Anderson C, Turner N (2003) Native and exotic tree species in the squares of the Magellan Region. An Inst Patagon 31:27–42

Rozzi R, Armesto JJ, Gutiérrez JR, Massardo F, Likens GE, Anderson CB, Poole A, Moses KP, Hargrove E, Mansilla AO, Kennedy JH, Willson M, Jax K, Jones CG, Callicott JB, Arroyo MTK (2012) Integrating ecology and environmental ethics: earth stewardship in the southern end of the Americas. Bioscience 62(3):226–236

Sola FJ, Peri PL, Huertas L, Martínez Pastur GJ, Lencinas MV (2016) Above-ground arthropod community structure and influence of structural-retention management in southern Patagonian scrublands, Argentina. J Insect Conserv 20:929–944

Sola FJ, Soler R, Huertas A, Lencinas MV (2018) Urbanización y diversidad de invertebrados en turbales de Tierra del Fuego. In: Proc XXVIII Reunión Argentina de Ecología, 29 Oct - 2 Nov, Mar del Plata, Argentina

Tallamy DW, Ballard M, D'Amico VD (2010) Can alien plants support generalist insect herbivores? Biol Invasions 12:2285–2292

Thomas K, Geller L (2013) Urban forestry: toward an ecosystem services research agenda. The National Academies Press, Washington

Tuhkanen S (1992) The climate of Tierra del Fuego from vegetation geographical point of view and its ecoclimatic counterpartes elsewhere. Acta Bot Fenn 145:1–64

Vergara PM, Schlatter RP (2006) Aggregate retention in two Tierra del Fuego *Nothofagus* forests: short-term effects on bird abundance. For Ecol Manag 225:213–224

Zerbe S, Maurer U, Schmitz S, Sukopp H (2003) Biodiversity in Berlin and its potential for nature conservation. Landsc Urban Plan 62:139–148

# Chapter 17
# Paradise Lot: A Temperate-Climate Urban Agroforestry Biodiversity Island

**Eric Toensmeier**

**Abstract** Paradise Lot is a garden and home site occupying 400 square meters in the cold temperate subhumid northeastern United States. Emerging from the permaculture movement, its agroforestry approach emphasizes perennial crops. The goal is to provide ecosystem services while producing food. Data are presented on plant biodiversity, including the numbers of species that are native and/or provide important functions including nectar for pollinators, wildlife cover, nitrogen fixation and groundcover. Anecdotal evidence is provided suggesting that the site is a haven for invertebrates and offers some benefits to vertebrates despite its small size. This case study demonstrates that food production and ecosystem function can go hand in hand, with the wide variety of plants incorporated here providing various ecological roles and human uses. Located in an urban area with a depauperate flora and fauna, the site serves as a biodiversity island.

**Keywords** Agroforestry · Food forest · Homegarden · Perennial crops · Permaculture · Urban agriculture

## 17.1 Introduction

Paradise Lot is an urban agroforestry garden, inspired by the permaculture movement, which serves as a biodiversity island. Designed and managed by the author and Jonathan Bates, the project has been documented in several books (Jacke and Toensmeier 2005; Toensmeier and Bates 2013).

Agroforestry systems incorporate woody plants into agriculture in a variety of ways. Paradise Lot practices both tree intercropping (growing annual crops and woody plants together) and multistrata agroforestry (perennial systems with multiple layers of crops) (Toensmeier 2016).

E. Toensmeier (✉)
Global Evergreening Alliance, Perennial Agriculture Institute, Yale University, Holyoke, MA, USA



F. Montagnini (ed.), *Biodiversity Islands: Strategies for Conservation in Human-Dominated Environments*, Topics in Biodiversity and Conservation 20,
https://doi.org/10.1007/978-3-030-92234-4_17

Toensmeier and Bates were very involved in the world of permaculture as they began the project (Toensmeier and Bates 2013). Permaculture is an international and regional movement that disseminates and practices a worldview, using a best practices framework and design systems (Ferguson and Lovell 2013). Ecosystem mimicry is one of permaculture's core practices, and it is at the heart of Jacke and Toensmeier (2005), which Toensmeier was writing in 2004 when the garden was established. This work lays a foundation for "edible forest gardens", a home-scale version of multistrata agroforestry drawing heavily on permaculture design and eastern forest ecosystem mimicry. This form of garden has become known as a "food forest" and has spread rather rapidly in the last two decades, due in part to appreciation of its multifunctional approach that provides ecosystem services while producing food (Clark and Nicholas 2013).

While this agroforestry practice is relatively new in temperate climate, it has been practiced for thousands of years in the tropics. The "tropical homegarden" produces highly biodiverse perennial and annual crops, and often small livestock, right around the home. Homegardens have been described by scientists as "the epitome of sustainability" for their impressive impacts on biodiversity and other ecosystem services (Nair 2006; Nair and Kumar 2006). Indeed, they have been ranked as having the highest biodiversity of any anthropogenic ecosystem (Nair 2006). These systems were a major, though little credited, inspiration and model for the early innovators of permaculture (Ferguson and Lovell 2014).

At the garden level, edible forest gardens strive to mimic the architecture (layers, density, patterns, and diversity), social structure (niches, relationships, and communities), soil ecology and fertility dynamics, and successional dynamics of natural ecosystems. In particular, it is aggrading mid-successional ecosystems that are the most important model, as in humid temperate climates these feature high rates of biomass growth and high diversity of perennial food plants (Jacke and Toensmeier 2005). This is reflected in the central importance of polycultures (growing more than one species in the same place, while attempting to minimize competition and maximize cooperation). Toensmeier and Bates (2013) observed wild species associations and replicated them, sometimes with slight modifications, in the garden. For example, elderberry (*Sambucus canadensis*) often serves as a trellis for groundnut (*Apios americana*) in the wild, and this association has been used as the basis for a polyculture in the garden, albeit with improved, higher-yielding forms of each.

## 17.2 Site History

The site has been documented in two books. Jacke and Toensmeier (2005) uses the site as a case study of ecological site design. Toensmeier and Bates (2013) is a garden memoir telling the story of the garden's inspiration and development.

### 17.2.1 Goals

The garden designers began the process of clarifying goals before identifying a site, and continued the process once moving in in 2004. These goals are published in Jacke and Toensmeier (2005), for which the site was a design process case study, and in Toensmeier and Bates (2013), which includes some revised goals.

The initial goal was to create a demonstration site to test the "edible forest garden" model in the northeastern United States. Specific goals included: "creating an intensively managed backyard foraging paradise;" "a mega-diverse living ark of useful and multifunctional plants from our own bioregion and around the world;" "to bring our dead and blighted backyard to life... creating a lush, semiprivate oasis that inspires our neighbors to plant their own;" and "to serve as a refuge in our biologically impoverished neighborhood" (Jacke and Toensmeier 2005). To expand a bit on the biodiversity oasis aspect of the goals, Toensmeier and Bates (2013) write: "It was too much to hope that moose and great blue herons would come to visit, but we did want to create a haven for smaller forms of wildlife, particularly those birds, insects and amphibians that help control pests." It was hoped that the garden would serve as an inspiration to gardeners in the neighborhood.

In terms of food production, the goal was to have as long a harvest season as possible for production of fruits and vegetables. Emphasis was on perennial crops in diverse polyculture systems. The design incorporated nitrogen-fixing plants to aid in restoration of degraded soils and maintain fertility, plants to attract pollinators and beneficial insects, groundcovers to serve as a living mulch and suppress weeds, and thicket-forming and evergreen plants for wildlife cover. Other elements included beds for annual crops, a greenhouse, a pond for aquatic vegetable production, and abiotic elements including patio, toolshed, compost piles, and vehicle access (Jacke and Toensmeier 2005).

In species selection, an emphasis was placed on multifunctionality. If a species performs more than one ecosystem function and/or provides more than one human use, it saves space – a key issue in this small, intensive garden. This kind of multifunctionality is a key principle of permaculture (Ferguson and Lovell 2014).

### 17.2.2 Site Analysis and Assessment

The Paradise Lot site is located in Holyoke, Massachusetts, in the northeastern United States. Climate is cold temperate subhumid with winter lows usually no lower than −12 °C and summer highs usually below 38 °C. Precipitation is around 1000 mm, with no dry season. The frost-free growing season is roughly 165 days (Toensmeier and Bates 2013).

Holyoke is a "rust belt" city, a former industrial center in the western part of the state. Most homes have lawns with some ornamental woody plants. Right across the street from the site is a brownfield, the abandoned and contaminated location of a

**Fig. 17.1** The neighborhood context. The site (flagged as 145 Brown Avenue) seen in context of a neighborhood comprised mostly of lawns with some trees. Brownfield complex across street to right, and steep wooded sloped above railroad tracks. (Image: Google Maps)

former dry-cleaning operation (contamination risk to Paradise Lot is likely minimal as the issue is buried tanks, and the brownfield is significantly lower elevation). Behind the brownfield is a steep wooded slope that provides a bit of a wildlife corridor between two densely settled neighborhoods. Little food is grown in the neighborhood and there are no visible plantings for wildlife and minimal landscaping with native plants (author personal observation) (Fig. 17.1).

The site for the garden was barren in 2004 when the project began (Fig. 17.2). The previous house had burned down. Fill (mineral soil with extremely low organic matter content) was brought in after construction to level the yard and parking lot. The property occupies 400 square meters including the house and parking, with the back yard where most of the garden is located occupying 195 square meters (Toensmeier and Bates 2013).

Testing determined that the site features three distinct soil types. The original soil, to the west end of the garden, is sandy and was pH 5.3 in 2004. It contained low levels of lead. The land between this area and the house consists of highly compacted clay with construction debris including bricks, concrete, and asphalt. The remainder of soils were gravel and sand fill (Toensmeier and Bates 2013).

Shade was another key limitation in the design. Buildings to the south shade much of the garden in winter, and trees overhanging from the property to the north provide shade in the summer. Only one small area offered sun year-round, which became the site of the greenhouse (Toensmeier and Bates 2013).

Del Tredici (2010) is a field guide to plants growing spontaneously in urban areas. It includes notes on the ecological conditions tolerated by these species and thus permits their use as indicator species to provide information about urban sites. Shading the northern end of the garden from across the fenceline were two mature

**Fig. 17.2** The site in 2004 before development of the garden. The Paradise Lot site was a typical degraded urban yard with major disturbances from recent construction. (Photo: E Toensmeier)

Norway maples (*Acer platanoides*), which produced many seedlings in the yard below. Del Tredici describes this species as tolerating compacted soils and road salt. The areas of recently arrived "fill" soil were virtually free of vegetation, with a few plants of the annual lamb's quarters (*Chenopodium album*), which del Tredeci notes is common in disturbed landscapes and tolerant of compacted soils. The original sandy acid soil was dominated by crabgrass (*Digitaria ischaemum*): "[a] disturbance-adapted colonizer of bare ground; tolerant of contaminated and compacted soil" (Del Tredici 2010). Fencelines were tangled with vines of bittersweet (*Celastrus orbicularis*), a noted aggressive species. Though these (mostly reviled) species were not desired by Bates and Toensmeier, they observed that the maples were the only place in the garden where birds congregated.

### *17.2.3 Design*

The design process used is laid out in Jacke and Toensmeier (2005). It involves goals clarification, site analysis and assessment, and the development of a design concept. This concept is fleshed out into greater levels of detail (schematic and detailed design), followed by more detailed designs for specific polycultures and infrastructure elements.

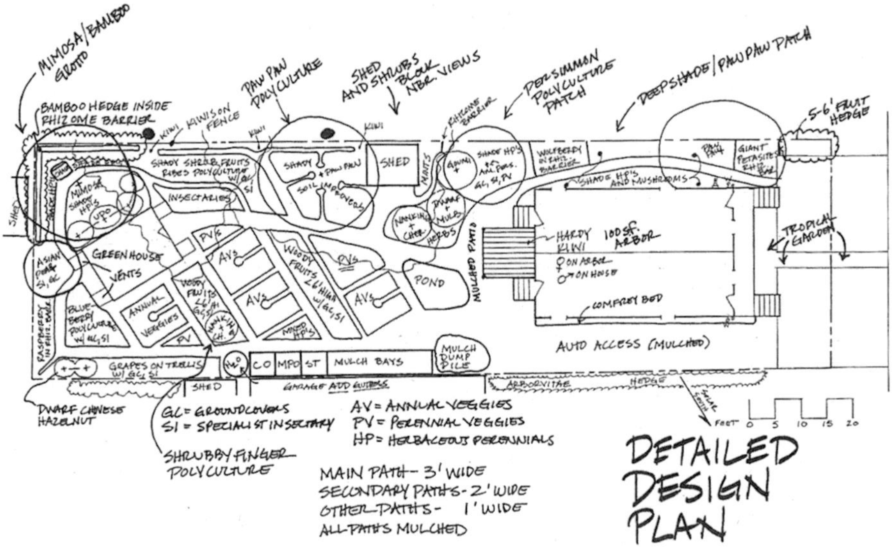


**Fig. 17.3** Map of the site. The initial design for the garden as shown in Jacke and Toensmeier (2005). (Image: Dave Jacke based on work by the author)

The design matches patches with particular conditions with suitable plants species or other design elements. For example, an area with full shade in summer and compacted clay soil is utilized for a toolshed. These patches must also fit into the overall design concept.

The Paradise Lot design is shown in Fig. 17.3. Along the north and west edges, taller trees are planted to mimic a forest edge. In sunnier areas, beds for annual crops alternate with lower-growing woody crop polycultures. Near the house, a kiwi trellis and pond are located, along with toolshed and compost and mulch piles.

### 17.2.4 Evolution and Management

Overall, the project has been successful in achieving its goals, with the exception of adoption of ecological practices by neighbors. Figure 17.4 illustrates the increase in biomass and habitat diversity since 2004. Fruit is available outdoors daily for about six months of the year (plus greenhouse citrus in winter), and vegetables are available for 9–10 months outdoors and throughout the winter in the greenhouse. Chickens (*Gallus gallus*), silkworms (*Bombyx mori*), worms (*Eisenia fetida*), and native black soldier fly larvae (*Hermetia illucens*) have all been raised at various times, with chickens providing eggs and the invertebrates providing compost and food for the chickens. Lead in the sandy soil type was remediated by buffering the pH, increasing organic matter, and mulching (in line with University recommendations), and has since fallen to safe levels according to ongoing testing.

**Fig. 17.4** The site in 2019 demonstrating increased biomass and biodiversity. Paradise Lot in fall 2019. Note 10-meter tall American persimmon at right. (Photo: E Toensmeier)

While the initial goals were largely met by 2009, the herbaceous perennial layer was poorly organized, with excessive competition from some aggressive plants. The initial enthusiasm for crop biodiversity for its own sake was dampened by the reality that some crops did not taste good, while others did not thrive in the site conditions (particularly the initially degraded soils). The site goals were amended with the following: "to grow the things we like to eat, that grow well for us, and assemble them in functioning polycultures" (Toensmeier and Bates 2013).

Over time, the goal to develop the collection of native plants became increasingly important. This extended to cultivating species that had been part of the Eastern Agricultural Complex, a suite of crops first domesticated several thousand years ago in what is today Missouri (Toensmeier and Bates 2013). As we found new resources on native plants as providers of food and habitat (Roth 2006, Tallamy 2007, Mäder et al. 2011, 2014), it was noted that the garden already included many of these species as a result of research carried out by Jacke and Toensmeier (2005). At the same time, additional species recommended in these references were incorporated into the garden.

In 2017, co-founding designer and gardener Jonathan Bates left the site to begin a farming operation. At this time the goals were again revised to simplify management. These included: further culling of undesirable or unproductive species;

planting more large perennial grasses for mulch production; expanding plantings for beneficial insects and pollinators; increasing rainwater capture and storage (city water is the main source of irrigation at moment); developing raised beds for annual crops; and testing new approaches to pathway management (author's personal experience). A quantitative assessment on hours of labor required (compared to lawn management) has not been performed but would be a useful contribution.

The initial expectation was that similar gardens would be planted throughout the neighborhood, as neighbors were inspired by the example. This has not occurred, perhaps because so many neighbors rent and do not have tenure agreements that permit gardening. Toensmeier and Bates speculate that this form of extreme "ecofunctional" gardening is perhaps too far from the norms of landscape management, and have advocated for "suburban landscape mimics" that incorporate the same elements but have an aesthetic more in line with American norms, even though this involves some reduction in ecosystem function. Perhaps other tenure models like leasing of yards might encourage adoption as well. At the same time, thousands of gardeners from all over the world have visited the site, and many have gone on to implement similar designs (Toensmeier and Bates 2013).

The project continues to present trade-offs. The high diversity and intensity create a high labor demand, as does the ongoing experimentation and "editing." Inclusion of non-native food plants, offering high yields of desirable crops, is in dynamic tension with the desire to maximize use of native multipurpose species. Native food plants are often at a lower level of domestication than exotic species, exhibiting lower yields and/or stronger flavors in some cases. For example, the native nut shrub chinquapin (*Castanea pumila*) was replaced with peaches (*Prunus persica*), which require more work but offer much higher and more palatable yields. Maintaining a high level of plant biodiversity is a priority, but presents an ongoing management challenge. Some species, both native and non-native, have become ongoing weed issues – yet new species are continually brought to the garden.

## 17.3 Plant Species

Data for this site are available on plant species but only anecdotal information is available on wildlife, so the focus here is on plants, both as biodiversity themselves and in their roles in providing food and habitat for wildlife. Data on uses of plants are also presented, to demonstrate that in this case habitat and human use need not be incompatible goals.

### *17.3.1 Plant Biodiversity*

The garden contains 59 botanical families, 130 genera, and 191 perennial or self-sowing species. In 2004, when planting began, there were perhaps five species present. This represents a substantial increase in botanical diversity, though not all of it is native.

Novel ecosystems, comprised of non-historic assemblages of native and non-native organisms, have become an area of study in recent decades (Hobbs et al. 2013). In some cases, novel ecosystems have been shown to provide similar and even improved ecosystem services when compared to historic ecosystems of the same region (Davis et al. 2011, Hallet 2013). Multistrata agroforestry systems like those practiced at Paradise Lot can be seen as anthropogenic novel ecosystems.

Table 17.1 shows the plant species present in 2019. Taxonomy, food uses, and ecological functions are noted. Additional uses occur in the garden (e.g., medicinal plants, bamboo for staking and trellising), but are not documented here. Almost all species now present were brought to the site intentionally, with the exceptions of *Lactuca canadensis* (spontaneous) and *Aster cordifolia* (present from the outset). Two species are present but undesired, currently in the process of eradication: *Thladiantha dubia* and *Houttyunia cordata.* More than a hundred other species have been attempted and either failed to succeed or were eradicated based on their poor flavor or other undesirable traits.

### *17.3.2 Ecological Functions*

The garden was designed with five ecological functions in mind (native plant species, nectary plants, wildlife habitat, nitrogen fixation, and groundcover), and species have been selected accordingly. Table 17.2 shows the number and percentage of plants in the garden providing these ecological functions, broken down by form (woody plants, vines, and herbaceous plants). Woody plants include trees, shrubs, and bamboos. Vines includes woody lianas as well as herbaceous perennial vines, and a few self-seeding annuals. Herbaceous plants are mostly perennial and include forbs, grasses, and more, along with a few self-sowing and broadcast annuals. The ecological functions of all species in the garden are shown in Table 17.1. In addition to these, some annual crops are produced in outdoor beds each year. Winter annuals and subtropical perennial crops including *Citrus* species are grown in the greenhouse.

The first category examined here is native plants, which are presumed more likely to have pre-existing relationships with native animals (Jacke and Toensmeier 2005). Overall, 46% of the species in the garden are native to the eastern United States, and many were collected from plants growing within a 15-min driving radius of the site.

**Table 17.1** Plant species present in 2019

| Latin name | Common name | Botanical family | Food uses | Ecological functions |
|---|---|---|---|---|
| **Woody Plants** | | | | |
| *Amelanchier spp.* | "Autumn brilliance" serviceberry | Rosaceae | Fruit | Native |
| *Amelanchier spp.* | "Regent" juneberry | Rosaceae | Fruit | Native, nectary, wildlife cover |
| *A. stolonifera* | Running serviceberry | Rosaceae | Fruit | Native, nectary, wildlife cover |
| *Amorpha fruticosa* | False indigo | Fabaceae | | Native, nitrogen fixation, nectary, |
| *Asimina triloba* | Pawpaw | Annonaceae | Fruit | Native |
| *Calycanthus floridus* | Carolina allspice | Calycanthaceae | Spice | Native, wildlife cover |
| *Castanea mollisima* | Chinese chestnut | Fagaceae | Nuts | |
| *Cercis canadensis* | Redbud | Fabaceae | Vegetable | Native |
| *Corylus spp.* | Hazelnut | Corylaceae | Nuts | Nectary, wildlife cover |
| *Cudrania tricuspidata* | Che fruit | Moraceae | Fruit | |
| *Diospyros virginiana* | American persimmon | Ebenaceae | Fruit | Native |
| *Elaeagnus multiflora* | Goumi | Elaeagnaceae | Fruit | Nitrogen fixation, nectary |
| *Hamamelis virginiana* | Witch hazel | Hamamelidaceae | | Native |
| *Indigofera kirilowii* | Kirilow's indigo | Fabaceae | | Nitrogen fixation, nectary, wildlife cover |
| *Lespedeza bicolor* | Bush clover | Fabaceae | Vegetable | Nitrogen fixation |
| *Lindera benzoin* | Spicebush | Lauraceae | Spice | Native |
| *Lycium chinense* | Edible-leaf goji | Solanaceae | Fruit, vegetable | wildlife cover |
| *Malus domestica* | Dwarf apple | Rosaceae | Fruit | Nectary |
| *Morus alba* | White mulberry | Moraceae | Fruit, vegetable | |
| *M. alba tartarica* | Tartarian mulberry | Moraceae | Vegetable | |
| *M. macrura* | Himalayan mulberry | Moraceae | Fruit | |
| *Phyllostachys aureosulcata* | Running bamboo | Poaceae | Vegetable | Wildlife cover |
| *P. Nigridolaria* | Running bamboo | Poaceae | Vegetable | Wildlife cover |
| *P. nuda* | Running bamboo | Poaceae | Vegetable | Wildlife cover |

(continued)

**Table 17.1** (continued)

| Latin name | Common name | Botanical family | Food uses | Ecological functions |
|---|---|---|---|---|
| *Prunus besseyi* | Sand cherry | Rosaceae | Fruit | Native, nectary, wildlife cover |
| *P. persica* | Peach | Rosaceae | Fruit | Nectary |
| *P. tomentosa* | Nanking cherry | Rosaceae | Fruit | Nectary |
| *Pyrus communis* | European pear | Rosaceae | Fruit | Nectary |
| *Pyrus spp.* | Asian pear | Rosaceae | Fruit | Nectary |
| *Ribes spp.* | Jostaberry | Grossulariaceae | Fruit | |
| *R. nigrum* | Black currant | Grossulariaceae | Fruit | |
| *R. odoratum* | Clove currant | Grossulariaceae | Fruit | Native, wildlife cover |
| *R. rubrum* | Red currant | Grossulariaceae | Fruit | |
| *R. uva-crispa* | Gooseberry | Grossulariaceae | Fruit | Wildlife cover |
| *Rubus idaeus* | Raspberry | Rosaceae | Fruit | Native, nectary, wildlife cover |
| *R. occidentalis* | Black raspberry | Rosaceae | Fruit | Native, nectary, wildlife cover |
| *Salix spp.* | Native willow | Salicaceae | | Native, nectary |
| *Sambucus canadensis* | Elderberry | Adoxaceae | Fruit, vegetable | Native, nectary |
| *Senna hebecarpa* | Wild senna | Fabaceae | | Native, nitrogen fixation, nectary, wildlife cover |
| *Tilia cordata* | Littleleaf linden | Malvaceae | Vegetable | |
| *Toona sinensis* | Chinese toon | Meliaceae | Vegetable | |
| *Vaccinium corymbosum* | Halfhigh blueberries | Ericaceae | Fruit | Native |
| *Viburnum opulus* | Highbush cranberry | Adoxaceae | Fruit | Native, nectary |
| *x Sorbopyrus* | Shipova | Rosaceae | Fruit | |
| *Zizyphus jujuba* | Jujube | Rhamnaceae | Fruit | |
| **Vines** | | | | |
| *Actinidia arguta* | Hardy kiwifruit | Actinidiaceae | Fruit | |
| *Amphicarpa bracteata* | Ground bean | Fabaceae | Beans, tubers | Groundcover, native |
| *Apios americana* | Groundnut | Fabaceae | Tubers | Native, nectary, nitrogen fixation |
| *A. priceana* | Price's groundnut | Fabaceae | Beans, tubers | Native, nectary, nitrogen fixation |
| *Dioscorea polystachya* | Chinese yam | Dioscoreaceae | Bulbils, tubers | |
| *Hablitzia tamnoides* | Caucasian spinach | Amaranthaceae | Vegetable | |
| *Humulus lupulus* | Hops | Cannabidaceae | Vegetable, tea | |
| *Melothria pendula* | Cucumber berry | Cucurbitaceae | Vegetable | Groundcover, native |

(continued)

**Table 17.1** (continued)

| Latin name | Common name | Botanical family | Food uses | Ecological functions |
|---|---|---|---|---|
| *Passiflora incarnata* | Maypop | Passifloraceae | Fruit, vegetable | Native, nectary |
| *Phaseolus polystachyos* | Thicket bean | Fabaceae | Beans | Native, nectary, nitrogen fixation |
| *Strophostyles helvola* | Woolly bean | Fabaceae | Beans | Native, nectary, nitrogen fixation |
| *S. umbellata* | Perennial woolly bean | Fabaceae | Beans | Native, nectary, nitrogen fixation |
| *Thladiantha dubia* | Manchu tubergourd | Cucurbitaceae | Fruit, vegetable | Nectary |
| *Vitis riparia* | Riverbank grape | Vitaceae | Fruit, vegetable | Native |
| *V. vinifera x labrusca* | Grape | Vitaceae | Fruit, vegetable | Native |
| **Herbs** | | | | |
| *Agastache foeniculum* | Anise hyssop | Lamiaceae | Tea | Native, nectary |
| *Allium ampeloprasum* | Perennial leek | Alliaceae | Culinary, vegetable | Nectary |
| *A. canadense* | Wild garlic | Alliaceae | Culinary, vegetable | Native |
| *A. cepa proliferum* | Walking onion | Alliaceae | Culinary, vegetable | |
| *A. cernuum* | Nodding wild onion | Alliaceae | Culinary, vegetable | Native, nectary |
| *A. fistulosum* | Welsh onion | Alliaceae | Culinary, vegetable | Nectary |
| *A. tricoccum* | Ramps | Alliaceae | Culinary, vegetable | Native |
| *A. tuberosum* | Garlic chives | Alliaceae | Culinary, vegetable | Groundcover, nectary |
| *A. ursinum* | Ramsons | Alliaceae | Culinary, vegetable | |
| *A. victorialis* | Victory onion | Alliaceae | Culinary, vegetable | |
| *A. vineale* | Field garlic | Alliaceae | Culinary, vegetable | |
| *Amorpha nana* | Dwarf false indigo | Fabaceae | | Native, nectary, nitrogen fixation |
| *Anthriscus sylvestris* | Woodland chervil | Apiaceae | Culinary, vegetable | Nectary |
| *Aquilegia canadensis* | Columbine | Ranunculaceae | Vegetable | Native |
| *Artemisia annuua* | Sweet Annie | Asteraceae | | |
| *Asarum canadense* | Wild ginger | Aristolochiaceae | | Groundcover, native |

(continued)

**Table 17.1** (continued)

| Latin name | Common name | Botanical family | Food uses | Ecological functions |
|---|---|---|---|---|
| *Asclepias exaltata* | Poke milkweed | Apocynaceae | Vegetable | Native, nectary |
| *A. incarnata* | Swamp milkweed | Apocynaceae | Vegetable | Native, nectary |
| *A. speciosa* | Showy milkweed | Apocynaceae | Vegetable | Native, nectary |
| *A. syriaca* | Common milkweed | Apocynaceae | Vegetable | Native, nectary |
| *A. tuberosa* | Butterfly weed | Apocynaceae | Vegetable | Native, nectary |
| *Asparagus officinalis* | Asparagus | Asparagaceae | Vegetable | |
| *Aster cordifolia* | Heartleaf aster | Asteraceae | | Native, nectary |
| *A. novae-angliae* | New England aster | Asteraceae | | Native, nectary |
| *A. scaber* | Chwinamul | Asteraceae | Vegetable | |
| *Astragalus glycyphyllos* | Licorice milkvetch | Fabaceae | Tea | Groundcover, nitrogen fixation |
| *Atriplex hortensis* | Orach | Amaranthaceae | Vegetable | |
| *Brassica napus* | Kale | Brassicaceae | Vegetable | |
| *B. rapa* | Mustard | Brassicaceae | Vegetable | |
| *Bunias orientalis* | Turkish rocket | Brassicaceae | Vegetable | Nectary |
| *Bunium bulbocastanum* | Earth chestnut | Apiaceae | Vegetable, root | Nectary |
| *Camassia quamash* | Camass | Asparagaceae | Root | |
| *Camassia scillioides* | Wild hyacinth | Asparagaceae | Root | Native |
| *Campanula porschariskayana* | Bluebells | Campanulaceae | Vegetable | Groundcover |
| *Campanula rapunculus* | Rampion | Campanulaceae | Root | |
| *Cardamine diphylla* | Toothwort | Brassicaceae | Culinary | Native |
| *Ceanothus americanus* | New Jersey tea | Rhamnaceae | Tea | Native, nectary, nitrogen fixation |
| *Chamaenerion angustifolium* | Fireweed | Onagraceae | Vegetable | Native, nectary |
| *Chasmanthium latifolium* | River oats | Poaceae | | Native, nectary, wildlife cover |
| *Chenopodium bonus-henricus* | Good king Henry | Amaranthaceae | Vegetable | |
| *Chrysogonum virginianum* | Green and gold | Asteraceae | | Groundcover, native, nectary |
| *Crambe maritima* | Sea kale | Brassicaceae | Vegetable | Nectary |
| *Crocus sativus* | Saffron crocus | Iridaceae | Culinary | |

(continued)

**Table 17.1** (continued)

| Latin name | Common name | Botanical family | Food uses | Ecological functions |
|---|---|---|---|---|
| *Cryptotaenia canadensis* | Honewort | Apiaceae | Vegetable | Native, nectary |
| *Dennestedia punctilobula* | Hay-scented fern | Dennestediaceae | | Groundcover, native |
| *Desmodium canadense* | Tick trefoil | Fabaceae | | Native, nitrogen fixation |
| *D. glutinosum* | Pointed-leaved tick trefoil | Fabaceae | | Groundcover, native, nitrogen fixation |
| *Diplotaxis muralis* | Sylvetta arugula | Brassicaceae | Vegetable | Nectary |
| *Disporum cantonense* | Chinese fairy bells | Liliaceae | Vegetable | |
| *Dryopteris expansa* | Spiny wood fern | Dryopteridaceae | Roots | |
| *Dystaenia takesimiana* | Korean celery | Apiaceae | Vegetable | Nectary |
| *Eurybia macrophylla* | Largeleaf aster | Asteraceae | Vegetable | Groundcover, native, nectary |
| *Festuca ovina* | sheep's fescue | Poaceae | | Groundcover |
| *Fragaria anasana* | Strawberry | Rosaceae | Fruit | Groundcover |
| *F. moschata* | Musk strawberry | Rosaceae | Fruit | Groundcover |
| *Fragaria vesca* | Wood strawberry | Rosaceae | Fruit | Groundcover, native |
| *Gaultheria procumbens* | Wintergreen | Ericaceae | Fruit, tea | Groundcover, native |
| *Helianthus tuberosus* | Sunchoke | Asteraceae | Root | Native, nectary, wild-life cover |
| *Hibiscus moscheotus* | Swamp hibiscus | Malvaceae | Vegetable | Native, nectary |
| *Hosta spp.* | Edible hosta | Asparagaceae | Vegetable | Groundcover |
| *Houyttunia cordata* | Houttyunia | Saururaceae | Vegetable | Groundcover |
| *Hydrophyllum spp.* | Waterleaf | Boraginaceae | Vegetable | Groundcover, native, nectary |
| *Impatiens capensis* | Touch me not | Balsaminaceae | | Native, nectary |
| *Kalimeris indica* | Yomena aster | Asteraceae | Vegetable | Nectary |
| *Lactuca canadensis* | Wild lettuce | Asteraceae | Vegetable | Native, nectary |
| *Ligularia fischeri* | Gomchee | Asteraceae | Vegetable | Nectary |
| *Lilium superbum* | Turk's cap lily | Liliaceae | Root | Native |
| *Lobelia cardinalis* | Cardinal flower | Lobeliaceae | | Native, nectary |

(continued)

**Table 17.1** (continued)

| Latin name | Common name | Botanical family | Food uses | Ecological functions |
|---|---|---|---|---|
| *L. siphilitica* | Blue lobelia | Lobeliaceae | | Native, nectary |
| *Matteuccia struthiopteris* | Ostrich fern | Onocleaceae | Vegetable | Native |
| *Melissa officinalis* | Lemon balm | Lamiaceae | Culinary, tea | Nectary |
| *Mentha piperita* | Peppermint | Lamiaceae | Culinary, tea | Nectary |
| *Mentha spp.* | Apple mint | Lamiaceae | Culinary, tea | Nectary |
| *Mertensia virginiana* | Oysterleaf | Boraginaceae | Vegetable | Native, nectary |
| *Miscanthus giganteus* | Giant miscanthus | Poaceae | Vegetable | Wildlife cover |
| *Monarda fistulosa* | Bergamot | Lamiaceae | Culinary | Native, nectary |
| *Musa basjoo* | Japanese fiber banana | Musaceae | Vegetable | |
| *Muscari botryoides* | Italian grape hyacinth | Asparagaceae | Vegetable | |
| *M. neglectum* | Musk hyacinth | Asparagaceae | Vegetable, root | |
| *Myrrhis odorata* | Sweet cicely | Apiaceae | Vegetable | Nectary |
| *Nelumbo nucifera* | Chinese lotus | | Root, nut | |
| *Oenanthe javanica* | Water celery | Apiaceae | Vegetable | Groundcover, nectary |
| *Osmunda cinnamomea* | Cinnamon fern | Osmundaceae | Vegetable | Native |
| *O. japonica* | Japanese royal fern | Osmundaceae | Vegetable | |
| *Pachysandra procumbens* | Allegheny spurge | Buxaceae | | Groundcover, native |
| *Phacelia bipinnatifida* | Purple phacelia | Boraginaceae | | Groundcover, native, nectary |
| *Physalis heterophylla* | Clammy groundcherry | Solanaceae | Fruit | Groundcover, native |
| *P. longifolia* | Longleaf groundcherry | Solanaceae | Fruit | Groundcover, native |
| *Phyalis spp.* | Perennial groundcherry | Solanaceae | Fruit | Groundcover |
| *Pleioblastus distichus* | Dwarf fernleaf bamboo | Poaceae | | Groundcover |
| *Podophyllum peltatum* | Mayapple | Berberidaceae | Fruit | Native |
| *Polygonatum commutatum* | Giant Solomon's seal | Asparagaceae | Vegetable, roots | Native |
| | Mountain mint | Lamiaceae | Tea | Native, nectary |

(continued)

**Table 17.1** (continued)

| Latin name | Common name | Botanical family | Food uses | Ecological functions |
|---|---|---|---|---|
| *Pycnanthemum hyssopifolium* | | | | |
| *P. muticum* | Mountain mint | Lamiaceae | Tea | Native, nectary |
| *Pycnanthemum spp* | Mountain mint | Lamiaceae | Tea | Native, nectary |
| *Rheum palmatum* | Turkish rhubarb | Polygonaceae | Vegetable | |
| *R. x cultorum* | Rhubarb | Polygonaceae | Vegetable | |
| *Rudebeckia laciniata* | Sochan | Asteraceae | Vegetable | Native, nectary |
| *Rumex acetosa* | Sorrel | Polygonaceae | Vegetable | |
| *Sagittaria latifolia* | Arrowhead | Alismataceae | Vegetable, root | Native |
| *Scilla siberica* | Siberian squill | Asparagaceae | Vegetable | |
| *Sedum ternatum* | Woodland stonecrop | Crassulaceae | Vegetable | Groundcover, native, nectary |
| *Sium suave* | Water parsnip | Apiaceae | Vegetable, root | Native, nectary |
| *Solanum ptycanthum* | Black nightshade | Solanaceae | Fruit, vegetable | Native |
| *Solidago spp.* | Goldenrod | Asteraceae | | Native, nectary |
| *Stachys affinis* | Chinese artichoke | Lamiaceae | Root | Groundcover |
| *Symphytum grandiflorum* | Large-flowered comfrey | Boraginaceae | | Groundcover, nectary |
| *Symphytum spp.* | "Hidcote blue" comfrey | Boraginaceae | | Groundcover, nectary |
| *S. x uplandicum* | Russian comfrey | Boraginaceae | | Nectary |
| *Trifolium repens* | White clover | Fabaceae | Vegetable | Groundcover, nectary, nitrogen fixation |
| *Tulipa clusiana* | Clusiana tulip | Liliaceae | Root | |
| *Tulipa spp.* | "Purissima" hybrid tulip | Liliaceae | Vegetable | |
| *Typha* spp | Cattail | Typhaceae | Vegetable, root | Native |
| *Vaccinium angustifolium* | Lowbush blueberry | Ericaceae | Fruit | Native |
| *Verbena hastata* | Blue vervain | Verbenaceae | Seeds, tea | Native, nectary |
| *Viola sororia* | Common blue violet | Violaceae | Vegetable | Groundcover, native |
| *Waldsteinia fragaroides* | Barren strawberry | Rosaceae | | Groundcover, native |
| *Zingiber mioga* | Mioga ginger | Zingiberaceae | Vegetable | |

**Table 17.2** Plant species performing ecological functions

| Function | # of Woody plants | % of Woody plants | # of Vines | % of Vines | # of Herbaceous plants | % of Herbaceous plants |
|---|---|---|---|---|---|---|
| Native plant species | 20 | 45% | 10 | 67% | 58 | 44% |
| Nectary plants | 19 | 43% | 7 | 47% | 54 | 41% |
| Wildlife cover | 17 | 39% | 0 | 0% | 2 | 2% |
| Nitrogen fixation | 5 | 11% | 6 | 40% | 6 | 5% |
| Groundcover | 0 | 0% | 2 | 13% | 31 | 23% |

Plants with flowers attractive to pollinators and beneficial insects were also prioritized, and a flowering calendar was created to ensure that pollen and nectar are available throughout the growing season. Species were selected using listings from sources including Jacke and Toensmeier (2005) and Mäder et al. (2011 and 2014). Nectary species of this kind constitute 42% of those grown in the garden.

Wildlife cover is defined here as being provided by vegetation that is at least 100 cm tall and either evergreen, forming multistemmed thickets, or both (Jacke and Toensmeier 2005). Wildlife cover is provided by 10% of the species in the garden, predominantly woody plants.

Nitrogen fixation is an important tool for restoring degraded sites and maintaining fertility (Jacke and Toensmeier 2005). Though only 9% of species in the garden fix nitrogen, they are densely planted to address nitrogen requirements.

Groundcovers provide a living mulch to suppress weed germination and maintain optimal conditions for healthy soil (Jacke and Toensmeier 2005). This function is provided by 17% of the species in the garden.

## *17.3.3 Human Uses*

Food production has been the primary human use for which plants have been selected at Paradise Lot. Table 17.3 shows the number and percentage of plants providing edible parts. These include: fruits (excluding those used as vegetables); vegetables including leaves, shoots, flowers, and fruits used as vegetables (Toensmeier et al. 2020); roots and tubers including bulbs, corms and rhizomes; nuts, seeds and dry beans; and culinary and spice plants and teas. The edible uses of all species in the garden are shown in Table 17.3. There are many non-edible uses for plants in the garden, such as medicinal plants and bamboo for garden stakes and trellises, but these are not analyzed here.

**Table 17.3** Number of plant species providing human uses

| Use | # of Woody plants | % of Woody plants | # of Vines | % of Vines | # of Herbaceous plants | % of Herbaceous plants |
|---|---|---|---|---|---|---|
| Fruit | 29 | 66% | 4 | 27% | 10 | 8% |
| Vegetable | 7 | 16% | 6 | 40% | 65 | 49% |
| Roots & tubers | 0 | 0% | 4 | 27% | 16 | 12% |
| Nuts, seeds & beans | 2 | 5% | 5 | 33% | 1 | 1% |
| Culinary, spice & tea | 2 | 5% | 0 | 0% | 31 | 23% |

Eighty-seven percent of the 191 plant species in the garden are edible. This includes 23% with edible fruit, 41% used as vegetables, 10% with edible roots and tubers, 4% with edible nuts, seeds, or beans, and 17% with culinary, spice, or tea uses. Some of these species produce more than one edible category, for example *Lycium chinese* has both edible leaves and fruit.

## 17.4 Anecdotal Observations on Animal Biodiversity

Several elements were designed into the garden to improve habitat for animals, primarily insects and birds. Access to water is provided via an artificial pond. Water is an important habitat element for arthropods and vertebrates (Jacke and Toensmeier 2005). Virtually no pesticides are used at all. The exception is kaolin clay powder, which is non-toxic and an important control for plum curculio (*Conotrachelus nenuphar*), which otherwise devastates pome and stone fruits (author personal experience). The garden is also designed with an irregular texture, with open areas, thickets, and forest-like areas to maximize habitat diversity. This "lumpy texture" has been shown to minimize pest damage by providing niches for many birds and other predators (Jacke and Toensmeier 2005).

Formal data have not been collected on animal diversity. However, some anecdotal observations will be shared here.

### *17.4.1 Arthropods*

Invertebrate density and diversity were low in 2004 and began to increase dramatically in 2005. Dead logs, leaf and wood chip mulch, and flowers were observed to be strong attractants of invertebrates. As of 2019, large and diverse populations of arthropods have been observed in the garden. Unlike birds and most other vertebrates, a small garden of this size appears to be able to provide sufficient habitat for many insects to live their entire life there. For example, Chinese mantis (*Tenodera aridifolia*) has been present in the garden for many years. Each year over a dozen oothecae (egg cases) are laid in the garden. The nymphs emerge in spring and appear to live their entire lives in the garden, which would seem to be large enough to provide all the space needed for multiple adult mantises.

### *17.4.2 Birds*

Initially the great majority of birds in the garden were urban species like starlings (*Sturnus vulgaris*) and house sparrows (*Passer domesticus*). As habitat (larger woody plants and the installation of the pond) and food availability (insects, seeds and fruit) increased over several years, bird diversity and abundance increased. Migrating birds now often stop to visit the garden including Baltimore oriole (*Icterus galbula*), multiple species of warbler (*Parulidae*), and wood thrush (*Hylocicha mustelina*). The presence of the wood thrush is notable as it is a forest understory species (Jacke and Toensmeier 2005). The garden alone would appear to be too small to provide sufficient territory for birds, though a flicker (*Colaptes auratus*) nested in a neighboring tree and fed in the garden daily one summer.

### *17.4.3 Mammals*

Mammals in the garden represent typical small to mid-sized urban residents. Grey squirrels (*Sciurus carolinensis*) are year-round residents. Frequent visitors, often only visible at night, are skunks (*Mephitis mephitis*) and opossums (*Didelphis virginiana*). Occasional visitors include red fox (*Vulpes vulpes*), star-nosed mole (*Condylura cristata*), and raccoon (*Procyon lotor*). The very high populations of neighborhood cats (*Felis catus*), which hunt in the garden severely limit populations of smaller rodents. Bats, likely *Eptesicus fuscus*, can be seen hunting at night over the garden in summer.

### *17.4.4 Herptiles*

Though introduction of tadpoles of the native American toad (*Anaxyrus americanus*) was a failure due to predation by cats, a lone wood frog (*Rana sylvatica*) has come to the pond in spring several years to unsuccessfully call for a mate. Several sightings of the red-backed salamander (*Plethodon cinereus*) may be the same individual or represent a larger population. It may have arrived in nursery pots. A DeKay's brownsnake (*Storeria dekayi*), a small slug-eating snake, was greeted enthusiastically by the gardeners but it was not seen again and may well have been consumed by cats.

## 17.5 Conclusions

This case study demonstrates that food production and ecosystem function can go hand in hand, with the wide variety of plants incorporated here providing various ecological roles and human uses. While the scale of the garden is too small to provide significant habitat for vertebrate wildlife, it is quite successful at serving the needs of arthropods and offers some benefits to neighborhood and migrating birds and mammals.

Researchers are invited to use the site to quantify the impacts on invertebrates and other wildlife, as only anecdotes are possible here. Research to compare this temperate site to tropical homegardens is also suggested. Its embracing of the novel ecosystem concept is controversial and also worth of study.

Development of similar, perhaps simpler models on public land in cities could increase biodiversity impacts. Demonstration and instruction in development of streamlined systems, for example modular systems for annual crops, small fruits and tree fruits, pollinator plantings, could increase adoption to the levels necessary to reintroduce some ecosystem services and habitat into biologically barren neighborhoods. Should this model be more widely adopted, even in simplified form, it could potentially offer benefits to people while providing important islands, and potentially corridors, of biodiversity in urban areas.

## References

Clark KH, Nicholas KA (2013) Introducing urban food forestry: a multifunctional approach to increase food security and provide ecosystem services. Landsc Ecol 28:1649–1669

Davis MA, Chew MK, Hobbs RJ, Lugo AE, Ewel JJ, Vermeij GJ, Brown KH (2011) Don't judge species on their origins. Nature 474:153–154

Del Tredici P (2010) Wild urban plants of the Northeast: a field guide. Comstock Publishing, Ithaca

Ferguson RS, Lovell ST (2013) Permaculture for agroecology: design, movement, practice, and worldview. A review. Agron Sustain Dev 34:251–274

Ferguson RS, Lovell ST (2014) Permaculture for agroecology: design, movement, practice, and worldview. Agron Sustain Dev 34:251–274.

Hallet LM (2013) Towards a conceptual framework for novel ecosystems. In: Hobbs RJ, Higgs ES, Hall CM (eds) Novel ecosystems: intervening in the new world order. John Wiley, Oxford, pp 16–28

Hobbs RJ, Higgs ES, Hall CM (2013) Introduction: why novel ecosystems? In: Hobbs RJ, Higgs ES, Hall CM (eds) Novel ecosystems: intervening in the new world order. John Wiley, Oxford, pp 1–8

Jacke D, Toensmeier E (2005) Edible forest gardens. Chelsea Green Publishing, White River Junction

Mäder E, Shepherd M, Vaughn M, Black SH, LeBuhn G (2011) The xerces society guide to attracting native pollinators: protecting North America's bees and butterflies. Storey Publishing, North Adams

Mäder E, Hopwood J, Morandin L, Vaughn M, Black SH (2014) The xerces society guide to farming with native beneficial insects: ecological pest control solutions. Storey Publishing, North Adams

Nair PK (2006) Whither homegardens? In: Nair PK, Kumar BM (eds) Tropical homegardens. Springer, Dordrecht

Nair PK, Kumar BM (2006) Introduction. In: Nair PK, Kumar BM (eds) Tropical homegardens. Springer, Dordrecht

Roth S (2006) Bird-by-bird gardening: the ultimate guide to bringing in your favorite birds, year after year. Rodale, Emmaus

Tallamy DW (2007) Bringing nature home: how you can sustain wildlife with native plants. Timber Press, Portland

Toensmeier E (2016) The carbon farming solution: a global toolkit of perennial crops and regenerative agriculture practices for climate change mitigation and food security. Chelsea Green Publishing, White River Junction

Toensmeier E, Bates J (2013) Paradise lot: two plant geeks, one-tenth of an acre, and the making of an edible garden oasis in the city. Chelsea Green Publishing, White River Junction

Toensmeier E, Ferguson R, Mehra M (2020) Perennial vegetables: a neglected resource for biodiversity, carbon sequestration, and nutrition. PLOS One, in press

# Chapter 18
# Contribution to the Domestication and Conservation of the Genetic Diversity of Two Native Multipurpose Species in the Yabotí Biosphere Reserve, Misiones, Argentina

**Fernando Niella, Patricia Rocha, Ariel M. Tuzinkievicz, Ricardo Buchweis, Christian Bulman Hartkopf, Peggy Thalmayr, José González, Florencia Montagnini, and Sandra Sharry**

**Abstract** In this chapter we describe a strategy for domestication and conservation of the genetic variability of two native species of the Interior Atlantic Forest, or Selva Paranaense, *Peltophorum dubium* and *Enterolobium contortosiliquum*. Both species are leguminous trees, commonly used for forest restoration projects. In tropical and subtropical forests, tree species are vulnerable to habitat fragmentation and population reductions. The resulting negative genetic effects, such as loss of genetic variability and inbreeding depression, can affect the long-term survival of forest species, leading to their further vulnerability or extinction. While *in situ* strategies such as protected areas and biodiversity islands may be an option for biodiversity conservation, in practice there are many challenges to these strategies in humid

F. Niella (✉) · P. Rocha · A. M. Tuzinkievicz · R. Buchweis · C. Bulman Hartkopf
School of Forest Science (FCF), Misiones National University (UNaM), Eldorado, Misiones, Argentina
e-mail: lpv@facfor.unam.edu.ar

P. Thalmayr
IBS-CONICET, Iguazú, Misiones, Argentina

J. González
Puerto Laharrague S. A, Montecarlo, Misiones, Argentina
e-mail: jgonzalez@laharrague.com

F. Montagnini
School of the Environment, The Forest School, Yale University, New Haven, CT, USA
e-mail: florencia.montagnini@yale.edu

S. Sharry
School of Agriculture and Forest Science, La Plata National University (UNLP), La Plata, Buenos Aires, Argentina

National University of Rio Negro, Viedma, Argentina

 
F. Montagnini (ed.), *Biodiversity Islands: Strategies for Conservation in Human-Dominated Environments*, Topics in Biodiversity and Conservation 20,
https://doi.org/10.1007/978-3-030-92234-4_18

subtropical ecosystems, where there is high tree species diversity and low abundance per species (0.1–1 trees/ha). Trees in remnants of forests, which can serve as biodiversity islands, must have high genetic diversity so that they can persist through time by adapting to disturbances. There is a great need to expand the genetic basis of collection of species' propagation material, so that the germplasm available for restoration programs includes the largest possible genetic diversity. To address this need, we established short- and long-term provenance and progeny trials and a vegetative reproduction methodology to produce propagation material to ensure the genetic diversity of these two species for domestication, including for restoration and enrichment. Our results lay the foundation for the conservation of genetic variability of *P. dubium* and *E. contortosiliquum* and contribute to the design of a possible biodiversity island strategy for these species.

**Keywords** Conservation strategy · Forest restoration · Genetic variability · Interior Atlantic forests · Native tree species · Selva paranaense · Vegetative propagation techniques

## 18.1 Introduction

The Yabotí Biosphere Reserve, with a total of 235,900 hectares, is located in the province of Misiones, Argentina and is part of the Atlantic Forest. The Atlantic Forest extends over the Atlantic coast of Brazil and is one of the most diverse ecosystems in the world, with about 15,000 species of vascular plants, of which at least 48% are endemic (Myers et al. 2000; Martini et al. 2007; Murray-Smith et al. 2009). The Atlantic Forest extends into the interior of Brazil and into the province of Misiones in the northeast of Argentina and eastern Paraguay, where it is known as Interior Atlantic Forest or *Selva Paranaense* (Paranaense forest) or *Selva Misionera.* The Interior Atlantic Forests, together with the northwest montane forest of the country known as the *Yungas* (mostly in the provinces of Salta, Jujuy and Tucumán), represent the most biodiverse regions of Argentina. The Atlantic Forest was included as one of the original biodiversity hotspots or critical zones in the widely cited publication of Ernst Mayr in 1988, along with 10 other regions of the globe, which have now been extended to 36 regions worldwide. Designation as a hotspot region should meet two strict criteria: a) it must contain at least 1500 endemic species of vascular plants (0.5% of the total vascular plants identified on Earth); and b) it must have experienced habitat loss of at least 70% of its original area. More than 50% of the world's plant species are endemic to the 36 biodiversity critical areas mentioned above and once covered 15.7% of the Earth's surface. Now, they are reduced to 2.3% of the Earth's surface and include many endemic species, which face a growing threat of extinction (Reed et al. 2011).

The growing population, the advance of the livestock agricultural frontier, and the extraction of timber trees have been the most frequent causes of the drastic decrease in surface area and consequent vulnerability of the 36 critical areas or hotspots. Given this situation, and as part of the commitments made on a global scale at the biodiversity convention in 1992 in Rio de Janeiro, the Global Strategy for Plant Conservation (GSPC 2010) agreement was signed in 2002 by more than 180 countries, with the aim of developing actionable policies to support plant conservation. GSPC Objective 8 requires that "at least 75% of threatened plant species are in *ex situ* collections and at least 20% available for recovery and restoration programs", as outlined by the United Nations Development Program (Reed et al. 2011).

The recognition of islands of biodiversity provides an important step for restoration and can contribute to reversing fragmentation and increasing biodiversity around the world (Montagnini et al. 2022). Several authors, such as Ratnam et al. (2014) and Thomas et al. (2014), have suggested strategies to increase resilience in forest restoration initiatives. Such measures include increasing tree population size and genetic diversity, maintaining forest cover in the landscape for genetic and geographical connectivity between tree populations, promoting genetic improvement of tree species, and enhancing the identification and protection of evolutionary shelters (Bhagwat et al. 2012; Pauls et al. 2013).

In tropical and subtropical forests, tree species are vulnerable to habitat fragmentation and population reductions. The resulting negative genetic effects, such as loss of genetic variability and inbreeding depression, affect the long-term survival of forest species, leading to their further vulnerability or extinction (Maina and Howe 2000). While *in situ* strategies such as protected areas and biodiversity islands may be an option for biodiversity conservation, in practice there are many challenges with these strategies in humid subtropical ecosystems, where there is high tree species diversity and low abundance (0.1–1 trees/ha of the same species).

In studies comparing fruits of the species *Enterolobium cyclocarpum* obtained from trees located in continuous forests and from isolated trees from pasture grazing areas, Rocha and Aguilar (2001) infer that habitat fragmentation alters the mechanisms through which plants regulate the quality of their offspring by increasing self-pollination rates in isolated trees and producing less fruit, which are then usually abortive. For this reason, isolated trees produce progeny that is less vigorous than the progeny of mother trees in continuous forests, which is another adverse effect of forest fragmentation. Therefore, intraspecific genetic variability is directly related to the resilience of the ecosystem and its ability to cope with abrupt changes in the climate and its habitat to ensure its long-term survival.

### 18.1.1 Importance of Genetic Diversity in Restoration and Conservation Programs

Genetic diversity is the basis of an organism's ability to adapt to changes in its environment through natural selection. Populations with little genetic variation are more vulnerable to the emergence of new pests or diseases, pollution, climate change, and habitat destruction due to human activities or other catastrophic events. The inability to adapt to changing conditions greatly increases the risk of extinction.

Reed and Frankham (2003) found that the correlation between genetic diversity at the population level and the adaptability of the population to changing environmental conditions was highly significant, concluding that the loss of heterozygosity[1] has a strong effect on the adaptability of the population. This is consistent with the provisions of the International Union of Conservation of Nature (IUCN) on the status of importance of the conservation of genetic diversity (McNeely et al. 1990). Genetic diversity is one of the three major components of biodiversity, but is still overlooked in most plans for conserving biodiversity. Therefore, the implementation of genetic criteria into the Red List assessment process will help to define more precisely the conservation status of the species (Bruford and Segelbacher 2018).

Meanwhile, there is also a commitment from the international community to restore hundreds of millions of hectares of degraded forest landscapes, following the Strategic Plan for Biodiversity 2011–2020, including the Aichi Biodiversity Targets (CBD 2018). However, the successes and failures of past restoration efforts remain in many cases undocumented and uncommunicated. Case studies show that failures may have been much more common than successes (Wuethrich 2007; Godefroid et al. 2011). The causes of restoration failures can be multiple. One reason, often ignored, is the inadequate consideration of the source and genetic quality of forest reproductive material (Godefroid et al. 2011; Le et al. 2012).

The impact of logging on the structure of forest species populations depends to a large extent on the degree of disruption and intensity of logging. The threat to genetic diversity posed by commercial logging correlates with the abundance of a species in each forest management unit. Natural regeneration, while allowing the transmission of genetic information to the next generation, does not guarantee adaptive and non-adaptive change of structures during the regeneration phase (Rajora and Pluhar 2003). When the number of seed trees left in the logged forest is low and dysgenic selection is heavily practiced according to the market demands, the tree phenotype left in the remaining population is skewed, leading to a gradual process of genetic erosion of the entire population. Without genetic diversity, evolution is impossible, and adaptation decreases, which can result in local extinctions. Processes such as natural selection, genetic drift,[2] and genetic flow[3] collectively affect the genetic diversity of populations and either promote or hinder local and wide-ranging adaptation.

Particularly in tropical and subtropical regions, the genetic diversity of tree species is a key component of the functioning of forest ecosystems (Ratnam et al. 2014). In tropical species, such as for example *Dipteryx odorata* (hermaphroditic and pollinated by insects), and *Bagassa guianensis* (dioecious and mainly wind-pollinated), studies of the impact of forest management have shown that selective

[1]The presence of different alleles at one or more loci are not the same (Acquaah 2012).

[2]The random fluctuations of gene frequencies in a population such that the genes amongst offspring are not a perfectly representative sampling of the parental genes (Schlegel 2010).

[3]Introduction of genetic material from one population of species to another, thereby, changing the composition of the gene pool of the receiving population (White et al. 2007).

logging induced an asynchrony in flowering, limited the flow of genes, and induced inbreeding, even in species that were cross-pollinated. In this manner, the regeneration of their offspring is affected with serious consequences, especially for tree species managed by selective logging. This is especially true in situations where the remaining forests have only few trees of reproductive age and lack pollinators, which contribute to the flow of genes (Ratnam et al. 2014).

### 18.1.2 Genetic Considerations for Restoration Programs for the Interior Atlantic Forest

The high intensity of some methods of logging, as mentioned above, can modify breeding patterns in residual trees and result in increasingly inbred seeds through self-fertilization or crossbreeding between closely related individuals (biparental inbreeding), compromising the role of a tree population as a source of seeds. The risk of inbreeding must be seriously considered in activities dealing with genetic resources, use of germplasm in practical forestry and tree improvement. In such cases, for restoration and/or enrichment programs, the use of germplasm obtained from similar ecological conditions, even if they are not from local sources, may be a better option than resorting to fragmented nearby forests or isolated trees (Navarro Pereyra 2002; Thomas et al. 2014).

Various general guidelines for seed collection that ensure a minimum of genetic diversity for restoration programs have been published, e.g., works by The Australian Network for Plant Conservation Inc. (Vallee et al. 2004), the University of California (Rogers and Montalvo 2004), the World Agroforestry Centre (ICRAF) (Kindt et al. 2006), and the Royal Botanic Gardens, Kew (2003), among others. It is essential that the collection of germplasm captures a representative sample of the genetic diversity of species to be used in restoration projects in a way that takes into consideration the natural distribution of the species as much as possible. A set of general guidelines for tree seed harvesting are intended to ensure a minimum level of genetic diversity. For example, as a general rule, a minimum of 30 randomly selected trees must be sampled for a completely cross-pollinated species (Rogers and Montalvo 2004).

In the province of Misiones, Argentina, the extraction of wood from forests and the harvesting and planting of yerba mate (*Ilex paraguariensis* A. St -Hil) began at the end of the nineteenth century. Forestry activities (mostly plantations of fast-growing exotic species such as *Pinus* spp) were consolidated by the end of the twentieth century (Ferrero 2005), along with plantations of tea, tobacco, and yerba mate monocultures, and more recently livestock production. In the area of study (Yabotí Biosphere Reserve) historically the selective cutting of timber species was done using low impact management practices, such as the Minimum Cutting Diameters (MCDs) that were established by provincial legislation.

The selective cutting method following the MCD requirement for forest harvesting is widely used in tropical and subtropical ecosystems. However, in many cases, this type of management is suited to industrial processing technologies and market demands, while not considering the reproductive biology of the individual trees, and the appropriate number of remaining seedlings, among other factors, to ensure the persistence in perpetuity of the harvested species. Therefore, a management program focused exclusively on the MCD does not result in ecologically sustainable forest management in the long term (Montagnini et al. 1998; Putz et al. 2000; Jennings et al. 2001; Sist et al. 2003).

Among the species that are vulnerable and/or threatened by excessive habitat fragmentation and extraction in the Interior Atlantic Forest, *Peltophorum dubium* (Spreng.) Taub and *Enterolobium contortosiliquum* (Vell.) Morong, are two important species that have been long used in restoration projects. These species are both hermaphrodites and pollinated by insects. Mori et al. (2013), who studied the reproductive system in natural populations of *P. dubium,* observed that the crossings were not random, and that the species was not self-incompatible, but presented a typical system of mixed mating, combining crosses with self-fertilization. Inbreeding was also detected in both the parental generation and in its offspring, although they observed evidence for natural selection between the juvenile and adult phase against inbred individuals.

There is a great need to expand the genetic basis of species propagation material, so that the germplasm available for restoration programs includes the largest possible genetic diversity. In this chapter, we propose a strategy for domestication and conservation of the genetic variability of two native species, *Peltophorum dubium* and *Entorolobium contortosiliquum*, to provide germplasm for enrichment and/or restoration projects of native forests. Our methods could also be used for other species, to produce propagation material with genetic diversity for domestication, enrichment, and/or restoration programs. This work in progress is expected to contribute to a strategy for establishing or enriching biodiversity islands.

## 18.2 Case Study: Provenance and Progeny Trials for the Domestication and Conservation of Genetic Diversity of *Peltophorum dubium* (Caña Fístola) and *Enterolobium contortosiliquum* (Timbó) in the Yabotí Biosphere Reserve, Misiones, Argentina

The species, provenance, and progeny trials aim to ascertain genetic variation at different levels: between species, between geographical regions, between stands and between individuals, and they are the first logical step of genetic improvement for any species (White et al. 2007). Therefore, as a case study, we proposed to assess and evaluate the genetic resources of two native species of the Interior Atlantic Forests, *P. dubium* and *E. contortosiliquum,* as a source of germplasm for enrichment and/or restoration projects of native forests. The species, *P. dubium* and *E. contortosiliquum,* both belonging to the legume family, are of great interest to the local forest industry due to the quality of their wood, which has led to high levels of extraction and in turn, genetic erosion. Currently, there is also an increasing interest in planting them in small- and medium-sized farms, in compliance with provincial regulations. These species can facilitate the production of honey, have good properties for restoration of degraded areas, can be used in agroforestry or silvicultural consociated systems, and have dendro-energy potential. Local towns have high demand for these species for use as ornamentals in landscaping projects in urban areas, since they are representative of the Misiones Forest (Eibl et al. 1998; Barth et al. 2008; Alexandre et al. 2009; Eibl et al. 2017). As seen, local communities consider both as multipurpose species.

Initial field trials to assess their behavior under different silvicultural conditions, have identified *P. dubium* as a heliophyte species, with rapid initial growth and good behavior in open monoculture plantations. In the west Chaco, field trials have shown that *P. dubium* can reach 45 cm in diameter at 20 years of age (Gómez and Cardozo 2003). In Misiones, plantation trials of *E. contortisiliquum* have shown that this species can reach 25 cm in diameter at 14 years of age (Montagnini et al. 2006), demonstrating fast growth even when genetic material with little or no domestication was used. Similar results have been obtained in forest enrichment experiments carried out under canopy of native forest San Pedro, Misiones (Eibl et al. 1993) and in San Ignacio in plantations under the canopy of *Pinus* spp. stands (Crechi et al. 2016).

The main goal of the present case study was to develop a working strategy (Fig. 18.1) for domestication and conservation of genetic variability of two native species of the Interior Atlantic Forests, *P. dubium* and *E. contortosiliquum,* which contemplated the following activities (short, medium and long term): 1) selection and marking of seed trees using the mother seed tree method (White et al. 2007) covering the ranges of both species in Argentina, and subsequent seed harvest from these mother trees; 2) study of morphometric characteristics of fruits and seeds of provenance and progeny of *P. dubium* and *E. contortosiliquum*; 3) establishment of provenance and progeny trials of *P. dubium* and *E. contortosiliquum* in the greenhouse (short-term) and in the field (long-term) to analyze the genetic diversity of their adaptive characteristics; and 4) development of mini-clonal gardens of provenance and progeny of *P. dubium* and *E. contortosiliquum* for the conservation and propagation of propagules with genetic diversity, for domestication, enrichment and/or restoration projects.

Below, we describe each of the activities[4] that were carried out.

[4]The studies developed in the activities described in this chapter were carried out using appropriate statistical designs and analyses, which will be described in future publications.

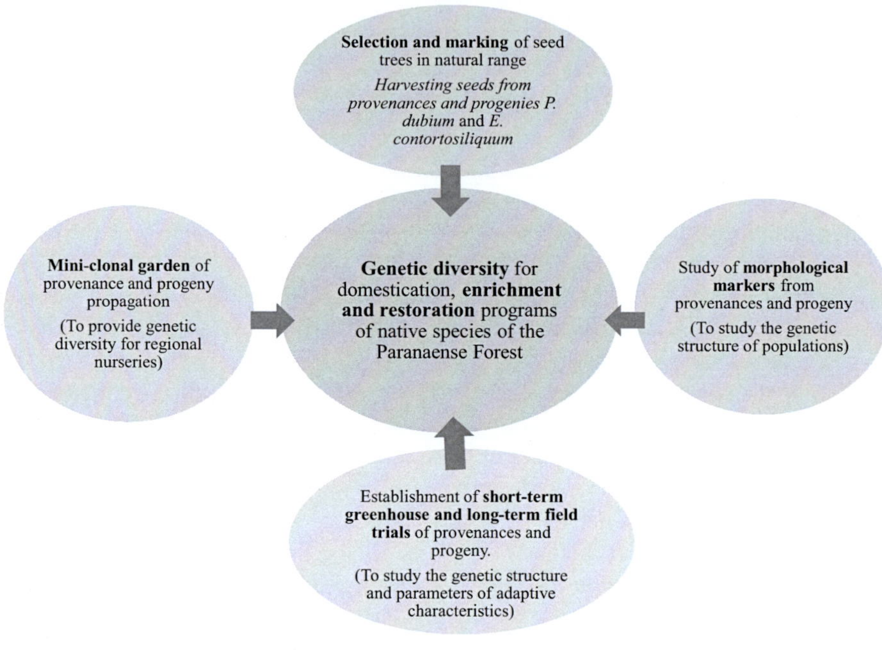


**Fig. 18.1** Proposed strategy for the domestication and conservation of genetic diversity of *P. dubium* (Caña fístola) and *E. contortosiliquum* (Timbó) in the Yabotí Biosphere Reserve, Misiones, Argentina (Own source)

### *18.2.1 Selection and Marking of Seed Trees and Harvest of Seeds from Provenances and Progenies of* **P. dubium** *and* **E. contortosiliquum**

A survey of seed trees, and seed harvests were carried out in much of the natural distribution range of these species in Argentina. Seeds were harvested from individuals in native populations of the northwestern and northeastern region of Argentina (latitudes 23° 28′ to 32° 08´ South, and longitude 54° 30′ to 65° 18´ West), over an area of 1,025,000 km$^2$. We were able to harvest seeds from 54 trees of *E. contortosiliquum* and 35 trees of *P. dubium* from the provinces of Misiones, Corrientes, Entre Rios, Formosa, Tucuman, Salta, and Jujuy.

The information survey in the field and subsequent processing of the data were carried out as follows: 1- Identification and contact with owners who possessed isolated individuals or remnant forest with the species under study. 2- Visit to the property to be able to identify, mark and take the following data from the seed trees: diameter at breast height (DBH) (cm), shaft height (meters), total height (meters), treetop diameter (meters), phenological state, sanitary state, environmental variables and other observations; photograph of the selected tree; information on roads to

**Fig. 18.2** Seed and fruits of one progeny of *P. dubium*. (Photo: Y. López)

reach the site and location coordinate data with GPS. The information was processed using a Geographic Information System (GIS).

### *18.2.2 Study of Morphometry of Fruits and Seeds, from the Provenances and Progenies of* P. dubium *and* E. contortosiliquum

A minimum of 2000 seeds per tree were harvested from each geo-referenced tree. The fruit and seeds morphometric variables of weight, width, length, number of seeds per fruit, number of seeds per kg of fruits, and number of fruits per kg were measured for *P. dubium,* from the material collected, as described in point 2.1.1 (Fig. 18.2).

Seeds of *P. dubium* and *E. contortosiliquum* (Fig. 18.3) were also used to study the germination power, seedling morphometry and survival in the nursery under controlled conditions of temperature and humidity, and they were also used for the field progeny trials. The collection sites, i.e. the provinces, are the *provenances* and the progeny from each tree registered in this study, are *progeny groups*. The morphometric characteristics of fruits and seeds, as well as germination and seedling growth, were evaluated and measured in the Vegetative Propagation Laboratory of the Faculty of Forestry Sciences (UNaM), in Eldorado (Misiones).

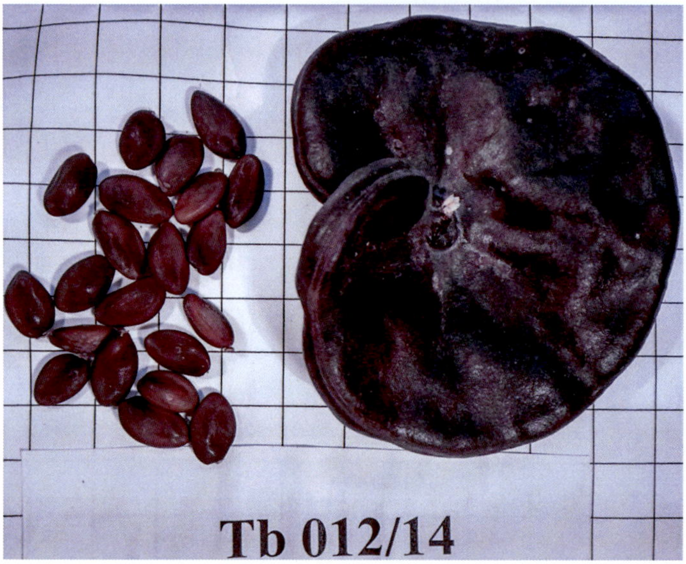


**Fig. 18.3** Seed and fruits of one progeny of *E. contortosiliquum*. (Photo: Y. Lopez)

## *18.2.3 Establishment of Short-Term Greenhouse, Long-Term Under-Canopy Enrichment Field Trials, and Mini Clonal Gardens of* **P. dubium** *and* **E. contortosiliquum**

### 18.2.3.1 Short-Term Greenhouse Trials

To perform germination power tests, the seeds of each provenance and progeny of each species were scarified with sandpaper (150 granulometry) and disinfected with sodium hypochlorite (NaClO) at 50% (v/v) for 15 minutes, with 4 subsequent rinses with distilled water. The seeds were germinated in 90 $cm^3$ HIKO™ trays, in winter with semi-controlled conditions of humidity (micro spray irrigation) and temperature (between 20 and 30 °C). The substrate used was composted pine bark, with application of Plantacote Plus slow release fertilizer™ (3 $kg/m^3$).

From each offspring, seed batches of between 25 to 40 seeds were taken, and germinations counts were made after 4, 6, 10, 24, 25 and 31 days. Within 10 days of planting the germination and emergency stabilized, therefore this was considered the period to assess the germination capacity. Seedling morphometry and survival were assessed at 60 and 120 days. Seedling height (cm), neck diameter (ND, mm), and number of internodes were measured.

#### 18.2.3.2 Long-Term Under Canopy Enrichment Field Trial of *P. dubium* and *E. contortosiliquum*

For the establishment of the under-canopy long-term field trial of provenances and progenies of *P. dubium* and *E. contortosiliquum,* the site selected is in the Municipality of El Soberbio, Province of Misiones, in the Yabotí Biosphere Reserve (latitude 27°08'32''South and longitude 53°58'57''West), belonging to the Company El Moconá SA and managed by Puerto Laharrague SA.

The site has a subtropical climate that corresponds to a Cfa classification according to Köppen (1936), without a dry season, with average temperatures ranging from 12 to 31 °C and with more than 2000 mm of annual precipitation. Soils are deep, red Ultisols classified as Kandiudults. For the preparation of the site, the strips were opened mechanically and manually, ranging from 3 to 3.5 m wide and 90 to 190 m long, with 20 m of undisturbed area between strips.

Seedling establishment was done in a completely randomized block design with 20 replicate blocks. For each 3 m-wide enrichment blocks on the opened strips of forest, the distance between blocks was 20 m and the seedlings were set in rows with 3 m between seedlings (Fig. 18.4). The following characteristics were evaluated:

**Fig. 18.4** Long term field trial establishment in enrichment blocks of *P. dubium.* (Photo: P. Thalmayr)

survival at the first, second and third year, diameter at breast height (DBH), total height, and trunk shape. The morpho-physiological variables (leaf size, stomatal density, photosynthesis, transpiration, phenology), will be evaluated at the fourth, seventh year and at half of the rotation age. The field experiments are being maintained up to 3 years post-establishment, with periodic control of weeds and ants, and replacement of failed seedlings.

#### 18.2.3.3 Mini-Clonal Gardens of Provenances and Progenies of *P. dubium* and *E. contortosiliquum*

To ensure the long-term conservation and production of provenances and progenies of *P. dubium,* a mini-clonal garden was established, using seedlings produced for the short-term greenhouse experiment, as described above. The methodology described by Niella et al. (2013, 2014, 2016a, b) was used for the mini-stumps management and rooting of mini-cuttings (Figs. 18.5 and 18.6).

**Fig. 18.5** Greenhouse mini-stumps garden (*P. dubium*). (Photo: F. Niella)

**Fig. 18.6** Rooted mini-cutting (*P. dubium*). (Photo: F. Niella)

## 18.3 Results and Discussion

### *18.3.1 Morphometry of Fruits and Seeds from Provenances and Progenies of* P. dubium *and* E. contortosiliquum

For *P. dubium,* the morphometric variables of weight, width, length, number of seeds per fruit, number of seeds per kg of fruits, and number of fruits per kg showed significant differences among provenances and progenies (Table 18.1, Figs. 18.2 and 18.3). These data agree with the results of a study of morphological variation of fruits and seeds of *Prunus nepaulensis* Steud. in Meghalaya, India (Shankar and Synrem 2012). The weight of the fruit showed the greatest variation (37.94%) among provenances. Positive correlations were found between fruit weight and length, and lower correlations among weight, fruit width, and number of seeds per fruit (data not shown).

Significant differences among provenances and progenies of *P. dubium* were observed for all seed morphometric variables (Table 18.2). Of these, seed length varied the most. Seed weight was positively correlated with germination percentage.

In this study, the association between the size of the *P. dubium* seeds, expressed in terms of length and width, with the size of the seedlings with 4 months of growth was verified. These data agree with other studies that demonstrate a correlation between seed size with seedling size, verifying that larger seeds produce more

**Table 18.1** Means of seed morphometric variables for *P. dubium* provenance and progenies

| Provenances | Progenies | REP | PF (g) | AF (mm) | LF (mm) | N° SEM | Sem $kg^{-1}$ | Fruits $kg^{-1}$ |
|---|---|---|---|---|---|---|---|---|
| MISIONES | CF31 | 100 | 0.23a | 15.70b | 71.72bc | 1.45bcde | 6.21 h | 4.76i |
| MISIONES | CF23 | 100 | 0.20abcd | 14.37 cd | 80.51a | 1.56bcd | 7.40fgh | 5.27ghi |
| MISIONES | CF32 | 100 | 0.20bcde | 14.87de | 68.84cde | 1.42cdef | 7.12fgh | 5.65fghi |
| MISIONES | CF25 | 100 | 0.17efg | 17.28a | 66.53def | 1.21efg | 7.81efgh | 6.40 cdfg |
| FORMOSA | CF22 | 100 | 0.22ab | 13.09 fg | 67.75cdef | 1.70ab | 7.84efgh | 5.13hi |
| FORMOSA | CF21 | 100 | 0.19bcde | 14.16 cd | 65.11efg | 1.60abcd | 9.62 cd | 6.18cdefg |
| FORMOSA | CF24 | 100 | 0.16fgh | 12.72 g | 64.18 fg | 9.91bc | 9.91bc | 7.46ab |
| FORMOSA | CF20 | 100 | 0.15gh | 12.84 g | 64.62efg | 11.39ab | 11.39ab | 7.08abc |
| FORMOSA | CF30 | 100 | 0.14 h | 14.66c | 70.58bcd | 1,16 g | 8.05defg | 7.73a |
| JUJUY | CF34 | 100 | 0.21abc | 15.88b | 67.84cdef | 1.67cbc | 7.71efgh | 5.27ghi |
| JUJUY | CF17 | 100 | 0.18cdef | 15.55b | 74.38b | 1.55abcd | 8.52cdef | 5.85efgh |
| TUCUMÁN | CF28 | 100 | 0.16fgh | 13.56ef | 71.01bc | 8.77cdef | 8.76 cdef | 6.72bcde |
| TUCUMÁN | CF33 | 100 | 0.17efg | 14.08cde | 70.38bcd | 1.8a | 11.68a | 6.18cdefg |
| SALTA | CF27 | 100 | 0.18cdefg | 15.88b | 67.71cdef | 1.17 fg | 6.77gh | 5.97defgh |
| CORRIENTES | CF29 | 100 | 0.17defg | 13.15 fg | 61.76 g | 1.37 cdefg | 9.24 cde | 6.85abcd |

Adapted from Tuzinkievicz (2019)

REP (replications). Means with a common letter are not significantly different ($p > 0.05$). Test: Tukey Alpha-0.05. PF (g) (fruit weight in grams), AF (fruit width in mm), LF (fruit length in mm), No. SEM (number of seeds per fruit), Sem $kg^{-1}$ (seeds per kg) and fruits $kg^{-1}$ (fruits per kg)

**Table 18.2** Means of seed morphometric variables for *P. dubium* provenance and progenies

| Provenance | Progenies | REP | PS 1000 (g) | Sem $kg^{-1}$ | AS (mm) | LS (mm) |
|---|---|---|---|---|---|---|
| MISIONES | CF31 | 100 | 60.20a | 16.79i | 4.49bc | 10.16a |
| MISIONES | CF23 | 100 | 51.40 cd | 20.06fgh | 3.98ef | 9.48bc |
| MISIONES | CF32 | 100 | 55.71b | 1824hi | 4.06def | 9.20 cd |
| MISIONES | CF25 | 100 | 54.21bc | 19.11ghi | 4.04def | 8.77ef |
| FORMOSA | CF22 | 100 | 51.22bcd | 19.42fghi | 4.52ab | 9.30bc |
| FORMOSA | CF21 | 100 | 50.46cde | 20.49fgh | 4.76a | 8.37 fg |
| FORMOSA | CF24 | 100 | 49.93de | 20.27fgh | 4.24 cd | 8.77ef |
| FORMOSA | CF20 | 100 | 33.14j | 32.30a | 3.56 h | 8.50efg |
| FORMOSA | CF30 | 100 | 35.44ij | 29.35b | 3.85 fg | 8.87de |
| JUJUY | CF34 | 100 | 46.58ef | 21.89ef | 4.02def | 9.29bcd |
| JUJUY | CF17 | 100 | 40.99gh | 24.76 cd | 4.02def | 8.13 g |
| TUCUMÁN | CF28 | 100 | 48.33def | 21.68efg | 3.88f | 9.68b |
| TUCUMÁN | CF33 | 100 | 37.85hi | 27.27bc | 4.18de | 9.61bc |
| SALTA | CF27 | 100 | 41.27gh | 24.95 cd | 4.15de | 8.76ef |
| CORRIENTES | CF29 | 100 | 44.43 fg | 23.16de | 3.60gh | 9.24 cd |

Adapted from Tuzinkievicz (2019)
PS 1000 (g) (weight of 1000 seeds in grams). Sem $kg^{-1}$ (seeds per kg). AS (mm) (seed width in mm). LS (mm) (seed length in mm). REP (number of replications). Means with a common letter are not significantly different ($p > 0.05$). Test: Tukey Alpha-0.05

vigorous seedlings, and with better survival. (Frazao et al. 1985; Haig and Westoby 1991; Pastorino and Gallo 2000; Ursulino Alves et al. 2005).

For the species *E. contortosiliquum*, the studies of fruit and seed morphometry are still in process.

## 18.3.2 Short-Term Greenhouse, Long-Term Under Canopy Trials, and Mini Clonal Gardens of Provenances and Progenies of P. dubium and E. contortosiliquum

### 18.3.2.1 Short-Term Greenhouse Trials

For *P. dubium* the results of germination capacity tests stabilized within 10 days of planting. Seeds from the Jujuy and Entre Ríos provinces had the highest germination rates. The greatest variation in germination was recorded for seeds from Misiones and the smallest for seeds from Jujuy. Germination power and survival rate presented high coefficients of variation both within and between progenies from different provenances, except progenies from Jujuy (Tuzinkievicz 2019). Seed weight was positively correlated with germination rate. The emergence speed in this study was an average of 6.67 AD (average days), compared to the results observed by Espínola Areco and Rodriguez Espínola (2010), where the *P. dubium* seeds registered an emergency speed of 6.85 AD.

At 60 to 120 days of growth, *P. dubium* seedlings showed significant differences among provenances and progenies for seedling height (HT cm) (P value: 0.001) and root collar diameter (RCD mm) (P-value: 0.001). Progenies CF10 and CF6 from the Corrientes provenance had the highest mean heights at 60 and 120 days of nursery growth. Progenies CF13 and CF15 from the Misiones provenance had the lowest mean heights at 60 and 120 days of nursery growth. The Misiones provenance had both the minimum and maximum mean RCD, with progenies CF26 having the maximum RCD at 60 days and CF31 at 120 days, while CF13 showed the minimum RDC at 60 and 120 days of growth. Overall, positive correlations were found between seedling height at 120 days and seed width and length, and between fruit width and seedling height of seedlings at 60 days, suggesting that the larger the fruits and seeds, the larger the seedlings.

While there were no significant differences in seedling survival among provenances, there were significant differences in survival among progenies (Table 18.3). Nevertheless, Misiones presented the highest number of progenies mortality. Survival was negatively correlated with seedling size at 120 days of growth, suggesting that as the plants grew, the survival in the nursery decreased if the seedling density in the tray was high. According to Villagra (2012), *P. dubium* seedlings are characterized by a high light requirement and fast growth. Therefore, since the tray was not thinned, competition could have been one of the problems. This also agrees with work on the evaluation of containers for the production of *Eucalyptus globulus* Labill, in which they concluded that in some cases, such as those obtained for *P. dubium*, the greenhouse low survival rate could be explained by the accelerated growth of the seedlings, which suppress the growth of other individual seedlings (Molina et al. 1992).

Significant differences were found among progeny for germination and seedling growth of *E. contortosiliquum*, with seedling survival per progeny ranging from 0 to 77% at 60 days of growth (Table 18.4). Preliminary results showed that on the third day after planting, more than 50% of progeny had at least one seedling germinated, and that the average germination rate was 23.3% after 120 days of growth (Buchweis 2019). The germination rate in general was low for the species (23%), which could be attributed to the fact that many of the seed trees from which the seeds were collected were isolated. This is in accordance with findings by Rocha and Aguilar (2001), who studied the reproductive biology of *E. cyclocarpum*, a species of the same genus, and concluded that trees that grow in continuous forests were almost six times more likely to produce more seeds than those from isolated trees. Moreover, the seeds from the continuous forests are more vigorous than those of trees that grow isolated, probably due to increasing self-pollination rates in isolated trees, thereby producing less fruit, which are usually abortive.

Significant differences among progenies were found, both for seedling height and RCD. The Misiones province had the progenies with the highest mean seedling heights and RCDs followed by progenies from the Corrientes and Tucumán provinces. The overall mean seedling height was 37.3 cm, mean RCD was 5.48 mm, and mean survival was 44.85% after 120 days of growth (Buchweis 2019).

**Table 18.3** Mean greenhouse (%SUVGH) and field (%SUVFIELD) survival capacity at 12 months for *P. dubium* provenances and progenies

| Provenances | Progenies | %SUVGH | %SUVFIELD |
|---|---|---|---|
| MISIONES | CF 12 | 30ab | 80ab |
| MISIONES | CF13 | 0b | NA |
| MISIONES | CF14 | 0b | NA |
| MISIONES | CF15 | 0b | 100a |
| MISIONES | CF16 | 0b | NA |
| MISIONES | CF 18 | 57.5ab | 90ab |
| MISIONES | CF 19 | 7.5b | 100a |
| MISIONES | CF 23 | 42.5ab | 71ab |
| MISIONES | CF 25 | 60ab | 75ab |
| MISIONES | CF 26 | 45.7ab | 81ab |
| MISIONES | CF 31 | 65ab | 86ab |
| MISIONES | CF 32 | 50ab | 86ab |
| MISIONES | CF 35 | 40ab | 75ab |
| CORRIENTES | CF 6 | 50ab | 80ab |
| CORRIENTES | CF 7 | 63ab | 85ab |
| CORRIENTES | CF8 | 0b | 100a |
| CORRIENTES | CF 9 | 11.4b | 82ab |
| CORRIENTES | CF 10 | 66.7ab | 90ab |
| CORRIENTES | CF 11 | 5ab | 100a |
| CORRIENTES | CF 29 | 30ab | 80ab |
| ENTRE RIOS | CF 1 | 14.3b | 92ab |
| ENTRE RIOS | CF 2 | 16b | 50b |
| ENTRE RIOS | CF 3 | 65ab | 85ab |
| ENTRE RIOS | CF 4 | 80a | 63ab |
| ENTRE RIOS | CF 5 | 62.5ab | 81ab |
| FORMOSA | CF 20 | 15b | 69ab |
| FORMOSA | CF 21 | 14b | 80ab |
| FORMOSA | CF 22 | 60ab | 80ab |
| FORMOSA | CF24 | 0b | 100a |
| FORMOSA | CF 30 | 5b | 100a |
| JUJUY | CF 17 | 42.5ab | 40b |
| JUJUY | CF 34 | 55ab | 60ab |
| TUCUMÁN | CF 28 | 12.5b | 65ab |
| TUCUMÁN | CF 33 | 62.5ab | 45b |
| SALTA | CF 27 | 65ab | 62ab |

Means with a common letter are not significantly different ($p > 0.05$). Test: Tukey Alpha-0.05

#### 18.3.2.2 Long-Term Under Canopy Enrichment Field Trials

One year after establishment of the field trial, the survival rates of *P. dubium* progenies differed significantly, with a mean of 79% (Table 18.3, Fig. 18.4). Progenies of *E. contortosiliquum* showed no significant differences in field survival

**Table 18.4** Mean greenhouse (%SUVGH) and field (%SUVFIELD) survival capacity at 12 months for *E. contortisiliquum* provenances and progenies

| Provenance | Progenies | %SUVGH | %SUVFIELD |
|---|---|---|---|
| Misiones | TB1 | 45 | 89ab |
| Misiones | TB2 | 22.5b | NA |
| Misiones | T28 | 75a | 80ab |
| Misiones | T29 | 77.5a | 70bc |
| Misiones | T30 | 55ab | 70bc |
| Misiones | T31 | 10b | NA |
| Misiones | T32 | 15b | NA |
| Misiones | TB8 | 7.5b | NA |
| Corrientes | TB3 | 0 | NA |
| Corrientes | TB4 | 0 | NA |
| Corrientes | TB5 | 35ab | 78ab |
| Corrientes | TB6 | 5 | NA |
| Corrientes | TB7 | 0 | NA |
| Tucumán | TB9 | 52.5ab | 67bc |
| Tucumán | TB10 | 25ab | 78ab |
| Tucumán | TB11 | 32.5ab | NA |
| Tucumán | TB12 | 17.5b | NA |
| Tucumán | TB13 | 0 | NA |
| Tucumán | TB14 | 0 | NA |
| Tucumán | T15 | 62.5ab | 100ab |
| Tucumán | T16 | 22.5b | NA |
| Tucumán | T17 | 5b | NA |
| Tucumán | T18 | 5b | NA |
| Tucumán | T19 | 22,5b | NA |
| Jujuy | T21 | 5b | NA |
| Jujuy | T22 | 5b | NA |
| Jujuy | T23 | 5b | NA |
| Jujuy | T24 | 47.5ab | 60bc |
| Jujuy | T25 | 12.5b | NA |
| Jujuy | T26 | 15b | NA |
| Salta | T20 | 45ab | 70bc |
| Formosa | T27 | 15b | NA |

Adapted from Buchweis (2019)

Means with a common letter are not significantly different ($p > 0.05$). Test: Tukey Alpha-0.05

capacity, with a 70% mean survival (Table 18.4, Fig. 18.7). According to Widiyatno et al. 2014, the variable of survival rate could be used to select the best family for breeding programs in the future because this variable showed the adaptation of the family toward the extreme condition in their early establishment.

**Fig. 18.7** Seedling at the first-year evaluation field trial (*E. contortosiliquum*). (Photo: R. Buchweis)

#### 18.3.2.3 Mini-Clonal Gardens of Provenances and Progenies of *P. dubium* and *E. contortosiliquum*

A mean of 927 mini-cuttings/m$^2$/year was produced for both species, based on three collections per year; with the mean survival and rooting capacity of mini-cuttings both above 85% after 120 days of growth in greenhouse conditions (Niella et al. 2014). This technique offers great opportunities for improving the quality of planting stock, through its capability to produce any quantities of plantation materials in homogenous quality at any time (Yasman and Natadiwirya 2001).

## 18.4 Conclusions

At this stage of the study, it can be concluded that:

1. In spite of the low density of individuals/ha, the number of seed trees selected for germplasm collection of the species *P. dubium* and *E. contortosiliquum* met the

minimum of 30 randomly selected trees to be sampled according to Rogers and Montalvo (2004).
2. Fruits and seeds of *P. dubium* showed differences among provenances and progenies, and positive correlations between morphometric variables of fruits and seeds and the height and survival of seedlings at 60 and 120 days.
3. Short-term greenhouse trials showed significant differences among provenances and progenies of *P. dubium* for height at 60 and 120 days, and positive correlations between fruit and seed size and height. Field survival, 12 months after establishment differ significantly among progenies of *P. dubium;* nevertheless, no difference was observed on *E. contortosiliquum.*
4. The strategies presented in this case study agree with the ex-situ conservation and population improvement methods proposed by Navarro Pereyra (2002), which include: vegetative propagation of mother trees; import of seeds from other populations outside the one to be improved; establishment of conservation gardens that include at least twenty non-related progenies, and development of low intensity breeding programs for the tropical species.
5. The presented results are preliminary and long-term studies are required for both species. However, this study lays the foundation for understanding genetic variability in *P. dubium* and *E. contortosiliquum* and provides methods for the propagation of germplasm to ensure genetic diversity of these species for domestication, enrichment, and/or restoration programs. It will also contribute to delineation of a proper design for a biodiversity island strategy for *P. dubium* and *E. contortosiliquum.*

**Acknowledgements** The authors thank the Sustainable Management of Natural Resources Project, Ministry of Agriculture, Livestock, Fisheries and Food (MAGPyA), the International Bank for Reconstruction and Development (BIRF 7520 AR)/Project PIA10031, the General Secretariat of Science and Technology and the School of Forest Science (FCF) of Misiones National University (UNaM), Misiones, Argentina, as well as the Puerto Laharrague Company, and the forest technician Federico Robledo for his contribution to field trials establishment.

## References

Acquaah J (2012) Principles of plant genetics and breeding. John Willey & Sons LTD, West Sussex

Alexandre RS, Gonçalves FG, Rocha AP, Pereira M, Lemes E (2009) Tratamentos físicos y químicos na superaçâo de dormência em sementes de *Enterolobium contortisiliquum* (Vell.) Morong. Agrária 4:156–159

Barth SR, Eibl B, Montagnini F (2008) Adaptability and species growth. Forestry and environmental technical conferences of the FCF-National University of Misiones and EEA Monte Carlo. INTA, Eldorado, Misiones, pp 1–16

Bhagwat SA, Nogué S, Willis KJ (2012) Resilience of an ancient tropical forest landscape to 7500 years of environmental change. Biol Conserv 153:108–117

Bruford M, Segelbacher G (2018) IUCN SSC Conservation Genetics Specialist Group. 2018 Report. https://www.iucn.org/sites/dev/files/2018_conservation_genetics_sg_report_-_publication.pdf

Buchweis R (2019) Study of growth and survival of seedlings in nursery and in the field of progeny of *Enterolobium contortisiliquum* from northeast Argentina. Undergraduate's thesis for the Forestry Engineering Degree, Faculty of Forest Sciences of the Misiones National University. www.facfor.unam.edu.ar

Crechi E, Keller A, Vera C, Domecq C, Henning A (2016) Growth of 3 native hardwood species in open field and under pine canopy at 16 years, in Misiones, Argentina (*Cordia trichotoma, Balfourodendron riedelianum, Enterolobium contortisiliquum*). XVII Jornadas Técnicas Forestales y Ambientales. Posadas, Misiones. Producción Forestal Primaria. Pp. 78–80

Eibl B, Morandi F, Muñoz D, Martinez L (1993) Enrichment in sashes with native forest species in San Pedro, Misiones, Argentina. VII Jornadas Técnicas: Ecosistemas Forestales Nativos, Uso, Manejo y Conservación. ISIF-FCF. Eldorado. Misiones Actas I:268–277

Eibl B, Morandi F, Muñoz D, Martínez L (1998) El potencial de las especies nativas en programas de plantación. En: VI Jornadas Técnicas, Serie Técnica No 6, Ecología de Especies Nativas de la Selva Subtropical Misionera. Universidad Nacional de Misiones, Facultad de Ciencias Forestales. Eldorado, Misiones, Argentina. Mayo 1998. Pp.19–26

Eibl B, Montagnini F, Lopez M, Lopez LN, Montechiesi R, Esterche E (2017) Organic yerba mate, *Ilex paraguariensis*, in association with native tree species: a sustainable production alternative. In: Montagnini F (ed) Integrating landscapes: agroforestry for biodiversity conservation on food sovereignty, advances in agroforestry 12. Springer, Cham, pp 261–281

Espínola Areco NL, Rodríguez Espínola HN (2010) Efecto de los incleendios forestales sobre la emergencia de semillas de tres especies forestales nativas en condiciones in vitro. Investigación Agraria 12(1):29–34

Ferrero B (2005) Study of territorial management and natural resources, of the rural population of the Area of Influence of the Yabotí-Argentina Biosphere Reserve. Looking for alternatives for sustainable local development around a Biosphere Reserve. Programa Man and Biosphere, UNESCO. www.unesco.org/MAB/doc/mys/2004/ferrero.pdf. (12 de diciembre 2012) 184 pp

Frazao DAC, Costa SD, Coral FS, Azevedo JA, Figueiredo FJC (1985) Influência do peso da semente no desenvolvimento e vigor de mudas de cacau (*Theobroma cacao* L.). Revista de Agricultura 60(1):2–16

Godefroid S, Piazza C, Rossi G, Buord S, Stevens AD, Aguraiuja R, Cowell C, Weekley CW, Vogg G, Iriondo J, Johnson I, Dixon B, Gordon D, Magnanon S, Valentin B, Bjureke K, Koopman R, Vicens M, Virevair M, Vanderborght T (2011) How successful are plant species reintroductions? Biol Conserv 144(2):672–682

Gómez C, Cardozo F (2003) Enrichment of the native forest with ibira puita guazu (*Peltophorum dubium* spreng). Informe interno INTA-Centro Regional Chaco-Formosa E.E.A Sáenz Peña. Campos Anexo Presidencia de la Plaza

GSPC (2010) Global Strategy for Plant Conservation (http://www.plants2020.net/)

Haig D, Westoby M (1991) Seed size, pollination costs and angiosperm success. Evol Ecol 5(3): 231–247. https://doi.org/10.1007/BF02214230

Jennings SB, Brown ND, Boshier DH, Whitmore TC, Lopes J (2001) Ecology provides a pragmatic solution to the maintenance of genetic diversity in sustainably managed tropical rain forests. For Ecol Manage 154:1–10

Kindt R, Lillesø JPB, Mbora A, Muriuki J, Wambugu C, Frost W, Beniest J, Aithal A, Awimbo J, Rao S, Holding-Anyonge C (2006) Tree seeds for farmers: a toolkit and reference source. World Agroforestry Centre, Nairobi, Kenya. http://www.worldagroforestry.org/treesandmarkets/tree%20seeds/Documents%20of%20Toolkit/PDF/Toolkit.pdf. (Accessed April 2014)

Köppen W (1936) Das geographische System der Klimate. In: Köppen W, Geiger R (eds) Handbuch der Klimato - logie. Gebrüder Borntraeger, Berlin, pp 1–44

Le HD, Smith C, Herbohn J, Harrison S (2012) More than just trees: assessing reforestation success in tropical developing countries. J Rur Stud 28(1):5–19

Maina G, Howe HF (2000) Inherent rarity in community restoration. Conserv Biol 14:1335–1340

Martini AMZ, Fiaschi P, Amorin AMA, Paixa˜o JL (2007) A hot-point within a hot-spot: a high diversity site in Brazil's Atlantic Forest. Biodivers Conserv 16:3111–3128

McNeely JA, Miller KR, Reid WV, Mittermeier RA, Werner TB (1990) Conserving the world's biological diversity. World Conservation Union, World Resources Institute, Conservation International, World Wildlife Fund–US and the World Bank, Washington, D.C.

Molina MPB, Barros DR, Ipinza RC (1992) Análisis de distintos contenedores para la producción de plantas de *Eucalyptus globulus* Labill. Ciencia e Investigación Forestal 8(2):169–193

Montagnini F, Eibl B, Woodward C, Szczipanski L, Ríos R (1998) Tree regeneration and species diversity following conventional and uniform spacing methods of selective cutting in a subtropical humid forest reserve. Biotropica 30(3):349–361

Montagnini F, Levin B, Berg KE (2022) Biodiversity Islands. Strategies for conservation in human dominated environments. In: Montagnini F (ed) Biodiversity islands: strategies for conservation in human dominated environments. Springer, Cham, pp 1–35

Montagnini F, Eibl B, Fernández R (2006) Rehabilitation of degraded lands in Misiones. Argentina Bois et Forets des Tropiques 288(2):51–65

Mori E, Sebbenn A, Tambarussi E, Gurie R (2013) Sistema de reprodução em populações naturais de *Peltophorum dubium*. Sci For, Piracicaba 41(99):307–317

Murray-Smith C, Brummitt NA, Oliveira-Filho AT, Bachman S, Moat J, Lughadha EMN, Lucas EJ (2009) Plant diversity hotspots in the Atlantic Forest in Brazil. Conserv Biol 23:151–163

Myers N, Mittermeier RA, Mittermeier CG, Fonseca GAB, Kent J (2000) Biodiversity hotspots for conservation priorities. Nature 403:853–858

Navarro Pereyra CM (2002) Genetic resources of *Cedrela odorata* L. and their efficient use in Mesoamerica. Academic Dissertation in Forest Tree Breeding. Faculty of Agriculture and Forestry of the University of Helsinki. ISBN: 952–10–0791-5 (PDF)

Niella F, Eibl B, Rocha P, Dummel C, Keller HA (2016a) Strategy for the conservation, domestication and rescue of plant genetic resources native to the Paranaense forest. XVII Jornadas Técnicas Forestales y Ambientales. FCFUNaM-INTA-COIFORM. Posadas, Misiones-Argentina. Pp. 302–304. www.jotefa.com.ar

Niella F, Rocha P, Eibl B, Bohren A, Ayala L, Conti P, Franco M (2016b) Development of clonal propagation techniques for the sustainable use of multipurpose native forest species and/or nitrogen fixators and non-wood forest products. In: Forest Research 2011–2015: Applied Research Projects. Propagation section. Buenos Aires: Ministerio de Agroindustria. Unidad para el cambio rural-UCAR. Pp. 386–389

Niella F, Rocha P, Eibl B, Bohren A, Ayala L, Conti P, Franco M, Radins M (2013) Development of clonal propagation techniques for the sustainable use of multipurpose native forest species and/or nitrogen fixers and non-wood forest products. 4to Congreso Forestal Argentino y Latinoamericano, Iguazú, Misiones, Argentina

Niella F, Rocha P, Eibl B, Schoffen C, Martinez M, Conti P, Franco M, Ayala L (2014) Clonal propagation of *Peltophorum dubium* (cane fistola), *Myrocarpus frondosus* (incense), and *Cordia trichotoma* (Peteribi) for conservation and domestication. Revista Forestal Yvyrareta 21(2014):43–50. http://www.yvyrareta.com.ar/index.php/ediciones-anteriores/diciembre-2014-n21

Pastorino MJ, Gallo LA (2000) Variación geográfica en peso de semilla en poblaciones naturales argentinas de "Ciprés de la Cordillera". Bosque (Valdivia) 21(2):95–109

Pauls SU, Nowak C, Bálint M, Pfenninger M (2013) The impact of global climate change on genetic diversity within populations and species. Mol Ecol 22(4):925–946

Putz FE, Redford KH, Robinson JG, Fimbel R, Blate GM (2000) Biodiversity conservation in the context of tropical forest management. In: The World Bank environment department, biodiversity series impact studies paper 75. The World Bank, Washington DC, USA

Rajora OP, Pluhar SA (2003) Genetic diversity impacts of forest fires, forest harvesting and alternative reforestation practices in black spruce (Picea mariana). Theor Appl Genet 106: 1203–1212

Ratnam W, Rajora OP, Finkeldey R, Aravanopoulos F, Bouvet JM, Vaillancourt RE, Kanashiro M, Fady B, Tomita M, Vinson C (2014) Genetic effects of forest management practices: global synthesis and perspectives. For Ecol Manage 333:52–65

Reed BM, Sarasan V, Kane M (2011) Biodiversity conservation and conservation biotechnology tools. In Vitro Cell Dev Biol Plant 47:1–4

Reed DH, Frankham R (2003) Correlation between fitness and genetic diversity. Conserv Biol 17(1):230–237

Rocha O, Aguilar G (2001) Reproductive biology of the dry forest tree *Enterolobium cyclocarpum* (Guanacaste) in Costa Rica: a comparison between trees left in pastures and trees in continuous forest. Amer J Bot 88(9):1607–1614

Rogers DL, Montalvo AM (2004) Genetically appropriate choices for plant materials to maintain biological diversity, report to the USDA Forest Service. University of California, Rocky Mountain Region, Lakewood

Royal Botanic Gardens, Kew (2003) A field manual for seed collectors. Seed Collecting for the Millennium Seed Bank Project, Royal Botanic Gardens, Kew

Schlegel RHJ (2010) Dictionary of plant breeding, 2nd edn. CRC Press Taylor & Francis Group, Boca Raton

Shankar U, Synrem YIL (2012) Variation in morphometric traits of fruits and seeds of prunus nepaulensis steud in Meghalaya, India. Trop Ecol 53(3):273–286

Sist P, Fimbel R, Douglas S, Nasi R, Chevallier A (2003) Towards sustainable management of mixed dipterocarp forests of Southeast Asia: moving beyond minimum diameter cutting limits. Environ Conserv 30(4):364–374

Thomas E, Jalonen R, Loo J, Boshier D, Gallo L, Cavers S, Bordacs S, Smith P, Bozzano M (2014) Genetic considerations in ecosystem restoration using native tree species. For Ecol Manage 333: 66–75

Tuzinkievicz A (2019) Geographical variation of morphometric characteristics of fruits, seeds and seedlings, germination power and survival in the origins and progeny of *Peltophorum dubium* (Spreng.) Taub. of the northwest and coastal region of Argentina. Master's thesis in Forestry Sciences, Faculty of Forestry Sciences of the Misiones National University. www.facfor.unam.edu.ar

Ursulino Alves E, de Lucena R, Alcântara B, Pereira de Oliveira A, Ursulino Alves A, de Paula DC (2005) Influência do tamanho e da procedência de sementes *Mimosa caesalpiniifolia* Benth. Sobre a germinação e vigor. Revista Árvore 29:877–885. https://doi.org/10.1590/S0100-67622005000600006

Vallee L, Hogbin T, Monks L, Makinson B, Matthes M, Rossetto M (2004) Guidelines for the translocation of threatened plants in Australia, 2nd edn. Australian Network for Plant Conservation, Canberra, Australia

Villagra M. (2012) Plasticidad morfológica y fisiológica de especies arbóreas del Bosque Atlántico en respuesta a cambios en la disponibilidad de luz y nutrientes. Tesis doctoral. Facultad de Ciencias Exactas y Naturales. Universidad de Buenos Aires. 139pp

White T, Adams W, Neale D (2007) Forest genetics. CABI Publishing, CAB International, Wallingford, UK, p 682

Wuethrich B (2007) Biodiversity. Reconstructing Brazil's Atlantic rainforest. Science 315:1070–1072

Yasman I, Natadiwirya M (2001) Dipterocarp plantation: the strategy and the approaches of PT Inhutsni I. In: Tielges BA, Sastrapradja SD, Rimbawanto A (eds) In-situ and ex-situ conservation of commercial tropical trees. ITTO-Gadjah Mada University, Yogiakarta, Indonesia, pp 407–412

# Part IV
# Safeguarding the Environmental, Economic, and Social Benefits of Biodiversity Islands

# Chapter 19
# How Community-Led Action Can Advance the Development of Biodiversity Islands

**Brett Levin**

**Abstract** Community-led action can contribute to the development of biodiversity islands. Biodiversity islands constitute areas of high biodiversity nested within human-dominated landscapes. Community action towards the development of biodiversity islands may be longstanding within cultures or learned and applied from ethical, philosophical, scientific, cultural, or economic motivations. Land access and long-term decision-making power provide the basis for communities to maintain areas of high biodiversity within degraded landscapes on private, public, and indigenous lands. The legal system and governing process of the presiding people and culture determine the tools that can aid in the establishment and protection of these areas. Some useful legal and financial tools may include land trusts, conservation easements, supportive zoning, novel financial resource pooling, and strong indigenous land rights. Community-developed biodiversity islands may be governed through a vast array of methods. Management of biodiversity islands requires methods that enhance or maintain biodiversity outcomes through time. This can be achieved through a broad array of community land uses and techniques. Examples of grassroots community action for the advancement of biodiversity conservation practices are numerous, diverse, and worldwide. Indigenous, religious, governmental, nonprofit, and for-profit organizations are capable of further expanding community-led action for the development of biodiversity islands. Examples of community-led biodiversity islands are described, including spiritual and religious sites, public areas, agricultural systems, and beyond. No matter the governing and organizational approach, considerations of social and environmental justice remain an important factor in how biodiversity islands are recognized, developed, and managed through time.

**Keywords** Advocacy · Conservation easements · Governance · Indigenous management · Land trusts · Sacred sites

---

B. Levin (✉)
Dietrick Institute for Applied Insect Ecology, Ventura, CA, USA
e-mail: Brett.Levin@aya.yale.edu



F. Montagnini (ed.), *Biodiversity Islands: Strategies for Conservation in Human-Dominated Environments*, Topics in Biodiversity and Conservation 20,
https://doi.org/10.1007/978-3-030-92234-4_19

## 19.1 Introduction

Biodiversity islands are areas of high biodiversity nested within ecologically degraded, human-dominated landscapes (Montagnini et al. 2022). Community-led action can advance the development and perpetuity of biodiversity islands. Through the empowerment of traditional biodiversity conservation practices, grassroots community action, and the utilization of legal and financial tools, community-led management can protect intact sections of land where plants and animals can thrive without ecologically degenerative interference from human activity (Montagnini et al. 2022). Such areas of high biodiversity within otherwise degraded landscapes can be found in rural areas, for example, in agroforestry systems (Montagnini and del Fierro 2022) as well as in suburban and urban contexts such as in homegardens, residential gardens, and urban green infrastructure (Toensmeier 2022; Negret et al. 2022; Soler et al. 2022).

In many instances, traditional community-based approaches to landscape management include biodiversity enhancing outcomes which can be considered biodiversity islands. Examples include sacred sites, sacred groves, community forests, church forests, and other sites that are protected and managed by local communities (Bhagwat and Rutte 2006; Ceperley et al. 2010; Moeliono et al. 2015; Baez Schon et al. 2022). The term "biodiversity island" in many cases is used to refer to these indigenous and traditional biodiversity conservation practices, highlighting results and understandings around procedures that advance appreciation and empowerment of such systems.

This chapter highlights the commons, which are the cultural and environmental resources accessible to all members of a society, including ecological assets such as air, water, and ecosystem health. These resources are held in common and not privately owned. Emphasis is placed on advances in community-based conservation, as governance approaches that tend to appreciate social complexities, utilize democratic and inclusive decision-making processes, and advance pluralistic frameworks. Grassroots community action and legal tools such as conservation easements through partnerships with non-profit Land Trusts can also aid in the development of biodiversity islands. Examples of grassroots efforts for the advancement of conservation practices are numerous, diverse, and worldwide. Legal frameworks, methods of enforcement, modes of implementation, and levels of community engagement are site and context dependent, but through these efforts of community-led action, biodiversity islands may be protected and managed through time.

### *19.1.1 Community Defined*

A community can be defined as a group of individuals living in the same area or who have a characteristic in common (Berkes 2004, 2017). The scale and scope of this definition can encompass a significant range of variables dependent on culture and perspective, from humancentric to biocentric and animistic (Harvey 2005). Such an inclusive definition of community may naturally lead to sustainable uses of land which protect biodiversity and provide for human habitation through time. One such example includes the aboriginal people of Australia, where worship of nature and the abiotic features of the landscape leads to a great respect for conservation and the value nature holds. In Oaxaca, Mexico, low intensity forest use and traditional rotational agriculture are spurred by deep connections to the land. This creates a more diverse forest-agriculture mosaic where traditional forest management harmonizes livelihoods and sustained biodiversity conservation (Berkes 2017). The Kenyah Dayak people of Indonesia manage subsistence orchards as gifted foresters. Their gardens look wild but are carefully cultivated (Peters 2018). Such morally and spiritually motivated biocentric approaches towards landscape management are present within many indigenous groups around the world. When considering community-led action towards biodiversity, such varied perspectives produce interesting considerations; the relinquishment of humans as the main component of community to humans as listeners or collaborators; the shift from anthropocentrism to biocentrism; and pathways towards rights of nature in the dialogue for community conservation (Prieto Méndez 2013).

From a less metaphysical and more human centric perspective, one cannot deny the abiotic factors of climate, topography, geology, soil, and precipitation in determining the species present and historical development of humans in an area (Dunson et al. 1991). Patterns of settlement are interrelated with the presence or absence of natural resources. Transportation corridors tend to follow ease through topographies. Agriculture and cultivation are inextricably linked to soils, water resources, and climate. From this vantage point, biological and physical components of a place are inextricably linked with community, whether centrally focusing on the human dimensions or expanding into the biophysical.

## 19.2 Motivations for Managing for Biodiversity Through Time

For numerous indigenous and traditional communities, the motivations for conserving and managing biodiversity through time are an integral part of the culture. Interactions with the landscape flora and fauna are learned and passed down over generations, which maintain or enhance biodiversity. These interactions form the basis of Traditional Ecological Knowledge (TEK) and become nested in culture (Gadgil et al. 1993; Inglis 1993). In ecologically degraded human-dominated landscapes, learning from, protecting, properly valuing, and/or restoring such community-led cultural practices may allow for areas of higher biodiversity, or biodiversity islands, to persist or emerge (Berkes 2017). This is not to advocate notions of the "noble savage" and attributes of romantic primitivism across all traditional or indigenous cultures (Raymond 2007). Instead, here a focus on scientific findings, supported by historical ecology, demonstrate longstanding ethnoecological knowledge leads to improved biodiversity outcomes in numerous communities worldwide (Berke et al. 1994; Balée 1998).

Several cases exist where improved biodiversity outcomes result from expanded perspectives of community, inclusive of genetic diversity, habitat diversity, landscape diversity, and temporal diversity (Salick 2012). One such example is evidenced in Maranhão, Brazil, with landscape management of increased biodiversity by indigenous people focusing on sustainable extraction of specific forest products, and the nurturance of specific forest plant species (Balée 1993). The loss of such long-standing land-based culture may result in biological degradation, even when newly implemented management objectives are advanced to achieve 'conservation' or 'biodiversity goals' (Crosby 2004). Another example relates to the traditional agricultural management of Pima farmers in the Sonoran Desert. There, many emerged volunteer plant species were left in the field for a broad range of reasons including more diverse food yields, to host edible insect larvae, for protective hedges, and to provide shade and cycle nutrients. Such intensive and intentional practices have become a rarity in the now conventionally cultivated fields of the region (Nabhan et al. 1989). These examples highlight cultural practices which maintain or enhance biodiversity over time, and suggest that protecting and empowering such practices can play a key role in the development and expansion of biodiversity islands.

For other cultures and societies, learned ethical, scientific, philosophical or economic motivations may drive actions for community-led biodiversity conservation within a landscape. From a western perspective, there is a broad range of literature and media that has helped advance such notions. Books such as Aldo Leopold's *A Sand County Almanac* (1949), Rachel Carson's *Silent Spring* (1962), James Lovelock's *Gaia Hypothesis* (1979), E.O. Wilson's *Biophilia* (1984), and Murray Bookchin's *Ecology of Freedom* (1982) are just a few highlighted examples that will be described below.

### 19.2.1 Philosophical and Ethical Motivations

Aldo Leopold's "land ethic" brought forth an advancement in ethics as a process of ecological evolution through a moral responsibility for western humanity to care for the natural world (Leopold 1949). From this vantage point, the moral sense of right and wrong and how community is defined expanded beyond humans to include the soils, trees, grasses, and all species and biophysical components of the environment. Leopold's "community concept" argues true community-led action must then be built upon consideration of these human and non-human pieces. From this standpoint, where humans alter a landscape in ways that diminish non-human life through time, an ethical violation has occurred. Bringing this vantage point into the realm of western consciousness provides a foundation for people to come together towards the protection of species where landscapes would otherwise become degraded. This learned ethical approach towards conservation, biodiversity, and their inherent values has helped shape the recommendations of community organizations, planning bodies, and governments. Throughout the United States, many recent conservation projects which may be recognized as biodiversity islands within a landscape can be attributed to the realization of the philosophical underpinnings of Aldo Leopolds' work. Today, his legacy continues through The Aldo Leopold Foundation (www.aldoleopold.org). Located on his home site in Baraboo, Wisconsin, the organization was founded in 1982 by his children as a non-profit organization to continue the practices incorporated by his land ethic. The foundation owns and manages his property to conduct education and outreach for land stewardship programs. There are also other academic and research institutions throughout the United States, run through the Forest Service, that function as centers of wilderness education, agriculture, and programs, celebrating Leopold's legacy and promoting his ideals and teachings towards a land ethic.

James Lovelock's Gaia Hypothesis, which proposed that the biosphere is a complex and self-regulating organism, advanced philosophical motivations towards conservation of biodiversity (Lovelock 1979). This hypothesis promulgated the notion that living matter on earth collectively determines and controls the material conditions necessary for the regulation of life. From this premise, philosophy regarding the role of humanity as part of Gaia permeated the realm of western environmental consciousness. Are the negative environmental impacts of humans a destabilizing force which will eventually be self-regulated? Can humans persist as part of a global community of living organisms in a fashion compatible and harmonious with self-regulation? Many of us are still seeking answers to these questions, though the notion of humans as a force within the planetary community towards improved biodiversity was advanced by Lovelock's work.

E.O. Wilson's *Biophilia* emphasized the philosophical underpinnings of the human compulsion and innate drive towards interactions with other forms of life. Wilson argued that "the connections that human beings subconsciously seek with the rest of life are inherent in our biology." It is argued that modern western society has disengaged the masses in recognizing and engaging with their love of life-giving systems. This mindset is responsible for ecological degradation and biodiversity loss (Wilson 1984). To reverse these outcomes, Wilson argues a philosophical motivation towards conservation of species through reconnecting with our inherent biophilic nature. Through this teaching and mindset, community groups and organizations have worked together to advance conservation projects, many of which may be considered biodiversity islands.

Murray Bookchin's book "Ecology of Freedom" also advanced philosophical motivations towards community-led action towards greater biodiversity through the message of social ecology, linking human relationships within society to human relationships towards the land and natural systems (Bookchin 1982). In societies where domination and exploitation are the norm, one may recognize similar patterns imposed on the landscape. These outcomes may also manifest in a downtrodden psyche of the individual. It is suggested that societies built upon foundations of mutual respect among people may materialize a more uplifted psyche of the individual and permeate into a more harmonious societal relationship with the landscape and natural resources. Though criticized as utopian by some, the greater message purveyed is that through such reciprocal positive mechanisms of right relationship between people to people and people to place, the individual spirit may be lifted, and greater freedom emerge. Through this philosophical concept, further motivations towards greater biodiversity outcomes have entered the western canon of environmental thought, in conjunction with advances in environmental ethics, as discussed in the previous section.

More recently, evolution in environmental philosophy and ethics advanced by Stephen Kellert, Mary Evelyn Tucker, and John Grimm, among others, has continued to shape a mindset towards a more interconnected worldview of community. In Mary Evelyn Tucker and Brian Swimme's "Journey of the Universe," the interconnection and interrelation of all beings brings a biophilic understanding towards the basis of community (Swimme and Tucker 2011). Stephen Kellert advanced an ethical imperative within design to incorporate both human and ecological considerations within any project (Kellert 2018). "The Value of Life" is an exploration of the actual and perceived importance of biological diversity for human beings and society (Kellert 1997). Stephen R. Kellert identifies ten basic values, which he describes as biologically based, inherent human tendencies that are greatly influenced and moderated by culture, learning, and experience. Drawing on 20 years of original research, he considers the universal basis for how humans value nature. Differences in those values vary by gender, age, ethnicity, occupation, and geographic location. He discusses how environment-related activities affect values, variation in values relating to different species, how values vary across cultures, and their policy and management implications. Throughout his book "Nature by design: The practice of biophilic design," Kellert argues that the preservation of biodiversity is fundamentally linked to human well-being in the largest sense, as he illustrates the importance of biological diversity to the human sociocultural and psychological condition (Kellert 2018).

### 19.2.2 Scientific Motivations

Rachel Carson advanced biological conservation and subsequent conservation projects through the scientific underpinnings of "Silent Spring" her famous book that pinpointed the environmental harms brought about by widespread and unregulated pesticide use (Carson 1962). Carson's effective and open communications of the loss of biological diversity brought regulation, ignited public discourse, bolstered the formation of the United States Environmental Protection Agency (EPA), and inspired a generation of environmentalists. This heightened awareness and profoundly shaped the western mindset regarding the environment.

Advances in conservation biology, regarding the measurement, distribution, abundance, and loss of species through time create another motivating factor for conserving biodiversity (Hawksworth 1995). As it becomes clear that more habitat is lost due to deforestation and habitat destruction, science-based motivations for community-led efforts for biodiversity conservation grow (Hawksworth 2010a, b). As such, knowledge regarding inventories for understanding baseline terrestrial biodiversity for estimating both local and global species diversity can allow for both rapid assessment and comparison of species diversity across geographies (Colwell and Coddington 1994). Numerous textbooks, journals, and publications have been created to further advance knowledge in these realms (Hawksworth 2012). As it becomes clear that species loss is greater, motivations to implement practices for conservation are bolstered. In many instances, communities may advance and aid in the scientific understanding of conservation or utilize scientific methodology to implement best practices in biodiversity conservation.

Overall, there can be a strong motivation towards the advancement of biodiversity islands from the academic and scientific realm. Biodiversity islands are the subject of study from many different angles, from edge effect to island biogeography (Montagnini et al. 2022). Deepening the scientific understanding of forest patch dynamics also provides a motivation for conservation of such landscape features.

### *19.2.3 Grassroots Motivations*

Other motivations for biodiversity conservation may arise from grassroots actions and through counterculture. In communities where, for a variety of reasons, institutional or governmental engagement has been ineffective, grassroots education and action may emerge to bring forth greater biological diversity through community efforts in degraded, human-dominated landscapes. This can be seen in various forms and is distinctly recognized through development of grassroots community gardening, urban agriculture, and in the permaculture movement worldwide (Veteto and Lockyer 2008; Ferguson and Lovell 2014).

Gardens are known to have the potential to foster significant diversity in cultural, biological, and agro-biological ways (Galluzzi et al. 2010; Goddard et al. 2010; Negret et al. 2022). In urban centers throughout the United States and beyond, community gardens have emerged, bringing forth both resilient food systems and positive biodiversity outcomes from grassroots formation (Clarke and Jenerette 2015; Di Pietro et al. 2018). Additionally, the permaculture movement is responsible for the development of thousands of projects throughout the globe which enhance biodiversity outcomes (Toensmeier 2022). Built upon ethical principles, permaculture and the training course known as a Permaculture Design Certificate (PDC) offer students a method for designing biologically diverse "permanent" agriculture systems and explore methods for sustained human habitation on a landscape (Mollison 1988). Similar local agroecological trainings, workshops, and community-led courses provide a grassroots approach to such training, especially in those cases where institutional and formal education fails to provide this information.

### 19.2.4 Economic Motivations

Economic motivations may also drive community-led action towards the development of biodiversity islands. Although historically economic motivations have often led to exploitation and landscape degradation, such motivations can also be a force for conservation of biodiversity.

Examples of profitable biodiversity enhancing community enterprise models span a diverse range of geographies, cultures, products, and services offered (Hay-Edie and Halverson 2006). This includes sustainable ecotourism, where visitors pay to stay and engage with local cultures and traditions. Though complex and sometimes difficult to obtain, the biodiversity of ecotourism sites can be significantly higher than that of the surrounding human-dominated and degraded landscape (Gossling 1999; Chung et al. 2018). Examples of other community enterprises which support similar biodiversity outcomes have been shown to produce a broad range of products such as basketry, incense, herbal medicines, and teas (Jarrett et al. 2017; Rocha et al. 2017).

Agricultural production may also play an important role in community enterprise development which supports biodiversity conservation within a broader degraded landscape (Badgley 2018). Sustainable agricultural models in which biodiversity islands are developed through collectives and community action include crops such as cacao, coffee, tea, yerba mate, dried fruits, and various other species (Erisman et al. 2016; Dudley and Alexander 2017; Hunter et al. 2017; Montagnini and del Fierro 2022). Sustainable timber production has also been a method of enterprise creation that can respect and support biodiversity conservation within degraded landscapes in addition to positive economic outcomes (Carey et al. 1999).

## 19.3 Land Access, Tenure, and Long-Term Control for Pockets of Biodiversity

For any community action towards the development of biodiversity islands to take place, long term control, tenure, and decision-making authority over the land base are paramount. Community land management without legal authority may prove difficult into posterity. This is documented in numerous case studies worldwide, from local government closure of community gardens on vacant urban land to the removal of indigenous people from ancestrally managed lands (Springer 2009; Holmes and Cavanagh 2016). Acquiring land title or the ability to manage lands in response to governmental territorial expansions may prove particularly challenging for societies without robust property laws or in cultures without concepts of land ownership. In societies where property rights are recognized and biodiversity outcomes are a community objective, people can find innovative ways to engage with legal frameworks to gain and maintain land access and tenure through time. Examples of such methods include use of land trusts and conservation easements, supportive zoning, inventive financial resource pooling, robust laws supporting indigenous land rights, and the use of public lands which encourage community-led biodiversity conservation. These methods are discussed in the following section.

### *19.3.1 Conservation Easements and Community Land Trusts*

Conservation easements are one tool that may be advocated for by a community for the development of biodiversity islands. Conservation easements are voluntary legal agreements between landowners and a land trust or government agency that protects conservation values on a property by permanently limiting uses of the land and offering tax incentives to the landowner. A land trust or community land trust is a non-profit organization that acquires land or conservation easements through support from donations or government funding. The opportunity for land trusts to maintain and protect biodiversity through such conservation easements is significant (Rissman et al. 2007; Wilson 2011). In many instances, a land trust acts as a conservation organization to help draft, implement, and ensure compliance with an easement while prioritizing community and conservation goals. Community land trusts are a subset of land trusts which specifically focus on the development and conservation of community assets for community benefit. Such a model may be particularly valuable towards the development of biodiversity islands. There are several examples around the world where conservation easements are used for the protection of biodiversity in the USA, Canada, Australia, Costa Rica, and other countries (Alexander and Hess 2012).

Conservation easements may include restrictions on development and land uses such as recreation, forestry, agroforestry, or agriculture. These easements, which provide conservation value in addition to opportunities for land-based revenue generation, are known as working land conservation easements. Throughout the world, communities can support and develop community land trusts to assist in the development of community-led biodiversity islands to transition private lands into long-term community conservation projects. In Chap. 1 (Montagnini et al. 2022) an example is shown of a Land Trust that received a donation from a family whose members preferred that their land was preserved and used for recreation instead of selling it. This is just one example where through land trusts and conservation easements community action provides protection of a biodiversity island and judicious management for posterity.

### 19.3.2 Innovative Financing for Biodiversity Islands

The use of innovative financing to acquire property is another method for long-term community-led land control which can support the development of biodiversity islands. A group of community members or a community organization may pool financial resources to acquire land outright through a cash purchase or a loan. This is seen in numerous countries throughout the world including lands and resources surrounding Juan Castro Blanco National Park in Costa Rica (Castro-Arce and Vanclay 2020). This community pooling of resources is also known as crowdsourcing. While there has been significant advancement and literature on the use of crowdsourced data collection in relation to biodiversity conservation, such as monitoring bird species and populations, there is much opportunity for the use of internet platforms as tools for further crowdsourced funding for the acquisition of property which may enhance biodiversity outcomes through time. The advent of multiple online platforms which allow for numerous smaller donations or investments to develop into projects, provides a groundwork for the advancement of crowdfunding towards biodiversity conservation (Gallo-Cajiao et al. 2018).

### 19.3.3 Indigenous Land Rights Protecting Biodiversity

Strong indigenous land rights can allow for biodiversity islands to emerge or persist throughout otherwise degraded landscapes. Dominion within current or historically inhabited and managed indigenous lands can support sovereignty, knowledge sharing, cultural empowerment, and positive biodiversity outcomes through time (Langton and Rhea 2005; Erikson 2008; Sobrevila 2008; Garnett 2018; Baldwin and Beazley 2019; Beller et al. 2020). Native title (Australia), Indian title (United States), Customary title (New Zealand) or Indigenous or Aboriginal title are common law doctrines which strengthen sovereignty of historical land rights through recognition by governments. While such rights may be powerful, many indigenous peoples emphasize indigenous rights that do not necessitate state sanctions to exist (Gilbert 2016).

Secure and long-term indigenous management of landscapes are worth highlighting in relation to community-led action toward the development of biodiversity islands. Research on this topic is steeped in complexity. Not all indigenously managed lands promote biodiversity, particularly in instances where resources were overexploited through time (Raymond 2007). Additionally, there are many examples where indigenous sovereignty has been withdrawn in the name of biodiversity conservation (Baldwin and Beazley 2019; Beller et al. 2020). With these considerations in mind, this section seeks to highlight the potential power and opportunity of indigenous land rights as a tool for community-led biodiversity islands to emerge and persist in otherwise degraded landscapes.

## 19.4 Governance and Management of Biodiversity Islands

Once land tenure is achieved, proper governance and subsequent management of biodiversity islands is essential for long term success. Such governance and management can increase biodiversity through time, allow community members to have a voice in decision-making, and avoid resource degradation, species loss, and a tragedy of the commons. Governance can take many varied forms across a range of scales depending on culture and context. Commonly, hierarchical decision making dominates organizational governance. While hierarchical governance is effective to achieve certain ends, alternative egalitarian governance structures may also be effective for community managed lands (Leventon et al. 2019). This chapter highlights three methods of land governance aligned with community-led action that may support high levels of biodiversity within a larger degraded landscape through commons, cooperatives, and community based conservation approaches.

Governance as commons, where land is held equally by all community members, allows for egalitarian ownership of a space. Common resources governed and adapted to local conditions with clear boundaries, measured outcomes, defined rules and mechanisms for conflict resolution, and self-determination prove resilient. Such governance allows for nested enterprises to emerge and common resource pools to persist through time. While some argue that community-owned resources degrade through time due to motivations of self-interest leading to overuse, deemed a tragedy of the commons, countless examples of long-term community managed resources refute such claims (Hardin 1968; Ostrom 1990). As applied to biodiversity islands, these governance strategies and other types of community governance are foundational support for management through time.

Cooperatives as an organizational approach can also provide a more egalitarian governance and management method to community-fostered biodiversity islands. In such approaches, ownership and decision making tend to be more equitably distributed throughout the organization, and outcomes, yields, and profits are then distributed accordingly. Globally, examples of agricultural cooperatives that value and incorporate agrobiodiversity into their cultivation practices are numerous (Méndez et al. 2007). Cooperative organizations which incorporate ethos of biodiversity conservation span many industries and both non-profit and for-profit ventures. Such systems can enhance social capital, which has shown to maintain positive biodiversity outcomes within degraded landscapes (Pretty and Smith 2004).

Advances in community-based conservation (CBC) may also provide the mechanisms toward effective development, governance, and management of biodiversity islands through time. CBC seeks to align development and conservation outcomes emergent from the community. Criticism of such methods of conservation have emerged due to the preponderance of linear thinking about development, often considering a single variable such as Gross Domestic Product (GDP) and situations where conservation and human habitation are seen as incompatible. Approaches that consider humans as part of the ecosystem and promote wide scale participation in ecosystem management provide a more feasible context from which CBC may emerge (Otto et al. 2013). The integration of concepts of common property, traditional ecological knowledge, environmental ethics, political ecology, and environmental history can further support the longevity of projects, providing knowledge of the past, reflection of the present resources, opportunities among stakeholders, and a context for visioning future goals.

The importance of fluid and adaptive collaborative approaches which recognize multiple stakeholders and development goals may also contribute significant value to CBC projects (Berkes 2004). Participatory consent-driven decision making as exemplified by the frameworks of Holacracy,[1] provides an example of participatory governance which may be applied to further mature CBC (Robertson 2015). The examples of CBC projects throughout the world that have increased biodiversity within degraded landscapes are extensive (Otto et al. 2013). For this reason, it is important to note that CBC can provide a strong avenue for community-led action towards the development of biodiversity islands.

## 19.5 Community Engagement and Advocacy Towards Biodiversity Islands

Community engagement and advocacy are other important forces towards the development of biodiversity islands. Intercultural and interpersonal communication, education and political engagement which support the appreciation and values of biodiversity within a landscape are foundational. Interpersonal communication through talking and engaging with neighbors and community members and fostering good relationships is an essential first step in building community coalitions for biodiversity conservations. Once achieved, intercultural communication, with different belief systems, histories, stories, and relationships to landscape and place, can then build the foundations from which a more biodiverse landscape may arise within a community (Pretty et al. 2009).

Environmental education focused around topics of biodiversity are shown to be effective to engage, empower, and bridge science and social issues (Van Weelie and Wals 2002). When this knowledge is imparted on youth, a next generation may prioritize issues of biodiversity conservation. When this knowledge is shared with politicians through actions of lobbying and political engagement, innovative legal tools and frameworks may be adopted or developed by governments. Scheduling meetings with local, regional, and national representatives, attending governmental meetings, writing letters, and coalition building are strategic tactics for empowering government action.

[1]Holacracy is a method of decentralized management and organizational governance, in which authority and decision-making are distributed throughout self-organizing teams rather than being vested in a management hierarchy.

## 19.6 Examples of Community-Led Conservation Strategies

Examples of grassroots community action for the advancement of conservation practices are numerous, diverse, and worldwide (de Boef et al. 2013; Otto et al. 2013). Ancient traditional cases of biodiversity islands within the landscape are exemplified by religious and sacred sites worldwide. Other examples span urban and rural settings, agrarian and nomadic cultures, and incorporate various organizational approaches.

Sacred groves, or natural sites that are dedicated to ancestral and spiritual deities, are found throughout the world. These sites often harbor and protect greater biodiversity than surrounding degraded landscapes (Bhagwat and Rutte 2006; Khan et al. 2008). Church forests, ancient temples, and dedicated sections of rivers and waterways have been protected and maintained by cultures throughout the globe for religious purposes for millennia (Baez Schon et al. 2022). Deeply held intergenerational importance of such sites provides strong community incentive for conservation through time. Other examples of sacred sites that serve as biodiversity islands include the Mizorom sacred groves in Northeastern India, sacred pools called Íbú ódó protected by Tchabè communities along the Ouèmé and Okpara Rivers of Central Benin (West Africa), sacred cacao groves of the Maya, and other sacred groves in Zimbabwe, Ghana, Thailand, China, and Nepal (Gómez-Pompa et al. 1990; Gadgil et al. 1993; Bhagwat and Rutte 2006; Ceperley et al. 2010). These biodiversity islands share common outcomes where community action enhances and protects biodiversity through time in human-dominated landscapes. Although the original purpose in most cases is not to create biodiversity islands, biodiversity islands are a consequence of their actions. Other community-protected sites include the village forests in Indonesia, known as Hutan Desa, which are legally recognized for the ecosystem services and benefits to society they provide. Their management and protection are guided by traditional communal governance as well (Moeliono et al. 2015).

Traditional methods of community-based biodiversity conservation which can maintain or build biodiversity islands may also include agrarian and nomadic cultures. The Yanesha at the headwaters of the Amazon basin in Central Peru have hunted, gathered, and farmed sustainably for thousands of years with little evidence of biodiversity degradation (Salick 1989). As deforestation and unsustainable development expand, community-led traditional landscape management provides the opportunity for biodiversity islands to emerge. In some instances, levels of biodiversity within traditional communities can increase through agricultural production as documented throughout Amazonia (Erikson 2008). Community-fostered swidden systems from Southwest China to the Ecuadorian Amazon to Madagascar can be managed in accordance with biodiversity enhancing outcomes (Rerkasem et al. 2009; Xu et al. 2009). In other examples, such as in the Southwestern-United States, species composition of agricultural plots is actively managed to embrace diversity and complexity in both production species and management of surrounding habitat (Nabhan 2000).

Moreover, community-led biodiversity islands exist in highly varied geographies catalyzed by diverse organizational structures. Rural community action from Zapatistas in southern Mexico to community restoration projects in Northern Ethiopia to the newly developing global concept of Ecosystem Restoration Camps, all seek to enhance biodiversity outcomes in rural settings through community efforts (https://www.ecosystemrestorationcamps.org/, Sigman 2022). Non-profit organizations such as Eco-agriculture Partners and The Forest Dialogue also aim to help engage community action for biodiversity in rural settings through methods of community engagement within integrated landscape management (https://ecoagriculture.org/, https://theforestsdialogue.org/). In suburban settings homegardens, community garden projects, and ecovillages act as biodiversity islands in the fragmented landscape (Negret et al. 2022; Toensmeier 2022). Enterprises such as Permaculture Artisans, a California-based landscaping company, develop properties into productive gardens, creating biodiversity islands on the suburban landscape (www.permacultureartisans.com). In the urban context, the rise of urban forestry, urban community gardens, and educational centers which support these ends continue to grow in popularity (Soler et al. 2022). These examples in varied scales of population density, diverse cultures, and various locations, provide excellent lessons for community-led action advancing the creation, establishment, and maintenance of biodiversity islands.

## 19.7 Conclusion

Community-led action towards the development of biodiversity islands demands looking beyond just the human factors of place to include all living beings and biophysical components that make up the community. Motivations for biodiversity conservation are diverse across cultures. In many instances, historical ecology and indigenous and traditional philosophies of the region may empower decision making and cultural practices towards the development and preservation of biodiversity islands. When empowered, communities across a range of contexts may effectively mobilize towards protecting, creating, and expanding islands of biodiversity. Land access and long-term decision-making authority become essential in the longevity of such pockets of biodiversity. Various legal and financial strategies such as conservation easements, community land trusts, and novel methods of crowdsourced funding can further aid in successful outcomes. Once biodiversity islands emerge within a landscape, egalitarian management and governance approaches have proven effective through time. Community advocacy may further bolster related positive outcomes. By employing these tactics, community-led biodiversity islands can rapidly multiply and continue to span the globe, as protected pockets of biodiversity.

## References


Alexander L, Hess GR (2012) Land trust evaluation of progress toward conservation goals. Conserv Biol 26(1):7–12

Badgley C (2018) Biodiversity and sustainable agriculture. Promoting biodiversity in food systems. In: Hawkins IW (ed) Promoting biodiversity in food systems. CRC Press, Boca Raton, pp 165–182

Baez Schon M, Woods CL, Cardelús CL (2022) Sacred church forests in northern Ethiopia: biodiversity and cultural Islands. In: Montagnini F (ed) Biodiversity islands: strategies for conservation in human-dominated environments. Springer, Cham, pp 531–549

Baldwin RF, Beazley KF (2019) Emerging paradigms for biodiversity and protected areas. Land 8(3). https://doi.org/10.3390/land8030043

Balée W (1993) Indigenous transformation of Amazonian forests: an example from Maranhão, Brazil. L'Homme 33:231–254

Balée W (1998) Advances in historical ecology. Columbia University Press, New York. 429pp

Beller EE, McClenachan L, Zavaleta ES, Larsen LG (2020) Past forward: recommendations from historical ecology for ecosystem management. Global Ecol Conserv 21. https://doi.org/10.1016/j.gecco.2019.e00836

Berkes F (2004) Rethinking community-based conservation. Conserv Biol 18(3):621–630

Berkes F (2017) Environmental governance for the Anthropocene? Social-ecological systems, resilience, and collaborative learning. Sustainability 9(7):1232. https://doi.org/10.3390/su9071232

Berkes F, Folke C, Gadgil M (1994) Traditional ecological knowledge, biodiversity, resilience and sustainability. In: Perrings CA, Mäler KG, Folke C, Holling CS, Jansson BO (eds) Biodiversity conservation. Springer, Switzerland AG, pp 269–287

Bhagwat SA, Rutte C (2006) Sacred groves: potential for biodiversity management. Front Ecol Environ 4(10):519–524

Bookchin M (1982) The ecology of freedom. Cheshire Books, Palo Alto, p 480

Carey AB, Lippke BR, Sessions J (1999) Intentional systems management: managing forests for biodiversity. J Sustain For 9(3–4):83–125

Carson R (1962) Silent Spring. Houghton Mifflin, Boston

Castro-Arce K, Vanclay F (2020) Transformative social innovation for sustainable rural development: an analytical framework to assist community-based initiatives. J Rural Studies 74:45–54

Ceperley N, Montagnini F, Natta A (2010) Significance of sacred sites for riparian forest conservation in Central Benin. Bois Forêts Tropiques 303:5–23

Chung MG, Dietz T, Liu J (2018) Global relationships between biodiversity and nature-based tourism in protected areas. Ecosyst Serv 34:11–23

Clarke LW, Jenerette G (2015) Biodiversity and direct ecosystem service regulation in the community gardens of Los Angeles, CA. Landsc Ecol 30(4):637–653

Colwell RK, Coddington JA (1994) Estimating terrestrial biodiversity through extrapolation. Philos Trans R Soc Lond B Biol Sci 345(1311):101–118

Crosby AW (2004) Ecological imperialism: the biological expansion of Europe, 900–1900. Cambridge University Press, Cambridge

de Boef WS, Subedi A, Peroni N, Thijssen M, O'Keeffe E (eds) (2013) Community biodiversity management: promoting resilience and the conservation of plant genetic resources. Routledge, Abington

Di Pietro F, Mehdi L, Brun M, Tanguay C (2018) Community gardens and their potential for urban biodiversity. In: Gladtron S, Granchamp L (eds) The urban Garden City. Springer, New York, pp 131–151

Dudley N, Alexander S (2017) Agriculture and biodiversity: a review. Biodiversity 18(2–3):45–49

Dunson WA, Travis J (1991) The role of abiotic factors in community organization. Amer Nat 138(5):1067–1091

Erickson CL (2008) Amazonia: the historical ecology of a domesticated landscape. In: Isbell W (ed) Silverman H. The Handbook of South American Archaeology Springer, New York, pp 157–183
Erisman JW, van Eekeren N, de Wit J, Koopmans C, Cuijpers W, Oerlemans N, Koks BJ (2016) Agriculture and biodiversity: a better balance benefits both. Agric & Food 1(2):157–174
Ferguson RS, Lovell ST (2014) Permaculture for agroecology: design, movement, practice, and worldview. A review. Agron Sustain Dev 34(2):251–274
Gadgil M, Berkes F, Folke C (1993) Indigenous knowledge for biodiversity conservation. Ambio:151–156
Gallo-Cajiao E, Archibald C, Friedman R, Steven R, Fuller RA, Game ET, Morrison TH, Ritchie EG (2018) Crowdfunding biodiversity conservation. Conserv Biol 32(6):1426–1435
Galluzzi G, Eyzaguirre P, Negri V (2010) Home gardens: neglected hotspots of agro-biodiversity and cultural diversity. Biodivers Conserv 19(13):3635–3654
Garnett ST, Burgess ND, Fa JE, Fernández-Llamazares Á, Molnár Z, Robinson CJ, Watson JE, Zander KK, Austin B, Brondizio ES, Collier NF (2018) A spatial overview of the global importance of indigenous lands for conservation. Nat Sustain 1(7):369
Gilbert J (2016) Indigenous peoples' land rights. International Law Brill Publishers, Leiden
Goddard MA, Dougill AJ, Benton TG (2010) Scaling up from gardens: biodiversity conservation in urban environments. Trends Ecol Evol 25(2):90–98
Gómez-Pompa A, Flores JS, Fernández MA (1990) The sacred cacao groves of the Maya. Lat Amer Antiq 1(3):247–257
Gössling S (1999) Ecotourism: a means to safeguard biodiversity and ecosystem functions? Ecol Econ 29(2):303–320
Hardin G (1968) The tragedy of the commons. Science 162(3859):1243–1248
Harvey G (2005) Animism: respecting the living world. Wakefield Press, Adelaide
Hawksworth DL (ed) (1995) Biodiversity: measurement and estimation. Chapman and Hall, London
Hawksworth DL (ed) (2010a) Management and the conservation of biodiversity. Springer, New York
Hawksworth (ed) (2010b) Methods and practice in biodiversity conservation. Springer, New York
Hawksworth DL (2012) Books on biodiversity and conservation. Biodivers Conserv 21(14): 3745–3769
Hay-Edie T, Halverson E (2006) Community action to conserve biodiversity: linking biodiversity conservation with poverty reduction. GEF Small Grants Programme, UNDP, New York. sgp.undp.org
Holmes G, Cavanagh CJ (2016) A review of the social impacts of neoliberal conservation: formations, inequalities, contestations. Geoforum 75:199–209
Hunter D, Guarino L, Spillane C, McKeown PC (eds) (2017) Routledge handbook of agricultural biodiversity. Routledge, Abington
Inglis J (ed) (1993) Traditional ecological knowledge: concepts and cases. IDRC, Ottawa
Jarrett C, Cummins I, Logan-Hines E (2017) Adapting indigenous agroforestry systems for integrative landscape management and sustainable supply chain development in Napo, Ecuador. In: Integrating landscapes: agroforestry for biodiversity conservation and food sovereignty. Springer, Cham, pp 283–309
Kellert SR (1997) The value of life: biological diversity and human society. Island Press, Washington DC
Kellert SR (2018) Nature by design: the practice of biophilic design. Yale University Press, New Haven
Khan ML, Khumbongmayum AD, Tripathi RS (2008) The sacred groves and their significance in conserving biodiversity: an overview. Int J Ecol Environ Sci 34(3):277–291
Langton M, Rhea ZM (2005) Traditional indigenous biodiversity-related knowledge. Aust Acad & Res Libr 36(2):45–69

Leopold A (1949) A Sand County almanac, and sketches here and there. Oxford University Press, New York

Leventon J, Schaal T, Velten S, Loos J, Fischer J, Newig J (2019) Landscape-scale biodiversity governance: scenarios for reshaping spaces of governance. Environ Policy Gov 29(3):170–184

Lovelock J (1979) Gaia: a new look at life on earth. Oxford University Press, Oxford

Méndez V, Gliessman SR, Gilbert GS (2007) Tree biodiversity in farmer cooperatives of a shade coffee landscape in western El Salvador. Agric Ecosyst Environ 119(1–2):145–159

Moeliono M, Mulyana A, Adnan H, Manalu P, Yuliani L (2015) Village forests (hutan desa): empowerment, business or burden? World Agroforestry Centre-ICRAF, Southeast Asia Regional Office, Bogor

Mollison B (1988) Permaculture: a Designer's manual. Tagari Publications, Sisters Creek

Montagnini F, del Fierro S (2022) Functions of agroforestry systems as biodiversity islands in productive landscapes. In: Montagnini F (ed) Biodiversity islands: strategies for conservation in human-dominated environments. Springer, Cham, pp 89–116

Montagnini F, Levin B, Berg KE (2022) Biodiversity Islands. Strategies for conservation in human-dominated environments. In: Montagnini F (ed) Biodiversity islands: strategies for conservation in human dominated environments. Springer, Cham, pp 1–35

Nabhan GP (2000) Native American management and conservation of biodiversity in the Sonoran Desert bioregion. In: Minnis PE, Elisens WJ (eds) Biodiversity and native America. University of Oklahoma Press, Norman, pp 29–44

Nabhan G, Hodgson W, Fellows F (1989) A meager living on lava and sand? Hia ced O'odham food resources and habitat diversity in oral and documentary histories. J Southwest:508–533

Negret HRC, Negret R, Montes-Londoño I (2022) Residential garden Design for Urban Biodiversity Conservation: Experience from Panama City, Panama. In: Montagnini F (ed) Biodiversity islands: strategies for conservation in human-dominated environments. Springer, Cham, pp 387–417

Ostrom E (1990) Governing the commons: the evolution of institutions for collective action. Cambridge University Press, Cambridge

Otto J, Zerner C, Robinson J, Donovan R, Lavelle M, Villarreal R, Salafsky N, Alcorn J, Seymour F, Kleyneyer C, Pearl M (2013) Natural connections: perspectives in community-based conservation. Island Press, Washington D.C

Peters CM (2018) Managing the wild: stories of people and plants and tropical forests. Yale University Press, New Haven

Pretty J, Smith D (2004) Social capital in biodiversity conservation and management. Conserv Biol 18(3):631–638

Pretty J, Adams B, Berkes F, de Athayde S, Dudley N, Hunn E, Maffi L, Milton K, Rapport D, Robbins P, Sterling E, Stolton S, Tsing A, Vintinnerk E, Pilgrim S (2009) The intersection of biological diversity and cultural diversity: towards intersection. Conserv Soc 7:100–112

Prieto Méndez JM (2013) Derechos de la naturaleza. Fundamento, contenido y exigibilidad jurisdiccional. Corte Constitucional del Ecuador, CEDEC, Quito. 280pp

Raymond H (2007) The ecologically noble savage debate. Ann Rev Anthropol 36:177–190

Rerkasem K, Lawrence D, Padoch C, Schmidt-Vogt D, Ziegler AD, Bruun TB (2009) Consequences of swidden transitions for crop and fallow biodiversity in Southeast Asia. Hum Ecol 37(3):347–360

Rissman AR, Lozier L, Comendant T, Kareiva P, Kiesecker JM, Shaw MR, Merenlender AM (2007) Conservation easements: biodiversity protection and private use. Conserv Biol 21(3): 709–718

Robertson BJ (2015) Holacracy: the new management system for a rapidly changing world. Henry Holt and Company, New York

Rocha P, Niella F, Keller H, Montagnini F, Metzel R, Eibl B, Kornel J, Romero F, López L, Araujo J, Barquinero J (2017) Ecological indigenous (EIK) and scientific (ESK) knowledge integration as tool for sustainable development in indigenous communities. Experience in Misiones, Argentina. In: Montagnini F (ed) Integrating landscapes: agroforestry for biodiversity conservation and food sovereignty, advances in agroforestry 12. Springer, Cham, pp 235–260

Salick J (1989) Ecological basis of Amuesha agriculture, Peruvian upper Amazon. Adv Econ Bot:189–212

Salick J (2012) Indigenous peoples conserving, managing, and creating biodiversity. In: Harlan JR, Gepts P, Famula TR, Bettinger RL, Brush SB, Damania AB, McGuire PE, Qualset CO (eds) Biodiversity in agriculture: domestication, evolution, and sustainability. Cambridge University Press, Cambridge, pp 426–445

Sigman E (2022) Safeguarding the benefits of biodiversity islands in northern Ethiopia in the midst of political change. In: Montagnini F (ed) Biodiversity islands: strategies for conservation in human-dominated environments. Springer, Cham, pp 647–674

Sobrevila C (2008) The role of indigenous peoples in biodiversity conservation World Bank, Washington D.C.

Soler R, Benítez J, Solá F, Lencinas MV (2022) Biodiversity islands at the world's southernmost city: plant, bird and insect conservation in urban forests and peatlands of Ushuaia, Argentina. In: Montagnini F (ed) Biodiversity islands: strategies for conservation in human-dominated environments. Springer, Cham, pp 419–437

Springer J (2009) Addressing the social impacts of conservation. Conserv Soc 7(1):26–29

Swimme B, Tucker ME (2011) Journey of the universe. Yale University Press, New Haven

Toensmeier E (2022) Paradise lot: a temperate urban multistrata agroforestry island of biodiversity. In: Montagnini F (ed) Biodiversity islands: strategies for conservation in human-dominated environments. Springer, Cham, pp 439–459

Van Weelie D, Wals A (2002) Making biodiversity meaningful through environmental education. Int JSci science Edu 24(11):1143–1156

Veteto JR, Lockyer J (2008) Environmental anthropology engaging permaculture: moving theory and practice toward sustainability. Cult Agric 30(1–2):47–58

Wilson EO (1984) Biophilia. Harvard University Press, Cambridge MA

Wilson R (2011) Documenting and protecting biodiversity on land trust projects. Land Trust Alliance, Washington DC

Xu J, Lebel L, Sturgeon J (2009) Functional links between biodiversity, livelihoods, and culture in a Hani swidden landscape in Southwest China. Ecol Soc 14(2). https://doi.org/10.5751/ES-02916-140220

# Chapter 20
# Priorities, Perspectives, and Use of a Community Forest by Surrounding Residents in Mayagüez, Puerto Rico: Protecting the Forest for Its Services

**Gabriela M. Morales-Nieves**

**Abstract** Community forests are important for protection of lands, reduction of deforestation, conservation of biodiversity, and carbon sequestration while providing socioeconomic benefits to those living around them. In the insular Caribbean, community forests do not seem to be used widely. Given that it is a region with high biodiversity and endemism, while being also the sub region in Latin America and the Caribbean with highest population density, protection of lands in fragmented landscapes is especially important. To expand the adoption of community forests by communities, it is important to understand the interest of the residents in managing them and the opportunity they see in community forests. This research used interviews with people living around a community forest in Mayagüez, Puerto Rico to understand local perceptions of and interactions with the forest. Interviewees were asked which activities they would be interested in participating in the forest, perceived ecosystem services and what should be the management priorities and services the forest should provide to the community. The most important ecosystem service for residents was air quality, followed by recreation. Motives that discouraged the use of the forest by residents were the presence of non-desirable species and lack of knowledge regarding the functioning of the forest. The forest is currently a meeting space for community activities such as celebrations, workshops and the selling of food and other products. Although conservation of biodiversity is just one motivation for managing the forest as a community, the values that people assign to the forest contribute to its preservation as a biodiversity island within a rural-urban landscape.

**Keywords** Caribbean · Community forestry · Conservation · Ecosystem services · Human perceptions · Participation · Urban expansion

G. M. Morales-Nieves (✉)
Yale University, School of the Environment, The Forest School, New Haven, CT, USA
e-mail: gabriela.morales-nieves@aya.yale.edu



F. Montagnini (ed.), *Biodiversity Islands: Strategies for Conservation in Human-Dominated Environments*, Topics in Biodiversity and Conservation 20,
https://doi.org/10.1007/978-3-030-92234-4_20

## 20.1 Introduction

Community forestry can be broadly defined as government-approved or recognized initiatives that have the objective of increasing the role of surrounding communities in governing and managing forest resources to provide social and economic benefits while promoting the sustainable management of said forest lands (Bowler et al. 2012; Gilmour 2016). Community forests have been considered as an essential component for the conservation of lands and reducing deforestation rates (Porter-Bolland et al. 2012; Lambrick et al. 2014), maintaining biodiversity (Souza et al. 2016), and sequestering carbon (Wood et al. 2019), while providing different socioeconomic benefits to surrounding residents (Bridhikitti and Khadka 2019). However, mixed results have been found in cases when biodiversity conservation goals may conflict with interest in using the forest resources. There are often some trade-offs between activities that enhance biodiversity conservation, like controlling wildlife hunting, and those that could have negative impacts on biodiversity and ecosystem functions, like for example, removing or selecting certain plant species and replacing for others of commercial value (Shrestha and Shrestha 2010; Rasolofoson et al. 2015). Even though the term "community forestry" is broad and used in varying contexts, most of them coincide in that they "aim to protect forests and benefit local livelihoods" (Rasolofoson et al. 2015).

In the Caribbean islands region, literature on participatory management of resources seems to focus more on those in coastal ecosystems (Geoghegan and Renard 2002; McConney and Pena 2012; Smith 2012; Dalton et al. 2012). No scientific literature has been found on community forests in the Caribbean islands, but there is some gray literature and reports on community-based forest management, such as a report by the Caribbean Natural Resources Institute (CANARI) and the Food and Agriculture Organization (FAO) (2012). As in other parts of the world, with climate change, the Caribbean islands are expected to experience changes in temperatures, precipitation patterns (Pulwarty et al. 2010) and increased intensity of hurricanes (Holland and Bruyère 2014; Knutson et al. 2015). This is especially important given their small physical dimensions, their high ranking in vulnerability to global disasters, and their high dependency on natural resources for the tourism industry, among other aspects (Pulwarty et al. 2010).

The Caribbean region is also a world biodiversity hotspot and it is important for biodiversity conservation due to its high levels of endemism and impending threats related to increases in human populations and encroachment on forests and other natural areas (Myers et al. 2000; Anadon-Irizarry et al. 2012). The Caribbean Islands comprise the most densely populated sub region in Latin America and the Caribbean, given 2017 data obtained from the World Bank (DataBank | The World Bank) found in Appendix A. Nine of ten of the most densely populated countries within Latin America and the Caribbean are from the insular Caribbean. Therefore, maintaining biodiversity islands in a human-dominated landscape is especially important for the conservation of high biodiversity in a region with limited territorial extent.

To expand the adoption of community forests as a tool for ecosystem and biodiversity protection and also for economic development in the Caribbean Islands, it is important to understand the interests of communities surrounding forested lands in managing them and also the opportunities seen in them. This chapter examines the use of a community-managed forest in Mayagüez, Puerto Rico as a case study. This research aimed to understand how the surrounding community used or would use the forest for, what ecosystem services or benefits they felt the forest provided to the community, and what the management priorities for the forest should be. It uses these findings to discuss the potential of community forests for conservation and offers recommendations for management in other similar contexts in the region and elsewhere.

## *20.1.1 Recent Trends in Land Use History of Puerto Rico and the Caribbean*

### 20.1.1.1 Puerto Rico

To understand the results of this case study, it is necessary to contextualize the land use history of Puerto Rico and the Caribbean. By the nineteenth century, with consistent increases in the population in Puerto Rico post colonization, agriculture, comprised mostly of sugarcane, coffee, plantain, cotton and rice, had spread throughout. In addition, by that time pasturelands covered about 55% of the land (Grau et al. 2003). By the 1930s, agriculture and cattle production represented about 45% of the gross national product (GNP). In the late 1940s, "post war efforts to promote industrialization shifted the economy from agriculture to light industry," with agriculture comprising less than 5% of the GNP by 1980 (Grau et al. 2003). Since agricultural land abandonment due to this transition in economy from mainly agricultural to industry and service-based in current times, Puerto Rico gained from 5–6% forest cover in the 1940s to 50–57% by 2009 (Álvarez-Berríos et al. 2013; Brandeis and Turner 2013), remaining in a steady state up to 2014 (Marcano-Vega 2017).

During the first decade of the 2000s, urban cover also increased. Urban expansion was the major cause for deforestation in Puerto Rico during this time and the previous decade, with the loss of agricultural lands as the major land use change, however with an overall increase in woody vegetation recovery (Álvarez-Berríos et al. 2013). Although urbanization contributed to woody regrowth in the past, recent patterns of urban expansion are the major driver of deforestation and forest fragmentation in the island. Urbanization seems to compete and replace the same lands as agriculture in Puerto Rico, since the appropriate characteristics for urbanization are very similar to those that are needed for intensive agriculture (e.g. lower elevations, flat topography, proximity to roads), with 60% of total development by 2003 occurring where the most productive lands for agriculture are located (López et al. 2001; Grau et al. 2003; Martinuzzi et al. 2007).

By 2016, 16.1% of Puerto Rico's land cover was under some form of protection, whether under state, federal (U.S.A.), private or non-profit organization supervision. Of these protected lands, 27% are considered forest lands and 56% are forested wetlands (Castro-Prieto et al. 2019).

#### 20.1.1.2 Caribbean Region

By 2012 in the Greater Antilles forest cover ranged from less than 4% for Haiti, to more than 64% for Puerto Rico (González and Scalley 2016). In the Lesser Antilles, most islands range between 36–61% forest cover, with Barbados having 19–31% and St. Lucia and St. Vincent and the Grenadines having more than 64%. Similarly, north of the Lesser Antilles, Bahamas and Turks and Caicos range mostly around 36–61% forest cover.

Another study has found that in St. Kitts, Nevis, Grenada and Barbados, forest cover has increased from 50% to 95% after the 1950s (Helmer et al. 2008). At the same time, cultivated lands have decreased by 59–99%, and developed land area has increased, especially at lower elevations, reflecting a very similar land use change pattern as in Puerto Rico, due to similar reasons. This study also found that these islands, though their humid forest, dry forests and mangroves cover has increased, are under very little protection and therefore are high conservation priorities.

Jamaica and Trinidad and Tobago were found to have net forest loss between 2001 and 2010 (Aide et al. 2013). The same authors also found that Haiti had an increase in woody vegetation during the same time and that overall, the Caribbean had experienced a net gain in woody vegetation between 2001 and 2010.

### *20.1.2 Community Forests*

Motivations behind forest conservation by a community can vary from considering that forests are sacred sites (Ceperley et al. 2010), an opportunity for the economic development of a community, or a vehicle for increased recognition of autonomy, and a sense of conservation commitment (Moeliono et al. 2015; Ruiz-Mallén et al. 2015). After centuries of colonization, and in response to failure of centralized government management of forested lands, during the 1970s and 1980s many countries in the global South and the New World re-adopted or supported small-holder and community-based forestry as it became increasingly recognized in international arenas (Gilmour 2016).

In the same report, Gilmour explains that community-based forestry regimes can range from participatory conservation, where stakeholders do not own the land but participate in some form of decision-making regarding a forest but with little authority, to private ownership by a community. In the private ownership regime form, most rights to access, management and even compensation belong to the households, groups or communities that own the land. Finally, Gilmour (2016) provides estimates of forest lands under some sort of community-based forestry regime globally, which range between 200 and 505 million hectares.

### *20.1.3 Community Forests in Latin America and the Caribbean*

Especially during the 1980s and 1990s, many efforts from indigenous communities to get their territorial autonomy recognized resulted in the establishment and legal recognition of community-forests throughout Latin America (Ruiz-Mallén et al. 2015). México is probably the country with the most internationally recognized community forestry and history, with around 80% of the country's forests under some type of legal jurisdiction by local communities (Gilmour 2016). More recent efforts include the Programa de Manejo Florestal Comunitário e Familiar, established in 2009 in Brazil.

Gilmour (2016) also reported that around 272 million hectares are under a community-based forestry regime in Latin America, although none of the countries listed are in the Caribbean islands. In collaboration with the FAO, the Caribbean Natural Resources Institute developed a synthesis of community-based forestry in the Caribbean (CANARI 2012). In this report, they use 14 case studies in countries or territories like Haiti, Cuba, St. Kitts and Nevis and Grenada. Many of these case studies seem to be in the form of a collaborative regime, where the community has some sort of participation in the management of forests, some type of extraction rights on state-owned forests, or permission of use within a state-owned forest, with some groups requesting autonomy or management control, or on privately-owned land.

All projects include livelihood improvement and capacity-building as primary or secondary objectives. Primary objectives range from watershed rehabilitation to plantation timber production. In Puerto Rico, the Department of Natural Resources has collaborative agreements with different organizations to perform activities related to "restoration, education, ecotourism, agroecology and scientific research" in different protected areas that the state owns (Acuerdos Colaborativos 2015). A very emblematic case of a co-managed forest is Bosque del Pueblo in Adjuntas, Puerto Rico, where the community organization Casa Pueblo successfully halted a mining project in the mountainous interior of the main island and proposed and established a protected forest instead (Massol González et al. 2006).

### 20.1.4 Río Hondo Community Forest: Establishment and History

The Río Hondo Community Forest (RHCF) is located in Barrio Río Hondo of the municipality of Mayagüez, in western Puerto Rico (18°.174455 N, −67°.132376 W). The Río Hondo Community Forest is 27.5 hectares in size. In its recently published land management plan, it is described as a community forest in the rural-urban interface. The land where the forest is now located had been used for growing sugar cane decades prior and later as pasture land for cattle and horses. By 1975, the Barrio Río Hondo land area only had 4% forest cover. By 2010, 95% of the land was covered by secondary forest that is no more than 40 years old after agricultural activity was abandoned. In the Barrio Río Hondo area, urban expansion has increased even though population has declined between 2000 and 2010 (Castro-Prieto et al. 2017).

This land was privately owned and was intended to be turned into a housing project in 2007. Upon learning about this, different community members organized to put a halt to said project. The community started renting the land from the landowners and organizing activities on the land. They originally had intentions of building a theater-museum and other infrastructure as well as dedicating land to serve as demonstration sites of different historical agricultural practices in Puerto Rico. However, that idea later turned into keeping the land as forest. Its main governing body is the Río Hondo Community Forest Board, which is composed of volunteers, mainly residents of the community, who have an interest in maintaining the forest, organizing activities and carrying out all proposed projects within their land management plan.

The RHCF land management plan was created under a proposed co-management agreement between the RHCF Community Board, the Mayagüez Municipality and the United States Forest Service. In 2018, the Mayagüez Municipality and the United States Forest Service, under the Community Forest and Open Space Conservation Program (How the Community Forest Program Works | US Forest Service), provided the funding to formally purchase the entire property (Fig. 20.1). The official title of the land belongs to the Mayagüez Municipality, protecting the land into perpetuity and ensuring that all activities are in line with the statutes established by the Community Forest and Open Space Conservation Program (Figueroa Vázquez, Program Manager for State and Private Forestry for Puerto Rico and the US Virgin Islands, US Forest Service, pers. comm.).

#### 20.1.4.1 Biodiversity in the Río Hondo Community Forest

As described in the land management plan, a total of 23 tree species have been identified in the RHCF (Rodríguez Candelaria et al. Working Document), eight of which are considered introduced and the rest native (Appendix B). Different stands within the forest have different dominant tree species, like *Albizia procera*

**Fig. 20.1** Walking along a trail that runs through an early-successional part of the property bordering households with members of the Community Forest Board and my research assistant. (Photo: Gabriela Morales-Nieves)

(non-native), *Spathodea campanulata* (non-native), *Senna siamea* (non-native), *Cupania americana* (native) and *Guarea guidonia* (native), with a distinct amount of native tree species currently in the understory of all these stands.

Thirty-four bird species have been identified in the forest, nine of which are endemic to Puerto Rico. In addition, 4 reptile species and 3 amphibians (one endemic) have also been identified in the Río Hondo Community Forest (Appendix C).

## 20.2 Methods

I first identified all houses that directly surrounded the forest (134 total) and selected a set of 70 houses randomly using ArcGIS software, aiming for 50 random houses, while making space for those that were possibly unoccupied or non-residential structures (Fig. 20.2). Either through previous agreement or by visiting, I conducted semi-structured interviews to record residents' interests, desires and values for the forest. I gathered basic household data on the size, ages and other social demographics of the household. I asked both closed and open-ended questions. Household residents were asked about different benefits they felt they received from the

**Fig. 20.2** Aerial image of the Río Hondo Community Forest. The green dots represent all potential households that directly border the forest. (Image: ArcMap)

adjacent forest, what opportunities they see in this forest, what activities they would like to be a part of, and what crops and tree species they would choose to plant in the forest. Finally, I used the six goals set in the recently published land management plan for the forest and asked residents to rank these goals from most important to least important for them. The complete goals can be found in Appendix D.

Using MAXQDA software, I categorized the responses to the questions (e.g. positive vs. negative perceptions) to be analyzed quantitatively, following Garen et al. (2009). After coding the responses, I analyzed the proportion of residents that touched on a specific topic in an individual question. Other relevant comments made outside the formal survey were recorded and used to supplement or support formal responses to questions in the survey.

I calculated the means for each goal that I asked residents to rank, using six as the most important and one as the least important goal. I used MINITAB software to perform a one-way ANOVA to analyze if there was a significant difference in how interviewees organized the goals in order of importance; in other words, if there was a common pattern among interviewees in the order of importance of the goals. The ANOVA was followed by a Tukey Pairwise Comparisons test to determine which means were significantly different from which.

## 20.3 Results

Out of the 70 houses I reached out to, I performed interviews in 37 households. Some of the 70 houses were not occupied, were commercial buildings rather than homes, did not have residents there at the moment, or had residents that did not want to be interviewed. Twenty-four of the interviewees identified as female, with the rest (13) identifying as male. Out of the 37 interviewees, 34 of them owned the house they lived in. The average age of interviewees was 59 years (±16, s.d.). The average age time of residence in the household was 32 years (± 21, s.d.). The amount of people living in the household ranged between 1 and 6 people. Including all members in each household surveyed, most people living around the community forest are above 40 years old (Fig. 20.3).

### *20.3.1 Participation and Visits to the Río Hondo Community Forest*

Of the 37 residents that were interviewed, 19 people had visited the forest and 18 had not. However, two of the 18 that had not visited the forest mentioned that another member of the household had visited or participated in activities hosted at the forest. Besides mentioning whether they had visited or not, at least 2 other interviewees expressed that they had spent time in the land before it was declared a protected area,

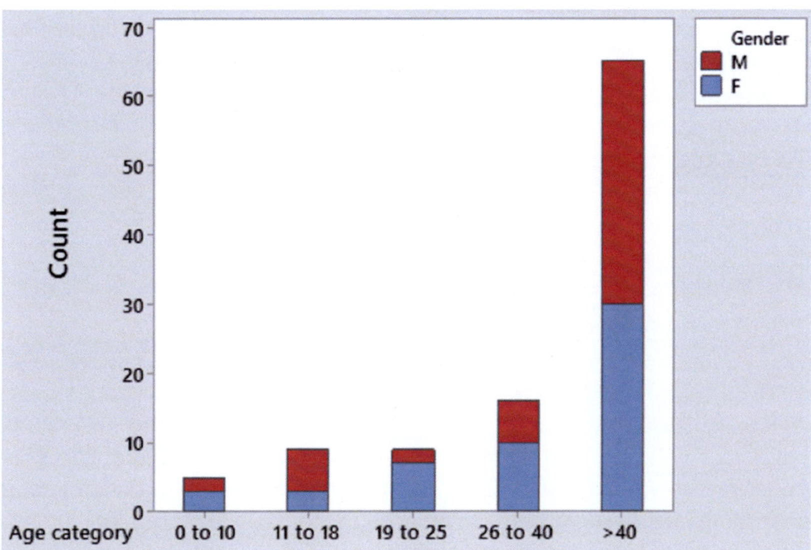


**Fig. 20.3** Age distribution by gender of residents in households surveyed around the Río Hondo Community Forest

since they had lived in the community for many years. There was no difference in responses between visiting or not having visited the forest when analyzed by gender.

In terms of what was the purpose of their visits to the forest, 17 responded it was to participate in specific activities. These responses were grouped into common categories in Table 20.1.

The most notable common activities were social gatherings and parties. These happen mainly in a common, more open area at the entrance to the property. Many residents specifically mentioned a community Christmas party that is thrown every year and added that they enjoyed it. For example, one resident mentioned: "When I most enjoy [going to the forest] is when they bring music and people have fun." It should be noted that some of these activities happen at the same time. Parties on occasion also include interpretative hikes in the trails, for example.

### *20.3.2 Interest in Activities and Other Recommendations*

Seventeen interviewees recommended activities beyond those mentioned among the ones that currently are organized in the forest, most of them related to recreation (Table 20.2). Within this topic, 3 people consistently mentioned the development of trails specifically for mountain biking: "Well I am a cyclist and I think it would be good if they would incentivize cycling in trails".

Within recreation, residents also recommended developing more infrastructure to promote it, like gazebos and resting areas. Residents also recommended social activities like bingos and barbeques. Four of the 5 educational activities recommended were meant for children: "[I would recommend] a summer camp where they show children the value of crops and where it is educational, subsidized by the [Mayagüez] municipality to promote agriculture". Finally, residents recommended to perform different commercial activities, mainly selling forest products like trees and added value products from harvested crops.

### *20.3.3 Benefits and Ecosystem Services*

The most common theme that arose in conversations with interviewees was air quality. Even more so, 5 people specifically referred to the forest as a "lung" to the community: "That forest is a lung for this area. If they didn't protect it, they (original landowners) were going to cut it down to build houses and put concrete." When asked what benefits they felt the forest provided to them or the community, residents, including some that had not actually visited the forest, mentioned a variety, including benefits they may not feel they receive currently, but more so as potential in the future: "It could be a source of employment for people that are unemployed."

**Table 20.1** Categories of responses to the open question regarding for what purpose residents have visited the Río Hondo Community Forest. Each interviewee might have mentioned more than one category

| Social gatherings and parties | Purchasing of products | Informational and Planning Meetings | Educational | Volunteering | Recreation | Arts and workshops |
|---|---|---|---|---|---|---|
| 8 | 7 | 4 | 3 | 3 | 3 | 2 |

**Table 20.2** Interest in activities that the Río Hondo Community currently hosts. Each interviewee might have mentioned more than one category.

| Related to agriculture | Sales and exhibition of crops and crafts | Educational walks or tours | Recreational activities (e.g. hiking, camping) | Related to crafts | Be part of the community forest board | Volunteering |
|---|---|---|---|---|---|---|
| 30 | 28 | 26 | 25 | 17 | 14 | 10 |

Others, though less frequently, mentioned the forest as serving aesthetic purposes ("The forest is beautiful") or as a source of tranquility ("Air purification, for the birds... peacefulness. That land serves for a lot of things"). Finally, 6 residents did not perceive any benefit or service that the forest provided to them or the community, the majority of them because they had not visited or were unfamiliar with the forest.

The different categories in which people mentioned perceived benefits or ecosystem services provided by the forest are summarized in Table 20.3. There is overlap between perceived benefits and observed positive impressions of the forest. However, there were certain comments that expressed general enthusiasm about the forest, like the following:

> "[I would be a] promoter so that the forest always serves a purpose."
>
> "[...] but the forest is wonderful."
>
> "[...] and I love that in there."

## 20.3.4 Negative Impressions and Concerns

Within the different questions and additional comments, concerns or conflicting desires regarding the forest's purpose were also captured. There was also some misinformation in the knowledge people had of the forest. For example, there was not much clarity around where funding for operations came from, who owned the forest and therefore, what activities were allowed or not.

People expressed concerns about certain activities or events being charged to the public and lack of participation of the community in events or meetings. Two main topics surfaced, that will be described in more detail below, which were first, about non-native species or those that are perceived as a nuisance and second, about the concept of cleanliness of the forest.

### 20.3.4.1 Perspectives on Non-native and Non-beneficial Plants Species

In conversation with residents, many of them associated the presence of the forest as housing different species of both plants and wildlife that they consider a nuisance for different reasons. For example, there is a vine that grows in the area, commonly called "picapica" (*Mucuna pruriens,* considered "probably native" which has hairs

**Table 20.3** General categories of perceived benefits or ecosystem services provided by the Río Hondo Community Forest. Interviewees might have mentioned more than one category

| Air quality | Recreation and tourism | Education | Don't know or perceive any | Food and medicine source | Wildlife and habitat | Aesthetics and relaxation | Employment | Physical protection |
|---|---|---|---|---|---|---|---|---|
| 11 | 9 | 6 | 6 | 6 | 5 | 2 | 2 | 1 |

that line the seed pod and that upon contact or release with the wind can attach and cause irritation and itchiness to the skin (Acevedo-Rodríguez 2005). Other species of concern were iguanas (*Iguana iguana*, non-native) and red-tailed boas (*Boa constrictor*, non-native), especially the former as a threat for getting into backyards and eating residents' crops.

Finally, residents mentioned the tree species *Albizia procera* and *Senna siamea* (both referred to as 'Acacia' and both non-native) as a species of little worth or as a nuisance. "Around there what there is is 'picapica', iguanas and 'acacia', which is not even useful for coal." Three residents mentioned that trees bordering the forest could potentially fall after a hurricane and could damage their houses or property due to how close they are to the border with their land. "I have a small ranch and I would like for [the people that manage the forest] to cut down some Acacia trees that may fall on the ranch."

#### 20.3.4.2 "Cleanliness" of the Forest

Another common theme in discussion with residents was that of the forest needing cleaning or fixing in order to fulfill its purpose:

> "If they clean it and they set it well then maybe it can serve as a tourism attraction or if they make more trails."

> "Well if they fix it, it will be a forest that serves for internal tourism. It is abandoned; it's been like that for more than 10 years."

Residents also expressed concern about some areas that have overgrown or that reduce mobility for walking. There was also a desire to reduce the presence of 'picapica' along the borders of the forest and concern about people dumping trash along the road portion that runs through the forest: "Residents have requested them to clean at least the borders [of the forest] to reduce the 'picapica', which is horrible."

### *20.3.5 Ranking of Goals in the Management Plan*

Three people refused to answer the section of ranking goals, two of them explaining that they thought all goals were equally important. On the other hand, there is a significant difference between what the remaining interviewees chose as the most and the least important goals (Table 20.4). When analyzing the Tukey Pairwise Comparisons, the first goal, which referred to maintaining the forest and planting trees for the purpose of improving air quality and reducing pollution, was the highest ranked and significantly different than 3 other goals: one that promoted participation of different sectors in the forest (18c), one that aims to strengthen cultural and social development of the community through the forest (18d), and one that aims to increase resiliency of the community in the face of climate change and economic crises (18e) (Table 20.5). The rest of the goals were not significantly different from each other.

**Table 20.4** Mean ranking for each of the six goals identified in the RHCF's Land Management Plan

| Goal | N | Mean | Standard deviation | 95% CI |
|---|---|---|---|---|
| 18a | 34 | 4.50 | 1.52 | (3.94, 5.06) |
| 18b | 34 | 3.68 | 1.99 | (3.12, 4.24) |
| 18c | 34 | 3.35 | 1.74 | (2.76, 3.89) |
| 18d | 34 | 3.32 | 1.60 | (2.53, 3.65) |
| 18e | 34 | 3.09 | 1.32 | (2.50, 3.62) |
| 18f | 34 | 3.06 | 1.70 | (2.79, 3.91) |
| *Pooled StDev = 1.66* | | | | |

The means are significantly different (see Appendix D for description of goals)

**Table 20.5** Tukey Pairwise Comparisons of all six goals using a 95% confidence interval

| Question | N | Mean | Grouping | |
|---|---|---|---|---|
| 18a | 34 | 4.50 | A | |
| 18b | 34 | 3.68 | A | B |
| 18f | 34 | 3.35 | A | B |
| 18c | 34 | 3.32 | | B |
| 18d | 34 | 3.09 | | B |
| 18e | 34 | 3.06 | | B |

The grouping column represents which goal is different from which. Means that do not share a letter are significantly different. In this case, goal 18a is different from goals 18c, 18d, 18e. Goals 18c, 18d and 18e, however, are not significantly different from each other (see Appendix D for description of goals)

## 20.4 Discussion

The Río Hondo Community Forest is a small protected forest that has undergone a similar pathway of land use history to many other parts of Puerto Rico and the rest of the Caribbean Islands. However, it is the first forest to receive funding from the United States Forest Service's Community Forest and Open Space Conservation Program in the Caribbean to purchase the property and maintain it as protected land, co-managed by its surrounding community. The surrounding community is one where many of their residents have been living for a large part of their lifetime. Therefore, the community forest has the opportunity to build long-lasting relationships with residents in a community that does not have fast turnover.

### 20.4.1 Demographic Trends

The results showed that there is a large number of residents that are above 40 years of age. This is perhaps a reflection of a more country-wide trend, where many Puerto Ricans, especially due to an ongoing economic crisis and consequent migratory wave since 2006 that has been accentuated after the passage of hurricane María, have migrated from Puerto Rico. Even though data by age for these dates are not yet available, between 2013 and 2017, 30.2% of the people that migrated were between 1–19 years old, 58.4% were 20–59 years old and 11.4% were above 60 years old (Junta de Planificación de Puerto Rico 2019).

The Center for Puerto Rican Studies in a research brief explains that "since Hurricane María, migration intensified, especially among families with children" and that "this pattern has shifted the Island's demographic structure". Finally, it reports that "population decline [...] was 129,848 in Puerto Rico immediately following the storm" (Center for Puerto Rican Studies 2019). Age distribution is important to know if the RHCF board wishes to develop activities and curriculum catered to the age of most likely visitors.

### 20.4.2 Addressing Negative Perceptions and Recommendations

Even though I did not ask interviewees why they had not visited the forest, some mentioned the following reasons: fear of wildlife (specifically snakes), not being interested, being busy with work and childcare, not knowing of the existence of the project, not knowing where to look for information, and finally, not perceiving the forest as functional or open to the public. Similarly, other research has found that around half of interviewed local residents have not visited protected areas that are adjacent to their homes (Pérez-Verdin et al. 2004; Moorman 2006).

Regardless of whether they had visited the forest or not, 33 of the 37 interviewees expressed interest in participating in at least one of the activities that the RHCF board already organizes. This implies that rather than lack of interest from community members, perhaps increased communication about the activities, setting weekly hours for visits and making other offerings available could increase participation of residents that surround the community forest. Increasing opportunities of surrounding residents to learn about the ecology of the forest and its establishment may increase their knowledge and valuation of the forest as a protected area (Moorman 2006). This effort of increased communication is a good opportunity since, beyond just participating in activities, 14 interviewees mentioned interest in forming part or getting involved in the development or management activities and events with the RHCF board.

The perception of the forest not being open or functional, or interviewees preferring to visit the forest if it was "clean" has to do with different values that people attribute to natural areas, and probably with aesthetics and what is visually pleasing, as well as with the presence of trash, especially on the sides of the road that crosses the forest. Other studies have demonstrated that attributes such as removal of dead trees, removal of understory, improved visibility, greater crown density, less leaf litter and improved visual distance contribute to perceived aesthetic improvements and landscape quality (Tyrväinen et al. 2003; Chen et al. 2015) and are usually related to factors such as accessibility and security of the area (Tyrväinen et al. 2003).

Ecological and aesthetic management goals can compete with each other, especially when some management for ecological purposes, like improved wildlife habitat, can be perceived as not aesthetically pleasing. The forest therefore might be perceived as not functional and as potentially -but not currently- optimal for tourism and recreation because the current forest structure is heterogeneous and has a dense understory, even more so due to the creation of canopy openings after hurricane María. For the same reason the forest also has many snags, and tree species that have associated negative perceptions, like *Albizia procera.*

Although not documented in the survey, informal conversations with residents shed light in some confusion in terms of the ownership of the forest and what management practices are allowed or prohibited. Some people referred to the forest as belonging to the U.S. federal government and some as belonging to the Mayagüez municipality, and overall they were not quite sure of the role of the community in making decisions on the management of the forest. This should be an important clarification to make in further outreach and communications coming from the RHCF board, as protected areas that are "perceived to have a level of mutual benefit and co-management generate more support toward conservation and recreational use goals" (Buta et al. 2014).

### 20.4.3 Exploring the Potential of Positive Perceptions and Values for Sustaining Conservation

The results on purpose of interviewee's visits to the forest highlight the importance of the forest to be used as a common place for social gatherings for the community. Meanwhile, the results on the activities that residents would go for, shows that many of the activities that the RHCF board already hosts would be attended by residents that surround the forest, especially those that are dedicated to agriculture, sales, education and recreation. Many of the recommendations by interviewees, like increased recreational infrastructure, are present in the goals of the forest's management plan. This highlights that the participatory process of developing this plan captured many interests of the community, which have been emphasized in this research project. However, the implementation of these goals is contingent on sourcing funding and the voluntary labor (for now) of members of the RHCF board and others.

In terms of air quality, it has been found that urban trees can remove air pollution, like particulate matter, ozone, nitrogen dioxide, sulfur dioxide and carbon monoxide, in amounts of around 80 pounds per acre per year (~90 kg/ha) at least in the United States (Nowak and Heisler 2010). Urban trees also have been found to lower local temperatures, which can also result in lower emission of pollutants and improved air quality. The reduction in local temperature also results in lower energy consumption from air conditioning from surrounding households. However, trees can also be a source of pollutants, like pollen and volatile organic compounds (Nowak and Heisler 2010; Leung et al. 2011; Baumgardner et al. 2012). Considering the hard edge of the forest with neighboring households, and in consideration with comments about the "cleanliness" of the forest, the RHCF could contemplate managing the understory at least in the borders of the forest to reduce the presence of species like the "picapica" (*Mucuna pruriens*).

This research also highlighted how much people value the forest for the perceived benefit of air quality improvement. As an interviewee mentioned: "If they didn't protect it, they (original landowners) were going to cut it down to build houses and put concrete", while another mentioned: "Because of the trees, for the air. If you notice, there are no trees in all of this area". This demonstrates that residents value the forest for its mere presence in the area. Different perhaps from many other community forests in Latin America, surrounding residents are not highly dependent on the forest resources for their livelihoods, like for fuelwood, food and medicine. They mostly value it or perceive benefits from more passive uses, like air quality and recreation.

Research in other community forests has shown that they provide opportunities for nature-based recreation and for income either through the direct charging of services associated with the forest or indirectly from visitor expenditure in surrounding businesses (Birch et al. 2014). The RHCF is in a great position to expand recreation as a service of the forest since many of its current activities are based on recreation and education and many residents in this research highlighted recreation, both as a perceived benefit of the forest and as activities that they would participate in.

Beyond these two main services, the forest can also provide other services to the community such as serving as a source of food. Among both perceived services and even more so among activities that residents would visit the forest for, were those that were related to food and medicine. Other community forests, even those that are of small scale such as the RHCF, have been proven to serve as a source of food and addressing food insecurity (Paudel 2018; Bridhikitti and Khadka 2019). In case studies from other community-based forestry projects in the Caribbean, results show that skills, knowledge and capacity building; strengthened organizational capacity and empowerment; and increased financial income, have been the main impacts and benefits (CANARI 2012).

The goal to maintain and improve the quality of life of the community through the conservation of the RHCF as the highest ranked goal highlighted once again the intrinsic value of the presence of the forest for surrounding residents. On the other hand, the competing rankings for the rest of the goals shows that all goals delineated in the management plan are equally important, and no pattern demonstrated other goals to be more important for people, made more evident with two interviewees declining to rank them at all.

Other research has found common attributes that contribute to successful management of community forests and influence surrounding residents to engage in conservation projects. These attributes I would recommend be considered by those managing the RHCF based on my findings and are summarized below:

- Interest and positive attitudes of surrounding residents toward forest management along with strong, open, inclusive, involvement and participation (Pagdee et al. 2006; CANARI 2012; Bridhikitti and Khadka 2019), with an emphasis on youth (CANARI 2012)
- Incentives for forest use for fulfilling basic community needs (Pagdee et al. 2006)
- Market and state economic incentives, human and financial resources to provide a new source of income (Pagdee et al. 2006; Ruiz-Mallén et al. 2015)
- Clear ownership to be able to use and manage the forest (Pagdee et al. 2006)
- Clearly defined forest boundary (Pagdee et al. 2006)
- Strong leadership and effective local organizations (Pagdee et al. 2006; CANARI 2012)
- Continuous transfer of knowledge and use of traditional techniques to harvest forest products (Pagdee et al. 2006)
- A developed sense of place, historical connection, belonging and cultural identity of community members as well as social cohesion (CANARI 2012; Ruiz-Mallén et al. 2015)
- Building trust and maintaining open and frequent communication with government partners for transparency and accountability (CANARI 2012)

The protection of this forest, besides being important for the ecosystem services it provides for people, can also provide benefits for biodiversity in the area. As mentioned by an interviewee, it is the most important ecosystem service of all: "But the most important thing for the community is to leave space for habitat for animals, and for bees to pollinate plants." This forest can be a refuge for wildlife and plant life, even when used for different purposes. Agroforests and managed secondary forests, like the RHCF, have been found to have high plant diversity, including threatened species and often have similar basal area than that of old-growth forests, all while improving local livelihoods (Souza et al. 2016). Perhaps a combined effort on education and demonstration of the uses of non-native species like *Albizia procera* for fuelwood (Lugo et al. 1990) for example, to be then replaced with assisted planting of native species and other desirable species by the community could increase forest use, while improving ecosystem function and services. As mentioned by Castro-Prieto et al. (2017) there seems to be increasing development pressures around the RHCF, therefore its protection as a biodiversity island could be considered increasingly relevant.

## 20.5 Conclusions

People living around the Río Hondo Community Forest in Mayagüez, Puerto Rico valued the forest for its mere presence: as a source of clean air, recreation or socializing, beauty and education. Even though nobody opposed the presence of the forest, some people had concerns about the status of the forest, especially in relation to the presence of wildlife and plant species that are perceived as a nuisance; dense understory, and lack of park-like infrastructure.

Given similar demographic trends of the community to other parts of Puerto Rico and ecologic traits of the forest itself, if this model of community forest were to be replicated in other areas, one could anticipate similar expectations, perceived benefits and potential uses by other communities. However, as the land is guaranteed to remain as forest into perpetuity, the RHCF within their efforts, according with what is already outlined in their management plan, and with additional resources and funding, could expand and teach about other potential or increased uses, like very small-scale timber harvesting, agroforestry, or other. They could also promote more active sources of income for surrounding community residents, like tourism and development of value-added products.

This project is a case study that can be comparable to other forests with similar land use history in many other parts of Puerto Rico and the Caribbean and could be used as an example to be replicated taking into account the findings in this research, along with attributes that make community-based forestry successful. Based on observed wildlife and tree species found, including endemic ones, this forest serves as a biodiversity island within a fragmented landscape in the rural-urban interface in western Puerto Rico.

**Acknowledgements** I would like to thank my advisors, Drs. Florencia Montagnini and Mark Ashton; Dr. Oscar Abelleira for connecting me with this wonderful forest; and Drs. Amity Doolittle, Eva Garen, Kathleen McGinley and Germán Ramos for their additional advice in the design of this project. I also thank Magaly Figueroa, from the US Forest Service's International Institute of Tropical Forestry; Betsy Acevedo and all the members of the RHCF Community Board for their advice, for their time to meet with me and for the great voluntary service they provide to the broader community in managing this forest. I also thank Yale University's Tropical Resources Institute and Sustainable Food Program for providing funding for this research project. Finally, I thank Camila Martínez for her assistance during data collection.

# Appendices

## *Appendix A: Population Density in 2017 of Countries in Latin America (LA) and the Caribbean ©*

| Country name | LA or C | Population density (people per sq. km of land area) |
|---|---|---|
| Saint Maarten (Dutch part) | C | 1193 |
| St. Martin (French part) | C | 672 |
| Barbados | C | 666 |
| Aruba | C | 585 |
| Haiti | C | 398 |
| Puerto Rico | C | 375 |
| Curacao | C | 361 |
| Grenada | C | 326 |
| El Salvador | LA | 308 |
| Virgin Islands (U.S.) | C | 306 |
| St. Lucia | C | 297 |
| St. Vincent and the Grenadines | C | 282 |
| Trinidad and Tobago | C | 270 |
| Jamaica | C | 270 |
| Cayman Islands | C | 264 |
| Dominican Republic | C | 218 |
| Antigua and Barbuda | C | 217 |
| St. Kitts and Nevis | C | 200 |
| British Virgin Islands | C | 197 |
| Guatemala | LA | 158 |
| Cuba | C | 109 |
| Costa Rica | LA | 97 |
| Dominica | C | 95 |
| Honduras | LA | 84 |
| Ecuador | LA | 68 |
| Mexico | LA | 64 |
| Panama | LA | 55 |
| Nicaragua | LA | 53 |
| Colombia | LA | 44 |
| Turks and Caicos Islands | C | 39 |
| The Bahamas | C | 38 |
| Venezuela | LA | 33 |
| Brazil | LA | 25 |
| Chile | LA | 25 |
| Peru | LA | 25 |
| Uruguay | LA | 20 |
| Paraguay | LA | 17 |

(continued)

| Country name | LA or C | Population density (people per sq. km of land area) |
|---|---|---|
| Belize | LA | 16 |
| Argentina | LA | 16 |
| Bolivia | LA | 10 |
| Guyana | LA | 4 |
| Suriname | LA | 4 |

## Appendix B: Tree Species Found in Río Hondo Community Forest (Obtained from Rodríguez Candelaria et al. Working Document)

| | | |
|---|---|---|
| *Albizia procera*[a] | *Ceiba pentandra* | *Randia acuelata* |
| *Andira inermis* | *Cinamomun montanum* | *Schefflera morotononi* |
| *Artocarpus altilis*[a] | *Cordia laevigata* | *Senna siamea*[a] |
| *Casearia aculeata* | *Cupania americana* | *Spathodea campanulata*[a] |
| *Casearia decandra* | *Delonix regia*[a] | *Swietenia macrophylla*[a] |
| *Casearia guianensis* | *Guarea guidonia* | *Terminalia catappa*[a] |
| *Casearia sylvestris* | *Inga laurina* | *Zanthoxylum martinicense* |
| *Cecropia peltata* | *Melicocus bijogatus*[a] | |

[a]Non-native species

## Appendix C: Wildlife Species Observed in the RHCF Between 2009 and 2017 (Obtained from Rodríguez Candelaria et al. Working Document)

| Bird Species | | |
|---|---|---|
| *Icterus dominicensis*[a] | *Turdus plumbeus* | *Mimus polyglottos* |
| *Melanerpes portorricencis*[a] | *Lonchura punctulata* | *Patagioenas squamosa* |
| *Spindalis portorricensis*[a] | *Butorides virescens* | *Coereba flaveola* |
| *Myiarchus antillarum*[a] | *Coccyzus vieilloti*[a] | *Anthracothorax dominicus* |
| *Todus mexicanus*[a] | *Vireo latimeri* | *Falco sparverius* |
| *Dendroica discolor* | *Loxigilla portorricencis*[a] | *Buteo jamaicencis* |
| *Mniotilta varia* | *Vireo altiloquus* | *Anthracothorax viridis*[a] |
| *Parula americana* | *Euplectes franciscanus* | *Tyrannus dominicensis* |
| *Dendroica tigrina* | *Columbina passerina* | *Brotogeris verscicolurus* |

(continued)

| Bird Species | | |
|---|---|---|
| *Icterus icterus* | *Margarops fuscatus* | *Crotophaga ani* |
| *Zenaida asiatica* | *Tiaris bicolor* | *Tyrannus caudifasciatus* |
| *Zenaida aurita* | | |
| Reptiles | | |
| *Ameiva exsul* | *Anolis* spp. | *Iguana iguana* |
| *Alsophis* spp. | | |
| Amphibians | | |
| *Bufo marinus* | *Eleutherodactylus* spp.[a] | *Leptodactylus albilabris* |

[a]Endemic species

## *Appendix D: The Six Goals Identified During the Creation of the RHCF Land Management Plan (Obtained from Rodríguez Candelaria et al. Working Document)*

18a. Maintain and improve the quality of life of the community through the conservation of the RHCF (for example, through the planting and maintenance of trees that improve air quality and reduce pollution).

18b. Provide economic benefits to communities through the generation of employment, maintenance of clean conditions of the forest and being self-sufficient by selling products, services and activities.

18c. Promote the participation of members of the community, community organizations, educational institutions, government entities and environmental conservation organizations.

18d. Strengthen the social development of the community by highlighting the customs and the cultural, historic, folkloric, generational and traditional elements that represent and integrate the community with the objective of creating and contributing to the cultural development.

18e. Increase the resiliency of the community to face the challenges of climate change and economic crises (for example, install a rainwater collection system to share in times of lack of water, etc.)

18f. Promote the forest as a recreational space for the individual, family and community well-being (for example, the maintenance and creation of new walking trails, maintenance of areas for camping, etc.)

## References

Acevedo-Rodríguez P (2005) Vines and climbing plants of Puerto Rico and the Virgin Islands. Smithsonian Inst. Contr US Nat Herb 51:1–483

Acuerdos colaborativos (2015) In: DRNA. http://www.drna.pr.gov/programas-y-proyectos/acuerdos-colaborativos/. Accessed 12 Feb 2020

Aide TM, Clark ML, Grau HR, López-Carr D, Levy MA, Redo D, Bonilla-Moheno M, Riner G, Andrade-Núñez MJ, Muñiz M (2013) Deforestation and reforestation of Latin America and the Caribbean (2001–2010). Biotropica 45:262–271. https://doi.org/10.1111/j.1744-7429.2012.00908.x

Álvarez-Berríos NL, Redo DJ, Aide TM, Clark ML, Grau RH (2013) Land change in the greater Antilles between 2001 and 2010. Land 2:81–107. https://doi.org/10.3390/land2020081

Anadon-Irizarry V, Wege DC, Upgren A, Young R, Boom B, Leon YM, Arias Y, Koenig K, Morales AL, Burke W (2012) Sites for priority biodiversity conservation in the Caribbean Islands biodiversity hotspot. J Threat Taxa:2806–2844. https://doi.org/10.11609/JoTT.o2996.2806-44

Baumgardner D, Varela S, Escobedo FJ, Chacalo A, Ochoa C (2012) The role of a peri-urban forest on air quality improvement in the Mexico City megalopolis. Environ Pollut 163:174–183. https://doi.org/10.1016/j.envpol.2011.12.016

Birch JC, Thapa I, Balmford A, Bradbury RB, Brown C, Butchart SHM, Gurung H, Hughes FMR, Mulligan M, Pandeya B, Peh KS-H, Stattersfield AJ, Walpole M, Thomas DHL (2014) What benefits do community forests provide, and to whom? A rapid assessment of ecosystem services from a Himalayan forest. Nepal Ecosyst Serv 8:118–127. https://doi.org/10.1016/j.ecoser.2014.03.005

Bowler DE, Buyung-Ali LM, Healey JR, Jones JP, Knight TM, Pullin AS (2012) Does community forest management provide global environmental benefits and improve local welfare? Front Ecol Environ 10:29–36. https://doi.org/10.1890/110040

Brandeis TJ, Turner JA (2013) Puerto Rico's forests, 2009. USDA Forest Service, Asheville

Bridhikitti A, Khadka B (2019) Assessing factors to successful management for small-scale community forest under threat of urban growth: in a case of Ban Na Kham Noi community forest, Mukdahan, Thailand. J Sustain For 0:1–17. https://doi.org/10.1080/10549811.2019.1631184

Buta N, Holland SM, Kaplanidou K (2014) Local communities and protected areas: the mediating role of place attachment for pro-environmental civic engagement. J Outdoor Recreat Tour 5–6: 1–10. https://doi.org/10.1016/j.jort.2014.01.001

CANARI (2012) Community forestry in the Caribbean: a regional synthesis. Caribbean Natural Resources Institute (CANARI). Laventille, Trinidad and Tobago

Castro-Prieto J, Gould W, Ortiz-Maldonado C, Soto-Bayó S, Llerandi-Román I, Gaztambide-Arandes S, Quinones M, Cañón M, Jacobs KR (2019) A comprehensive inventory of protected areas and other land conservation mechanisms in Puerto Rico. IITF-Gen Tech Rep 50. IITF, Rio Piedras, Puerto Rico

Castro-Prieto J, Martinuzzi S, Radeloff VC, Helmers DP, Quiñones M, Gould WA (2017) Declining human population but increasing residential development around protected areas in Puerto Rico. Biol Conserv 209:473–481. https://doi.org/10.1016/j.biocon.2017.02.037

Center for Puerto Rican Studies (2019) Puerto Rico two years after hurricane María. Hunter College CUNY, New York

Ceperley N, Montagnini F, Natta A (2010) Significance of sacred sites for riparian forest conservation in Central Benin. Bois Forets Trop 303(1):5–23

Chen Y, Sun B, Liao S, Chen L, Luo S (2015) Landscape perception based on personal attributes in determining the scenic beauty of in-stand natural secondary forests. Ann For Res 59:91–103. https://doi.org/10.15287/afr.2015.440

Dalton T, Forrester G, Pollnac R (2012) Participation, process quality, and performance of marine protected areas in the wider Caribbean. Environ Manag 49:1224–1237. https://doi.org/10.1007/s00267-012-9855-0

DataBank | The World Bank. https://databank.worldbank.org/home. Accessed 22 Jan 2020

Garen EJ, Saltonstall K, Slusser J, Mathias S, Ashton MS, Hall JS (2009) An evaluation of farmers' experiences planting native trees in rural Panama: implications for reforestation with native species in agricultural landscapes. Agrofor Syst 76(1):219–236

Geoghegan T, Renard Y (2002) Beyond community involvement: lessons from the insular Caribbean. Parks 12(2)

Gilmour DA (2016) Forty years of community-based forestry: a review of its extent and effectiveness. FAO, Rome

Gonzalez G, Scalley TH (2016) Building a collaborative network to understand regional forest dynamics and advance forestry initiatives in the Caribbean. Caribb Nat 2016 Spec Issue 1:245–256

Grau HR, Aide TM, Zimmerman JK, Thomlinson JR, Helmer E, Zou X (2003) The ecological consequences of socioeconomic and land-use changes in postagriculture Puerto Rico. Bio Science 53:1159–1168. https://doi.org/10.1641/0006-3568(2003)053[1159:TECOSA]2.0.CO;2

Helmer EH, Kennaway TA, Pedreros DH, Clark ML, Marcano-Vega H, Tieszen LL, Schill SR, Carrington CMS (2008) Land cover and forest formation distributions for St. Kitts, Nevis, St. Eustatius, Grenada and Barbados from decision tree classification of cloud-cleared satellite imagery. Caribb J Sci 44:175–198

Holland G, Bruyère CL (2014) Recent intense hurricane response to global climate change. Clim Dyn 42:617–627. https://doi.org/10.1007/s00382-013-1713-0

How the Community Forest Program Works | US Forest Service. https://www.fs.usda.gov/managing-land/private-land/community-forest/program. Accessed 10 Feb 2020

Junta de Planificación de Puerto Rico (2019) Resumen Económico De Puerto Rico. Junta de Planificación de Puerto Rico, San Juan. jp.gbierno.pr

Knutson TR, Sirutis JJ, Zhao M, Tuleya RE, Bender M, Vecchi GA, Villarini G, Chavas D (2015) Global projections of intense tropical cyclone activity for the late twenty-first century from dynamical downscaling of CMIP5/RCP4.5 scenarios. J Clim 28:7203–7224. https://doi.org/10.1175/JCLI-D-15-0129.1

Lambrick FH, Brown ND, Lawrence A, Bebber DP (2014) Effectiveness of community forestry in prey long forest, Cambodia. Conserv Biol 28:372–381. https://doi.org/10.1111/cobi.12217

Leung DYC, Tsui JKY, Chen F, Yip W-K, Vrijmoed LLP, Liu C-H (2011) Effects of urban vegetation on urban air quality. Landsc Res 36:173–188. https://doi.org/10.1080/01426397.2010.547570

López T d M, Aide TM, Thomlinson JR (2001) Urban expansion and the loss of prime agricultural lands in Puerto Rico. Ambio 30:49–54

Lugo AE, Wang D, Herbert Bormann F (1990) A comparative analysis of biomass production in five tropical tree species. For Ecol Manag 31:153–166. https://doi.org/10.1016/0378-1127(90)90158-8

Marcano-Vega H (2017) Forests of Puerto Rico, 2014. USDA Forest Service, Asheville

Martinuzzi S, Gould WA, Ramos González OM (2007) Land development, land use, and urban sprawl in Puerto Rico integrating remote sensing and population census data. Landsc Urban Plan 79:288–297. https://doi.org/10.1016/j.landurbplan.2006.02.014

Massol González A, González E, Massol Deyá A, Deyá Díaz T, Geoghegan T (2006) Bosque del Pueblo, Puerto Rico: Cómo la lucha antiminera cambió la política forestal desde la base comunitaria. International Institute for Environment and Development, London

McConney P, Pena M (2012) Capacity for (co)management of marine protected areas in the Caribbean. Coast Manag 40:268–278. https://doi.org/10.1080/08920753.2012.677632

Moeliono A, Mulyana M, Adnan A, Manalu H, Yuliani P, Balang L (2015) Village forests (hutan desa): empowerment, business or burden? World Agroforestry Centre (ICRAF), Bogor

Moorman RS (2006) Benefits of local residents visiting La Selva Biological Station, Costa Rica. Environ Conserv 33:89–99. https://doi.org/10.1017/S0376892906002955

Myers N, Mittermeier RA, Mittermeier CG, Fonseca GAB d, Kent J (2000) Biodiversity hotspots for conservation priorities. Nature 403:853–858. https://doi.org/10.1038/35002501

Nowak D, Heisler G (2010) Air quality effects of urban trees and parks. Res Ser Monogr Ashburn VA Natl Recreat Parks Assoc Res Ser Monogr 44:1–44

Pagdee A, Kim Y, Daugherty PJ (2006) What makes community forest management successful: a meta-study from community forests throughout the world. Soc Nat Resour 19:33–52. https://doi.org/10.1080/08941920500323260

Paudel J (2018) Community-managed forests, household fuelwood use and food consumption. Ecol Econ 147:62–73. https://doi.org/10.1016/j.ecolecon.2018.01.003

Pérez-Verdin G, Lee ME, Chávez DJ (2004) Outdoor recreation in a protected area in southern Durango, Mexico: analysis of local residents' perceptions. Soc Nat Resour 17:897–910. https://doi.org/10.1080/08941920490505310

Porter-Bolland L, Ellis EA, Guariguata MR, Ruiz-Mallén I, Negrete-Yankelevich S, Reyes-García V (2012) Community managed forests and forest protected areas: an assessment of their conservation effectiveness across the tropics. For Ecol Manag 268:6–17. https://doi.org/10.1016/j.foreco.2011.05.034

Pulwarty RS, Nurse LA, Trotz UO (2010) Caribbean Islands in a changing climate. Environ Sci Policy Sustain Dev 52:16–27. https://doi.org/10.1080/00139157.2010.522460

Rasolofoson RA, Ferraro PJ, Jenkins CN, Jones JPG (2015) Effectiveness of community forest management at reducing deforestation in Madagascar. Biol Conserv 184:271–277. https://doi.org/10.1016/j.biocon.2015.01.027

Rodríguez Candelaria I, López Méndez C, Rivera Sanantonio J, García Santiago D, Crespo Vélez S, Pérez Méndez A, Abelleira Martínez O, Ramos Cartagena G (n.d.) (Working Document) Plan de Co-Manejo del Bosque Comunitario de Río Hondo en Mayagüez, Puerto Rico: Empresa Comunitaria con Propósito Social, Ambiental, Cultural y Recreativo. Mayagüez, Puerto Rico. Bosque Comunitario de Río Hondo-Mayagüez-Facebook

Ruiz-Mallén I, Schunko C, Corbera E, Rös M, Reyes-García V (2015) Meanings, drivers, and motivations for community-based conservation in Latin America. Ecol Soc 20:3. https://doi.org/10.5751/ES-07733-200333

Shrestha B, Shrestha S (2010) Biodiversity conservation in community forests of Nepal: rhetoric and reality. Int J Biodivers Conserv 2:98–104

Smith SL (2012) Toward inclusive co-management: factors influencing stakeholder participation. Coast Manag 40:327–337. https://doi.org/10.1080/08920753.2012.677642

de Souza SEXF, Vidal E, Chagas G d F, Elgar AT, Brancalion PHS (2016) Ecological outcomes and livelihood benefits of community-managed agroforests and second growth forests in Southeast Brazil. Biotropica 48:868–881. https://doi.org/10.1111/btp.12388

Tyrväinen L, Silvennoinen H, Kolehmainen O (2003) Ecological and aesthetic values in urban forest management. Urban For Urban Green 1:135–149. https://doi.org/10.1078/1618-8667-00014

Wood A, Tolera M, Snell M, O'Hara P, Hailu A (2019) Community forest management (CFM) in south-west Ethiopia: maintaining forests, biodiversity and carbon stocks to support wild coffee conservation. Glob Environ Change 59:1–11. https://doi.org/10.1016/j.gloenvcha.2019.101980

# Chapter 21
# Sacred Church Forests in Northern Ethiopia: Biodiversity and Cultural Islands

**Mabel Baez Schon, Carrie L. Woods, and Catherine L. Cardelús**

**Abstract** Sacred forests, protected due to their religious importance, form a vast network of informal and often inadvertent (shadow) conservation sites worldwide. Despite socioeconomic and political pressures increasing deforestation worldwide, these shadow conservation sites are remarkable in their ability to resist these pressures through their cultural, religious, and social significance. In this chapter we discuss our research on the sacred forests of the Ethiopian Orthodox Tewahido Church (EOTC), the dominant religion in Ethiopia. Ethiopia has 11% forest cover distributed in a patchwork of forest fragments which are, in many regions, the only repository of biodiversity, including many endemic and endangered species. In the South Gondar Region (14,607 km$^2$) of northern Ethiopia, around 1022 of these forest fragments are sacred forests that surround churches of the EOTC. Despite their small size, on average 5.2 hectares (range 1.7–148.9 ha), they constitute 100% of the forests in the region. High levels of biodiversity, including endangered and endemic taxa, endow these forests with great value. They provide essential ecosystem services to the surrounding community and are integral to the rituals and culture of the EOTC. Far from being static cultural and religious relics scattered throughout the Ethiopian landscape, these sacred church forests are complex and dynamic socio-ecological systems. We discuss EOTC forests as islands of biocultural diversity, their central role in local communities, the ecosystem services they provide, and the threats they face. We base our observations on a 10-year (2010–2020) interdisciplinary research project where we explore the mechanisms of the religious management of EOTC church forests.

M. Baez Schon (✉)
School of Forest, Fisheries, and Geomatics Sciences, University of Florida, Gainesville, FL, USA
e-mail: mabelbaez@ufl.edu

C. L. Woods
Department of Biology, University of Puget Sound, Tacoma, WA, USA

C. L. Cardelús
Department of Biology, Colgate University, Hamilton, NY, USA

 
F. Montagnini (ed.), *Biodiversity Islands: Strategies for Conservation in Human-Dominated Environments*, Topics in Biodiversity and Conservation 20,
https://doi.org/10.1007/978-3-030-92234-4_21

**Keywords** Biocultural conservation · Church forests · Ethiopian Orthodox Tewahido Church · Sacred forests · Sacred natural sites · Shadow conservation

# 21.1 Introduction

## *21.1.1 Sacred Natural Sites*

Sacred natural sites (SNS) are defined as "areas of land or water having special spiritual significance to peoples and communities" (Oviedo et al. 2005, p. 3). Sacredness can be interpreted in different ways, and is attributed to natural sites by people across different religious backgrounds and faiths—from indigenous and traditional to institutionalized religions (Verschuuren et al. 2010; Pungetti 2012). As such, the definition of SNS is an open working definition, acknowledging the fact that interpretation of what areas are sacred, and why, is not fixed (Verschuuren et al. 2010). In some indigenous traditions, for example, all nature is sacred and animals and people are related (Harmon and Putney 2003; Pungetti 2012). Sacred natural sites can be rock formations (e.g. Uluru, Australia), mountain ranges (e.g. Mt. Hakusan, Japan), or whole ecosystems (e.g. Tanchara Wetlands, Ghana). Water is an essential resource that is often depicted as sacred (Altman 2002). Sacred water sites include sacred pools (e.g. Benin) (Ceperley et al. 2010), rivers (e.g. the Ganges River, India), lakes, wells, and springs (e.g. sacred fertility springs in Kyrgyzstan) (Altman 2002; Verschuuren and Wild 2012). Natural areas may also have sacred energies, sacred species, sacred ancestral burial grounds, contain individual sacred trees, encompass entire forest groves (e.g. Indian sacred groves), and/or be sacred at specific times of the year (Bhagwat and Rutte 2006; Delgado et al. 2012; Pungetti 2012).

Sacred natural sites are ubiquitous and found on every continent except Antarctica (Bhagwat and Rutte 2006; Rutte 2011). Sacred forests, a type of SNS, are often small in size but together make up an important network of forest conservation areas. These are often referred to as "shadow" conservation areas. Shadow conservation areas are protected indirectly through community use rather than directly through legal protection designation. Notably, they are not protected for the explicit conservation of biodiversity (Bhagwat and Rutte 2006; Cardelús et al. 2017; Kent and Orlowska 2018). Sacred forests are some of the oldest forms of conserved areas and are important for biodiversity at the local, regional, and national scales comparable to, and complementary to, protected areas (PA) (Bhagwat and Rutte 2006; Tengö et al. 2007; Verschuuren et al. 2010). The religious, cultural, and biological importance of sacred forests are well known and documented (Verschuuren et al. 2010; Pungetti et al. 2012).

The origins of sacred forests are as diverse as their various examples. In areas with extensive agricultural development, sacred forests are likely as old as agriculture (Bhagwat and Rutte 2006). As land was cleared for farming, humans set aside parcels of forested land that were considered aesthetically pleasing or were religiously valuable. As populations grew, and agricultural practices intensified, economic and sociocultural pressures increased to use these fragments in a more utilitarian manner. As communities began to use these land areas as resources, benefitting from the ecosystem services that they provided, the cultural services the land could provide were increasingly threatened (Bhagwat and Rutte 2006).

Ecosystem services can be divided into 4 categories—(1) provisioning services (food, water, timber), (2) regulating services (regulation of climate, floods, water quality), (3) cultural services (recreation, aesthetic enjoyment, and spiritual fulfillment), and (4) supporting services (soil formation, photosynthesis, and nutrient cycling) (Millenium Ecosystem Assessment 2010). To prevent the overexploitation of ecosystem services and natural resources that the forest fragments provided, "social fences" grew around them (Bhagwat and Rutte 2006). Social fences are described as informal community conservation mechanisms that prevent overexploitation of forest resources and ensure their perpetuity (Colding and Folke 2001; Bhagwat and Rutte 2006). Thus, while every ecosystem service may not have been the primary purpose for protecting these sacred forests, their very presence had utilitarian value when the landscape changed. This argument is supported by research on the landscape characteristics of protected areas (PAs) and SNS in Italy, in which the network of PAs and SNS showed little overlap. Sacred natural sites were found more often in extensively agricultural and peri-urban settings, while PAs where located farther from human settlements (Frascaroli et al. 2019).

While sacred forests are not officially considered conservation mechanisms in the International Union for Conservation of Nature (IUCN) Protected Area network, sacred forests are a form of community forest management, which currently play, and have traditionally played, an integral role in the conservation of biodiversity and should be considered in conservation initiatives (Bhagwat and Rutte 2006; Dudley and Higgins-Zogib 2012; Montagnini et al. 2022). Their role in conservation of biocultural (biological and cultural) diversity was acknowledged in the 2005 United Nations Educational, Scientific and Cultural Organization (UNESCO) international symposium in Tokyo which resulted in the joint IUCN-UNESCO guidelines for management of SNS (Wild and McLeod 2008). Sacred natural sites are currently recognized under UNESCO'S World Heritage Sites (Wild and McLeod 2008). Ultimately sacred forests are protected due to their spiritual, historical, and cultural value (Bhagwat and Rutte 2006), also providing benefits to the surrounding area by serving as biodiversity islands.

In this chapter we discuss the sacred forests of the Ethiopian Orthodox Tewahido Church (EOTC) (henceforth: EOTC church forests), the dominant religion in Ethiopia. EOTC church forests are islands of biocultural diversity due to their central role in local communities, the ecosystem services they provide, and the threats the otherwise denuded landscape faces. Despite their designation as SNS, they have received comparatively little attention relative to other well-known SNS, such as sacred groves in India (Ormsby and Bhagwat 2010; Notermans et al. 2016) or western Africa (e.g., SNS in Ghana, Ormsby 2012). Our observations are based on extensive ecological and ethnographical field research conducted by an interdisciplinary research team in the South Gondar Administrative Zone (SGAZ), Amhara National Regional State, Ethiopia. We researched 46 church forests over 10 years

(2010–2020) to document and evaluate the relationship of the church forests and the communities that inhabit and depend on them (Cardelús et al. 2013; Klepeis et al. 2016; Scull et al. 2017; Woods et al. 2017; Kent and Orlowska 2018; Orlowska and Klepeis 2018; Cardelús et al. 2019; Woods et al. 2020; Cardelús et al. 2020).

## 21.2 History of Sacred Natural Sites in Ethiopia and Their Shadow Conservation

Sacred church forests of northern Ethiopia have a fascinating history believed to have originated in the fourth century AD (Wassie 2002; Karbo 2013). Community-driven "shadow conservation" through traditional institutions has allowed these sacred forests to persist to today (Dudley et al. 2009; Cardelús et al. 2017). This contrasts with the negative perception that exaggerates historical anthropogenic degradation and deforestation in Ethiopia. The most pervasive story is that forest cover decreased from 40% to 4% in the late nineteenth and early twentieth centuries as detailed by McCann (1997). The first published reference to this dramatic forest loss was published, uncited, in the Food and Agriculture Organization's "Agriculture in Ethiopia" (1961) (McCann 1997). This narrative took hold abroad, as well as within Ethiopia. In 1985, the Ethiopian Relief and Rehabilitation Commission officially connected the Ethiopian famine to the 36% reduction in forest cover from 1885 to 1985. The story of Ethiopia's degradation was directly attributed to Ethiopians' mismanagement of their natural resources. While there is strong evidence of human reduction of forest cover, many of the current forests exist because of human intervention (McCann 1997; Sigman 2022).

The earliest known landscape descriptions of the Ethiopian highlands from the 1830s highlight a lack of forest cover. The presence of trees was not an indication of remnant trees spared from deforestation, but rather of careful management. Today, across the Ethiopian highlands, forests are almost exclusively found with an Ethiopian Orthodox Tewahido Church (EOTC) at their center (Wassie 2002). The persistence of these sacred church forests in the region is because of their shadow conservation by the community (Cardelús et al. 2017; Kent and Orlowska 2018). A precise number of EOTC church forests is unknown; however estimates place them at around 35,000 throughout Ethiopia (McCann 1997; Berhane-Selassie 2008). While Ethiopia is ethnically, culturally, and linguistically diverse, there persists a remarkable sacred connection between the Ethiopian peoples and their forests, a cultural phenomenon in which there is EOTC Christian, non-Christian, and indigenous veneration of sacred trees and forests (Berhane-Selassie 2008; Doffana 2019).

EOTC church forests have proven to be resilient within the landscape. A recent study compared changes in the northern Ethiopian highlands sacred forests and surrounding forest cover using aerial photographs taken during the Italian occupation of Ethiopia (1935–1941) against modern (2014–2016) aerial photographs. While they found a decrease in the forest cover surrounding the sacred forests and adjacent agricultural lands, they did not find a decrease in EOTC church forest size (Scull et al. 2017). During this time period, only four EOTC church forests disappeared. These findings corroborate those of McCann (1997) and show remarkable persistence of forest patches due to local stewardship. Specifically, the gradual deforestation of buffer zones surrounding sacred forests, but not of the sacred forests themselves, highlights the religious value of SNS and their importance to the EOTC community (Scull et al. 2017). The elimination of buffer zones also presents a new threat to these islands of biodiversity as they are more susceptible to "edge effects," which are characterized by increased air temperature, vapor pressure deficits, UV radiation, wind penetration, evapotranspiration, and tree mortality, along with decreased soil moisture and changes in nutrient availability that diminish the productivity and integrity of forests (Kapos 1989; Malcolm 1994; Scariot 2000).

To examine the resilience of EOTC church forests in Ethiopia, it is important to understand the significant social and political changes within Ethiopia since the 1930s. Following the return of emperor Haile Salassie in 1941, marking the end of the Italian occupation (1935–1941), the EOTC experienced a process of centralization. Under Salassie's rule, the church and state were united. In 1974, the Marxist Derg regime overthrew emperor Haile Salassie. While the Derg regime maintained ties to the Ethiopian Church, largely due to its importance to Ethiopians, in 1974 the church and state were officially separated. The Public Ownership of Rural Land Proclamation on 1975 further weakened the EOTC, by ending the landlord system through the nationalization of all rural land (Keeley and Scoones 2000; Ancel and Ficquet 2015; Zerga 2016). This had significant implications for sacred lands throughout Ethiopia, and specifically for the EOTC lands. The proclamation led to the redistribution of non-forested church land to landless individuals, though ultimately land ownership remained in the hands of the State (Keeley and Scoones 2000; Ancel and Ficquet 2015; Zerga 2016).

The restructuring of land tenure systems was a significant blow to EOTC income and land security as 5–50% of productive land used for agriculture and economic activities was lost (Yigremew 2002). Historically, the narrative of degradation reflected and supported a broader movement of land centralization in the name of preventing the overuse of the common resources (Zerga 2016).

The Derg Regime was overthrown by the current government in 1991. Despite promises to change land-tenure systems throughout Ethiopia, under the current government, land tenure has remained essentially the same. The post-revolutionary government introduced an ethnic territoriality system known as "Kilil", which often ignores local land claims, such as those held by the EOTC (Berhane-Selassie 2008). Currently, the EOTC only has usufruct rights to their lands, and these sacred church sites are not officially recognized by the state (Berhane-Selassie 2008). While the church forests *de facto* belong to the church, by law, they belong to the state.

Despite the 1974 separation of church and state, Orthodox Christianity is the most popular religion in Ethiopia (41%), followed by Islam, (35%), Protestantism (20%) and traditional/indigenous religions followed by a minority of the population (3%) (FDRE Population Census Commission 2008). The distribution of religion throughout Ethiopia is clumped, with many regions overwhelmingly Muslim or Orthodox

(FDRE Population Census Commission 2008). Across northern Ethiopia, where our research is based, the majority (83%) of the population is Orthodox Christian and EOTC church forests are at the center of social and religious life; essential for both religious and secular outreach (e.g., FDRE Population Census Commission 2008; Klepeis et al. 2016; Cardelús et al. 2017; Orlowska and Klepeis 2018). We chose Northern Ethiopia as our focus of study for three main reasons: (1) Orthodox Christianity is dominant in the region, (2) all of the region's forests are EOTC church forests (Cardelús et al. 2017) and (3) northern Ethiopia is home to the country's second largest city, Bahir Dar, which finds itself at the center of much of the country's rapid economic and infrastructure development plans (Fig. 21.1).

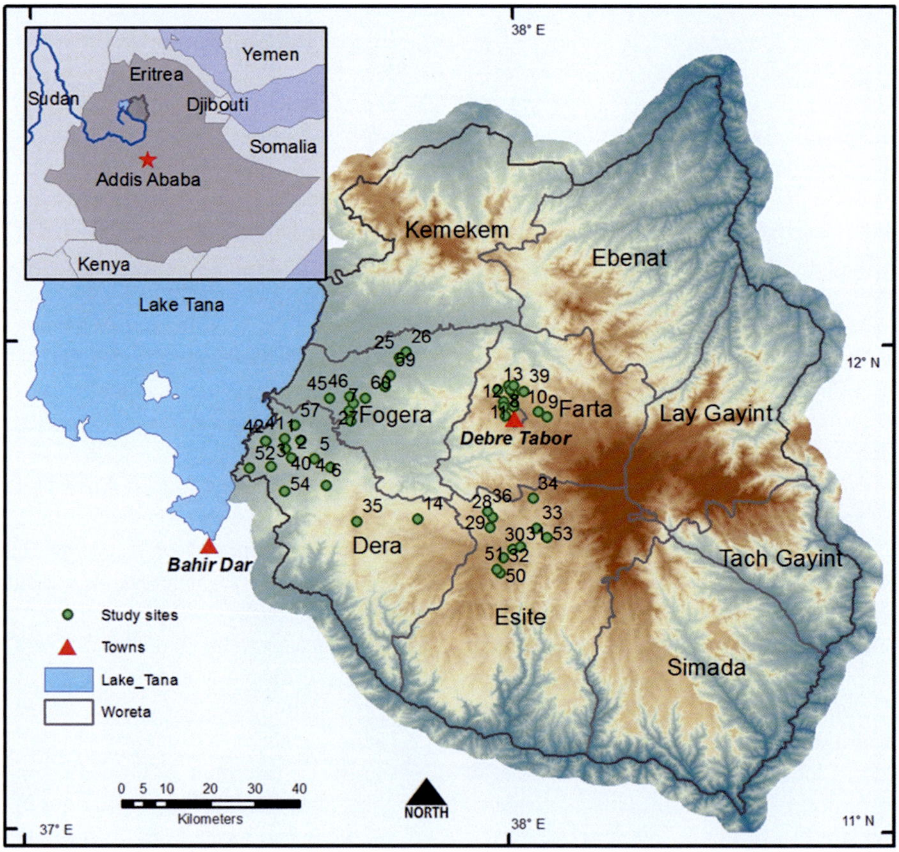


**Fig. 21.1** Map of the study area in South Gondar Administrative Zone (SGAZ), Amhara National Regional State, Ethiopia. (Reproduced from Cardelús et al. (2019))

## 21.3 Community Use of EOTC Church Forests in Northern Ethiopia

### *21.3.1 Religious Functions*

Sacred forests, of the EOTC and other religions, are an integral part of Ethiopian culture and are considered the interface between the divine and the mortal (Berhane-Selassie 2008; Kent and Orlowska 2018). Each sacred forest is believed to be located in a divinely chosen place (Kent and Orlowska 2018). Some stories proclaim that when Jesus was crucified in Jerusalem and his blood dispersed throughout the world, areas anointed with his blood were deemed suitable for the establishment of EOTC churches. Other stories state that the founding priest of each church, while wandering felt a stone or wooden replica of the *tabot*, the Ark of the Covenant given to Moses at Mount Sinai, grow incredibly heavy as they reached the sacred spot upon which the church should be constructed (Klepeis et al. 2016). The main function of the church at the center of the forest is to house the sacred *tabot*.

For the EOTC community, the forests themselves are not the objects of worship, nor are they protected to maintain biodiversity. Rather, their main function is to act as a shield to the church and the *tabot*, which is generally found in the center of the church in the middle of the forest. The forest becomes more sacred as one nears the *tabot* and acts as a shroud to protect it and the church within which it is contained (Klepeis et al. 2013). Thus, in EOTC tradition, the forests themselves may not have the intrinsic religious value often associated with sacred groves in other contexts where individual trees are sacred (Verschuuren et al. 2010).

EOTC church forests are also the resting place for the dead, both men and women (Kent and Orlowska 2018). Within the EOTC tradition, the souls of the dead are not properly at rest until they are buried within a sacred forest (Klepeis et al. 2016). Church forests also have springs with holy water, either inside the forest itself, or on adjacent land. This sacred water is used by community members to cure ailments, from headaches to complex neurological and psychological problems. The strength, and thus capacity for curing ailments, varies among holy springs.

EOTC church forests are also training sites for young priests and provide shelters for monks and nuns (Kent and Orlowska 2018). In the traditional EOTC church schools, young boys are housed in wooden huts within the church forests. They are trained to reach a high level of asceticism, or avoidance of all forms of indulgence by spending hours daily learning Ge'ez, an ancient Semitic language used in EOTC liturgy, practicing EOTC rituals, and begging for food (Kent and Orlowska 2018). Learning the teachings of the EOTC, regardless of whether a boy becomes a priest, is a way to increase their social standing in the community (Kent and Orlowska 2018). EOTC church forests also provide shelter for monks and nuns who renounce all worldly things, fast extensively, and practice sexual abstinence in order to achieve a level of asceticism and purity that makes them closer to the divine (Kent and Orlowska 2018; Orlowska and Klepeis 2018). Women can become nuns only after menopause as menstruation is considered impure (Kent and Orlowska 2018). Monks and nuns are often elders whom turn to the church for shelter and food. In the church, however, as monks and nuns, they maintain their respectability (Orlowska and Klepeis 2018, pg 6).

### 21.3.2 Financial and Social Functions of EOTC Church Forests

The Public Ownership of Rural Lands Proclamation of 1975 decreased the abilities of churches to perform income-generating activities such as selling grass on non-forested lands. This significantly reduced church income, requiring a diversification of church revenue production. In particular, priests became more reliant on donations from the community and sacred forests became supplemental sources of income through the cultivation of fast growing and economically profitable plants such as *Eucalyptus*, *Cupressus*, and coffee (Klepeis et al. 2016; Cardelús et al. 2019). The church forests also provide community members with other products such as honey, medicinal plants, and wood for fuel (Cardelús et al. 2017).

Church forests are critical areas for community interactions. The forest has a series of paths and clearings where community members congregate throughout the week to meet in "Mehabirs", or social subgroups of the congregation. Areas where Mehabirs meet are called "Mehabir clearings". These gathering spaces are often the only shaded meeting spots in the community. Mehabir meetings vary in purpose from planning burials to hosting feasts (Klepeis et al. 2016). Throughout the week, community members use the space to seek peace, preach, and volunteer for community service. Due to the centrality of church forests in communities, government stakeholders also use these areas for community outreach, including for health campaigns (Fig. 21.2) (Orlowska and Klepeis 2018) and even the local HIV positive support group (Authors' Pers. Obs.). Overall, Klepeis et al. (2016) describe these forests as the centers of the community, a place of interaction between humans and the divine, a meeting place for the community, and a place that facilitates interactions among community members and secular authorities. EOTC church forests are therefore important for the conservation of forests, as well as for the role they play as anchors that sustain a community's cultural identity, making them biocultural islands (Bhagwat and Rutte 2006; Cardelús et al. 2012). As such, despite the top-down nature of the EOTC in Ethiopia, each EOTC church forest is unique due to its specific uses by the surrounding community.

### 21.3.3 Governance

Forest use is regulated in three main ways—through "wugz" or "wugzet" (a curse), by hired guards, or by community enforcement (Klepeis et al. 2016; Kent and

**Fig. 21.2** An EOTC church forest near Bahir Dar illustrates how sacred forests can function as the centers of the community. (Photo: C. Cardelús)

Orlowska 2018). The "wugzet" is a curse executed by the priests which stipulates that any transgressor of forest use would face the wrath of God (Klepeis et al. 2016). Protection mechanisms also include fines, time in jail and excommunication, with punishments escalating. Other deterrents come in the form of official fines for cutting trees, allowing cattle into the forest, etc. These rules are not evenly enforced among forests, illustrating the strong effect of community on forest management (Klepeis et al. 2016; Reynolds et al. 2017). In fact, variation in plant species composition was also found to vary across "Woreda" (or district) regardless of size, which indicates a strong and direct connection between the local community management of their church forests, their species composition, and the health and long-term persistence of these forests (Woods et al. 2020).

Given the historical structuring of governance of the EOTC, in which the Patriarch of the church holds the most power, there is a hybrid top-down and bottom-up forest governance system that relies on both the Patriarch and the locals (Klepeis et al. 2016). EOTC church forests have continued to serve as biodiversity refuges integral to the Ethiopian cultural identity (Berhane-Selassie 2008). It is evident that religion in this case acts as an effective "social fence" (Bhagwat and

Rutte 2006), and the combination of religious and utilitarian worth makes them valuable to the community, ensuring their preservation.

## 21.4 Landscape Analysis Status

### *21.4.1 EOTC Church Forest Status and Edge Effects*

Across the South Gondar Administrative Zone (SGAZ) of northern Ethiopia (14,607 km$^2$), there are around 1022 EOTC forests; surprisingly, these forests are almost uniformly distributed at around 2 km intervals (Cardelús et al. 2013). While these forests are protected, their small average size (approximately 5 ha) and high level of human activity has resulted in high disturbance; on average forests have 56% disturbance, characterized by trails, structures, graves, plantations (e.g., coffee, *Eucalyptus*), etc. (Cardelús et al. 2019). Across 46 sacred forests, disturbance was not found to be correlated with proximity to population centers but was correlated with forest size: the smaller the forest, the greater the disturbance. This disturbance negatively impacts plant species richness, forest biomass, and forest regeneration (Cardelús et al. 2019). Despite the high levels of disturbance, EOTC church forests are valuable repositories of biodiversity in an otherwise denuded landscape (Aerts et al. 2016; Woods et al. 2020). The resilience of these forests is compromised by the slow decay of the surrounding matrix of shrubs and remnant trees, not by deforestation within the forest itself (Fig. 21.3; Scull et al. 2017).

The resilience of forests in South Gondar is also seen in other EOTC church forests in South Western highlands where maintenance of the church forests has continued to be demonstrated despite forest area decreasing by 37% concurrently with agriculture increasing by 109% (1995–2010) (Daye and Healey 2015). The loss of the landscape matrix, which acted as limited buffers, exposes these forests to greater edge effects, threatening their integrity and regeneration capacity. Furthermore, it eliminates refugia for pollinators and dispersers, such as perch sites for birds, and reduces movement of dispersers between forests, limiting the dispersal of fruits and seeds, compromising forest regeneration (Cordeiro et al. 2009; Woods et al. 2020). While until now sacredness was enough to protect these forests, the loss of the matrix is compromising the forests and the communities that rely on them.

### *21.4.2 Forest Regeneration*

These church forests are the remaining "Noah's arks" for many indigenous species in the region. Up to 168 woody plant species have been recorded in these forests, 95% of which are endemic (Wassie et al. 2010). However, the presence of large

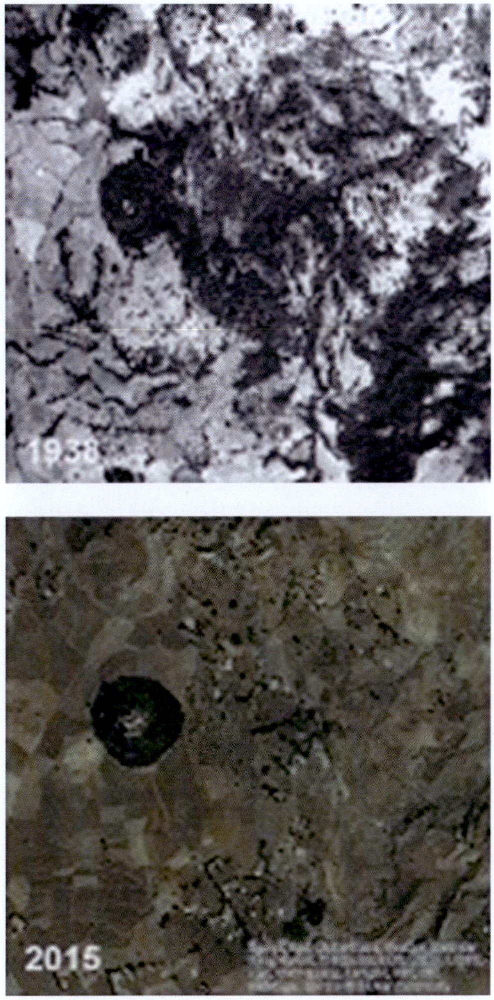


**Fig. 21.3** An example of an EOTC sacred forest in the South Gondar Administrative Zone (SGAZ) (38.028°N, 11.903°E) shown in 1938 and 2015 (DigitalGlobe, by way of ESRI). Note there is minimal decline of woody biomass within the church forest compared to a significant decline in woody biomass in the area surrounding the church forest. Scale is 1:10,000. (Source: Scull et al. 2017)

canopy trees could be a mirage of a healthy forest if there are no seedlings of old-growth trees in the understory.

Tree regeneration in these forests is compromised by human activity and cattle entering the forests off the main trails (Cardelús et al. 2019; Wassie et al. 2010), as well as by the limited number of tree species represented in the seed bank (Teketay and Granström 1995; Wassie and Teketay 2006). In response to these disturbances, and to better demarcate church land, some EOTC church forests have built a small wall around their church, which has resulted in a marginally healthier seedling community (Woods et al. 2017).

## 21.5 Threats to EOTC Church Forests: Ecological, Social, and Cultural

Ethiopian Orthodox Tewahido church forests are important examples of how SNS can be islands of biocultural diversity in human modified landscapes. As with many SNS that have been historically protected, shifts in the social, cultural, and economic landscapes can challenge the very traditional institutions which have ensured their longstanding protection. Ethiopia has one of the fastest growing economies in the world (Shiferaw 2017) with significant investment from foreign governments, particularly China (Chakrabarty 2016; US Department of State 2019). This has led to an increase in infrastructure projects throughout the country, including roads as well as a planned highway that would connect Djibouti with Bahir Dar in the South Gondar Administrative Zone, where our research takes place (Chakrabarty 2016; Cheru 2016).

These shifts in economic development have the potential to impact sacred natural sites in Ethiopia through their effect on the local population (Chandrakanth et al. 2004; Daye and Healey 2015; Reynolds et al. 2017). For example, in a chronological study (2002 and 2014) of four EOTC church forests in the South Gondar Administrative Zone, Reynolds et al. (2017) found that increasing wealth is often associated with more elaborate cement burial tombs and larger churches. These changes increase disturbance within the already small church forests (Klepeis et al. 2016; Reynolds et al. 2017). The replacement of native trees with economically beneficial *Eucalyptus* around and within church forests has reduced soil quality and compromised future forest expansion (Cardelús et al. 2019; Cardelús et al. 2020). We also found native species planted within the forests for economic benefit, including *Juniperus*, used as building material, coffee (*Coffea*), and *Rhammus* which is used for making the traditional beer called "Tella" (Cardelús et al. 2019). However, due to the small size of EOTC church forests, even low levels of disturbance have significant effect on their ecological integrity and regenerative capacity.

Alongside rapid economic growth are social and cultural changes (Reynolds et al. 2017). In the EOTC church forest context, the most notable example of weakening community protection of church forests is the shift away from the expectations that local communities and church leaders are responsible for forest protection to the expectancy that church leaders and secular authorities are in charge (Reynolds et al. 2017). Due to their isolation, high levels of disturbance, and low seedling recruitment, EOTC church forests may be ecologically "living dead" such that when the last large trees die, so will the forests (Cardelús et al. 2019). Despite this, EOTC church forests are still at the center of social and religious life. Church forests still have several factors aiding in their protection. They have "grassroots participation, have sociocultural legitimacy, and have long demonstrated ecological efficacy" (Sheridan 2008, pg 10), which are all essential for sustainable conservation.

## 21.6 Global SNS: Social, Economic, and Cultural Threats and Opportunities

Sacred natural sites (SNS) historically have served as successful forms of local biocultural conservation, their protection rooted in local beliefs and the strength of their religious leaders and guardians. However, SNS are not static and have adapted and changed responding to economic, social and ecological pressures (Sheridan and Nyamweru 2008; Ormsby 2012; Bossou et al. 2020). For some SNS these changes have brought increased protection (See: Ormsby 2012), while many SNS are threatened by social and religious change through the displacement of traditional beliefs and power as well as by ambiguous property rights (Chandrakanth et al. 2004; Tengö et al. 2007; Ormsby 2012; Bossou et al. 2020).

The dynamic nature of SNS can provide sources of adaptation that allow for continued or even increased protection, despite ongoing social and cultural changes that may threaten traditional beliefs and power. In some cases, external support may strengthen local customs (Ormsby and Edelman 2010; Ormsby 2012). An example of this is the Tafi Atome Monkey Sanctuary, in Ghana (Ormsby and Edelman 2010; Ormsby 2012). According to the locals, upon migrating to Tafi Atome they brought with them their fetish, a local god, which they placed within the forest. This forest immediately became sacred and protected by the community. Soon thereafter villagers noticed monkeys had followed them to their new home, a subspecies of the mona monkey (*Cercopithecus mona mona*). The villagers in Tafi Atome believed that the monkeys were representatives of the gods and therefore considered them sacred and taboo to kill them. The sacred forest was protected for more than 200 years; however, the arrival of Christianity directly contradicted these local customs. In an attempt to directly challenge traditional religious beliefs, local priests encouraged the killing of the sacred monkeys and exploitation of the forest for personal economic gains. In 1996, in order to provide a counter economic incentive to protect the forest, the Taffi Atome community in collaboration with the Ghana Tourist board and the Nature Conservation Research Centre—a leader in developing rural ecotourism and community protected areas— established the Tafi Atome Monkey Sanctuary. The ecotourism project reaffirms local traditions, protects the sacred forest and the monkeys, and is also an economic boom for the town (Ormsby and Edelman 2010; Ormsby 2012). This is an example of an SNS which, with external help, stopped the erosion of local traditions and increased protection of their sacred forest.

Sacred natural sites have the capacity to adapt and persist on the landscape; it is important to note, however, that physical persistence does not guarantee ecological integrity. A well-known example of this is the Ganges River of India. The Ganges River is one of the most sacred rivers in Hinduism and considered the physical manifestation of the divine goddess, Ganga, also providing water to 40% of India's population (Das and Tamminga 2012; Sachdeva 2017). The Ganges Basin is one of the most populated areas of the world, and lack of widespread and adequate sanitation infrastructure makes the Ganges one of the most heavily polluted rivers in the world and a potential health hazard to users (Das and Tamminga 2012). Despite the high levels of pollution, the Ganges remains sacred—bathing in it, for example, is thought to cleanse the soul (Sachdeva 2017). The juxtaposition of pollution and sacredness can prove challenging to conservation when the ecological integrity of the SNS is disconnected from their sanctity in a way that can challenge their very existence.

## 21.7 The Role of SNS in Preserving Biodiversity: Conservation and Management

Ethiopian Orthodox Tewahido church forests are central to the religious, ecological, and social fabric of the Ethopian Orthodox Tewahido church. Far from being fixed cultural and religious relics spread throughout the Ethiopian landscape, these sacred forests are complex and dynamic socio-ecological systems (Sheridan 2008). The role of these forests within Ethiopia reflects larger global patterns whereby SNS have successfully preserved biocultural diversity through protection by and for their communities (Verschuuren et al. 2010).

The role of SNS in preserving biodiversity through culturally legitimate and successful community forest management practices make them attractive to those who seek to increase the network of protected habitats (Sheridan 2008; Wild and McLeod 2008; Verschuuren et al. 2010). However, SNS are often inadvertent (shadow) conservation sites which may limit their ability to protect biodiversity and ecosystem services. Conservation initiatives involving SNS need to be collaborative, anchored within the community, involve all relevant stakeholders, and follow guidelines such as IUCN'S *Sacred Natural Sites: Guidelines for Protected Areas Managers* (Wild and McLeod 2008, (Bossou et al. 2020). This approach would ensure that conservationists take into account the traditional beliefs that have historically protected these forests and are important for the local cultural identity.

Sacred groves across Africa maintain unique and diverse communities of flora and fauna. In Guinea-Bissau in western Africa, sacred groves hold a unique avifauna of forest specialists and insectivorous birds relative to other forest types in the region (Kühnert et al. 2019) and in Benin, riparian forests around sacred pools contain a diversity of plant species endemic to western Africa (Ceperley et al. 2010). However, sacred groves are currently under threat from increased demand for arable land due to increasing human populations, changes to cultural traditions and beliefs, and climate change (Ormsby and Bhagwat 2010). Agricultural expansion remains the largest threat to the conservation of sacred groves. In Benin, a replacement of shifting cultivation in forest patches to cashew as a cash crop has resulted in a reduction in biodiversity and increased food insecurity (Temudo and Abrantes 2014).

The maintenance of local-level resource management through continued traditional beliefs and social cohesion rather than the reliance on external development or conservation initiatives are essential to the conservation of sacred groves. Examples of successful local-level resource management include, for example, the sacred forest of Nvangkpe on the border of Nigeria and Cameroon which has the largest and rarest trees in the region and has survived because it is valued for its spiritual, ancestral and magical worth amidst the pressures of agricultural expansion (Kamanda et al. 2003). In Benin, the appreciation of sacred pools by local communities and the continued education of newly arrived migrants about social norms has conserved the riparian forests around these pools from overfishing and cattle grazing (Ceperley et al. 2010). These conservation successes reinforce the importance of local custom and local reinforcement of traditions.

However, conservation and forest management in the absence of community input and inclusion of indigenous narratives of protection will likely not succeed (Bongers et al. 2006; Orlowska and Klepeis 2018; Temudo 2012). There are many cautionary tales in which good intentions were not successful. For example in Guinea-Bissau in western Africa, the intention of Non-Governmental Organization (NGO) conservationists to create a National Park (known as "fortress conservation") led to a weakening of local institutions and social cohesion because local conservation management practices were denied and local stakeholders fought each other to secure their livelihoods (Temudo 2012). Thus, the continued practices of local communities for sacred forest protection must be supported.

## 21.8 Conclusion

Sacred natural sites may not widely be considered as important contributors to biodiversity conservation, but in some regions where religion occupies a central role, such as the South Gondar Region of northern Ethiopia, they are biodiversity islands. Due to their spiritual status, sacred forests in South Gondar are readily protected by the Ethiopian Orthodox Tewahido Church and the surrounding communities, which has allowed forest cover to remain at relatively constant levels as nearby land has been cleared for a growing agriculture economy. While SNS are not primarily protected because of ecological concerns, they do offer a potentially successful alternative to more traditionally recognized conservation strategies, particularly in regions where there may be few options. However, the successful conservation requires cooperation between the church, nearby communities, and any other relevant groups such as government actors who have ties to the SNS. Without involvement from all parties that rely on or use the land and ensure its protection, the rising economic and sociocultural pressures of the region will lead to the degradation of the few remaining islands of biodiversity in Ethiopia, potentially having serious implications to the land's environmental and spiritual significance.

**Acknowledgements** We are grateful to the respected Ethiopian priests and monks who allowed us access to their sacred church forests. We would like to acknowledge the students of the Cardelús and Scull labs at Colgate University, the Woods lab at the University of Puget Sound, and the

Tsegay lab at the University of Bahir Dar, Ethiopia, for their contributions to the lab and field work. We want to thank Ben Everett-Lane for his editorial work and feedback on our revisions. We also want to thank our reviewers Dr. Alison Ormsby and Dr. Keith Kirby. This work was funded through the Dynamics of Coupled Natural Human Ecosystems Program of the National Science Foundation [Award No. 1518501] and by the National Science Foundation Graduate Research Fellowship [Award No. DGE 1842473]. Initial work was funded by The Picker Interdisciplinary Science Institute at Colgate University: http://www.colgate.edu/centers-and-institutes/picker-interdisciplinary-science-institute.

## References

Aerts R, Van Overtveld K, November E, Wassie A, Abiyu A, Demissew S, Daye DD, Giday K, Haile M, TewoldeBerhan S, Teketay D, Teklehaimanot Z, Binggeli P, Deckers J, Friis I, Gratzer G, Hermy M, Heyn M, Honnay O, Paris M, Sterck FJ, Muys B, Bongers F, Healey JR (2016) Conservation of the Ethiopian church forests: threats, opportunities and implications for their management. Sci Total Environ 551–552:404–414. https://doi.org/10.1016/j.scitotenv.2016.02.034

Altman N (2002) Sacred water: the spiritual source of life. Paulist Press

Ancel S, Ficquet É (2015) The Ethiopian orthodox Tewahido church (EOTC) and the challenges of modernity. In: Prunier G, Ficquet E (eds) Understanding contemporary Ethiopia: monarchy, revolution, and the legacy of Meles Zenawi. Hurst and Company, London, pp 65–91

Berhane-Selassie T (2008) The socio-politics of Ethiopian sacred groves. In: Sheridan MJ, Nyamweru C (eds) African sacred groves: ecological dynamics and social change. Ohio University Press, Athens

Bhagwat SA, Rutte C (2006) Sacred groves: potential for biodiversity management. Front Ecol Environ 4:519–524. https://doi.org/10.1890/1540-9295(2006)4[519:SGPFBM]2.0.CO;2

Bongers F, Wassie A, Sterck F, Bekele T, Teketay D (2006) Ecological restoration and church forests in northern Ethiopia. J Dry Lands 1:35–44

Bossou B, Lokossou O, Assongba M, Agbangla M (2020) Saving the sacred. Yokohama

Cardelús CL, Lowman MD, Eshete AW (2012) Uniting church and science for conservation. Science 80(335):2006–2008

Cardelús CL, Scull P, Hair J, Baimas-George M, Lowman MD, Eshete AW (2013) A preliminary assessment of ethiopian sacred grove status at the landscape and ecosystem scales. Diversity 5: 320–334. https://doi.org/10.3390/d5020320

Cardelús CL, Scull P, Wassie Eshete A, Woods CL, Klepeis P, Kent E, Orlowska I (2017) Shadow conservation and the persistence of sacred church forests in northern Ethiopia. Biotropica 49: 726–733. https://doi.org/10.1111/btp.12431

Cardelús CL, Woods CL, Mekonnen AB, Dexter S, Scull P, Tsegay BA (2019) Human disturbance impacts the integrity of sacred church forests, Ethiopia. PLoS One 14:1–14. https://doi.org/10.1371/journal.pone.0212430

Cardelús CL, Mekonnen AB, Jensen KH, Woods CL, Baez MC, Montufar M, Bazany K, Tsegay BA, Scull P, Peck WH (2020) Edge effects and human disturbance influence soil physical and chemical properties in sacred church forests in Ethiopia. Plant Soil 329–342. https://doi.org/10.1007/s11104-020-04595-0

Ceperley N, Montagnini F, Natta A (2010) Significance of sacred sites for riparian forest conservation in Central Benin. Bois Forets des Trop 303(1):5–23. https://doi.org/10.19182/bft2010.303.a20450

Chakrabarty M (2016) Ethiopia – China economic relations. A Classic Win – Win Situation? 7: 226–248

Chandrakanth MG, Bhat MG, Accavva MS (2004) Socio-economic changes and sacred groves in South India: protecting a community-based resource management institution. Nat Resour Forum 28:102–111. https://doi.org/10.1111/j.1477-8947.2004.00077.x

Cheru F (2016) Emerging southern powers and new forms of South–South cooperation: Ethiopia's strategic engagement with China and India. Third World Q 37:592–610. https://doi.org/10.1080/01436597.2015.1116368

Colding J, Folke C (2001) Social taboos: "invisible" systems of local resource management and biological conservation. Ecol Appl 11:584–600. https://doi.org/10.1890/1051-0761(2001)011[0584:STISOL]2.0.CO;2

Cordeiro NJ, Ndangalasi HJ, McEntee JP, Howe HF (2009) Disperser limitation and recruitment of an endemic African tree in a fragmented landscape. Ecology 90:1030–1041. https://doi.org/10.1890/07-1208.1

Das P, Tamminga KR (2012) The ganges and the GAP: an assessment of efforts to clean a sacred river. Sustainability 4:1647–1668. https://doi.org/10.3390/su4081647

Daye DD, Healey JR (2015) Impacts of land-use change on sacred forests at the landscape scale. Glob Ecol Conserv 3:349–358. https://doi.org/10.1016/j.gecco.2014.12.009

Delgado F, Escobar C, Verschuuren B, Hiemstra W (2012) Sacred natural sites, biodiversity and Well-being: the role of sacred sites in endogenous development in the COMPAS network. Sacred Nat Sites Conserv Nat Cult 188–197. https://doi.org/10.4324/9781849776639

Doffana ZD (2019) Sacred sites and ancestor veneration in Sidama, Southwest Ethiopia: a socio-ecological perspective. Cogent Soc Sci 5. https://doi.org/10.1080/23311886.2019.1704600

Dudley N, Higgins-Zogib L (2012) Protected areas and sacred nature: a convergence of beliefs. In: Pungetti G, Oviedo G, Hooke D (eds) Sacred species and sites: advances in biocultural conservation. Cambridge University Press, New York, pp 36–45

Dudley N, Higgins-Zogib L, Mansourian S (2009) The links between protected areas, faiths, and sacred natural sites. Conserv Biol 23:568–577. https://doi.org/10.1111/j.1523-1739.2009.01201.x

FDRE Population Census Commission (2008) Summary and statistical report of the 2007 population and housing census: population size by age and sex. Addis Ababa

Frascaroli F, Zannini P, Acosta ATR, Chiarucci A, d'Agostino M, Nascimbene J (2019) Sacred natural sites in Italy have landscape characteristics complementary to protected areas: implications for policy and planning. Appl Geogr 113:102100. https://doi.org/10.1016/j.apgeog.2019.102100

Harmon D, Putney A (2003) The full value of parks: from the economics to the intangible. Rowman & Littlefield Publishing Group, Lanham

Kamanda B, Angu AK, Nguinguiri J-C (2003) The social value of the Nyangkpe scared forests of south West Province of Cameroon. In: Harmon D, Putney AD (eds) The full value of parks: from economics to intangible. Rowman & Littlefield Publishing Group, Lanham, pp 77–89

Kapos V (1989) Effects of isolation on the water status of Forest patches in the Brazilian Amazon. J Trop Ecol 5:173–185

Karbo T (2013) Religion and social cohesion in Ethiopia. Int J Peace Dev Stud 4:43–52. https://doi.org/10.5897/ijpds2013.0164

Keeley J, Scoones I (2000) Knowledge, power and politics: the environmental policy-making process in Ethiopia. J Mod Afr Stud 38:89–120. https://doi.org/10.1017/S0022278X99003262

Kent EF, Orlowska I (2018) Accidental environmentalists. Worldviews Environ Cult Relig 22:113–142. https://doi.org/10.1163/15685357-02201101

Klepeis P, Scull P, Lalonde T, Svajlenka N, Gill N (2013) Changing forest recovery dynamics in the northeastern United States. Area 45:239–248. https://doi.org/10.1111/area.12016

Klepeis P, Orlowska IA, Kent EF, Cardelús CL, Scull P, Wassie Eshete A, Woods C (2016) Ethiopian church forests: a hybrid model of protection. Hum Ecol 44:715–730. https://doi.org/10.1007/s10745-016-9868-z

Kühnert K, Grass I, Waltert M (2019) Sacred groves hold distinct bird assemblages within an Afrotropical savanna. Glob Ecol Conserv 18:e00656. https://doi.org/10.1016/j.gecco.2019.e00656

Malcolm JR (1994) Edge effects in central Amazonian Forest fragments author. Ecol Soc Am 75: 2438–2445

McCann JC (1997) The plow and the Forest: narratives of deforestation in Ethiopia, 1840-1992. Environ Hist 2:138–159

Millenium Ecosystem Assessment (2010) Ecosystems and human Well-being: biodiversity synthesis. Ecosystems 285. https://doi.org/10.1057/9780230625600

Montagnini F, Levin B, Berg KE (2022) Introduction. Biodiversity Islands: strategies for conservation in human-dominated environments. In: Montagnini F (ed) Biodiversity islands: strategies for conservation in human-dominated environments, Series: Topics in Biodiversity and Conservation. Springer, Cham, pp 1–35

Notermans C, Nugteren A, Sunny S (2016) The changing landscape of sacred groves in Kerala (India): a critical view on the role of religion in nature conservation. Religions 7. https://doi.org/10.3390/REL7040038

Orlowska I, Klepeis P (2018) Ethiopian church forests: a socio-religious conservation model under change. J East African Stud 12:674–695. https://doi.org/10.1080/17531055.2018.1519659

Ormsby A (2012) Cultural and conservation values of sacred forests in Ghana. In: Sacred species and sites: advances in biocultural conservation. Cambridge Univ Press, Cambridge, pp 335–350

Ormsby AA, Bhagwat SA (2010) Sacred forests of India: a strong tradition of community-based natural resource management. Environ Conserv 37:320–326. https://doi.org/10.1017/S0376892910000561

Ormsby A, Edelman C (2010) Community-based ecotourism at Tafi atome monkey sanctuary, a sacred natural site in Ghana. In: Pungetti G, Oviedo G, Hooke D (eds) Sacred natural sites: conserving nature and culture. EarthScan, London, pp 233–243

Oviedo G, Jeanrenaud S, Otegui M (2005) Protecting sacred natural sites of indigenous and traditional peoples: an IUCN perspective

Pungetti G (2012) Sacred species and sites: dichotomies, concepts and new directions in biocultural diversity conservation. In: Pungetti G, Oviedo G, Hooke D (eds) Sacred species and sites: advances in biocultural conservation. Cambridge University Press, New York, pp 13–27

Pungetti G, Oviedo G, Hooke D (eds) (2012) Sacred species and sites advances in biocultural conservation. Cambridge University Press, New York

Reynolds TW, Stave KA, Sisay TS, Eshete AW (2017) Changes in community perspectives on the roles and rules of church forests in northern Ethiopia: evidence from a panel survey of four Ethiopian orthodox communities. Int J Commons 11(355–387). https://doi.org/10.18352/ijc.707

Rutte C (2011) The sacred commons: conflicts and solutions of resource management in sacred natural sites. Biol Conserv 144:2387–2394. https://doi.org/10.1016/j.biocon.2011.06.017

Sachdeva S (2017) The influence of sacred beliefs in environmental risk perception and attitudes. Environ Behav 49:583–600. https://doi.org/10.1177/0013916516649413

Scariot A (2000) Seedling mortality by litterfall in Amazonian forest fragments. Biotropica 32:662–669. https://doi.org/10.1111/j.1744-7429.2000.tb00513.x

Scull P, Cardelús CL, Klepeis P, Woods CL, Frankl A, Nyssen J (2017) The resilience of Ethiopian church forests: interpreting aerial photographs, 1938–2015. L Degrad Dev 28:450–458. https://doi.org/10.1002/ldr.2633

Sheridan MJ (2008) The dynamics of African sacred groves: ecological, social & symbolic processes. In: Sheridan MJ, Nyamweru C (eds) African sacred groves: ecological dynamics and social change. Ohio University Press, Athens

Sheridan M, Nyamweru C (2008) African sacred groves: ecological dynamics & social change. James Currey Publisher, Oxford

Shiferaw A (2017) Productive capacity and economic growth in Ethiopia. CDP Backgr Pap No 34 ST/ESA/2017/CDP/34 6:1–17

Sigman E (2022) Safeguarding the benefits of biodiversity islands in northern Ethiopia in the midst of political change. In: Montagnini F (ed) Biodiversity islands: strategies for conservation in human-dominated environments, Series: Topics in Biodiversity and Conservation. Springer, Cham, pp 647–674

Teketay D, Granström A (1995) Soil seed banks in dry Afromontane forests of Ethiopia. J Veg Sci 6:777–786. https://doi.org/10.2307/3236391

Temudo MP (2012) "The white men bought the forests": conservation and contestation in Guinea-Bissau, Western Africa. Conserv Soc 10:354–366. https://doi.org/10.4103/0972-4923.105563

Temudo MP, Abrantes M (2014) The cashew frontier in Guinea-Bissau, West Africa: changing landscapes and livelihoods. Hum Ecol 42:217–230. https://doi.org/10.1007/s10745-014-9641-0

Tengö M, Johansson K, Rakotondrasoa F, Lundberg J, Andriamaherilala JA, Rakotoarisoa JA, Elmqvist T (2007) Taboos and forest governance: informal protection of hot spot dry forest in southern Madagascar. Ambio 36:683–691. https://doi.org/10.1579/0044-7447(2007)36[683:TAFGIP]2.0.CO;2

US Department of State (2019) 2019 investment climate statements: Ethiopia. state.gov

Verschuuren B, Wild R (2012) Sacred natural sites- sources of biocultural diversity

Verschuuren B, Wild R, McNeely J, Oviedo G (2010) Sacred natural sites: conserving nature and culture. Earth Scan, Routledge

Wassie A (2002) Opportunities, constraints and prospects of Ethiopian orthodox tewahido churches in conserving forest resources: the case of churches in South Gonder, Northern Ethiopia. Swedish Univeristy of Agricultural Sciences, Uppsala

Wassie A, Teketay D (2006) Soil seed banks in church forests of northern Ethiopia: implications for the conservation of woody plants. Flora Morphol Distrib Funct Ecol Plants 201:32–43. https://doi.org/10.1016/j.flora.2005.04.002

Wassie A, Sterck FJ, Bongers F (2010) Species and structural diversity of church forests in a fragmented Ethiopian Highland landscape. J Veg Sci 21:938–948. https://doi.org/10.1111/j.1654-1103.2010.01202.x

Wild R, McLeod C (2008) Sacred natural sites: guidelines for protected area managers. IUCN, Gland. iucn.org

Woods CL, Cardelús CL, Scull P, Wassie A, Baez M, Klepeis P (2017) Stone walls and sacred forest conservation in Ethiopia. Biodivers Conserv 26:209–221. https://doi.org/10.1007/s10531-016-1239-y

Woods CL, Bitew Mekonnen A, Baez-Schon M, Thomas R, Scull P, Abraha Tsegay B, Cardelús CL (2020) Tree community composition and dispersal syndrome vary with human disturbance in sacred church forests in Ethiopia. Forests 11:1082. https://doi.org/10.3390/f11101082

Yigremew A (2002) Review of landholding systems and policies in Ethiopia under the different regimes. Ethiopian economic association,/Ethiopian economic policy research institute, Addis Ababa, Ethiopia. fao.org

Zerga B (2016) Land resource, uses, and ownership in Ethiopia: past, present and future. Int J Sci Res Eng Technol 2:2395–2566

# Chapter 22
# Beyond the Island: Integrated Approaches to Conserving Biodiversity Islands with Local Communities

**Michael S. Esbach, Mahi Puri, Robinson Botero-Arias, and Bette A. Loiselle**

**Abstract** This chapter highlights how graduate students can work with local communities and stakeholders to improve biodiversity conservation and rural development outcomes both in and outside of biodiversity islands with examples from Ecuador, India, and Brazil. In the first case study, Michael Esbach draws on his long-term work with the Indigenous Cofán people in the Ecuadorian Amazon to describe the creation of a biodiversity island within the Cofán territory of Zábalo. In the second case study, Mahi Puri focuses on a human-dominated landscape in India with a high occurrence of carnivores. Finally, Robinson Botero-Arias tells the story of the development of sustainable use plans and conservation of wildlife in an aquatic ecosystem in the Amazon. These case studies demonstrate the "learning and action" platform of the interdisciplinary Tropical Conservation and Development (TCD) program at the University of Florida (UF), which stresses the importance of working with partners and local communities to design problem-oriented research to address challenges at the intersection of conservation and development. Through formal and informal methods, TCD embodies the lessons learned from the examples presented in this chapter, which include the linkage between biodiversity conservation and human well-being; the necessity of understanding local needs, cultural values, and motivations; and engaging stakeholders in partnerships that center participatory research targeted at finding solutions that make the goals of local actors and biodiversity conservation more compatible.

---

M. S. Esbach
Tropical Conservation and Development Program, Center for Latin American Studies, University of Florida, Gainesville, FL, USA

High Meadows Environmental Institute, Princeton University, Princeton, NJ, USA

M. Puri · R. Botero-Arias · B. A. Loiselle (✉)
Tropical Conservation and Development Program, Center for Latin American Studies, University of Florida, Gainesville, FL, USA

Department of Wildlife Ecology and Conservation, University of Florida, Gainesville, FL, USA
e-mail: bloiselle@latam.ufl.edu



F. Montagnini (ed.), *Biodiversity Islands: Strategies for Conservation in Human-Dominated Environments*, Topics in Biodiversity and Conservation 20,
https://doi.org/10.1007/978-3-030-92234-4_22

**Keywords** Interdisciplinary approaches · Partnership · Participatory research · Stewardship

## 22.1 Introduction

The past several decades have seen significant changes in the spatial distribution of intact forests and the patterns of land use surrounding these landscapes (Hansen et al. 2013; Kim et al. 2015; Grantham et al. 2020). These processes have resulted in numerous biodiversity islands: biologically diverse, ecologically intact refugia that are often surrounded by a matrix of human dominated and transformed land. Conserving biodiversity islands, including the ecological functions and services found within, becomes increasingly challenging as they become smaller and more isolated from each other (as per MacArthur and Wilson 1967; Damschen et al. 2019; Montagnini et al. 2022). Scientific studies have often focused on island characteristics and the ways in which the matrix can both positively and negatively impact these islands (e.g., Renjifo 2001). These studies, however, often fail to integrate important social, economic, and cultural contexts, which may drive conservation outcomes within and surrounding biodiversity islands. Ignoring these human dimensions also makes bridging the gap between theory and practice extremely challenging. In this chapter, we argue that scientific research needs to move beyond the island in order to support conservation within the island. By embracing multi-stakeholder participatory processes and partnerships, researchers can develop studies that support conservation science and local action (Duchelle et al. 2009; Kainer et al. 2019). We focus on the role graduate (doctoral) students can play in strengthening such work.

Designing and conducting problem-oriented dissertation studies that respond to local concerns, meet requirements of academic advisors and committees, and bridge theory and practice is exceedingly difficult to achieve. Success will depend on a number of underlying mechanisms that empower and prepare graduate students to collaborate with local actors in research processes, integrate social and biological sciences, and communicate research results to a number of audiences, including peer-reviewed journals and local stakeholders. The interdisciplinary learning and action framework of the Tropical Conservation and Development (TCD) program at the University of Florida (UF) was developed to catalyze and enable a generation of young scientists to embrace new approaches to biodiversity conservation research (Kainer et al. 2019). This framework focuses on the development of theoretical foundations and skills necessary to work across disciplines and effectively collaborate and communicate with local stakeholders. Several TCD courses focus on the development and application of key skills (e.g., communication, facilitation, conflict management), which empower students to effectively listen, learn, and act with key local actors (Duchelle et al. 2009).

Dissertation studies offer an opportunity to put these skills to use. Student researchers are encouraged to establish partnerships with local communities and governmental or non-governmental organizations to better understand the social-ecological context and to more effectively generate and exchange knowledge. Below we identify important principles for effective partnerships. The case studies that follow highlight these principles in practice.

- *Build Social Capital:* In order to build strong relationships, it is important to spend time with your partners, often outside the context of a specific research agenda. By focusing on issues with a clear motivation for partners (e.g., access to certain kinds of information or training), students can establish trust and a record of engagement prior to initiating a research project.
- *Acknowledge Diversity:* It is important to recognize that different parties will have different motives, objectives, and values. Student research should consider mechanisms for incorporating this diversity, assessing trade-offs, and creating equitable conflict resolution mechanisms (Gavin et al. 2018).
- *Share Power:* Effective partnerships require the sharing of power between parties. In addition to utilizing diverse participatory methods (e.g., Reason and Bradbury 2001; Reed 2008), students are encouraged to co-design and implement research with local partners. Sharing power in this way encourages joint responsibility and active relationship management (Mead 2013).
- *Encourage Long-Term Commitments:* It is important to recognize that effective partnerships are built upon long-term commitments that can be resource-intensive and transcend the timeline of graduate research. We encourage students to carefully consider this prior to initiating new relationships. At the same time, students may be able to build upon relationships they already have or work in collaboration with ongoing opportunities. This allows them to capitalize on previous work and provide continuity to that progress.

The remainder of this chapter uses case studies to highlight how graduate students can leverage diverse skills and partnerships to conduct research in support of biodiversity islands. The three case studies are diverse and include (1) efforts by an Amazonian Indigenous community to simultaneously maintain subsistence livelihoods and wildlife populations in Ecuador, (2) efforts to establish a sustainable working landscape surrounding a tiger reserve which benefits small landowners while increasing suitable habitat for large predators in India, and (3) management and sustainable harvest of a threatened top predator by local communities in a sustainable development reserve in the Brazilian Amazon. While different, each case study highlights common themes: leveraging existing local partnerships, working across diverse stakeholders with different objectives, and student researchers utilizing their unique professional skills, experience, and history in their particular study area.

## 22.2 Case Studies

We present three case studies in the following section. In the first case study, Michael Esbach draws on his long-term work with the Indigenous Cofán people in the Ecuadorian Amazon to describe the creation of a biodiversity island within the Cofán territory of Zábalo. The emergence and persistence of this reserve area represents a social process whereby the Cofán developed coordinated, collective arrangements to maintain resources within their territory from generation to generation. Using ethnographic research and common property theory, he examines Cofán common property institutions, critical components of Indigenous tenure regimes that shape conservation outcomes like biodiversity islands. In the second case study, Mahi Puri focuses on a human-dominated landscape in India with a high occurrence of carnivores. Limited research on carnivore ecology outside protected areas (PAs) has hindered conservation efforts. Concurrently, for marginalized farmers, losses due to fluctuating weather patterns, conflict with wild animals, and minimal government support reinforces issues of food insecurity. Combining ecological models on species distribution with concepts from environmental economics, Mahi explores potential solutions that could contribute to biodiversity conservation in partnership with private landowners in shared landscapes. The final case study tells the story of the development of sustainable use plans and conservation of wildlife in an aquatic ecosystem in the Amazon. Here, Robinson (Robin) Botero-Arias has taken on a leadership role as part of the staff of the Mamiraua Institute for Sustainable Development in Brazil. As part of his dissertation research and building on his and other long-term population ecology studies of black caiman, a large crocodilian species, he describes the social, ecological, and political context of the Mamiraua Sustainable Development Reserve and the efforts to develop a caiman harvesting program with local communities.

### *22.2.1 Indigenous Stewardship of a Biodiversity Island in Ecuador*

The Amazon Basin is the largest tropical rainforest in the world, widely recognized for its biological and cultural diversity (Jenkins et al. 2013). This cultural diversity includes over 370 Indigenous ethnicities that have inhabited and managed the region for millennia (Denevan 1992). Despite significant challenges, including colonization, disease, and dispossession, Indigenous peoples have successfully established property rights to over 25 percent of the Amazon (RAISG 2016). Within these territories, Indigenous peoples have created biodiversity islands that host high biodiversity (Pretty et al. 2009; Stevens 2014), prevent deforestation and wildfires (Gray and Coates 2010; Blackman et al. 2017, Schleicher et al. 2017), and maintain the culture and livelihoods of Indigenous peoples (Garnett et al. 2018).

The social processes that create biodiversity islands within Indigenous territories are complex and understudied (Bremner and Lu 2006). In response, the case study presented here focuses on the emergence and maintenance of a biodiversity island within the Cofán territory of Zábalo in the Ecuadorian Amazon. This research was made possible through a long-term partnership between Michael Esbach and the Cofán people. His approach to partnership was informed by previous experiences working with Indigenous peoples across the globe, as well as through diverse learning experiences provided through UF's TCD program. This approach, as noted in the introduction, focuses on building social capital, acknowledging diverse motives and objectives, power sharing, and a long-term engagement.

Prior to initiating his research, Michael regularly visited with Cofán people in order to build social capital and learn more about the history of Zábalo. During this time he focused on supporting Cofán initiatives not directly related to his dissertation research. By focusing on Cofán needs and interests, such as securing funds for their ongoing initiatives, he was able to establish a record of engagement and build trust with the community. Over time, Michael developed a plan of research in partnership with the Cofán that responded to their interests and future goals. Throughout this process, he managed expectations by recognizing that each party should benefit, but in potentially different ways. For instance, while Michael's research was aligned with Cofán interests, he also recognized the importance of paying Cofán residents for their time and support and seeking out training opportunities to further their local capacity. At the same time, Michael benefitted from the Cofán's expertise and participation in his dissertation. Ultimately, Michael and his Cofán partners were able to explore local resource management institutions, ecological knowledge, and trends in animal populations over time. This work directly supports the Cofán's desire to improve self-determination over resource management within their territory. This is important because it shows the effectiveness of local institutions in maintaining animal populations over time. Beyond these accomplishments, Michael continues to work with the Cofán on a number of initiatives, deepening his partnership with the Cofán.

The remainder of this case study is dedicated to results from this research partnership, which explores how the Cofán created institutions to conserve and protect a biodiversity island within their territory of Zábalo. One of five Cofán territories with legal title, Zábalo is located in the northeastern Ecuadorian Amazon within the Cuyabeno Wildlife Reserve (Fig. 22.1). Like the majority of Indigenous territories in the Ecuadorian Amazon, Zábalo is owned collectively by the Cofán. This communal tenure regime signals that resources within the territory are managed through formal and informal rules known as common property institutions (Bremner and Lu 2006). These institutions govern how resources are managed in Zábalo, allowing Cofán residents to enforce rules among themselves and exclude external individuals from using resources found within the territory (Ostrom 1990).

Zábalo was founded by residents from Dureno, another Cofán community, in the early 1980s in response to agrarian reform initiated by the Ecuadorian state which opened the Cofán's ancestral lands to colonization. This process supported the emergence of the Cofán common property institutions that regulate resource use

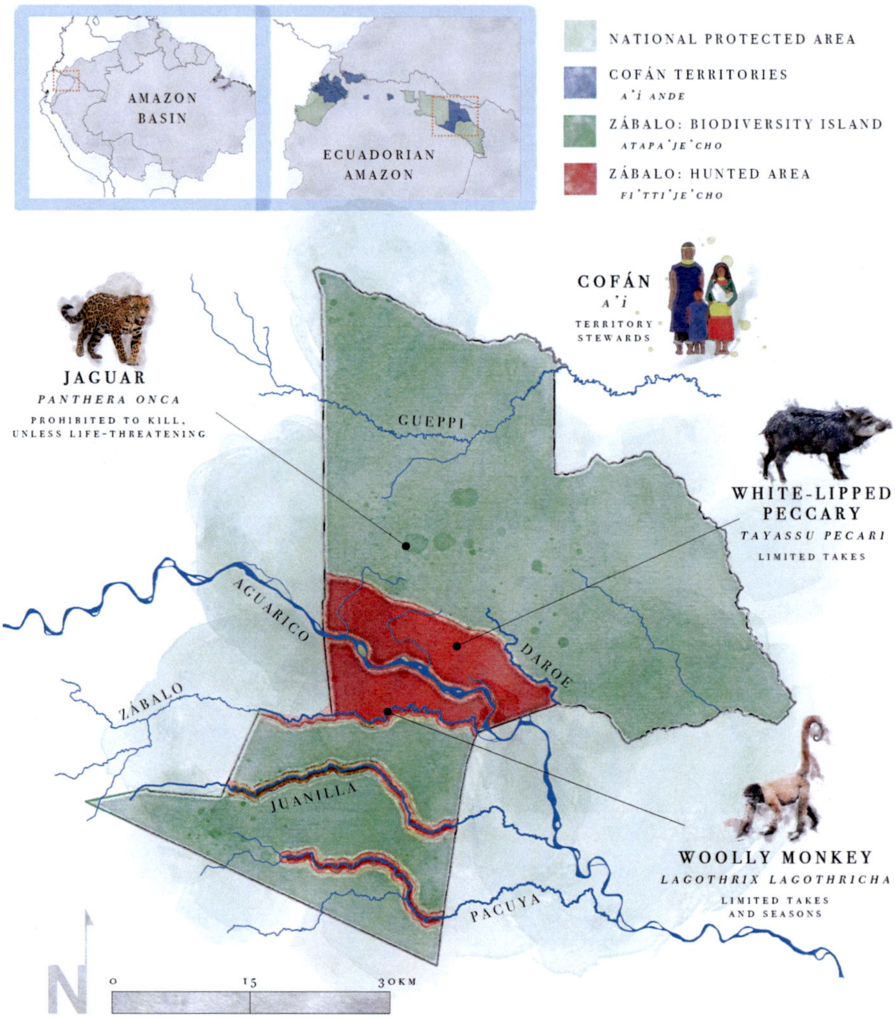


**Fig. 22.1** Map of Zábalo in the Ecuadorian Amazon depicting the biodiversity island stewarded by the Cofán people. Resource use activities are prohibited within this area, and additional, specific rules have been institutionalized within the hunted area. Map and illustration created by Michael Esbach using freely available data and images

and promote conservation. Research suggests that for these institutions to emerge, people need to recognize that resources important to their livelihoods are becoming scarce and that they are capable of reducing the pressures that create this scarcity (Bromley and Cernea 1989; Lu 2001). As colonists increasingly encroached on their traditional territory and degraded the surrounding environment, the Cofán realized that their forest-based livelihood was dependent upon an intact environment and fought to preserve it through three primary methods: (1) a secure land title, (2) the ability to exclude outsiders, and (3) the importance of managing resource use.

Across the Amazon, one of the primary drivers for Indigenous peoples to secure legal property rights are external pressures on important resources (Lu 2001). Zábalo residents first filed for legal property rights in 1981. It was not until 1992, however, with the expansion of the Cuyabeno Wildlife Reserve, that the Cofán were able to come to an agreement with the Ministry of Environment (MAE). This co-management agreement established the Cofán's legal title to 85,000 hectares in exchange for their management of the area. The Cofán would go on to sign additional agreements with the MAE, increasing their territorial rights to approximately 145,000 hectares. Today there are about 180 people living in Zábalo, representing nearly 40 families. While some of these families live in more isolated areas, most are clustered into four locations along the Aguarico River. Zábalo residents hunt across their territory, fish within both white-water and black-water rivers, and practice agriculture near their homes and within individual plots scattered along the Aguarico River.

With their land title secure, Cofán leaders crafted approaches to defend their territory from outside threats. This included the creation of the Cofán park guard program, which focuses on patrolling Zábalo's boundaries, clearing trails, and manning guard stations in order to keep outsiders from extracting resources. Cofán leaders have demonstrated the effectiveness of this program, and as a result, the national government signed two official documents (*Acuerdo No. 138* in 2002 and *Registro Official No. 5* in 2003) in support of the Cofán park guard program. Currently, all Zábalo residents participate in the guard program, conducting patrols and clearing boundaries on a yearly basis. Group of men and women are deployed to different regions of the territory to conduct these responsibilities, and different families are assigned to live in guard stations for months at a time.

In addition to a secure land title and the ability to exclude outsiders, the Cofán also developed a sophisticated resource management system within Zábalo called *se'pi'cho* (see Cepek 2012). Over the past 30 years, the community of Zábalo has experimented with *se'pi'cho* in order to manage their subsistence activities in a way that sustains communal resources. These efforts have resulted in an effective system of adaptive management whereby daily subsistence acts as a type of monitoring. Residents, individually and collectively, develop detailed knowledge related to changes in resources, including the distribution of animal populations and their abundance across the territory. This knowledge is shared annually during a community meeting when Zábalo residents discuss their observations. During this meeting, they have the opportunity to create and/or revise rules based on their knowledge. Proposed changes will be voted on by the community and take written form if the majority approve. The rule is then applied to all community members and enforced strictly through different fine structures and public acknowledgement when a resident breaks a rule. The interaction of these components—where observation represents a form of monitoring and knowledge production that informs rules with revisions over time based on additional observations—is the crux of adaptive management.

*Se'pi'cho* rules currently take diverse forms, including users, activities, and species-specific restrictions (see Cepek 2012). The Cofán have also developed spatial boundaries which recognize areas for human use, primarily along the Aguarico River (e.g., houses, gardens) and subsistence activities where hunting, fishing, and gathering are permitted, as well as a reserve area which acts as a source for wildlife populations (Fig. 22.1). This reserve area functions as a biodiversity island where animals and plants are protected from human use, but tourism and scientific research activities are allowed. Most of the land base in the territory of Zábalo comprises a biodiversity island whereby community members utilize diverse strategies, from adaptive management to the exclusion of outsiders, to sustain biodiversity that is important to Cofán livelihoods and identity. The creation and sustainability of biodiversity islands within Indigenous territories, therefore, is a social process whereby Indigenous peoples secure and enforce communal property rights and develop institutions to manage resources in culturally appropriate ways.

### *22.2.2 Buffering Impacts to a Biodiversity Island in India*

India is a megadiverse country with approximately 20% of its land forested (Mittermeier et al. 2004). Less than five percent of its land is protected, however. Most of India's biodiversity islands, or PAs, were created from princely hunting grounds or reserve forests established during the colonial era (Rangarajan 1996). As a result, the approximately 800 PAs scattered across the country are small, fragmented, and isolated, making them less than optimal in terms of their ecological capacity and potential (Nayak et al. 2020). These PAs are also vulnerable to expanding infrastructure, agricultural development, and economic activities such as mining, dams, and commercial logging (Jayadevan et al. 2020).

The foundation of PAs in India is the Wildlife Protection Act (1972), which spurred the creation of many biodiversity islands through the relocation of human settlements (Rangarajan 2005; Rangarajan and Shahabuddin 2006). Limited science-based conservation planning, along with the exclusion of local communities from decision-making (Kothari et al. 1995), has contributed to conflicts over the management of resources. Suspension of the traditional rights of local communities over forests and restrictions on access has resulted in resentment towards the government and conservation policies (Maikhuri et al. 2001). In addition to these challenges, the high density of people and livestock around PAs has resulted in numerous human-wildlife conflicts (Bagchi and Mishra 2006; Karanth et al. 2012). Unsustainable resource use has also contributed to forest degradation and exacerbated negative impacts on wildlife populations (Shankar et al. 1998; Kumar and Shahabuddin 2005; Davidar et al. 2010).

Given their mixed success in achieving ecological objectives, exclusion of local peoples, and mounting development threats, government and non-government organizations across India have started experimenting with alternative conservation strategies (Saberwal and Rangarajan 2003; Sinha et al. 2012). In this case study,

Mahi Puri first explores the conservation potential of legally sanctioned community reserves and other voluntary community-led initiatives, and presents results from her own research working directly with private landowners living in a buffer around a biodiversity island.

In 2003, an amendment to the Wildlife Protection Act allowed for the establishment of community reserves. Today, there are over 100 community reserves where local users have greater access to resources which are administered by local people and elected village institutions (Kothari 2006). While limited research has been conducted on these reserves, results show that they significantly outperform open-access areas in terms of biodiversity conservation and can play a complementary role to strict PAs (Shahabuddin and Rao 2010). In terms of governance, community reserves have had varying degrees of success due to bureaucratic inefficiency (Pathak 2006). Despite being included in the country's PA system, several reserves continue to face economic and political constraints.

Other forms of communally managed forests exist in India that do not have the same legal status as community reserves. Across the country, community forest resource (CFR) rights have been recognized under the Forest Rights Act (2006), allowing communities to conserve and manage forest resources. Management plans developed by communities are diverse, responding to local environments and needs. Communities have initiated measures for soil and water conservation, carried out fire management, imposed restrictions on grazing activities in areas under assisted natural regeneration, removed invasive species, and even set aside areas to allow wildlife presence. Communities have experimented with short-rotation species (including bamboo and various native fruiting trees) that provide annual economic returns and have been able to generate substantial incomes through the sustainable harvest of non-timber forest products. These initiatives have contributed to employment creation (especially for the landless) and empowerment of local village-level institutions and women's collectives (for detailed case studies see Agarwal and Saxena 2018). While some CFR areas have seen tremendous success, many others continue to be challenged by a lack of tenure rights, power imbalances with the forest authorities, conflicts with neighboring villages, and crop damage by wild animals.

In addition to community-led initiatives, there are also private landowners and organizations assisting in forest regeneration and restoration. In the Kodagu district of Karnataka, Save Animals Initiative (SAI) Sanctuary is a private forest managed by a non-profit trust. Elsewhere, ecotourism-based initiatives such as Kumaon Maati and Gorukana channel their revenues back into conservation or for the benefit of the rural poor (Puri et al. 2019). However, most of these projects have been independent initiatives, conducted at small spatial scales and with limited involvement of local people. The need to expand wildlife habitat, and increased effectiveness of the existing PA network must be reconciled with the fact that the land outside PAs is largely privately owned. This mandates incorporating the social context in ecological studies in order to reduce the gap between conservation goals and their current feasibility.

Natural regeneration on degraded or modified lands is a slow and long-term process. The process can be sped up through active restoration and intensive management. Adoption of agroforestry systems that support food production and provide other resources (such as fodder, firewood) is a potential strategy that can benefit local livelihoods and concurrently enhance the biodiversity value of the hostile landscape surrounding PAs by maintaining connectivity across landscapes and reducing pressures on PAs (Montagnini et al. 2022). However, the success of such an intervention is contingent upon the decisions and choices of individual landowners (Knight et al. 2010). Within this context, Mahi Puri undertook a study rooted in a social-ecological systems approach (Liu et al. 2007; Carter et al. 2014) to examine the (1) habitat use patterns of a carnivore assemblage (including tigers, leopards, and wild dogs) in a human-dominated landscape, and (2) willingness of farmers to modify existing land-use by adopting agroforestry, if provided with a monetary incentive.

The study was conducted in the administrative buffer of Pench Tiger Reserve, a biodiversity island in central India (Fig. 22.2). The site was selected based on Mahi's long-term work in the region, specifically her research on landscape connectivity and human-carnivore conflicts. Prior discussions with the local Forest Department and the partnering NGO also helped in identifying specific objectives for the region. The field studies were comprised of camera trap surveys in approximately 500 square kilometers of multi-use forests, and interviews with over 600 farmers. At the data collection stage, Mahi employed members from the local communities. Engagement of local people in the research project helped in building trust with other local farmers who participated in the surveys. The larger goal of this research project was to prioritize restoration in the landscape and establish an incentive-based conservation initiative, which integrates ecological values with the preferences of key decision-makers (i.e., private landowners).

The results from the study demonstrate that at fine spatial scales, habitat use by carnivores was positively influenced by the composition and configuration of the landscape. With regard to modifying current land-use, farmer decision-making was positively influenced by their education and current income from agriculture, and negatively influenced by household size. In addition, the percentage of forest cover around the landholding and recurring incidents of crop damage by wild herbivores contributed to people's willingness to adopt agroforestry. Variables such as people's perceptions of benefits from agroforestry, perceived risks, as well as social norms were also important driving factors. Interdisciplinary studies such as this could be useful in exploring the potential for carnivore conservation and collaboration with local stakeholders within a forest-agriculture matrix. This could provide valuable information for identifying areas where conservation interventions might be pursued and succeed. Ultimately, it could aid in landscape level planning that integrates agricultural lands with natural areas, concurrently contributing to farmer incomes through an incentive-based program.

Moving forward, bridging the gap between scientific research and conservation action will require a combination of both bottom-up and top-down approaches. Participatory methods that consider local needs and preferences are an essential component for supporting local stewardship. To overcome economic barriers of

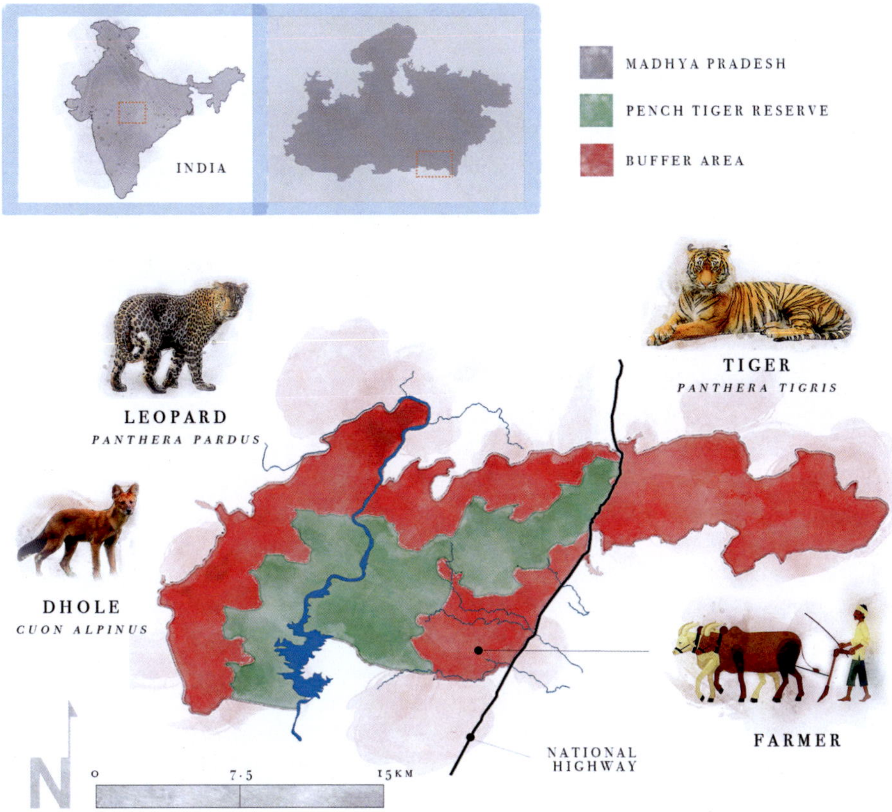


**Fig. 22.2** Pench Tiger Reserve in central India is surrounded by a buffer area that supports the conservation of key carnivore species and the livelihoods of local farmers. Map and illustration created by Michael Esbach using freely available data and images

restoration, government incentives, private investments, and support from NGOs will be crucial. An important tool in this regard is payment for ecosystem services (PES). Unlike many countries around the world where government support exists, a formal PES system is currently absent in India. Another important consideration for the program's effectiveness is the need for structured mechanisms such as insurance and compensation schemes. These mechanisms may reduce and/or mitigate conflicts as people practice agroforestry and their interactions with wildlife increase. In the long-term, improved habitat quality that encourages wildlife presence can create opportunities to promote alternative forms of tourism in such landscapes. This could help offset tourism pressure on PAs, diversify existing tourism activities, and contribute to income generation. Ultimately, integrating restored lands through community-led initiatives and wildlife-friendly agriculture on private lands with the country's extant PA network has the potential to minimize some developmental challenges, thereby accomplishing the dual objectives of improving landscape conservation and securing local livelihoods.

### *22.2.3 Sustainable Management of an Aquatic Biodiversity Island in Brazil*

The Brazilian Amazon supports high biodiversity and immense reservoirs of carbon and water (Mittermeier et al. 2005; Ritter et al. 2017). The aquatic wildlife and river systems found across this region, which can be viewed as "islands" in forested landscapes, are extremely important from both social and ecological perspectives (Antunes et al. 2016). To safeguard these resources, Brazil has developed an extensive PA network which covers approximately 28% of the Amazon region. This network includes different categories of PAs which utilize different approaches to conserve biodiversity, including strict nature conservation, sustainable development, tourism, scientific research, and education (Fearnside 2003; Silva 2005; de Queiroz 2010). Tailored to unique social-ecological contexts, this system is able to include local people in PA governance and support the sustainable use of natural resources through management planning and zoning systems (e.g., strictly protected zones and harvest areas) (Gillingham 2001; Rylands and Brandon 2005; de Queiroz 2010). This case study focuses on Sustainable Development Reserves (SDR), a Brazilian PA category that integrates local stakeholders into conservation priorities. In this context, Robinson Botero-Arias discusses his work with black caiman (*Melanosuchus niger*) and local communities within the Mamiraua SDR.

Brazil's SDRs utilize both scientific and local knowledge to create conservation priorities and management strategies. By facilitating the involvement of local people and utilizing their knowledge base, SDRs integrate diverse knowledge systems to establish guidelines for natural resource use and zoning of activities within a specific area. The zoning plan, for example, must take into consideration the subsistence needs of local communities, together with ecological requirements for sustainable use and biodiversity conservation (Castello 2008; Castello et al. 2009; Marioni et al. 2013a, b). The Mamiraua SDR has implemented a successful pirarucu (*Arapaima gigas*) sustainable management program with local stakeholders. In response to overfishing by external parties, this program focused on building a participatory monitoring system of the fishery (Castello et al. 2009). This system has effectively incorporated local communities in the conservation of this large fish species. Recognizing the success of the pirarucu participatory management program, caimans were identified as an ideal target species for further participatory action given local needs and conservation threats. Given Robin's leadership role in international crocodilian conservation and knowledge of local systems, his dissertation research in collaboration with Mamiraua SDR was designed to advance this sustainable management goal.

Populations of black caimans have suffered from past overexploitation. Intense hunting began between 1940 and 1950 and then continued, though less intensively, through the 1970s. Until the 1990s, the black caiman was considered one of the most threatened Neotropical crocodilians due to unsustainable use, habitat loss and degradation, and over-hunting by local fishermen (Da Silveira and Thorbjarnarson 1999; Thorbjarnarson and Da Silveira 2000; Marioni et al. 2013a, b). Given the importance of black caiman to local livelihoods, as both a source of food and income, the Brazilian government sought ways to sustainably harvest this species. A key first step was the down-listing of the black caiman from CITES (Convention on International Trade in Endangered Species) Appendix I to Appendix II in 1999; this change signaled international support for establishing a caiman management system. Shortly thereafter, the Brazilian State of Amazonas initiated an event in partnership with scientific and local communities that aimed to develop a caiman conservation, monitoring, and harvest management program. The Mamiraua SDR was chosen as the site for a pilot program, and was chosen for two main reasons: (1) the existence of ongoing monitoring actions with evidence of a high abundance of black caimans, and (2) a demonstrated capacity of local communities to engage in effective, sustainable use of natural resources (such as with the pirarucu management program) (Botero-Arias et al. 2010).

Five years after the creation of Mamiraua SDR in 1996, local populations of black caiman began to recover, suggesting that supervised harvests might successfully generate income for local residents without negative, long-term impacts on the wild population. Robinson Botero-Arias has been collaborating with the Caiman Conservation and Management Research Program at the Mamiraua Institute for Sustainable Development since 2007. There he has supported processes to establish a framework for legal caiman harvesting. This legal framework was strengthened via two resolutions in 2011 that established (1) technical procedures for caiman management in PAs, and (2) criteria for the harvest and processing of crocodilians, both in the State of Amazonas. The foundation of this framework is a participatory process that integrates conservation principles with local knowledge and needs. Specifically, adaptive management principles help maintain stable and healthy caiman populations over time by incorporating feedbacks from caiman monitoring, various sustainability indicators, and local participation. Through this system, black caiman populations have demonstrated significant resilience to harvest (Botero-Arias and Regatieri 2013; Marioni et al. 2013a, b). The sustainable use of Amazonian caimans has become an important management strategy to conserve the species and preserve its habitats.

By working with local communities, government institutions, and other researchers, Robin has supported the sustainability of caiman harvests within the Mamiraua SDR. Participatory processes have proven critical to these efforts. By working directly with local people that have historically benefited from the caiman harvest to develop different zones (e.g., harvesting areas, reserve areas; Fig. 22.3), his work has supported local decision-making that directly supports local needs. His work also points to the importance of adaptive management performed through partnerships between local communities and researchers that draw on different knowledge systems and values. This partnership has supported the integration of biological research on harvested species with the social and economic needs of local peoples through a constant participatory evaluation process that combines sustainability indicators and caiman population monitoring. Robin's work has also sought to support local communities by connecting them with regional supply chains so that they can develop a fair market for caiman with competitive pricing. This way, local

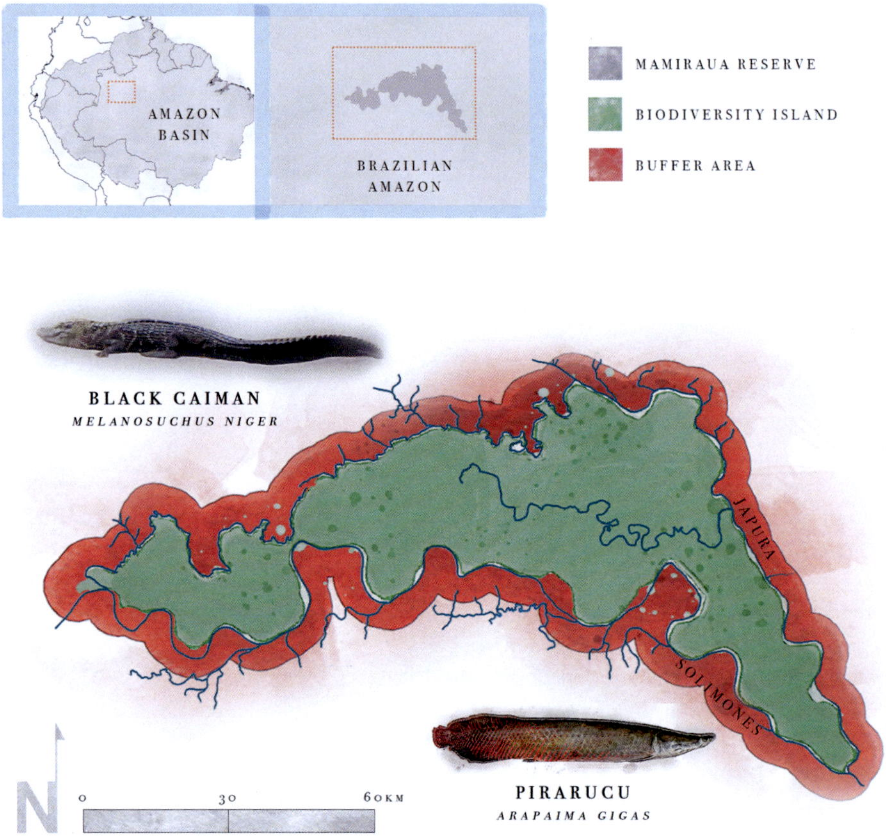


**Fig. 22.3** The Mamiraua Sustainable Development Reserve functions as a biodiversity island focused on the conservation and sustainable management of black caiman and pirarucu. Map and illustration created by Michael Esbach using freely available data and images

people can get the best prices for their harvested meat, which simultaneously reinforces local benefits and motivation to sustainably manage caimans and their habitat.

Management systems for the use of the natural resources within SDRs have positioned local socio-cultural values at the center of the decision-making process, with the complementary aim of conserving biodiversity and decreasing illegal activities threatening wildlife populations. Despite its success over the past decade, caiman management in the State of Amazonas is still considered experimental. As such, Robin will continue to work with his partners in Mamiraua SDR through his affiliation with the Mamiraua Institute for Sustainable Development to conduct further collaborative research focused on producing information about the populations to be exploited, the long-term sustainability of their harvest, and various social and organizational challenges that arise when working in a complex social-ecological system.

## 22.3 Conclusions

Conserving biodiversity islands, like those described in this chapter, requires the development of approaches that facilitate multi-stakeholder inclusion and participation in order to address underlying research needs and integrate research findings into participatory decision-making processes. Graduate students can play an important role in developing and advancing such interdisciplinary research like that described here which responds to local concerns while contributing to finding viable solutions for biodiversity conservation in complex social-ecological systems that characterize "islands" of native habitat.

The case studies presented here represent the dissertation work of three graduate students with considerable knowledge of local systems in which they conducted their research. Although the contexts are different, each student took on research questions that addressed local needs in partnership with local communities, conservation NGOs, and/or government actors. The case studies highlight (1) the importance of existing partnerships with local communities and NGOs to identify issues threatening biodiversity, (2) application of interdisciplinary problem-oriented research to better understand threats and potential solutions, and (3) current or proposed mechanisms to share results with local stakeholders to advance solutions appropriate to particular social-ecological contexts. Outcomes from their research included enhanced understanding of mechanisms by which Indigenous peoples successfully manage wildlife populations in community reserves in Ecuador, identification of opportunities for increasing habitat for large carnivores in rural India that generate alternative income sources via more wildlife-friendly agriculture, and the development of participatory management systems for the sustainable use of aquatic wildlife populations in Amazonian PAs. This work was supported by the learning and action platform emphasized through the TCD program at UF. TCD graduate students like Michael, Mahi, and Robin were provided opportunities to enrich their professional skills by working across disciplines and engaging diverse stakeholders to develop problem-oriented research questions in partnership with local peoples, NGOs, and other partners and with whom they could jointly develop more holistic solutions to the conservation of biodiversity islands.

## References

Agarwal S, Saxena AK (2018) People's forests: is community forest resource governance the future of India's jungles? Centre for Science and Environment, New Delhi

Antunes AP, Fewster RM, Venticinque EM, Peres CA, Levi T, Rohe F, Shepard GH (2016) Empty forest or empty rivers? A century of commercial hunting in Amazonia. Sci Adv 2(10):e1600936

Bagchi S, Mishra C (2006) Living with large carnivores: predation on livestock by the snow leopard (*Uncia uncia*). J Zool 268(3):217–224

Blackman A, Corral L, Lima ES, Asner GP (2017) Titling indigenous communities protects forests in the Peruvian Amazon. Proc Natl Acad Sci 114(16):4123–4128

Botero-Arias R, Regatieri SA (2013) Construindo as bases para um Sistema de Manejo Participativo dos Jacarés Amazônicos. Série Prot, Tefé, Amazonas, Brazil
Botero-Arias R, Marmontel M, Queiroz HL (2010) Projeto de manejo experimental de Jacarés no Estado Do Amazonas: Abate de Jacarés no setor Jarauá - Reserva de Desenvolvimento Sustentável Mamirauá, Dezembro de 2008. Scientific Magazine UAKARI 5(2):49–57
Bremner J, Lu F (2006) Common property among Indigenous peoples of the Ecuadorian Amazon. Conserv Soc 4(4):499–521.
Bromley DW, Cernea MM (1989) The management of common property natural resources: some conceptual and operational fallacies. World Bank discussion paper 57, World Bank, Washington DC
Carter NH, Viña A, Hull V, McConnell WJ, Axinn W, Ghimire D, Liu J (2014) Coupled human and natural systems approach to wildlife research and conservation. Ecol Soc 19(3):43
Castello L (2008) A socio-ecological synthesis on the conservation of the pirarucu (*Arapaima*) in floodplains of the Amazon. PhD dissertation, State University of New York, Syracuse
Castello L, Viana JP, Watkins G, Pinedo-Vasquez M, Luzadis VA (2009) Lessons from integrating fishers of arapaima in small-scale fisheries management at the Mamirauá Reserve. Amazon Environ Manage 43(2):197–209
Cepek ML (2012) A future for Amazonia: Randy Borman and Cofán environmental politics. University of Texas Press, Texas.
Da Silveira R, Thorbjarnarson JB (1999) Conservation implications of commercial hunting of black and spectacled caiman in the Mamiraua sustainable development reserve, Brazil. Biol Conserv 88:103–109
Damschen EI, Brudvig LA, Burt MA, Fletcher RJ Jr, Haddad NM, Levey DJ, Orrock JL, Resasco J, Tewksbury JJ (2019) Ongoing accumulation of plant diversity through habitat connectivity in an 18-year experiment. Science 365(6460):1478–1480
Davidar P, Sahoo S, Mammen PC, Acharya P, Puyravaud JP, Arjunan M, Garrigues JP, Roessingh K (2010) Assessing the extent and causes of forest degradation in India: where do we stand? Biol Conserv 143(12):2937–2944
de Queiroz HL (2010) Protected areas in the Amazonian Várzea and their role in its conservation: the case of Mamirauá Sustainable Development Reserve (MSDR). In: Junk W, Piedade M, Wittmann F, Schöngart J, Parolin P (eds) Amazonian floodplain forests. Ecological studies (analysis and synthesis). Springer, Dordrecht, pp 465–483
Denevan WM (1992) The pristine myth: the landscape of the Americas in 1492. Ann Assoc Am Geogr 82(3):369–385
Duchelle AE, Biedenweg K, Lucas C, Virapongse A, Radachowsky J, Wojcik DJ, Kainer KA (2009) Graduate students and knowledge exchange with local stakeholders: possibilities and preparation. Biotropica 41(5):578–585
Fearnside PM (2003) Conservation policy in Brazilian Amazonia: understanding the dilemmas. World Dev 31(5):757–779
Garnett ST, Burgess ND, Fa JE, Fernández-Llamazares Á, Molnár Z, Robinson CJ, Leiper I (2018) A spatial overview of the global importance of indigenous lands for conservation. Nat Sustain 1(7):369–374
Gavin MC, McCarter J, Berkes F, Mead ATP, Sterling EJ, Tang R, Turner NJ (2018) Effective biodiversity conservation requires dynamic, pluralistic, partnership-based approaches. Sustainability 10:1846
Gillingham S (2001) Social organization and participatory resource management in Brazilian Ribeirinho communities: a case study of the Mamirauá sustainable development reserve. Amazonas Soc Nat Res 14(9):803–814
Grantham HS, Duncan A, Evans TD, Jones KR, Beyer HL, Schuster J, Walston JC, Robinson JG, Callow M, Clements T, Costa HM, DeGemmis A, Elsen PR, Ervin J, Franco P, Goldman E, Goetz S, Hansen A, Hofsvang E, Jantz P, Jupiter S, Kang S, Langhammer P, Laurance WF, Lieberman S, Linkie M, Malhi Y, Maxwell S, Mendez M, Mittermeier R, Murray NJ, Possingham H, Radachowsky J, Saatchi S, Samper C, Silverman J, Shapiro A, Strassburg B,

Stevens T, Stokes E, Taylor R, Tear T, Tizard R, Venter O, Visconti P, Wang S, Watson JEM (2020) Anthropogenic modification of forests means only 40% of remaining forests have high ecosystem integrity. Nat Commun 11:5978
Gray M, Coates J (2010) 'Indigenization'and knowledge development: extending the debate. Int Soc Work 53(5):613–627
Hansen MC, Potapov PV, Moore R, Hancher M, Turubanova SA, Tyukavina A, Thau D, Stehman SV, Goetz SJ, Loveland TR, Kommareddy A, Egorov A, Chini L, Justice CO, Townshend JRG (2013) High-resolution global maps of 21$^{st}$-century forest cover change. Science 342:850–853
Jayadevan A, Nayak R, Karanth KK, Krishnaswamy J, DeFries R, Karanth KU, Vaidyanathan S (2020) Navigating paved paradise: evaluating landscape permeability to movement for large mammals in two conservation priority landscapes in India. Biol Conserv 247:108613
Jenkins CN, Pimm SL, Joppa LN (2013) Global patterns of terrestrial vertebrate diversity and conservation. Proc Nat Acad Scie USA 110(28):2602–2610
Kainer KA, Binnqüist CL, Dain JL, Jaimes BC, Castillo PN, Basulto RG, Contreras FM (2019) Leading by listening, learning by doing: modeling democratic approaches to conservation leadership in graduate education. J Environ Stud Sci 9(2):206–217
Karanth KK, Gopalaswamy AM, DeFries R, Ballal N (2012) Assessing patterns of human-wildlife conflicts and compensation around a Central Indian protected area. PLoS One 7:e50433
Kim D, Sexton JO, Townshend JR (2015) Accelerated deforestation in the humid tropics from the 1990s to the 2000s. Geophys Res Lett 42:1e7
Knight AT, Cowling RM, Difford M, Campbell BM (2010) Mapping human and social dimensions of conservation opportunity for the scheduling of conservation action on private land. Conserv Biol 24:1348–1358
Kothari A (2006) Community conserved areas: towards ecological and livelihood security. Parks 16(1):3–13
Kothari A, Suri S, Singh N (1995) Conservation in India: a new direction. Econ Polit Wkly 30: 2755–2766
Kumar R, Shahabuddin G (2005) Effects of biomass extraction on vegetation structure, diversity and composition of forests in Sariska Tiger Reserve. India Environ Conserv 32(3):248–259
Liu J, Dietz T, Carpenter SR, Folke C, Alberti M, Redman CL, Schneider SH, Ostrom E, Pell AN, Lubchenco J, Taylor WW (2007) Coupled human and natural systems. Ambio 36(8):639–649
Lu F (2001) The common property regime of the Huaorani Indians of Ecuador: implications and challenges to conservation. Human Ecol 29(4):425–447
MacArthur RH, Wilson EO (1967) Island biogeography. Princeton University Press, Princeton
Maikhuri RK, Nautiyal S, Rao KS, Saxena KG (2001) Conservation policy–people conflicts: a case study from Nanda Devi Biosphere Reserve (a world heritage site). India For Pol Econ 2(3): 355–365
Marioni B, Botero-Arias R, Fonseca-Junior SF (2013a) Local community involvement as a basis for sustainable crocodilian management in protected areas of Central Amazonia: problem or solution? Tropical Conserv Sci 6(4):484–492
Marioni B, Farias I, Verdade LM, Bassetti L, Coutinho ME, de Mendonça SH, Vieira TQ, Magnusson WE, Campos Z (2013b) Avaliação do risco de extinção do jacaré-açu Melanosuchus niger (Spix, 1825) no Brasil. Biodiv Brasil 3(1):31–39
Mead ATP (2013) Sharing power: Māori and protected areas. In: Katene S, Mulholland M (eds) Governance and leadership in future challenges for Māori: He Kōrero Anamata. Huia Publishers, Wellington, pp 195–204
Mittermeier RA, Gil PR, Hoffmann M, Pilgrim J, Brooks T, Lamoreux CM, Da Fonseca GAB (2004) Hotspots revisted: Earth's biologically wealthiest and most threatened ecosystems. CEMEX, México, pp 99–103
Mittermeier RA, Da Fonseca GA, Rylands AB, Brandon K (2005) A brief history of biodiversity conservation in Brazil. Conserv Biol 19:601–607
Montagnini F, Levin B, Berg KE (2022) Introduction. Biodiversity Islands: strategies for conservation in human-dominated environments. In: Montagnini F (ed) Biodiversity islands: strategies

for conservation in human-dominated environments, Topics in Biodiversity and Conservation. Springer, Cham, pp 1–35
Nayak R, Karanth KK, Dutta T, Defries R, Karanth KU, Vaidyanathan S (2020) Bits and pieces: Forest fragmentation by linear intrusions in India. Land use policy:104619
Ostrom E (1990) Governing the commons: the evolution of institutions for collective action. Cambridge University Press, Cambridge
Pathak N (2006) Community conserved areas in South Asia. Protected Areas Programme 56
Pretty J, Adams B, Berkes F, De Athayde SF, Dudley N, Hunn E, Pilgrim S (2009) The intersections of biological diversity and cultural diversity: towards integration. Conserv Soc 7(2):100–112
Puri M, Karanth KK, Thapa B (2019) Trends and pathways for ecotourism research in India. J Ecotour 18(2):122–141
RAISG (Rede Amazônica de Informação Socioambiental Georreferenciada) (2016) Amazonia 2016: protected areas and indigenous territories. Amazon Network of Geo-Referenced Socio-Environmental Information. https://www.amazoniasocioambiental.org
Rangarajan M (1996) The politics of ecology: the debate on wildlife and people in India, 1970-95. Econ Polit Wkly:2391–2409
Rangarajan M (2005) India's wildlife history: an introduction. Orient Blackswan, Hyderabad
Rangarajan M, Shahabuddin G (2006) Displacement and relocation from protected areas: towards a biological and historical synthesis. Conserv Soc 4(3):359–378
Reason P, Bradbury H (2001) Handbook of action research: participative inquiry and practice. Sage Publications, London
Reed MS (2008) Stakeholder participation for environmental management: a literature review. Biol Conserv 141:2417–2431
Renjifo LM (2001) Effect of natural and anthropogenic landscape matrices on the abundance of SubAndean bird species. Ecol Appl 11:14–31
Ritter CD, McCrate G, Nilsson RH, Fearnside PM, Palme U, Antonelli A (2017) Environmental impact assessment in Brazilian Amazonia: challenges and prospects to assess biodiversity. Biol Conserv 206:161–168
Rylands AB, Brandon K (2005) Brazilian protected areas. Conserv Biol 19(3):612–618
Saberwal VK, Rangarajan M (2003) Battles over nature: science and the politics of conservation. Orient Blackswan, Hyderabad
Schleicher J, Peres CA, Amano T, Llactayo W, Leader-Williams N (2017) Conservation performance of different conservation governance regimes in the Peruvian Amazon. Sci Rep 7(1): 1–10
Shahabuddin G, Rao M (2010) Do community-conserved areas effectively conserve biological diversity? Global insights and the Indian context. Biol Conserv 143(12):2926–2936
Shankar U, Murali KS, Shaanker RU, Ganeshaiah KN, Bawa KS (1998) Extraction of non-timber forest products in the forests of Biligiri Rangan Hills, India. 4. Impact on floristic diversity and population structure in a thorn scrub forest. Econ Bot 52(3):302–315
Silva M (2005) The Brazilian protected areas program. Conserv Biol 19(3):608–611
Sinha BC, Qureshi Q, Uniyal VK, Sen S (2012) Economics of wildlife tourism – contribution to livelihoods of communities around Kanha tiger reserve, India. J Ecotour 11:207–218
Stevens S (ed) (2014) Indigenous peoples, national parks, and protected areas: a new paradigm linking conservation, culture, and rights. University of Arizona Press, Tucson
Thorbjarnarson J, Da Silveira R (2000) Secrets of the flooded Forest. Nat Hist 109(2):70–79

# Chapter 23
# Agroecology and Forest Conservation in Three Types of Land Reform Communities in the Cacao Region of Bahia, Brazil

**Kathleen R. Painter, Robert Buschbacher, Luiz Carlos Souto Silva, and Emerentina Costa e Silva**

**Abstract** To address conflicts between human occupation and forest conservation in the cacao region of southern Bahia, Brazil, the Jupará Agroecological Movement, in partnership with WWF, promotes an agroecological model of land reform. This chapter presents results of this initiative, and assesses its contribution to two kinds of biodiversity islands: diversified agroforestry systems and remnants of natural forest. The chapter utilized interviews and visits to 60 families in three communities with different forms of land occupation: "contested" and "negotiated" models of land reform, and a "traditional community" created by escaped slaves (*quilombo*). Participants and non-participants in the Jupará program are compared, with forest cover changes documented using remote sensing imagery. Jupará promotes agroecological practices, community organization and commercialization. Jupará participants adopted three key agroecological practices: substituting organic fertilizer and pest control for agrochemicals, investing in contour erosion strips, and implementing agroforestry systems that incorporate biodiversity and ecological services. Maintenance of native forest varies by community type. Land reform communities on former cacao estates have fertile soil and can be highly productive while maintaining a diverse native forest overstory (*cabruca*), also protecting remnants of native forest.

K. R. Painter
Institute for Social and Economic Development (ISED) Solutions, Garden City, ID, USA
https://isedsolutions.net

R. Buschbacher (✉)
School of Forest Fisheries and Geomatics Sciences and Tropical Conservation and Development Program, University of Florida, Gainesville, FL, USA
e-mail: rbusch@ufl.edu

L. C. Souto Silva
Jupará Agroecological Movement and COOPASB Cooperative, Bahia, Brazil

E. Costa e Silva
Associação Agricola do Projeto de Assentamento Frei Ventuy, Bahia, Brazil

 
F. Montagnini (ed.), *Biodiversity Islands: Strategies for Conservation in Human-Dominated Environments*, Topics in Biodiversity and Conservation 20,
https://doi.org/10.1007/978-3-030-92234-4_23

Land reform communities on less fertile soils require initial forest clearing. As these communities develop, agroforestry systems increase; native forest cover initially declines, but there is potential to recover and reconnect some remnants. These results draw attention to contextual factors beyond the local scale of project implementation. While diversified agroforestry biodiversity islands were viable in all communities, protection of native forest remnants depends on public policies and consideration of broader historical, geographic and biophysical drivers of land use dynamics.

**Keywords** Agroforestry · Biodiversity islands · Cabruca · Forest conservation · Jupará agroecological movement · Land reform · Quilombo · WWF

## 23.1 Introduction

### *23.1.1 Competing Conservation Paradigms*

According to a prevalent conservation paradigm, large blocks of undisturbed habitat are the most effective approach to achieve conservation goals (Kramer et al. 1997; Noss et al. 1999; Wuerthner et al. 2015; Dinerstein et al. 2017). Yet the social and biological assumptions and implications of this approach are complex, and its normative character is often unacknowledged. Biologically, one consequence of the "large, intact, contiguous blocks" paradigm is that protected areas tend to be in remote, steep, infertile and inaccessible regions that are less suitable for human occupation (Hunter and Yonzon 1993; Scott et al. 2001; Joppa and Pfaff 2009, 2011; Barnes et al. 2018; Putz 2020). The species composition of such regions is not likely to be as rich as, nor representative of, the more anthropocized flat, lowland, fertile habitats. Furthermore, the ecology of so-called "intact" natural areas may in fact be highly influenced by past and present large and small-scale human activities (Denevan 1992; Cronon 1996; Heckenberger et al. 2008; Heckenberger and Neves 2009; Pivello 2011; Ross 2011; Clement et al. 2015; Boivin et al. 2016; Levis et al. 2017, 2020; de Souza et al. 2018; Franco-Moraes et al. 2019; Shepard et al. 2020). To illustrate the normative aspect of the large-block protection paradigm, we can consider that while contemporary conservationists generally abhor fragmentation, not very long ago "edge" habitats and ecotones were considered particularly valuable because they incorporate a diversity of habitats and consequently high local species richness (Leopold 1933 cited in Silbernagel 2003; Odum 1971). In part, this shift in emphasis is a normative change from valuing game species to seeking to protect rare species that require a large range and are particularly vulnerable to extinction (Laurance 2008; Laurance et al. 2018). Yet a recent review of fragmentation research (Fahrig 2017) found that biodiversity effects of habitat fragmentation are usually confounded with effects of habitat loss and that 76% of significant effects of fragmentation were positive. Arroyo-Rodríguez et al. (2022) carried out a multi-taxonomic assessment of small patches of Lacandona rainforest in Mexico and found that species density was as great in patches as in continuous forest, including for forest-specialist species. They conclude that preservation of all forests is important to biodiversity, especially preservation of small patches in human-dominated areas.

Besides these biological considerations, many human populations live in high biodiversity areas, and an exclusionary conservation paradigm creates an inherent conflict between conservation advocates and local people who depend on use of land and management of natural resources for their livelihoods. The potential power and wealth imbalances between external conservation advocates and local communities raises ethical concerns and calls into question the long-term political viability of conservation interventions (Chapin 2004; West et al. 2006; Brockington et al. 2008; Brockington and Wilkie 2015; Holmes and Cavanagh 2016; Kopnina 2016; Kohler and Brondizio 2017; Ybarra 2018).

An alternative conservation paradigm of community-based conservation has been developed since the 1980s. There is strong empirical evidence that community control of resources and habitat leads to equal or better conservation outcomes than state-tenure fully protected areas (Charnley and Poe 2007; Bray et al. 2008; Lele et al. 2010; Soares-Filho et al. 2010; Nelson and Chomitz 2011; Porter-Bolland et al. 2012; Radachowsky et al. 2012; Wood et al. 2019; Elleason et al. 2020; Vélez et al. 2020). Nevertheless, integrating biodiversity conservation with socioeconomic development is challenging and context-specific. Furthermore, the alternative paradigms or approaches may very well be complementary rather than competitive. The concept of biodiversity islands provides ample room for both fully protected areas and biodiversity protection within areas of human occupation (Montagnini et al. 2022).

### *23.1.2 Key Questions and Overview of the Chapter*

The purpose of this chapter is to present the experience, results and lessons from the Jupará Agroecological Movement, which worked with land reform communities to address the apparent conflict between forest protection and human occupation in the cacao region of southern Bahia, Brazil. Using the perspective of biodiversity islands, we analyze whether the initiative was effective in terms of land sharing and land sparing (i.e. creation and maintenance of biodiversity islands through agroforestry production systems, and maintenance of remnants of intact native forest within the occupied landscape).

The implementation and effectiveness of any community-based conservation initiative is highly context specific. Therefore, following this introduction Sect. 23.2 presents a historical overview of human occupation of the cacao region in southern Bahia, and then describes the land reform process occurring in this region. This overview provides the geographical, biological and social context of the apparent conflict between forest conservation and human occupation in this region.

Section 23.3 presents the Jupará Agroecological Movement (hereafter shortened to Jupará), which emerged, with the support of World Wildlife Fund (WWF), to promote an alternative, agroecological model of land reform in 35 land reform and traditional agricultural communities throughout the cacao region of southern Bahia. Jupará's approach and results varied according to the characteristics of each community, and this study focuses on three communities representative of different types of land occupation, which are also presented in Sect. 23.3.

Section 23.4 provides an overview of the research methodology, Sect. 23.5 presents the results of Jupara's work in terms of adoption of agroecological practices, and Sect. 23.6 presents the dynamics of native forest and agroforestry cover in the different communities. These results are presented both as a comparison between participants and non-participants in Jupará activities from within the same communities, and by comparison among the three types of community.

We conclude with reflections on the viability of Jupará's approach for creating the two types of biodiversity islands within land reform communities in southern Bahia (Sect. 23.7). Does the Jupará Agroecological Movement present a viable "proof-of-concept" for agroecological production, community organization and commercialization to achieve forest conservation? How do the historical, ecological and geographic differences between different types of communities enable or limit the adoption of agroecological practices and the protection of forest remnants? And what are the implications of these findings for mobilizing public policy to produce positive outcomes for ecosystems and communities?

## 23.2 Regional Context: Human Occupation and Land Reform in the Cacao Region

### *23.2.1 Cacao and Human Occupation of Brazil's Atlantic Forest*

Brazil's Atlantic Forest is considered a global "hot spot" for biodiversity conservation (Ribeiro et al. 2011). While overall only 12% of this biome maintains its forest cover (including regenerating areas and degraded forest, Ribeiro et al. 2011), it covers a vast region with varied topography, vegetation, human occupation and conservation status. WWF divides the Atlantic Forest into nine ecoregions (Olson and Dinerstein 1998). Two of these, the Serra do Mar Coastal Forest and the Coastal Forests of Bahia, maintain relatively higher levels of forest coverage, the former because of relatively low human occupation, and the latter because of how humans occupied and used the forest. In the Serra do Mar Coastal Forest ecoregion, steep topography drove early colonization of southern Brazil to "leapfrog" from the heavily occupied coast (e.g. Santos and Rio de Janeiro) to the inland plateau that includes major cities such as São Paulo and Curitiba (Morse 1974). In contrast, the Coastal Forests of Bahia ecoregion, in particular in the cacao region of southern Bahia, has maintained large blocks of forest because of an occupation model based on cacao agroforestry with an overstory of native forest (*cabruca* system).

While the port city of Salvador was a major colonial city founded in 1564, indigenous resistance, steep topography and inhospitable climate kept southern Bahia's forests relatively intact as late as 1820 (Adonias Filho 1978). At that point, cacao cultivation took off, and by 1895 cacao was the economic basis for the entire state of Bahia. Between 1895 and 1930, cacao was consolidated as the second greatest source of wealth in all of Brazil. Adonias Filho (1978) refers to the first period as the pioneer phase (*desbravadores*), while the second period was dominated by the "cacao barons" (*coronéis*) who controlled large estates. The latter period was immortalized by Jorge Amado in novels such as Terras do Sem Fim (1942, translated as The Violent Land, 1945).

While cacao cultivation produced great wealth for some, it required a large mass of salaried workers who lived in precarious social and economic conditions. According to Mascarenhas et al. (1999), the Human Development Index in the cacao region of Southern Bahia (5.1) was worse than in three other cacao producing regions of Brazil (5.8–6.5 in Rondonia, Pará and Espiritu Santo), much worse than Indonesia (7.0), Ecuador (8.0) and Malaysia (9.0), and only slightly better than the Ivory Coast (4.0) and Ghana (5.0). Brazil has one of the most unequal distributions of both wealth and land in the world. As of 1979, the region's Gini coefficient was 0.62, with 9% of rural properties being larger than 200 hectares and occupying 49% of the region, while 57% of the properties were less than 50 hectares and occupied 16% of the region (Trevizan 1982). However, productivity was inversely proportional to property size – smaller properties produce relatively more cacao (Trevizan 1982).

Traditionally, cacao was cultivated exclusively in the *cabruca* system. Cacao (*Theobroma cacao*) is an understory tree native to the Amazon which requires shade. *Cabruca* is the system of planting cacao underneath the canopy of native old-growth forest supplemented by some cultivated forest species such as jackfruit (*Artocarpus heterophyllus*) and cajá (*Spondias mondim*), which leaves most of the native forest biomass and ecological services intact. From the 1930s to the 1950s, cacao production declined for multiple economic and biological reasons, including the Great Depression, lack of investment, soil exhaustion and cacao diseases (Adonias Filho 1978). In 1957, a federal agency was created to provide research, credit, rural electrification and other services to strengthen the cacao economy (CEPLAC – Comissão Executiva do Plano da Lavoura Cacaueira, Executive Commission for Cacao Farming Plan). CEPLAC developed a technical package that included homogenous plantations of cacao with a monoculture overstory of either rubber or the leguminous tree *Erythrina*, and included use of chemical fertilizers, herbicides and insecticides.

According to Almeida et al. (2001), this high-input technical package offered a 160% increase in private profit to the landowner, but a lower "social return" measured as the broader distribution of benefits throughout society. Notably, the traditional, low-input system provided greater employment and thus Almeida et al. (2001) calculated that in this case the social return was three times greater than the private profit. Mascarenhas et al. (1999) indicate that labor is 67% of the production cost of cacao (varying from 54 to 84%, depending on which level of technology is used).

In the 1990s, cacao production in southern Bahia went into crisis. High global prices in the 1970s stimulated increasing production in other tropical countries, so that by 1992 prices were at a low point. At the same time, the Brazilian government overvalued its currency as a strategy to control inflation (*Plano Real*), further lowering the price in Reais. Low prices meant less investment in maintaining cacao plantations, and when witches broom, a fungal disease (*Crinipellis perniciosa*), invaded the region in 1989, it quickly spread out of control (Alger and Caldas 1994). Large landowners saw their income decline, government revenue decreased, and masses of salaried rural workers became unemployed. Eventually, land prices of cacao estates declined, and landowners were eager to sell out. Today much of the cacao produced in Bahia is produced by small and medium-sized landholders, because of the labor intensity required to control the disease (Rice and Greenberg 2000).

### 23.2.2 Land Reform and the Pressure on Forests

Land reform (also known as agrarian reform) has been on Brazil's political agenda as a mechanism of addressing poverty and inequality since the era of industrialization following World War II. The Catholic church has been active with this issue since the 1960s (Ecclesiastic Base Communities, CEB) and 1970s (Pastoral Land Commission, CPT, created in 1975). Land reform social movements, such as the Movement of Landless Rural Workers (MST), the Worker's Party (PT), and the Rural Worker's department within the national union umbrella organization CUT (Central Única dos Trabalhadores or Unified Workers' Central), all emerged between 1979 and 1985 (Deere and León 1999). In the 1990s, hundreds of thousands of landless poor engaged in hundreds of land occupations throughout the country. The Fernando Henrique Cardoso administration (1995–2002) settled 525,000 families on 18 million hectares. This was a rather chaotic process with a focus on numerical targets, with limited support for credit or technical assistance, the key factors for success identified by a global analysis of agrarian reform experiences throughout the world (Smith 2001).

The land reform process in the cacao region of Southern Bahia had a very violent trajectory until 1998, when conditions changed dramatically. Key to understanding this is the fact that, in Brazilian land reform generally, only 5% of projects were initiated by the government (i.e. by INCRA, the National Agency for Colonization and Land Reform), with the other 95% initiated by land reform movements such as MST, who organized a group of settlers and petitioned INCRA's cooperation (Cullen et al. 2005). The process of staking a claim and establishing a settlement involved occupation of a suitable property (*acampamento*), either directly or on the side of the road adjacent to it. Such occupations could take months to years to be regularized by INCRA, who would expropriate the land and compensate the landowner if it could be shown to be "unproductive" and/or if enough political pressure was applied. During occupation, settlers could be forcibly removed (*despejo*), either by landowners' "hired guns" and/or by the police; in many cases settlements suffered multiple removals but persisted for months and years until land was expropriated.

During this "contested" land reform, settlers had a strong motivation to occupy areas of forest, protected areas, remote and infertile areas, because in all of these, landowner resistance would be less. In the case of protected areas, settlers might be peaceably removed by offering an alternative piece of land. In the case of large cacao estates, landowners might be open to negotiation if occupation occurred on an uncultivated forest reserve versus established plantations. Inevitably, converting newly occupied forest into productive farms required extensive forest clearing.

Around 1998, the cacao crisis (as explained in Sect. 23.2.1) permitted a transition from this "contested" situation to what we call a "negotiated" form of land reform. At that point, the price of cacao had crashed and productivity had declined as well (due to what some settlers sardonically referred to as "Saint Witches Broom"). Landowners were willing to cede land to expropriation, in which case they might even negotiate compensation (paid by INCRA) far in excess of the market price. This was a tipping point, and agrarian reform rapidly switched from painfully violent occupation of marginal lands to peaceful distribution of the region's best lands, since the major cacao estates were on the best soils and had received infrastructure investments. In some cases, such as that of Fazenda Liberdade in the municipality of Maraú, former paid laborers making less than minimum wage abruptly became collective owners of lands that they had lived and worked on for decades. Converting laborers to owners produced a virtuous cycle of positive incentives for them to invest in the labor- and managerially-intensive control of Witches Broom, thus increasing productivity and quality of cacao.

## 23.3 Study Context: Jupará Approach and Types of Communities

### *23.3.1 The Jupará Agroecological Movement*

The Jupará Agroecological Movement originated in a "collective" of church and rural worker union organizations that supported land reform in the cacao region in the 1980s. In the same region and time period, WWF had been successfully working to create and consolidate the 11,400-hectare Una Biological Reserve, the only legally protected habitat of the endangered golden-headed lion tamarin (*Leontopithecus chrysomelas*) (Saracura 1997). Subsequent WWF-sponsored research showed that the Reserve was not large enough to support a minimum viable population of this species (Dietz et al. 1996), and in 1995 WWF provided an initial grant to Jupará to explore alternatives to forest clearing in land reform communities near the Una Reserve.[1] Jupara's goal was to incorporate conservation in a productive agricultural landscape, while simultaneously seeking to improve social and economic well-being.

[1] WWF's initial funding went from 1995 to 1998, with an 18-month funding gap and a second phase of support from 1999 to 2003 (Buschbacher 2008).

WWF and Jupará set the following program goals: implementation of agroecological practices, organic certification, increased family incomes, and maintenance of 30% of community areas in forests and 40% in agroforestry systems (Buschbacher 2008). The overall goal was to substitute slash and burn agriculture, prevalent among smallholders in areas of tropical, moist forest, with a more resilient and sustainable form of family agriculture. Jupará's agroecological production system substituted natural products and ecological techniques such as compost, green manure, contour erosion barriers, and multicropping for the use of fire and agrochemicals. A total of 10 agroecological practices were promoted (Table 23.1).

Jupará's conceptual model, developed jointly with WWF, integrated production, community organization, commercialization and forest conservation (Fig. 23.1). Jupará took a holistic, agroecological approach. On-the-ground activities integrated technical assistance on agroecological production with a methodology based on empowerment, autonomy and solidarity, including a strong focus on gender equity.

Jupará recognized that economic viability would depend on improving the terms of trade for inputs and products (commercialization in Fig. 23.1). A cooperative (COOPASB – Cooperativa dos Pequenos Produtores e Produtoras Agroecologistas do Sul da Bahia Ltda.) was formed to negotiate better prices at scale. In addition, Jupará proposed small agroindustries to enhance market value and access for diverse crops, such as cacao, fruits, guaraná (*Paullinia cupana*) and manioc. This included processing (i.e. fermentation and drying of cacao and guaranaa seeds and dehydration of fruits), packaging, and value-added products (tapioca cakes from manioc, liqueurs, and jellies and frozen pulp from many kinds of fruit, and eventually chocolate).

Community organization and leadership was supported by Jupará to strengthen cooperation within communities on common production practices, management of agroindustrial enterprises, and articulation with government and other partners.

Finally, the conceptual model considered forest conservation as an inherent characteristic of the agroecological system being promoted. Organizations that wished to participate in Jupará agreed to eliminate forest clearing, prioritize organic agriculture, and commit to quantitative targets for diversified agroforestry production systems and intact forest (40% and 30% of total area, respectively). While the project did not target specific species nor measure biodiversity, reducing forest clearing was expected to conserve biodiversity and ecological services. In addition, relative to either of the prevalent alternatives of slash and burn subsistence agriculture or high-chemical-input cacao cultivation, organic and agroforestry-based production systems are inherently biodiverse and have conservation value (Schroth et al. 2004; McNeely and Schroth 2006; Bhagwat et al. 2008; Jose 2009, 2012; Palacios Bucheli and Bokelmann 2017; Haggar et al. 2019; Santos et al. 2019; Udawatta et al. 2019; Marsden et al. 2020; Montagnini et al. 2022; Montagnini and del Fierro 2022). Protected natural forests and agroforestry-based production systems are the two types of biodiversity island addressed by this chapter.

**Table 23.1** Description of agroecological production practices promoted by Jupará

| |
|---|
| **1. Compost.** Composting of organic materials such as manure, cacao shells and food scraps to produce organic fertilizer |
| **2. Homemade liquid organic fertilizer.** A type of compost tea, using manure and other organic ingredients, that Jupará extensionists taught farmers to make for themselves |
| **3. Green manures.** Nitrogen fixing plants that are planted either between crops or during a fallow period and allowed to decay on site. Can be used to improve soil quality or control erosion; farmers also report using green manures to control ants. Species commonly used were Canavalia (*Canavalia enisformis*) and pigeon pea (*Cajanus cajan*) |
| **4. Commercial organic fertilizer.** Jupará produced this fertilizer at a central location and sold it to farmers at cost |
| **5. Mulching.** Use of mulches such as leaves or cacao shells, usually around the base of trees, to add nutrients and control erosion |
| **6. Multicropping.** More than one species is planted in each field. Tree crops might be planted with annual crops or several tree crops planted together |
| **7. Crop rotation.** Rotation of the location of crops from year to year. An annual crop should be followed by a crop from a different family. In agroforestry systems, annual crops are gradually replaced by perennial and tree crops |
| **8. Contour erosion barriers.** Planting along the contour, with the creation of some kind of barrier to control erosion and gradually form terraces perpendicular to the slope. The barrier is usually created by piling up the organic residue (weeds and fallow vegetation hand cleared in preparation for planting) in contour rows. Typically, annual crops are planted in the cleared area between the rows of residue, and tree crops are planted in the rows. This way of treating the organic residue is an alternative to burning; it is labor-intensive but promotes long-term soil productivity. |
| **9. Elimination of agrochemicals.** Agrochemicals commonly used in Bahia include chemical fertilizers and pesticides to eliminate leafcutter ants. Jupará hoped to replace these with organic products |
| **10. Elimination of fire.** Farmers in Bahia commonly burn to clear forest or clear fallowed fields for planting. Jupará hoped to replace burning with intensive, permanent agroforestry systems and the use of contour barriers |

Source: Painter (2006)

### *23.3.2 Three Types of Settlements in the Cacao Region of Bahia, Brazil*

The Jupará-WWF project worked in 35 communities in 17 municipalities throughout the cacao region of southern Bahia (Fig. 23.2). The historical origins, and consequently the social, biological and geographical characteristics of these communities fall into three clearly differentiated categories. This study looks at one representative community of each type (Table 23.2): a traditional *quilombo* community (Lagoa Santa); a "contested" land reform settlement (Fortaleza); and a "negotiated" land reform settlement (Cascata).

Jupará received WWF support from 1995 to 2003; the dynamics of land reform in this region changed around 1998, and Jupará adapted accordingly. Initially they worked with settlements from "contested" land reform. These were mainly forest

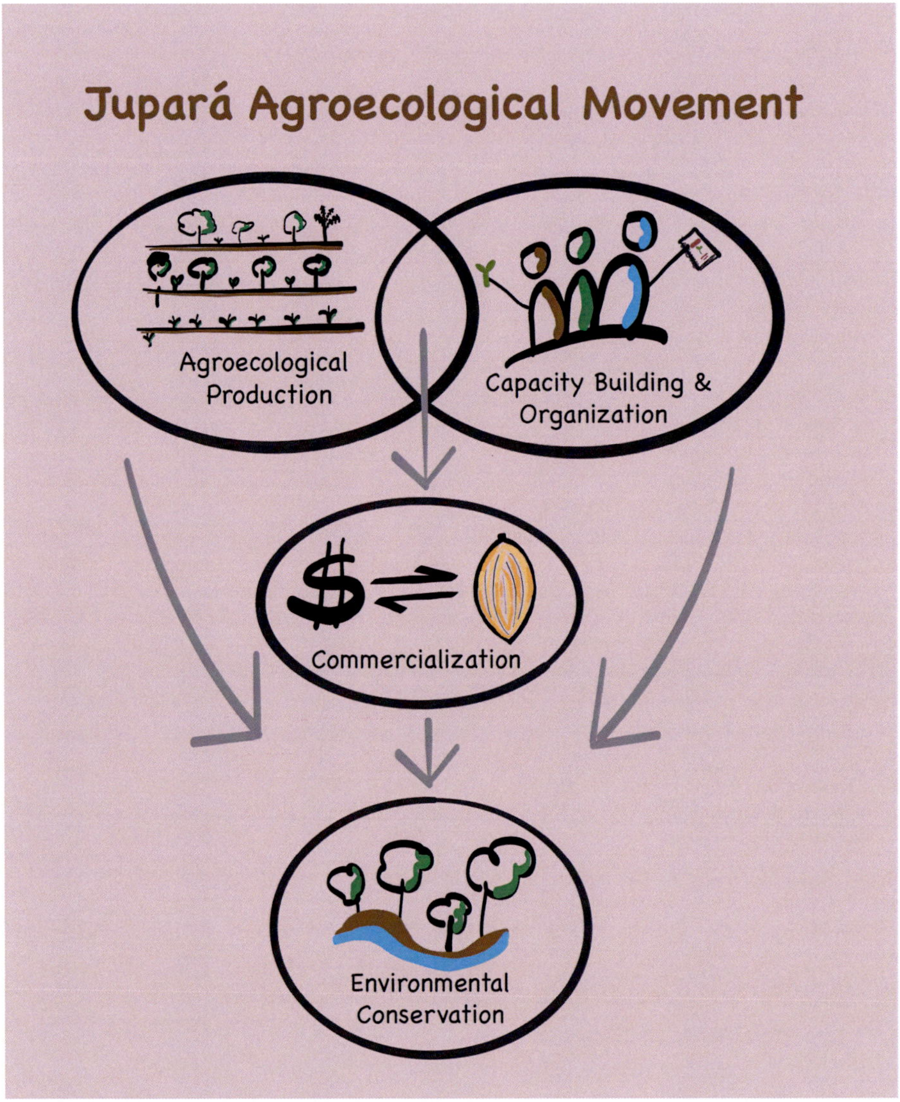


**Fig. 23.1** Conceptual model jointly developed by WWF and the Jupará Agroecology Movement. (Figure elaborated by Carolina Jordão)

remnants that had been passed over by cacao-driven occupation in the nineteenth and early twentieth centuries because of disadvantageous soils and geographies. Fortaleza is typical of this pattern and was selected as a priority area by Jupará and WWF because it is in the same municipality as the Una Biological Reserve.

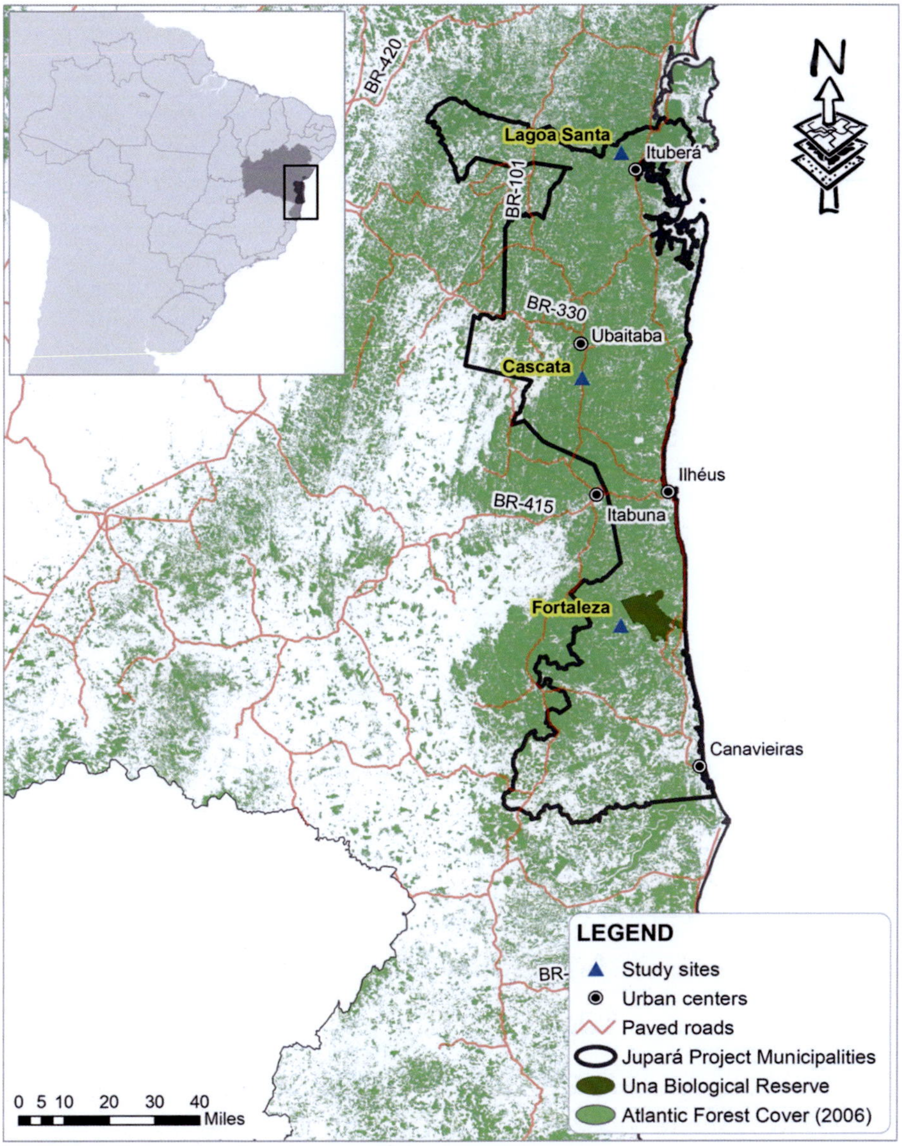


**Fig. 23.2** Location of three agricultural communities within the Cacao Region of Southern Bahia, Brazil. (Map elaborated by Felipe Veluk Gutierrez using MapBiomas)

After 1998, Jupará continued to work with many of these communities but added several others that were located on cacao estates, such as Cascata, that became accessible to "negotiated" land reform after the cacao crisis around 1998.

**Table 23.2** Main social, historical and biophysical characteristics of three family agriculture communities in the cacao region of southern Bahia, Brazil

| Community | Lagoa Santa | Fortaleza | Cascata |
|---|---|---|---|
| Origin history | Traditional; quilombo | Agrarian reform (contested) | Agrarian reform (negotiated) |
| Origin date | 1800s | Mid-1970s[a] | 1998 |
| Average landholding | 9 ha | 22 ha[b] | 11 ha[c] |
| Accessibility | Isolated | Isolated | On major highway |
| Soils | Moderate; Oxisols; Colônia formation | Moderate; Oxisols; Colônia formation | Fertile; Alfisols; Cepec formation |
| Production system | Highly diversified | Diversified | Cacao |
| % of income from farming | 45 | 67 | 75 |
| Percent natural forest | 8% | 36% | 40% |
| Percent agroforestry | 75% | 39% | 46% |
| Forest tendency | Maintain and expand agroforestry | Increase forest on private lands; loss of community forest | Maintain agroforestry and forest reserve |

Source: Based on data from Painter (2006)
[a]The area was occupied starting in the 1970s and expropriated by INCRA in 1986 which gave the 50 occupying families collective land rights. However, it was not until 1998 that INCRA began the process of sub-dividing the area into defined lots for each family
[b]This includes 17 hectare per family lot + 5 hectare per family in the community forest reserve
[c]All the land is held collectively, but each family is assigned 4–5 hectare of cacao plantation to work

The third type of community was not explicitly involved in official land reform, but is a historical remnant of a kind of spontaneous land reform that was self-generated by slaves who escaped from plantations, Jesuit missions, and the slave trade in the nineteenth century. Lagoa Santa was one community of this type where Jupará worked intensively because of the protagonism of a dynamic local leader, André Jesus de Conceição (known as Catixa). These "traditional communities" make an intriguing comparison because they can be conceived as an extrapolation into the future of the contested land reform type of community (e.g. Fortaleza) because they were also settled in remote remnants of forest with minimal capital or support, and have now been occupied for generations.

#### 23.3.2.1 Lagoa Santa

Lagoa Santa is a traditional agricultural community located in the municipality of Ituberá. Lagoa Santa is a *quilombo* community, meaning a community that was originally founded by a group of escaped slaves in the nineteenth century. Resident families have farmed this land for several generations, resulting in numerous land divisions among farmers' descendants and, consequently, small landholdings. Landholding size ranges from 2 to 25 hectares, with an average size of 9 hectares. Many of the larger landholdings are shared by extended families. Little native forest remains in this area. Soil quality is moderate. Terrain is very hilly. While fairly close to the municipal headquarter city of Ituberá, access to Lagoa Santa is difficult.[2] Farmers in Lagoa Santa maintain highly diverse agroforestry systems, producing mainly rubber, cloves, piassava (*Attalea funifera*) palm fiber for roofing material, and small amounts of cupuaçu (*Theobroma grandiflorum*), açaí palm (*Euterpe oleracea)*, guaraná, cacao, black pepper, and manioc flour. On average, 67% of family income comes from agriculture and the remaining 33% comes from off-farm sources, including employment in Ituberá and on nearby plantations.

#### 23.3.2.2 Fortaleza

Fortaleza is a land reform settlement in the municipality of Una and a neighbor of the Una Biological Reserve. Fifty families were settled on 1102 hectares, with an average lot size of 17 hectares. Approximately 250 hectares were designated community areas including a small forest reserve and a community cacao plantation. Families began settling the area in the mid-1970s when the area was completely forested. Most settlers grew up on their parents' farms, usually also in Southern Bahia, or worked on nearby rubber plantations before settling in Fortaleza. INCRA (Instituto Nacional de Colonização e Reforma Agraria, the government agency responsible for land reform) officially divided the land and gave titles to these landholders only in 1997. At that time, and as part of the titling process, some families were moved from their original plot of land to a different plot: in some cases, a new, completely forested plot. In other places, agroforestry plantations that had been cultivated by one family were allocated to a different family; any compensation had to be arranged informally between these families. At the time of this research, many of the original farmers in Fortaleza were getting older and had children who were married and beginning their own families. In some cases, the land has been unofficially divided among several sons. In one case, 23 people were surviving on the production of one 17 hectare parcel. In other cases, the children of these farmers worked off-farm in Una or on nearby plantations. Farmers in Fortaleza produced mainly rubber and cacao as cash crops, as well as some manioc, corn, fruit and vegetable crops for consumption. Few families owned livestock beyond one mule for hauling cacao and a few chickens. On average, 47% of family income came from agricultural production and 53% came from off-farm sources, usually employment in the nearby town of Una, or retirement pensions. Most of the agricultural income came from rubber.

[2] Because of their origin as refuges of escaped slaves, quilombos occur in areas of difficult access. While Lagoa Santa is only about 5 miles from Ituberá, due to very hilly terrain it currently takes about 30 min by car and a good part of a day by foot.

#### 23.3.2.3 Cascata

Cascata, in the municipality of Aurelino Leal, was an abandoned cacao plantation that was sold for land reform, settling 40 families in 1998. The land titles were granted not to individuals but to the community association, although each family was assigned an area of 4–5 hectares of the plantation to maintain and harvest. In addition, about 180 hectares of Atlantic Forest were designated as a community forest reserve. While at the time of this research Lagoa Santa and Fortaleza lacked electricity, potable water, basic sanitation, and regular public transportation services, Cascata is located along a major highway, allowing for easy access to public transportation, local markets, secondary education and health services. Because Cascata was established on a former plantation, much infrastructure was already in place, including a dozen workers' houses occupied by resident families, a large building for meetings and events, a church and school, production areas for processing cacao, and a large cacao dryer. Families enjoy well-constructed brick homes, electricity, running water, modern sanitary facilities, and a public telephone. Cascata has fertile soils and produces cacao almost exclusively; the cacao plantation was well established when families arrived, but heavily disease-infested, and cultural practices focused on recuperation of these plantations.

## 23.4 Methods Used to Assess Agroecological Adoption and Forest Conservation

In order to evaluate the long-term efficacy of the Jupará Agroecological Movement's model of land reform, and to assess the extent to which it is able to create and maintain two kinds of biodiversity islands (diversified agroforestry systems and remnants of natural forest), in 2005 a University of Florida (UF) Master's student and participant of UF's Tropical Conservation and Development program conducted research in collaboration with Jupará to address the following questions:[3] Are families that participated in the Jupará project meeting the quantitative goals of 40% of their area in diversified agroforestry production systems and 30% in intact forest? How do family dynamics and size of landholding influence project implementation? Are project results consistent in different types of communities? (Painter 2006).

[3]Painter collected and analyzed all data presented in this chapter. Buschbacher was her thesis advisor. He worked at WWF until 2000, was the main WWF collaborator with Jupará from 1994 to 2000, and wrote an unpublished case study on the Jupará project which is the basis for much of the regional description in this chapter (Buschbacher 2008). Luiz Carlos Souto Silva and Emerentina Costa e Silva are the leaders of Jupará, and facilitated the design and implementation of Painter's field work. Souto provided information on Jupará agroindustry and commercialization activities post-2008.

Painter (2006) interviewed and made site visits to 60 families in the 3 representative communities described above (10 families in Cascata, 20 in Lagoa Santa, and 30 in Fortaleza). In Cascata, where management practices are collective and uniform among families, data were analyzed for the settlement as a whole. For Lagoa Santa and Fortaleza, an index of participation was used in order to differentiate and compare results between families who were active participants in the Jupará agroecological movement and those who were not (Painter 2006). In Fortaleza, land use change over time was assessed using Landsat imagery (1986 and 2001) analyzed in ArcGIS; participatory mapping with GPS reference points and a digital map provided by INCRA were used to align remote sensing imagery to properties (Painter 2006).

## 23.5 Adoption of Agroecological Production Practices in Different Settlement Types

The Jupará project promoted a set of 10 agroecological production practices (Table 23.1). The results of the interviews conducted by Painter in 2005 show that these 10 practices could be partitioned into 3 groups: three that were widely implemented by both participants and non-participants in the Jupará program; three that were rarely implemented by any of the farmers; and three practices that were differentially implemented by Jupará project participants and not by non-participants (Fig. 23.3).

In 2005, 10 years after initiation of the Jupará project, multi-cropping, crop rotation and mulching were implemented by 70–100% of all farmers, without clear differentiation between Jupará participants and non-participants. These can be considered well established practices in this region (Painter 2006).

On the other hand, relatively few farmers used green manure, homemade liquid organic fertilizer, or organic compost, whether they were Jupará participants or not. Green manures were used by less than 20% of both participants and non-participants in both communities. This practice was not as vigorously promoted as others (many farmers had never even heard of the practice), and seeds were not readily available. The few farmers who did report using green manures reported using them as a pest control method against leaf-cutter ants, rather than as a source of fertilizer. Both the homemade liquid organic fertilizer and organic compost require animal manure as a key ingredient; very few of the farmers in these communities have livestock, and "importing" the manure proved infeasible.

In this context, Jupará changed their strategy to produce their own "organomineral fertilizer" in the industrial zone of Ilhéus, a central city in the cacao region (See Fig. 23.2) where appropriate inputs (ground rock rich in phosphorus and micronutrients, ash, and composted organic matter) could be obtained (Buschbacher 2008). This fertilizer can build up soil organic matter content and contribute to long-

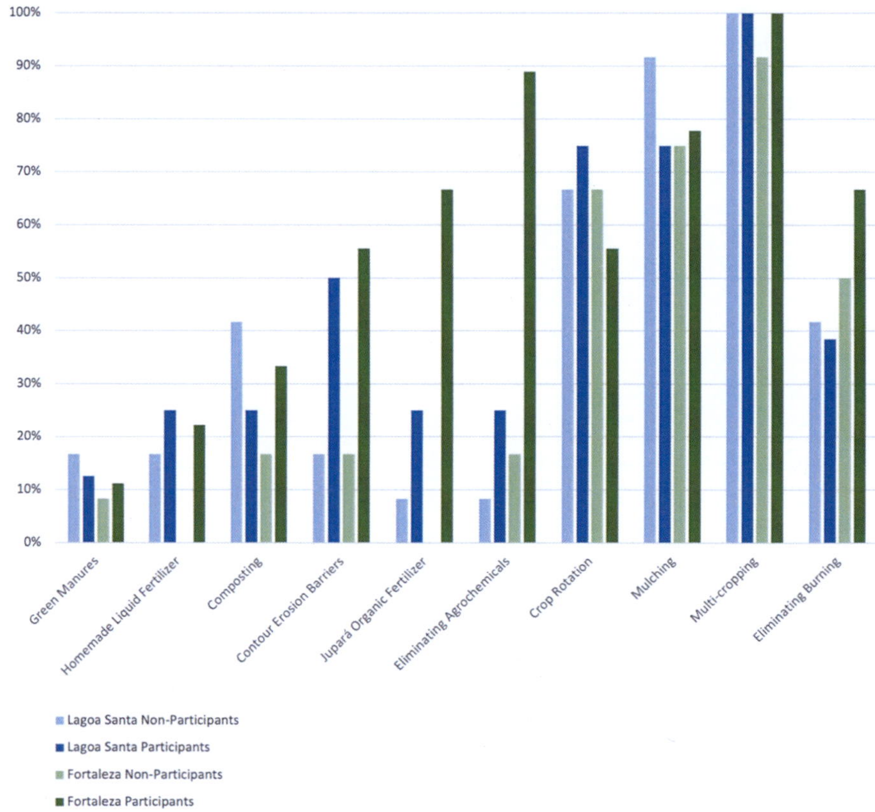


**Fig. 23.3** Adoption of 10 targeted agroecological practices in 2005 by participants and non-participants of Jupará Agroecology movement in two communities of southern Bahia, Brazil

term soil fertility, but it was slightly more expensive than commercial chemical fertilizers and also less concentrated (greater weight and volume), which made it more difficult to transport and apply. These disadvantages were cited by several farmers who chose not to use it, but nevertheless the use of Jupará's organic fertilizer was one of the agroecological practices that were differentially implemented by Jupará participants versus non-participants. There was considerable overlap between the use of Jupará's organic fertilizer and the practice of total elimination of agrochemicals including pesticides, herbicides and chemical fertilizer.

The third practice that was much more widely adopted by Jupará participants than non-participants was the construction of contour erosion barriers. Contour erosion barriers are labor intensive and require some technical expertise to create. The fact that Jupará participants widely adopted this practice indicates that Jupará was successful in providing technical expertise and motivating farmers to invest in a labor-intensive but ecologically beneficial process that enhances long-term productivity. In Fortaleza, around 55% of Jupará participants used contour erosion barriers, 65% used Jupará's organic fertilizer, and nearly 90% reported no use of agrochemicals; in each case less than 20% of non-participants adopted these practices (Fig. 23.3). A similar difference of 50% of participants versus less than 20% of non-participants created contour erosion barriers in Lagoa Santa, whereas only about 25% of Lagoa Santa participants used Jupará's organic fertilizer and totally eliminated agrochemicals (versus less than 10% of non-participants).

The use of fire to prepare fields, generally considered ecologically undesirable, was persistent among Jupará participants and non-participants alike. Because interviews considered the use of fire as a binary yes/no variable, the data do not distinguish how fire is used or how extensively. Certain agroecological practices are incompatible with the use of fire, such as the contour erosion strips which form a vegetative barrier and is a labor-intensive investment that would be destroyed by fire. In addition, agroforestry systems are long-term investments and once established not likely to be intentionally burned. Thus, it is most likely that burning was used only (or mainly) for the initial establishment of agricultural fields, whether from native forest or successional regrowth forest (i.e. previously abandoned farmland). The key to eliminating this practice is continuous cropping once fields are established (i.e. agroforestry systems). Thus, the data on maintenance of forest cover and extent of agroforestry systems (Sect. 23.6) are probably much better indicators of ecological and biodiversity values than whether or not fire is completely eliminated from the repertoire of agricultural practices.

Both biophysical factors and production systems are quite different in Cascata, the former cacao plantation (Table 23.2). The cacao agroforestry system (*cabruca*, with native forest overstory) was already established but heavily attacked by Witches Broom disease; thus, efforts were focused on pruning diseased branches, managing shade and humidity with overstory thinning and repair of drainage ditches, weeding, organic fertilization, replanting of overstory gaps, and grafting disease-resistant clones of cacao (Buschbacher 2008). All Cascata farmers were Jupará participants and the community received a collective organic certification in 2000. All used Jupará's organic fertilizer and none used agrochemicals. Crop rotation and contour erosion barriers were less utilized in the *cabruca* system, but 50% of farmers did use the homemade liquid organic fertilizer for foliar application (Painter 2006). Cascata does have some livestock used to transport cacao from the fields to the processing center for fermentation and roasting, and in 2004 the community built a biogas facility that converted manure to gas for the dryer and other machinery processing cacao and fruits, with the residue used as fertilizer. Only one farmer reported using fire (probably for a small area of subsistence crops, not for the cacao plantation).

## 23.6 Forest Conservation in Different Settlement Types

Although Jupará participants implemented a different subset of agroecological practices when compared with non-participants, differences in forest cover maintenance between these groups were small and not statistically significant (Painter 2006). However, there were major differences between communities, and these are illustrative of how history, land tenure, biophysical characteristics and production systems influence forest dynamics and conservation.

In both Lagoa Santa and Fortaleza, much of the forest cover was converted to agricultural use long before the Jupará project began. The traditional quilombo community of Lagoa Santa has had long occupation, with landholdings subdivided across generations, such that average landholdings are a relatively small 9 hectare per family. Only 8% of this area remains in native forest, far lower than in either of the other communities; yet far more of this community is in agroforestry systems, which cover 75% of the entire community. These agroforestry systems are highly diverse, with rubber, cloves, and piassava palm fiber (*Attalea funifera*) as major crops, often intercropped with cacao, banana, plantain, açai, cupuaçu, black pepper, guaraná, passionfruit, pineapple, manioc, timber species and others. Anecdotally, Jupará participants may have more diverse agroforestry systems than non-participants; certainly, the more prevalent use of contour erosion strips is indicative of more intensive management.

Fortaleza was a typical land reform settlement of the "contested" period when cacao estates were heavily defended and land occupations occurred opportunistically in remote, unsettled areas of forest. When Fortaleza was occupied in the mid-1970s, it was fully forested. By 2005, it had 36% of forest cover, plus 39% in agroforestry (Painter 2006). Agroforestry systems in Fortaleza are diverse but less diverse than in Lagoa Santa. Rubber and cacao are the main cash products; manioc and other short-lived crops are produced during the early stages of agroforestry system establishment, and piassava, açai, cupuaçu and other species are also present.

From the time of occupation until 1998 when the federal land reform agency INCRA actually demarcated individual lots, the community had set aside 20% of its area as a community forest reserve. Rather than formalize the land division that had been established during the first 20 years of occupation, INCRA decided unilaterally to allocate the territory, such that eight of the 50 families were given lots in fully forested areas within the community forest reserve. The families that were relocated into the forest reserve had to clear part of that forest to produce. In spite of this, forest cover from remote sensing images showed that total community forest cover remained almost constant between 1986 and 2001 (the latest cloud-free image available in 2005): natural forest cover declined by 2% while agroforestry and successional forest (which could not be distinguished in the remote sensing imagery) increased by 3% (Painter 2006, Fig. 23.4). In fact, while forest cover decreased on the "new lots" of community members who were relocated to the forest reserve, forest cover increased on the older lots by 20% among Jupará participants and nearly 10% among non-participants. Much of this increase in forest cover occurred along

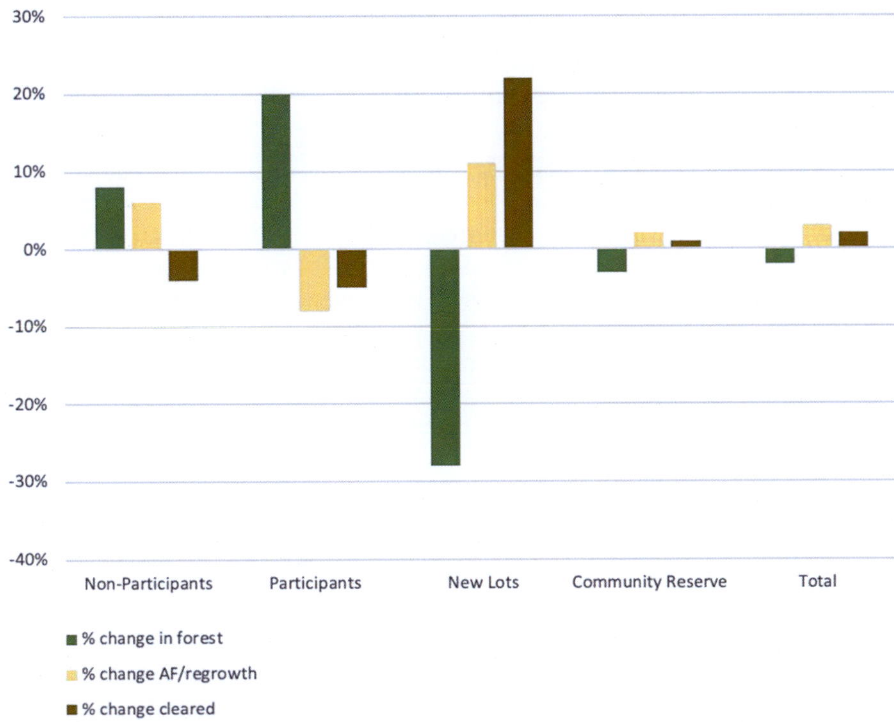


**Fig. 23.4** Percent change in forest cover, agroforestry, and cleared areas among participants and non-participants of the Jupará Agroecology movement, in new lots established within the forest reserve, and in the remaining forest reserve of Fortaleza Land Reform Settlement, Bahia, Brazil, 1986–2001

two riparian corridors in the center of the community, consistent with Jupará recommendations. Furthermore, the community forest reserve was never fully intact; about 5% had been cleared by 1986, which increased to 6% in 2001 (as opposed to 28% cleared in the areas that were transformed into family lots).

The case of Cascata, which was an established cacao plantation that was converted into a land reform settlement, is quite different. Upon establishment as a land reform community, the estate already had 46% of its area in agroforestry systems, specifically *cabruca* cacao. In addition, 40% of the settlement area was intact forest. This is rather typical for cacao estates, in which landowners claimed and controlled as much land as they could and gradually implemented cacao plantations in the most accessible and least steep portions. Although no remote sensing analysis was performed, there was no major observable change to the coverage of either the native forest or agroforest (Painter 2006).

## 23.7 The Effectiveness of Jupara's Agroecological Land Reform Model and Its Applicability for Regional Conservation and Development

This final reflection will look at Jupará's impact in terms of agroecological production practices, the social and economic viability of this approach to land reform, and the larger implications for regional forest conservation. Does the Jupará Agroecological Movement present a viable "proof-of-concept" for agroecological production, community organization and commercialization to achieve forest conservation? How do the historical, ecological and geographic differences between different types of communities enable or limit the adoption of agroecological practices and the protection of forest remnants? And what are the implications of these findings for mobilizing public policy to produce positive outcomes for ecosystems and communities?

### *23.7.1 Viability of Jupará's Agroecological Model*

In terms of agroecological production practices, Jupará was able to introduce and implement diversified cropping systems with organically enriched soils that will be productive for decades in all three communities. The adoption of organic practices (Jupará organic fertilizer and elimination of agrochemicals) and long-term investment in diversified agroforestry (contour erosion barriers) by Jupará participants demonstrates a significant shift towards more sustainable agriculture.

Furthermore, Jupará was able to capitalize on the agroecological production and community organization to implement a series of commercialization activities between 2002 and 2012. Jupará's Cooperative COOPASB exported one container of organic cacao directly to Europe with WWF's support in 2002 and continued selling certified organic cacao to regional cacao trading companies thereafter. In addition, food processing centers ("*Casas de Agroaproveitamento*") in six different communities produced dehydrated fruit and fruit pulp for juices, including jackfruit (*Artocarpus heterophyllus*), cajá (*Spondias mombin*), banana, plantain, pineapple and jenipapo (*Genipa americana*). Sales agreements for these products were negotiated with government programs for food acquisition from family agriculture.[4] Jupará, COOPASB and participating communities managed resources from federal agencies INCRA, CONAB[5] and the Secretariat for Territorial Development, and

[4]PAA – Programa de Aquisição de Alimentos (Food Acquisition Program), destined for poor populations and national food stocks and PNAE – Programa Nacional de Alimentação Escolar (National Program of School Nutrition).

[5]CONAB – Companhia Nacional de Abastecimento, implements the PAA and executed a purchase agreement with COOPASB which included advanced payment that was used for capital investment in the food processing centers.

attained production levels of 80–100 tons per year of dehydrated fruits and 4–5 tons per month of fruit pulp, providing revenue of 3.5 million Brazilian Reais per year and partial employment to 400 community members, with priority given to women and youth.[6]

Finally, environmental conservation through the land sharing type of biodiversity island is evident in the agroecological practices. These increase productivity and long-term sustainability of the production system (Sandhu et al. 2010; Powlson et al. 2011; Palm et al. 2014; Reganold and Wachter 2016), while enhancing biodiversity and ecological services both on and off site. The practices decrease manufacturing, transport and sale of toxic chemicals, as well as toxic exposure of both producers and consumers. They also increase the quality of water runoff, by reducing erosion, leaching of nutrients and toxic chemical load (Sebesvari et al. 2012; Cambardella et al. 2015). Organic matter and soil organisms are noticeably more abundant in contour erosion strips where weeds and other organic material are piled to decay, when compared to adjacent land between the strips, and especially when compared to sites prepared by burning (authors' personal observation). The use of organic pest control maintains the biodiversity of natural pest predators and such systems are inherently bird-friendly (Bengtsson et al. 2005; Crowder et al. 2010).

Taken together, these positive results for agroecological production, community organization, commercialization and land sharing clearly demonstrate the viability of Jupará's model of land use in agricultural communities of the cacao region of southern Bahia. Jupará's accomplishments, especially the integrated model including the biodigestor implemented in the community of Cascata, received national and global recognition: the Social Technology award from the Bank of Brazil Foundation in 2005, "Best Practice for Local Management" in the category of "Environmental management, combating poverty, changing patterns of production and consumption" from the Caixa Economica Federal bank, and "Best Practice in Conservation of the Natural Environment" from UNESCO in 2006.[7]

### *23.7.2 Application of Jupará's Model for Forest Conservation at Landscape Scale*

To understand the broader implications of Jupará's model of agroecological land reform for regional forest conservation (i.e. the land sparing type of biodiversity island), the comparison among the three different communities is decisive. When given the opportunity to occupy a productive part of the landscape (i.e. the cacao estate at Cascata), there was no necessity to clear forest and the large forest remnant covering 40% of this settlement was maintained. The 46% of the settlement that is in

[6] Each center employed about 20 workers, which was rotated among 80 community members, i.e. 1 week per month each.

[7] https://mirror.unhabitat.org/bp/bp.list.details.aspx?bp_id=1580

*cabruca* cacao absorbed the community's labor and provided the economic basis for a prosperous quality of life. This result could be undermined in the future by population growth or increasing economic aspirations (e.g. Jevons paradox – see Alcott 2005; Polimeni et al. 2015; York and McGee 2016). Nevertheless, Jupará's actions to increase the productivity of the forest overstory by processing the widespread jackfruit and cajá, diversifying this overstory with other economically valuable species, and converting animal manure into biogas, as well as proposals to produce artisanal chocolate and develop agritourism with access to native forest and a waterfall, would all be beneficial to the sustainability of this system and its biodiversity and ecological services.

The case of Fortaleza is in some ways opposite to Cascata. Landless poor occupied an area of forest in the 1970s and had to clear the forest to produce. Poor soils made it more difficult to produce the more nutrient-demanding cacao as intensively as in Cascata, and poor logistics and limited markets made it difficult to commercialize the full array of potential agroforestry crops.[8] The slash and burn model of tropical forest agriculture would lead to continuation of this initial forest loss, but by basing production on diversified agroforestry (primary income sources are rubber and cacao) these farmers achieved an equilibrium with a mix of agroforestry, remnant forest, and some cleared areas (Fig. 23.4). In fact, our data show that forest cover on some parts of the settlement is increasing, more so among Jupará participants than non-participants (Fig. 23.4). We observed that community members gave priority to reforesting connected riparian areas (Painter 2006). Unfortunately, the public policy implemented by INCRA relocated some families onto the community forest reserve, and the cycle of initial forest clearing began again.

Lagoa Santa may represent the future tendency of Fortaleza extrapolated through several generations. The biophysical and geographical situation of these two communities is similar (Table 23.2), but Lagoa Santa has had much longer occupation. Subdivision of landholdings across generations has led to more intensively managed agroforestry systems that are more diverse in Lagoa Santa than in Fortaleza, but very little remaining forest (8% in Lagoa Santa compared to 36% in Fortaleza).

Notably, these differences in socioeconomic and conservation outcomes across communities are not a function of Jupará efforts, but rather due to the broader historical, geographic and biophysical characteristics of each community. Cascata had the most successful economic and conservation outcomes because of favorable soils, location, and pre-existing land use. In addition, Cascata maintained a collective organization, with all community members participating in Jupará, versus about half of community members from other communities participating in Jupará. Unlike Lagoa Santa and Fortaleza, where farmers work individually on their own widely-separated landholdings, Cascata Land Reform Settlement was created on an existing plantation, with an established, continuous plantation of cacao and a centralized

[8] Cascata is on a main road with access to public transport, whereas the other two sites are much more difficult to access. This matters both for access to markets and access to other jobs for off-farm income, and potentially less pressure for the land to support everyone in the family.

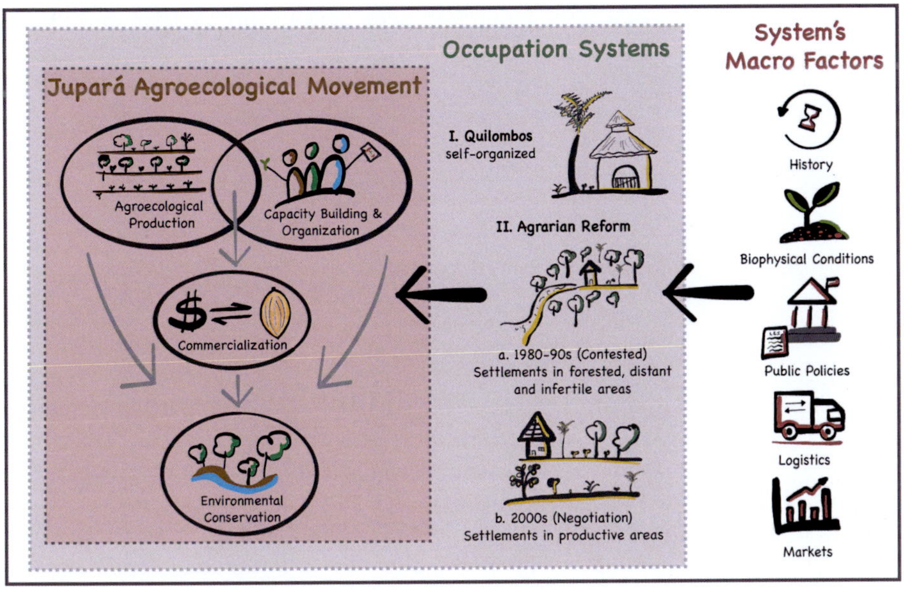


**Fig. 23.5** Conceptual model modified to consider the type of land occupation and macro-systemic factors. (Figure elaborated by Carolina Jordão)

hamlet of housing and processing facilities. This enabled implementation of the biogas digester, centralized processing of cacao to ensure high quality, and collective commercialization.

These findings lead us to propose a revised version of Jupará's conceptual model, expanding it to look beyond the specific actions within the community that Jupará can influence directly (Fig. 23.1), to incorporate the larger-scale system within which Jupará's efforts are embedded (Fig. 23.5). Agroecological practices, from production systems to community organization and commercialization, vary dramatically between settlements in productive areas (i.e. former cacao estates such as Cascata) and settlements within forested, inaccessible and less fertile areas. Among the latter, there are significant differences in landholding size, demographics, and production systems between long-established "traditional communities" such as quilombos and newly established "contested land reform communities" (i.e. Lagoa Santa versus Fortaleza). All of these differences reflect the cacao region's history, biophysical variability, and geographic implications for markets and logistics. Finally, the establishment, implementation and viability of all of these settlement types is strongly affected by public policies.

The broader systemic perspective on regional development and land reform (Fig. 23.5) calls attention to the complexity of the challenges of conservation and development. One critical yet under-studied aspect of sustainability and livelihoods of these family farmers is the complex set of linkages between rural and urban areas. For example, 25–55% of family income in these communities comes from sources other than agriculture, whether off-site labor or retirement or other payments

(Table 23.2). The successful production and commercialization system described below (André Catixa) is based on access to the weekly municipal market. These are some of the complexities that come into focus when we broaden our perspective to a larger systemic scale.

### 23.7.3 Public Policy Influence on Implementation of Agroecological Practices and Forest Conservation

The potential for maintenance of native forest within agricultural settlements in southern Bahia varies from the forest rich cacao estates to the traditional communities which are rich in agroforestry but have little native forest. Within this panorama, public policy can be critical. The pre-1998 approach to land reform – reactive legalization following occupation of economically marginal forest rather than proactive distribution of existing productive lands – catalyzed forest clearing. Various INCRA policies exacerbated this outcome, such as considering private lands with more than 50% forest as "unproductive" and thus subject to occupation and expropriation, and demarcating lands that exceed the legally required minimum of 20% forest as areas to be farmed.[9] The outrageous case of Fortaleza where 8 of the 50 families were moved internally within the settlement by INCRA onto lots that were wholly natural forest illustrates the negative impact of these policies. While in one part of the community families were increasing the area of forest along riparian corridors, in another part INCRA resettled people with no option but to clear forest.

But with the transition of the cacao economy that occurred in 1998, land reform began to occur in the most productive part of southern Bahia's landscape. As a result, it is now possible to reconcile land reform with conservation in both kinds of biodiversity islands, agroforestry production systems and forest remnants. Cascata exemplifies this, and at one point Jupará incorporated several such settlements throughout the region. These communities contain important forest remnants, with the potential to amplify their impact by taking a regional perspective. Just as farmers within Fortaleza protected a corridor of forest in a riparian area, the geography and topography of the entire cacao region is amenable to creating regional corridors. Cacao estates open onto roads and back up into steep slopes; *cabruca* agroforestry is concentrated near the entrance and considerable forest remnants are left in the interior. These forest remnants could be linked to those of nearby landholdings with similar layouts and designated as regional corridors/biodiversity islands.

The current situation of André Catixa, the community leader from Lagoa Santa, is an exemplary case of how government support can be integrated with the Jupará agroecological approach. Through the agency of community members, the state

[9] Even though this is contrary to Atlantic Forest Decree 750/93. "*Ficam proibidos o corte, a exploração e a supressão de vegetação primária ou nos estágios avançado e médio de regeneração da Mata Atlântica*"

**Fig. 23.6** Catixa (André Jesus de Conceição) in his family's irrigated production system for edible greens, with a small portion of his agroforestry plot immediately behind, Lagoa Santa community. Far background is neighboring land that does not participate in Jupará. (Photo: Robert Buschbacher)

government has formally recognized Lagoa Santa as a *quilombo* community, with accompanying land rights; some non-quilombo landowners are still present but over time these should be expropriated and removed. Public policies have provided support for housing, and Catixa's extended family lives in a community of six houses; initial construction of these was sub-standard but community members have repaired and expanded these and currently live in dignified conditions. Credit to *quilombo* members enabled the family to install an irrigation system, and they currently have a small, terraced plot that produces a variety of edible greens (Fig. 23.6) that the family harvests weekly, cleans, packages, and sells at the Ituberá Saturday market. The family has access to education, internet, a car, and motorcycles. The remainder of their land is in diversified agroforestry systems, with multiple plots divided among different family members. Planting is responsive to market opportunities, with recent increase in açai planting and diversification into high-value exotic fruits such as mangosteen and rambutan.

André Catixa was able to leverage government support to achieve exemplary results, but unfortunately the Jupará extension model has not yet been widely adopted by the government. Rather than being supportive or even proactive, government extension agents who prepare family agriculture loan projects, and the banks who process them, have to be coaxed and guided to enable agroforestry-based projects. The case of land demarcation in Fortaleza shows INCRA working directly contrary to forest conservation. The Jupará Agroecological Movement has demonstrated social, economic and production feasibility, but until this approach is mainstreamed in existing governmental programs, its implementation will depend on necessarily-limited external funding and favorable macro conditions.

## 23.8 Conclusions

The goal of the Jupará Agroecological Movement was to demonstrate and implement a socially, environmentally and economically sustainable model of land reform adapted to the historical, biophysical and socioeconomic conditions in the cacao region of southern Bahia. WWF supported this initiative in the belief that it could reduce the rate of forest clearing and conserve biologically important habitat. Their partnership was guided by a conceptual model (Fig. 23.1) which integrated agroecological production practices, community organization, and commercialization of multiple, value-added products, to achieve both land sharing in the organic agroforestry plots and land sparing in protected remnants of native forest.

Jupará was able to introduce and implement diversified cropping systems with organically enriched soils that will be productive for decades in all three communities. The adoption of organic practices (Jupará organic fertilizer and elimination of agrochemicals) and long-term investment in diversified agroforestry (contour erosion barriers) by Jupará participants demonstrates a significant transition. In addition, Jupará was able to demonstrate economic viability in the context of government investments in land reform communities and policies for food acquisition from family agriculture. Finally, the organic, agroforestry-based production systems were clearly compatible with the land-sharing type of biodiversity island. These positive results for agroecological production, community organization, commercialization and land sharing clearly demonstrate the viability of Jupará's model of land use in agricultural communities of the cacao region of southern Bahia.

In terms of the regional applicability of this model and its implications for the land sparing type of biodiversity island, our analysis shows that larger-scale contextual factors – ranging from the type of land occupation to broader systemic variables – determine the type of productive system, how land can be utilized, and thus the feasibility of maintaining forest cover. We thus propose a revised conceptual model (Fig. 23.5) that recognizes the different types of community land-use occupation and considers broader systemic social, political, economic and biophysical factors. These driving variables were not taken into account when WWF and Jupará participants agreed on quantitative targets for 30% forest[10] and 40% agroforestry

[10]The 30% target was conceived as an incremental improvement on the Brazilian Forest Code's requirement, albeit with limited to no enforcement, that all landholdings in the Atlantic Forest maintain 20% forest cover.

coverage. There was never an enforcement mechanism related to these targets, nor even a measured baseline or monitoring of forest coverage prior to Painter (2006). This was because Jupará conceived standing forest as a valued asset, not as an opportunity cost (Buschbacher 2008). Results show that the agroforestry targets were achievable on their economic merits, but that protection of forest remnants depends on enabling conditions, as illustrated by the three types of community analyzed in this chapter. One implication is that Jupará could develop more viable targets for forest protection by taking existing conditions into account at the level of communities or individual landholdings (i.e. the fact that Lagoa Santa has only 8% forest cover or that some farmers in Fortaleza were moved to landholdings with 100% forest). More broadly, the key finding of this study with regard to the land sparing type of biodiversity island within land reform communities of southern Bahia is that it does occur in some areas, but that supportive public policies are essential to achieve large-scale maintenance of natural forest within the broader landscape.

Public policies can support diversified production systems, maintenance of forest remnants, and socioeconomic well-being of family farmers in two ways: by providing more support (i.e. financing, technical assistance, and public purchasing) and by ensuring that existing forms of support align with agroecological land use. The example of André Catixa shows the positive outcomes from well-aligned policies, capacity and motivation for agroecological land use. The case of land demarcation in Fortaleza shows INCRA working directly contrary to forest conservation. The Jupará Agroecological Movement has demonstrated social, economic and production feasibility, but until this approach is mainstreamed in existing governmental programs, its implementation will depend on necessarily-limited external funding and favorable macro conditions.

## References

Alcott B (2005) Jevons' paradox. Ecol Econ 54(1):9–21

Alger K, Caldas M (1994) The declining cocoa economy and the Atlantic Forest of southern Bahia, Brazil: conservation attitudes of cocoa planters. Environmentalist 14(2):107–119

Almeida FA, Mascarenhas GCC, Midlej RR (2001) Estudo da cadeia agroindustrial do cacau. In: RCMT V, Teixeira Filho A (eds) Cadeias Produtivas no Brasil: Análise de Competitividade, vol 1, 1st edn. Embrapa, Brasília, pp 110–135

Amado J (1942) Terras do Sem Fim. Livraria Martins, São Paulo

Arroyo-Rodríguez V, Arasa-Gisbert R, Arce-Peña N, Cervantes-López MJ, Cudney-Valenzuela SJ, Galán-Acedo C, Hernández-Ruedas MA, San-José M, Fahrig L (2022) The importance of small rainforest patches for biodiversity conservation: a multi-taxonomic assessment in the Lacandona region, Mexico. In: Montagnini F (ed) Biodiversity islands: strategies for conservation in human-dominated environments, Topics in Biodiversity and Conservation. Springer, Cham, pp 41–60

Barnes MD, Glew L, Wyborn C, Craigie ID (2018) Prevent perverse outcomes from global protected area policy. Nat Ecol Evol 2(5):759–762

Bengtsson J, Ahnström J, Weibull AC (2005) The effects of organic agriculture on biodiversity and abundance: a meta-analysis. J Appl Ecol 42(2):261–269

Bhagwat SA, Willis KJ, Birks HJB, Whittaker RJ (2008) Agroforestry: a refuge for tropical biodiversity? Trends Ecol Evol 23(5):261–267

Boivin N, Zeder MA, Fuller DQ, Crowther A, Larson G, Erlandson JM, Denham T, Petraglia MD (2016) Ecological consequences of human niche construction: examining long-term anthropogenic shaping of global species distributions. Proc Natl Acad Sci 113(23):6388–6396

Bray DB, Duran E, Ramos VH, Mas JF, Velazquez A, McNab RB, Barry D, Radachowsky J (2008) Tropical deforestation, community forests, and protected areas in the Maya Forest. Ecol Soc 13(2):56

Brockington D, Wilkie D (2015) Protected areas and poverty. Philos Trans R Soc B Biol Sci 370(1681):20140271

Brockington D, Duffy R, Igoe J (2008) Nature unbound: conservation, capitalism and the future of protected areas. Earthscan, London/Sterling

Buschbacher R (2008) Reforma Agrária Ecológica na Região Cacaueira do Sul da Bahia: O Projeto Jupará de Conservação e Desenvolvimento. Unpublished manuscript, 92pp

Cambardella CA, Delate K, Jaynes DB (2015) Water quality in organic systems. Sust Agric Res 4(3):60–69

Chapin M (2004) A challenge to conservationists. World Watch 17(6):17–31

Charnley S, Poe MR (2007) Community forestry in theory and practice: where are we now? Annu Rev Anthropol 36:301–336

Clement CR, Denevan WM, Heckenberger MJ, Junqueira AB, Neve EG, Teixeira WG, Woods WI (2015) The domestication of Amazonia before European conquest. Proc R Soc B Biol Sci 282(1812):20150813

Cronon W (1996) The trouble with wilderness: or, getting back to the wrong nature. Environ Hist 1(1):7–28

Crowder DW, Northfield TD, Strand MR, Snyder WE (2010) Organic agriculture promotes evenness and natural pest control. Nature 466(7302):109–112

Cullen L Jr, Alger K, Rambaldi DM (2005) Land reform and biodiversity conservation in Brazil in the 1990s: conflict and the articulation of mutual interests. Conserv Biol 19(3):747–755

de Souza JG, Schaan DP, Robinson M, Barbosa AD, Aragão LE, Marimon BH Jr, Marimon BS, da Silva IB, Khan SS, Nakahara FR, Iriarte J (2018) Pre-Columbian earth-builders settled along the entire southern rim of the Amazon. Nat Commun 9(1):1–10

Deere CD, León M (1999) Towards a gendered analysis of the Brazilian agrarian reform. Occasional Paper #16, Latin American Studies Consortium of New England. University of Connecticut, Storrs

Denevan WM (1992) The pristine myth: the landscape of the Americas in 1492. Ann Assoc Am Geogr 82(3):369–385

Dietz JM, de Sousa SN, Billerbeck R (1996) Population dynamics of golden-headed lion tamarins *Leontopithecus chrysomelas* in Una Reserve, Brazil. Dodo J Wildl Preserv Trusts 32:115–122

Dinerstein E, Olson D, Joshi A, Vynne C, Burgess ND, Wikramanayake E et al (2017) An ecoregion-based approach to protecting half the terrestrial realm. Bioscience 67(6):534–545

Elleason M, Gua Z, Deng Y, Jiang A, Goodale E, Mammides C (2020) Strictly protected areas are not necessarily more effective than areas in which multiple human uses are permitted. Ambio 2020:1–16

Fahrig L (2017) Forty years of bias in habitat fragmentation research. In: Kareiva P, Marvier M, Silliman B (eds) Effective conservation science. Oxford University Press, Oxford

Filho AA (1978) Sul da Bahia: Chão de Cacau (Uma Civilização Regional). Editora Civilização Brasileira, Rio de Janeiro

Franco-Moraes J, Baniwa AF, Costa FR, Lima HP, Clement CR, Shepard GH Jr (2019) Historical landscape domestication in ancestral forests with nutrient-poor soils in northwestern Amazonia. For Ecol Manag 446:317–330

Haggar J, Pons D, Saenz L, Vides M (2019) Contribution of agroforestry systems to sustaining biodiversity in fragmented forest landscapes. Agric Ecosyst Environ 283:106567

Heckenberger M, Neves EG (2009) Amazonian archaeology. Annu Rev Anthropol 38:251–266

Heckenberger MJ, Russell JC, Fausto C, Toney JR, Schmidt MJ, Pereira E, Franchetto B, Kuikuro A (2008) Pre-Columbian urbanism, anthropogenic landscapes, and the future of the Amazon. Science 321(5893):1214–1217

Holmes G, Cavanagh CJ (2016) A review of the social impacts of neoliberal conservation: formations, inequalities, contestations. Geoforum 75:199–209

Hunter ML, Yonzon P (1993) Altitudinal distributions of birds, mammals, people, forests, and parks in Nepal. Conserv Biol 7(2):420–423

Joppa LN, Pfaff A (2009) High and far: biases in the location of protected areas. PLoS One 4(12): e8273. https://doi.org/10.1371/journal.pone.0008273

Joppa LN, Pfaff A (2011) Global protected area impacts. Proc R Soc B Biol Sci 278(1712): 1633–1638

Jose S (2009) Agroforestry for ecosystem services and environmental benefits: an overview. Agrofor Syst 76(1):1–10

Jose S (2012) Agroforestry for conserving and enhancing biodiversity. Agrofor Syst 85(1):1–8

Kohler F, Brondizio ES (2017) Considering the needs of indigenous and local populations in conservation programs. Conserv Biol 31(2):245–251

Kopnina H (2016) Half the earth for people (or more)? Addressing ethical questions in conservation. Biol Conserv 203:176–185

Kramer R, van Schaik C, Johnson J (eds) (1997) Last stand: protected areas and the defense of tropical biodiversity. Oxford University Press, Oxford

Laurance WF (2008) Theory meets reality: how habitat fragmentation research has transcended island biogeographic theory. Biol Conserv 141(7):1731–1744

Laurance WF, Camargo JL, Fearnside PM, Lovejoy TE, Williamson GB, Mesquita RC, Meyer CF, Bobrowiec PE, Laurance SG (2018) An Amazonian rainforest and its fragments as a laboratory of global change. Biol Rev 93(1):223–247

Lele S, Wilshusen P, Brockington D, Seidler R, Bawa K (2010) Beyond exclusion: alternative approaches to biodiversity conservation in the developing tropics. Curr Opin Environ Sustain 2(1–2):94–100

Leopold A (1933) Game management, 1986 edn. University of Wisconsin Press, Madison, 481pp

Levis C, Costa FR, Bongers F, Peña-Claros M, Clement CR, Junqueira AB, Neves EG, Tamanaha EK, Figueiredo FO, Salomão RP, Castilho CV (2017) Persistent effects of pre-Columbian plant domestication on Amazonian forest composition. Science 355(6328):925–931

Levis C, Peña-Claros M, Clement CR, Costa FR, Alves RP, Ferreira MJ, Figueiredo CG, Bongers F (2020) Pre-Columbian soil fertilization and current management maintain food resource availability in old-growth Amazonian forests. Plant Soil 450:29–48

Marsden C, Martin-Chave A, Cortet J, Hedde M, Capowiez Y (2020) How agroforestry systems influence soil fauna and their functions – a review. Plant Soil 453(9):29–44

Mascarenhas GCC, Midlej RR, Trevizan SP, Afonso JM, Andrade Filho EN, de Medeiros JX (1999) O cluster do cacau no sul da Bahia. In: Haddad PR (ed) A Competitividade do Agronegócio e o Desenvolvimento Regional no Brasil: Estudo de Clusters. CNPq/EMBRAPA, Brasilia, pp 59–123

McNeely JA, Schroth G (2006) Agroforestry and biodiversity conservation–traditional practices, present dynamics, and lessons for the future. Biodivers Conserv 15(2):549–554

Montagnini F, del Fierro S (2022) Functions of agroforestry systems as biodiversity islands in productive landscapes. In: Montagnini F (ed) Biodiversity islands: strategies for conservation in human-dominated environments, Topics in Biodiversity and Conservation. Springer, Cham, pp 89–116

Montagnini F, Levin B, Berg KE (2022) Introduction. Biodiversity islands: strategies for conservation in human-dominated environments. In: Montagnini F (ed) Biodiversity Islands: strategies for conservation in human-dominated environments, Topics in Biodiversity and Conservation. Springer, Cham, pp 1–35

Morse RM (1974) From community to metropolis: a biography of São Paulo, Brazil. Octagon Press, London

Nelson A, Chomitz KM (2011) Effectiveness of strict vs. multiple use protected areas in reducing tropical forest fires: a global analysis using matching methods. PLoS One 6(8):e22722
Noss RF, Dinerstein E, Gilbert B, Gilpin M, Miller BJ, Terborgh J, Trombulak S (1999) Core areas: where nature reigns. In: Soulé ME, Terborgh J (eds) Continental conservation: scientific foundations of regional reserve networks. Island Press, Washington, DC, pp 99–128
Odum EP (1971) Fundamentals of ecology, 3rd edn. Saunders, Philadelphia
Olson DM, Dinerstein E (1998) The Global 200: a representation approach to conserving the Earth's most biologically valuable ecoregions. Conserv Biol 12(3):502–551
Painter KR (2006) Effectiveness of a non-governmental organization (NGO) extension program for forest conservation and sustainable agroforestry in southern Bahia, Brazil. Master's thesis, University of Florida, Gainesville, FL
Palacios Bucheli VJ, Bokelmann W (2017) Agroforestry systems for biodiversity and ecosystem services: the case of the Sibundoy Valley in the Colombian province of Putumayo. Int J Biodivers Sci Ecosys Serv Manag 13(1):380–397
Palm C, Blanco-Canqui H, DeClerck F, Gatere L, Grace P (2014) Conservation agriculture and ecosystem services: an overview. Agric Ecosyst Environ 187:87–105
Pivello VR (2011) The use of fire in the Cerrado and Amazonian rainforests of Brazil: past and present. Fire Ecol 7:24–39
Polimeni JM, Mayumi K, Giampietro M, Alcott B (2015) The Jevons paradox and the myth of resource efficiency improvements. Routledge, Oxfordshire/New York
Porter-Bolland L, Ellis EA, Guariguata MR, Ruiz-Mallé I, Negrete-Yankelevich S, Reyes-García V (2012) Community managed forests and forest protected areas: an assessment of their conservation effectiveness across the tropics. For Ecol Manag 268:6–17
Powlson DS, Gregory PJ, Whalley WR, Quinton JN, Hopkins DW, Whitmore AP, Hirsch PR, Goulding KW (2011) Soil management in relation to sustainable agriculture and ecosystem services. Food Policy 36(Supplement 1):S72–S87
Putz FE (2020) Science-to-conservation disconnections in Borneo and British Columbia. For Chron 96(1):20–26
Radachowsky J, Ramos VH, McNab R, Baur EH, Kazakov N (2012) Forest concessions in the Maya Biosphere Reserve, Guatemala: a decade later. For Ecol Manag 268:18–28
Reganold JP, Wachter JM (2016) Organic agriculture in the twenty-first century. Nat Plants 2(2): 15221
Ribeiro MC, Martensen AC, Metzger JP, Tabarelli M, Scarano F, Fortin MJ (2011) The Brazilian Atlantic forest: a shrinking biodiversity hotspot. In: Zachos FE, Habel JC (eds) Biodiversity hotspots: distribution and protection of conservation priority areas. Springer, Berlin/Heidelberg, pp 405–434
Rice RA, Greenberg R (2000) Cacao cultivation and the conservation of biological diversity. AMBIO 29(3):167–173
Ross NJ (2011) Modern tree species composition reflects ancient Maya "forest gardens" in northwest Belize. Ecol Appl 21(1):75–84
Sandhu HS, Wratten SD, Cullen R (2010) Organic agriculture and ecosystem services. Environ Sci Pol 13(1):1–7
Santos PZF, Crouzeilles R, Sansevero JBB (2019) Can agroforestry systems enhance biodiversity and ecosystem service provision in agricultural landscapes? A meta-analysis for the Brazilian Atlantic forest. For Ecol Manag 433:140–145
Saracura VF (1997) Reserva Biológica de Una: Plano de Manejo. ICMBio, Brasilia. https://www.icmbio.gov.br/portal/images/stories/imgs-unidades-coservacao/REBIO%20Una.pdf
Schroth G, da Fonseca GAB, Harvey CA, Gascon C, Vasconcelos HL, Izac AMN (eds) (2004) Agroforestry and biodiversity conservation in tropical landscapes. Island Press, Washington, DC/Covelo/London
Scott JM, Davis FW, McGhie RG, Wright RG, Groves C, Estes J (2001) Nature reserves: do they capture the full range of America's biological diversity? Ecol Appl 11(4):999–1007

Sebesvari Z, Le HTT, Van Toan P, Arnold U, Renaud FG (2012) Agriculture and water quality in the Vietnamese Mekong Delta. In: Renaud FG, Künzer C (eds) The Mekong Delta system. Springer, Dordrecht, pp 331–361

Shepard GH Jr, Neves E, Clement CR, Lima H, Moraes C, dos Santos GM (2020) Ancient and traditional agriculture in South America: tropical lowlands. In: Oxford research encyclopedia of environmental science. Oxford University Press, New York

Silbernagel J (2003) Spatial theory in early conservation design: examples from Aldo Leopold's work. Landsc Ecol 18:635–646

Smith T (2001) Beyond land rights. Newsweek International. January 21, 2002, p. 26

Soares-Filho B, Moutinho P, Nepstad D, Anderson A, Rodrigues H, Garcia R, Dietzsch L, Merry F, Bowman M, Hissa L, Silvestrini R (2010) Role of Brazilian Amazon protected areas in climate change mitigation. Proc Natl Acad Sci 107(24):10821–10826

Trevizan SP (1982) Estrutura fundiária e produtividade na região cacaueira da Bahia. Boletim Técnico 103. Centro de Pesquisas do Cacau (CEPLAC), Bahia, Brasil. pp 3–20

Udawatta PR, Rankoth L, Jose S (2019) Agroforestry and biodiversity. Sustainability 11(10):2879

Vélez MA, Robalino J, Cardenas JC, Paz A, Pacay E (2020) Is collective titling enough to protect forests? Evidence from Afro-descendant communities in the Colombian Pacific region. World Dev 128:104837

West P, Igoe J, Brockington D (2006) Parks and peoples: the social impact of protected areas. Annu Rev Anthropol 35:251–277

Wood A, Tolera M, Snell M, O'Hara P, Hailu A (2019) Community forest management (CFM) in south-west Ethiopia: maintaining forests, biodiversity and carbon stocks to support wild coffee conservation. Glob Environ Chang 59:101980

Wuerthner G, Crist E, Butler T (eds) (2015) Protecting the wild: parks and wilderness, the foundation for conservation. Island Press, Washington, DC/Covelo/London

Ybarra M (2018) Green wars: conservation and decolonization in the Maya forest. University of California Press, Oakland

York R, McGee JA (2016) Understanding the Jevons paradox. Environ Soc 2(1):77–87

# Chapter 24
# Preserving Biodiversity in Appalachian Mixed Mesophytic Forests Through the Permit-Based Harvest of American Ginseng and Other Forest Botanicals

**Karam Sheban**

**Abstract** Appalachian mixed mesophytic forests are home to some of the world's most diverse temperate forests; 90% of their plant biodiversity is in the understory. Many of these understory plants are of great cultural and/or commercial significance, making their protection vital from ecosystem, sociocultural, and economic perspectives. Due to development and forest fragmentation across the Eastern United States, large, relatively undisturbed tracts of Appalachian mixed mesophytic forests exist primarily in federally protected national parks and forests. Protected national forests are managed under a multiple-use mandate; they exist to protect native forest and also to provide services to the public such as recreation and harvestable forest products. A limited number of national forests have permit-based harvest programs for medicinal and culinary understory plants, some of which, like American ginseng, are commercially valuable. Demand for these plants and concern over their conservation status, however, is resulting in national forests shrinking their permit-based harvest programs or eliminating them altogether. Contrary to this trend, I argue that a renewal of the relationship between people and culturally significant understory plants—through the expansion of permit-based harvest programs on national forestland as well as through the intentional cultivation of forest herbs (referred to as forest farming)—presents the best way forward to reestablish viable populations of understory herbs in the mature second growth forests of the Appalachian region. In conjunction with forest farming of understory herbs, permit-based harvesting is an effective mechanism for safeguarding the benefits of biodiversity islands in mixed mesophytic forests.

**Keywords** Deciduous forests · Eastern USA · Forest botanicals · Forest farming · Herbs · Understory

---

K. Sheban (✉)
School of Forestry and Environmental Studies, Yale University, New Haven, CT, USA
e-mail: karam.sheban@yale.edu



F. Montagnini (ed.), *Biodiversity Islands: Strategies for Conservation in Human-Dominated Environments*, Topics in Biodiversity and Conservation 20,
https://doi.org/10.1007/978-3-030-92234-4_24

## 24.1 Introduction

The Appalachian mixed mesophytic forest is an ecoregion within the more extensive deciduous forests ranging across the Eastern United States. This forest type represents the epicenter of temperate forest biodiversity in the U.S. and is characterized by the wealth of species occupying each stratum of the forest, from the canopy down into the understory. The canopy of Appalachian Mixed Mesophytic Forests is incredibly diverse, with a single site often supporting more than 30 tree species (Loucks et al. n.d.). The understory is the main source of floral biodiversity in these forest systems, however, accounting for 90% of the plant diversity in the region (Gillium 2007).

While mixed mesophytic forest used to cover a much greater extent across North America, its range has been reduced considerably, and the remaining tracts have experienced widespread degradation; it is estimated that 95% or more of this habitat has been converted or degraded at some point over the past 200 years (Loucks et al. n.d.). Forest cover generally has increased across the Eastern U.S. in this timeframe, with the abandonment of cropland, pasture, and other cleared lands, although the trend has begun to reverse in the twenty-first century (Drummond and Loveland 2010); however, these second-growth forests do not contain the same species associations or diversity as are seen in primary forests (Braun 1951; Gilliam 2014). In particular, populations of slow-growing herbaceous understory species take many years to reach reproductive maturity, struggle to disperse widely, and in most cases do not recover fully in secondary forests (Duffy and Meier 1992).

Herbaceous understory plants provide not only tremendous ecological value, but sociocultural value as well. The understory of Appalachian mixed mesophytic forests is home to over 50 species that have been harvested for hundreds or thousands of years for use in indigenous, traditional, and folk medicine, cooking, and spiritual practices (Burkhart and Jacobson 2009; Chamberlain 2005). A number of these plants—such as American ginseng (*Panax quinquefolius*), goldenseal (*Hydrastis canadensis*) and ramps (*Allium triccocum*)—have significant commercial value; prices paid for ginseng plants harvested in the wild can reach over $1,000 per pound (Hamilton 2017).

The remaining intact tracts of Appalachian mixed mesophytic forest can be considered biodiversity islands, "area[s] of high biodiversity within ecologically degraded, human dominated landscapes" (Montagnini et al. 2022, this volume). These tracts are largely contained within state and federally controlled public lands, such as national parks and national forests across the Appalachian region (Loucks et al. n.d.). Of these remaining tracts, only the national forests allow varying degrees of permit-based understory plant harvesting. This is in keeping with the principles laid out in the Multiple-Use Sustained-Yield Act of 1960 passed by the United States Congress, a guiding piece of legislation that dictates that national forests are managed for a variety of uses, including timber harvesting, livestock grazing, watershed protection, fish and wildlife, and recreation. Along with the harvest of firewood and the hunting of animals such as deer, squirrels, and turkeys, the permitted harvesting of understory plants on select national forests takes place during a designated season. These permits are open to the public for a nominal fee and are limited to one permit per person per season. Permit-based plant harvesting confers many advantages: it creates a legal pathway for plant resource access and use; generates funds for the monitoring of plant harvest levels and the allocation of personnel; engages the public in the conservation and management of forest biodiversity; and elevates the profile of understory plants within the national forest system.

In this chapter I will argue that the permit-based harvest of understory plants in national forests represents a strategic and just way to conserve and restore populations of dispersal-limited understory plants and to protect these threatened islands of mixed mesophytic biodiversity, while simultaneously preserving Appalachian traditions and traditional livelihoods. Critically, I will suggest that this strategy is contingent on the parallel development of markets for understory herbs through the expansion of intentional cultivation operations using the agroforestry practice of forest farming. This chapter will start with an overview of mixed-mesophytic forests and the unique biodiversity they house. It will cover some of the threats facing the biodiversity of these forests. The chapter will briefly touch on the history of understory plant harvesting in the Appalachians, focusing on American ginseng as a case study in how biodiversity, the management of public lands, market forces, and cultural traditions can collide. And lastly, the chapter will consider how shifting the supply chain balance of commercially valuable herbs from wild-harvested toward sustainably cultivated products could support a permit-based allowable harvest on public lands, achieving multiple social and ecological benefits.

## 24.2 The Appalachian Mixed Mesophytic Forest

The Appalachian mixed mesophytic forest of the Eastern United States was first described by the forest ecologist and University of Cincinnati professor E. Lucy Braun in 1916 (Braun 1916). Using Braun's nomenclature, deciduous forests in the Eastern U.S. are organized into forest associations, defined by the dominant tree species found within them. Mixed mesophytic forests represent the oldest and most complex of these forest associations; "the epicenter of highest development of the Eastern deciduous forest" (Rosson 2008). They are situated in the geographic center of Eastern U.S. deciduous forests. From this mixed forest or the preceding forests of the Tertiary period, all other forest associations within eastern deciduous forests have arisen (Braun 1951).

Appalachian mixed mesophytic forest ranges from Northwest Alabama and East-Central Tennessee through Eastern Kentucky, Western North Carolina, the majority of West Virginia, up to Southeast Ohio and Southwest Pennsylvania (Fig. 24.1). Within this range, the fullest expression of this forest type is seen in ravines, gorges, coves, and valleys where the topography is varied enough to support a multitude of species (Braun 1951; Rosson 2008) (Fig. 24.2). While the majority of forest associations in the U.S. are dominated by two or three primary canopy species,

**Fig. 24.1** Appalachian mixed mesophytic forests range map. (Source: USGS Ecoregions Index Map, https://pubs.usgs.gov/pp/p1650-e/wwf_maps.html)

ecological complexity is the defining feature of mixed mesophytic forests, with up to 30 tree species on a single site.[1] In addition to floral diversity, these forests house incredibly diverse fungi, songbird, salamander, land snail, and beetle communities. Their freshwater communities "are the richest temperate freshwater ecosystems in the world, with high species richness and endemism in mussels, fish, crayfish, and other invertebrates" (Loucks et al. n.d.).

[1]For an incomplete list of tree, shrub, and herbaceous species found in Appalachian Mixed Mesophytic Forests see appendix at the end of this chapter.

**Fig. 24.2** The Red River Gorge in the Daniel Boone National Forest, a national forest in the heart of the Appalachian mixed mesophytic range. (Photo: blog.campnative.com)

Mixed mesophytic forests once covered much of the temperate regions of the Northern Hemisphere, across what is now North America and the deciduous regions of China, though their range has shrunk considerably due to the pressures of colonization and development. In the Eastern United States, it is estimated that half or more of the present forest cover has been cut some time in the past three centuries as a result of agricultural expansion, development, and harvesting of fuelwood and other resources (Drummond and Loveland 2010). Beginning in the early 1900s as settlement and agriculture moved westward, the Eastern United States experienced a gain in forest cover, including natural forest regeneration across the range of mixed mesophytic forest. But that trend has reversed, as logging, urbanization, and the urban sprawl of low-density settlement have begun affecting forest regrowth. The Eastern U.S. experienced a 2.3% decline in forest cover between 1973 and 2000—a total loss of over 165,000,000 hectares—with regions of mixed mesophytic forest, such as the Piedmont and the Central and Southwestern Appalachians experiencing declines in forest cover at an even faster rate (Drummond and Loveland 2010). Since 2000, the trend has largely continued and projections suggest forest cover across the Eastern U.S. will continue to decline in the coming decades (Adams et al. 2019).

The contemporary trend of forest decline indicates the urgency of preserving the remaining forests, despite many of these being the result of secondary growth. Mature second growth forest can contain high levels of biodiversity, as well as eventually achieve an old-growth structure. E. Lucy Braun observed that, on the best sites within the mixed mesophytic forest association, the regrowth may return to a similar species composition of an old growth forest if the forest was only cut once and not utilized for other purposes such as agriculture (Braun 1951). More chronic disturbance, such as a century of use as pasture or for agriculture, can leave behind a significantly transformed landscape, more susceptible to colonization by invasive species (Webb et al. 2000), in turn suppressing growth and reproduction of native species (Miller and Gorchov 2004).

## 24.3 The Herbaceous Layer: Challenges and Opportunities

Only recently have scientists began to appreciate that the herbaceous layer in deciduous forests—their greatest source of biodiversity—is much more sensitive to disturbance than the canopy layer (McCarthy 2014). In a landmark paper, Duffy and Meier (1992) concluded that herb populations in mixed mesophytic forests would take several centuries to recover after a major disturbance. Additional research has documented the difficulty understory species face in recolonizing former forest habitat after a disturbance such as logging or agriculture (Holmes and Matlack 2018).

The challenges to recolonization are myriad. Understory herbs are limited in their sexual reproduction due to small crops of seed, which are often poorly dispersed by gravity or insects, and generally have low germination rates (Duffy and Meier 1992). For species that reproduce asexually, the ability to move via vegetative sprouting is limited to less than 1 m per decade (Sobey 1977), and it takes years for an individual plant to reach sexual maturity for many species (Bierzychudek 1982). On top of these challenges, predation by deer and other animals can present a serious challenge to population establishment (Alverson et al. 1988; McGraw and Furedi 2005).

The restoration of understory populations and the preservation of existing wild populations provide two critical approaches to protecting understory biodiversity in a landscape facing the dual pressures of deforestation and fragmentation (Drummond and Loveland 2010; Loucks et al. n.d.). While important, the first approach—restoring populations of threatened understory herbs—presents some challenges. Restoration plantings require funding, personnel, and botanical expertise, all of which are particularly limited when it comes to restoration of understory herb species. Additionally, the reintroduction of native understory plants into the wild requires a seed or transplant source, which for many understory species simply is not available. For species such as American ginseng, for which planting stock is available, the introduction of ginseng seed with an industrial provenance (primarily produced on large industrial ginseng farms in Wisconsin and Quebec) into geographic regions across the plant's native range raises concerns about maladaptation and the corruption of local genetics (Schluter and Punja 2002). There is a critical need for native plant nurseries to provide planting stock for restoration plantings, private forest farming operations, and local botanical gardens and sanctuaries.

The second approach to preserving existing wild populations of understory plants—what McCoy et al. (2019) refers to as the in-situ conservation of medicinal and social-use understory taxa—represents the most cost-effective path forward. While expertise and funding are required for this approach as well, starting with an existing botanical resource eliminates some of the initial challenges facing protection efforts. It is critical that in-situ conservation and restoration take place on private forestland, and this is occurring through a variety of mechanisms which we will cover later in detail. United Plant Savers, a nonprofit organization headquartered in Rutland, Ohio, has established a network of "botanical sanctuaries" on private lands, providing a framework for preserving habitat and populations of threatened native species with an emphasis on medicinal understory plants (www.unitedplantsavers.org). The Appalachian Beginning Forest Farmers Coalition, a consortium of universities, government agencies, and nonprofit organizations, promotes the cultivation of at-risk understory species in forest farming systems. These programs are based on the premise of "Conservation Through Cultivation," which posits that by cultivating and selling commercially valuable understory plants, pressure can be taken off wild populations as the supply chain transitions towards sustainably grown products (www.appalachianforestfarmers.org) (Burkhart and Jacobson 2009; Burkhart 2011). This approach has additional benefits such as maintaining private lands as forest, generating land-based income, and strengthening the local forest and farming sectors by diversifying the array of crops producers can cultivate. Given contemporary land use trends, these methods of plant conservation on private lands are becoming more important than ever and adoption rates of both practices are growing.

## 24.4 Human Use and Harvest of Understory Plants in the Appalachians

Protecting public lands is critical. It would be misguided, however, to suggest that the only way to preserve biodiversity islands within public lands is to create reserves where access to plant material is sealed off from the public. This is a well-established pattern on public lands in the U.S.: adopting a protectionist approach toward natural resources in the name of saving them. While the U.S. is frequently lauded internationally for its public land conservation, the protectionist model it employs has been criticized. Embodied by the National Park Service (NPS), this approach to public lands conservation is sometimes referred to as the "Yellowstone model," and has been exported around the world. A wealth of academic literature, however, suggests that this model of closing off of public land has been harmful to vulnerable rural populations and provides national economic benefits at the expense of local communities while incurring high costs for park protection and failing to effectively conserve biodiversity (Schelhas 2001). Even if this approach to land management was resulting in irrefutably positive changes in biodiversity—which, in the case of understory plants, is dubious—there is an increasing consensus among academics, policy makers, and communities around the world that considering simply the ecological effects of a particular policy divorced from its broader social, economic, and political implications is not enough (Ban et al. 2013).

It is important to note that the distribution of medicinal and culinary understory plants across the Eastern United States is inextricably linked to human use over millennia (Abrams and Johnson 2017). This is important for several reasons, but particularly in the context of how and to whom we allow or disallow access to plant material in the name of saving it. Historically, the harvest of understory plants for personal use and for income generation has taken place across a great forest-commons, with place-based, plant-specific knowledge passed down generationally. This was true prior to the colonization of North America and has remained true across the Appalachian region, throughout the mixing of traditional plant-based medicinal knowledge from old-world cultures—such as Spanish, African, British, and French—with indigenous practices (Light 2008). These traditions persist throughout the Appalachian region, where the biological diversity of mixed mesophytic forests and relative isolation kept plant-based medicinal and cultural traditions alive (Alwhaibi et al. 2017). Across the range of mixed mesophytic forest, the knowledge of how to harvest and use plants, along with practices of good stewardship, was transferred intergenerationally. For understory herbs, this included practices such as only harvesting mature plants with ripe berries, planting seeds before harvesting the plant, and leaving behind the most productive plants ("mother" plants) to replenish local populations (Chittum et al. 2019).

Considering the historical and contemporary interactions between people and understory plants in the mixed mesophytic forests of the Appalachian region, it is important to acknowledge that humans have and continue to function as dispersal agents for the understory biodiversity of the region. Humans have been harvesting and replanting medicinal and culinary herbs for millennia; this stewardship relationship is one of the main ecological dynamics responsible for the "natural" distribution of many understory plants. In fact, given the long history of human-plant interactions, some have made the argument that it is impossible to disentangle anthropogenic influence from the observed ecology of understory plants in the Appalachian region and beyond, and that it would be misguided to predict the future of plants without considering the role humans may play (Turner and McGraw 2015; Albuquerque et al. 2018).

When considering historical human-plant interactions in forests, the question arises as to whether these interactions have positive or negative impacts on biodiversity. There is evidence that direct seeding of understory plants improves germination rates (Filyaw and Sheban 2015) and that plant reintroduction into mature second growth forests provides an avenue for circumventing the aforementioned dispersal limitations that plague understory herbs. It may be that a renewal of this relationship with understory plants presents the best way forward to reestablish viable populations of understory herbs in the mature second growth forests of the Appalachian region. The time to act is now—these traditional and indigenous economies and this way of life are disappearing (Cavender and Beck 1995) along with the primary forests on which they have historically depended. Shifting patterns

**Fig. 24.3** A mature American Ginseng plant. (Photo by Karam Sheban)

of land ownership, the closing off of public space, the unsustainable harvest of plant resources, and the reduction of undisturbed forest habitat are moving the baseline on both forest health and forest livelihoods.

## 24.5 American Ginseng as a Bellwether: The Future of Understory Plants in Appalachian Mixed Mesophytic Forests

### *24.5.1 American Ginseng Biology and Brief History*

When considering the cultural, economic, and ecological forces acting upon understory biodiversity across mixed mesophytic forests, American ginseng provides a particularly useful case study in understanding both threats and opportunities confronting forest herbs. American ginseng (*Panax quinquefolius,* Araliaceae) is a long-lived perennial herb, reaching a maximum height of 50 cm (Anderson et al. 2002) (Fig. 24.3).

A mature ginseng plant has three to four palmately lobed leaves, each with three or five leaflets, radiating from a central aerial stem, which itself grows out of the underground root. Ginseng diggers commonly refer to the leaves of a ginseng plant as its "prongs," and the number of prongs provide an indication of the plant's

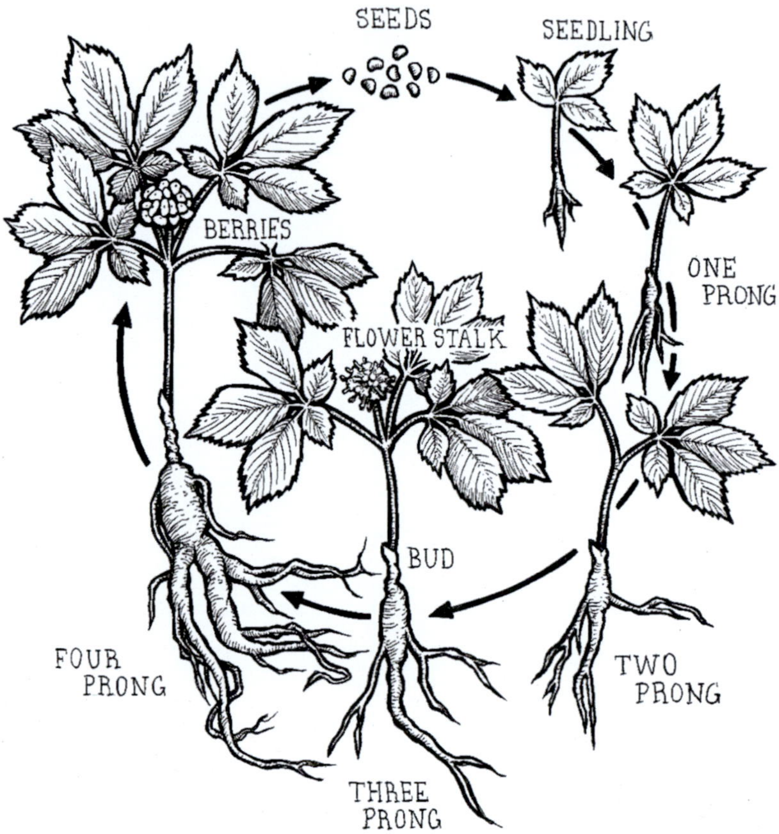


**Fig. 24.4** The lifecycle of American ginseng. (Source: Rural Action (www.ruralaction.org))

maturity (Anderson et al. 2002). The plant develops leaves in early spring, and its flower emerges midsummer. The fruit ripens around late August, producing a cluster of red berries, each carrying one or two seeds, and the leaves die back in late summer to early fall (Fig. 24.4). Ginseng roots form arbuscular mycorrhizal relationships, which help the plant access soil nutrients (Anderson et al. 2002).

Ginseng naturally occurs from Southern Quebec and Ontario, to South Dakota, throughout the Midwest, and south to Georgia and Oklahoma (Anderson et al. 2002) (Fig. 24.5). It tends to occur in moderately to highly shaded hardwood forests on slopes under a canopy of mesic trees species, such as sugar maple (*Acer saccharum*), red oak (*Quercus rubra*), American beech (*Fagus grandifolium*), shagbark and mockernut hickory (*Carya ovata* and *C. tomentosa*), and white ash (*Fraxinus americana*). These trees are characteristic of the ravines, coves, gorges, and valleys that Braun identified as home to the fullest diversity of mixed mesophytic forest (Braun 1951). Ginseng prefers moist but well-drained sites and can grow under a range of edaphic conditions, though research has indicated that high calcium levels are particularly important in promoting plant growth and development (Davis and Persons 2014).

**Fig. 24.5** Native range of American ginseng (*Panax quinquefolius*), from the USDA Plants Database

American ginseng has been used culturally and medicinally for thousands of years across the globe. In indigenous and folk medicine in North America and in traditional medicine across Eastern Asia, American ginseng (along with Korean ginseng, *Panax ginseng*, and other ginseng species) has been used to treat a wide range of health conditions and ailments, from the common cold to erectile disfunction, as well as being used generally to combat stress and boost energy. Ginseng's purported ability to mitigate stress led to its categorization—in the parlance of modern herbal medicine—as an adaptogen, a substance "with a relatively high degree of safety that help[s]...organism[s] adapt to various types of stress" (Dharmananda 2002). These traditional uses have inspired contemporary examinations of the plant's medicinal potential. Modern pharmacology has attributed a range of medicinal properties to American ginseng, the majority of which are thought to derive from a family of bioactive phytochemical compounds, known as ginsenosides, which are abundant in the plant's leaves, stems, and roots (Szczuka et al. 2019). These compounds are actively being studied for their potential anti-diabetic, anti-obesity, anti-cancer, anti-aging, and antimicrobial properties, as well as their ability to alleviate the symptoms of a wide range of other conditions (Szczuka et al. 2019).

Ginseng is unique among understory plants as "the most economically important wild-harvested native medicinal plant in the United States" (Division of Scientific

Authority 2018). It has long been harvested and used in North America, but it is also the centerpiece of major international trade between North America and China, the primary importer of American ginseng. This trade drives the significant harvest of ginseng plants from the wild; nearly 8.5 million ginseng plants were harvested from the wild in 2017, according to data reported by states and Tribes with export programs (Division of Scientific Authority 2018).

The extensive harvesting of wild ginseng to supply the international market has been ongoing for over 300 years, since the first shipment of ginseng roots from North America to China in 1717 (Parsons 2016). Due to concerns around the plant's conservation, ginseng was listed as a controlled species under Appendix II of the Convention on International Trade in Endangered Species (CITES) in 1975, signifying an acknowledgement that ginseng was "not necessarily now threatened with extinction but that [it] may become so unless trade is closely controlled" (https://www.cites.org/eng/app/index.php).

The challenge presented by forest botanical species, but especially American ginseng—as its commercial value is orders of magnitude greater than other merchantable temperate herb species—is a strong incentive to harvest the plant when it is encountered. For a potential harvester, leaving the plant in the ground passes the opportunity on to another harvester. And unlike some other non-timber forest products (NTFPs)—such as collecting maple sap to produce syrup or gathering berries or tree nuts—harvesting a ginseng root, the merchantable part of the plant, is lethal. Considering that American ginseng is dispersal-limited and typically requires a minimum of 5 years to reach reproductive maturity, its high commercial value and the lethality of harvest has created conservation challenges across ginseng's entire range. And yet, as put by Curt Freese in his book "*Harvesting Wild Species*," "[r] egardless of what conservationists think, much of the world depends on wild species for an array of products, whether for food, fiber, or medicine. Thus, in many cases, the question is not *whether* to use wild species, but rather how to move from a system of use that is clearly not sustainable toward one that is better" (Freese 1997).

### *24.5.2 Permit-Based Harvesting of Ginseng*

Currently, the harvest of wild ginseng plants is allowed in 19 states—Alabama, Arkansas, Georgia, Illinois, Indiana, Iowa, Kentucky, Maryland, Minnesota, Missouri, New York, North Carolina, Ohio, Pennsylvania, Tennessee, Vermont, Virginia, West Virginia, and Wisconsin—and also in the territory of the Menominee Indian Tribe of Wisconsin. While the monitoring and regulation of American ginseng falls under the federal purview of the U.S. Fish and Wildlife Service, the states bear the bulk of actual oversight, regulating the harvest, sale, and certification of wild ginseng roots within their jurisdiction. The purpose of these regulations is to promote the regeneration of plant populations. To achieve this end, the states and tribes enforce a harvest season, a minimum harvest size (3 or 4 leaves) and/or age (5 to 10-year-old plants), and certain sustainable practices such as planting the seeds of harvested plants on site (Division of Scientific Authority 2018).

The proliferation of private property across the United States and the increasing ownership of forest and farm lands by absentee landowners (Petrzelka et al. 2013) has shifted control over and access to plant material across large swaths of the Eastern United States. Now more than ever, public lands provide one of the only avenues for legal plant harvest for landless Americans. The trend of restricting access to plant resources that have traditionally existed in a commons continues today. Most states (17 out of 19) do not allow harvest on state lands, and only five states allow the harvesting of plants on national forest land. Within these states, the number of national forests that allow the harvesting of plants is shrinking. In 2016, the Daniel Boone National Forest in Kentucky shut down its permit-based harvest program due to conservation concerns (Division of Scientific Authority 2018).

Since 2013, in national forests where ginseng harvesting is still permitted, the number of permits issued, the allowable harvest quantity, and the length of the harvest season have all been reduced (Division of Scientific Authority 2018). This is not without some scientific justification. Understory plant populations are being affected by the harvest of wild populations for commercial gain, but without baseline data on population numbers for various merchantable understory plants, it is difficult to say exactly how significant that effect is. Regarding ginseng, there is research showing that on lands that have an allowable ginseng harvest there is a higher proportion of juvenile compared to adult (i.e., sexually mature) plants than on lands where harvest is not allowed (Cruse-Sanders and Hamrick 2004; Young et al. 2009; McGraw et al. 2013). And beyond ginseng, regular harvests of large quantities of other understory medicinal and edible plants—such as goldenseal (*Hydrastis canadensis*), ramps (*Allium triccocum*), and black cohosh (*Actaea racemosa*)—raise questions of sustainability. A recently developed initiative out of Virginia Tech University, RootReport, has begun to track the harvest quantities for 14 of the more commonly traded native forest medicinal species (https://www.rootreport.frec.vt.edu/). Data collection efforts such as this will be necessary for understanding the long-term ecological trends of forest understory species going forward.

While additional research on forest herbs other than ginseng is certainly needed, in the meantime, McGraw et al. (2013) suggest that because we know more about ginseng than other understory herbs, it can be used as a "phytometer"—a species that can be used as a measure of physiological responses to various environmental factors—to "better understand how environmental changes are affecting many lesser-known species that constitute the diverse temperate flora of Eastern North America" (McGraw et al. 2013). Ginseng's economic arc may provide an indication of the commercial future of other understory species as well. As the herbal products market continues to grow—the international market for herbal medicine was valued at $60 billion dollars in 2019 (Ahmad Khan and Ahmad 2019)—and as wild plant populations become increasingly scarce, demand and prices can be expected to rise broadly for all merchantable plants. Indeed, there has been a recent spike in prices paid for goldenseal, resulting in shifting demand for the herb and likely having an effect on wild populations (Tanner Filyaw, personal communication). Thus, although ginseng's history, cross-cultural significance, and high value make it unique among understory plants, we can likely expect other species to follow suit, both ecologically and commercially.

In the face of these trends—increasing demand for plant material and decreasing access of formerly communal plant resources—and in considering how best to protect biodiversity islands within the forests of the Eastern United States, the U.S. is reaching an inflection point. If decreasing forest biodiversity in Appalachian mixed mesophytic forests can be understood broadly as the consequence of colonization, development, and climate change, who should bear the costs associated with its protection? Are there ways to preserve and even enhance this biodiversity while also deviating from the historic trend of restricting the access to resources by local communities and landless Americans? Facilitating a targeted set of social, economic, and ecological changes in order to maintain and even expand the permit-based harvest of plant material on public lands represents a way to achieve this.

It has been shown that allowing ginseng harvest on national forestland can impact the proportion of adult to juvenile wild plants (McGraw et al. 2013), but could it be possible that an allowable harvest also results in a greater number of juvenile plants overall? If best practices are required and then followed by legal harvesters on public land—only harvesting mature plants with ripe berries, replanting those berries on site—this would be the expected result. Such a consideration has the potential to subvert the narrative of harvest necessarily undermining long-term population survival.

### *24.5.3 Factors to Consider in Allowing Plant Harvest on Public Lands*

In light of this scientific and management ambiguity in how harvest affects wild American ginseng populations, several factors should be considered when weighing the value of an allowable harvest on public land. First, the establishment of a permit-based harvest on national forestland generates revenue which could be used to fund critical long-term population monitoring efforts. This research is essential for understanding how understory plant populations are responding to myriad changes in their environment, including climate change, herbivory, and harvest pressure.

Population monitoring is taking place in the Nantahala and Pisgah National Forests in North Carolina, Cherokee National Forest in Tennessee, and Wayne National Forest in Ohio (Division of Scientific Authority 2018; Tanner Filyaw, personal communication). In the Wayne National Forest, Forest Product Permits are separated into one permit for ginseng and one permit for six other species: Blue Cohosh (*Caulophyllum thalictroides*), Black Cohosh (*Actaea racemosa*), White Snakeroot (*Ageratina altissima*), Goldenseal (*Hydrastis canadensis*), Bloodroot (*Sanguinaria canadensis*), and Wild Ginger (*Asarum canadense*). Permits cost $20 each and, with between 60 and 100 permits issued each year, the program generates between $1,000 and $2,000 annually (Tanner Filyaw, personal communication).

**Fig. 24.6** Forest farming ramps; a demonstration led by Rural Action, a nonprofit in Southeast Ohio. (Photo by Karam Sheban)

While this is not much, it is enough to fund monitoring efforts on a portion of the Wayne National Forest. Funds generated from understory plant harvest permits in Wayne National Forest are also directed towards education efforts in partnership with Rural Action, a local nonprofit with a history of working with commercially valuable forest herbs (www.ruralaction.org), among other groups. This partnership funds workshops that educate local landowners and community members about the sustainable harvest and management of at-risk forest herbs. Promoting "Conservation through Cultivation," landowners are taught how to forest farm ginseng and other commercial species such as ramps (Fig. 24.6). Programs such as these raise the profile of understory diversity, which may otherwise go largely unnoticed and unfunded in a national forest.

Illegal harvest or "poaching" of ginseng and other wild plants is a major issue across the Appalachian region, and it takes place on both public and private lands. There are many factors responsible for rates of ginseng poaching across the region, from poverty and drug use (Pokladnik 2008; Young et al. 2011) to false perceptions that stealing ginseng will lead to a quick windfall, promoted by television shows such as the History Channel's *Appalachian Outlaws.* This television show and others like it (National Geographic's *Smokey Mountain Money*) rely on derogatory stereotypes and depict a lawless, get-rich-quick ginseng economy across the Appalachian region. There are indications that these shows may have been partly responsible for an uptick in rates of ginseng poaching on private lands after their debut in 2014 (Arnold 2016; Hamilton 2017).

The establishment of an allowable harvest on public lands creates a legal avenue for harvesting ginseng and other medicinals for people who may not own or have access to private lands for plant material. Given the reality of poaching, the establishment of plant monitoring efforts represents an effective approach to evaluating the extent to which poaching is occurring on national forestland. Such efforts could be partially or fully funded through harvest permits. Additionally, education about the ecology and sustainable harvest of understory plants—also promoted through permit-based harvesting—represents a systemic way to combat the theft of valuable understory herbs.

Currently permit-based harvesting of understory plants is generating only a small amount of revenue due to the limited number of permitting programs and the prices paid for understory plants. As demand for herbal products continues to grow, it is important to consider how permit-based harvesting might evolve to support harvesters and plant populations in tandem.

## 24.6 Forest Farming as a Means to Support and Grow Permit-Based Harvesting Programs

The majority of traded American botanical products are supplied to markets through wild-harvesting (Dentali 2007; Dentali and Zimmerman 2012), with some of the most prominent species being harvested from forests. Scholars, farmers, herbalists, conservation groups, government agencies, and nonprofit organizations have called for the intentional cultivation of native North American forest plants, particularly through systems such as forest farming. Agroforestry practices such as forest farming confer advantages over field-based cultivation (Bannerman 1997; Robbins 1998, 1999; Gladstar and Hirsch 2000; Burkhart and Jacobson 2009), including fewer costs and the potentials for higher product quality, reduced disease pressure, and income generation while preserving or even increasing forest diversity (Teel and Buck 1998; Hill and Buck 2000; Rao et al. 2004).

I propose that an additional advantage of the intentional cultivation of medicinal and culinary forest plants in forest farming systems is its potential to support the permit-based harvest of these same plants on public lands. This is achieved by stabilizing supply chains, taking pressure off of wild plant populations, and increasing prices for forest-based medicinal and culinary plants. Improving the price points for a range of understory species would facilitate greater revenue generation on public lands, allowing for enhanced investment in long-term monitoring, plant protection, and education and outreach efforts around native biodiversity. Thus, increasing the support for, as well as the overall viability of, forest farms on private lands is a necessary step in the protection of remaining islands of native plant biodiversity while allowing for continued access to plant resource commons. This could be accomplished by ensuring that forest farmers are receiving prices that support the cultivation of multiple understory species, rather than a select few.

However, in order for the transition from wild-harvesting to sustainable cultivation of understory herbs to equitably impact people who rely on native forest biodiversity for livelihoods or cultural expression, it must provide opportunities for the most economically disadvantaged. Increasing prices for understory plants must be translated into greater investment in NTFP permit-harvesting programs on public lands, rather than higher prices and greater harvest pressure leading to these programs being shut down.

Currently, prices paid for American ginseng are extremely high when compared to other wild-harvested understory species. Research suggests that ginseng and goldenseal are the only medicinal herb species in the Eastern U.S. that can profitably be forest farmed, while the price points for other medicinals remain too low to justify the expense (Burkhart and Jacobson 2009). This dramatic imbalance in price creates uneven pressures on wild plant populations. Freese (1997) suggests this phenomenon can be observed in wild-harvested natural commodities around the world. He cites the example of Latin American mahogany, which commands a very high price compared to the lumber of species which are considered "lesser-known", placing inordinate demand on one species while undervaluing others. In this scenario, one strategic conservation response would be to grow markets and demand for these lesser-known species to spread harvest pressure evenly across a range of species (Freese 1997). While industrial buyers understand that relying on wild-collected herbal material creates sustainability challenges (as well as issues with adulteration due to a lack of supply chain transparency and traceability (Jiang et al. 2011), many are not willing or able to pay prices high enough to encourage the cultivation of additional understory species.

Recently, however, this has begun to change. United Plant Savers (www.unitedplantsavers.org) administers a third-party certification system, called "Forest Grown Verification" (FGV), designating that a given forest farmer is intentionally and sustainably cultivating plants rather than harvesting them from wild populations. This certification structure is the result of years of meetings between forest farmers, Technical Service Providers, and industry groups seeking solutions to supply chain challenges. Designed in the spirit of Organic certification, the program seeks similar beneficial outcomes: increased profitability, reduced inputs, greater economic and ecological resilience, and increased employment potential (Jouzi et al. 2017). Forest Grown Verification is supported by herbal products companies, such as Oregon-based Mountain Rose Herbs (www.mountainroseherbs.com), through the introduction of FGV product lines for multiple herb species, including ginseng, goldenseal, and black cohosh. The program is financed partly through consultation and annual renewal fees, while verified growers are paid a premium for their sustainably-cultivated product. Industry buyers like Mountain Rose Herbs that establish product lines for FGV material are critical partners in establishing an herbal products supply chain based increasingly on cultivated products.

Programs such as FGV have the potential to make forest farming increasingly profitable, as industry buyers begin to pay forest farmers prices that more accurately reflect what their cultivated plants are worth. In turn, this increased value is supported by consumers, willing to pay a premium for a product they believe is higher quality and produced in a way that is worth supporting. This is the model of ethical consumption that has helped sustain the growth of organically certified foods across the globe (Johnston et al. 2011; Sebastiani et al. 2013). Public-lands harvesting is supported as a consequence, through boosting regional plant populations and raising overall awareness of at-risk understory biodiversity. At the same time, higher price-points for lesser-known species could allow for greater permit fees on public lands, generating more revenue, which could feed into long-term monitoring and educational programs.

## 24.7 Discussion

New scientific research has reinvigorated the call to better understand, monitor, and support understory herb populations (Spicer et al. 2020). Forest farming and public-harvest programs both represent approaches to raise the profile of understory herbs and to advance conservation, cultural use, and positive socioeconomic development. Transitioning the supply of forest botanicals from wild-harvested toward sustainably-cultivated plants represents a promising opportunity to preserve public access to plants on public lands while supporting biodiversity islands in mixed mesophytic forests. This entails formalizing the supply chains (or value chains) for commercial understory species. Belcher and Schreckenberg (2007) describe this process in their analysis of the commercialization of non-timber forest products. They provide a useful framework for thinking about the characteristics of NTFPs, drawing a distinction between NTFPs and smallholder agricultural products (Table 24.1). By bringing medicinal and culinary herbs into cultivation, landowners are moving these plants away from the authors' categorization of NTFPs and towards that of smallholder agricultural products (see the authors' note in Figure 7 on the nature of a spectrum between these two categories).

According to Belcher and Schreckenberg, bringing NTFPs under cultivation can introduce novel challenges, both ecological and socioeconomic. They suggest that "[i]ntensively managed NTFP production systems may completely displace the natural vegetation within the management unit," though acknowledge that "the impacts at the landscape or plot level are less clear for some of the less intensively managed cases" and that "[i]n most cases ecological and biodiversity impacts have not been measured or even estimated" (Belcher and Schreckenberg 2007). These concerns, however, may be less germane to understory medicinal and culinary herb forest farming systems in Eastern North America. Few forest farming operations are intense enough to require the removal of competing vegetation. The majority of small landowners introduce native understory plants into their forestland and simply allow them to grow—an approach to forest farming referred to as "wild-simulated production" (Carroll and Apsley 2013). These systems are often mixed and diverse, both spatially and temporally, containing introduced native plants within an assemblage of other native species, producing a range of products and value throughout the year and over the years. Multistrata agroforestry is often seen as a practice restricted

**Table 24.1** Key differences in the value chains of NTFPs and smallholder agricultural products[a]

| Factor | NTFPs | Smallholder agricultural products |
|---|---|---|
| Resource biology | Collection areas for wild harvested NTFPs often distant from the home | Fields usually close to or in walking distance of home |
| Resource biology | Low-density production means bulking-up becomes very important | Cultivation leads to higher density; usually many producers in one area |
| Resource biology | Usually "wild" or relatively unimproved leading to problems of inconsistent quality, sometimes highly dependent on the vagaries of weather | Known varieties and availability of inputs allow for more uniform production |
| Resource tenure | Insecure tenure over collection areas leads to risk of over-exploitation; inability to manage the resource (to improve quality and/or quantity) | Individual tenure, therefore ability to exclude others, provides incentive to invest in the resource |
| Resource – Knowledge Base | Traditional knowledge only, little formal research | Many staple and minor agricultural products subject of agricultural research and extension programs |
| Policy issues | Little relevant policy in support of commercialization; usually restricts harvest and/or transport and sale of NTFPs | Supportive policies in place, including credit provision, extension, research |
| Market structure | "Thin markets"– Often few buyers for the total product from a production area | Many different buyers at different scales; producers have more options for trading |
| Market information | Very little available; channeled through intermediaries | Often widely available |
| Production volumes | Often a supplementary activity, therefore production varies as producers choose between different livelihood opportunities | Usually a more consistent part of livelihoods, leading to more predictable production volumes |
| Destination markets | Very diverse, faddish, frequently "luxury" goods and niche markets | Better known markets and more predictable |
| Intellectual property rights issues | May be critical for medicinal products and, if active ingredients are synthesized away from original source, requiring negotiation of benefit-sharing agreements | Can be an issue with respect to propagation of improved varieties |

**Note:** For the purpose of the discussion, the table polarizes the two extremes, although some NTFPs are not cultivated as smallholder crops, and some smallholder crops are produced for niche markets, giving them similar characteristics to NTFP value chains

[a]Adapted from Belcher and Schreckenberg in Development Policy Review, Issue 25, Volume 3, 2007, page 363

to the tropics, but it is important to recognize that a practice such as the wild-simulated approach to forest farming described in this chapter can serve as an example of multistrata agroforestry within Eastern temperate forests. Aside from herbs, other products that can be integrated into these multistrata systems include tree syrups, tree bark, firewood, timber, and mushrooms.

Belcher and Schreckenberg identify land tenure, grower organization, and a clear understanding of the biology and ecology of specific understory species as constraints to the successful cultivation of historically wild-harvested species (Belcher and Schreckenberg 2007). The authors caution that "domestication of NTFPs is constrained by the need for secure tenure (usually at the individual level) and, depending on the requirements of the plant, some technical skills and investment capital" and that "[t]hese conditions can prevent landless and other poor people from participating" (Belcher and Schreckenberg 2007). Increasing quality and consistency of cultivated products can also disadvantage those who continue to wild harvest. They also point out that "many NTFPs represent important sources of 'safety-net' (a resource that households can turn to in times of need), subsistence (for households' own consumption) or "cash" and that increased commercialization of the resource can lead to overexploitation or shifting control of property rights such that local people who have historically used the resource are ultimately worse off (Belcher and Schreckenberg 2007).

These considerations need to be front-of-mind as we transition botanical supply chains away from depending primarily on the wild-harvest of herbal material from mixed mesophytic forests and Eastern deciduous forests in general. There is a need for more research into understory species aside from American ginseng, which is relatively well-studied. This research should be connected to organizational efforts such as the Appalachian Beginning Forest Farmers Coalition, as well as University Extension, local nonprofits, conservation districts, and other Technical Service Providers, as these organizations provide an effective structure to disseminate information and organize forest farmers around the production, verification, and marketing of their products. For landless Americans, innovative opportunities for forest farming are being explored through contract growing and leasing arrangements on private family and corporate-owned lands (for example, on a coal company's land in West Virginia) (Tanner Filyaw, personal communication).

The continued access to plant material through permit-based harvesting on public lands also remains a critical way to ensure equality of access. A more dynamic, robust, and resilient harvest program on public lands should be integrated into the overall strategic plan of national forests in the region to support conservation of the remarkable biodiversity of remaining Appalachian mixed mesophytic forests. Further research into and education around novel and sustainable harvest practices will help address overexploitation. The development of product lines of dried ginseng leaf represents an opportunity for non-lethal harvest based on research indicating that ginseng stems and leaves may have higher concentrations of ginsenosides than the root itself (Kang and Kim 2016; Zhang et al. 2020). Research indicating that the harvest of ramp leaves is not lethal to the plant and represents a sustainable form of harvesting that still allows for enjoyment of the resource is similarly promising (Dion et al. 2016).

Expanding markets and access to plant material on public lands should be accompanied by stricter regulations to protect plant biodiversity. This requires better education for law enforcement, adaptive policy that supports forest farmers, and certification structures that support cultivated plant material in the herbal products market. Shifting the minimum harvest age from 5 to 10 years for wild ginseng has been proposed, as well as the implementation of new harvest regulations supporting overall population growth, prohibiting the harvest of not just the youngest ginseng plants, but the most reproductive plants as well, supporting overall population growth (McGraw et al. 2013). Training law enforcement to recognize the value of forest plants and understand that their theft is not a trivial matter is also a critical component of protecting native biodiversity and supporting forest farmers. Ginseng growers routinely have thousands, even hundreds of thousands of dollars of herbal material stolen from them, and often no legal recourse to recover their financial loses (Arnold 2016; Johannsen 2002). Law enforcement can provide important protection for both wild and cultivated plants but must understand their financial value. At the policy level, forest farmers could be supported through the creation of a designation under CITES for the export of "wild-simulated" ginseng. This would allow ginseng growers in states without an export program to certify their product as cultivated, while still maintaining the high value of ginseng roots grown under natural conditions.

It is also crucial that, along with direct investment in private forest farming systems and botanical markets, we have policies in place that support the overall viability of diverse farming livelihoods and of forestland ownership. As the number of farms in the U.S. shrinks, the average age of farmers grows, and the average farm size increases (MacDonald 2020), we need policies that support small-scale agricultural operations growing a diverse array of products. Programs that allow forestland owners to gain economic livelihoods from the sustainable management and retention of forests are critical as well—programs such as The Nature Conservancy's new Family Forest Carbon Program, seeking to allow small forestland owners access to carbon markets. Currently, a very small number of private forestland owners receive Payments for Ecosystem Services (PES). Only 6% of private forestland owners in the U.S. have participated in a cost-share program, 2% have conservation easements on their land, and a similar number receive payments for hunting leases or entrance fees. Payments for wetland mitigation, representing 38% of all ecosystem services payments applied to forest ecosystems, were accessed by .00002% of private forest landowners in the United States (Mercer et al. 2011). Increasing access to PES programs would allow for more diverse and resilient forest ownership, of which forest farming could be a part, helping to avoid the overharvest of commercially valuable understory plants.

Additionally, from a cultural perspective there is a need for a shift toward a conservation ethic concerning understory plants. For many Americans, the most significant reference point they may have for American ginseng is the reality television show *Appalachian Outlaws*, which attracted nearly three million viewers in 2014 (Hufford 2014). This show has been criticized for projecting cultural stereotypes suggesting the harvest of ginseng from the mountainous Appalachians is a lawless endeavor, where clandestine treachery and the evasion of law enforcement is endemic to local culture (Hufford 2014). As a counterbalance to this negative imagery, there is a need for a comparable monetary investment in education around the conservation and sustainable harvest of ginseng and other understory plants, as well as a reinvigoration of the stewardship practices that have accompanied plant harvesting for generations. This cultural shift will require continued access to and engagement with these plants with permit-based harvesting providing an equitable point of contact through our national forest system.

In the context of the COVID-19 pandemic, this work has taken on increased urgency. Early research has suggested that markets for nutraceutical products, including medicinal herbs, have been growing rapidly in response to the crisis as consumers seek out traditional and alternative forms of medicine and as medical researchers search for new plant-based compounds to treat COVID infections (Ayseli et al. 2020; Tahir et al. 2020). Natural products companies found themselves overwhelmed by orders for alternative medicines and herbal products (see Mountain Rose Herbs' temporary suspension of new orders, https://mountainroseherbs.com/about/new-order-closure).

## 24.8 Conclusion

With projections for the future of forest cover in the U.S. suggesting it may continue to decline, and with climate change resulting in large-scale plant species migrations, the protection of remaining native biodiversity islands has never been more important. As the primary source of biodiversity in Eastern deciduous forests, understory plants should be a focal point for preservation efforts (Spicer et al. 2020). Development and implementation of an effective management strategy for American ginseng could establish precedents for similar programs directed toward other increasingly valuable and threatened forest understory species. There is a great opportunity to enlist the public in the protection of these plants through the creation of policies and incentive structures that support the livelihoods of those who live in and around the biodiverse Appalachian mixed mesophytic Forests. Permit-based harvesting of understory plants represents one important way to engage the public in forest conservation programs. This can only work, however, if harvest on public lands is strengthened simultaneously with the intentional cultivation, or forest farming, of valuable medicinal and culinary plants using an approach of "Conservation through Cultivation" to ensure that plant populations expand rather than dwindle.

Recent years have demonstrated how public health and geopolitical events—exemplified by COVID-19—can directly impact patterns of natural resource use. If the uncertainty of the future leads more people to traditional foods and medicines, and thereby to the plants that support these traditions, we must ensure we have laid the groundwork to support an equitable and sustainable future for both plants and people. Protecting and expanding our existing islands of plant biodiversity, as an investment in the plants themselves, the people that use them, and the generations to come, is a project worthy of a national effort.

## Appendix: Some Common Species of Appalachian Mixed Mesophytic Forests

### *Canopy*

| Scientific name | Common name |
|---|---|
| *Fagus grandifolia* | American beech |
| *Liriodendron tulipifera* | Tulip poplar/yellow poplar/tulip tree |
| *Tilia americana* | American basswood |
| *Acer saccharum* | Sugar maple |
| *Aesculus flava* | Yellow buckeye/common buckeye, sweet buckeye |
| *Quercus rubra* | Red oak |
| *Quercus alba* | White oak |
| *Tsuga canadensis* | Eastern hemlock |
| *Betula lenta* | Black birch |
| *Prunus serotina* | Black cherry |
| *Magnolia acuminata* | Cucumber magnolia |
| *Fraxinus americana* | White ash |
| *Acer rubrum* | Red maple |
| *Nyssa sylvatica* | Black gum/sour gum/tupelo |
| *Juglans nigra* | Black walnut |
| *Carya spp.* | Hickory species |

### *Midstory*

| Scientific name | Common name |
|---|---|
| *Cornus florida* | Flowering dogwood |
| *Magnolia spp.* | Magnolia species |
| *Oxydendrum arboreum* | Sourwood/sorrel tree |
| *Acer pensylvanicum* | Striped maple |
| *Cercis canadensis* | Eastern redbud |
| *Carpinus caroliniana* | Ironwood/American hornbeam |
| *Ostrya virginiana* | American Hophornbeam |
| *Ilex opaca* | American holly |
| *Amelanchier arborea* | Service berry/shadbush |
| *Lindera benzoin* | Spicebush |
| *Hamamelis virginiana* | Witch hazel |
| *Asimina trilobal* | American pawpaw |
| *Hydrangea arborescens* | Smooth hydrangea/wild hydrangea |
| *Cornus alternifolia* | Alternative-leaved dogwood/green osier |
| *Rhododendron maximum* | American rhododendron/great laurel |

(continued)

| Scientific name | Common name |
|---|---|
| *Viburnum acerifolium* | Maple-leaved dogwood |
| *Ribes cynosbati* | Eastern prickly gooseberry |
| *Pyrularia pubera* | Buffalo nut |
| *Stewartia ovata* | Mountain camellia |
| *Sambucus canadensis* | American black elderberry/common elderberry |
| *Euonymus americanus* | Strawberry bush/bursting-heart |
| *Clethra acuminata* | Mountain pepper bush |
| *Aralia spinosa* | Prickly ash/devil's walking stick |

## *Understory*

| Scientific name | Common name |
|---|---|
| *Trillium grandiflorum* | White trillium |
| *Trillium erectum* | Red trillium/wake robin/bethroot |
| *Erythronium americanum* | Trout lily |
| *Cypripedium calceolus* | Lady's slipper orchid |
| *Viola spp.* | Violet species |
| *Sanguinaria canadensis* | Bloodroot |
| *Stylophorum diphyllum* | Celandine poppy/woods poppy |
| *Panax quinquefolius* | American ginseng |
| *Delphinium tricorne* | Dwarf larkspur |
| *Hydrophyllum spp.* | Waterleaf species |
| *Phacelia bipinnatifida* | Fernleaf phacelia/spotted phacelia |
| *Phlox divaricata* | Wild blue phlox/wild sweet William |
| *Synandra hispidula* | Guyandotte beauty |
| *Hydrastis canadensis* | Goldenseal/yellowroot |
| *Allium triccocum* | Ramps/wild leeks |
| *Anemone lancifolia* | Lanceleaf anemone/mountain thimbleweed |
| *Anemone quinquefolia* | Wood anemone |
| *Anemonella thalictroides* | Rue anemone |
| *Actaea pachypoda* | White baneberry/doll's eyes |
| *Actaea racemosa* | Black cohosh/black snakeroot |
| *Caulophyllum thalictroides* | Blue cohosh/squaw root |
| *Claytonia virginica* | Virginia springbeauty |
| *Claytonia caroliniana* | Carolina springbeauty |
| *Dicentra canadensis* | Squirrel corn |
| *Dicentra cucullaria* | Dutchman's breeches |
| *Cardamine douglassii* | Limestone bittercress/purple cress |
| *Tiarella cordifolia* | Heartleaf foamflower/false miterwort |
| *Dryopteris goldieana* | Giant wood fern |

(continued)

| Scientific name | Common name |
|---|---|
| *Phegopteris hexagonoptera* | Broad beech fern |
| *Adiantum pedatum* | Northern maidenhair fern |
| *Athyrium pycnocarpos* | Narrow-leaved-spleenwort/glade fern |
| *Athyrium thelypteroides* | Silvery glade fern/silvery spleenwort |
| *Osmunda claytoniana* | Interrupted fern |
| *Symphyotrichum cordifolius* | Common blue wood aster/heartleaf aster |
| *Eurybia divaricatus* | White wood aster |
| *Solidago caesia* | Blue-stemmed/wooden goldenrod |
| *Solidago latifolia* | Broadleaved/zigzag goldenrod |
| *Eupatorium rugosum* | White snakeroot/white sanicle/richweed |

# References


Abrams MD, Johnson SE (2017) The impact of native American activity on vegetation and soil charcoal in the eastern US. Glob J Archaeol Anthropol 1. https://doi.org/10.19080/GJAA.2017.01.555554

Adams AB, Pontius J, Galford G, Gudex-Cross D (2019) Simulating forest cover change in the northeastern U.S.: decreasing forest area and increasing fragmentation. Landsc Ecol 34:2401–2419. https://doi.org/10.1007/s10980-019-00896-7

Ahmad Khan MS, Ahmad I (2019) Herbal medicine: current trends and future prospects. In: Ahmad Khan MS, Ahmad I, Chattopadhyay D (eds) New look to phytomedicine. Academic, Cambridge, MA, pp 3–13. https://doi.org/10.1016/B978-0-12-814619-4.00001-X

Albuquerque UP, Gonçalves PHS, Ferreira Júnior WS, Chaves LS, da Silva Oliveira RC, da Silva TLL, dos Santos GC, de Limra Araújo E (2018) Humans as niche constructors: revisiting the concept of chronic anthropogenic disturbances in ecology. Perspect Ecol Conserv 16:1–11. https://doi.org/10.1016/j.pecon.2017.08.006

Alverson WS, Waller DM, Solheim SL (1988) Forests too deer: edge effects in northern Wisconsin. Conserv Biol 2:348–358

Alwhaibi M, Goyat R, Kelly KM (2017) The use of herbal remedies among mothers of young children living in the Central Appalachian region. Evid Based Complement Alternat Med. https://doi.org/10.1155/2017/1739740

Anderson RC, Anderson MR, Houseman G (2002) Wild American Ginseng. Native Plants J 3:93–105

Arnold C (2016) As demand for ginseng soars, poachers threaten its survival. Natl Geogr. URL https://www.nationalgeographic.com/culture/food/the-plate/2016/october/with-demand-for-ginseng-high%2D%2Dpoachers-threaten-its-survival/. Accessed 2 Feb 2020

Ayseli YI, Aytekin N, Buyukkayhan D, Aslan I, Ayseli MT (2020) Food policy, nutrition and nutraceuticals in the prevention and management of COVID-19: advice for healthcare professionals. Trends Food Sci Technol 105:186–199. https://doi.org/10.1016/j.tifs.2020.09.001

Ban NC, Mills M, Tam J, Hicks CC, Klain S, Stoeckl N, Bottrill MC, Levine J, Pressey RL, Satterfield T, Chan KM (2013) A social–ecological approach to conservation planning: embedding social considerations. Front Ecol Environ 11:194–202. https://doi.org/10.1890/110205

Bannerman JE (1997) Goldenseal in world trade: pressures and potentials. HerbalGram J Am Bot Counc 41:51

Belcher B, Schreckenberg K (2007) Commercialisation of non-timber forest products: a reality check. Dev Policy Rev 25:355–377. https://doi.org/10.1111/j.1467-7679.2007.00374.x

Bierzychudek P (1982) Life histories and demography of shade-tolerant temperate forest herbs: a review. New Phytol 90:757–776
Braun EL (1916) The physiographic ecology of the Cincinnati region. Ohio State University, Columbus
Braun EL (1951) Deciduous forests of eastern North America. Blakiston Press, Philadelphia
Burkhart EP (2011) "Conservation through cultivation:" economic, socio-political and ecological considerations regarding the adoption of ginseng forest farming in Pennsylvania [doctoral dissertation]. Penn State University
Burkhart EP, Jacobson MG (2009) Transitioning from wild collection to forest cultivation of indigenous medicinal forest plants in eastern North America is constrained by lack of profitability. Agrofor Syst 76:437–453. https://doi.org/10.1007/s10457-008-9173-y
Carroll C, Apsley D (2013) Growing American ginseng in Ohio: an introduction (Fact Sheet). The Ohio State University, Columbus
Cavender AP, Beck SH (1995) Generational change, folk medicine, and medical self-care in a rural Appalachian community. Hum Organ 54:129–142
Chamberlain JL (2005) Conserving the Appalachian medicinal plant industry. In: Proceedings of the Appalachian opportunities medicinal and aromatic plants conference. Presented at the Appalachian opportunities medicinal and aromatic plants: producing, using, and marketing herbs and non-timber forest products, Beckley, West Virginia, p 13
Chittum HK, Burkhart EP, Munsell JF, Kruger SD (2019) A pathway to a sustainable supply of forest herbs in the eastern United States. HerbalGram J Am Bot Counc 124:60–77
Cruse-Sanders JM, Hamrick JL (2004) Genetic diversity in harvested and protected populations of wild American ginseng, Panax quinquefolius L. (Araliaceae). Am J Bot 91:540–548. https://doi.org/10.3732/ajb.91.4.540
Davis J, Persons S (2014) Growing and marketing ginseng, goldenseal and other woodland medicinals, 2nd edn. New Society Publishers. newsociety.com
Dentali S (2007) Tonnage survey of select North American wild-harvested plants, 2004–2005. American Herbal Products Association, Silver Spring
Dentali S, Zimmerman M (2012) Tonnage survey of select North American wild-harvested plants, 2006–2010. American Herbal Products Association, Silver Spring
Dharmananda S (2002) The nature of ginseng: traditional use, modern research, and the question of dosage. HerbalGram J Am Bot Counc 54:34–51
Dion P-P, Bussières J, Lapointe L (2016) Sustainable leaf harvesting and effects of plant density on wild leek cultivation plots and natural stands in southern Quebec, Canada. Agrofor Syst 90:979–995. https://doi.org/10.1007/s10457-015-9878-7
Division of Scientific Authority (2018) General advice for the export of wild and wild-simulated American ginseng (Panax quinquefolius) legally harvested during the 2018 harvest season in 19 states and one tribe with a CITES Export Program for American ginseng. U.S. Department of the Interior, Fish and Wildlife Service, Marion
Drummond MA, Loveland TR (2010) Land-use pressure and a transition to forest-cover loss in the eastern United States. Bioscience 60:286–298. https://doi.org/10.1525/bio.2010.60.4.7
Duffy DC, Meier AJ (1992) Do Appalachian herbaceous understories ever recover from clearcutting? Conserv Biol 6:196–201
Filyaw T, Sheban K (2015) Restoring populations of American ginseng, goldenseal, and ramps on the Wayne National Forest – round two (final grant report to the National Forest Foundation No. AH-903) Rural Action, Plains, OH, ruralaction.org
Freese C (ed) (1997) Harvesting wild species: implications for biodiversity conservation. Johns Hopkins University Press, Baltimore. jhupooks.press.jhu.edu
Gilliam FS (ed) (2014) The herbaceous layer in forests of eastern North America. Oxford University Press, Oxford
Gillium F (2007) The ecological significance of the herbaceous layer in forest ecosystems. Bioscience 57:845–858
Gladstar R, Hirsch P (eds) (2000) Planting the future: saving our medicinal herbs. Healing Arts Press, Rochester

Hamilton J (2017) Spreading the ginseng gospel: case study in ginseng production and promotion from Watauga County Cooperative Extension. In: Leopold S, Ormsby A (eds) Proceedings: the future of ginseng and forest botanicals. Presented at the future of ginseng and forest botanicals symposium, Morgantown, West Virginia, pp 109–118

Hill D, Buck L (2000) Forest farming practices. In: Garrett H, Rietveld W, Fisher R (eds) North American agro-forestry: an integrated science and practice. American Society of Agronomy, Madison

Holmes MA, Matlack GR (2018) Assembling the forest herb community after abandonment from agriculture: long-term successional dynamics differ with land-use history. J Ecol 106:2121–2131. https://doi.org/10.1111/1365-2745.12970

Hufford M (2014) Review of Appalachian outlaws. J Appalach Stud 20:234–239. https://doi.org/10.5406/jappastud.20.2.0234

Jiang B, Ma C, Motley T, Kronenberg F, Kennelly EJ (2011) Phytochemical fingerprinting to thwart black cohosh adulteration: a 15 Actaea species analysis. Phytochem Anal PCA 22:339–351. https://doi.org/10.1002/pca.1285

Johannsen KL (2002) Root Rustlers. Mother Jones. URL https://www.motherjones.com/politics/2002/07/root-rustlers/. Accessed 23 Apr 2020

Johnston J, Szabo M, Rodney A (2011) Good food, good people: understanding the cultural repertoire of ethical eating. J Consum Cult 11:293–318. https://doi.org/10.1177/1469540511417996

Jouzi Z, Azadi H, Taheri F, Zarafshani K, Gebrehiwot K, Van Passel S, Lebailly P (2017) Organic farming and small-scale farmers: main opportunities and challenges. Ecol Econ 132:144–154. https://doi.org/10.1016/j.ecolecon.2016.10.016

Kang O-J, Kim J-S (2016) Comparison of ginsenoside contents in different parts of Korean ginseng (Panax ginseng C.A. Meyer). Prev Nutr Food Sci 21:389–392. https://doi.org/10.3746/pnf.2016.21.4.389

Light PD (2008) A history of southern and Appalachian folk medicine. J Am Herbalists Guild 8:27–38

Loucks C, Olson D, Dinerstein E, Weakley A, Noss R, Stritholt J, Wolfe K (n.d.) Appalachian mixed mesophytic forests | ecoregions | WWF. World Wildl Fund. URL https://www.worldwildlife.org/ecoregions/na0402. Accessed 19 Jan 2020

MacDonald JM (2020) Tracking the consolidation of U.S. agriculture. Appl Econ Perspect Policy 42:361–379. https://doi.org/10.1002/aepp.13056

McCarthy BC (2014) The herbaceous layer of eastern old-growth deciduous forests. In: Gilliam FS (ed) The herbaceous layer in forests of eastern North America. Oxford University Press, Oxford

McCoy J-A, Young JH, Nifong JM, Hummer K, DeNoma J, Avendaño-Arrazate CH, Greene SL, Kantar MB (2019) Species for medicinal and social use with an emphasis on Theobroma cacao L. (Cacao), Nicotiana tabacum L. (Tobacco), Actaea racemosa L. (Black Cohosh), and Humulus lupulus L. (Hops). In: Greene SL, Williams KA, Khoury CK, Kantar MB, Marek LF (eds) North American crop wild relatives, volume 2: important species. Springer, Cham, pp 645–692. https://doi.org/10.1007/978-3-319-97121-6_19

McGraw JB, Furedi MA (2005) Deer browsing and population viability of a forest understory plant. Science 307:920–922

McGraw JB, Lubbers AE, der Voort MV, Mooney EH, Furedi MA, Souther S, Turner JB, Chandler J (2013) Ecology and conservation of ginseng (Panax quinquefolius) in a changing world. Ann N Y Acad Sci 1286:62–91. https://doi.org/10.1111/nyas.12032

Mercer DE, Cooley D, Hamilton K (2011) Taking stock: payments for forest ecosystem services in the United States. Ecosystem Marketplace. Forest Trends, forest-trends.org

Miller KE, Gorchov DL (2004) The invasive shrub, Lonicera maackii, reduces growth and fecundity of perennial forest herbs. Oecologia 139:359–375. https://doi.org/10.1007/s00442-004-1518-2

Montagnini F, Levin B, Berg KE (2022) Introduction. Biodiversity islands. Strategies for conservation in human-dominated environments. In: Montagnini F (ed) Biodiversity islands. Strategies for conservation in human-dominated environments. Series: Topics in Biodiversity and Conservation, Springer, Cham, pp.

Parsons CM (2016) The natural history of colonial science: Joseph-François Lafitau's discovery of ginseng and its afterlives. William Mary Q 73:37–72

Petrzelka P, Ma Z, Malin S (2013) The elephant in the room: absentee landowner issues in conservation and land management. Land Use Policy 30:157–166. https://doi.org/10.1016/j.landusepol.2012.03.015

Pokladnik RJ (2008) Roots and remedies of ginseng poaching in Central Appalachia. Antioch University, Keene

Rao MR, Palada MC, Becker BN (2004) Medicinal and aromatic plants in agroforestry systems. Agrofor Syst 61:107–122. https://doi.org/10.1023/B:AGFO.0000028993.83007.4b

Robbins C (1998) American ginseng: the root of North America's medicinal herb trade. TRAFFIC 52–54. Washington

Robbins C (1999) Medicine from US wildlands: an assessment of native plant species harvested in the United States for medicinal use and trade and evaluation of the conservation and management implications. TRAFFIC North America (traffic.org)

Rosson JF (2008) Current forest conditions of older stands of the mixed mesophytic forest region on the Appalachian plateaus province of eastern Kentucky. In: Proceedings of the 16th central hardwoods forest conference, pp 213–222

Schelhas J (2001) The USA national parks in international perspective: have we learned the wrong lesson? Environ Conserv 28:300–304. https://doi.org/10.1017/S0376892901000327

Schluter C, Punja ZK (2002) Genetic diversity among natural and cultivated field populations and seed lots of American ginseng (Panax quinquefolius L.) in Canada. Int J Plant Sci 163:427–439. https://doi.org/10.1086/339512

Sebastiani R, Montagnini F, Dalli D (2013) Ethical consumption and new business models in the food industry: evidence from the Eataly case. J Bus Ethics 114:473–488. https://doi.org/10.1007/s10551-012-1343-1

Sobey DG (1977) The structure and rate of growth of the rhizomes of some forest herbs and dwarf shrubs of the New Brunswick-Nova Scotia border region. Can Field-Nat 91:377–383

Szczuka D, Nowak A, Zakłos-Szyda M, Kochan E, Szymańska G, Motyl I, Blasiak J (2019) American ginseng (Panax quinquefolium L.) as a source of bioactive phytochemicals with pro-health properties. Nutrients 11. https://doi.org/10.3390/nu11051041

Tahir AH, Javed MM, Hussain Z (2020) Nutraceuticals and herbal extracts: a ray of hope for COVID-19 and related infections (review). Int J Funct Nutr 1:1. https://doi.org/10.3892/ijfn.2020.6

Teel WS, Buck LE (1998) From wildcrafting to intentional cultivation: The potential for producing specialty forest products in agroforestry systems in temperate North America. North American Conference on Enterprise Development Through Agroforestry: Farming the Agroforest for Specialty Products. Minneapolis, Minnesota

Turner JB, McGraw JB (2015) Can putative indicator species predict habitat quality for American ginseng? Ecol Indic 57:110–117

Webb SL, Dwyer M, Kaunzinger CK, Wyckoff PH (2000) The myth of the resilient forest: case study of the invasive Norway Maple (Acer platanoides). Rhodora 102:332–354

Young JA, King TL, van Manen FT, Eackles MS (2009) Habitat characterization, genetic diversity, and population abundance of American ginseng. U.S. Geological Survey final report to the U.S. Fish and Wildlife Service, Division of Scientific Authority, Arlington, Virginia

Young JA, van Manen FT, Thatcher CA (2011) Geographic profiling to assess the risk of rare plant poaching in natural areas. Environ Manag 48:577–587. https://doi.org/10.1007/s00267-011-9687-3

Zhang F, Tang S, Zhao L, Yang X, Yao Y, Hou Z, Xue P (2020) Stem-leaves of Panax as a rich and sustainable source of less-polar ginsenosides: comparison of ginsenosides from Panax ginseng, American ginseng and Panax notoginseng prepared by heating and acid treatment. J Ginseng Res. https://doi.org/10.1016/j.jgr.2020.01.003

# Chapter 25
# Farmer Perceptions of Tropical Dry Forest Restoration Practices on the Azuero Peninsula of Panama – Implications for Increasing Biodiversity in a Human-Dominated Landscape

**Vicente Vásquez, Cristina Barber, Yassine Dguidegue, T. Trevor Caughlin, Roxana García, and Ruth Metzel**

**Abstract** Patches of second-growth forest are biodiversity islands with critical importance in agricultural landscapes, providing ecosystem services that support human livelihoods and promote biodiversity, no matter the size. Natural regeneration presents an opportunity for forest landscape restoration (FLR) at low cost. However, this is not always the chosen strategy for the smallholder farmers that control about 80% of farms worldwide and are the primary decision-makers for large land areas with restoration potential. Understanding the perceptions of farmers on restoration strategies can increase our capacity to implement restoration projects. We surveyed 64 Panamanian land managers in Los Santos province, where 53% of farmers are smallholders, to better understand how they perceive restoration strategies (reforestation, regeneration, and assisted natural regeneration) and their preferences. Participants were most confident defining reforestation, associating it with tree planting. Farmers associated natural regeneration with independently growing vegetation, and assisted natural regeneration with human intervention and tree planting. Farmers were polarized in their preference for land clearing, with 36% preferring it as a first option and 42% preferring it as a last option. Over half of participants ranked letting vegetation grow and tree planting among their first and second options. High percentages of tree cover were associated with low preference for land clearing. Large farm size was associated with higher preference for natural regeneration. Expanding natural regeneration requires clearly defining and standardizing this

V. Vásquez (✉) · Y. Dguidegue
University of the Ozarks, Clarksville, AR, USA
e-mail: vvsqu937@ozarks.edu

C. Barber · T. T. Caughlin
Boise State University, Boise, ID, USA

R. García · R. Metzel
Fundación Pro Eco Azuero, Pedasí, Los Santos, Panamá

 
F. Montagnini (ed.), *Biodiversity Islands: Strategies for Conservation in Human-Dominated Environments*, Topics in Biodiversity and Conservation 20,
https://doi.org/10.1007/978-3-030-92234-4_25

practice. Understanding farmers' preferences and knowledge of restoration practices will allow the best FLR strategies to be implemented in each specific case to achieve desired restoration goals.

**Abstract (Spanish)** Los parches de bosque secundario son islas de biodiversidad con importancia crítica para paisajes agrícolas, porque proveen servicios ecosistémicos que mantienen los medios de vida humanos y promueven la biodiversidad. La regeneración natural ofrece una oportunidad de restauración del paisaje forestal (FLR, por sus siglas en inglés) a bajo costo. No obstante, esta no es siempre la estrategia de restauración preferida por los productores que manejan fincas de pequeña escala que representan el 80% de las fincas en el mundo. Dichos productores son los principales actores en la toma de decisiones sobre una proporción significativa de tierra con potencial para restauración. Entender cómo los pequeños productores perciben la restauración podría ayudar a mejorar la eficiencia en la implementación de proyectos de FLR. Realizamos 64 encuestas a productores panameños en la provincia de Los Santos, donde 53% de los productores manejan fincas de pequeña escala. El objetivo de estas encuestas fue comprender cómo perciben diferentes estrategias de restauración (reforestación, regeneración natural y regeneración natural asistida), y sus preferencias. Los participantes definieron con más confianza reforestación, que asociaron con plantar árboles, mientras que asociaron regeneración natural con vegetación que crece por sí misma, y regeneración natural asistida con intervención humana y plantar árboles. Los productores tenían opiniones polarizadas respecto a limpiar toda la parcela: el 36% la evaluaron como su estrategia preferida y un 42% como menos preferida. Más de la mitad de los encuestados clasificaron plantar árboles y dejar que la vegetación crezca como su primera o segunda opción. Altos porcentajes de cobertura arbórea fueron asociados a una menor preferencia por limpiar toda la parcela, y parcelas más grandes fueron asociadas a una mayor preferencia por regeneración natural. Expandir la regeneración natural como estrategia de restauración depende en gran medida de alcanzar una definición común de esta práctica. Esto ayudará a integrar los conocimientos y necesidades de los ganaderos con los objetivos de restauración, e implementar la mejor estrategia de restauración en cada caso.

**Keywords** Cattle · Community-based research · Forest landscape restoration (FLR) · Participatory · Reforestation · Regeneration · Tree planting

## 25.1 Introduction

Patches of second-growth forest are biodiversity islands with critical importance in agricultural landscapes. Even small patches of second-growth forest can provide ecosystem services that support human livelihoods and promote biodiversity (Souza et al. 2016, Acevedo-Charry and Aide 2019). At the same time, they represent a low-cost natural climate solution with great potential to sequester carbon (Chazdon et al. 2016a, b, Griscom et al. 2017).

Due to the numerous benefits of forest recovery, land managers and policymakers across the tropics are promoting forest and landscape restoration (FLR) as a way to increase tree cover in working landscapes while supporting multiple needs of local stakeholders (Stanturf et al. 2019). Large scale landscape restoration can contribute to biodiversity conservation of critically endangered species. For example, in Los Santos province, the Azuero spider monkey (*Ateles geoffroyi azuerensis*) is known to need contiguous forest corridors with large width and tall canopy height, which are in danger if forest restoration practices are not instituted (Schelegle 2011).

The focus of this chapter is on understanding farmer definitions and preferences on how to restore the tree cover of large contiguous patches on their farm. Understanding their perceptions on restoration strategies can increase our capacity to implement restoration projects, helping to integrate farmers' knowledge and needs with forest restoration goals, and matching each specific case to the best restoration strategy.

### 25.1.1 Choice of Restoration Strategy: Natural Regeneration, Tree Planting

A very important decision for forest restoration is whether to pursue strategies that involve a high level of human intervention, like tree-planting, or to enable forest to recover with minimal intervention via strategies like natural regeneration (Holl and Aide 2011). Restoration, reforestation, assisted natural regeneration, and natural regeneration are all terms used often and with wide ranging definitions (Chazdon et al. 2016a, b). Global, academic, or policy conceptualizations of what it means to restore, reforest, or regenerate an area can differ from locally used definitions of these concepts, as it will be discussed in this chapter. In this chapter we generally use the term reforestation, unless otherwise stated, to mean "re-establishment of forest through planting and/or deliberate seeding on land classified as forest" as defined by the FAO's Forest Resources Assessment (FAO 2020). This definition excludes natural regeneration, which is defined as the "establishment of new forest by self-sown seed, coppice shoots or root suckers" (Grebner et al. 2013). Assisted natural regeneration is natural regeneration with human intervention, such as fencing, weed competition removal, enrichment planting, and protection from fire and other threats (Shono et al. 2007).

### 25.1.2 Advantages of Natural Regeneration

While tree planting has long served as a default restoration strategy, recent ecological research has revealed numerous benefits of natural regeneration, particularly when regeneration sites are located near existing forest fragments (Griscom 2020). When compared using meta-analyses, natural regeneration was found to, on average, provide greater improvements in ecosystem function than other restoration practices (Crouzeilles et al. 2017; Jones et al. 2018; Reid et al. 2018). The financial costs of natural regeneration can also be significantly lower than maintaining nurseries or buying tree saplings for planting (Elliott et al. 2013). Other researchers have suggested implementing tree-planting practices only in land areas where natural regeneration would otherwise be slow (Strassburg et al. 2019).

Despite its potential effectiveness, natural regeneration remains underutilized by decision-makers, farmers, and restoration practitioners (Chazdon and Uriarte 2016). Recent research concluded that natural regeneration should be especially adopted in areas adjacent to riparian zones (Griscom 2020). From a cost-benefit analysis standpoint, natural regeneration should be the first restoration approach attempted before resorting to "intensive restoration methods" (Brancalion and Holl 2020; Holl and Brancalion 2020).

Recent work has examined the political and economic reasons why natural regeneration is rarely a major component of restoration planning. These reasons include a lack of economic policy to monetize naturally-regenerating forests, and agrarian reform laws that penalize landowners for uncultivated land (Chazdon et al. 2020). Explanations for the slow adoption of natural regeneration are largely institutional. In the context of individual motivations to naturally regenerate vegetation, the different ways in which farmers in Azuero, Panama protect valuable species for livelihood purposes on their land has been thoroughly studied (Garen et al. 2011; Metzel and Montagnini 2014). However, it is important to better understand land managers' perceptions and preference of natural regeneration compared to reforestation in order to implement effective restoration projects.

### *25.1.3 Understanding Farmers' Preferences on Restoration Strategies*

As emphasized in major FAO reports, smallholding farmers own over 80% of farms worldwide (Schneider 2016); therefore, they are the primary decision-makers for large portions of land with restoration potential. Key research and policy organizations have emphasized the indispensable need to include smallholding farmers in restoration initiatives (FAO 2017). Implementing sustainable restoration practices requires understanding the different factors that contribute to farmers' preferences for restoration strategies. This includes understanding smallholding farmers' attitudes towards differing forest restoration practices, and examining existing types of collaboration among farmers, governmental stakeholders, and environmental organizations (Thacher et al. 1996; Schneider 2016).

Understanding farmers' preferences on restoration strategies, including perceptions of natural regeneration versus reforestation, can improve the capacity to achieve FLR goals. In some cases, spontaneous tree growth in silvopastoral systems may already be a central tenet of farm management, which could be leveraged to support natural regeneration. Farmers maintain trees in pastoral landscapes for diverse reasons, including cultural, economic, and agronomic motivations (Garen et al. 2011). For example, cattle ranchers in the Ecuadorian Amazon promote recruitment and survival of native tree species in pastures to sustain the viability of their cattle-focused farming activities (Lerner et al. 2015). Understanding that farmers maintain tree cover for a variety of reasons, some of which may relate to natural regeneration, is crucial to customizing FLR in a way that responds to specific needs of farming communities (Garen et al. 2009). However, in some cases farmers perceive natural regeneration as "messy" and undesirable and may be more averse to implementing it in their farmland (Zahawi et al. 2014). Participatory research with farmers could provide ideas for implementing natural regeneration in ways that improve aesthetic value, and ultimately longevity, of second-growth forests, providing more incentives for farmers to use natural regeneration strategies, including those who may not have originally appreciated them.

We studied farmer preferences for restoration strategies, including reforestation, natural regeneration, and assisted natural regeneration in a tropical landscape undergoing secondary succession. Our study area, Los Santos province in Panama, is representative of many Latin American pastoral systems that were once extensively deforested for cattle ranching and now present opportunities for forest recovery (Heckadon-Moreno 2009) (Fig. 25.1). In Los Santos, 53% of farms are managed by smallholders, as defined by the two-hectare classification used in Schneider's global study (2016, INEC 2010). Similar to many areas with relatively sparse tree cover, estimates based on forest production (e.g. Hansen et al. 2013, cited by Fagan 2020) are likely to provide potentially inaccurate assessments of tree cover change in Azuero (Caughlin et al. 2016). Although some research showed a regional forest loss of 220,000 ha from 2001 to 2018, other satellite remote sensing provides some evidence that tree cover is actually increasing in Los Santos (Metzel 2010; Hansen et al. 2013; Sloan 2015). Remote sensing detection devices with very high-resolution imagery provide additional support for regional increases in tree cover, albeit with high spatial variability (Tarbox et al. 2018; Caughlin et al. 2019). One explanation for high spatial variability in tree cover change in Los Santos is heterogeneity in farmer decision-making across a landscape that consists of smallholder farms.

In Los Santos the Panamanian government and NGOs are working to promote large-scale restoration of land, a majority of which is privately owned. These current restoration efforts are often centered around reforestation and require purchasing seedlings and other materials, which limits the scope of restoration practices due to costs. Farmer preferences are central to understanding potential techniques to increase forest cover and to successfully implement restoration programs in the region.

## 25.2 Methodology

### *25.2.1 Study Site: Azuero Peninsula, Panama*

This study was primarily conducted in Los Santos province on the Azuero peninsula of Panama (Figs. 25.1 and 25.2). Los Santos province forms part of the eastern half

**Fig. 25.1** Pasture and secondary regrowth in the Los Santos Province, Azuero Peninsula, Panama. (Photo: Cristina Barber)

of the peninsula, which falls within the tropical dry forest biome, defined as "a vegetation type typically dominated by deciduous trees where at least 50% of trees present are drought deciduous, the mean annual temperature is > 25 °C, total annual precipitation ranges between 700 and 2,000 mm, and there are three or more dry months every year (precipitation < 100 mm)" (Sánchez-Azofeifa et al. 2005). This study region experiences a 5-month dry season from December through April, and a bimodal rainy season from May through November, with a short dry period in June (Love and Spaner 2005; Heckadon-Moreno 2009).

### *25.2.2 Survey of Farm Managers*

We developed a survey to better understand how farmers perceive restoration strategies (reforestation, regeneration, and assisted natural regeneration), their preferences for these strategies for increasing tree cover on the landscape, and the factors that determine these preferences. The semi-structured, voluntary, anonymous survey consisted of 19 questions about farmer characteristics, land management practices, and knowledge and perceptions of regeneration and reforestation. This survey was

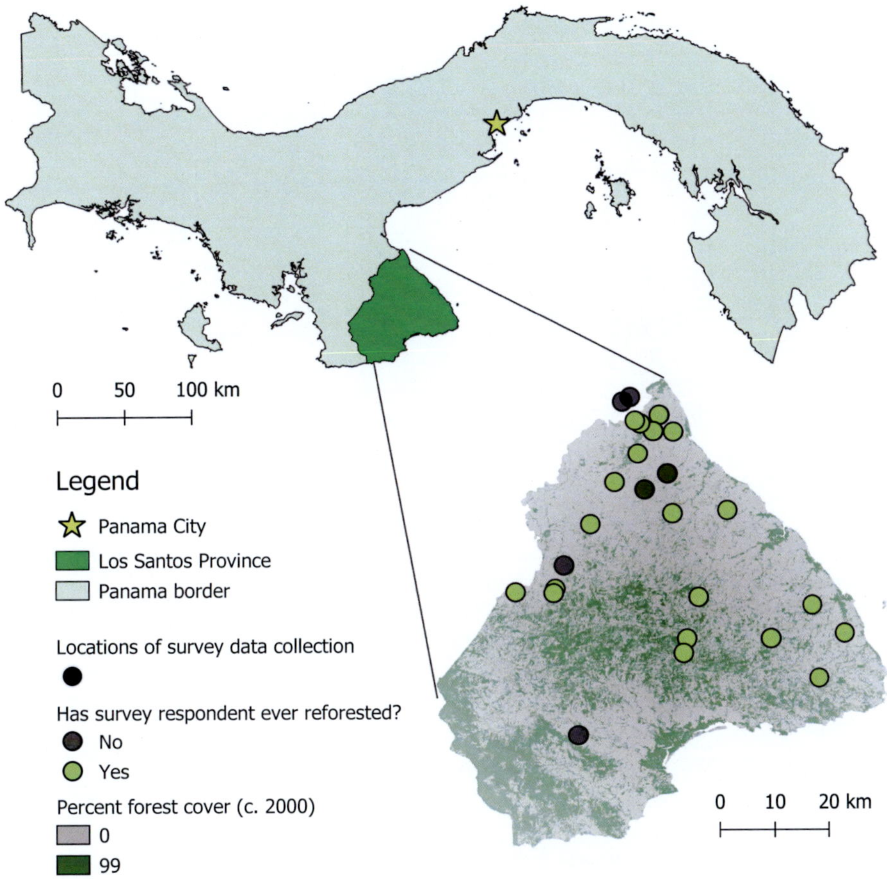


**Fig. 25.2** Map of survey locations. Symbol coloring represents percentage of farmers at each location who indicated that they had reforested their land in the past

developed by one of the authors from the Azuero peninsula, and it was revised through conversations with different landowners and after being used in a pilot survey group from July 2019–January 2020. We wanted to ensure a high level of comprehension of the survey by using non-technical terms when appropriate.

The survey was administered to 64 Panamanian "farm managers", a term which encompasses farm owners and farm administrators, from Herrera (5) and Los Santos (59) provinces in February 2020. Survey participants were selected by going door to door (58 surveys; 1 PM – 8:30 PM) and visiting agricultural supply stores (6 surveys; 9 AM – 1 PM) in sixteen rural communities (Fig. 25.3). Surveys varied in duration from 10–30 minutes and were conducted orally by one of the authors and a colleague who was born in the Azuero peninsula, using the Survey123 smartphone application to record participant answers.

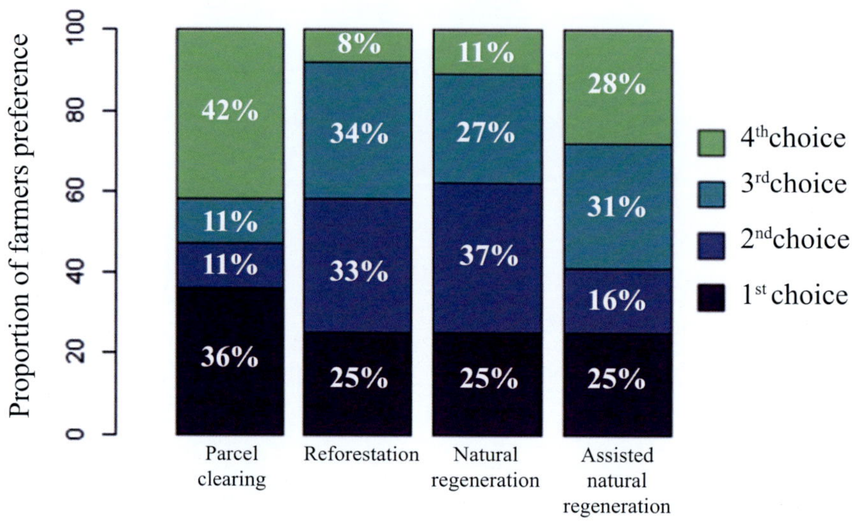


**Fig. 25.3** Percentage of farmers ranking parcel clearing, reforestation, natural regeneration (letting vegetation grow) and assisted natural regeneration (letting vegetation grow with fencing) as their first (most preferred), second, third or fourth (least preferred) land management strategy

## *25.2.3 Research Questions*

### 25.2.3.1 Farmers' Understanding of Reforestation and Natural Regeneration

To gain insight into how farmers understand restoration techniques that increase tree cover, we asked farmers to define concepts like reforestation, natural regeneration and assisted natural regeneration. We used the tidytext (Silge and Robinson 2016) tool to analyze their answers, including tokenizing linked words that repeat across participant answers, and classifying these into the actions, connotations and alternative definitions associated with the concept, as well as answers indicating lack of comprehension or response. The linked words repeated across participants to define reforestation, natural regeneration and assisted natural regeneration were counted and converted into percentages.

### 25.2.3.2 Farmers' Preferences for Reforestation or Natural Regeneration

To determine farmer preferences of land management options, we asked farmers to rank their preferences for the following four strategies: (1) clear the whole plot for cattle, (2) reforest the parcel, (3) let vegetation grow in the parcel (unassisted natural regeneration), and (4) let the vegetation grow in specific areas with the use of fences (assisted natural regeneration). The options in the farmers' native language

(Spanish) were: (1) "limpiar toda su parcela para ganado", (2) "reforestar su parcela", (3) "dejar que el monte se crezca" and (4) "dejar que el monte se crezca delimitando el área con cercas". This phrasing was adopted for the general survey to adapt these concepts to local terminology (i.e. "monte" is commonly used to refer to vegetation). In pre-survey consultations, "letting vegetation grow" was the phrase most evocative of natural regeneration for the farmers consulted, and "letting vegetation grow with fencing" was most evocative of an assisted natural regeneration practice that excludes cattle from the system through human labor to accelerate succession.

No questions on past or future land use for the forested land were asked in the survey, and talking about it was purposefully left to the preference of the survey participants. Survey participants ranked the land management option strategies from 1 (most preferred) to 4 (least preferred), and we then calculated the percentage of times that each strategy was mentioned in each preference category.

To identify which economic costs might drive farmers' decisions, we asked survey participants to choose from a list of resources which they would prefer to obtain during a reforestation or natural regeneration program. The possible options for reforestation were fencing materials, native tree seedlings, volunteer labor, and there was an option of not answering the question. The possible options for assisted natural regeneration were fencing materials, volunteers to build the fence, and the option of not answering the question. Answers were counted and shown in percentages.

#### 25.2.3.3 Factors Driving Farmer Preferences

We also quantified the relative importance of factors driving farmer land management preferences. We created four models including the four following management strategies: parcel clearing, reforestation, letting vegetation grow, and letting vegetation grow with fencing. The preferences of survey participants were tested against survey variables (Table 25.1). All variables were treated as fixed effects. We used a Bayesian modeling framework with Hamiltonian Monte Carlo sampling to analyze survey answers using the BRMS package (Bürkner 2017), and modeled the answers using a cumulative regression. This kind of distribution is used for categorical data in which the order of the categories is meaningful (McElreath 2020).

## 25.3 Results

### *25.3.1 Farmers' Understanding and Definition of Reforestation, Natural Regeneration and Assisted Natural Regeneration*

The number of farmers confident in defining reforestation was much greater than those confident in defining natural regeneration and assisted natural regeneration. The percentage of survey participants that stated not knowing or understanding

**Table 25.1** Variables included in the models to explain survey participants' preference for different restoration strategies: parcel clearing, reforestation, natural regeneration, and fenced natural regeneration

| Variable name | Type | Explanation |
|---|---|---|
| Farm size | Continuous | Farm size |
| Age | Continuous | Farmer age |
| Previous ref | Categorical | If the farmer has already reforested on their farm (yes/no) |
| Education | Ordered categorical | Farmers educational level (primary/ basic education/ high school/ undergraduate / masters) |
| Self-reported tree cover percentage | Continuous | Percentage of land reserved for trees reported by survey participants. |
| Future ref | Categorical | If the farmers would consider future reforestation on their farm (yes/no) |
| Ref C | Categorical | Preferred reforestation inputs (fencing wire/ native tree seedlings/ volunteers for fencing) |
| Future NR | Categorical | If the farmers would consider letting the vegetation to grow on their farm in the future (yes/no) |
| Know ref | Categorical | If the farmer knows someone that has reforested (yes/no) |

**Notes:** Continuous, categorical, and ordered categorical variables used to explain survey participant's preference for restoration: parcel clearing, reforestation, natural regeneration, and fenced natural regeneration. Future Ref stands for reforestation in the future, Ref C stands for the reforestation input categories, Future NR stands for Natural regeneration in the future, and Known ref. stands for knowing people that reforest

when asked to define the different land use alternatives was 3% for reforestation, 13% for natural regeneration and 19% for assisted natural regeneration. Of all survey participants, 77% included the planting of trees and seedlings in their definition of reforestation. When defining natural regeneration, 59% of survey participants included trees, other vegetation or forest growing by itself, and 8% described it as a process without human intervention. There were 8% of survey participants that included tree-planting in their definition of natural regeneration. Nearly half of survey participants defined assisted natural regeneration as a process that involves reforesting and planting trees (48%) and nearly one fifth (17%) defined it as a process involving human intervention. Fewer than 5% of participants mentioned letting trees, vegetation and forest grow on their own in their definitions of assisted natural regeneration. While 17% of survey participants defined reforestation as a process that happens on land previously deforested, fewer than 5% of participants defined natural regeneration or assisted natural regeneration as a process that happens on land previously deforested.

### 25.3.2 Farmers' Preferences for Reforestation or Natural Regeneration

Farmers had strong opinions about land clearing. Most farmers' answers were polarized between choosing parcel clearing as their most preferred (first; 36%) or least preferred (fourth; 42%) land management option (Fig. 25.3). Letting vegetation grow (natural regeneration) and reforestation were equally preferred as the top choice by one fourth of participants. Over half of survey participants (58%) preferred reforestation as their first or second option, while few (8%) chose it as their least preferred option. Over half of survey participants (62%) chose to let the vegetation grow (natural regeneration) as their first or second option, while few (11%) said it was their least preferred option. Letting vegetation grow with fencing (assisted natural regeneration) was the least preferred restoration strategy. More than half of survey participants (59%) chose it as their third or fourth option and fewer participants put it as their first or second choice compared to each of the three alternative land management options (Fig. 25.3).

Some survey participants expressed the reasons they consider reforesting or letting vegetation grow on part of their land. For reforestation reasons, they mentioned watershed protection, timber and fruit for their personal use, and to contribute to the environment. Natural regeneration reasons included contributing to solving climate change issues, protecting flora and fauna, watershed protection and allowing the land to recover.

### 25.3.3 Farmers' Preference of Inputs for Reforestation and Natural Regeneration

Results show that for participating in a reforestation project, 72% of farmers preferred to receive native tree seedlings, 14% preferred volunteer labor, 8% preferred fencing materials, and 6% did not answer. When asked about preference for inputs to support an assisted natural regeneration project, 61% of farmers preferred to receive volunteer labor, 28% preferred fencing materials, and 11% did not answer.

### 25.3.4 The Factors Driving Farmer Preferences for Land Management Strategies

Our analysis of the factors driving survey participant preferences for the four different management strategies found that a higher tree cover percentage on the farm and greater farm size increased the probability of ranking parcel clearing in 3rd or 4th place. The coefficients illustrate the relative effect size of the variables (i.e. how strong is the effect of each variable compared with the others). A positive

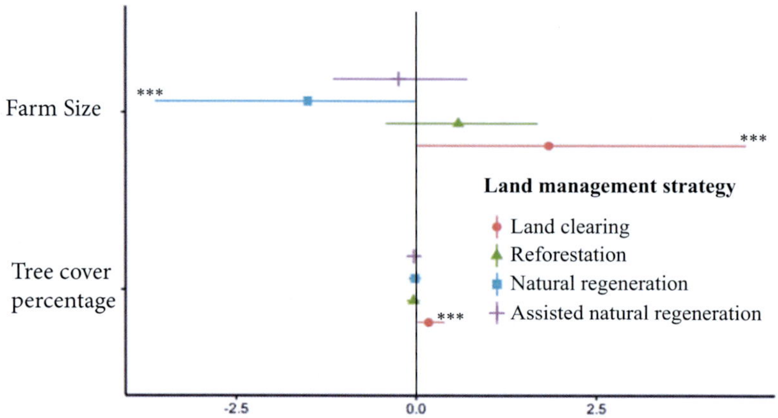


**Fig. 25.4** Coefficients plot for all land management strategies. The variables that had an effect on land management strategy preference are on the y axis. We consider that a variable has an effect on land management preference when the 95% CI does not cross zero. The vertical line marks the zero value. *** means that the variable has an effect on land management preference. In our preference rankings, fourth is least preferred and first is most preferred. Therefore, a lower parameter value (i.e. negative) indicates that the variable (farm size, tree area percentage) has a positive effect on preferred land management strategy

coefficient indicates more probability of grading the response variable as less desirable (3rd or 4th) and a negative coefficient indicates more probability of grading the response variable as more desirable (1st or 2nd). The 95% credibility intervals (CI) indicates the range of values containing the 95% of the posterior distribution and provides an uncertainty measure. The positive sign of the effect of self-reported tree cover percentage (relative effect size = 0.17; 95% CI: 0.03 to 0.40) and farm size (relative effect size = 1.84; 95% CI: 0.00 to 4.59) indicates that more tree cover on a farm is associated with less land clearing preference.

Larger farm size increases the probability of ranking natural regeneration in 1st or 2nd place. The negative sign of the effect of farm size (relative effect size = −1.50; 95% CI: −3.65 to 0.00) on natural regeneration indicates that as farm size increases, natural regeneration is a more attractive option. None of the variables shown in Table 25.1 seemed to influence farmers' choice of reforestation or assisted natural regeneration preference as land management strategy (Fig. 25.4).

## 25.4 Discussion

### *25.4.1 Reasons Why Farmers May Prefer Reforestation to Natural Regeneration*

Including smallholder farmers in restoration initiatives requires finding a common language and conceptualization of restoration strategies. If farmers and institutions do not share a common understanding and vocabulary about natural regeneration, this could impede its use as a landscape restoration strategy. Our results show that most farmers understand natural regeneration as vegetation growing without human intervention. If cattle farmers perceive natural regeneration as a restoration strategy devoid of human action, they are more likely to associate it with abandoned and unused land (Zahawi et al. 2014). Incorporating a human element to natural regeneration, such as fencing a parcel, fire prevention, or monitoring vegetation growth for scientific, societal or ecosystemic purposes may allow farmers to have a greater feeling of control and intervention in natural regeneration strategies, while at the same time saving them the high costs of tree-planting.

The percentage of farmers that appeared unfamiliar with the term natural regeneration was significantly higher than expected, given that many farmers (62%) chose letting vegetation grow as their first or second land management option, and also considering the recent land use changes in the Azuero Peninsula (Metzel 2010; Sloan 2015; Caughlin et al. 2016). One explanation for this discrepancy is that farmers may not perceive natural forest recovery on land that is underutilized for agriculture as part of a restoration strategy. At the regional-scale, agricultural labor scarcity has been proposed as a primary driver of forest cover increases, as Los Santos farmers increasingly are employed in off-farm jobs (Sloan 2015). The combination of lack of preference for natural forest growth relative to tree planting, combined with the labor needed to put up fences, could explain why assisted natural regeneration ranked lowest among management strategies in the survey of individual farmers. Despite the cost-effectiveness of assisted natural regeneration relative to tree planting and its potential to speed forest recovery relative to unassisted natural regeneration, there still exists a challenge of implementing this strategy on a widespread scale in working landscapes, including cattle pastures in Los Santos.

Reforestation may be a concept more widely understood by Azuero farmers because most of the previous institutional restoration efforts have focused on reforestation, and more specifically on tree-planting. Decades of forest restoration programs, such as the Alliance for a Million Hectares (Panama's Bonn Challenge commitment), the Smithsonian-Yale Native Species Reforestation Project (PRORENA, Hall et al. 2011), Yale's Environmental Leadership and Training Initiative (ELTI), Fundación Pro Eco Azuero's ecological corridor restoration program, government incentive programs to plant exotic species like teak and other initiatives by community based organizations have all played a role in informing farmers about reforestation and the benefits of tree planting.

### *25.4.2 Potential for Increased Use of Natural Regeneration*

Bridging the divide between farmer and institutional definitions and understandings of natural regeneration and assisted natural regeneration may help to increase implementation of these strategies. Increasing use of techniques related to natural regeneration will require facilitating the integration of different strategies in landscape restoration projects, as well as implementing donor/NGO success indicators to monitor progress of natural regeneration projects (Chazdon and Uriarte 2016; Chazdon et al. 2020). Fundación Pro Eco Azuero has implemented strategies like monitoring vegetative growth together with farmers to emphasize the value of natural regeneration and decreasing the density of planting to allow regeneration between planted trees while still retaining some of the benefits of a reforestation parcel. Other suggestions to promote natural regeneration may include: (1) leveraging natural regeneration at the boundary of existing forest reserves to expand secondary forest patches, (2) implementing assisted natural regeneration strategies as part of existing agroforestry and silvopastoral systems, and (3) shifting government and donor reporting frameworks away from use of indicators such as number of trees planted, and toward the total area of land restored.

Expanding the implementation of natural regeneration will also largely depend on the compatibility of individual farmers' needs and the benefits of natural regeneration strategies. Several studies have highlighted the uses and values that farmers attribute to the presence of trees on their parcels (Griscom et al. 2009; Metzel and Montagnini 2014; Jakovac et al. 2017). This indicates that farmers value restoration strategies where they are able to select the species and design of vegetation on their land. For example, farmers might prefer to have fruit trees close to their houses (Metzel and Montagnini 2014). By contrast, plants growing in a natural regenerating patch depend highly upon the species pool of neighboring forest patches due to dispersal limitation, thus restraining farmer species options. Farmers often prefer to protect rare species on their land (Lenkeek 2003), so young recruits of common species arriving from nearby patches might be perceived as less valuable. Although natural regeneration is a low-cost option, its implementation might be underutilized because of the disconnect between natural regeneration outcomes and farmer expectations.

### *25.4.3 Strategic Planning of Restoration Strategies: Understanding Farmers' Preferences*

Farmers evaluate the advantages and disadvantages of a) investing in labor to clear a plot for cattle, b) allowing vegetation to grow despite perceptions of abandonment, or c) opting for reforestation projects that provide a perceived higher degree of control, as well as tangible economic benefits from timber and fruit (Metzel and Montagnini 2014; Zahawi et al. 2014; Chazdon et al. 2020).

Understanding why some farmers are willing to restore their land using reforestation (tree planting) and/or natural regeneration, while others prefer parcel clearing as a predominant strategy, will assist strategic planning for where to target restoration efforts. We found that two attributes related to the landscape context of farms had strong impacts on management strategies: farm size and tree cover percentage. Parcel clearing, in particular, was affected most strongly by farm size and tree cover percentage. In contrast, the education level and age of survey participants had minimal impact on land management preferences.

Our finding that higher self-reported farm tree cover percentage is associated with a lower preference for land clearing has implications for land cover dynamics in Los Santos. This suggests that maps of farm tree cover and size could be used to predict willingness to forego land clearing and to contribute to strategic planning for restoration activities. Relationships between tree cover and farmer preferences are likely to change the rate of forest transformation, including rapid transitions between treeless and silvopastoral management techniques (Valencia Mestre et al. 2019). An increased willingness to support tree-intensive land management strategies when forest is present represents an example of a positive feedback loop that could promote regional increases in tree cover.

## 25.5 Conclusion

In a human dominated landscape such as the Azuero peninsula, Forest Landscape Restoration (FLR) strategies must align with the human preferences for land uses. In this chapter, we highlighted farmer definitions and preferences for restoration strategies like reforestation, natural regeneration and assisted natural regeneration. Smallholder farmers are crucial to the protection of the Azuero peninsula because they own a majority of farmland, which is why their participation and knowledge on restoration is indispensable to improve FLR capacity.

Farmers were more confident in defining reforestation, which they associated with tree planting, and were less confident in defining natural regeneration and assisted natural regeneration. They associated assisted natural regeneration with definitions of reforestation (human intervention and tree planting) more than with definitions of natural regeneration. Participants were more likely to define reforestation as a process occurring on previously deforested land.

Our findings indicate that the preferred management strategy among farmers was to clear the land for cattle. However, a similar proportion of farmers would choose clearing land for cattle as their least preferred option. This polarization may indicate that many farmers are looking for alternatives to treeless pasture systems. Over half of farmers chose natural regeneration and reforestation as their first and second restoration options, but assisted forest regeneration was the least preferred option. Assisted natural regeneration may be the least preferred option because the cost of labor or materials to fence off parcels is too high for a practice with seemingly limited economic benefits, especially given the exclusion of cattle from the parcel through fencing.

Farmers may prefer natural regeneration because letting vegetation grow is perceived as a land management strategy for unused land that involves a low investment of time and resources. While farmers associate both reforestation and assisted natural regeneration with human intervention, they may prefer to reforest because of a greater familiarity with the methods, as many restoration programs established on the peninsula to date have focused on tree planting, or because they can choose which species to plant, especially beneficial rare species that would not grow in the parcel otherwise. To make natural regeneration a viable option will require creating a broader understanding of restoration approaches among farmers in order to fine tune their implementation within each land area based on their benefits and obstacles to farmers.

Through the results of this research we gained insights into farmer's understanding, preferences, and drivers regarding restoration strategies to bridge their local knowledge and practices with academic discussions and debates around restoration alternatives. The results can be useful in the design and implementation of programs geared to accelerate forest landscape restoration at the global scale, which is needed to achieve the success of ambitious global agendas such as the Bonn Challenge, and the Sustainable Development Goals: climate action (SDG 13), and the restoration and sustainable use of terrestrial ecosystems (SDG 15).

**Acknowledgments** The authors would like to thank Oscar Vega of the ESRI Panama office for GIS assistance in creating the field surveys, and Sandra Vásquez and Jesus Batista for field implementation support.

## References

Acevedo-Charry O, Aide TM (2019) Recovery of amphibian, reptile, bird and mammal diversity during secondary forest succession in the tropics. Oikos 128:1065–1078

Brancalion PHS, Holl KD (2020) Guidance for successful tree planting initiatives. J Appl Ecol. https://doi.org/10.1111/1365-2664.13725

Bürkner PC (2017) Brms: an R package for Bayesian multilevel models using Stan. J Stat Softw 80(1):1–28

Caughlin TT, Rifai SW, Graves SJ, Asner GP, Bohlman SA (2016) Integrating LiDAR-derived tree height and Landsat satellite reflectance to estimate forest regrowth in a tropical agricultural landscape. Remote Sens Ecol Conserv 2:190–203

Caughlin TT, Graves SJ, Asner GP, Tarbox BC, Bohlman SA (2019) High-resolution remote sensing data as a boundary object to facilitate interdisciplinary collaboration. In: Perz SG (ed) Collaboration across boundaries for social-ecological systems science: experiences around the world. Springer International Publishing, Cham, pp 295–326

Chazdon RL, Uriarte M (2016) Natural regeneration in the context of large-scale forest and landscape restoration in the tropics. Biotropica 48:709–715

Chazdon RL, Broadbent EN, Rozendaal DMA, Bongers F, Zambrano AMA, Aide TM, Balvanera P, Becknell JM, Boukili V, Brancalion PHS, Craven D, Almeida-Cortez JS, Cabral GAL, de Jong B, Denslow JS, Dent DH, DeWalt SJ, Dupuy JM, Durán SM, Espírito-Santo MM, Fandino MC, César RG, Hall JS, Hernández-Stefanoni JL, Jakovac CC, Junqueira AB, Kennard D, Letcher SG, Lohbeck M, Martínez-Ramos M, Massoca P, Meave JA, Mesquita R, Mora F, Muñoz R, Muscarella R, Nunes YRF, Ochoa-Gaona S, Orihuela-Belmonte E, Peña-Claros M, Pérez-García EA, Piotto D, Powers JS, Rodríguez-Velazquez J, Romero-Pérez IE, Ruíz J, Saldarriaga JG, Sanchez-Azofeifa A, Schwartz NB, Steininger MK, Swenson NG, Uriarte M, van Breugel M, van der Wal H, Veloso MDM, Vester H, Vieira ICG, Bentos TV, Williamson GB, Poorter L (2016a) Carbon sequestration potential of second-growth forest regeneration in the Latin American tropics. Sci Adv 2(5):e1501639

Chazdon RL, Brancalion PHS, Laestadius L, Bennett-Curry A, Buckingham K, Kumar C, Moll-Rocek J, Guimarães Vieira IC, Wilson SJ (2016b) When is a forest a forest? Forest concepts and definitions in the era of forest and landscape restoration. Ambio:538–550

Chazdon RL, Lindenmayer DB, Guariguata M, Crouzeilles R, Benayas JMR, Chavero EL (2020) Fostering natural forest regeneration on former agricultural land through economic and policy interventions. Environ Res Lett 15(4):043002

Crouzeilles R, Ferreira MS, Chazdon RL, Lindenmayer DB, Sansevero JBB, Monteiro L, Iribarrem A, Latawiec AE, Strassburg BBN (2017) Ecological restoration success is higher for natural regeneration than for active restoration in tropical forests. Sci Adv 3:e1701345

de Souza SEXF, Vidal E, Chagas Gde F, Elgar AT, Brancalion PHS (2016) Ecological outcomes and livelihood benefits of community-managed agroforests and second growth forests in Southeast Brazil. Biotropica 48:868–881

Elliott S, Blakesley D, Hardwick K, Plukham S, Saengkam D, Agbor JN (2013) Restoring tropical forests: a practical guide. Royal Botanic Gardens, Kew

Fagan ME (2020) A lesson unlearned? Underestimating tree cover in drylands biases global restoration maps. Glob Chang Biol 26(9):4679–4690

FAO (2017) Smallholder Forest Producer Organizations in a Changing Climate, Rome. http://www.fao.org/3/a-i7404e.pdf

FAO (2020) Forest Resources Assessment Terms and Definitions. http://www.fao.org/3/I8661EN/i8661en.pdf

Garen EJ, Saltonstall K, Slusser JL, Mathias S, Ashton MS, Hall JS (2009) An evaluation of farmers' experiences planting native trees in rural Panama: implications for reforestation with native species in agricultural landscapes. Agr Syst 76(1):219–236

Garen EJ, Saltonstall K, Ashton MS, Slusser JL, Mathias S, Hall JS (2011) The tree planting and protecting culture of cattle ranchers and small-scale agriculturalists in rural Panama: opportunities for reforestation and land restoration. For Ecol Manage 261(10):1684–1695

Grebner DL, Bettinger P, Sirey JP (2013) Chapter 11 – common forestry practices. In: Grebner DL, Bettinger P, Sirey JP (eds) Introduction to forestry and natural resources. Academic, Cambridge, pp 255–285

Griscom HP (2020) The long-term effects of active management and landscape characteristics on carbon accumulation and diversity within a seasonal dry tropical ecosystem. For Ecol Manage 473:118296

Griscom HP, Griscom BW, Ashton MS (2009) Forest regeneration from pasture in the dry tropics of Panama: effects of cattle, exotic grass, and forested Riparia. Restor Ecol 17:117–126

Griscom BW, Adams J, Ellis PW, Houghton RA, Lomax G, Miteva DA, Schlesinger WH, Shoch D, Siikamäki JV, Smith P, Woodbury P, Zganjar C, Blackman A, Campari J, Conant RT, Delgado C, Elias P, Gopalakrishna T, Hamsik MR, Herrero M, Kiesecker J, Landis E, Laestadius L, Leavitt SM, Minnemeyer S, Polasky S, Potapov P, Putz FE, Sanderman J, Silvius M, Wollenberg E, Fargione J (2017) Natural climate solutions. Proc Natl Acad Sci 114(44):11645–11650

Hall JS, Love BE, Garen EJ, Slusser JL, Saltonstall K, Mathias S, van Breugel M, Ibarra D, Bork EW, Spaner D, Wishnie MH (2011) Tree plantations on farms: evaluating growth and potential for success. For Ecol Manage 261(10):1675–1683

Hansen MC, Potapov PV, Moore R, Hancher M, Turubanova SA, Tyukavina A, Thau D, Stehman SV, Goetz SJ, Loveland TR, Kommareddy A, Egorov A, Chini L, Justice CO, Townshend JRG (2013) High-resolution global maps of 21st-century forest cover change. Science 342(6160): 850–853

Heckadon-Moreno S (2009) De Selvas a Potreros: La Colonización Santeña en Panamá: 1850–1980. ExedraBooks, Panamá

Holl KD, Aide TM (2011) When and where to actively restore ecosystems? For Ecol Manage 261: 1558–1563

Holl KD, Brancalion PHS (2020) Tree planting is not a simple solution. Science 368(6491): 580–581

Instituto Nacional de Estadística y Censo (2011) VII Censo Nacional Agropecuario. Volumen II: Características de los Productores y las Explotaciones Agrarias. Cuadro 1. Productores agropecuarios en la República, por grupos de edad, empresas y organizaciones comunales, según provincia, comarca indígena y tamaño de la explotación: 24 de abril de 2011, Panamá
Jakovac CC, Peña-Claros M, Kuyper TW, Bongers F (2017) Loss of secondary-forest resilience by land-use intensification in the Amazon. J Ecol:67–77
Jones HP, Jones PC, Barbier EB, Blackburn RC, Benayas JMR, Holl KD, McCrackin M, Meli P, Montoya D, Mateos DM (2018) Restoration and repair of Earth's damaged ecosystems. Proc R Soc B 285:2017–2577
Lenkeek AG (2003) Diversity makes a difference: farmers managing inter- and intra-specific tree species diversity in Meru Kenya. PhD Diss, Wageningen University, Wageningen
Lerner AM, Rudel TK, Schneider LC, McGroddy M, Burbano DV, Mena CF (2015) The spontaneous emergence of silvo-pastoral landscapes in the Ecuadorian Amazon: patterns and processes. Regional Environ Change 15(7):1421–1431
Love B, Spaner D (2005) A survey of small-scale farmers using trees in pastures in Herrera Province, Panama. J Sustain For 20(3):37–65
McElreath R (2020) Statistical rethinking: a Bayesian course with examples in R and Stan. CRC Press, Boca Raton
Metzel RNB (2010) From "Finca" to forest: forest cover change and land management in Los Santos, Panama. Thesis, Princeton University, Princeton, New Jersey
Metzel R, Montagnini F (2014) From farm to forest: factors associated with protecting and planting trees in a Panamanian agricultural landscape. Bois Et Forets Des Tropiques 322:3–15
Reid JL, Fagan ME, Zahawi RA (2018) Positive site selection bias in meta-analyses comparing natural regeneration to active forest restoration. Sci Adv 4:eaas9143
Sánchez-Azofeifa GA, Quesada M, Rodriguez JP, Nassar JM, Stoner KE, Castillo A, Garvin T, Zent EL, Calvo-Alvarado JC, Kalacska MER, Fajardo L, Gamon JA, Cuevas-Reyes P (2005) Research priorities for neotropical dry forests. Biotropica 37:477–485
Schelegle S (2011) Connections in a fragmented landscape: Azuero spider monkey corridors in the Azuero Peninsula, Panama. Thesis, Princeton University, Princeton, New Jersey
Schneider S (2016) Family farming in Latin America and the Caribbean: looking for new paths of rural development and food security. Pap 137, FAO, http://www.fao.org/3/a-i5534e.pdf
Shono K, Cadaweng E, Durst P (2007) Application of assisted natural regeneration to restore degraded tropical forestlands. Restor Ecol 15:620–626
Silge J, Robinson D (2016) Tidytext: text mining and analysis using tidy data principles in R. JOSS 1(3). https://doi.org/10.21105/joss.00037
Sloan S (2015) The development-driven forest transition and its utility for REDD+. Ecol Econ 116: 1–11
Stanturf JA, Kleine M, Mansourian S, Parrotta J, Madsen P, Kant P, Burns J, Bolte A (2019) Implementing forest landscape restoration under the Bonn challenge: a systematic approach. Annals For Sci 76:50
Strassburg BBN, Beyer HL, Crouzeilles R, Iribarrem A, Barros F, de Siqueira MF, Sanchez-Tapia A, Balmford A, Sansevero JBB, Brancalion PHS, Broadbent EN, Chazdon RL, Filho AO, Gardner TA, Gordon A, Latawiec A, Loyola R, Metzger JP, Mills M, Possingham HP, Rodrigues RR, Scaramuzza CAM, Scarano FR, Tambosi L, Uriarte M (2019) Strategic approaches to restoring ecosystems can triple conservation gains and halve costs. Nature Ecol Evol 3:62
Tarbox BC, Fiestas C, Caughlin TT (2018) Divergent rates of change between tree cover types in a tropical pastoral region. Landsc Ecol 33:2153–2167
Thacher T, Lee DR, Schelhas JW (1996) Farmer participation in reforestation incentive programs in Costa Rica. Agr Syst 35(3):269–289
Valencia Mestre MC, Ferguson BG, Vandermeer J (2019) Syndromes of production and tree-cover dynamics of Neotropical grazing land. Agroecol Sustain Food Syst 43(4):362–385
Zahawi RA, Reid JL, Holl KD (2014) Hidden costs of passive restoration. Restor Ecol 22:284–287

# Chapter 26
# Safeguarding Biodiversity Islands in Northern Ethiopia Amidst Political Change

**Emily Sigman**

**Abstract** This chapter concerns the safeguarding of biodiversity islands established through community-based restoration in Northern Ethiopia, and centers particularly on the case study of a kebele (or village, also spelled q'ebele) called Abreha we Atsbeha. Severely degraded landscapes in the Tigray region of Northern Ethiopia have enjoyed extensive restoration efforts in recent decades, and these efforts have intensified further since Ethiopia joined the African Resilient Landscapes Initiative (ARLI) and became a member of the Bonn Challenge. The resultant restored areas serve as biodiversity islands in an otherwise highly degraded area. While praise abounds for the restoration successes and benefits to communities like Abreha we Atsbeha, there is scant research exploring how the restoration takes place, and insufficient attention has been paid to the capacity of these restoration efforts—and the biodiversity islands they enable—to survive in the midst of political change. This chapter therefore investigates the communal labor activities that enliven successful ecological restoration and relies on a political ecology framework to discuss governance and restoration in this context. Overall, the chapter demonstrates that restoration-based biodiversity islands in Abreha we Atsbeha are created and safeguarded through a robust single-party political system. This system at once mandates universal participation in restoration activities while limiting the participation of many community constituents in the planning of restoration activities. The author argues that this social-ecological system needs to be adapted to ensure that rural, restoration-based biodiversity islands in the region can continue to flourish alongside more pluralistic and democratic political norms and institutions.

**Keywords** Communal labor · Dryland · Political ecology · Restoration · Social-ecological systems · Watershed

---

E. Sigman (✉)
Yale School of Forestry and Environmental Studies, Yale Jackson Institute for Global Affairs, New Haven, CT, USA
e-mail: Emily.sigman@aya.yale.edu

 
F. Montagnini (ed.), *Biodiversity Islands: Strategies for Conservation in Human-Dominated Environments*, Topics in Biodiversity and Conservation 20,
https://doi.org/10.1007/978-3-030-92234-4_26

## 26.1 Introduction

The term 'biodiversity islands' tends to conjure up a certain image, one of pristine ecosystems kept away from the destructive hand of human civilization. High levels of degradation and destruction wrought on the planet—careless disregard for precious soils, rapacious plundering of the world's forests, flagrant pollution of life-giving air and water systems—do much to cement this idea. Environmental degradation is a powerful driver, motivating conservationists to do all they can to safeguard what precious few biodiversity islands have been spared from the clutches of human shortsightedness.

Yet environmental degradation also motivates a different, concurrent response: the push to restore. Increased awareness of humanity's mutual dependency on intact, connected, and diverse ecosystems has arisen in tandem with improved understandings of the role that human beings have played in the stewardship of magnificent ecosystems, from the Amazon rainforest (Levis et al. 2017) to the Great Plains of North America (Krech 2000). Moreover, we as a global community are beginning to appreciate and acknowledge the role that we can play in reversing the tides of environmental degradation. We are increasingly aware of our collective capacity to usher in a new era of flourishing. Harnessing the power of restoration, we can do more than simply protect biodiversity islands: we can also create them.

In rural areas across Ethiopia, especially in the Northern province of Tigray, many local communities have already long been exercising this capability. This chapter explores biodiversity islands in the context of one community in the Tigray Region of Northern Ethiopia, called Abreha we Atsbeha. There, rural farmers have successfully implemented impressive soil and water conservation techniques and pioneered new social norms that have effectively restored watersheds, catalyzed natural plant regeneration, and over time, provisioned biodiversity islands in landscapes that were otherwise so severely degraded that they were deemed uninhabitable by humans. These novel socio-ecosystems—which exist only because of extensive and direct human mediation—pose an interesting set of questions for those seeking strategies to safeguard biodiversity islands.

The following sections will synthesize recent developments in socio-ecological restoration, parse out the case study, and offer commentary on the often-understudied attributes of these kinds of restored landscapes, in particular local labor and political dynamics. The chapter swivels around the concept of a Restoration Dilemma, described in 2.3, with the goal of assuaging the tensions that arise as restoration ambitions and initiatives progress in scale. Overall, this chapter aims to highlight the active role that local communities can, do, and must play in creating biodiversity islands. It imagines a future where the fruits of restoration labor can be effectively safeguarded—and equitably savored—by all.

## 26.2 Restoration, Local Actors, and Biodiversity Islands

### *26.2.1 Global Impetus for Restoration*

Recent decades have witnessed an increasing global awareness of the need to restore landscapes on a mass scale. Deforestation and degradation have advanced so thoroughly that it is no longer enough to conserve what vestiges of intact forest remain; the task of safeguarding the world's ecosystems must also be fulfilled by efforts to actively restore what's been destroyed.

Globally, more than 2 billion hectares of deforested and degraded lands hold potential for restoration (Minnemeyer et al. 2014). The benefits of restoration are multitudinous, and are often encapsulated as 'ecosystem services' in the field's contemporary literature (Daily 1997; Costanza et al. 2014). Increased biodiversity ranks chiefly among the ecosystem services sought by restoration advocates (Mace et al. 2012). In foundational island biogeography theory (MacArthur and Wilson 1967), biodiversity islands—which in these early studies referred to physical islands surrounded by water—comprise areas in a landscape with biodiversity measures significantly higher than those of the surrounding landscape. Extending this concept, restored areas generally possess higher biodiversity metrics relative to their surrounding degraded environments, that is, restored areas serve as biodiversity islands in and of themselves in the context of degraded landscapes (Benayas et al. 2009). Biodiversity gains in these spaces can take the form of increased plant, wildlife, insect, and soil diversity—among many other measures—and many of these gains are understood to be of both direct and indirect human benefit (Aerts and Honnay 2011).

'Restored' biodiversity islands geared primarily towards human use (i.e. agroforests, timberlands, foraging areas, or other such zones intended to be regularly utilized) can likewise enhance other, most restrictive biodiversity islands (such as national parks or pristine zones). These areas can, for example, provision connective corridors between conservation areas, serve as reservoirs for pollinators and dispersers, contribute to increased genetic diversity needed to support healthy species reproduction, and help stabilize large-scale soil and water dynamics, which improve mosaic landscape function as a whole (Boesing et al. 2018). Moreover, with careful planning and proper governance, these areas can alleviate pressure to extract from protected zones, serving as sources of food, fuel, timber, fibers and other materials while acting as areas of intermediate biodiversity between degraded landscapes and zones of extreme conservation importance (Schroth et al. 2004; Kumar 2010).

A number of restoration initiatives have emerged from this understanding, and chief among them is the Bonn Challenge. The Bonn Challenge is a global, multi-stakeholder effort that was launched in 2011 with the goal of restoring 150 million hectares of degraded and deforested lands by the year 2020 and 350 million hectares by 2030.[1] It offers institutional, policy, and technical support to developing countries, which in turn have pledged to restore a certain number of hectares. The challenge promotes regional collaboration platforms, such as the African Resilient Landscape Initiative (ARLI) (World Bank 2015). ARLI, in turn, supports initiatives like the African Forest Landscape Restoration Initiative (AFR100), a country-led effort to bring 100 million hectares of land in Africa into restoration by 2030.[2]

Twenty-nine countries have so far made a commitment of more than 125 million hectares to the initiative, with Ethiopia pledging 15 million hectares, by far the region's largest single contribution. In exchange, development banks have contributed more than US$1 billion, with an additional $481 million pledged by the private sector (Anderson and Piembert 2019).

### *26.2.2 Increasing Focus on Local-Level Community Engagement*

Early attempts in the last decades to formalize restoration studies, projects, and initiatives worldwide tended to focus primarily on biophysical dynamics, without incorporating—or even acknowledging—the complex political, economic, and social forces at play in restoration projects (Brudvig 2011). More recent years have witnessed an outpouring of literature seeking to correct the course and expand the restoration lens to better capture and understand the intricacies of multipart cultural-environmental landscapes (Rovere 2015).

As ecologists have progressed in their understanding of forest stand dynamics—no longer viewing forest succession as advancing towards a 'climax' state, but rather existing in a process of perpetual evolution—so too have restoration scholars relaxed the tendency to view restoration as a reversion 'back' to some previous ecosystem (Trigger et al. 2008). Such a view allows 'restoration' to focus on key attributes of ecosystems—such as soil dynamics, watershed health, and biodiversity measures—while permitting and even encouraging the emergence of novel socio-ecosystems (Hobbs et al. 2009). At the same time, the intersecting fields of anthropology, environmental history, and political ecology have substantially rewritten the conservation script in recent years, convincingly illuminating the key role that human societies have historically played in the stewardship and even creation of ecosystems once considered unmediated or primary forests (Fernandez-Manjarres et al. 2018).

These streams of knowledge have started to merge, each contributing new dimensions to a growing, interdisciplinary inquiry that seeks to align the development of human society with environmental conservation and ecosystem regeneration. A Social-Ecological Systems (SoESs) framework was first proposed by Berkes and Folke (1998) to analyze resilience in local resource management systems; since

[1] Bonn Challenge. In: Bonn Challenge. https://www.bonnchallenge.org/. Accessed 25 Jan 2020

[2] AFR100 In: Afr100.org. https://afr100.org. Accessed 25 Jan 2020

then, the term has come to refer to a number of frameworks that expressly recognize the complex linkages between human and 'natural' systems (Martin 2017), especially in restoration and resilience contexts (López et al. 2017). Socio-ecological restoration, in turn, has emerged at the confluence of these schools of thought, as a practical discipline working to address interconnected social and environmental challenges (Colding and Barthel 2019).

This understanding—while stressing the need to restore ecosystems on a landscape-scale—has emphasized the importance of local community dynamics. This emphasis recognizes that local communities are often the direct beneficiaries and/or agents of restoration (Brancalion et al. 2014). It highlights the value of local ecological knowledge (Uprety et al. 2012), and promotes community participation (Maynard 2013), improved understanding of local power dynamics (Habtezion et al. 2015), long-term community monitoring and evaluation (Wortley et al. 2013), and an intentional awareness and anticipation of potential unintended consequences and pernicious outcomes (Daily and Maston 2008).

Scholars, public agencies, and non-government institutions have increasingly adopted the perspectives and practices advanced by the social-ecological restoration agenda, some of which now operates under the global framework promulgated by the UN Sustainable Development Goals (SDGs). This self-professed 'blueprint to achieve a better and more sustainable future for all' unambiguously understands the interconnected nature of the world's most pressing social and environmental problems (United Nations 2020). Many of the targets and indicators are expressly concerned with community integration and engagement. Despite this improved understanding, effective community-scale work remains difficult in practice, and coordination among actors across different scales has likewise proven challenging.

### *26.2.3 The 'Restoration Dilemma'*

On the whole, literature from socio-ecological restoration tends to suggest that the most effective and durable restoration projects are idiosyncratic, take place at a local level, are culturally and ecologically site-specific, and therefore tend to resist scaling. Yet irrefutable evidence from studies concerned with climate change, mass extinction and biodiversity loss, and reliable resource provisioning—bolstered by the imperatives of the international development agenda—clearly articulates the need for restoration at a global scale.

Restoration practitioners therefore face a profound dilemma. A *dilemma* is defined as 'a problem offering two possibilities, neither of which is unambiguously acceptable or preferable'(Garner 2009). Focusing on global-scale restoration often has pernicious outcomes at the local scale (which is not acceptable). Focusing on local-scale restoration often stymies practitioners attempting to transform "a thousand random acts of restoration" into a "coherent strategy" that can realistically meet global needs (Covelli-Metcalf et al. 2015; Budiharta et al. 2016) (which is not preferable).

Escaping through the horns of this dilemma requires careful, consistent and adaptive feedback about how national, regional, and global restoration directives reverberate at the local level and vice versa. A firm grasp on *how* restoration actually happens—the human as well as the non-human agencies involved—offers key insights into project dynamics and vulnerabilities, and in so doing, improves prospects for multi-scalar success.

## 26.3 Restoration in Ethiopia – The Case of Abreha We Atsbeha

### *26.3.1 The Case Study in Context*

Ethiopia offers fertile ground for those seeking to explore the kinds of biodiversity islands that emerge from restoration, and in particular, for studies examining how to safeguard such biodiversity islands in the midst of political change. Ethiopia today comprises Africa's second most populous country[3]—the most populated landlocked country to exist anywhere in the history of the planet—and by many accounts is home to the fastest growing economy in the world. Yet this growth occurs in the context of a largely rural society, comprised of a roughly 80% agrarian population.[4] Despite the fact that Ethiopia is essentially governed by a one-party state with a planned economy, the country's vast population is spread out over more than a million square kilometers and divided into nine politically autonomous ethnic states, each possessing unique socio-ecological features and agronomic patterns. The country as a whole is thus best conceived as a quilted patchwork stitched of an extraordinary diversity of landscapes, languages, and life-ways.

Restoration and conservation in this context are therefore highly suited to an ethno-regional focus. The country often garners substantial coverage in public media outlets and has earned a reputation among international agencies for its eye-catching national restoration activities, such as, for example, the recent planting of 350 million trees across the country over a period of 12 hours (UNEP 2019). Yet the attention paid to country-level restoration blitzes such as these tends to obfuscate an understanding of the heterogeneity of long-term restoration activities at the federation- and even regional-level. It is at these smaller scales where connections between restoration, livelihoods, and biodiversity are made apparent, and where prospects for safeguarding biodiversity islands are most accessible. This study is therefore

[3]The population of Ethiopia is roughly 112,000,000. The most populous country in Africa is Nigeria, with a population of roughly 201,000,000. From Ethiopia Overview. In: The World Bank: Where we Work. (2019) https://www.worldbank.org/en/country/ethiopia/overview. Accessed 25 Jan 2020

[4]Rural population (% of total population) – Ethiopia. In: The World Bank Data Indicators 2018 revision. https://data.worldbank.org/indicator/SP.RUR.TOTL.ZS?locations=ET. Accessed 25 Jan 2020

restricted to the northern Tigray region, and focuses on the dynamics of one kebele (or community, village, also spelled ‘q’ebele’) within this context. The case study helps concretize the ‘Restoration Dilemma’ outlined in Sect. 26.2.3, and draws out the multi-scalar dynamics that bind participatory restoration, biodiversity provisioning, and socio-political change.

### *26.3.2 Physical, Ecological, and Social Characteristics of Tigray*

The wrinkles of humanity are etched in the broken, volcanic plateaus of Tigray. The heart of the ancient Aksumite kingdom, one of the cradles of modern civilization, once pumped through this region, drained by the Tekeze and Gash rivers which ran to the Red Sea, along with the Upper Nile. This made of the empire a vast marine trading power unlike the territorial constraints that today render the whole of Ethiopia landlocked (McKenna 2019). The region has remained dominated by an Orthodox Christian population (95.6% in 2007 – IHSN 2019), which has a longstanding history as a seat of political power for the country as a whole, though Ethiopia’s capital is in Addis Ababa, located some thousand kilometers away in the Amhara region (Van Veen 2016). This political history—which also includes a notable lack of prolonged European colonialism, the survival of Haile Selassie’s Imperial Kingdom well into the twentieth century, and a punctuated, brutal communist interlude throughout the 1970s and 1980s—reverberates in contemporary Ethiopia and has important implications for understanding and promoting restoration and biodiversity initiatives in Tigray and beyond.

Tigray’s topography is characterized by its relatively high elevations, ranging from 1000 to > 3500 m in altitude, and semi-arid temperate climate, with a mean annual precipitation of 700–1200 mm (Bard et al. 2000). Temperatures range between 15 and 25° C, while soils range in agricultural quality, from Vertic Cambisols in the upper reaches, to Vertic Calcisols, Vertic Cambisols, Calcaric Phaeozems and Calcaric Regosols[5] in the lower regions (Rabia et al. 2013). Despite the difficult conditions, the highland regions served as a center of African plant domestication and agricultural innovation, and extensive human manipulation of the landscape has taken place for millennia (Bard et al. 2000). The Tigrean Plateau was once vegetated with dry evergreen, montane forests and deciduous wooded grasslands, yet today it is mostly montane grassland (Pankhurst and Ingrams 1988).

[5] A **Cambisol** is a soil with little or no profile differentiation. They are typically found in landscapes with high rates of erosion, and are exploited for agriculture. A **Calcisol** is a soil with a layer of migrated calcium carbonate in the soil profile. They are typically found in arid zones, and their chief use is for animal grazing. **Phaeozem** is a dark soil with high base status typically exploited for intensive agriculture. **Regosols** are poorly developed mineral soils in unconsolidated materials, extensive in eroding lands in arid areas.

This loss of forest cover is the result both of natural disasters such as drought and landslides, and human activity, including intensive deforestation, agriculture, and livestock grazing (Nyssen et al. 2000).

It is important to recognize that the barren landscape of present-day Tigray is not a phenomenon of recent decades, but likely dates back to at least the seventeenth century (Pankurst 1988). Few, if any, remnant forests exist that are not the direct result of human intervention in the region. Many such forests are the direct result of coordinated restoration activities undertaken in the absence of the edaphic and hydraulic conditions that would facilitate natural regeneration. As such, the great majority of biodiversity islands that exist in Tigray today should be understood as the result of intentional restoration and management, which enabled the growth of forests in an otherwise highly degraded landscape – a point which deserves great emphasis.

### *26.3.3 Contemporary Restoration in Ethiopia*

While the history of restoration in Ethiopia is not well studied, attempts at better documentation of present-day restoration activities furnish a cursory understanding of contemporary restoration dynamics. Much of the restoration is driven by a desire to deliver ecosystem services, notably the renewal of watersheds to provision water for agricultural systems (Gebregziabher et al. 2016). Over time, the lack of woody vegetation on hillsides has destabilized watersheds; instead of percolating slowly through the rocky soils, episodic rainfall rushes off the slopes, carrying with it precious topsoil and seedbanks. This not only leads to periodic inundation, sedimentation and gully formation in lowland settlements, but also to the gradual impoverishment of valley groundwater supplies. The persistent threat of drought was—and continues to be—compounded by this chronic incapacitation of groundwater, which was one of many factors that led to the perilous famines witnessed during the late twentieth century (Keller 1992). Today more than a quarter of the population of Tigray remains chronically food insecure, requiring the continued provision of international food aid (Alemu et al. 2014). This precariousness is further exacerbated by the growing incidence of erratic precipitation and rising temperatures associated with climate change (Teshome and Zhang 2019).

To reverse this trend and build resilience, international agencies and local communities recognize the need to restore watersheds, which in turn requires afforestation of degraded hillsides. As a signatory of the Bonn Challenge, supported by the AFR100, Ethiopia has committed to restore 15 million hectares, or roughly one-sixth of its total land area, by 2025 (African Resilient Landscapes Initiative, ARLI, https://afr100.org). Tremendous efforts have been taking place across the country under the banner of the Sustainable Land Management Programme (SLMP), a multi-stakeholder project commissioned by the World Bank and GIZ (GIZ 2020), which contributes to these substantial re-greening goals through programs broadly referred to as 'Soil and Water Conservation' (SWC) (Hurni et al. 2016), constituting

**Fig. 26.1** Free Labour Contribution Period (FLCP) activities. On left: two women use a repurposed food sack to carry stones from the excavation site to the terracing site. On right: A mixed group of men and women construct a stone terrace. These activities happen side by side continuously. (Photos: E. Sigman, with permission from community members photographed)

watershed-level interventions on agricultural landscapes. Numerous additional NGOs, research, and foreign government aid organizations have likewise made substantial investments towards conservation and reforestation in Ethiopia. While these are nation-wide efforts, a special emphasis has been placed on the Tigray region.

With such severe levels of degradation, natural regeneration typically does not take place even with the exclusion of agricultural and grazing activities on hillsides (Aynekulu et al. 2009). Moreover, in addition to proving cost-prohibitive, reforestation through active tree planting has shown to have dubious benefits, as—in the absence of underlying functional water and nutrient cycles—many trees do not survive to catalyze forest succession (Shono et al. 2007). Extensive efforts have therefore focused on the repair of watersheds, which—in the absence of financial and technical capital—has been conceived through mass hillside stabilization programs. Such programs are carried out by large groups of rural residents, often equipped with little but their own hand tools (see Fig. 26.1), and are followed by 'exclosure' systems that restrict access to these areas while natural or assisted regeneration takes places over several decades (Mekuria et al. 2017). The resultant forests, constituting newly created biodiversity islands, in addition to provisioning water and other ecosystem services, support populations of native wildlife and insects and facilitate connections to other conservation areas.

The greatest contribution to restoration and the creation of biodiversity islands in Northern Ethiopia therefore comes from rural communities. Farmers and pastoralists are the agents that ultimately contribute the labor required to stabilize hillsides, reverse watershed degradation, restore nutrient cycles, and catalyze functional regeneration. They are also the agents that must significantly alter their land practices to accommodate these programs, and consent to continued land-use restrictions which generate, maintain, and safeguard the newly-created biodiversity islands.

### 26.3.4 *Abreha we Atsbeha as a Model Community for Restoration in Ethiopia*

As noted, Ethiopia has gained recognition within the international restoration community for its recent activities and successes. Since the launch of a series of nationwide programs in 2010 (MOFED 2010), together with substantial financial investment from interested international agencies, Ethiopia has invested more than US$1.2 billion annually in restoration activities across the country, rehabilitating more than 12 million hectares of land and over 3000 watersheds, and supporting more than 1.6 million hectares under active 'Sustainable Land Management' (Seyoum 2016). Recent research published by the International Center for Tropical Agriculture (CIAT) confirms that these investments have significantly reduced runoff and soil erosion, increased crop yields, and enhanced soil organic carbon (Tamene et al. 2018). Meanwhile, other studies have demonstrated the benefits of coupling active restoration with community exclosures (Mekuria et al. 2018), particularly in Tigray, which has led to increased vegetation cover and biodiversity (Asefa et al. 2003; Mengistu et al. 2005; Mekuria et al. 2012), along with enhanced soil fertility (Mekuria et al. 2017), water flows (Dessalew et al. 2016), and groundwater recharge (Anwar et al. 2016).

According to some researchers (Nyssen et al. 2014), as a result of this blend of national initiatives, foreign assistance, and local participation, "Ethiopia is now greener than it has ever been during the last 145 years. . .human investments have overridden the impacts of climate change." In the midst of this movement—what some (e.g. Dodd 2015) refer to as Ethiopia's 'Green Revolution'—a community called Abreha we Atsbeha has garnered substantial recognition for its achievements in both implementation and innovation in the intersecting fields of restoration and conservation.

Abreha we Atsbeha—a village of about 5030 people in the Tigray Region of Northern Ethiopia (see Fig. 26.2)—has been widely promoted by both the Ethiopian federal government and a bevy of international research and aid organizations as a key example of restoration best practices, both from a biophysical and socioeconomic perspective (Lamond 2012).

Several journal articles and reports have been published specifically studying Abreha we Atsbeha. Some of these detail historical, demographic, ecological, and hydrologic conditions of the site and summarize water harvesting and restoration techniques (e.g. Tadesse et al. 2015), while others measure labor inputs to restoration (e.g. Hachoofwe 2012). Other reports have focused on livelihoods and economic stability, detailing the positive impacts of restoration activities on biodiversity, socioeconomics, production, and policy, while outlining some pathways forward for sustainability and replication (UNDP 2013).

The majority of information on Abreha we Atsbeha, however, comes from grey literature, much of which centers on the community's charismatic chairman Gibremechel Giday, better known as 'Aba Hawi'. In 2012, Aba Hawi traveled to Brazil to attend the Rio + 20 summit and accept the UN Equator Prize, an award

**Fig. 26.2** A day at the market in Abreha we Atsbeha. In the background is the Abreha we Atsbeha church for which the community is named. This rock-hewn edifice has been in continuous use since at least the tenth century. (Photo: E. Sigman)

bestowed by the UN Development Program (UNDP) upon 25 initiatives in recognition of their "outstanding projects working to advance sustainable development solutions for people, nature and resilient communities". Aba Hawi also features prominently in a documentary called *Ethiopia Rising: From Red Terror to Green Revolution,* which has won several international film awards, and has enjoyed thousands of public screenings throughout the globe. Stories of Abreha we Atsbeha's success have been published in BBC (Haslam 2015), Reuters (Win 2019), The Guardian (Watson 2016), and many other national and international news publications.[6] Blog posts about Abreha we Atsbeha are common among the websites of several major international organizations working in Ethiopia, including the World Food Program,[7] The World Agroforestry Center (Kuria et al. 2016), World Vision, World Resources Institute (Rejj 2015) and many others. There are also a number of short films about Aba Hawi—focusing on his role as a charismatic leader who has championed for restoration and inspired his constituents to undertake volunteer restoration work—and about Abreha we Atsbeha, created by visitors and international organizations on YouTube and Vimeo.

Ethiopia's regional and federal governments, as well as a number of international research and development organizations, are actively working to try to 'scale' Abreha we Atsbeha's success. Yet precisely how this 'scaling' will take place remains unclear. Most efforts thus far have centered on replicating specific ecological

[6] See "Ethiopia, 30 Years on from Famine and Live Aid." EthiopiaOnline.

[7] Ethiopian Village Recognized At Rio + 20 For Innovative Hunger Solution. In: United Nations World Food Program Blog. Accessed 25 Jan 2020

interventions, spreading practices through 'farmer-to-farmer' trainings, or provisioning particular resources such as materials for check-dam construction or nurseries for on-farm agroforestry intensification. Despite the significant attention paid to this community, critical gaps in research remain to understand both why the community has been so successful and how those successes could be most effectively scaled. Filling these research gaps is key to assuaging the 'Restoration Dilemma' at play in this case study, and safeguarding the biodiversity islands that have been created in communities like Abreha we Atsbeha.

## 26.4 The Free Labor Contribution Period (FLCP)

### *26.4.1 Contributions of Coordinated Group Labor to Restoration Programs and Biodiversity Outcomes*

Concomitant with rising global awareness and promotion of participatory restoration and conservation, coordinating bodies—particularly those emanating from international agencies—have celebrated Ethiopia's 'community-based' environmental activities. Widespread praise abounds for what are conceived as democratic, cooperative, and innovative restoration initiatives like those underway in Abreha we Atsbeha. Yet few international agencies, even those heavily vested in restoration programs, have made systemic inquiries into how participation in restoration is motivated and coordinated in practice, and existing studies examining cooperative structures have so far proven simplistic.

'Scaling' restoration in the Ethiopian context is not simply a matter of identifying and promoting technical landscape interventions. The agrarian nature of the county's overwhelmingly rural population, coupled with severe levels of environmental degradation, renders scaling a highly social and political undertaking. Moreover, though Ethiopia's economy is growing swiftly, and despite significant monetary investments in restoration, levels of financial and technical capital in these vast rural landscapes remain low. In this setting, unpaid communal labor provides an effective means to achieve needed restoration outcomes. For all of these reasons, residents of local communities are almost invariably the agents of restoration and the stewards of resultant restored landscapes. The way such residents are motivated, organized, and sustained in this work should consequently be of paramount interest to those researchers and agencies concerned with scaling restoration and safeguarding biodiversity islands.

It is therefore surprising to see how little attention has been paid to the coordinated group labor systems that animate restoration in practice. Who organizes people? How are they organized? What incentives motivate this participation? Few have sought answers to these questions, despite the vital role coordinated labor plays in achieving the highly praised restoration outcomes in Ethiopia, and despite the central function such coordinated systems must therefore occupy in attempts to scale existing successes.

**Fig. 26.3** Newly constructed check dams in an outwash area in Abreha we Atsbeha. (Photo: E. Sigman)

In particular, little attention has been paid to a widespread phenomenon called in Tigrinyan "Israin Tshanta"—translating to "Twenty (20) Days"— and referred to by others in the region as the "Free Labor Contribution Period" (FLCP). The FLCP is a central feature of rural life in the Tigray region; it is a period lasting between 20–60 days, typically carried out during the driest season (January–March). During this time, all able-bodied members of a community are expected to volunteer on community projects. Communities work on a wide range of tasks during this time, but most of them revolve around landscape restoration. Often, this includes labor-intensive undertakings such as terracing, check-dam construction (Fig. 26.3), and other SWC activities, organized and managed with high levels of oversight and coordination by local and regional officials (Sigman 2019a). The origins and evolution of the FLCP remain unclear, making it difficult to parse if and how perceptions about the FLCP have changed over time (Sigman 2019b).

It is estimated that between 2010 and 2015 more than 15 million rural dwellers in the region contributed the unpaid labor equivalent of US $750 million annually (Seyoum 2016). Initial assessments report that Tigray—a mountainous area approximating the size of Italy—has been almost entirely terraced. Laid end-to-end the terraces of Tigray would be longer than the Great Wall of China (Dodd 2015). All of this has occurred in the last 20–30 years, and much of it through the apparatus of the FLCP (see Fig. 26.4).

**Fig. 26.4** A mountainside in Abreha we Atsbeha undergoing active restoration via terracing and 'exclosure' programs. Vegetation and soil can cover evidence of terracing over time, obscuring from view the human labor required to catalyze regeneration. The two images are identical; on the right image the author has added lines to show where terracing has taken place. Human shapes in the photo give a sense of the scale of the undertaking. (Photo and rendering: E. Sigman)

The FLCP has been central to the success of restoration in Abreha we Atsbeha and elsewhere, and should be understood as the direct human mechanism through which restoration activities take place. Yet despite the obvious centrality of the FLCP to Ethiopia's many large-scale restoration initiatives, the FLCP as a social and political institution is rarely acknowledged. Though it is a critical feature of rural life in Tigray, and plays a key role in translating international, federal, and regional land management strategies to the local level, the FLCP is virtually absent from the discussions by the international development and restoration community surrounding how to best design, manage, evaluate and scale landscape changes, and how to safeguard resultant biodiversity islands.

The FLCP is a massive social and political institution, through which initiatives like the Bonn Challenge and AFR100 reverberate, and a substantial portion of Ethiopia's pledged 15 million hectares of restoration will likely be facilitated through the apparatus of this established group labor system. Studying the FLCP furnishes a keen understanding of how large-scale projects are implemented at the local level, making it an instructive lens through which to try to parse the 'Restoration Dilemma'. Such an understanding can help develop projects that are compatible with local systems, assess how the responsibilities and benefits of such projects flow through a community, and highlight possible synergies or incongruences that emerge at the intersection of landscape restoration and political power. These insights in turn improve prospects for scaling up restoration projects and minimizing threats to the region's newly established biodiversity islands.

## *26.4.2 Mechanisms of the Free Labor Contribution Period*

Theories and methods from political ecology supported the author's immersive field visits to Ethiopia in 2017 and 2018, oriented with the express purpose of documenting the FLCP and illuminating its centrality in restoration and conservation (Sigman 2019b). These visits focused on deepening an understanding of the labor dynamics at play during the FLCP in Abreha we Atsbeha, hypothesizing that this knowledge would provide a more accurate picture of what restoration actually entails, and therefore improve prospects for linking local level successes to global goals through scaling efforts (i.e., to assuage the 'Restoration Dilemma').

Participant observation, purposive sampling, interviews with key informants, and other qualitative methods undertaken during visits coinciding with the local FLCP, led to the creation of a local governance map (Fig. 26.5). This map provided a

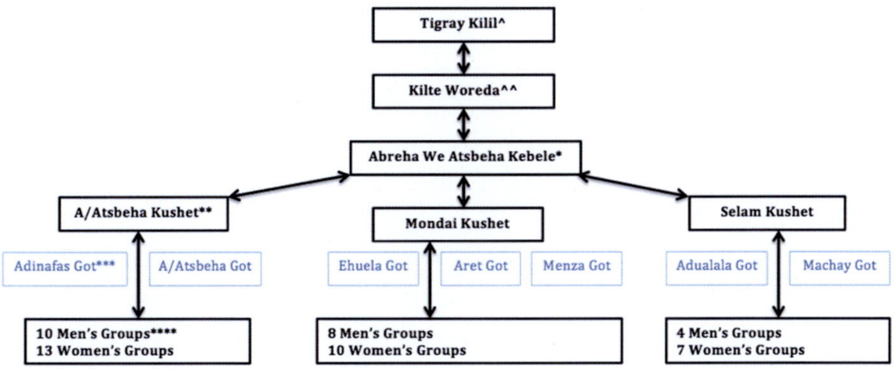


**Fig. 26.5** ^The Kilil is a region, or state, of Ethiopia. There are 9 Kilils in Ethiopia, divided ethno-linguistically

^^Woredas are districts within Kilils. There is an assigned woreda liaison responsible for A/Atsbeha, who occasionally comes to kebele meetings and also assigns and manages A/Atsbeha's 3 on-site development agents

*The Kebele is "the village" over which the chairman, Aba Hawi, presides. Aba Hawi is elected by a 208-person parliament (made up of the wahayo or "cells), of which all are members of the TPLF. Aba Hawi then appoints a 14-person cabinet: 3 wudaba members (people responsible for coordinating with the Woreda), 3 propaganda ministers, 2 finance ministers, 3 ministers of men's associations, and 3 ministers of women's associations. He also selects from this same pool various committees (including watershed committee), and kushet leaders. Everyone in the cabinet and committees are TPLF members

**The Kushet is the sub-village unit, determined geographically. Each Kushet is assigned a leader from the 14-person cabinet. All kushet leaders are party members. Kushet leaders meet with group leaders every Friday, and with committee members every Saturday

***The 'Got' is a neighborhood unit, determined geographically. Gots do not have assigned representatives, but rather serve as organization units, mostly for the watershed committee when determining where to do restoration activities

****Groups consist of about 25–30 people and can be made up of both party and non-party members. Each contain a 'cell' which is made up of 4 party members: cell leader, representative of the cell leader, secretary, and finance. The cell leader represents the group at Friday meetings

conceptual documentation of the decision-making structures that govern the community as a whole, and which of those are adapted to facilitate the FLCP and other related programs. It likewise furnished an understanding of important cleavages in the community, chief among them political affiliation, kushet residence (i.e., geographic partitioning) and gender.

Figure 26.5 represents a strong hierarchical system exercising power from the regional government to the household level. Figure captions explain how households in Abreha we Atsbeha, and other Tigrinyan kebele, are partitioned into gendered 'groups', which send representatives both to a local parliament (which elects the community chairman) and to regular weekly meetings, where decisions concerning the community as a whole are regularly made. It is during these regular weekly meetings, and through directives from the parliamentary-elected community chairman's chosen cabinet, that agendas concerning land use and the FLCP are set. While the meetings themselves are participatory and democratic from the perspective of the elected and appointed members, access to these meetings turned out to be highly restrictive.

One of the key insights that emerged from the conceptual mapping exercise was that, while in theory, residents of any political denomination could serve as cell members (i.e., as parliamentary and weekly meeting representatives), in practice, only recognized members of the ruling party (The Ethiopian People's Revolutionary Democratic Front – EPRDF) had ever held these positions. This homogeneity occurs despite the fact that less than a sixth of the population in Abreha we Atsbeha are recognized as EPRDF members.

Given Ethiopia's recent political history, this single-party hegemony is hardly surprising. However, recognizing that the supremacy of the party reaches all the way down to the kebele and even household level is important, especially given the fact that such a small portion of the population possesses access to representation and agency. This recognition contends with the realities of 'community-driven' restoration. What the conceptual map revealed is that restoration activities, carried out through the FLCP, are conducted by a rural citizenry that remains systematically removed from the decision-making processes that govern them, and possesses little to no recourse for meaningful participation in anything other than the FLCP itself.

This is not to say that local actors do not understand the value of their work, or appreciate the ways in which the restoration activities they undertake—and the biodiversity islands they collectively safeguard through exclosures—benefit them directly. Local actors benefit mostly via the increased provisioning of water, and through 'cut and carry' systems, wherein citizens are allowed to enter the exclosures on foot and hand harvest grasses and other fodder to bring back to their on-farm livestock. Ecological literacy and favorable perceptions of the 'water banks' created by restoration was high among virtually all surveyed respondents in Abreha we Atsbeha. Rather, it is to suggest that outside perceptions of restoration in Ethiopia—particularly those of international agencies lavishing praise on the participatory nature of the restoration activities—typically do not comprehend the stratification of power in local communities, and therefore fail to grasp the degree to which participation in restoration relies on a system of mass labor mandated by an elite political minority.

This realization raises a number of thorny questions for those interested in promoting and scaling participatory restoration initiatives. Most germane to this chapter is the question of longevity; such an asymmetrical system has the potential to breed discontent, and therefore threatens to destabilize restoration programs in the long-term. Safeguarding biodiversity islands in this context therefore requires an explicit understanding of the governance structures that sustain them, and a wholesale accounting of the myriad ways international agencies and initiatives reverberate through local configurations.

### *26.4.3 The Free Labor Contribution Period in a Pluralizing Context*

In early 2018, Ethiopia declared a national state of emergency following a mounting series of protests which culminated in the resignation of Hailemariam Dessalegn, the Prime Minister of Ethiopia and Chairman of the EPRDF.[8] These events coincided with the FLCP in Abreha we Atsbeha, whose regularly planned activities were interrupted by a series of emergency meetings held in kebeles at the behest of concerned district and regional party leaders. At various meetings, ranging in size and topic, registered party members were asked to voice their grievances, and share their ideas and opinions. A palpable aura of discontent—much of it directed at the FLCP and related programs—permeated throughout the course of these intensive, multi-day hearings. Decades of suppressed criticisms were unleashed as the fear of sharing long-held negative views dissolved.

Grasping the significance of these meetings requires some understanding of Ethiopia's recent political history. Starting in 1974, following the famine-induced collapse of the empire of Haile Selassie, Ethiopia was governed by a brutal military junta known as The Derg, followed by a dictatorial regime administered by a communist state party called People's Democratic Republic of Ethiopia (PDRE). Opposition to the Derg came primarily from groups based in the North, in Tigray. These opposition movements coalesced in the late 1980s, forming a powerful shadow government known as the Ethiopian People's Revolutionary Democratic Front (EPRDF) in 1988 (Rahmato 2009).

The EPRDF inflicted a total military defeat on the communist regime and established in 1991 a new government that has ruled Ethiopia ever since. This government has presided over nearly three decades of essentially single-party rule, exercising considerable control over the entire federation (Feltor 2018). The EPRDF is both the political party that committed Ethiopia to the Bonn Challenge and related restoration targets, and the political party that oversees and administers the FLCP, one key method through which these targets are currently met in Tigray and expected to be met across the country. This is especially true in the Tigray region, where Abreha we Atsbeha is located, and where the hegemony of the EPRDF is particularly strong.

[8]Ethiopia declares national state of emergency. BBC News (2018)

The EPRDF has suppressed many critical groups movements throughout its tenure, but mass protests and unrest beginning in 2016 have proven—so far—politically transformative. In April 2018, shortly after the conclusion of the FLCP in Abreha we Atsbeha, notable reformist Abiy Ahmed was sworn in as the new Prime Minister, becoming Ethiopia's first ethnically Oromo leader, thereby weakening the stronghold of Tigrinyan power. Ahmed swiftly launched a wide program of internationally-acclaimed political and economic reforms, and was awarded the 2019 Nobel Peace Prize for his work in brokering reconciliation with Eritrea, Tigray's tumultuous northern neighbor. Most recently, Sahle-Work Zewde was installed as Ethiopia's first female president, and Africa's only serving female head of state. The international community has generally lauded Ethiopia's recent progress towards a more open, pluralistic society and improved international relations.

Yet the single-party, restrictive government of Ahmed and Sahle-Work's predecessors remains central in places like Abreha we Atsbeha, and still resides at the core of rural institutions like the FLCP. During the emergency meetings held in lieu of 3 days of FLCP activities in Abreha we Atsbeha, many community members levied significant complaints and accusations against the party: its structure, its leaders, and specifically, the restoration and conservation programs it engenders. These events, along with the governance hierarchies described in Fig. 26.5, made clear that this autocratic system of government, especially at the local level—is the force that is overwhelmingly responsible for current restoration programs. Likewise, it is the same authority that sets norms and bylaws around continued safeguarding of the community's hard-earned biodiversity islands.

Scaling Abreha we Atsbeha's ecological success currently involves replicating these obscured socio-political dynamics. At the same time, there is little guarantee that such programs—invariably bound up in the politics of the last several decades—will prove compatible with the evolution of the country's governance systems as a whole. Safeguarding biodiversity islands therefore requires an explicit recognition of the FLCP, the socio-political structures that animate it, the misalignment between these existing structures and the advancement of Ethiopia's social, political and economic systems as a whole.

## 26.5 Sustaining Restoration and Safeguarding Biodiversity Islands Amidst Social and Political Change

### *26.5.1 Acknowledging Labor Realities in Restoration and Conservation Programs*

This chapter has repeatedly stressed the central role that coordinated group labor plays in the rural regions of Tigray. This centrality is a reflection of the region's manifest socio-ecological conditions: the landscape is semi-arid, mountainous, and highly degraded and the communities living there chronically lack financial capital and equipment. Establishing biodiversity islands where prolonged environmental degradation limits the ability for ecosystems to regenerate naturally requires substantial acts of human intervention, and this intervention must fit within the existing limits of community capacities.

These human interventions, moreover, must be thoughtfully planned and executed. Restoration of watersheds and the creation of biodiversity islands is a multifarious undertaking, requiring not only ecological but also sociological expertise and coordination. If human labor is the ultimate and direct mechanism through which restoration is achieved, then those concerned with promulgating successful restoration must be expressly concerned with the dynamics of that labor. The paucity of attention paid to this critical aspect of restoration in the region suggests that international and federal agencies alike largely take labor—including mass coordinated labor like the FLCP—for granted.

The previous section detailed the operations of a substantial local hierarchy, supported by a hegemonic regional authority that has been historically empowered by a single-party federal government. This hierarchy coordinates rigorous restoration activities among large segments of rural society, organizing, incentivizing, and monitoring individual and group performance in a calculated, iterative, and—from many perspectives—effective manner. Owing to this hierarchy, inhabitants of local communities abide by social norms dictating that their contributions be made reliably, compulsorily, and without pay. Acceptance of these bylaws thus substantially reduces the costs associated with restoration and thereby makes it possible in this otherwise limited context. Furthermore, the maintenance of this hierarchy and system of labor control is presently critical to the continued creation and protection of biodiversity islands in Tigray.

It is difficult to overstate the point that biodiversity islands in this region are generally *not* forest remnants to be protected against human encroachment, but rather are sites reflecting extraordinary feats of intentional restoration. The continued provisioning of such exclosure systems requires enormous community buy-in, strong bylaws against harvesting infractions, and high levels of ecological literacy that entail an understanding of the connections between hillside vegetation and lowland agriculture.

At the same time, these exclosure systems are not akin to national parks or protected areas; they typically remain the communal property of the kebele. Such ecosystems therefore depend not only on the regeneration of woody species and the encouragement of native animals and insects, but also on the continued renewal of community consent and respect for use-norms. The landscape of Tigray is a cautionary tale: systems don't always regenerate automatically. This is as true of the ecosystems as of the socio-political systems that are bound in the biodiversity islands of Abreha we Atsbeha. Paying attention to community labor programs is key for pinpointing critical vulnerabilities and building capacity around them before the system reaches a tipping point.

### 26.5.2 Recognizing Complexity in Local-Level Motivations

Precisely because people are not compensated financially for their contributions in the FLCP, scholars and practitioners must carefully consider the other motivations at play in these spaces. The author conducted interviews in Abreha we Atsbeha, where people generally displayed an impressive understanding of the complex watershed dynamics in the area, and were motivated by the understanding that restoring vegetation to hillsides would secure water supplies in their agricultural fields (Sigman 2019a). They also appreciated other benefits of restoration, such as increased biodiversity—especially wild animals—and access to cooler microclimates. At the same time, individuals also reported that they participated in the FLCP because they feared punishment if they withheld their labor or complained (Sigman 2019b).

This punishment could come in the form of social castigation or perceived favoritism of others by those in power, suggesting strong local norms around group participation. Additionally, participation in the FLCP is a prerequisite for membership in Tigray's welfare system, called the Productive Safety Net Program (PSNP). Through the PSNP, vulnerable members of society can receive subsistence amounts of food or money in exchange for an extension of their labor contribution to restoration activities. However, those members who are physically able to work must still contribute their labor without compensation during the FLCP in order to be eligible for the program. There is thus a complexity of compulsion-based motivations at play, which affect different segments of society asymmetrically, and may constitute a major obstacle to effective scaling.

There is a wide range of perceptions and motivations surrounding restoration and biodiversity islands in the area. These constitute not only different perceptions between different people, but individual people can have blended motivations—both positive and negative—for participating. Continued adherence to exclosure zones (i.e., the safeguarding of biodiversity islands) is driven by both positive and negative motivations; people can appreciate biodiversity islands (or exclosures) and the work it takes to create and maintain them while still holding negative views about the process. As demonstrated in the emergency meetings, these negative views are linked with feelings of disempowerment. Empowerment of stakeholders in the community therefore plays a key role in safeguarding biodiversity islands against these negative perceptions.

### 26.5.3 Understanding 'Political Monoculture' as a Threat to Biodiversity

Best practices in socio-ecological restoration demand that all stakeholders be able to actively influence and make decisions about the restoration activities that impact their daily lives. Yet in Abreha we Atsbeha—perhaps Ethiopia's most famous socio-ecological restoration case study—the majority of people who dedicate substantial time, labor, and resources to the work of restoration have very little access to the decision-making structures that would afford them a meaningful voice in the deliberation process. The nature, duration, location, and timing of restoration activities in Abreha we Atsbeha's FLCP, for example, are largely dictated by decisions made by Aba Hawi (the community chairperson) and his cabinet. The cabinet—relying on information gleaned from the kushet[9] meetings and taking into account directives from the woreda—will come up with a plan for the FLCP, which is then approved by the parliament. This entire process remains confined to registered party members, though the work of the FLCP necessarily involves many people who do not have a voice in the party system. Participatory restoration projects call for exactly the opposite structure, with full participation of all involved community members at each stage of the planning process.

Knowledge of this local political process is critical for understanding how the FLCP operates, and for thinking about how agency, attitudes, and voluntary labor are bound up in the restoration activities the FLCP advances. Within Abreha we Atsbeha's participatory restoration program, representation and agency among local stakeholders is essentially limited to EPRDF party members, despite the fact that registered members comprise only a sixth of the population. This asymmetry in representation, and its implications for long-term management of restoration programs, is therefore concerning.

The administrative system which has historically enabled restoration and safeguarded biodiversity islands in the region can be thought of as a political 'monoculture'. There exists one elite party, which despite comprising only a fraction of the population, dictates all of the activities of that population, in particular the labor-intensive activities associated with restoration, and the restrictive norms governing community exclosure from newly established biodiversity islands. This concentrated local political power may be seen as a consequence of the region's unique political history, and its extant behaviors perhaps justified to many by the exigencies of extreme land degradation coupled with the looming threat of famine. For these reasons and others, this system has been able to operate on the margins of the rights concerning representation and labor conditions which are otherwise typically espoused by the international development agenda.

However, those seeking to replicate this case study and safeguard the biodiversity islands in the region should recognize that intensive manual labor, ironclad bylaws, and political hegemony have played a central role in its evolution. The task of organizing thousands of people to give up communal grazing lands, volunteer up to two months of hard labor, and wait decades to reap the benefits of their work has been won in part by charisma and solidarity, and also in part by intense social pressures, by fear of castigation, and by a local political system that restricts much meaningful participation to all but a hand-selected elite.

[9] See Figure 26.1 caption for description of 'kushet' and 'woreda' units

This elite is bound up in what—for three decades—has been a recalcitrant, 'political monoculture' across the country. Just as diversity is a critical component of ecological resilience, so too is diversity a critical component of socio-political resilience. Restoration and conservation systems that rely in large part on political hegemony may be vulnerable to political changes, whether as a result of acute emergencies or long-term stressors.

While political hegemony may in fact be central to the success Abreha we Atsbeha has enjoyed in the past, Ethiopia's political system is becoming increasingly pluralistic at the national level, and—as emergency meetings held during the author's fieldwork activities in 2018 demonstrated—this shift is having growing impacts at the local level. Those interested in continuing Ethiopia's restoration success must therefore begin to think about how such large-scale and labor-intensive projects can continue or be supplemented if local political institutions start to divest power to a wider range of voices and agendas.

### *26.5.4 Need for Reassessment and Realignment Among International Coordinating Organizations*

Abreha we Atsbeha is undoubtedly a case study in successful dryland restoration across a number of metrics. Through decades of thoughtful, ecologically-sound and socio-economically sensitive interventions and innovations, the community of Abreha we Atsbeha has effectively stewarded a resilient biophysical ecosystem in their kebele. This effort has provided substantial social and economic benefit to the community as a whole.

At the same time, Abreha we Atsbeha's ecological and economic success is linked to the perpetual reifying of its social and political systems. The international community may be inadvertently bolstering these systems in Abreha we Atsbeha through its unexamined praise of the community's restored biodiversity islands. Moreover, it may be unwittingly promoting the replication of such systems in other areas in the attempts to scale Abreha we Atsbeha's successes in order to meet the demands of the Bonn Challenge and other such initiatives. A healthy debate on whether or not international sustainability and development organizations want to be associated with these kinds of political systems cannot even begin until there is a greater degree of reflection on the existence of these systems—and indeed their centrality—in the work of restoration in the first place.

Thorny questions on alignment between potentially conflicting development priorities aside, there is a pressing issue of practicality. With the country's recent national reforms, civil society in Ethiopia seems to be moving in a number of positive directions, including provisioning a more democratic, open and pluralistic public sphere. Could this increased plurality and openness pose an unrecognized threat to restoration and conservation programs built on hegemony? Can FLCP-mediated restoration be modified to be more inclusive in this new political landscape, perhaps by embracing more voices in the decision-making process? If not, what replaces the FLCP? What safeguards the exclosure zones? In short, how will restoration weather democratization? (NB: this chapter was authored prior to the surge in political violence that began sweeping the Tigray region in 2021 and, unfortunately, has reversed the country's progression towards peaceful pluralism. Recent events pose an even more urgent question: how will restoration weather another civil war?)

Those concerned with safeguarding biodiversity islands like those in Abreha we Atsbeha would do better to stay ahead of these questions than behind them. This can start with acknowledgement of local labor realities, recognition of complexity in local-level motivations, an understanding of relationships between 'political monoculture' and landscape-level biodiversity, and better alignment between international coordinating organizations, national initiatives, and local actors.

## 26.6 Conclusion

High levels of environmental degradation worldwide have spurred an increased appreciation for the contributions restoration can and must make towards ameliorating a number of interconnected global challenges. Restored areas serve as biodiversity islands in otherwise degraded landscapes, improving ecosystem connectivity and function, and offering a number of irreplaceable services to human societies around the globe. Recent years have witnessed mounting enthusiasm for restoration and an expanding consciousness surrounding the intricacies of socio-ecosystems. Substantial initiatives like the Bonn Challenge and the UN Sustainable Development Goals have elevated this awareness to the international arena, and have catalyzed mass-scale restoration movements while provoking research on local-level behaviors and impacts.

While this upsurge in awareness and activity has promised great benefits—such as improved food and water security, increased rural economic development, climate change adaptation and mitigation, and biodiversity protection—the realities of restoration in practice are still unfolding. Chief among the challenges faced by restoration practitioners is a phenomenon introduced here as the 'Restoration Dilemma'. The dilemma posits that restoration is highly site-specific and therefore resists scaling, and yet, degradation tends to be so pronounced that restoration usually *must* be scaled in order to realize its promised benefits (Sigman 2021).

To explore possibilities for assuaging the 'Restoration Dilemma', the present chapter focused on a case study from Northern Ethiopia, in the community of Abreha we Atsbeha. Here, a local community successfully re-vegetated substantial areas of highly degraded landscapes through integrative watershed management and hillside exclosures. This local-level restoration occurs in the context of a significant nationwide restoration platform called the Sustainable Land Management Program (SLMP)—the largest of its kind in Africa—and has been supported by a number of multinational restoration initiatives, including The Bonn Challenge.

The restoration activities of the village, and the biodiversity islands they enabled, have been widely lauded by the international community, and many have offered Abreha we Atsbeha as an example of how effective, coordinated, multi-scale restoration and conservation might proceed throughout the region and across the globe. Yet, as this analysis demonstrates, an insufficient amount of attention had been paid to the local-level dynamics that enable these restored landscapes. In particular, there has been little to no appreciation of the substantial 'Free Labor Contribution Periods' that were key to the creation of current biodiversity islands and remain central to the restoration programs in the region.

The expectation of substantial amounts of uncompensated rural labor exists in the context of a hegemonic, single-party 'political monoculture'. This stands at odds with the present-day realities of a country that is rapidly becoming politically pluralized. Sustaining restoration and safeguarding biodiversity islands amidst social and political change in Ethiopia therefore requires: (1) acknowledgement of local labor realities; (2) recognition of complexity in local-level motivations; (3) understanding relationships between 'political monoculture' and landscape-level biodiversity, and; (4) better alignment between international coordinating organizations, national initiatives, and local actors.

As the international community looks to the future of restoration, and together shapes the aims and aspirations of the global restoration movement, we must pay attention to governance and to power. Just as we work to promote diversity in our ecosystems, so too should we explore questions of political plurality and governance. This exploration should begin with an appreciation for the substantial and complex human labor dynamics that enable biodiversity islands, and should motivate us to probe deeper into the political structures that motivate, organize, and conscript them. In this way, we may begin to untangle the 'Restoration Dilemma' and, with care, enable diverse political, social, and ecological systems to thrive.

**Acknowledgements** This work would not have been possible without the generosity, assistance, knowledge, and hospitality offered by Aba Hawi, his family, and the many gracious citizens of Abreha we Atsbeha. Additionally, Cathy Watson, Niguse Hagazi, and Dr. Kiros Hagdu of the World Agroforestry Centre (ICRAF) supported the initial stages of this project, and were instrumental in facilitating later independent research. Abrha Berhane (Mekelle University) and Zufan Tsadik provided key translation and additional field analysis. Dr. Gordon Geballe (Yale FES) provided advice and resources at the planning stages, and Dr. Florencia Montagnini (Yale FES) served as advisor throughout the research process, and offered constructive editing. This work was supported by the National Science Foundation Graduate Research Fellowship Program under Grant No. DGE1122492, as well as by the Yale Tropical Resources Institute, the Yale Jackson Institute for Global Affairs, and Yale FES. The author also gratefully acknowledges the financial support of the CGIAR Research Program on Forests, Trees and Agroforestry and the CGIAR Trust Fund Donors.

Any opinions, findings, and conclusions or recommendations expressed in this material are those of the author's and do not necessarily reflect the views of supporters.

## References

Aerts R, Honnay O (2011) Forest restoration, biodiversity and ecosystem functioning. BMC Ecol 11:29. https://doi.org/10.1186/1472-6785-11-29

Alemu A, Getahun M, Bezabih T (2014) Comprehensive Food Security and Vulnerability Analysis (CFSVA). Ethiopia Central Statistical Agency/World Food Program

Anderson W, Pembert N (2019) Across Africa, restoring land is climate action. AFR100 Blog

Anwar AA, Dessalew WA, Seifu AT, Mekuria W, Mulugeta A, Tammo SS (2016) Effectiveness of land and water management interventions in reducing runoff and soil erosion in the northwestern highlands of Ethiopia. Paper presented at First African ESP Conference; Nairobi, Kenya

Asefa D, Oba G, Weladji R, Colman J (2003) An assessment of restoration of biodiversity in degraded high mountain grazing lands in northern Ethiopia. Land Deg & Dev 14:25–38. https://doi.org/10.1002/ldr.505

Aynekulu E, Denich M, Tsegaye D (2009) Regeneration response of Juniperus procera and Olea europaea subsp cuspidata to exclosure in a dry afromontane forest in northern Ethiopia. Mount Res and Dev 29:143–152. https://doi.org/10.1659/mrd.1076

Bard K, Coltorti M, DiBlasi CM, Dramis F, Fattovich F (2000) The environmental history of Tigray (northern Ethiopia) in the middle and late Holocene: a preliminary outline. Afr Archaeol Rev 17: 65–86. https://doi.org/10.1023/a:1006630609041

Benayas J, Newton A, Diaz A, Bullock J (2009) Enhancement of biodiversity and ecosystem services by ecological restoration: a meta-analysis. Science 325:1121–1124. https://doi.org/10.1126/science.1172460

Berkes F, Folke C (1998) Linking social and ecological systems: management practices and social mechanisms for building resilience. Cambridge University Press, Cambridge

Boesing AL, Nichols E, Metzger JP (2018) Biodiversity extinction thresholds are modulated by matrix type. Ecography 41(9):1520–1533

Brancalion P, Cardozo I, Camatta A (2014) Cultural ecosystem services and popular perceptions of the benefits of an ecological restoration project in the Brazilian Atlantic forest. Restor Ecol 22: 65–71. https://doi.org/10.1111/rec.12025

Brudvig L (2011) The restoration of biodiversity: where has research been and where does it need to go? Am J Bot 98:549–558. https://doi.org/10.3732/ajb.1000285

Budiharta S, Meijaard E, Wells J (2016) Enhancing feasibility: incorporating a socio-ecological systems framework into restoration planning. Environ Sci Policy 64:83–92. https://doi.org/10.1016/j.envsci.2016.06.014

Colding J, Barthel S (2019) Exploring the social-ecological systems discourse 20 years later. Ecol Soc. https://doi.org/10.5751/es-10598-240102

Costanza R, de Groot R, Sutton P, van der Ploeg S, Anderson SJ, Kubiszewski I, Farber S, Turner RK (2014) Changes in the global value of ecosystem services. Glob Environ Chang 26:152–158

Covelli-Metcalf E, Mohr J, Yung L (2015) The role of trust in restoration success: public engagement and temporal and spatial scale in a complex social-ecological system. Restor Ecol 23:315–324. https://doi.org/10.1111/rec.12188

Daily G (1997) Nature's services. Island Press, Washington, DC

Daily G, Matson P (2008) From theory to implementation. PNAS 105:9455–9456. https://doi.org/10.1073/pnas.0804960105

Dessalew WA, Anwar AA, Seifu AT, Mekuria W, Charles FN, Tammo SS (2016) Enhancing ecosystem services in the upper Blue Nile Basin, Ethiopia through the implementation of soil and water conservation measures. Presented at First African ESP Conference, Nairobi

Dodd M (2015) Ethiopia rising: Red terror to Green Revolution. 1080 Films, Abreha we Atsbeha, Ethiopia

Felter C (2018) Ethiopia: East Africa's emerging giant. Council on Foreign Relations Backgrounder

Fernández-Manjarrés JF, Roturier S, Bilhaut AG (2018) The emergence of the social-ecological restoration concept. Restor Ecol 26(3):404–410

Garner B (2009) Garner's modern American usage. Oxford University Press, USA, Oxford

Gebregziabher G, Abera DA, Gebresamuel G, Giordano M, Langan S (2016) An assessment of integrated watershed management in Ethiopia. Colombo, Sri Lanka: International Water Management Institute (IWMI). 28p. (IWMI Working Paper 170). https://doi.org/10.5337/2016.214
GIZ (2020) Sustainable use of rehabilitated land for economic development (SURED). In: Giz.de. https://www.giz.de/en/worldwide/18912.html. Accessed 25 Jan 2020
Habtezion S, Adelekan I, Aiyede E (2015) Earth system governance in Africa: knowledge and capacity needs. Curr Op in Envi Sust 14:198–205. https://doi.org/10.1016/j.cosust.2015.06.009
Hachoofwe, EM (2012) Local ecological knowledge of trees on farms, constraints and opportunities for further integration in Tigray Region, northern Ethiopia: A case study of smallholder farmers in Abreha Wa Atsbeha and Adi gudom. MSc thesis in Agroforestry. Gwynedd, UK: Bangor University
Haslan C (2015) Turning Ethiopia's desert green. BBC News
Hobbs R, Higgs E, Harris J (2009) Novel ecosystems: implications for conservation and restoration. Trends Ecol Evol 24:599–605. https://doi.org/10.1016/j.tree.2009.05.012
Hurni H, Berhe WA, Chadhokar P, Daniel D, Gete Z, Grunder M, Kassaye G (2016) Soil and water conservation in Ethiopia: guidelines for development agents. Second revised edition. Bern, Switzerland: Centre for Development and Environment (CDE), University of Bern, with Bern Open Publishing (BOP). 134 pp
Keller E (1992) Drought, war, and the politics of famine in Ethiopia and Eritrea. J Mod Afr Stu 30: 609–624. https://doi.org/10.1017/s0022278x00011071
Krech S (2000) The ecological Indian. W.W. Norton, New York
Kumar P (ed) (2010) The economics of ecosystems and biodiversity ecological and economic foundations. Earthscan, London and Washington, 422pp
Kuria A, Winowiecki, L, Ouya D (2016) Fresh water, the reward of land restoration, flows in Ethiopia's dry zone. In: Agroforestry World. http://blog.worldagroforestry.org/index.php/2016/11/10/water-reward-land-restoration-flows-ethiopias-dry-zone/. Accessed 29 Jan 2020
Lamond, G (2012) Local knowledge training using the AKT5 software and methodology at Mekelle University, funded by the AfricaRISING project: a report on a two week research study in the village of Abreha We Atsbeha in Tigray Region, northern Ethiopia. Nairobi, Kenya: World Agroforestry Centre
Levis C, Costa F, Bongers F (2017) Persistent effects of pre-Columbian plant domestication on Amazonian forest composition. Science 355:925–931. https://doi.org/10.1126/science.aal0157
López DR, Cavallero L, Easdale MH, Carranza CA, Ledesma M, Peri PL (2017) Resilience management at landscape level: an approach to tackle social-ecological vulnerability of agroforestry systems. In: Montagnini F (ed) Integrating landscapes: agroforestry for biodiversity conservation and food sovereignty, advances in agroforestry 12. Springer, Cham, pp 127–148
MacArthur RH, Wilson EO (1967) The theory of island biogeography. Princeton University Press, Princeton, 203 pp
Mace G, Norris K, Fitter A (2012) Biodiversity and ecosystem services: a multilayered relationship. Trends Ecol Evol 27:19–26. https://doi.org/10.1016/j.tree.2011.08.006
Martin D (2017) Ecological restoration should be redefined for the twenty-first century. Restor Ecol 25:668–673. https://doi.org/10.1111/rec.12554
Maynard C (2013) How public participation in river management improvements is affected by scale. Area 45:230–238. https://doi.org/10.1111/area.12015
McKenna A (2019) Tigray, historical region, Ethiopia. Encyclopedia Brittanica Online, https://www.britannica.com/place/Tigray-historical-region-Ethiopia
Mekuria W, Veldkamp E (2012) Restoration of native vegetation following exclosure establishment on communal grazing lands in Tigray, Ethiopia. App Veg Sci 15:71–83. https://doi.org/10.1111/j.1654-109x.2011.01145.x
Mekuria W, Langan S, Noble A, Johnston R (2017) Soil restoration after seven years of exclosure management in northwestern Ethiopia. Land Deg & Dev 28:1287–1297. https://doi.org/10.1002/ldr.2527

Mekuria W, Wondie M, Amare T (2018) Restoration of degraded landscapes for ecosystem services in North-Western Ethiopia. Heliyon 4:e00764. https://doi.org/10.1016/j.heliyon.2018.e00764

Mengistu T, Teketay D, Hulten H, Yemshaw Y (2005) The role of enclosures in the recovery of woody vegetation in degraded dryland hillsides of central and northern Ethiopia. J Arid Env 60: 259–281. https://doi.org/10.1016/j.jaridenv.2004.03.014

Ministry of Finance and Economic Development (MoFED) (2015) The Federal Democratic Republic of Ethiopia Growth and Transformation Plan (GTP). Ministry of Finance and Economic Development (MoFED), Addis Ababa

Minnemeyer S, Laestadius L, Sizer N (2014) Atlas of forest and landscape restoration opportunities. In: World Resources Institute. https://www.wri.org/resources/maps/atlas-forest-and-landscape-restoration-opportunities. Accessed 29 Jan 2020

Nyssen J, Haile M, Moeyersons J (2000) Soil and water conservation in Tigray (Northern Ethiopia): the traditional dagat technique and its integration with introduced techniques. Land Deg & Dev 11:199–208. https://doi.org/10.1002/1099-145x(200005/06)11:3<199::aid-ldr376>3.0.co;2-y

Nyssen J, Frankl A, Haile M (2014) Environmental conditions and human drivers for changes to north Ethiopian mountain landscapes over 145 years. Sci Total Environ 485-486:164–179. https://doi.org/10.1016/j.scitotenv.2014.03.052

Pankhurst R, Ingrams L (1988) Ethiopia engraved. Kegan Paul International, London

Rabia AH, Afifi RR, Gelaw AM, Bianchi S (2013) Soil mapping and classification: a case study in the Tigray Region, Ethiopia. J Agri Env Int Dev 107(1):73–99. https://doi.org/10.12895/jaeid.20131.81

Rahmato D (2009) 2 peasants and agrarian reforms: the unfinished quest for secure land rights in Ethiopia. Land Rights 33

Reij C (2015) How Ethiopia went from famine crisis to Green Revolution. In: World Resources Institute Blog. Accessed 25 Jan 2020

Rovere A (2015) Review of the science and practice of restoration in Argentina: increasing awareness of the discipline. Restor Ecol 23:508–512. https://doi.org/10.1111/rec.12240

Schroth G, Fonseca G, Harvey C (2004) Agroforestry and biodiversity conservation in tropical landscapes. Island Press

Seyoum Y (2016) Forest landscape restoration experiences in Ethiopia. AFR100 Conference, Addis Ababa

Shono K, Cadaweng E, Durst P (2007) Application of assisted natural regeneration to restore degraded tropical forestlands. Restor Ecol 15:620–626. https://doi.org/10.1111/j.1526-100x.2007.00274.x

Sigman, E (2019a) The role of group labor in socioecological restoration: perspectives from Abrha we Atsbeha, Fourth Annual World Agroforestry Congress, Montpellier, France

Sigman E (2019b) Who decides? Restoration policies, local governance, and group labor: a case study from Ethiopia. Yale Environmental Leadership and Training Initiative. Yale University, New Haven

Sigman E (2021) The Dilemma of Scale: competing imperatives for global restoration. Restor Ecol, 29: e13408. https://doi.org/10.1111/rec.13408

Tadesse A, Gebrelibranos T, Geberehiwot M, Mezgebu A (2015) Characterization and impact assessment of water harvesting techniques: a case study of Abreha Atsbeha watershed, Tigray, Ethiopia. Open Agricultural Resources (OER) Africa

Tamene L, Abera W, Woldearegay K, Tibebe D, Tadesse M, Admassu Z, Yaekob T, Tefere H, Alene T, Dubale W, Amede Y, Hailu H, Demeke G, Mekonnen K, Thome P (2018) Land restoration initiatives and their performances in Ethiopia: a systematic assessment based on meta-data analysis. AfricaRISING

Teshome A, Zhang J (2019) Increase of extreme drought over Ethiopia under climate warming. Adv Met 2019:1–18. https://doi.org/10.1155/2019/5235429

Trigger D, Mulcock J, Gaynor A, Toussaint Y (2008) Ecological restoration, cultural preferences and the negotiation of 'nativeness' in Australia. Geoforum 39:1273–1283. https://doi.org/10.1016/j.geoforum.2007.05.010

United Nations (2020) About the sustainable development goals. In: United Nations Sustainable Development. https://www.un.org/sustainabledevelopment/sustainable-development-goals/. Accessed 25 Jan 2020

United Nations Development Program (2013) Abrha Weatsbha Community, Ethiopia. NY, New York

United Nations Environment Program: News and Stories (2019) Ethiopia plants over 350 million trees in a day, setting new world record

Uprety Y, Asselin H, Bergeron Y (2012) Contribution of traditional knowledge to ecological restoration: practices and applications. Écoscience 19:225–237. https://doi.org/10.2980/19-3-3530

Van Veen E (2016) Perpetuating power: Ethiopia's political settlement and the organization of security. Chapter 1: dynamics of political power in Ethiopia: Past and Present

Watson C (2016) No one leaves anymore': Ethiopia's restored drylands offer new hope. The Guardian

Win T L (2019) With fruits and green shoots, Tigray swaps image of famine for resilience. Reuters

World Bank (2015) NEPAD launches initiative for the resilience and restoration of African landscapes

Wortley L, Hero J, Howes M (2013) Evaluating ecological restoration success: a review of the literature. Restor Ecol 21:537–543. https://doi.org/10.1111/rec.12028

# Part V
# Conclusions

# Chapter 27
# Conclusions: Challenges and Opportunities in Implementing Biodiversity Islands

**Florencia Montagnini, Ryan T. Smith, Benjamin Everett-Lane, Sara del Fierro, and Dara Albrecht**

**Abstract** This concluding chapter presents the lessons learned from the chapters in the four previous parts of this book: (I) Introduction; (II) Biodiversity Islands Establishment and Management: Challenges and Alternatives; (III) Biodiversity Islands Across the Globe: Case Studies; and (IV) Safeguarding the Environmental, Social, and Economic Benefits of Biodiversity Islands. Constraints limiting the adoption of Biodiversity Islands (BI), include conceptual, biophysical, economic, political, social, and cultural factors. Opportunities for increasing the implementation of BI are presented, particularly chances for working with groups from private conservation initiatives, such as those representing local communities, indigenous peoples, and conservation organizations. Examples of policies promoting agroecology are discussed, as well as current trends in conservation which support the BI concept. Despite the challenges posed to BI, local motivation, political will, and the right educational campaigns, can allow economically prosperous human communities and biodiversity to thrive harmoniously within shared landscapes. Many international efforts are currently underway, creating sustainable and dynamic BI within human-dominated environments. BI are a critical strategy for conservation in the twenty-first century while having the added benefit of contributing to climate adaptation and resiliency solutions. This book serves as a tool for policy makers, practitioners, and researchers interested in increasing the implementation of Biodiversity Islands.

F. Montagnini (✉) · R. T. Smith
School of the Environment, The Forest School, Yale University, New Haven, CT, USA
e-mail: florencia.montagnini@yale.edu

B. Everett-Lane · D. Albrecht
Yale University, Yale College, New Haven, CT, USA

S. del Fierro
United States Department of Agriculture (USDA), Washington, DC, USA



F. Montagnini (ed.), *Biodiversity Islands: Strategies for Conservation in Human-Dominated Environments*, Topics in Biodiversity and Conservation 20,
https://doi.org/10.1007/978-3-030-92234-4_27

**Keywords** Agroforestry · Biodiversity conservation · Community action · Ecosystem services · Indigenous communities · Private reserves · Social-ecological systems

## 27.1 Introduction

There are common claims of a "sixth mass extinction" happening today with the loss of species currently at a rate at least 1000 times the background rate (Pimm et al. 2014; Pimm 2021). The direct causes include the loss of habitats, the introduction of invasive exotic species, over-harvesting of biodiversity resources, and homogenization of species in agriculture. The common factor of all these elements is that they are mostly human-driven. The economic and social root causes behind biodiversity loss, include demographic changes, overconsumption and production patterns, economic growth, macro-economic policies and structures, social change, and development (https://www.greenfacts.org/en/global-biodiversity-outlook/l-3/6-threat-biodiversity.htm., https://www.cbd.int/).

Worldwide, industrialized agriculture grows at the expense of natural areas such as forests and savannas, for example through "invading" forests to grow soybeans in Latin America and palm oil in Indonesia. Commercial agriculture generated nearly 70% of deforestation in Latin America between 2000 and 2010 (FAO 2015a; Martin 2020). Southeast Asia, the region in the world that has suffered the greatest rate of deforestation, lost 30% of its forest cover over the last 40 years (Afelt et al. 2018). Tropical cloud forests, one of the world's most species- and endemism-rich terrestrial ecosystems, are threatened by direct human pressures and climate change, with substantial species losses worldwide, especially in readily accessible places (Karger et al. 2021, Newcomer et al. 2022).

Human population as well as per-capita resource consumption are expected to continue to rise, driving expanded urbanization, land use change, increased demand for agricultural land, and deforestation related activities such as unsustainable logging and mining. Industrial agriculture, focused on maximizing production through monocultural cropping engineered to provide maximum yields, continues to expand. However, it is dependent on advanced plant breeding, specialized (and costly) machinery, and agrochemicals for fertilization and control of weeds and pests. This approach contaminates and depletes agricultural soils, resulting often in soil degradation (Bern 2018). The planet is losing ~0.3% per year of its capacity to produce food due to soil degradation, currently jeopardizing the food and nutrition security of one third of the world's population (FAO 2015a).

Deforestation is linked to increased agricultural areas and poorly managed urban growth (Afelt et al. 2018). As forest ecosystems and their habitats are lost, displaced organisms along with their pathogens move from forests to anthropic environments and from animals to humans, thus creating breeding grounds for the emergence of new diseases. The COVID-19 pandemic is a call for attention to the fact that production models prevalent today are contributing, in multiple ways, to this health emergency (Martin 2020).

The "One Health" concept recognizes that human health is connected to animal health and to the environment (Afelt et al. 2018). The modern incursion of humans into the forest, involving changes to natural ecosystems, causes imbalances and frequently leads to the appearance and spread of zoonotic diseases and even pandemics like COVID-19. The main routes of transmission of zoonosis differ widely according to the specific underlying factors leading to emerging infectious diseases (Loh et al. 2015). This knowledge can be used to develop more effective strategies for controlling newly emerged diseases, taking into account the different underlying pressures leading to land use change.

Anthropized rural environments are characterized by a wide diversity of landscapes comprising houses, barns, fields, orchards, and woodlands of differing density, and can provide an acceptable habitat for a large range of small animal species which can carry a variety of pathogens next to human dwellings (Afelt et al. 2018). Anthropized rural environments are also likely to increase human exposure and risk of contracting insect-borne diseases, particularly as changing climate conditions facilitate expanded ranges for disease carrying insects in the tropics. Thus, it is crucial to educate and raise awareness about the risks associated with anthropized environments (Afelt et al. 2018).

The advance of industrialized agriculture also threatens the traditional production systems of indigenous peoples, local communities, and small and medium-sized producers that underpin agriculture. These traditional systems and smaller producers provide food for 70–80% of the world's population, as well as the collection of ancestrally associated knowledge, the preservation of genetic diversity and its territorial management systems (FAO 2015b, 2019; Gadgil et al. 2021; Montagnini and Berg 2019; Pimm 2021). Growing concern about the food and nutritional insecurity of much of the world's population, together with the impacts of the COVID-19 pandemic, have led to the inclusion of this problem in the international political agenda. To this end, it is important to promote alternative ways to produce food, as have been presented in several chapters in this book, e.g. Calle et al., González et al., Levin (a), Montagnini et al., Montagnini and del Fierro, Montes-Londoño et al., Painter et al., and Toensmeier. We must focus on more sustainable agricultural techniques, including regenerative agriculture, to produce food without depleting soil and damaging the climate (Bern 2018). We need a revaluation of traditional cultivation practices which for hundreds of years have given sustainment to local peoples (and continue to do so) with diverse, nutritious, and culturally appropriate foods (Gadgil et al. 2021; Pimm 2021). Likewise, it is important to promote the use of biological controls that can gradually replace the toxic chemicals that are now used in industrial agricultural systems.

It is possible to reconcile agricultural production with biodiversity conservation, when nature is part of human-dominated landscapes, truly sharing space by virtue of coexistence (Calle et al. 2022; Crespin and Simonetti 2019; Levin 2022a; Montagnini and del Fierro 2022; and other chapters mentioned above). Sustainable agricultural management techniques geared toward harmonizing ecosystem productivity and biodiversity conservation can contribute to mitigating or reversing detrimental effects of human impacts on landscapes while ensuring that agricultural productivity can meet the needs of human inhabitants for generations (Montagnini and Berg 2019; Montagnini et al. 2022). To face these issues, today's environmental, social, economic, and political circumstances require innovative responses that are appropriate to the emerging conditions (Berlyn 2021; Mc Neeley 2021).

Biodiversity islands (BI)—ecological refuges where plants and animals thrive without major interference from human activity—can contribute to the provision of ecological, economic, and social benefits at the ecosystem, landscape, and global levels. They can exist in an assortment of human-dominated landscapes (e.g., agricultural, wetland, urban) ranging in size from square meters to many square kilometers. In the following sections we summarize examples presented in this book of BI from throughout the world, discuss their challenges, and suggest viable alternatives in their implementation and management at each scale.

## 27.2 Key Messages from This Book

### *27.2.1 Part I, Introduction*

The first part of the book establishes the framework for understanding the complexities of biodiversity islands and the variety of strategies that can be used to establish them. The Introduction defines the term "biodiversity island" (BI) as a unique type of ecological refuge whose design depends on its purpose, as well as on the spatial distribution of reserves throughout the landscape, degree of landscape degradation, species present, and location within the urban-rural spectrum. BI can contribute to the ecological strength of a land area and make local agricultural areas more resilient, for example by converting them into agroforestry systems (AFS) using various agroecological strategies. Land use systems in a forestry matrix, as in many indigenous sacred sites and AFS, can also be part of BI.

### *27.2.2 Part II, Biodiversity Islands Establishment and Management: Challenges and Alternatives*

Design strategies for BI depend on landscape use within the matrix of habitat fragmentation. Integrated landscape management (ILM), including sustainable agriculture, agroforestry and community led action, may provide a framework for implementation of BI in complex landscape matrices. An experiment evaluating edge effects by Arroyo-Rodríguez et al. (2022) shows that natural forest patches of all sizes can benefit the ecosystem, and even small patches are valuable for conservation of forest-specialist species. The next chapters discuss AFS strategies, including regenerative agriculture, the integration of agricultural productivity and biodiversity conservation, silvopastoral systems (SPS), incorporating reforestation into livestock farms, and riparian buffers, which safeguard aquatic and riparian environments from harmful agricultural practices. These techniques can lead to agricultural systems containing significantly higher biodiversity than the surrounding area, showing that BI can exist in protected areas as well as in human-managed landscapes. However, the right species should be reintroduced to a landscape during restoration. The section concludes with a chapter by Eibl et al. (2022) stressing the necessity of seed source certification to ensure the quality of genetic material and species reintroduced to a landscape during restoration.

### *27.2.3 Part III, Biodiversity Islands Across the Globe: Case Studies*

The third part presents a total of 11 case studies where varied agroecological strategies were applied in the formation or conservation of BI in human-dominated landscapes. These case studies include (1) forest islands surrounded by flood-prone savannah-dominated landscapes of Paraguayan Chaco utilized for livestock production; (2) secondary forests that have persisted in the highly deforested landscape of the Ucayali region of the Peruvian Amazon; (3) the integrated network of conserved areas in Monteverde, Costa Rica which facilitate species movement across BI; (4) El Hatico Nature Reserve, a model of restoration and utilization of agricultural practices for sustainable production surrounded by a largely monoculture Colombian landscape; (5) Hacienda Pinzacuá, a restored, regenerative agriculture farm in the surrounding treeless central Andes of Colombia; (6) small persisting biodiverse land patches in the British countryside; and (7) the resilient islands of Las Rosas in the Argentinean humid pampas which represent an opportunity to propose diverse agroecosystems and develop local productive and economic strategies. Case studies focused on conservation in urban landscapes include: (8) residential gardens of Panama City; (9) the urban forests and peatlands of Ushuaia, Argentina; and (10) the perennial garden of Paradise Lot in the U.S.A. (Toensmeier 2022). The section closes with a case study (11) describing experiments to attain high genetic diversity in BI in Misiones, Argentina (Niella et al. 2022).

Techniques are discussed for raising livestock in landscapes vulnerable to risks from extreme weather or human-caused encroachment. Through extensive or semi-extensive livestock production or use of live fences, livestock's impact on the landscapes were minimized to allow for the conservation of BI. There may be benefits but also limitations in the use of BI for promoting species migration, leading to efforts to connect islands and allow species to travel greater distances. Where BI are surrounded by monoculture farms or treeless landscapes there may be ecological and economic benefits that they can provide to those regions. Even in urban landscapes BI can be havens for species and nodes in an interconnected network of land patches that allow wildlife to travel and prosper.

The variety of case studies from different types of landscapes from several regions of the world reveals the role BI play in conserving local flora and fauna that has been largely diminished by anthropogenic activities. In addition, these case studies show how these BI are able to strengthen or increase the genetic diversity of a human-dominated landscape, as shown by Niella et al. Furthermore, there are human-centered benefits in BI, from providing a deeper cultural connection to nature to supplying ecosystem services that make BI profitable to farmers and nearby communities.

### *27.2.4 Part IV, Safeguarding the Environmental, Economic, and Social Benefits of Biodiversity Islands, and Part V, Conclusions*

These final two parts of the book further detail the economic, social, political, and cultural aspects of the establishment and persistence of BI in anthropogenic landscapes. A variety of strategies can be used to establish BI, including local laws and legal tools, monetary aid and other financial resources, and local culture, particularly of indigenous communities. All these strategies rely on community-led action to contribute to the development and subsequent management of BI. Different community members' perspectives towards their local ecosystem provide further insights into the deciding factors or various motivations for conservation. For example, the priorities, perspectives and use of a community forest by the people living around Mayagüez, Puerto Rico, led them to be willing to protect the natural forest for its services. In general, community forests are important for protection of lands, reduction of deforestation, conservation of biodiversity, and carbon sequestration while providing socioeconomic well-being to those living around them (Morales-Nieves 2022).

The values that people assign to the forest contribute to its preservation as a BI within a rural-urban landscape, even if biodiversity is not the prime benefit. For example, sacred forests in Ethiopia have survived despite socioeconomic and political pressures increasing deforestation in the adjacent land area. These forests provide vital spaces for religious practice, as well as ecosystem services that contribute valuable resources to the community, further reinforcing the relationship between the community and the forest. Other case studies in Ecuador, Brazil and India underscore the importance of outside partners working directly with local communities when implementing conservation practices, as opposed to leaving local voices and knowledge out of the design process. The role of the community in conservation is further demonstrated through strategies for community-managed AFS that allow land reform to occur in a more sustainable way, maximizing social and economic returns while minimizing forest clearance in the cacao region of Bahia, Brazil.

Biodiversity islands can provide a variety of ecosystem services through the food and resources grown within them, with overharvesting prevented via proper regulation, as shown in the permit-based harvest of ginseng in Appalachian Mixed Mesophytic Forests in the U.S.A. The attitudes of farmers towards various agroecological approaches can determine what strategies farmers are willing to use in order to continue to benefit from their land, as shown in the research focusing on natural regeneration, reforestation and assisted natural regeneration as strategies to establish and maintain silvopastoral farms in the Azuero Peninsula of Panama. Moreover there may be tensions between the restrictions that government control may place on the relationship between the BI and the local community, as shown in Northern Ethiopia, and the positive role local community members can play in establishing and maintaining BI. Social-ecological systems need to be adapted to ensure that rural, restoration-based BI in the region can continue to flourish alongside more pluralistic and democratic political norms and institutions.

Biodiversity islands can provide valuable ecosystem services to the communities or farmers who choose to establish them, helping to maintain or improve productivity while also conserving local flora and fauna. BI should always be tailored specifically to the landscape, needs, and resources of the ecosystem to ensure they are effective at protecting native species and their genetic diversity. To ensure BI are enduring, however, the local community members must be allowed and encouraged to contribute to its design and maintenance. This leads to help people develop a more sustainable relationship with nature. In the remainder of this Conclusions chapter, the lessons learned are presented along with alternatives and suggestions for addressing some of the challenges to establishing or maintaining BI.

## 27.3 Barriers to Implementation of Biodiversity Islands

While BI offer a promising and practical option for conserving and restoring biodiversity across human-dominated landscapes, they are not without challenges. Several barriers to establishing and proliferating BI at scale have been recognized within the chapters of this book. Some of these are conceptual in nature, relating to theoretical pitfalls of this particular framework, while others have been gleaned from the specific challenges encountered within the case studies examined.

### *27.3.1 Conceptual and Biophysical*

A key challenge in designing and managing BI, articulated first in this book by Montagnini et al., and later in several other chapters (Arroyo-Rodríguez et al. 2022, Clavo Peralta et al. 2022, Esbach et al. 2022, Kirby 2022, Laino et al. 2022, Negret et al. 2022, Santos-Gally and Boege 2022, to name a few), is the question of priorities and tradeoffs, many of which are inherent in any conservation approach.

Designing a BI for the protection of one target species may come at the expense of other species with different habitat requirements. Prescribing a specific BI design without considering its relative priority within the broader landscape or the value of alternative land uses that may be at stake can similarly lead to misguided planning or even undermine wider conservation objectives. BI must be designed and undertaken with careful attention to both broad and local contexts and objectives.

As a land-sharing approach, BI managers may feel pressure or criticism from both sides of the conservation-production spectrum: from one side that they do not do enough to support biodiversity requirements, while from the other side that there is too much conservation at the expense of production and human needs. A similar balancing challenge that applies to BI is what Sigman (2022) calls in her chapter the "Restoration Dilemma". That is, the need for restoration efforts, such as BI, to be highly site-specific and therefore resistant to scaling, while at the same time needing to be scaled up because of the magnitude of the challenge and the need for widespread adoption to realize their full potential benefits. These types of tradeoffs must be reckoned with in endeavoring to establish BI.

When planning BI, design and expectations may not always match with reality. For example with ancient woodland islands in the British countryside, as described by Kirby 2022, the extent and pattern of patches as perceived by researchers—i.e. what is mapped as woodland—may be smaller or larger than the actual patch size used by the species in reality. This may in turn influence the success of the BI.

Certain BI designs may involve specific ecological and physical parameters that prevent them from being replicated elsewhere. For instance, the urban residential gardens in Panama City detailed in Negret et al. occupy a unique proximity to a native forest patch that allowed for the gardens, though small, to function as BI. Efforts to establish similar residential gardens in other urban settings may not be as successful if they do not similarly benefit from a nearby native forest patch.

Designing and managing BI becomes even more difficult in the face of climate change. Newcomer et al. point out that there are many unknowns as conditions alter in the context of climate change, from whether biological corridors will or will not support species migration, to how climate impacts will affect the region's socioeconomic conditions, all of which have implications for the long-term sustainability of a BI.

### *27.3.2 Economic and Political*

Beyond these conceptual and biophysical issues, real economic and political constraints also make BI implementation and management a challenge. As with other ecosystem services, while many of the benefits of BI are enjoyed broadly, the costs are private, potentially making it a less appealing option for land managers without an additional source of funding to compensate for opportunity costs or otherwise incentivize conservation over other land uses. The BI approach seeks to promote integrated land management, but it still faces challenges, similar to other conservation efforts, in competing with more productive, and often less biodiversity friendly, alternatives that may offer faster and, at least initially, more tangible returns.

Establishing BI may have high start-up and operational costs, especially where they involve more labor intensive approaches or ones with specific objectives like maintaining phylogenetic diversity, such as the projects described by Santos-Gally and Boege in their chapter on native tree islands within neotropical silvopastoral systems. The cumulative costs related to seed collection, germinating and transplanting seedlings, establishing a nursery, and then managing the tree islands for competition and protection from cattle were significant in the study and these costs may be prohibitive to other producers, especially those with limited resources. Some of these costs may be recouped in the long run from additional benefits from ecosystem enhancements, but the startup capital required for the initial transition may be a barrier.

Incentive programs themselves may require a minimum level of conservation before becoming sustained, viable approaches. The case study of American ginseng by Sheban is a good example of a promising regulatory conservation tool of permit-based harvesting. However, it can only be effective in supporting the understory ginseng in the BI if supplemented by simultaneous forest farming to sufficiently reestablish populations. Maintaining a BI may require a set of additional mechanisms to be successful, which may be a challenging configuration of approaches to orchestrate.

At a more macro level, the establishment of BI is often in competition with more intensive production systems that have the potential to influence prices or major infrastructure decisions, which in turn create path dependencies that may lock in these less sustainable alternatives. In Laino et al.'s case study of livestock production systems in Paraguay, major roads were being planned in anticipation of higher yielding commodity production in the region, further jeopardizing the prospect of BI establishment and maintenance. Once these types of large-scale investments are made that enable more intensive production systems, it can become even more challenging for a producer to change to a different type of production system not dependent on and often in conflict with such aggressive and intrusive incursions on the landscape.

Conservation decisions are inherently political, and in certain contexts, the political landscape may be even more complex than the biophysical ones. Sigman draws attention to the dynamics in Ethiopia's Tigray region where large scale government-led restoration projects resulting in unique BI enjoyed support from a political "monoculture", i.e., single party leadership, that is now losing favor as the country becomes politically pluralized. The success of the BI is complicated by the complex communal labor realities and motivations that enabled these projects and which may not be available going forward to support restoration, at least in their current form, as the political landscape changes.

Similarly, although the Jupará Agroecological Movement in Brazil demonstrated the feasibility of a unique agroecological model of land reform, its success may be limited to only its local context unless broader historical, geographic and biophysical drivers are accounted for and integrated into supportive public policies (Painter et al. 2022). If the model is not supported by a wider enabling environment, it may continue to only rely on limited and potentially unstable external funding and its potential to scale up to support large-scale conservation of natural forest will be constrained.

### 27.3.3 Social and Cultural

In addition to these economic and political barriers, cultural and social preferences and norms may sometimes create resistance to BI. Farmers may have different attitudes or preferences that hinder their willingness to try new approaches like BI, including risk aversion, unfamiliarity, or pressure from existing social trends. In Vásquez et al.'s chapter on farmer perceptions of forest restoration practices in Panama, the main management strategy preferred among farmers was land clearing for cattle. Farmers tended to prefer familiar options or ones over which they have more control, such as tree-planting, which was a well-known practice in the region due to its predominance among various restoration programs across the Azuero peninsula.

A related barrier is the lack of necessary information and knowledge dissemination. This applies not only to knowledge of new and innovative approaches or models of BI, but also of past practices. The latter was the emphasis in the chapter by Clavo Peralta et al., where subsequent waves of migration from various parts of the country resulted in varied land uses over time in the Ucayali region of the Peruvian Amazon. While earlier local communities passed along knowledge of different uses of the diverse species maintained in the remaining forest fragments, more recent settlers did not have that same knowledge and therefore undervalued conservation and opted for practices that were more reflective of their originating regions. Several chapters in the book emphasize the importance of indigenous knowledge in supporting BI (Levin 2022a, b; Esbach et al. 2022), but without proper record and dissemination of that knowledge, its relevance risks being lost.

González et al. (2022) have noted in their chapter that the challenge is not simply to influence individual farmers' preferences, but rather to facilitate a deeper structural transformation that would replace the existing paradigm of competition and economic profit with one centered on cooperation and relationships based on mutual solidarity and concern. Switching from conventional systems to more agroecological approaches requires "a completely different mindset" that surrenders control of nature in favor of learning from nature, as Montes-Londoño et al. (2022) describe in their chapter on a silvopastoral case study in the Colombian Andes.

Furthermore, the success of BI is often contingent on a variety of stakeholders coming together around shared, or at least congruous, goals. Baez Schon et al., referring to the sacred natural sites that serve as BI in northern Ethiopia, point out the need for support from different involved/interested groups (e.g. the church, nearby communities, government) if they are going to be viable alternatives in the face of increasing economic and sociocultural pressures. Newcomer et al.'s case study of the Monteverde Reserve Complex in Costa Rica similarly describes the storied history of the multiple local reserves and the various sets of actors who came to be involved in their establishment and management, including Quaker settlers, community-based organizations, NGOs, schools, international research scientists, farmers, and eco-tourism organizations. While the unique constellation of actors in these case studies created just the right context for establishing this BI, the reliance on diverse stakeholders may pose a challenge in sustaining the BI into the future, especially if priorities and needs shift in the face of new pressures.

Many of these barriers presented throughout the book are not unique to BI, but are in fact issues that have challenged various approaches to conservation and are present in any undertaking that seeks to bring together sometimes divergent ecological, social, economic, and political goals and dynamics. Nonetheless, these barriers are not insurmountable. As many of the case studies have shown, given the right incentives, support and enabling policies, barriers can be overcome and BI can be scaled up to support conservation and production.

## 27.4 Viable Alternatives and Opportunities for Establishment of Biodiversity Islands

### *27.4.1 Community-Based Opportunities*

Biodiversity islands designed in partnership with local communities or indigenous groups can show the greatest potential for long-term success (Reyes-García et al. 2019). Social factors have been identified by restoration practitioners as having a far greater influence on the longevity of restoration projects than ecological factors, with multi-stakeholder engagement being the greatest challenge (Nerfa et al. 2021). For this reason, improving communication between policy makers, practitioners, and local communities is critical for improving BI implementation. Esbach et al. show that multi-stakeholder participation can be facilitated by intentionally integrating local partnerships and participatory research into conservation and development strategies. Participatory research can be fine-tuned to meet local needs while empowering communities to play active roles in developing solutions, demonstrating that the goals of local actors and BI are compatible.

Community based natural resource management can incentivize local communities to sustainably manage resources for their long-term availability, serving as an alternative to degradative cycles of exploitation. While community-based forestry generally has positive environmental and income related outcomes, it can sometimes inadvertently restrict the rights of communities to access forest resources (Hajjar et al. 2021). Structured engagement with local communities can help design projects that are more biodiverse, and also meet the needs of their multiple users (Dumont et al. 2019).

Forest and Farmer Producer Organizations such as community forestry user groups or producer cooperatives can help generate local support for sustainably managed BI. González et al. suggest that producer and consumer, or "prosumer", cooperatives, working across stakeholder levels, can help localized biodiverse agri-food systems. Community actors (both producers and consumers, among others) are the key to building locally managed agroecological systems. Levin (b) notes that cooperative business structures can also improve social outcomes such as farm-worker health and empowerment, core tenets of regenerative agriculture.

Education and capacity building are needed to further the implementation of BI globally. The chapter by Vásquez et al. (2022) shows that many rural farmers in Panama do not see assisted natural regeneration (ANR) as a restoration practice and therefore training farmers on the benefits of ANR could help scale up the restoration of degraded lands. Demonstration farms such as Hacienda Pinzacuá and El Hatico in Colombia, and Paradise Lot in Massachusetts, U.S.A. serve as examples of heightened productivity and ecological functioning, which can inspire other farms to adopt similar practices. Urban BI are also educational centers which can connect people to land and local ecosystems, from tropical Panama to the temperate Northeastern United States and Ushuaia, Argentina (Soler et al. 2022).

When working with local stakeholders, it is important to understand a community's diverse motivations for engaging in restoration and conservation, as demonstrated in different chapters of this book. Levin (b) notes that community-led action can be motivated by ethical, philosophical, scientific, cultural, and economic values. Morales-Nieves reveals that air quality and recreation were the highest priorities in an urban community forest in Puerto Rico. In Ethiopia, church forests are managed for their spiritual use (Baez Shon et al. 2022) but, as Sigman points out, the political "monoculture" may present a risk to restoration efforts sponsored by the political party in power. In Monteverde, Costa Rica, the local conservation movement is made more resilient by its many different motivations driving conservation (Newcomer et al. 2022). Painter et al.'s work in the Atlantic forest of Brazil shows that outside support can help BI meet community needs for production, conservation, and socioeconomic well-being when coupled with understandings of the motivations of community members, nuances of land tenure, and appropriate enabling conditions.

### *27.4.2 Current Restoration/Conservation Efforts Favoring Biodiversity Islands*

In spite of financial and other constraints faced by restoration and conservation projects, many examples have recently been brought to our attention showing evidence of a current and increasing trend of BI implementation. Several of these efforts are nurtured along by non-government organizations (NGOs) with limited funding and by private individuals and entities whose sole motivation relies on their own vocation and desire to preserve nature. A recent trend for the resurgence of agroecology strategies and experiences, as mentioned by Levin (a, b) and in the chapters by González et al. and by Painter et al., contributes to harmonizing production with conservation and promotes the spread of BI. In this subsection, a few examples are presented which illustrate that the implementation of new BI is already happening and gaining momentum. Further support and guidelines to assist their management and persistence are needed.

#### 27.4.2.1 Examples of Recently Launched NGOs and Private Restoration/Conservation Projects

Several business enterprises located in critical areas in need of solid conservation efforts, such as the Amazon, place a strong emphasis on biodiversity conservation while promoting restoration and sustainable agriculture practices, including agroforestry, to improve livelihoods of local people. For example, in the Ecuadorian Amazon, "Aliados" was formed in 2018 to build resilient community businesses based on supporting biodiversity in the Andes and the Amazon and to connect them to markets across the globe. Aliados restores and conserves landscapes in the Amazon rainforest along with local communities and in partnership with NGOs, private sectors and other key partners through promoting agroforestry and strengthening value chains (https://www.losaliados.org/). Their projects are located in the upper Napo Watershed in the Andes-Amazon foothills, in two critical ecological corridors at the crossroads of four National Parks composed of 150,000 hectares of megadiverse cloud and rain forest. They have designed their own Theory of Change, aiming towards reaching impact at the landscape level by combining their experience and network of community, corporate and philanthropic partners to build a regenerative agroforestry and restoration landscape for a fundamentally new model of doing business in the Amazon. Their actions contribute to the creation and maintenance of BI financed by their enterprise profits as well as from charitable donations and international funding.

Several recent efforts and initiatives worldwide to advance conservation have surged with the support of local conservation organizations and people. For example, the NRDC (Natural Resources Defense Council, https://www.nrdc.org), founded in 1970 by a group of law students and attorneys at the forefront of the environmental movement, is actively working on the 30x30 initiative: to prevent mass extinctions and bolster resilience to climate change, we must protect at least 30 percent of our lands, rivers, lakes, and wetlands by 2030. The 30x30 targets "...will help maintain global biodiversity, preserve the integrity of ecosystems on which we all depend, provide safe havens to help wildlife adapt to climate change, and sustain natural systems that store carbon, such as forests, mangroves, seagrasses, wetlands, and grasslands." NRDC is focused on achieving these goals while protecting diverse habitats, improving access for all people, and identifying and managing these areas in collaboration with indigenous communities (https://www.nrdc.org/30x30-nrdcs-commitment-protect-nature-and-life-earth). Their multifaceted strategy involves key areas of work in the United States and abroad. The Sierra Club and other international conservation organizations are joining this effort as well (Brett Levin, personal communication, May 2021).

A number of other recent private conservation projects have some alignment with the BI concept that we are promoting through this book. For example, Homegrown National Park is a grassroots organization in the U.S.A. calling for action to restore biodiversity and ecosystem function by planting native species and creating new ecological networks. Their mission is to restore biodiversity and ecosystem function, stemming from the realization that every human being on this planet needs diverse, highly productive ecosystems to survive. They are catalyzing a collective effort of individual homeowners, land managers, farmers, and "anyone with some soil to plant in...to start a new habitat by planting native plants and removing most invasive plants," claiming this represents the largest cooperative conservation project ever conceived or attempted in the country. Their goal is to reach 20 million acres (8,093,713 hectares) of native plantings in the U.S.A., an area that represents approximately half of the green lawns of privately-owned properties in the country (https://homegrownnationalpark.org/resources).

### *27.4.3 Private Protected Areas (PPAs)*

Private Protected Areas (PPAs) are areas of land or water that fulfill the conditions to be considered Natural Protected Areas (NPA) by the IUCN and that are managed by private governance (Mitchell et al. 2018, https://www.iucn.org/). All over the world there are families, communities and organizations that have decided to do something to change the current loss of natural areas and biodiversity and have begun to protect watershed headers and habitats of threatened species, restore degraded areas, develop education strategies, and promote positive contact with nature, among other initiatives that transform the way we relate to nature. Some of them have used tools of voluntary conservation, such as PPAs, with a view to getting greater formality and legal security to their ventures (Monteferri 2019). These areas hold BI that need management guidelines to ensure their efficacy and persistence.

In recent decades, the voluntary conservation movement on private lands has grown in different parts of the world, contributing to address the loss of biodiversity (Roldán et al. 2010). For example, a study in South Africa showed that, if PPAs were considered within the protected area system, results on estimations of species diversity would almost triple (Gallo et al. 2009). The level of consolidation and growth of voluntary conservation movements vary depending on each region. The private conservation movement in Latin America has been growing in recent years. At the 2018 Private Conservation Areas Congress for Latin America, this increase was made evident with a total of 4152 protected areas covering 4,618,042 hectares (Monteferri 2019). In Peru, there are a total of 1.5 million hectares of PPA in the whole country, including different forms of private conservation: PPAs as well as conservation and ecotourism concessions. Around 70% of these areas are located in the Peruvian Amazon, combining different legal tools for private conservation (Carolina Butrich, NGO "Conservamos por Naturaleza", personal communication, April 2021, https://spda.org.pe/wpfb-file/acp-en-peru-301-pdf/).

Owners and managers of PPAs often suffer from financial, logistical and other difficulties which pose a threat to their conservation efforts, as manifested by Víctor Zambrano, personal communication, April 2021. He owns a PPA in the Peruvian Amazon and works in the "Comité de Gestión de la Reserva Nacional Tambopata" (Management Committee of the Tambopata National Reserve), located in Madre de Dios, Peru. For Víctor, the main challenge is for landowners to find ways to generate long-term value without having to decrease land productivity and ecosystem services.

Voluntary conservation plays a key role as it creates a culture of conservation and makes it more accessible to all citizens. When managing land with conservation as a major purpose, agroforestry, agrobiodiversity and silvopasture gain space in the face of monocultures; permaculture and agricultural biodynamics become more important than chemical fertilizers; organic farming eliminates pesticide use that is generating drastic decreases in insects in the world (Monteferri 2019). Owners of PPAs explore ways of managing from a perspective of custodians, seeing nature as an ally rather than an obstacle.

Given the multitude of complex social and ecological challenges, there is an urgency to take action, and conservation at local and regional levels takes on unique importance (Morton 2013). Leadership at the local level will play a key role in the decades to come, as interconnectedness facilitates the replication of local initiatives, with communities becoming increasingly more informed and eager to see change. Voluntary conservation allows volunteers, companies, farmers, families, schools and universities to be recognized, participate and collaborate in preservation, and facilitates respect of biodiversity at all scales. To ensure their persistence and successful management, tax benefits could be provided to PPAs along with some measure of compensation from other commercial interests that benefit from the carbon sequestration potential and other ecosystem services they provide.

#### 27.4.3.1 Biodiversity Islands in Indigenous Territories

Land use systems made up of complex assemblages embedded in a forestry matrix, as is the case in many traditional indigenous sacred sites, agroforestry systems (Baez Schon et al. 2022; Gadgil et al. 2021), and indigenous territories can be considered BI in themselves, as explained in the Introductory chapter of this book (Montagnini et al. 2022). Known also as "Islands of Nature," they can be many square kilometers in size, generally use native forest species for sustainable food production and biodiversity, and are integrated with the natural forest. Several examples located in indigenous territories worldwide were well documented in a recent report by The Intergovernmental Science-Policy Platform on Biodiversity and

Ecosystem Services (IPBES) (United Nations IPBES 2019). The protection of these Islands of Nature is ensured as long as the indigenous peoples' territories and rights to use the land are respected, which often conflicts with development goals of other sectors. For example, protected territories inhabited by indigenous peoples in the Peruvian Amazon are threatened by road development, oil extraction, and other industries (Joseph Zárate, personal communication, April 2021, Zárate 2021). As oil exploration moves from industrialized countries to other locations such as in Peru and Ecuador, indigenous territories and their biodiversity face increasing threats.

Indigenous movements and their supporters, including the legal system in each affected country/territory, however, are actively seeking justice and winning court battles. For example, the Union of People Affected by Texaco (UDAPT, www.udapt.org) in Ecuador won a legal battle against the oil company, found to be guilty of egregious pollution and irresponsibility in the Ecuadorian rainforest. Ten years later, they are still fighting for the settlement that would benefit the tens of thousands of indigenous people who have been impacted (Julio Prieto, personal communication, May 2021). The case is explained in detail in a recent Forces for Nature podcast (https://forcesfornature.com/podcast/advocating-for-environmental-justice-in-the-ecuadorian-rainforest/).

In Ecuador, the Confederation of Indigenous Nationalities of the Ecuadorian Amazon (La Confederación de las Nacionalidades Indígenas de la Amazonia Ecuatoriana), or CONFENIAE, is the regional organization of indigenous peoples in the Oriente region of the Ecuadorian Amazon. Nine indigenous peoples living in the region—Quichua, Shuar, Achuar, Huaorani, Siona, Secoya, Shiwiar, Záparo, and Cofán—are represented politically by the Confederation (http://www.confeniae.org.ec/). CONFENIAE is one of three major regional groupings that constitute the Confederation of Indigenous Nationalities of Ecuador (CONAIE). It is also part of the Amazon Basin indigenous organization, COICA (http://COICA.org.ec). While these organizations get some technical and financial assistance from several NGOs and other groups, and the indigenous people appreciate their help, they often would prefer greater autonomy in their decision making (Efren Nango, CONFENIAE, personal communication, February 2021). The indigenous peoples and their allies are challenged to not allow external forces to divide and disrespect their organization as they advance efforts to exert their rights to the land and biodiversity, including the embedded BI within their territories.

#### 27.4.3.2 Examples of Ongoing Agroecological Initiatives Supporting Biodiversity Islands

New ways of practicing agroecology to harmonize food production with conservation often lead to the spread of BI, as described in both chapters by Levin, and by González et al. (2022). These practices and experiences carried along by different groups of people are spreading, managing to overcome financial, technical training and other constraints. For example, recent presentations on agroecology and biodiversity by local biologists and agronomists were offered to local farmers who are transitioning to agroecological production systems, in an event taking place at the agroecological farm "La Dorita" located in Basavilbaso, Entre Ríos province, Argentina (Libertario González, personal communication, May 2021). These are small BI which are delineated and protected to fulfill local needs.

Individual farmers worldwide who practice agroecology in a variety of ways, including agroforestry, contribute to the creation of BI through their individual efforts and often using their own financial resources, as described in the chapters mentioned above as well as in other chapters in this volume from Calle et al., Esbach et al., Montes-Londoño et al., Painter et al., and Toensmeier. An agricultural producer in the state of Zulia, Venezuela who manages a 1000-hectare farm including silvopastoral systems with buffalo, many saman (*Samanea saman*) trees that produce feed for cattle, secondary forests of different ages, as well as areas with bamboo and oil palm, was seeking our advice on how to design and manage BI after watching a program on CNN[1] (Wilmer Morán, personal communication, April 2021). Thus, our book hopefully will provide information to help individual efforts like this one to design, protect and manage BI.

#### 27.4.3.3 Recent Government Policies Promoting Agroecological Practices

In some cases agroecological transitions from conventional agriculture are already being supported by government initiatives that go along with local agroecology movements, even in countries like Argentina and the U.S.A., where industrial agriculture for large scale production of grains is the norm. These initiatives are a response to the environmental issues created by monoculture agriculture as mentioned earlier in this chapter.

In Argentina, where export taxes on soybean production comprise a large portion of the Gross Domestic Product, the Argentine Society of Agroecology (SAAE) was created in 2018 and has been pivotal in the consolidation of agroecology in the country (https://es-la.facebook.com/pages/category/Interest/SAAE-Sociedad-Argentina-de-Agroecolog%C3%ADa-126879274899065/). Brazil and Argentina are the only two Latin American countries that have constituted a National Society of Agroecology. The First Argentine Congress of Agroecology was held in 2019 (https://fcagr.unr.edu.ar/?p=14040). Training events to lead the transition to agroecology aimed at producers and technical personnel have been organized by the National Institute of Agricultural Technology, INTA (INTA Procadis, https://inta.gob.ar/acerca-de-procadis). In the Buenos Aires province, by April 2021 there were 350 farmers registered as agroecological, comprising a total of about 40,000 hectares

[1] Interview, CNN en Español, from CNN studio in Atlanta, GA, U.S.A., aired March 27–28, 2021, "Pandemia, Biodiversidad y Futuro" ("Pandemic, Biodiversity and the Future") in GloboEconomía, a CNN program with José Antonio Montenegro, a Warner Media production, https://www.youtube.com/watch?v=f68tzy65zq4, or search YouTube:"Cómo afectó el ser humano a la biodiversidad"

distributed all over the province, including large and small producers (Germán Lanzer, Director of Agroecology for the Buenos Aires province, Transition to Agroecology course offered by INTA Procadis, April 2021).

In 2020, the Ministry of Agriculture, Fisheries and Livestock of Argentina created the National Direction of Agroecology, within the Secretariat of Food, Bioeconomy and Regional Development (https://www.revistainternos.com.ar/2020/08/=). Its primary objective is "to intervene in the design and implementation of policies, programs and projects that promote intensive and extensive agroecologically-based primary production at all levels." Interacting with producers, agricultural organizations and municipal and provincial governments, they are executing a Strategic Productive Transition Plan for agroecological implementation, providing technical assistance, and establishing credits or tax tools for its promotion.

Also in Argentina, the National Network of Municipalities and Communities that promote Agroecology (Renama) works with some thirty municipalities in the different productive regions of the country (www.renama.org). Renama has incorporated about 170 producers so far, with an estimated 86,000 hectares under agroecological management. "This is not an alternative practice. On the contrary, it is the agriculture of the coming years" (Eduardo Cerdá, Director of the Division of Agroecology, and also president of Renama, 2020 interview with InterNos, local media from Buenos Aires, Argentina). In the same interview he added: "There's a strong paradigm shift. The current production model was important at the time, when new fertilizers, herbicides and strong technological innovations resulted in good production increases. But the continuity of this model has brought us many problems in terms of soil losses (according to INTA, more than 50% of the country's soil organic matter has been lost) and increases in herbicide-resistant weeds. This has brought along the use of more agrochemicals to control them, which in turn has increased costs above profits, leaving a lot of producers out of the game. In addition, producers realize what it means to be in contact with these substances which were claimed not to generate acute poisoning. But it turns out that the problem was their chronic toxicity, that is, toxicity generated from long exposure to the products. This is a needed change of paradigm from the perspective of improving nutrition and increasing resilience in the face of the current pandemic" (http://www.revistainternos.com.ar/2020/03/actual-modelo-de-produccion-es-drogodependiente/).

Dissemination of agroecological practices and regenerative agriculture in all their forms that are environmentally friendly, harmonizing productivity and environmental goods and services, is driving the promotion and persistence of BI in human dominated environments. There is already enough traditional and scientific knowledge among farmers and international and local institutions as we have described in several chapters of this book (e.g., Baez Schon et al., Clavo Peralta et al., Levin (a), Montagnini et al., Montagnini and del Fierro). These practices need to be promoted using incentives, especially in the early years of implementation, until adequate production levels are reached. These incentives can take many forms such as soft loans, materials, and tools, and should include education and technical assistance at all levels.

In the United States, the Department of Agriculture (USDA) has long provided technical and financial assistance and other resources to farmers and ranchers via agencies including the Natural Resources Conservation Service and the National Agroforestry Center. These agencies and their programs help landowners implement practices that conserve and restore the natural resources on their land and in their operations. This assistance is provided through a variety of different incentive programs including the Environmental Quality Incentives Program (EQIP), which offers financial resources and one-on-one help to plan and implement improvements that address natural resource concerns. The Conservation Reserve Program provides an annual payment for removing environmentally sensitive land from agricultural production and planting species that will improve environmental health and quality (https://www.fsa.usda.gov/programs-and-services/conservation-programs/conservation-reserve-program/; https://www.nrcs.usda.gov/wps/portal/nrcs/main/national/programs/financial/eqip/).

With the United States re-entering the Paris Agreement of the United Nations Framework Convention on Climate Change in 2021, renewed attention on its international climate mitigation commitments may help bolster these working lands conservation programs in recognition of the climate mitigation opportunities they could potentially provide. Via executive order, the Biden-Harris administration recently directed the USDA to develop a strategy to encourage the voluntary adoption of climate-smart agricultural and forestry practices that result in additional, measurable, and verifiable carbon reductions and sequestration. While the contours of this strategy are still being developed, the strategy will be designed to accommodate all farmers, and will seek to integrate climate-smart practices into existing programs or create new programs that strengthen markets for agriculture and forestry products generated through climate-smart agriculture and forestry practices, including via support of private voluntary carbon markets (https://www.usda.gov/sites/default/files/documents/climate-smart-ag-forestry-strategy-90-day-progress-report.pdf). Agroforestry practices, including windbreaks, buffers, alley cropping, silvopasture and multi-story cropping, as well as other agroecological practices such as cover crops and no-till, are among the practices being considered for inclusion in a potential new climate-smart agriculture and forestry program, which may facilitate the establishment of BI on working lands in the U.S.A.

### *27.4.4 Valuing and Financing Biodiversity Conservation*

A large disparity exists between the resources needed and the available funds for financing conservation, and this has created a dire situation for the future of biodiversity conservation, as discussed in a recent interview published by the Nature Conservancy: it has been estimated that by 2030 the world needs as much as an additional US$598 billion to $824 billion annually to close the financing gap (Solberg 2021). Financing biodiversity on its own is a difficult task, considering that the effects and payoffs of such conservation are not always tangible, and that conservation of biodiversity is a multidimensional global issue. Financing biodiversity can become increasingly easier, however, when tangible and valued resources such as carbon or water are incorporated into conservation strategies (Sheban 2022).

The first step towards building these tangible and effective conservation strategies is the proper assessment and valuation of ecosystem services. Proper valuation is essential in developing new initiatives, and programs such as Payments for Ecosystem Services (PES) may be efficient in encouraging future biodiversity conservation through compensation. PES programs have considerably increased in recent years, although biodiversity-focused PES programs have remained limited. Compensatory mitigation banking is also growing, while voluntary biodiversity offsets are a recent policy development and have yet to experience large amounts of growth (Salzman et al. 2018). Valuation systems that bundle multiple services could better encourage sustainable land uses to farmers and protect biodiversity, even if the PES is only provided for a single service such as water or carbon (Montagnini and Finney 2011). Biodiversity objectives might also be incorporated into future ecological action programs.

Another opportunity for funding conservation is through taxing certain activities that are detrimental to worldwide ecosystems, such as imposing emissions taxes on airline travel to discourage fossil fuel consumption or implementing a variety of "polluter pays" taxation programs. These funds could in turn be used to finance conservation initiatives such as the implementation and maintenance of BI. If the financial penalties were established at sufficient levels, perhaps industries would self-regulate to address environmental degradation and avoid paying such taxes. Such an approach might accelerate progress for sustainability by directing pollution tax funding toward implementing BI and carbon sequestration projects. As Newcomer et al. described in their chapter in this volume, such an approach has seen moderate success in Costa Rica with gas taxes being used to fund the national PES program.

As we have described, the effort to integrate biodiversity into the market is already growing, with investments worldwide coupled with positive environmental outcomes gaining in popularity. Agricultural and consumer markets are increasingly shifting products and supply chains to align with conservation goals. Because BI establishment generally happens on fragmented landscapes, funding and acquiring these small patches of land through government action and policy could satisfy both land development and biodiversity conservation goals in the future, and may be more effective than the approach of designating large amounts of land solely for conservation initiatives. Preserving numerous small areas of land for BI conservation rather than rendering large swaths of land economically unusable seems mutually beneficial for governments and for landowners and can contribute to conservation goals as well.

Voluntary conservation actions from companies, farmers, families, schools and universities should be financially recognized as they participate and collaborate in creating BI at all scales. Financial incentives to ensure successful management and persistence of privately established BI may take many forms such as loans, soft credits, and the provision of materials and tools. In particular, tax benefits could be provided to Private Protected Areas as well as to landowners who protect sensitive areas within their properties, for example, riparian forests that contribute to landscape connectivity and other ecosystem services as shown by Giraldo et al. (2022). This type of incentive can be calculated based on the area covered by the BI, or by valuing the ecological importance of a particular area for conservation purposes. Other compensation could be derived from other commercial interests such as companies or enterprises that benefit from the carbon sequestration potential and other ecosystem services provided by BI.

### *27.4.5 The Framework: Current Trends in Biodiversity Conservation Aligning with the Biodiversity Islands Concept*

There have been several trends and useful perspectives on how protected area design issues could be addressed more effectively (McNeely 2021). Bengtsson et al. (2003) highlighted some of the limitations of protected areas, which covered just over 11% of the land at that time. While the landscape perspective and the integration of resilience theory and biodiversity conservation is now much more reflected in biodiversity policy (e.g., see IPBES 2019), improvements in conservation practice and landscape management have recently been quite small (Bengtsson et al. 2021). The CBD's Strategic Plan for Biodiversity 2011–2020 was highly relevant to protected areas as a major conservation tool (Pimm 2021; Sayer et al. 2021). Its target 11 called for establishing at least 17% of terrestrial and inland water biomes as protected areas, along with 10% of coastal and marine biomes. These were to be effectively and equitably managed, ecologically representative, and well connected as parts of systems of effective area-based conservation measures that are integrated into wider landscapes and seascapes.

At the time, however, protected areas were considered too static when they need to be more dynamic (Bengtsson et al. 2003, 2021; Mc Neely 2021). A dynamic landscape approach that mimics natural disturbance regimes could include, for example, successional lands that are recovering from over-exploitation. These resemble the territories managed by indigenous peoples such as the shifting cultivation practices described by Gadgil et al. (2021). Indigenous territories or parts of them where sustainable multistrata agroforestry and forest systems are practiced are considered BI as defined in this book. The importance of landscape connectivity is a current hot subject in the conservation mainstream (McNeely 2021), with detailed guidelines prepared by an IUCN international team (Hilty et al. 2020). Landscape connectivity is discussed in this book in relation to the effectiveness of BI.

The Aichi targets set in 2010 have led to the expansion of systems of national parks and other categories of protected areas, and over 90 of the signatory countries have attained the 17% target. However, some governments have established protected areas in degraded and marginal areas that have limited conservation value, often because those protected areas simply were not needed for anything else. There is still not enough evidence of the effectiveness of protected landscapes in delivering biodiversity outcomes (Sayer 2021). The value of landscape approaches to conserve biodiversity through management of the broader landscape within which conventional protected areas are located is expected to increase (Sayer et al. 2021). This concept aligns with the integrated landscape management approaches suggested in this book.

The fifteenth meeting of the Convention on Biological Diversity (CBD) Conference of the Parties (COP), Kunming, China, October 2021 has established seven thematic programs of work corresponding to some of the major biomes on the planet. Each program establishes a vision and basic principles to guide future work. They also set out key issues for consideration, identify potential outputs, and suggest a timetable and means for achieving them. Implementation of these programs depends on contributions from signatory parties, the Secretariat, and relevant intergovernmental and other organizations (https://www.cdb.int). CBD COP-15 is likely to adopt relatively simple, aspirational, and politically attractive targets for biodiversity conservation (Sayer et al. 2021). Civil society is likely to prefer enhanced conservation measures, but people are often reluctant to accept actions that restrict their material wellbeing, thus conservation strategies need to be scientifically sound and aligned with the cultures and economies of local societies such as is described for BI in this book. Research, training, and capacity building are needed to manage programs of biodiversity conservation (Sayer et al. 2021). This also pertains to BI, and this book has set the basis for implementing and successfully managing BI using an inclusive, landscape oriented and integrated approach.

It has been broadly recognized that solutions to the problems of biodiversity conservation come down to working with people, their lives, aspirations, fears, and social complexity. Although people and nature will not always peacefully coexist, rigorous science and intelligent technology can help (Berlyn 2021; Pimm 2021). Evidence from chapters of this book shows that BI may constitute such a solution when properly designed and managed within the social milieu where they are embedded.

### *27.4.6 The Context: Biodiversity Islands in Changing Environments*

Climate change is increasingly a threat to both natural and human systems. The success of BI as effective conservation solutions is vulnerable to changing climate conditions, including via impacts on habitat requirements and species ranges, as well as the cascading effects of shifting socioeconomic demands. At the same time, however, BI have an important role to play in contributing to climate change solutions. Land use, including agriculture and forestry, accounts for an estimated 23% of total anthropogenic greenhouse gas emissions (Shukla et al. 2019).

Reducing deforestation and introducing, or reintroducing, trees and herbaceous cover into working lands not only benefit biodiversity, but can have outsized impacts on greenhouse gas emissions via conserved carbon stocks, reduced emissions, and increased carbon sequestration. These so-called natural climate solutions, which include conservation, restoration and improved agricultural practices, can contribute up to one-third of the cost-effective climate mitigation we need with the potential to remove as much as 23.8 petagrams of $CO_2$ equivalent (Griscom et al. 2017).

Co-benefits of conservation, including the livelihood support they provide and ecosystem services, such as watershed regulation, can also help increase the resilience of communities impacted by climate change (Griscom et al. 2017). Given the magnitude and urgency of this challenge, curbing climate change and limiting global temperature rise to below the 2 °C degree threshold set by the Paris Agreement will require ambitious and radical changes in our production systems. As an integrated, flexible conservation approach with great potential, BI can be an important piece of the nature-based solutions necessary to address the climate crisis.

## 27.5 Conclusions: Biodiversity Islands in Action

The lessons learned from the chapters of this book form a collection of experiences showing positive and promising results of how BI can be implemented and managed. Through a series of informative case studies and detailed explanations of the intricacies of BI creation, the contributions to this volume provide a comprehensive context for the impediments and opportunities of BI implementation amidst various societal and environmental factors. Several chapters in the book discuss the challenges that BI development face, including a lack of conceptual understanding of BI function, incompatibility with current local societal practices or government priorities, and competing economic/productive land use.

Many practical alternatives are presented that address these challenges through the creation of BI. Several of these opportunities relate to communities' motivations, which is why addressing social issues and improving communication are critical for establishing effective and lasting BI. The book presents a variety of examples of BI created in differing circumstances, from NGO action to AFS implementation, where involvement of local and indigenous community members plays a vital role.

BI present an opportunity for sustainable, dynamic, productive conservation, which is why they are becoming increasingly desirable in the global conservation movement. As societies look for alternatives to maintain economic prosperity, provide culturally-important community spaces, and conserve local ecosystem biodiversity, BI are sure to become more widely used, which underscores the importance of the information described in this book as a tool for planning and implementing BI.

## References

Afelt A, Frutos R, Devaux C (2018) Bats, coronaviruses, and deforestation: toward the emergence of novel infectious diseases? Front Microbiol 9:Article 702. https://doi.org/10.3389/fmicb.2018.00702

Arroyo-Rodríguez V, Arasa-Gisbert R, Arce-Peña N, Cervantes-López MJ, Cudney-Valenzuela SJ, Galán-Acedo C, Hernández-Ruedas MA, San-José M, Fahrig, L (2022) The importance of small rainforest patches for biodiversity conservation: a multi-taxonomic assessment in the Lacandona region, Mexico. In: Montagnini F (ed) Biodiversity islands: strategies for conservation in human-dominated environments. Series: Topics in Biodiversity and Conservation. Springer, Cham, pp 41–60

Baez Schon M, Woods CL, Cardelús CL (2022) Sacred Church forests in northern Ethiopia: biodiversity and cultural islands. In: Montagnini F (ed) Biodiversity islands: strategies for conservation in human-dominated environments. Series: Topics in Biodiversity and Conservation, Springer, Cham, pp 531–549

Bengtsson J, Angelstam P, Elmqvist T, Emanuelsson U, Folke C, Hise M, Moberg F, Nyström M (2003) Reserves, resilience and dynamic landscapes. Ambio 32:389–396. https://doi.org/10.1579/0044-7447-32.6.389

Bengtsson J, Angelstam P, Elmqvist T, Emanuelsson U, Folke C, Ihse M, Moberg F, Nyström M (2021) Reserves, resilience and dynamic landscapes 20 years later. 50th anniversary collection: biodiversity conservation. Ambio 50:962–966. https://doi.org/10.1007/s13280-020-01477-8

Berlyn GP (2021) Some thoughts on biodiversity and sustainability. J Environ Sci Public Health 5(3):371–373. https://doi.org/10.26502/jesph.96120136

Bern L (2018) Den metabola pandemin: det största hotet mot en hållbar framtid. Recito Publishers, Sweden, 481 pp https://www.bokus.com/bok/9789177653059/den-metabola-pandemin-det-storsta-hotet-mot-en-hallbar-framtid/

Calle Z, Molina CCH, Molina DCH, Molina DEJ, Molina EJJ, Murgueitio B, Murgueitio A, Murgueitio E (2022) A highly productive biodiversity island within a monoculture landscape: El Hatico nature reserve (Valle del Cauca, Colombia). In: Montagnini F (ed) Biodiversity islands: strategies for conservation in human-dominated environments. Series: Topics in Biodiversity and Conservation. Springer, Cham, pp 279–304

Clavo Peralta ZM, Vela Alvarado JW, Alvez-Valles CM (2022) Biodiversity islands and dominant species in agricultural landscapes of the Western South Amazon, Peru. In: Montagnini F (ed) Biodiversity islands: strategies for conservation in human-dominated environments. Series: Topics in Biodiversity and Conservation. Springer, Cham, pp 207–235

Crespin SJ, Simonetti JA (2019) Reconciling farming and wild nature: integrating human–wildlife coexistence into the land-sharing and land-sparing framework. Ambio 48:131–138. https://doi.org/10.1007/s13280-018-1059-2

Dumont ES, Bonhomme S, Pagella TF, Sinclair FL (2019) Structured stakeholder engagement leads to development of more diverse and inclusive agroforestry options. Exp Agric 55(S1): 252–274. https://doi.org/10.1017/S0014479716000788

Eibl B, Montagnini F, López LN, Romero HF, Dummel CJ, Küppers G, López MA, Lavignolle P, Cortes CJ, De La Vega MA (2022) Conservation, connectivity and registration of seed areas in remnants of natural reserves in the province of Misiones, Argentina. In: Montagnini F (ed) Biodiversity islands: strategies for conservation in human-dominated environments. Series: Topics in Biodiversity and Conservation. Springer, Cham, pp 157–181

Esbach MS, Puri M, Botero-Arias R, Loiselle BA (2022) Beyond the island: safeguarding biodiversity requires integrated approaches. In: Montagnini F (ed) Biodiversity islands: strategies for conservation in human dominated-environments. Series: Topics in Biodiversity and Conservation. Springer, Cham, pp 551–568

FAO (2015a) Status of the World's Soil Resources. Main report. Prepared by Intergovernmental Technical Panel on Soils (ITPS) http://www.fao.org/3/a-i5199e.pdf

FAO (2015b) Los Pueblos Indígenas y las Políticas Públicas de Seguridad Alimentaria y Nutricional en América Latina y el Caribe. Una visión regional. http://www.fao.org/3/i4678s/i4678s.pdf

FAO (2019) Informe de Seguridad Alimentaria. http://www.fao.org/3/ca6030es/ca6030es.pdf

Gadgil M, Berkes F, Folke C (2021) Indigenous knowledge: from local to global. 50th anniversary collection: biodiversity conservation. Ambio 50:967–969. https://doi.org/10.1007/s13280-020-01478-7

Gallo J, Pasquin L, Reyers B, Cowling RM (2009) The role of private conservation areas in biodiversity representation and target achievement within the little Karoo region, South Africa. Biol Conserv 142(2):446–454

Giraldo LP, Chará J, Calle Z, Chará-Serna AM (2022) Riparian corridors: longitudinal biodiversity islands in farming landscapes. In: Montagnini F (ed) Biodiversity islands: strategies for conservation in human-dominated environments. Series: Topics in Biodiversity and Conservation. Springer, Cham, pp 139–156

González LH, Neffeng G, Benedetto V, Sánchez Miñarro M, García A, Biasatti R, Rimoldi P, Alesio C, Paiz D (2022) The anthropization of the natural landscape in the Pampa region in Santa Fe province, Argentina. Resilience of the environment, opportunity for a change of the agri-food paradigm. In: Montagnini F (ed) Biodiversity islands: strategies for conservation in human-dominated environments. Springer, Cham, pp 353–386

Griscom BW, Adams J, Ellis PW, Houghton RA, Lomax G, Miteva DA, Schlesinger WH, Shoch D, Siikamäki JV, Smith P, Woodbury P, Zganjar C, Blackman A, Campari J, Conant RT, Delgado C, Elias P, Gopalakrishna T, Hamsik MR, Herrero M, Kiesecker J, Landis E, Laestadius L, Leavitt SM, Minnemeyer S, Polasky S, Potapov P, Putz FE, Sanderman J, Silvius M, Wollenberg E, Fargione J (2017) Natural Climate Solutions. Proc Natl Acad Sci 114(44):11645–11650. https://doi.org/10.1073/pnas.1710465114

Hajjar R, Oldekop JA, Cronkleton P, Newton P, Russell AJ, Zhou W (2021) A global analysis of the social and environmental outcomes of community forests. Nat Sustain 4(3):216–224

Hilty J, Worboys G, Keely A, Woodley S, Lauche B, Locke H, Carr M, Pulsford I, Pittock J, White JW, Theobald DM, Levine J, Reuling M, Watson JEM, Ament R, Tabor GM (2020) Guidelines for conserving connectivity through ecological networks and corridors. Best Practice Protected Area Guidelines Series No. 30. Gland, Switzerland: IUCN

IPBES (2019) Summary for policymakers of the global assessment report on biodiversity and ecosystem services of the Intergovernmental Science-Policy Platform on Biodiversity and Ecosystem Services. IPBES Secretariat, Bonn. https://ipbes.net/global-assessment

Karger DN, Kessler M, Lehnert M, Jetz W (2021) Limited protection and ongoing loss of tropical cloud forest biodiversity and ecosystems worldwide. Nat Ecol Evol 5:854–862. https://doi.org/10.1038/s41559-021-01450-y

Kirby K (2022) Islands of trees in long-fragmented landscapes in Great Britain. In: Montagnini F (ed) Biodiversity islands: strategies for conservation in human-dominated environments. Series: Topics in Biodiversity and Conservation. Springer, Cham, pp 337–352

Laino R, Musalem K, Laino L, Caballero A, Bueno D, Aranda L, Esquivel A, Ferreira Riveros M, Romero Nardelli L, Cantero N, Irala R (2022) Island forests among savannahs: conservation of biodiversity in rangelands of the Paraguayan humid Chaco. In: Montagnini F (ed) Biodiversity islands: strategies for conservation in human-dominated environments. Series: Topics in Biodiversity and Conservation. Springer, Cham, pp 185–205

Levin B (2022a) Regenerative agricultural systems as biodiversity islands. In: Montagnini F (ed) Biodiversity islands: strategies for conservation in human-dominated environments. Series: Topics in Biodiversity and Conservation. Springer, Cham, pp 61–88

Levin B (2022b) The role of community led action for the creation of biodiversity islands. In: Montagnini F (ed) Biodiversity islands: strategies for conservation in human-dominated environments. Series: Topics in Biodiversity and Conservation. Springer, Cham, pp 487–504

Loh EH, Zambrana-Torrelio C, Olival KJ, Bogich TL, Johnson CK, Mazet JAK, Karesh W, Daszak1 P (2015) Targeting transmission pathways for emerging zoonotic disease surveillance

and control. Vector-Borne and Zoonotic Diseases 15 (7):432–437, https://doi.org/10.1089/vbz.2013.1563
Martín E (2020) Antología Toxicológica del Glifosato. Naturaleza de derechos. Argentina. 5th ed., 269 pp. Available at: https://imagenagropecuaria.com/revista/wp-content/uploads/2020/05/antologia5.pdf
McNeeley JA (2021) How to conserve biological diversity: perspectives from Ambio. 50th anniversary collection: biodiversity conservation. Ambio 50:957–961. https://doi.org/10.1007/s13280-020-01479-6
Mitchell BA, Stolton S, Bezaury-Creel J, Bingham HC, Cumming TL, Dudley N, Fitzsimons JA, Malleret-King D, Redford KH, Solano P (2018) Guidelines for privately protected areas. Best Practice Protected Area Guidelines Series No. 29. UICN, Gland
Montagnini F, Berg KE (2019) Feeding the world. In: Carver F, Gossington H, Manuel C (eds) Sustainable Development Goals. Transforming our world. Witan Media Ltd, Painswick, UK, United Nations Association – UK, London, pp 98–100. https://www.sustainablegoals.org.uk/feeding-the-world/
Montagnini F, del Fierro S (2022) Functions of agroforestry systems as biodiversity islands in productive landscapes. In: Montagnini F (ed) Biodiversity islands: strategies for conservation in human-dominated environments. Series: Topics in Biodiversity and Conservation. Springer, Cham, pp 89–116
Montagnini F, Finney C (2011) Payments for environmental services in Latin America as a tool for restoration and rural development. Ambio 40:285–297
Montagnini F, Levin B, Berg KE (2022) Introduction. Biodiversity islands: strategies for conservation in human-dominated environments. In: Montagnini F (ed) Biodiversity islands: strategies for conservation in human dominated environments. Series: Topics in Biodiversity and Conservation. Springer, Cham, pp 1–35
Monteferri B (Ed) (2019) Áreas de conservación privada en el Perú: avances y propuestas a 20 años de su creación. Sociedad Peruana de Derecho Ambiental, Lima, Peru
Montes-Londoño I, Calle A, Montes-Calderon A, Montes-Botero O (2022) Hacienda Pinzacuá: an island of regenerative agriculture in the piedmont of the Colombian Central Andes. In: Montagnini F (ed) Biodiversity islands: strategies for conservation in human-dominated environments. Series: Topics in Biodiversity and Conservation, Springer, Cham, pp 305–335
Morales-Nieves GM (2022) Priorities, perspectives and use of a community forest by surrounding residents in Mayagüez, Puerto Rico: protecting the forest for its services. In: Montagnini F (ed) Biodiversity islands: strategies for conservation in human-dominated environments. Series: Topics in Biodiversity and Conservation, Springer, Cham, pp 505–530
Morton T (2013) Hyperobjects: philosophy and ecology after the end of the world. University of Minnesota Press, Minneapolis
Negret HRC, Negret R, Montes-Londoño I (2022) Residential garden design for urban biodiversity conservation: experience from Panamá City, Panama. In: Montagnini F (ed) Biodiversity islands: strategies for conservation in human-dominated environments. Springer, Cham, pp 387–417
Nerfa L, Wilson SJ, Reid JL, Rhemtulla JM (2021) Practitioner views on the determinants of tropical forest restoration longevity. Restor Ecol 29(3):e13345. https://doi.org/10.1111/rec.13345
Newcomer Q, Camacho Céspedes F, Stallcup L (2022) The Monteverde Cloud Forest: evolution of a biodiversity island in Costa Rica. In: Montagnini F (ed) Biodiversity islands: strategies for conservation in human-dominated environments. Series: Topics in Biodiversity and Conservation, Springer, Cham, pp 237–278
Niella F, Rocha P, Tuzinkievicz A, Buchweis R, Bulman C, Thalmayr P, González J, Montagnini F, Sharry S (2022) Native multipurpose species in the Yaboti Biosphere Reserve, Misiones-Argentina. In: Montagnini F (ed) Biodiversity islands: strategies for conservation in human-dominated environments. Series: Topics in Biodiversity and Conservation, Springer, Cham, pp 461–483
Painter K, Buschbacher R, Souto Silva LC, Costa e Silva E (2022) Agroecology and Forest Conservation in Three Types of Land Reform Community in the Cacao Region of Bahia,

Brazil. In: Montagnini F (ed) Biodiversity islands: strategies for conservation in human-dominated environments. Series: Topics in Biodiversity and Conservation, Springer, Cham, pp 569–599

Pimm S (2021) What is biodiversity conservation? 50th anniversary collection: biodiversity conservation. Ambio 50:976–980. https://doi.org/10.1007/s13280-020-01399-5

Pimm SL, Jenkins CN, Abell R, Brooks TM, Gittleman JL, Joppa LN, Raven PH, Roberts CM, Sexton JO (2014) The biodiversity of species and their rates of extinction, distribution, and protection. Science 344:1246752

Reyes-García V, Fernández-Llamazares Á, McElwee P, Molnár Z, Öllerer K, Wilson SJ, Brondizio ES (2019) The contributions of indigenous peoples and local communities to ecological restoration. Restor Ecol 27(1):3–8. https://doi.org/10.1111/rec.12894

Roldán M, Carminati A, Biganzoli F, Paruelo JM (2010) Las reservas privadas ¿son efectivas para conservar las propiedades de los ecosistemas? Ecol Austral 20(2):89–221

Salzman J, Bennett G, Carroll N, Goldstein A, Jenkins M (2018) The global status and trends of payments for ecosystem services. Nat Sustain 1:136–144

Santos-Gally R, Boege K (2022) Biodiversity islands. Native tree islands within silvopastoral systems in a neotropical region. In: Montagnini F (ed) Biodiversity islands: strategies for conservation in human-dominated environments. Series: Topics in Biodiversity and Conservation, Springer, Cham, pp 117–138

Sayer J, Margules C, McNeely JA (2021) People and biodiversity in the 21st century. 50th anniversary collection: biodiversity conservation. Ambio 50:970–975. https://doi.org/10.1007/s13280-020-01476-9

Sheban K (2022) Preserving biodiversity in Appalachian mixed mesophytic forest through permit-based harvest of ginseng and other forest botanicals. In Montagnini F (ed) Biodiversity islands: strategies for conservation in human-dominated environments. Series: Topics in Biodiversity and Conservation, Springer, Cham, pp 601–628

Shukla PR, Skea J, Calvo Buendia E, Masson-Delmotte V, Pörtner HO, Roberts DC, Zhai P, Slade R, Connors S, Van Diemen R, Ferrat M (2019) Climate change and land: an IPCC special report on climate change, desertification, land degradation, sustainable land management, food security, and greenhouse gas fluxes in terrestrial ecosystems. IPCC

Sigman E (2022) Safeguarding the benefits of biodiversity islands in northern Ethiopia in the midst of political change. In: Montagnini F (ed) Biodiversity islands: strategies for conservation in human-dominated environments. Series: Topics in Biodiversity and Conservation, Springer, Cham, pp 647–674

Solberg D (2021) Closing the gap. Nature Conservancy Magazine Summer 2021: 38–40, Nature.org/Magazine

Soler R, Benítez J, Solá F, Lencinas MV (2022) Biodiversity islands at the world's southernmost city: plant, bird and insect conservation in urban forests and peatlands of Ushuaia, Argentina In: Montagnini F (ed) Biodiversity islands: strategies for conservation in human-dominated environments. Series: Topics in Biodiversity and Conservation, Springer, Cham, pp 419–437

Toensmeier E (2022) Paradise Lot: A Temperate-Climate Urban Agroforestry Biodiversity Island. In: Montagnini F (ed) Biodiversity islands: strategies for conservation in human-dominated environments. Series: Topics in Biodiversity and Conservation, Springer, Cham, pp 439–459

United Nations IPBES (2019) Global assessment report on biodiversity and ecosystem services. https://sdg.iisd.org/news/ipbes-previews-2019-global-assessment-report-on-biodiversity/

Vásquez V, Barber C, Dguidegue Y, Caughlin TT, García R, Metzel R (2022) Farmer perceptions of tropical dry forest restoration practices on the Azuero Peninsula of Panama – implications for increasing biodiversity in a human-dominated landscape. In: Montagnini F (ed) Biodiversity islands: strategies for conservation in human dominated environments. Series: Topics in Biodiversity and Conservation, Springer, Cham, pp 629–646

Zárate J (2021) Wars of the interior. Granta Publications Ltd., Cambridge. https://www.amazon.co.uk/Wars-Interior-Joseph-Zarate/dp/1783786159

# Index




F. Montagnini (ed.), *Biodiversity Islands: Strategies for Conservation in Human-Dominated Environments*, Topics in Biodiversity and Conservation 20,
https://doi.org/10.1007/978-3-030-92234-4

The manufacturer's authorised representative in the EU is Springer Nature Customer Service Centre GmbH, Europaplatz 3, 69115 Heidelberg, Germany. If you have any concerns regarding our products, please contact ProductSafety@springernature.com

Printed and bound by CPI Group (UK) Ltd, Croydon, CR0 4YY
30/09/2026
02224059-0013